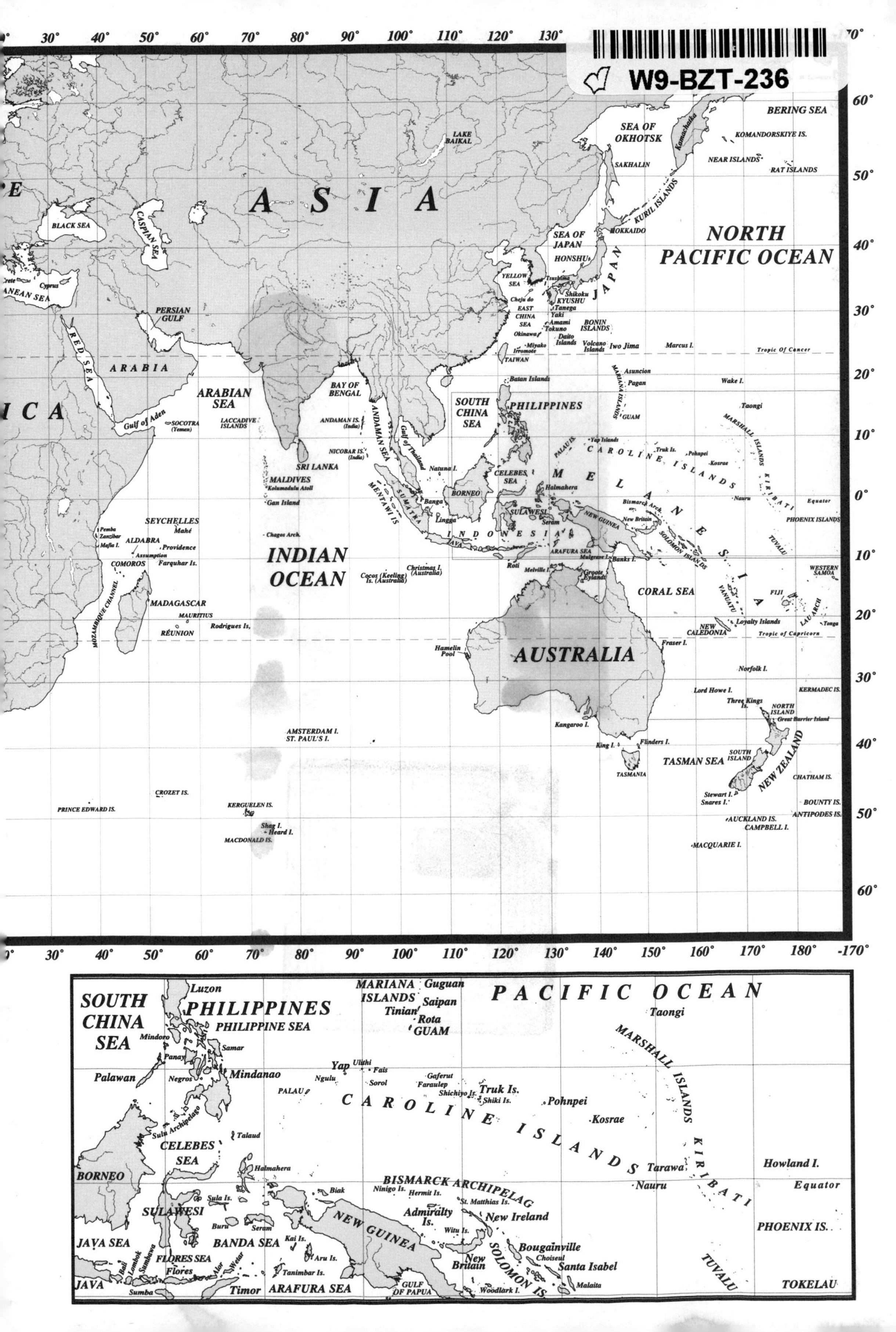

W9-BZT-236
30° 40° 50° 60° 70° 80° 90° 100° 110° 120° 130°
60° 50° 40° 30° 20° 10° 0° 10° 20° 30° 40° 50° 60°
A S I A
LAKE BAIKAL
BLACK SEA
CASPIAN SEA
Crete
Cyprus
PERSIAN GULF
RED SEA
ARABIA
ARABIAN SEA
Gulf of Aden
SOCOTRA (Yemen)
LACCADIVE ISLANDS
BAY OF BENGAL
ANDAMAN IS. (India)
NICOBAR IS. (India)
ANDAMAN SEA
Gulf of Thailand
SRI LANKA
MALDIVES
Kolumadulu Atoll
Gan Island
Chagos Arch.
SEYCHELLES
Mahé
Providence
Pemba
Zanzibar
Mafia I.
ALDABRA
Assumption
COMOROS
Farquhar Is.
INDIAN OCEAN
MOZAMBIQUE CHANNEL
MADAGASCAR
MAURITIUS
RÉUNION
Rodrigues Is.
Cocos (Keeling) Is. (Australia)
Christmas I. (Australia)
AMSTERDAM I.
ST. PAUL'S I.
CROZET IS.
PRINCE EDWARD IS.
KERGUELEN IS.
Shag I.
Heard I.
MACDONALD IS.
MENTAWI IS.
SUMATRA
Banga
Lingga
JAVA
Natuna I.
BORNEO
SOUTH CHINA SEA
PHILIPPINES
CELEBES SEA
SULAWESI
Halmahera
Seram
INDONESIA
Roti
Melville I.
ARAFURA SEA
Mulgrave I.
Groote Eylandt
Banks I.
Batan Islands
TAIWAN
Tropic of Cancer
SEA OF OKHOTSK
SAKHALIN
Kamchatka
KURIL ISLANDS
HOKKAIDO
SEA OF JAPAN
HONSHU
JAPAN
YELLOW SEA
Tsushima
Shikoku
KYUSHU
Tanega
Cheju do
EAST CHINA SEA
Yaki
Amami
Tokuno
Okinawa
Daito Islands
Miyako
Iriomote
BONIN ISLANDS
Volcano Islands
Iwo Jima
Marcus I.
BERING SEA
KOMANDORSKIYE IS.
NEAR ISLANDS
RAT ISLANDS
NORTH PACIFIC OCEAN
MARIANA ISLANDS
Asuncion
Pagan
GUAM
Wake I.
Taongi
MARSHALL ISLANDS
Yap Islands
PALAU IS.
Truk Is.
Pohnpei
Kosrae
CAROLINE ISLANDS
MELANESIA
Nauru
KIRIBATI
Equator
PHOENIX ISLANDS
Bismarck Arch.
New Britain
NEW GUINEA
SOLOMON ISLANDS
TUVALU
WESTERN SAMOA
CORAL SEA
VANUATU
FIJI
LAU ARCH.
Tonga
NEW CALEDONIA
Loyalty Islands
Tropic of Capricorn
Hamelin Pool
AUSTRALIA
Fraser I.
Norfolk I.
Lord Howe I.
Three Kings Is.
KERMADEC IS.
NORTH ISLAND
Great Barrier Island
Kangaroo I.
King I.
Flinders I.
TASMANIA
TASMAN SEA
SOUTH ISLAND
NEW ZEALAND
CHATHAM IS.
Stewart I.
Snares I.
BOUNTY IS.
ANTIPODES IS.
AUCKLAND IS.
CAMPBELL I.
MACQUARIE I.
30° 40° 50° 60° 70° 80° 90° 100° 110° 120° 130° 140° 150° 160° 170° 180° -170°
SOUTH CHINA SEA
Luzon
PHILIPPINES
PHILIPPINE SEA
Mindoro
Samar
Panay
Palawan
Negros
Mindanao
Sulu Archipelago
Talaud
CELEBES SEA
BORNEO
Halmahera
Sula Is.
SULAWESI
Buru
Seram
JAVA SEA
BANDA SEA
Kai Is.
Aru Is.
Bali
Lombok
Sumbawa
FLORES SEA
Flores
Alor
Wetar
Tanimbar Is.
JAVA
Sumba
Timor
ARAFURA SEA
GULF OF PAPUA
MARIANA ISLANDS
Guguan
Saipan
Tinian
Rota
GUAM
PACIFIC OCEAN
Taongi
Yap
Ulithi
Fais
Ngulu
Sorol
PALAU
Gaferut
Faraulep
Shichiyo Is.
Truk Is.
Shiki Is.
Pohnpei
Kosrae
CAROLINE ISLANDS
MARSHALL ISLANDS
KIRIBATI
Tarawa
Nauru
Howland I.
Equator
BISMARCK ARCHIPELAG
Ninigo Is.
Hermit Is.
St. Matthias Is.
Admiralty Is.
New Ireland
Witu Is.
NEW GUINEA
New Britain
SOLOMON IS.
Bougainville
Choiseul
Santa Isabel
Malaita
Woodlark I.
PHOENIX IS.
TUVALU
TOKELAU

SeaLife

A Complete Guide to the Marine Environment

SeaLife

A Complete Guide to the Marine Environment

Edited by
Geoffrey Waller

Principal Contributors
Michael Burchett and Marc Dando

Illustrated by
Marc Dando and Richard Hull

Smithsonian Institution Press
Washington, D.C.

Published in the United States of America
by Smithsonian Institution Press

ISBN 1-56098-633-6

Library of Congress Catalog Number is 95-72463

First published in the United Kingdom
by Pica Press, East Sussex, England

Typeset by Fluke Art, East Sussex, England
Colour printing by Clifford Press, Coventry, England
Printed and bound in Cornwall, England
by Hartnolls Limited

00 99 98 97 96   5 4 3 2 1

Contents

FOREWORD

The oceans cover over two thirds of the world's surface and form the largest habitat on Earth. This book is concerned with the animal and plant life that lives in and just above the surface waters of the sea and at its fringes. It describes the bio-oceanographic processes that define the nature of the seas and control the life rhythms of all marine animal and plant species. The aquatic world is the evolutionary 'cradle' from which the first fish-like amphibians emerged onto land in a new phase in the history of evolution of life on Earth, a phase that marked the dawn of the long vertebrate conquest of the terrestrial environment. The shoreline interface between sea and land was the evolutionary bridgehead for the rise of land animals which led eventually to the appearance of the ancestors of modern humans. The land-living ancestors of marine mammals also crossed this interface to re-enter their ancestral aquatic home.

Awareness about life in the oceans is central to our understanding of the way in which the seas impact on our lives. The oceans are not only our larder but are also our source of freshwater, distilled many times through evaporation and precipitation from the sea surface. The plant life of the oceans with its floating phytoplankton 'forests' forms a global oxygen 'factory' producing the air we breathe. Since humans have lived near the seas and crossed the oceans, the waters have not only yielded useful materials and food resources, but have provided a convenient dumping ground for human waste including sewage, hard materials, chemicals and even radioactive pollutants. Careless marine accidents and the notion of 'out of sight and out of mind' must become internationally and socially unacceptable if the oceans are to remain healthy for future generations. It is already a sad fact that stranded whales and dolphins have been classified as 'toxic waste' due to the high levels of PCBs that have accumulated in their bodies. Protection, conservation and sustainability must become the key issues to safeguard the future of our oceans.

The conflict between over-exploitation of local fish and shellfish resources and the need to maintain breeding stocks now makes frequent headline news. In the North Atlantic region alone, the Western Approaches, the Bay of Biscay, the North Sea and the Grand Banks of Newfoundland have all been areas of recent disputes between fishermen. Tunas are amongst the world's most valuable fishes and are the subject of dispute in both the Northern and Southern Hemisphere fisheries. A single adult Southern Bluefin Tuna from the southern oceans can be worth tens-of-thousands of dollars at market.

Only five species of bony marine fishes are currently listed by the Convention on International Trade in Endangered Species of Wild Flora and Fauna (CITES) which provides a framework for regulating trade in species threatened with extinction. It is difficult to assess the populations of such wide-ranging fishes as tunas, but the burgeoning oriental demand for shark fins and shark products generally, which consumes as many as an estimated 100 million sharks annually, has led to a CITES investigation of world shark fisheries. The nomadic Blue Shark which is found throughout the surface waters of the world is probably one of several shark species most heavily exploited by this trade. Blue Sharks are the most wide-ranging of all sharks and are caught in large numbers by tuna longliners.

Since the International Whaling Commission moratorium on commercial whaling started in 1986, there have been comparatively limited catches of a few species of whales, generally for 'scientific' or 'cultural' purposes. The small Minke Whale is, for example, currently hunted by several nations in both hemispheres for meat and oil as are Gray Whales and Bowhead Whales in the Arctic. However, even 'cultural' catches of whales are questionable as the products from these animals are seldom necessary for the survival of their human hunters. The work of international conservation and regulatory bodies is particularly important because some Great Whale species have been hunted almost to the point of extinction. Whether some of these near extinct populations that consist of widely dispersed individuals or groups can survive in the future remains to be seen.

This book provides a new way of approaching the complex subject of the marine environment from a scientific viewpoint. For the first time, a single text describes the astonishing diversity of life in the oceans. It is a highly illustrated work that uses much recent scientific knowledge for its classification, identification and descriptions of sea life. The book includes a comprehensive treatment of the zoology of the seas with numerous new and unique observations and descriptions. Its accessible style and wealth of illustration make it an ideal book for students of marine science, the seafarer and the general reader wishing to learn about the marine environment and sea life.

ACKNOWLEDGEMENTS

We are very grateful to the many individuals who have helped to make this book a reality; through their contributions, which we gratefully acknowledge, this book has been made possible.

Advice on many marine groups was kindly provided by specialists who were generous with their valuable time: Dr P. Cornelius, Dr G. Paterson and Dr G. Boxshall FRS commented on an early draft of the invertebrate chapter. Paul Clark gave invaluable advice on commensal crustaceans, and Dr I. Dixon and Christine Howson are thanked for their constructive comments on the final draft of the invertebrate chapter. Dr Q. Bone FRS read and commented on early drafts of two vertebrate chapters. Callan Duck reviewed the pinniped chapter and provided many helpful suggestions on the pinniped plate captions. Dr. B. McConnel also gave much invaluable advice on pinniped biology. Dr M. White, Dr A. North and Dr I. Boyd are thanked for assistance with locating obscure reference material and many helpful suggestions. Dr R. Luxmore helped with information on marine turtles. Paula Jenkins is thanked for providing access to the sirenian collections in the Natural History Museum, London and Richard Sabin for curatorial assistance.

Preparation of the text was greatly assisted by the technical expertise of Julie Reynolds (Fluke Art) who worked tirelessly on the many stages of the book from its inception, and Nigel Redman who skilfully edited the final text. We also thank Christopher Helm for his belief in and commitment to the project and for his encouragement and patience, especially during the early phase of the work.

Michael Burchett would like to thank Kenneth Burchett and Dr J. Burchett for their help and encouragement in writing the book.

Marc Dando would like to thank Julie Reynolds for her encouragement, advice and belief in this project which seemed to grow larger by the day; Kevin Morgan for providing much early source material; Anne Davis for her help in locating reference material; Dean Harmer and Tracy Partridge for their assistance and patience in reproducing the many early sketches; and finally thanks are due to Nigel Redman who provided the initial spark, and Christopher Helm who kept the torch burning.

List of Contributors

Michael Burchett
Biological Consultant

Marc Dando
Wildlife Illustrator

Sara Heimlich-Boran
Sea Mammal Research Unit, c/o British Antarctic Survey

Richard Hull
Bird Illustrator

Colin McCarthy
Zoology Department, The Natural History Museum

Kevin Morgan
Environmental Consultant

Nigel Redman
Series Editor

Iain Robertson
Ornithologist

Geoffrey Waller
Zoology Department, The Natural History Museum

EXPLANATION OF THE PLATES

On every plate a number is appended to each illustrated species. This number refers to a corresponding number in the caption text. The caption begins with the species' common name and then its scientific name. A bold number on the right indicates the page number of a distribution map for the species concerned. The description which follows contains information about the appearance and, in many cases, biology of the species. On the seabird plates, breeding adults are illustrated unless otherwise stated. The range is given by ocean or a specific geographical or hydrothermal region. World hydrothermal regions used here are shown on the map below. Length measurement is usually recorded as maximum known total length. In animal groups where particular types of length measurement are used, such as in certain invertebrates, fishes and marine mammals, these are noted separately in the captions. For seabirds, two measurements are generally given; average length (bill tip to tail tip) and average wingspan (wing tip to wing tip). In a few instances where wingspan is not known, only one measurement is given (e.g. 27/?cm). In some species, maximum adult body weight is recorded. For certain very large species where accurate body weight is difficult to measure, an approximate figure is given (e.g. 500kg+). Remarks about the species (given in appropriate cases) complete each caption. Population numbers are given where known, and endangered species classification follows the IUCN Red Data Book listing prepared by the World Conservation Union. Species classified as endangered are considered likely to become extinct unless specific protection means are taken.

Abbreviations:

AO	Atlantic Ocean
IO	Indian Ocean
PO	Pacific Ocean
MS	Mediterranean
IP	Indo-Pacific

These are usually qualified as sections of an ocean, e.g. nwAO refers to northwestern Atlantic Ocean.

Seabird colour plates and line drawings by Richard Hull. All other colour plates and drawings by Marc Dando. Plate captions written by Marc Dando (invertebrates), Iain Robertson (seabirds) and Geoffrey Waller (all other vertebrates).

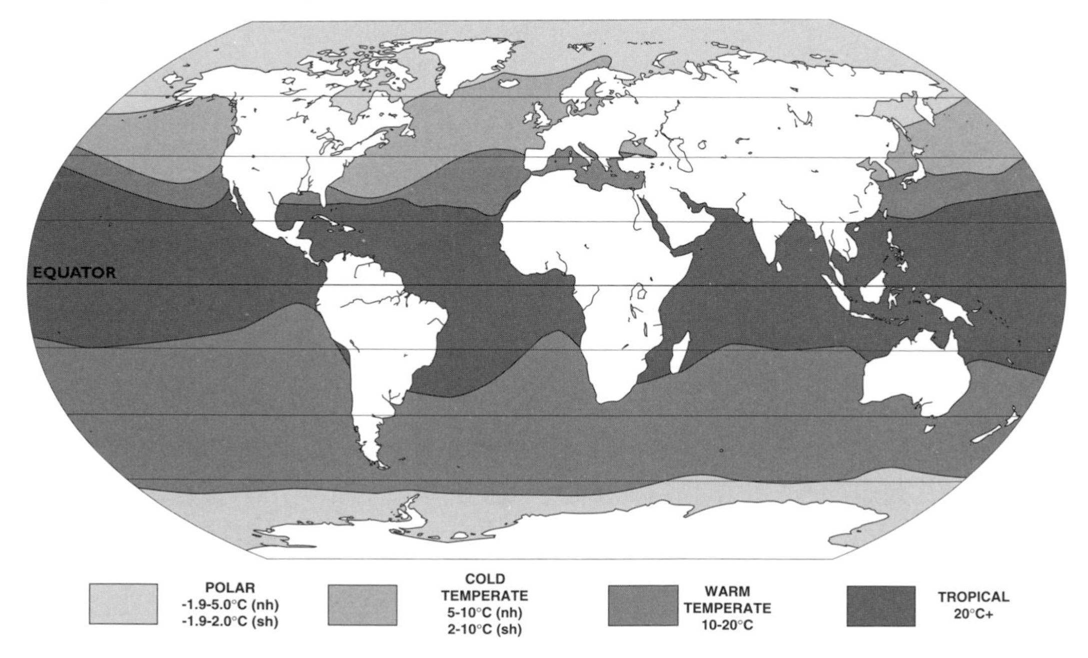

Major marine biogeographical areas of the world's oceans based on temperature (°C). (nh) = northern hemisphere, (sh) = southern hemisphere.

List of Plates

Dedicated to the memory of Sir Alister Hardy FRS, 1896-1985

Oceanography
and
Marine Biology

OCEANOGRAPHY AND MARINE BIOLOGY

by

Michael Burchett

HISTORICAL INTRODUCTION

VOYAGES OF MARINE DISCOVERY (1492-1870)

The beginnings of a systematic study of the oceans can be traced back to the era following the 'Age of Discovery' which occurred over a relatively short 30 year period between 1492 and 1522. It was during this period that Christopher Columbus discovered the 'New World' (12 October 1492) and Ferdinand Magellan's ship, *Victoria*, circumnavigated the world and returned to Spain on 6 September 1522 (fig.1).

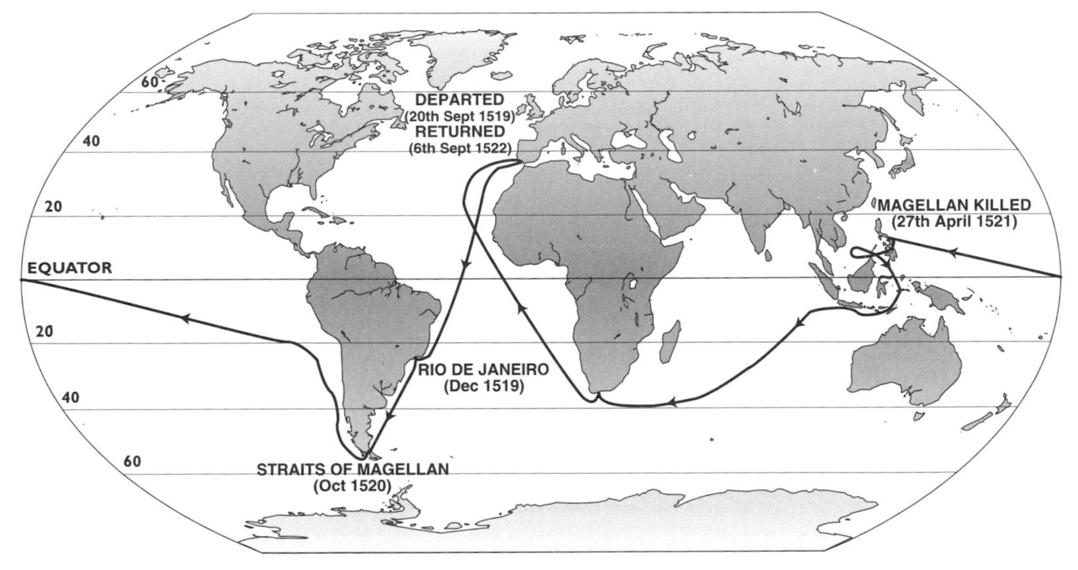

Figure 1. The Voyage of Magellan.

Spain maintained its dominance of the high seas until 1588, when the British fleet led by Sir Francis Drake, defeated the Spanish Armada. To help maintain British supremacy and an expanding empire, ships were sent on voyages to find and map new lands. The British navigator James Cook (1728-79) undertook three major seafaring voyages to map and scientifically explore new regions. The major objective of the first two voyages was to seek out a large southern continent which French navigators said they had previously sighted. On his first voyage in 1768 Captain Cook commanded *HMS Endeavour* and during this voyage he discovered and mapped much of the coastline of New Zealand and the east coast of Australia. Accurate chart making and navigation was only made possible by the invention of the John Harrison chronometer, which was the first accurate timepiece. Without accurate time keeping it is not possible to calculate the true position of a ship through latitude and longitude determinations. On the second voyage (1772-75) Cook commanded *HMS Adventure* and *HMS Resolution.* Using the prevailing westerly winds, he sailed round the Cape of Good Hope and set a course close to the 60° line of southern latitude and on 17 January 1773 he went beyond the Antarctic Circle in the hope of discovering the southern continent. Instead, the important discoveries of South Georgia and the South Sandwich Islands were made. He returned to England having circumnavigated the southern oceans, but did not find the southern continent because of thick ice. Cook's last voyage (1788-89) once again led him into the Pacific Ocean and to the discovery of the Hawaiian islands. His adventure also led him northwards into the Bering Sea until pack ice barred the way at a northern latitude of 70°44'. On 14 February 1779, Captain Cook, became the first navigator to sail both polar seas. On his return journey he passed once more through the Hawaiian islands and it was here that he was killed after a dispute over a stolen ship's boat. His legacies were the meticulous charts and coastal maps, and the observations on geology, biology and local native populations made by the naturalists that accompanied his voyages. These findings paved the way for further exploration and the start of marine resource exploitation. Commercial exploitation in the waters of the southern oceans began soon after the reports of Captain Cook's discoveries were published. Within 10 years of their discovery, the subantarctic islands were being regularly visited for the exploitation of elephant seals and fur seals for their oil and skins. The potential wealth of ocean resources highlighted by previous expeditions, and the improvements to ships and ship instrumentation, allowed the pace of global marine trade to increase rapidly in the fifty years that followed Cook's death.

The various maritime trading nations soon began to realise that knowledge of the physical aspects of oceans would help in the safe and speedy passage of ships. Knowledge of winds, tides and currents would give a trading nation the edge over its competitors. The American naval officer, Matthew Fontaine Maury (1806-73), director of the US Naval Department of Charts and Instruments, realised the need for international cooperation in gathering ocean data if a more comprehensive understanding of ocean currents and prevailing winds was to be achieved. In 1855 he published an important book titled *The Physical Geography of the Seas* which was of immediate value to all seafarers. Through his endeavours Maury came to the conclusion that the ocean was a "dynamic whole... as perfect and harmonious as that of the atmosphere or the blood". This book laid the foundations of modern oceanography and Maury is known as the 'father of oceanography'.

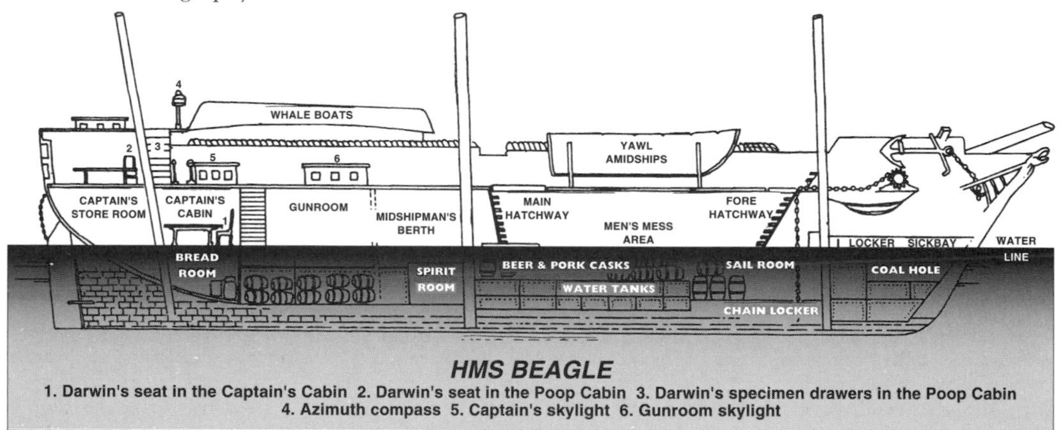

Figure 2. Cross section of *HMS Beagle.*

Early in the nineteenth century Charles Darwin (1809-82), a young naturalist of great insight, signed on as naturalist on *HMS Beagle* (fig.2), commanded by captain Robert Fitzroy (1805-65). The voyage of the Beagle (fig. 3) was to last nearly five years (1831-36) and during this time Darwin became fascinated with the beauty and diversity of coral reefs. After his return Darwin published *The Structure and Distribution of Coral Reefs* (1843) in which he correctly described the formation of coral atolls. In 1859 he published *On the Origin of Species* in which he described his theory of the evolution of biological organisms through natural selection. The observations made by Darwin during the voyage of the *Beagle* were fundamental to the development of his theory. His understanding revolutionised biological thinking and its contradiction of the Christian belief of creation caused much debate in its time.

Edward Forbes (1815-54) was another key figure among British naturalists. Forbes undertook dredging operations in the Aegean region of the Mediterranean and other European seas. By charting the distribution of certain organisms he was able to recognise major faunal areas and distinct faunal zones in relation to water depth (bathymetric zones). He was the first person to distinguish the 'littoral zone' (between tide marks), the 'Laminaria zone' (named after the seaweed *Laminaria*), the 'coralline zone' (50m or more) and the 'deep-sea coral zone'. The names of some of these zones have now changed having been incorporated into a better scheme of zonation. Forbes' understanding of the vertical distribution of life in the oceans led him to conclude that plant life was limited to the sunlit surface waters, and that the abundance of animal life decreased with depth of water. He believed there was little or no light or oxygen in the deep ocean environments and that life ceased to exist below a depth of about 500m. However, pioneering marine research carried out by Sir John Ross (1777-1856) in Baffin Bay, Canada in 1817-18 disproved this hypothesis. He used a 'deep-sea clam' dredge to haul up seabed samples from depths of 1.8km, in which living organisms were found. The nephew of Sir John Ross, Sir James Clark Ross (1800-62), also undertook similar research in Antarctic waters. During the British Antarctic Expeditions, dredging operations brought to the surface an abundance of seabed life from depths of over 7,000m. The expedition's naturalist Dr Joseph Hooker (1817-1911) made another fundamental discovery. He realised the importance of microscopic plant organisms called diatoms and he was sure they were a vital component in marine food chains. His ideas on this subject were later found to be correct.

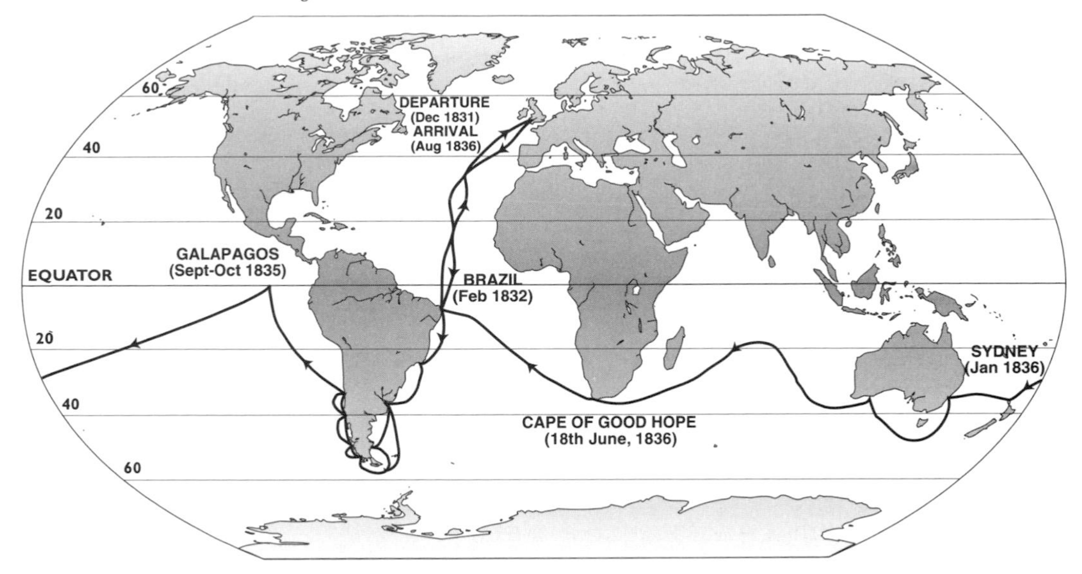

Figure 3. The voyage of *HMS Beagle*, commanded by Captain Robert Fitzroy.

THE DEVELOPMENT OF MARINE SCIENCE 1870-1900

By 1870 there was enough evidence to establish that life existed in the oceans to depths of at least 2,000m. The quest to find deep-sea 'living fossils' and the request for further hydrographical information sparked off the first era of marine research (1870-1900). In 1871 the Royal Society of London requested the British government to raise funds for a major expedition to investigate the world's oceans. The purpose was to ascertain the chemical composition of seawater, the distribution of marine organisms at all depths, the nature of coastal and ocean currents and to examine the sea floor. This global approach to ocean studies was a revolutionary step forward and for this purpose the British government released a spar-decked naval corvette of 2,360 tons named *HMS Challenger* which was to be fitted out for deep-sea explorations (fig.4). The Challenger Expedition was led by Captain Charles Wyville Thomson and included a complement of six scientists. Between 1872 and 1876 the crew and scientists of *HMS Challenger* completed 362 deep-sea soundings covering all three major oceans (fig.5).

Other operations carried out included dredging, trawling, tow-netting, water sampling, temperature measurements, meteorological observations and the charting of ocean currents. In addition more than 7,000 sealife specimens were collected, some from depths in excess of 6,000m. Each specimen was described, catalogued and preserved for later laboratory analysis. The information gathered from the Challenger Expedition took more than 23 years to analyse and the findings were published in a series of 50 volumes. In all, 5,000 new species of marine organisms were identified and described. Although no living fossils were caught, life was found to exist at all levels in the ocean, including the sea floor. The Challenger Expedition was the precursor of modern oceanographic research and it set many of the standard techniques of sampling and investigation.

The Challenger Expedition was followed by several other deep-sea explorations from the various maritime nations. Through the collation of information from many of these expeditions, scientists began to gain an overall picture of the physical and biological relationships which existed in the marine environment. Variations in the seasonal abundance of marine organisms, and the migration of animals in the water column (vertical migration) and across the oceans were some of the major discoveries made.

Figure 4. *HMS Challenger.*

By the time of the Challenger Expedition it was understood that the surface layer of the sea contained a 'soup' of small organisms consisting of tiny plants and animals. Some of these organisms could only be observed under a microscope, while others were clearly visible under a hand lens or with the naked eye. In the 1820s an army surgeon, J. Vaughan Thompson, spent much of his free time towing fine meshed nets off the coast of Cork (Ireland). He thought that the tiny organisms caught by the netting operations must all be adults. However, on closer inspection, he concluded they included the young stages of much larger animals. Through these endeavours he later showed that these organisms were indeed the larval stages of animals such as the crab's 'zoea' and 'nauplius', and the 'cypris' stages of barnacles.

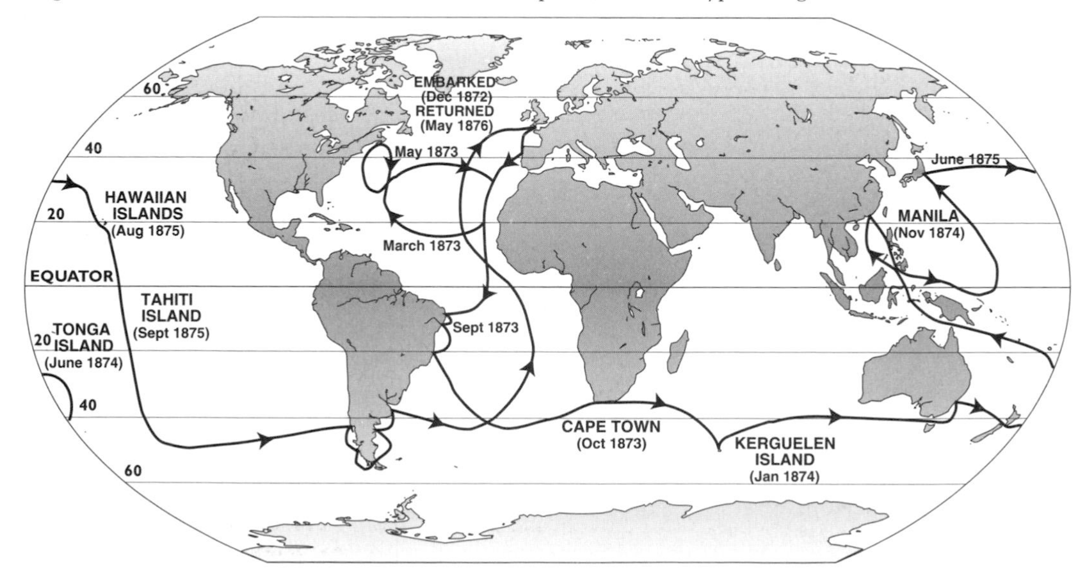

Figure 5. The voyage of *HMS Challenger.*

By the late nineteenth century the eminent German scientist Victor Hensen (1835-1910) had also turned his talents to the investigation of the small, floating creatures of the sea, known as plankton. However Hensen took a much broader approach to the study of ocean life. He was one of the first 'marine ecologists' who looked at the relationships between the various groups of marine plants and animals. Hensen carried out plankton studies aboard the ship, *National*, and the main objective was to study plankton distribution at depth and on the surface of the sea. Using his own design of plankton net and new sampling methods, he was the first scientist to carry out proper quantitative (large scale) plankton studies. From his findings he concluded that plankton was just as abundant in the cold waters of polar regions as it was in tropical waters. He also assumed that there were no breaks in the distribution of plankton across the wide expanses of oceans. However, this theory of even plankton distribution was later disproved and it is now known that plankton is often patchy in its distribution.

At the end of the nineteenth century, polar oceanography began with the unusual and remarkable voyage of the ship *Fram* (fig.6). The Norwegian explorer Fridtjof Nansen (1861-1930) designed and built a 38m, three-masted schooner with a hull thickness of 1.2m. The design of this sturdy little ship enabled it to endure being locked into pack ice and to resist its crushing forces. In September 1893 the *Fram* with a crew of 13 men and provisions for five years sailed into the sea ice north of Siberia. Here the ship remained trapped, as it drifted with the pack ice at an average speed of 2km per day. At one point it came to within 400km of the North Pole. On 13 August 1896 the *Fram* was released from the ice near the island of Spitsbergen, having drifted a total of 1658km during her three year entrapment (fig.7).

Figure 6. The *Fram* locked in Arctic Ocean ice.

The drift of the *Fram* proved that, unlike Antarctica, there was no equivalent northern continent but instead, a shifting mass of frozen water - the Arctic Sea. During the little ship's drift the crew carried out many oceanographic and meteorological observations. They found that the depth of the Arctic Sea sometimes exceeded 3,000m and that the temperature of deep Arctic water was warmer at a depth between 150m and 900m, compared to Antarctic waters. Nansen correctly deduced that this was caused by Atlantic water slipping below the less saline Arctic water. His observations of the direction of ice drift relative to the wind direction were used by V. Walfrid Ekman (1874-1954) to develop a mathematical explanation linking wind direction and ocean current flow. This became known as 'Ekman's Spiral'.

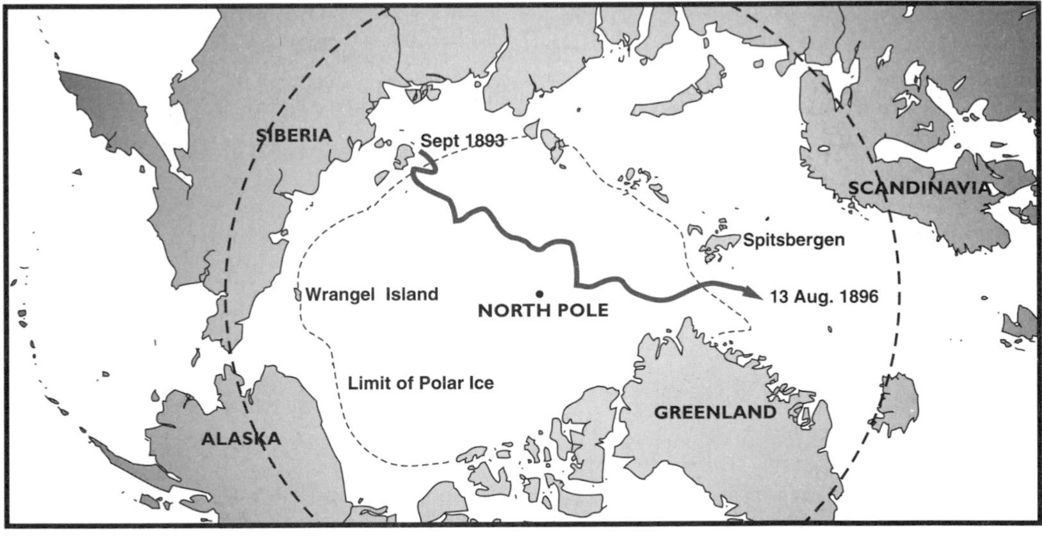

Figure 7. The Arctic voyage of the *Fram*.

MODERN OCEANOGRAPHY AND MARINE BIOLOGY

Modern oceanography is considered to stem from the beginning of the twentieth century. At the turn of the century no large scale deep-sea expeditions or explorations were being undertaken. The first major oceanographic expedition of the century was carried out by a Dane, Johannes Schmidt, between 1908 and 1910. In 1910 a Norwegian, Johan Hjort, studied the biology and hydrography of the Mediterranean and North Atlantic. His findings formed the basis of a classic marine book called *The Depths of the Ocean* which was published in 1912. After World War I (1914-18) Schmidt led a series of expeditions in the sister ships *Dana I* and *Dana II* and these expeditions continued throughout the 1920s. These explorations covered three major oceans and immense collections of oceanic organisms were made. The vast amounts of information formed the foundation of the *Dana Reports* which are still useful today.

Between 1925 and 1927 German scientific expeditions in the ship *Meteor* increased our understanding of the physics, chemistry, biology and topography of the southern Atlantic Ocean. Using highly advanced equipment, continuous monitoring and sampling were undertaken for a period of 25 months. The ship covered 310 sampling stations at which 9,383 pairs of temperature and salinity values were recorded from the surface to the ocean floor. Acoustic soundings were also taken every two to three nautical miles and the information gathered on such topographical surveys was invaluable in the understanding of oceanic water circulation. The chief scientist, George Wurst, put forward a theory of a four-layered structure of water circulation which later proved to be correct and is still recognised today. From the biological information collected, it became clear that there was a firm relationship between the quantity of ocean plant life (phytoplankton) and the amount of nutrients in the surrounding waters.

During the 1920s, British research centred on whales and the whaling industry in Antarctic waters. The Discovery Investigations began in 1925 and the research was initially carried out from Captain Scott's original ship the *Discovery*. Later surveys were undertaken from a 32 ton whale-catching vessel, the *RRS William Scoresby*. In 1929 the 1,036 ton purpose-built scientific research ship *Discovery II* replaced Scott's old ship. Many *Discovery* expeditions were carried out before and after World War II (1939-45). The research was extended to cover the entire southern oceans and certain parts of the Atlantic, Indian and Pacific Oceans. Much of the data collected from the various trips were put together as the *Discovery Reports* and these volumes established modern Antarctic marine research.

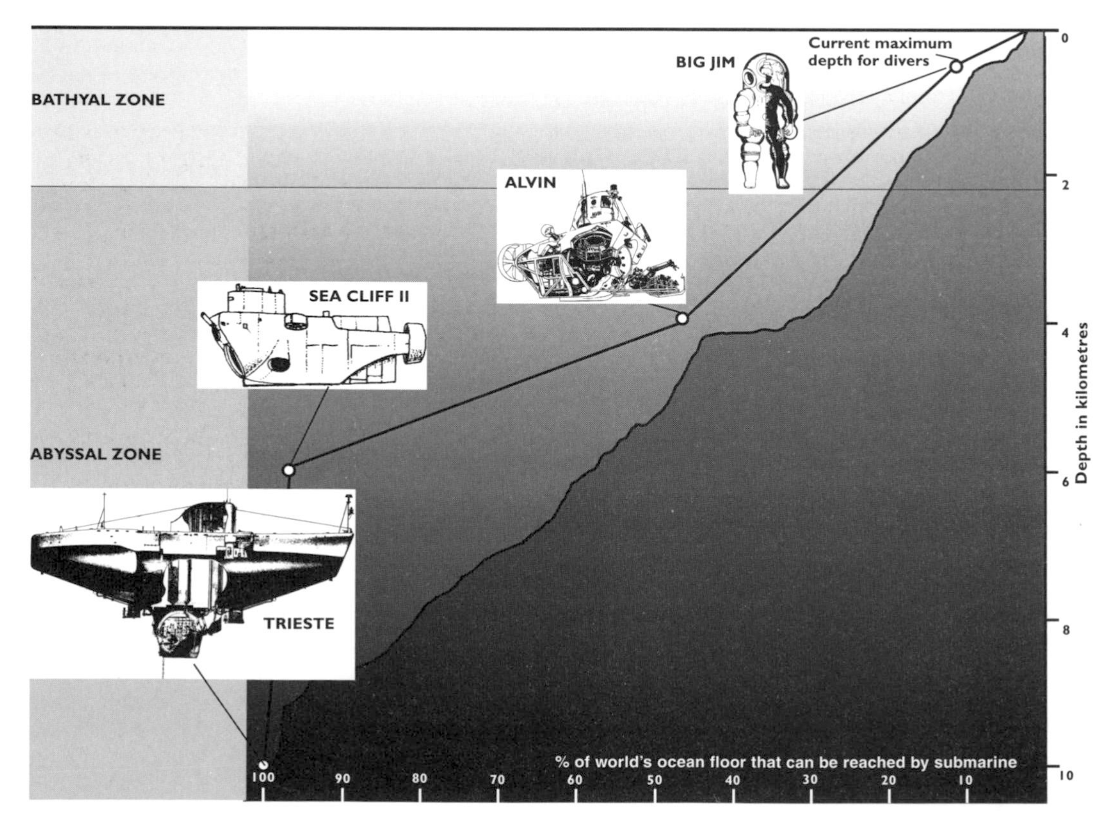

Figure 8. Deep-sea exploration by different submersible craft.

After World War II, there followed a great expansion of oceanographic and marine biological research and many countries began to develop their own institutions dedicated to marine studies. Notable examples include the Scripps Institution of Oceanography (USA), established in 1908, and the Woods Hole Oceanographic Institution (USA) which began in 1930. Over the past 40 years, technology has helped marine scientists gain access to all parts of the deepest oceans. Scuba diving and manned submersibles have enabled humans to descend to greater depths in relative safety. In 1934 the zoologist William Beebe descended to a depth of 923m (3,027ft) in a tethered 'bathysphere' to observe deep-sea life. The Swiss scientist Auguste Piccard designed the 'bathyscaphe' (deep boat) named *Trieste.* This underwater vessel allowed man to reach the deepest parts of any of the world's oceans and on 23 January 1960 it recorded a depth of 10,915m (35,801ft) in the Mariana Trench off the island of Guam in the Pacific (fig.8). Two other manned submersibles that are widely used in deep-sea operations are *Alvin* which has a working depth of 4,000m, and *Sea Cliff II* with a working depth of 6,000m.

Today, most seafaring nations have their own oceanographic research vessels which are capable of carrying out a variety of marine biological and oceanographical work. Seagoing research is often linked to shore-based laboratories which are essential for studies that require time and facilities not available at sea. Much of the collected data and specimens are returned to shore-based facilities for subsequent analyses. Many more laboratories have now been established to look primarily at nearshore marine and coastal environments. The Marine Biological Association (England) was founded in 1884 for just such a purpose. Many of the developments in marine ecology (the study of biological relationships in the marine environment) have depended largely on research carried out in coastal waters and the use of shore-based facilities. By its very nature, much fisheries research is confined to the shallow seas of continental shelf areas where many of the major fish food species live and spawn.

The main thrust of oceanographic science since the early 1960s has been linked with mineral resource exploration and exploitation for oil and gas. Marine geophysics, marine geology, hydrophysics and palaeoceanography have become important branches of marine science. Quite often, the amount of capital investment and human resources required to undertake large oceanographic ventures requires international cooperation. This may involve collaboration among marine scientists from many disciplines, universities and research institutes. Notable programmes of this type have included the 1957-8 International Geophysical Year (IGY), the 1959-65 International Indian Ocean Expedition, the International Decade of Ocean Exploration (IDOE) of the 1970s, the 1972 Geochemical Ocean Sections Study (GEOSECS) and the 1975 International Programme of Ocean Drilling (IPOD).

FUTURE MARINE RESEARCH

With modern methods and technology, future oceanographic and biological research will continue in its many forms. Improvements in technology will allow the pace of marine research to be accelerated. Remote unmanned underwater vehicles bristling with sensor equipment enable exploration of the depths without exposing human life to its dangers. Microchip technology and modern computers rapidly increase the collection, collation and interpretation of data. Remote sensing from satellites is becoming an important tool in the study of oceans and its sealife. Satellite instrumentation can measure surface sea temperatures, ice cover, atmospheric and weather conditions, and provide information for the production of geographical and topographical maps. Satellite communication can also relay information on the movements of whales, seals and birds by using modern electronic tagging methods. The information can be relayed by satellite to laboratory computers which analyse it within seconds. Satellites also provide a picture of large areas of surface fish and plankton distribution and pollution. Underwater scanning equipment has also become cheaper and more reliable over the past decade. Proton magnetometers which detect large underwater objects and underwater cameras allow rapid remote scanning and image exploration of marine environments.

Modern marine science has become far more applied, with many governments encouraging links between research and industry. Pure science (science for its own sake) has been on the decline as competition for funding has intensified. Independent environmental groups including the Worldwide Fund for Nature and Greenpeace have become the global 'watchdogs' of over-exploitation and pollution. Let us hope that oceanographic and marine biological research provides the necessary information to promote sensible use of the Earth's precious marine resources and to keep it in a fit and healthy state for future generations.

A CHRONOLOGY OF OCEANIC EXPLORATIONS AND MARINE RESEARCH

4,000BC	The Egyptians developed shipbuilding and ocean pilot skills.
1,000–600BC	The Phoenicians explored the whole of the Mediterranean Sea, reached England and circumnavigated Africa by keeping to the coastline.
384–322BC	Aristotle described how sharks produced live young and recognised that dolphins are not fish in his book *Historia Animalium.*
276–192BC	The Greek, Eratosthenes, accurately determined the circumference of the Earth using trigonometry.
54BC–AD30	The Roman, Seneca, put forward his hypothesis on the 'hydrological cycle' of water.
AD150	The Greek, Ptolemy, compiled the *Roman World Map* which incorporated basic lines of longitude and latitude.
300–800	The Polynesians spread east across the Pacific reaching New Zealand, Hawaii and Easter Island.
673–735	The English monk, Bede, described the lunar influences on tidal cycles and recognised the monthly tidal variations.
982	The Norseman, Eric the Red, discovered North America when he landed on Baffin Island (Canada).
995	Leif Ericson (son of Eric the Red) established the settlement of Vinland on the North American coast of Newfoundland.
1420s–30s	The Portuguese, Henry the Navigator, directed major voyages in European waters and several important discoveries included Madeira, the Azores and Cape Verde Islands.
1452–1519	The Italian, Leonardo da Vinci, observed and recorded the movements of waves and currents. From his observations of fossils he deduced that the sea levels must have been higher in the past.
1492	Christopher Columbus rediscovered the 'New World' of the Americas by way of the West Indies.
1498	Vasco da Gama reached the Orient and successfully returned.
1513–18	Vasco Nunez de Balboa crossed the Isthmus of Panama and sailed into the Pacific Ocean.
1519–22	The Spaniard, Ferdinand Magellan, set off to circumnavigate the world in 1519 and this was completed in 1522 by Sebastian del Cano after Magellan's death in the Philippines
1569	Gerardus Mercator constructed the *Mercator Projection Map* and this was adapted to marine charts, which greatly improved navigation.
1674	Robert Boyle carried out pioneering oceanographic measurements on temperature, salinity, pressure and depth. The findings were reported in *Observations and Experiments on the Saltiness of the Sea.*
1675	The Royal Observatory at Greenwich (England) was established and set the line of longitude at 0° (the Greenwich Meridian).
1725	Luigi Marsigli produced a book incorporating the whole of marine science as it was then known, which was titled *Histoire Physique de la Mer.*
1769–70	Benjamin Franklin published the first ocean charts of the Gulf Stream to help the passage of ships across the Atlantic.
1768–79	Captain James Cook (1728-79) undertook three major voyages of discovery and attempted to find the southern continent of Terra Antarctica. He did not discover Antarctica but he was the first navigator to cross the Antarctic Circle on 17 January 1773. His meticulous map-making pioneered the way for the success of future explorations.
1817–18	Sir John Ross discovered living organisms at a depth of 1.8km, near Baffin Island in the Arctic Ocean.
1831–36	Charles Darwin (1809-82), the naturalist aboard *HMS Beagle*, correctly deduced the formation of coral atolls.
1839–43	Sir James Clark Ross (1800-62), nephew of Sir John Ross, and the naturalist Dr Joseph Hooker (1817-1911), discovered life at depths of 7km in Antarctic waters.

1835–1910	Victor Hensen made important discoveries on the nature of plankton. He pioneered many new techniques for plankton study.
1854	Sir Edward Forbes (1815-54) published an influential book entitled *Distribution of Marine Life.*
1855	The American, Matthew Fontaine Maury (1806-73), compiled wind and current data to help shipping, and published his findings in *The Physical Geography of the Seas.* Maury became known as the 'Father of Oceanography'.
1872–76	The naturalist, Charles Wyville Thomson, conducted worldwide scientific investigations aboard *HMS Challenger.* The findings were published as the Challenger Reports and in 1873 he published the oceanography book *The Depths of the Sea.*
1884	The Marine Biological Association (England) was established to study coastal ecology.
1892	The Norwegian polar explorer, Fridtjof Nansen (1861-1930), had the ship *Fram* specially built to withstand ice-entrapment. The drift of the ship confirmed there was no Arctic continent.
1895–98	The American, Joshua Slocum, became the first man to circumnavigate the world single-handed, in his yacht *Spray.*
1902	The Danes set up ICES (International Council for the Exploration of the Seas) to investigate the oceanography and fisheries of the North Sea.
1902	V. Walfrid Ekman (1874-1954) developed a mathematical explanation linking wind direction and current flow (Ekman's spiral).
1908	The Scripps Institution of Oceanography (USA) was established.
1912	Johan Hjort published the classic marine book *The Depths of the Ocean.*
1912	The German scientist, Alfred Wegener, proposed his theory of 'continental drift'.
1920s	The Dane, Johannes Schmidt, carried out worldwide oceanographic studies in the ships *Dana I* and *Dana II.* The findings were published in the extensive volumes of the Dana Reports.
1920s–40s	The British-led 'Discovery Investigations' carried out extensive work in many regions of the southern oceans. The results were published in the Discovery Reports and established modern Antarctic marine research.
1925–27	A German expedition directed by George Wurst, aboard the ship *Meteor,* made detailed studies of the oceanography of the Atlantic Ocean. Many modern oceanographic techniques were first used on these voyages.
1930	The Woods Hole Oceanographic Institution (USA) was established at Cape Cod, Massachusetts.
1932	The International Whaling Commission (IWC) was set up to collect whaling information and to regulate the shore-based whaling industry and subsequently the pelagic whaling fleets.
1934	The zoologists William Beebe and Otis Barton descended to a depth of 923m (3,072ft) in a tethered 'bathysphere' to observe deep-sea life.
1942	H. Sverdrup, R. Flemming and M. Johnson published the classic marine science book *The Oceans.*
1943	Jacques Cousteau and Emile Gagnan developed the automatic demand valve system for use by scuba divers. Cousteau undertook many filming voyages in the ship *Calypso.*
1947	Thor Heyerdahl sailed from Peru across the Pacific on the balsa raft *Kon Tiki.*
1957–58	The International Geophysical Year (IGY) was organised to investigate marine and terrestrial geophysics. It established a pattern for international scientific cooperation.
1958	The nuclear submarine *USS Nautilus,* commanded by Cdr. Andersen, reached the North Pole under the sea ice.
1960	The Swiss scientist, Auguste Piccard, developed the untethered 'bathyscaphe' *Trieste.* Jacques Piccard and Donald Walsh descended to the deepest part of the world's oceans to a depth of 10,915m in the Mariana Trench, off the island of Guam in the Pacific Ocean.

1959–65	The International Indian Ocean Expedition (IIOE) was established to investigate the oceanography of the Indian Ocean.
1966	The first supertanker disaster, *Torrey Canyon*, occurred off the coast of SW England, where large quantities of spilled oil caused a major ecological disaster. Since then other major oil-related disasters in the marine environment have included the *Amoco Cadiz*, *Exxon Valdez* and oil spills in the Persian Gulf resulting from the Gulf War.
1968–1975	The US National Science Foundation organised the Deep Sea Drilling Project (DSDP) to investigate the ocean crust. It confirmed the theory of 'sea floor spreading'.
1970s	The United Nations initiated the International Decade of Ocean Exploration (IDOE).
1972	The Geochemical Ocean Sections Study (GEOSECS) was organised to study ocean chemistry and biochemical recycling of chemical substances.
1977	Manned American submersibles discovered remarkable deep-sea communities associated with hydrothermal vents along sea floor ridges.
1978	The first 'remote-sensing' oceanographic satellite (Seasat-A) was launched to study the oceans.
1980s	The coordinated Ocean Research and Exploration Sections programme (CORES) was established by the UN to continue scientific work from the IDOE investigations of the 1970s.

Evolution of the Planet, Life and Oceanographic Processes

STRUCTURE OF THE EARTH

The Earth is about 4,500 million years old and is the only planet that is known to support organic life. The Earth is round in shape but bulges slightly at the equator and narrows towards the poles. It has a circumference of about 40,000km (24,800 miles) and a diameter of approximately 12,700km (7,900 miles). The internal composition of the planet consists of a number of concentric rings which have been laid down according to the densities of the various materials within each ring. There is a solid 'core' of heavy materials that have sunk to the centre of the planet, around which an envelope of lighter molten rock has formed (fig.9). The molten core is surrounded by the 'mantle' which is an intermediate layer of dense, hot rocks comprising 84% of the Earth's volume and 68% of it's mass. The main chemical constituents of the mantle include silicon, magnesium, iron and oxygen.

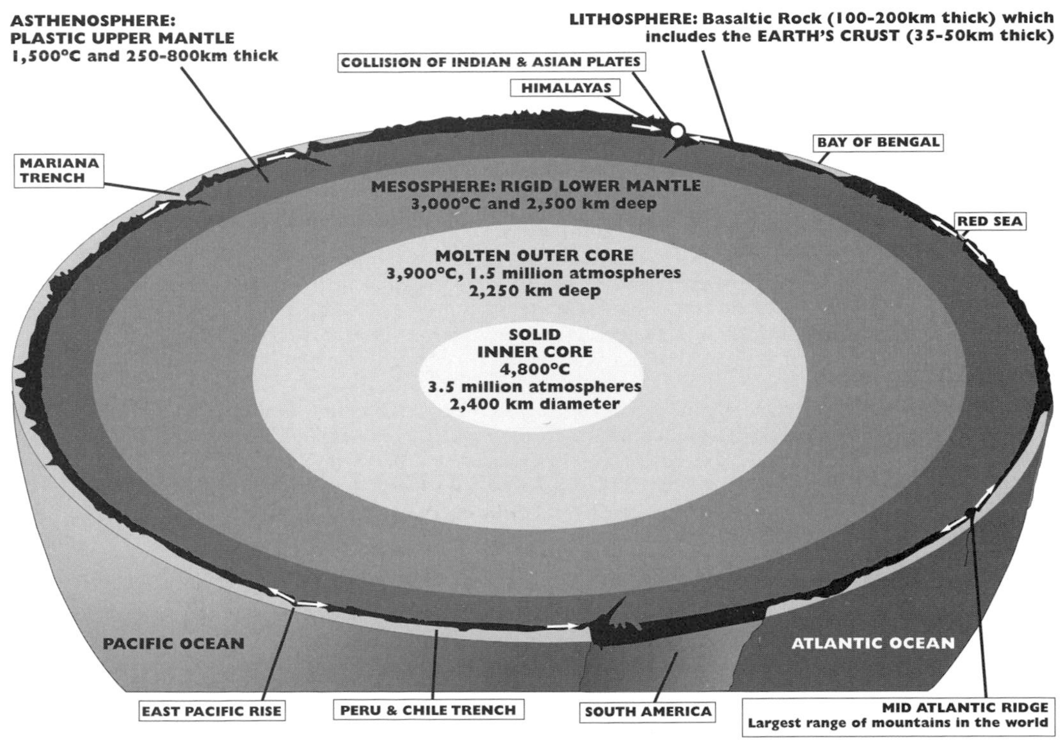

Figure 9. The structure of the interior of the Earth.

The planet remained in a molten state for nearly 1,000 million years until it cooled sufficiently to allow the formation of a thin, solid skin called the 'crust'. The crust is a surface layer of only 35-50km in thickness. It is part of a much deeper layer called the 'lithosphere', which extends down from the surface to a depth of 100-200km. Below the lithosphere and separating it from the more solid mantle (the 'mesosphere') is the 'asthenosphere'. This layer is made of partially melted upper mantle and has a thickness of about 250-800km (average 350km).

FORMATION OF THE ATMOSPHERE AND OCEANS

About 4,000 million years ago conditions on Earth were not suitable to support organic life. The gaseous envelope (atmosphere) at that time was composed mostly of hydrogen, methane, ammonia, carbon dioxide, water vapour and a little free oxygen. The ultraviolet rays of the sun bombarded the Earth and the intense heat of the surface would have caused most of the gaseous atmosphere to be lost into space. Eventually the surface did cool sufficiently to allow a 'crust' to form and the atmospheric gases to be retained. At this time most of the gases lost to the atmosphere were replaced through gaseous volcanic activities. The cooling gases eventually formed a vast envelope of clouds which completely surrounded the planet and 60% of all incoming sunlight was reflected back into space. This allowed further cooling and water vapour to condense into rain. The first rains hitting the crust would have rapidly boiled away but over a long period of time, permanent accumulations of water began to fill the irregular depressions of the surface on the Earth. The planet now had a 'hydrosphere' (areas covered by water) as well as an atmosphere. Today water dominates the Earth's surface and the oceans cover approximately 70% of the surface with about 1.3 million cubic kilometres of water. Seawater makes up about three quarters of all water stocks by volume, and therefore freshwater is a relatively rare commodity by comparison. However, the vast amounts of surface water only account for 10% of the entire planet's reservoir. The bulk of the water is chemically locked into the mineral matter of the Earth's mantle.

The present atmosphere of the planet now comprises about 78% nitrogen, 21% oxygen and 1% other gases. The atmosphere has gradually changed from its primeval form through the combination of dynamic environmental activity and biological processes.

THE ORIGIN OF LIFE IN THE SEA

Life could not evolve until conditions on the surface of the Earth became favourable and the right chemical substances were available. The hostile atmosphere of hydrogen, methane, ammonia and water vapour evolved into a life-supporting atmosphere of oxygen (O_2), carbon dioxide (CO_2) and nitrogen (N_2). The mechanism of the change has been a subject of much debate among scientists. However, it is probable that by about 3.6 billion years ago (bya) the seas had cooled enough to form a marine environment sufficiently protected from the harmful rays of the sun to provide favourable conditions for complex chemical reactions to take place. Although no 'true' life existed then, carbohydrates, amino acids, proteins, nucleic acids and fatty acids formed and accumulated in the early seas. Central to this chemical evolution was the production of nucleic acids (adenine, guanine, thymine and cytosine) which have the capacity to replicate themselves. This 'self-replicating' or 'reproductive' capability is essential if life is to continue from one generation to the next. Energy required to power many of the chemical reactions for the construction of the complex molecules came from a variety of sources; these include heat from the sun, lightning, heat from volcanic activity and the intense pressures and temperatures generated by large meteorites hitting the Earth. The mechanisms leading to formation of living organisms are not clearly understood. However, for a 'primordial soup' of chemical substances to be transformed into organic life, it would be necessary for the complex molecules to fuse together in an orderly fashion.

A true 'living organism' can be described as one that is able to grow and replicate itself. Organisms that are alive grow by the process of 'biosynthesis'. In this process, raw materials are obtained, broken down into simpler units and then reconstructed into more complex, more useful substances that can be used by the organism. It is reasonable to assume that the first living organisms must have appeared in the oceans and fed on the organic molecules, breaking them down and reconstructing them without the use of oxygen (anaerobes). Similar organisms still exist today and they include certain types of anaerobic 'bacteria' which have been known to exist for at least 3.6 billion years. By about 2.3bya another significant step forward in the evolution of life occurred when the important green pigment 'chlorophyll' first appeared. This enabled some organisms to obtain food substances from water and carbon dioxide using solar radiation as an energy source. The biochemical reaction which took place then still happens in all plants today and is known as 'photosynthesis' or 'phototropism'. All organisms using this method of organic production became known as the 'primary producers' or 'fixers' of solar energy. Only the plants (flora) can produce sugars and nearly all other forms of life depend upon them for food in one way or another (see page 54).

The simplest of all living creatures consist of a single living unit called the 'cell'. More complex organisms, be they plants or animals, are made up of many hundreds or even millions of such cells. Complex creatures will have a variety of cells performing different functions. However, all living organisms are characterised by their ability to respond to the environment around them, reproduce and be able to harness energy sources for their own use. This is also true for single celled organisms. Despite the immense variety and diversity of creatures both past and present, they were, and all are, fashioned from the same types of molecular building blocks of proteins, carbohydrates, nucleic acids and fats. Nucleic acids can not only replicate themselves but also carry the 'genetic code' or instructions (DNA) that are passed down from parent to offspring. These instructions determine what the organism is to be, but with some slight but significant variations between subsequent generations. Proteins are built from amino acids and are used to make and repair tissues, protect the body from disease, carry oxygen in haemoglobin, maintain body structures and act as catalysts (biological enzymes) in biochemical reactions that keep an organism alive. Carbohydrates and fats are the main energy providers and energy stores.

The first living organisms to use photosynthesis for the manufacture of complex carbon compounds were probably primitive, single-celled, plant-like algae similar to the blue-green algae that exist today. These simple plant-like structures are known to have existed for the past 1.7 billion years. About 1.2bya more complex organisms evolved and these protists also have contemporary relatives, the single-celled protozoans of today. By about 900 million years ago (mya) sexual reproduction (reproductive material coming from more than one source) started to appear in some organisms and this mechanism enabled the pace of biological evolution of more complex organisms to increase. One such method that still exists today is the formation of a colony by a loose group of single-celled organisms to form a much larger structure e.g. the sponges (Porifera). The jellyfish (Cnidaria) were probably the first true multi-cellular organisms with nervous, muscular and reproductive tissues, producing slight alterations of DNA instructions between successive generations. The combination of sexual reproduction and multi-cellular tissue formation resulted in another acceleration of biological evolution. By the beginning of the Cambrian period (570mya) 'organs' were becoming more widespread in the bodies of animals. An organ can be described as a number of different types of tissues which are arranged together to perform at least one major body function. Early organs would have included reproductive and light-sensitive organs. At the most complex end of the evolutionary spectrum, highly advanced animals are characterised by their well defined organs which include eyes, heart, lungs, kidneys, gonads and a brain.

BIOLOGICAL EVOLUTION BY NATURAL SELECTION

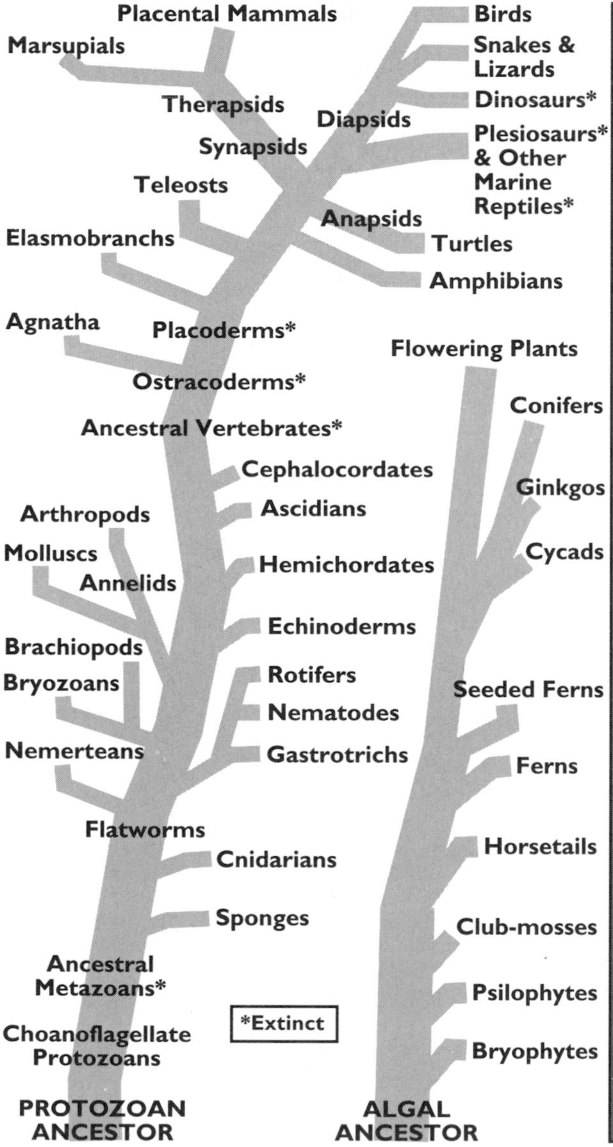

Placental Mammals
Marsupials
Therapsids
Synapsids
Diapsids
Birds
Snakes & Lizards
Dinosaurs*
Plesiosaurs* & Other Marine Reptiles*
Teleosts
Anapsids
Elasmobranchs
Turtles
Amphibians
Agnatha
Placoderms*
Ostracoderms*
Ancestral Vertebrates*
Cephalocordates
Ascidians
Hemichordates
Echinoderms
Rotifers
Nematodes
Gastrotrichs
Arthropods
Molluscs
Annelids
Brachiopods
Bryozoans
Nemerteans
Flatworms
Cnidarians
Sponges
Ancestral Metazoans*
Choanoflagellate Protozoans
PROTOZOAN ANCESTOR
*Extinct
Flowering Plants
Conifers
Ginkgos
Cycads
Seeded Ferns
Ferns
Horsetails
Club-mosses
Psilophytes
Bryophytes
ALGAL ANCESTOR

Era	Period	Began	Lasted	Development of Life
CENOZOIC	Quaternary	2	2	Development and spread of modern man. Marine life much as it is today
	Tertiary	65	63	Flowering plants dominant. Primates & hoofed mammals appear. Adaptation of whales & sea cows to marine life.
MESOZOIC	Cretaceous	136	70	Flowering plants appear. Mammals & birds become numerous. Fishes evolving. Reptiles still dominate seas.
	Jurassic	195	60	Appearance of primitive birds. Widespread coniferous forests. The age of the reptiles.
	Triassic	225	30	Deserts worldwide. Dominant reptiles. First mammals. First ichthyosaurs.
PALEOZOIC	Permian	280	55	Appearance of modern insects. Sea & freshwater life abundant.
	Carboniferous	345	65	First reptiles. Appearance of winged insects. Ferns & horsetails common.
	Devonian	395	50	Fishes abundant. First amphibians. Ancestors of all modern fishes evolve.
	Silurian	440	45	Sea scorpions & jawed fishes common. Seaweeds abundant. First land plants.
	Ordovician	500	60	Corals & trilobites common. All life restricted to water. First vertebrates appear.
	Cambrian	570	70	First abundant fossils. Sea urchins common. All life restricted to seas. All major invertebrate groups evolve.
	Precambrian	4600	4030	Earliest traces of life: algae & bacteria in warm seas.

Figure 10. a) Evolutionary tree of life b) Geological time scale and the evolution of life on earth.

The evidence that plant and animal life evolved by small but continuous changes in body form over many successive generations has been provided from the study of fossil remains (palaeontology). Fossils of plants and animals from successive geological eras show that a distinct progression of complexity has taken place by small but important mutations (fig.10a). These changes are related to the success of biological adaptations which are the result of changing environmental conditions on the Earth at that time. Further evidence for evolution has come from the study of the comparative anatomy of ancient and modern forms of plants and animals.

In 1858 the combined ideas of Charles Darwin (1809-82) and Alfred Russel Wallace (1823-1913) were published as the hypothesis *Origin of species by the process of evolution through natural selection.* They concluded that successful organisms survived at the expense of the unsuccessful ones. In addition, all new individuals begin life with small, but important differences so that no two individuals are exactly alike when produced sexually. In the battle for survival, the competition for scarce resources tends to eliminate the weakest in favour of stronger individuals or those that are most adaptable. It was also thought that 'evolution by natural selection' is essentially sensible. In a species (an organism that produces viable offspring) well adapted to a stable environment, most mutations are not favourable and the individuals carrying the mutations quickly die out. However, in a situation that favours a mutation, the organism carrying the altered genes is more likely to survive and pass on the favourable genetic changes to future generations.

Throughout the evolution of life there have been successive waves of dominance and extinctions of various plant and animal groups (fig.10b). Nobody really knows the reasons why some groups survived at the expense of others. However, one possible explanation is that sudden or large environmental changes do not give some organisms time to adjust to the new conditions and others that are more closely suited to the new conditions will take over. As biological evolution progressed through geological time, organisms not only became more diverse but also more complex. At many stages during the evolutionary process there must have been strong competition for space and resources between similar groups of organisms. Those best suited to the prevailing conditions survived. Today the arthropods and molluscs among the invertebrates and the fish, birds and mammals among the vertebrates (animals with a backbone), are some of the most successful groups on the planet in terms of species diversity.

CONTINENTAL DRIFT AND EVOLUTION

From their studies of the Earth's crust, scientists have distinguished three broad geological eras dating from the Precambrian times. These comprise the Palaeozoic (Greek for ancient life), Mesozoic (middle life) and Cenozoic (recent life) eras. Each of these eras is subdivided into units called periods; the Tertiary period, which spans some 65 million years, is further subdivided into epochs.

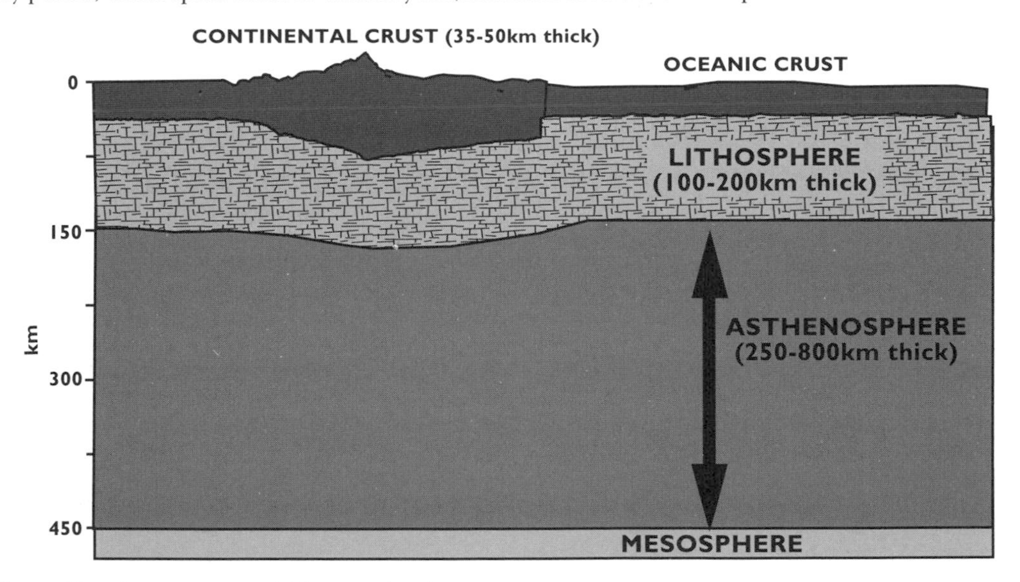

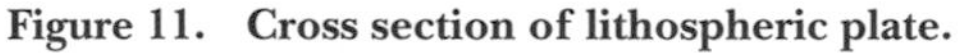
Figure 11. Cross section of lithospheric plate.

Fossils preserved in rocks provide an incomplete record of life on Earth and the climate. The rocks and fossils provide evidence that huge supercontinental shifts have taken place over the face of the planet. Distributions of plants and animals can readily be explained if continental areas which are now separated by deep oceans were once joined. Evidence to support this can be found in the Upper Carboniferous rocks from both sides of the Atlantic. Rocks from North America and Europe have yielded similar fossils which tend to support an ancient continental link. There is also a remarkably close match between the shapes of the Atlantic continental shelves of Africa and South America. In 1915 this led Alfred Wegener (1880-1930) to propose his 'match' theory of continental drift. However, it was not until the early 1970s that his theory was confirmed when geophysical events of sea-floor spreading were proved correct. More evidence for this hypothesis came from the Deep Sea Drilling Project of the *Glomar Challenger* (1968) which sampled the sediments and crusts beneath the oceans.

The theory of global plate tectonics explains the process by which continents drift apart or collide, and ocean basins are formed. It is now thought that the crust is made up of about 12 distinct plates which slip and move on the Earth's mantle. The hard continental blocks and ocean crust (lithosphere) are thought to float on the denser asthenosphere mantle beneath it. This concept is similar to ice floating and moving around on the surface of water (fig.11). Using this analogy, it is clear that some ice cubes move apart form each other (divergence) and others will collide (convergence). When continental plates spread apart (divergent boundaries) new materials are added to some of the plate margins through lines of volcanoes. One such line of volcanic activity is the mid-Atlantic Ridge which is causing the Atlantic Ocean to widen at a rate of about two to three centimetres per year. At other plate boundaries, continents come together (convergent boundaries) and material is lost as one continent slides under the other (fig.12). Many of the plate boundaries are regions of high volcanic activity and earthquakes.

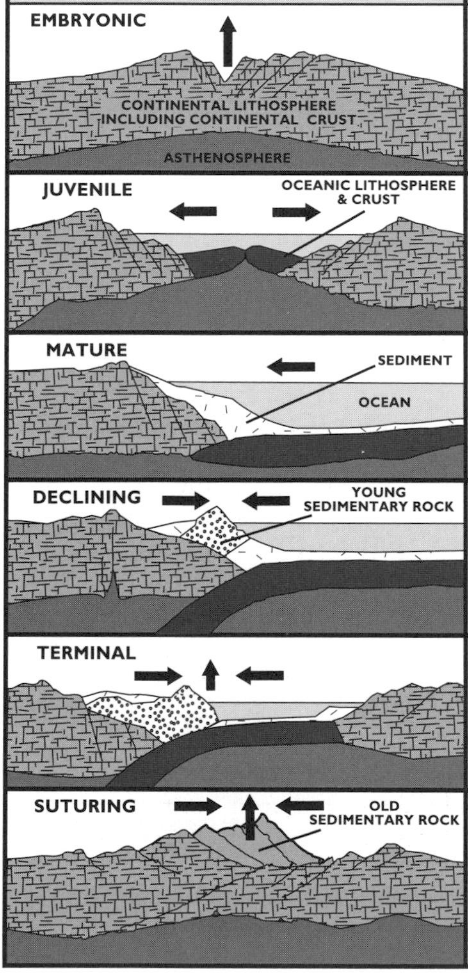

STAGE	MOTION	PHYSIOGRAPHY	EXAMPLE
EMBRYONIC (CONTINENTAL LITHOSPHERE INCLUDING CONTINENTAL CRUST; ASTHENOSPHERE)	Uplift	Complex system of linear rift valleys on continent	East African rift valleys
JUVENILE (OCEANIC LITHOSPHERE & CRUST)	Divergence (spreading)	Narrow seas with matching coast lines	Red Sea
MATURE (SEDIMENT; OCEAN)	Divergence (spreading)	Ocean basins with continental margins	Arctic Ocean Atlantic Ocean
DECLINING (YOUNG SEDIMENTARY ROCK)	Convergence (subduction)	Island arcs and trenches around basin edge	Pacific Ocean
TERMINAL	Convergence (collision) and uplift	Narrow, irregular seas with young mountains	Mediterranean Sea
SUTURING (OLD SEDIMENTARY ROCK)	Convergence and uplift	Young to mature mountain ranges	Himalayas

Figure 12. The Wilson's cycle of ocean basin evolution and the effect on associated landmasses (based on Pinet 1992).

FORMATION OF OCEAN BASINS

About 250mya it is believed that the surface of the Earth had only one large landmass, the supercontinent called Pangaea which was surrounded by a large ocean named Panthalassa. This ocean was the precursor of the Pacific Ocean of today. By the end of the Jurassic period (130mya) fossil records provide the first signs of a break up of Pangaea. The land mass divided itself into Laurasia to the north (which comprised North America, Europe and Asia) and Gondwanaland to the south (which comprised South America, Africa, India, Australia and Antarctica). Separating these northern and southern continents was the shallow Sea of Tethys. Although the north–south division of the two continents continued, an east–west movement was also taking place at the same time. These movements have continued to the present day creating the layout of the continents and oceans as we know them (fig.13).

The break up of Pangaea had several important consequences. Global temperatures would have fluctuated as the continents moved round relative to the angle of the sun. Also, the geographical areas of coastlines would have increased with the formation of continental shelves around the margins of the continents. These fringing coastal waters provided a more stable, conducive environment for the evolution of species. Separation of the continents may have promoted faunal (animal) and floral (plant) isolation where different populations could adapt in their own way to the new habitats and environmental conditions. Such isolation would eventually lead to the evolution of new species. By the middle of the late Jurassic era there were two well-defined faunal realms, and today over 30 marine provinces are recognised in the world's oceans.

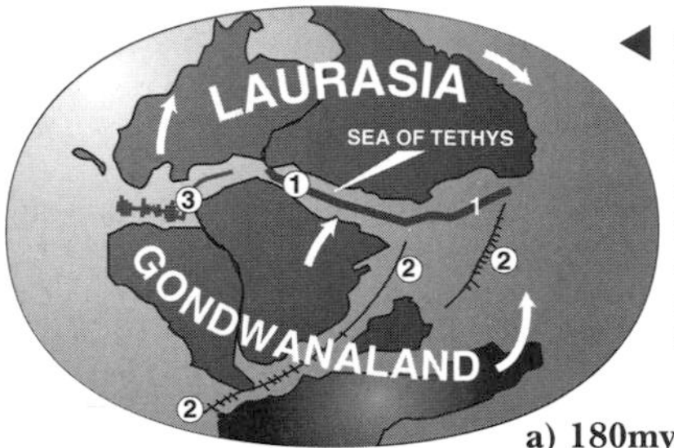

◀ 200 million years ago there was one large land mass, the supercontinent Pangaea. Shown here is the theoretical first break-up of the land mass about 180 million years ago. This formed the new land masses of Laurasia and Gondwanaland which were separated by the Sea of Tethys. The surrounding ocean was named Panthalassa. Over the next 140 million years this fragmentation continued with Gondwanaland and Laurasia subdividing, establishing the major continents as we know them today.

a) 180mya

At about 65-60 million years ago the eastern and western continents drifted northwards splitting Greenland from North America, while India collided with Asia to form the Himalayas. ▶

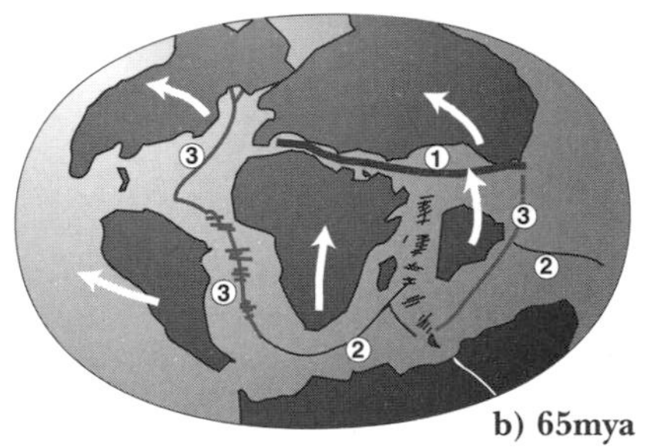

b) 65mya

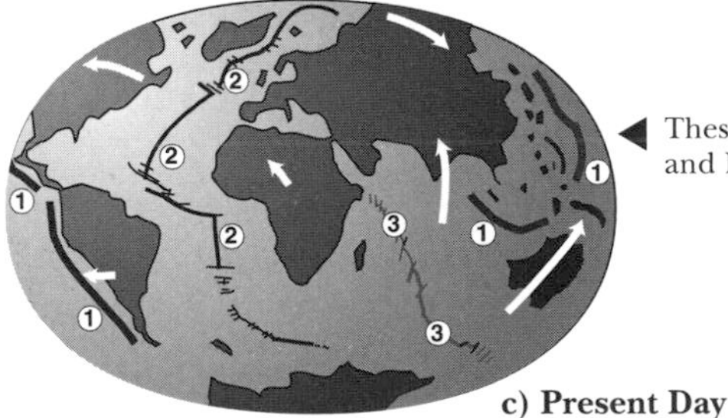

◀ These movements continue to the present day, with Australia, India and North America moving at a faster rate.

c) Present Day

KEY
1. Trenches
2. Mid-ocean ridges
3. Transverse faults
⇨ Direction of drift

Figure 13. Pangaea and the shifting of the continents.

THE PHYSICAL AND CHEMICAL NATURE OF OCEANS

For a full appreciation of marine ecology it is useful to have some understanding of the physical and chemical constraints of the marine environment. Some of the constraints may be imposed by the fluid nature of the seawater and the chemicals dissolved in it. Chemical variables which are known to affect marine organisms include salinity, nutrients, dissolved gases and the pH (acidity). Other environmental features affecting organisms include sunlight, temperature, air and water movements. These factors are controlled and altered as a result of the dynamic interaction between the sea surface and the Earth's atmosphere.

SUNLIGHT

Solar radiation, or sunlight is essential to life in the seas and on land. As the radiation enters the Earth's atmosphere only about 50% of it will reach the ocean surface. The rest is scattered, absorbed or is reflected back into space by the clouds. The remainder of the radiation striking the surface of the ocean is either absorbed or is reflected back into the atmosphere. The amount of reflection will be determined by the angle of the incoming rays, the time of the year, the time of day and the latitude. At the equator the total daily radiation is fairly constant throughout the year. However, at 50° latitude distinct summer and winter variations will occur. Above the Polar Circles there is a period of continuous daylight during the summer and a period of continuous darkness in winter.

Sunlight that does manage to enter the ocean will be scattered and absorbed. The different wavelengths of light in the visible spectrum will be absorbed at different depths in the water column. Red light is quickly absorbed; only one percent remains at a depth of 10m in the clearest ocean waters. Blue light penetrates the furthest and only one percent of this remains at 150m in clear water. Therefore to the human eye, most things appear to be a blue colour underwater. In the turbid coastal waters and river estuaries, high levels of suspended sediment or organic matter will greatly reduce the depth to which any light can penetrate. Under these conditions it is possible for most of the light to be absorbed in the first few metres, while at the other extreme the clearest ocean waters may still have visible light down to a depth of 1,000m. However, any animals that can detect these minimal light levels are extremely sensitive to it.

Sunlight affects marine plants in several different ways. Some fraction of the solar radiation which penetrates the surface waters is absorbed by plants (e.g. microscopic algae and large seaweeds). The sun's energy is used in the production of organic material from inorganic substances by plants using photosynthesis. The collective mass of all the new organic material produced by the plants is called 'primary production' and plants are therefore the primary producers on this planet. Most forms of life depend, directly or indirectly, on organic plant material as a food source. Light intensity also controls the depth to which plants can survive and produce enough new organic material to sustain themselves. The 'euphotic zone' describes the region in which there is sufficient light to support the growth and reproduction of new plant tissue by photosynthesis. The net gain or newly produced organic material by the plants, must equal or exceed that lost by the plants through tissue respiration. The bottom of the euphotic zone (about 150m) is the depth at which production gain equals respiratory loss and below this point a plant will eventually die. The vertical extent of the euphotic zone will vary according to the turbidity of the water. Below the euphotic zone is the 'disphotic zone' or the 'dim light zone' (bottom limit about 1,000m). The deepest regions of the open oceans form the 'aphotic zone' and this extends down to the sea-floor (fig.23). At these depths no light can penetrate and the only available light is produced by animals themselves (bioluminescence or cold light).

The energy of sunlight can also be absorbed by the water and converted into heat, therefore raising the temperature of the seawater. Temperature plays an important role as it helps drive some of the water movements of the oceans and is responsible for the global distribution of many marine organisms. Global temperature changes of surface water are generally gradual. However, it is often useful to describe regions according to the temperature regimes that prevail e.g. tropical, temperate and polar regions. Light intensity also has an indirect effect on the distribution of animal life within the water column, e.g. where there is light there is plant life, and where there is plant life there will be animal life which feeds upon it (herbivores). Animal distribution is therefore often dependent on the distribution of plants, and this in turn is controlled by the light levels at which photosynthesis can occur. Herbivores in the water column actively seek the depth of maximum plant abundance for consumption. The vision of many animals is also dependent on visible light and this may control certain behavioural and biological rhythms such as migration and breeding cycles. The control factor may be something as simple as light intensity or periodic light changes such as day length.

SALINITY

Salinity refers to the salt (sodium chloride) content of seawater, and in oceanic water salinity is about 35 parts of salt per 1,000 parts of seawater (35 ‰). Although sodium chloride (NaCl) makes up 85% by weight of the salts there are other minor constituents called nutrient salts which are vital for life (Table 1).

Components	(g per kg)	%
Chloride	19.0	55.0
Sodium	10.6	30.6
Sulphate	2.7	7.7
Magnesium	1.3	3.7
Calcium + Potassium + others	1.0	3.0

Table 1. The chemical constituents of seawater by approximate weight.

Variation in the salinity of ocean water is linked primarily to climatic conditions. The salinity of surface water is increased by the removal of water through evaporation. It decreases through dilution with the addition of freshwater in the form of rain which can pour directly onto the sea or flow into it from the land. At higher polar latitudes the salinity is also reduced by ice and melting snow. In the open oceans, salinity readings tend to be higher at lower latitudes where values are linked to the rate of evaporation and precipitation. The highest salinity values tend to occur between latitudes 20 and 30° north and south of the equator where there is least rain and the highest evaporation rate (fig.14). Table 2 shows typical values of salinity for various environmental regions.

Environmental type	Salinity (‰)
Open oceans	32-38 (mean 35)
Shallow coastal areas	27-30 (brackish)
Estuaries	0-30
Semi-enclosed seas (e.g. Baltic Sea)	25 or less
Hypersaline environments (e.g. rock pools, tropical lagoons, Red Sea)	40 or more

Table 2. Typical salinity values for various environmental regions.

Salinity variations are at their highest in the surface layer of oceanic water. At depths greater than 1,000m there is little or no variation and readings remain remarkably stable at about 34.5 to 35.0 ‰ in all regions of the world.

The salinity of water and the dissolved nutrient salt concentrations have profound effects on all marine living organisms. The primary productivity of the oceans depends on the quantities of nutrients that are available to the living plants. Coastal waters and estuaries generally have higher concentrations of nutrient nitrates and phosphates, which are washed into the seas by the freshwater run-off from the land. During storm periods nutrient levels of the seawater are high due to mixing by wind and waves. In temperate latitudes, high nutrient levels and increasing water temperatures encourage the planktonic plants to flourish. Coastal algal blooms first appear in the spring and the water becomes greenish-brown in colour. As nutrients are used up, the bloom subsides but it can reappear in the late summer when autumn storms once again replenish the nutrients by mixing the deeper water layers. The second algal bloom diminishes with the oncoming winter as light levels and temperatures fall. The rise and fall of algal blooms is closely followed by the rise and fall of zooplankton numbers (tiny animal plankton) which feed on the algae (see page 56).

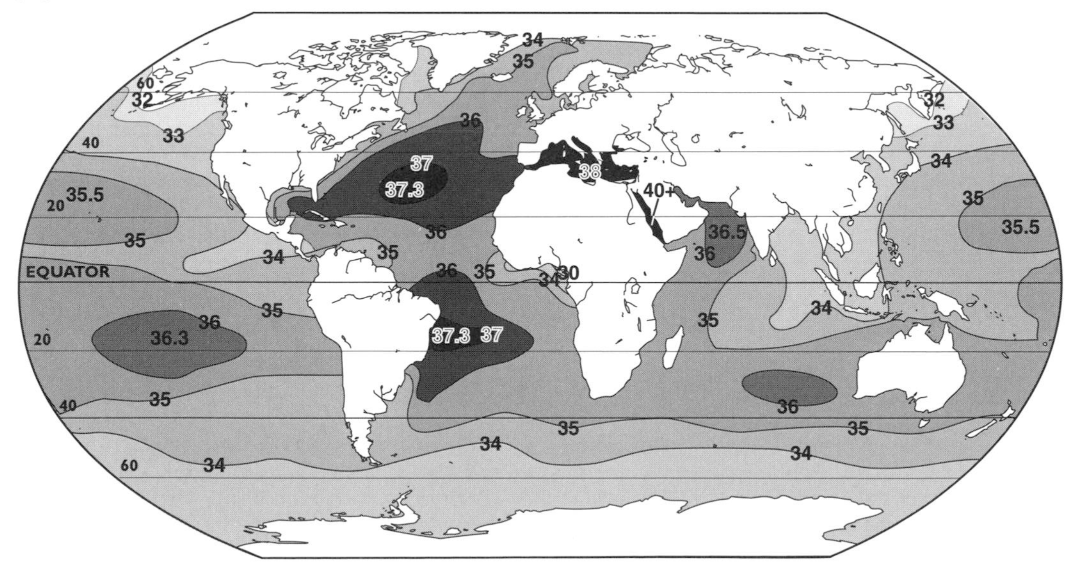

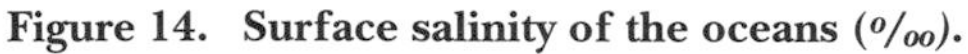
Figure 14. Surface salinity of the oceans (‰).

Seawater salinity also has a profound effect on the concentration of salts in the tissues and body fluids of organisms. Slight shifts of salt concentrations in the bodies of animals can have stressful or even fatal consequences. Therefore animals have either evolved mechanisms to control body salt levels, or they let them rise and fall with the levels of the seawater around them. In most marine invertebrates the salt content of the body fluids is held at about the same level as the surrounding water. However, the bony fishes (teleosts) maintain body salt concentrations at about 30-50% of the seawater's concentration. Water tends to move from areas of low salt concentration to areas of high salt concentration in order to equalise the salt concentrations by the process of osmosis. Bony fishes are continually getting rid of salt by pumping it out through the gill membranes. The seabirds, turtles and marine reptiles, which have terrestrial ancestry, also have body fluids which are less concentrated than seawater but, unlike humans,these groups can drink seawater without upsetting their body metabolism. To do this these animal groups have special salt excretory glands which can produce a five percent salt solution that is twice as salty as seawater. The salt is removed in different ways from different groups of animals. In marine iguanas and seabirds the salt is excreted from the nostrils while marine turtles 'cry' salty tears. Some seabirds have channels in their bills to help the salty fluid drain away. The 'tubenose' bird species (petrels and shearwaters)that spend most of their time flying, will forcibly eject the salt from the nostrils.

Animals that inhabit areas of rapidly changing salinity such as coastal regions and estuaries have a constant 'osmotic' (diffusion) balance to maintain. Those species which migrate between freshwater and seawater environments (e.g. salmon, sea trout and eels) have to spend an acclimatisation period in brackish water areas such as estuaries. Animal species that can tolerate a wide range of salinity are termed 'euryhaline' and those that can only tolerate a narrow salinity band are known as 'stenohaline' organisms.

Salinity has another indirect effect on animals because of its capacity to depress the freezing point of water to a lower temperature. In polar regions seawater freezes at -1.9°C instead of the normal 0°C for freshwater. Creatures living in this supercooled environment may freeze if a part of their body comes into contact with floating ice. However, even this problem has been overcome by some marine animals (e.g. some bony fishes) which produce blood and body fluids with an antifreeze additive or supercool the body fluids to the temperature of the surrounding water.

DISSOLVED GASES

Oxygen (O_2) is not readily soluble in seawater and the level of solubility of gases will decrease with rising water temperatures and higher salinities (Table 3). The amount of oxygen present in seawater will vary with the rate of production by plants, the consumption by animals and plants, bacterial decomposition and degree of mixing at the air-sea interface by wind and wave actions. Surface water which sinks to deeper levels will take with it a store of oxygen that will sustain animal life even in the deepest oceans.

Factors	Effects
Wave and current turbulence	Increases the exchange of seawater gases with the atmosphere.
Difference in gas concentration	Gases diffuse across the air-sea interface from high to low areas of concentration until chemical equilibrium is attained.
Temperature	A drop in water temperature increases the solubility of gases.
Salinity	A rise in salinity increases the solubility of gases.
Pressure	A rise in pressure increases the solubility of gases.
Photosynthesis	Increases concentration of O_2; decreases concentration of CO_2.
Respiration	Increases concentration of CO_2; decreases concentration of O_2.
Decomposition	Increases concentration of CO_2; decreases concentration of O_2.
pH	Controls the relative concentrations of the CO_2 in water

Table 3. Factors regulating concentration of gases in water (based on Pinet 1992).

Carbon dioxide (CO_2) is an essential gas which is required by plants for photosynthetic production of new organic matter (see page 54). Carbon dioxide is 60 times more concentrated in seawater that it is in the atmosphere. Within seawater it forms carbonic acid (H_2CO_3), bicarbonate (HCO_3), carbonate (CO_3) and also dissolved CO_2. Marine plants use bicarbonate to supply their CO_2 needs as dissolved quantities of the gas are not sufficient to supply this fully. Regions of the Earth with low water temperatures have higher levels of dissolved gas in a given volume of water compared to seawater of tropical regions. The Antarctic Icefish (*Chaenocephalus aceratus*) has lost its red blood cells which normally carry the oxygen around the body. Instead it leads a sluggish lifestyle in oxygen-rich water and oxygen is taken in through the skin. The fish also carries small amounts of dissolved oxygen in the plasma fluid. Higher levels of CO_2 also fuel massive phytoplankton blooms in polar regions at certain times of the year.

ACIDITY (pH)

Acidity or pH is a measure of the concentration of hydrogen ions in the water. pH has a scale of 1 to 14, and pH7 is the neutral value. Values either side of neutral means either an increase in acidity (moving towards pH1) or an increase in alkalinity (moving towards pH14). Surface seawater often has a pH between 8.1 and 8.3, but generally the acidity of ocean water is remarkably stable with a neutral pH. In shallow seas and coastal areas with slow flushing times, the pH can be altered by plant and animal activities, and by pollution from the terrestrial environments.

WATER PRESSURE

Hydrostatic pressure is the relationship between a given depth and the weight of the overlying water column per unit area. The relationship between depth and pressure is essentially a linear one; a given volume of gas is compressed to half its volume for every 10m increase in depth. Therefore at a depth of 30m the volume of gas will be a quarter of its surface value. The body tissues of organisms consist mostly of liquid or solid material which is virtually incompressible by water. However, many fishes and cephalopods adjust their buoyancy by filling a body cavity with gas. This is similar to inflating a life-jacket to keep a person afloat. Teleost fishes use a 'swim-bladder' to achieve neutral buoyancy while the cephalopod *Nautilus* uses buoyancy chambers inside its shell. As the animals ascend or descend in the water column the buoyancy must be continually adjusted, as a rapid ascent to the surface may cause over-inflation and damage. Some of the air-breathing animals that inhabit the marine environment can dive to great depths in search of food. However, whales, penguins, seals and marine otters rely on the natural buoyancy of fatty tissue, fur or feathers to keep them afloat. Unlike humans, they have evolved a variety of physiological mechanisms to prevent gas bubbles appearing in the body as they rise swiftly to the surface. To the human scuba diver a rapid return to the surface from depths greater than 10m can cause physiological damage, paralysis and in the worst cases, death from the 'bends' as gas bubbles expand to cause tissue damage.

DENSITY

Density (mass per unit volume) of seawater is altered mostly by changes in temperature and salinity. As salinity increases, the density will increase and as temperature increases the density will decrease. It is evident to anyone who can swim that the human body floats slightly higher in seawater than it does in freshwater. This is taken to extreme in the highly saline Dead Sea where it is difficult to submerge a swimming body.

Water of uniform density can be thought of as a body of water which moves together as a total mass. A water mass with uniform density, salinity and temperature will demonstrate different movement patterns as it comes into contact with other water masses of different densities. Water masses that are less dense will float over water masses of higher density and they will therefore probably remain at the surface. If a surface water mass increases its density through falling temperatures or increased salinity, it may sink to a different depth. The oceans can be thought of as having a three-layered system of water masses, i.e. the surface layer (0-550m), an intermediate layer (550-1,500m) and a deep-water layer (below 1,500m to the sea floor).

SEA TEMPERATURE

Water temperature is one of the most important physical factors of the marine environment and it has a major impact on many biological, chemical and physical processes. Temperature controls the rate at which many chemical reactions and biological processes occur. Density of seawater is determined by a combination of temperature and salinity and it is density which is the major driving force of vertical water movements. Water temperature partly affects the amount of gases which can be dissolved in water. Water temperature is also the single most important factor influencing the biogeographical distribution of marine species.

The surface temperature of the sea is influenced by the amount of solar radiation penetrating the water. The level of sunlight it receives will vary according to day length and cloud cover. However, the latitude and time of year governs the angle of the sun and therefore the strength of the sun's rays. Surface sea water temperature is highest at the equator where it can sometimes reach 40°C in the shallow lagoons of tropical reefs. At the other extreme the temperature of the seawater can fall as low as -1.9°C in the high polar regions. The heating effect of sunlight is generally confined to the immediate surface of the ocean with 98% of the infrared radiation being absorbed in the first metre of water. However, water has the physical property of absorbing large amounts of heat with little change to its temperature. The large volumes of seawater and evaporation from the surface will also help to reduce water temperatures, and therefore ocean water tends to heat up and cool down slowly. For most marine life this slow change is vital as they need to acclimatise to the range of temperatures to which they are exposed. The vast majority of organisms will not survive rapid changes or abnormal temperature conditions. Many of the life processes of living organisms such as the rate of respiration, rate of growth, feeding and behaviour

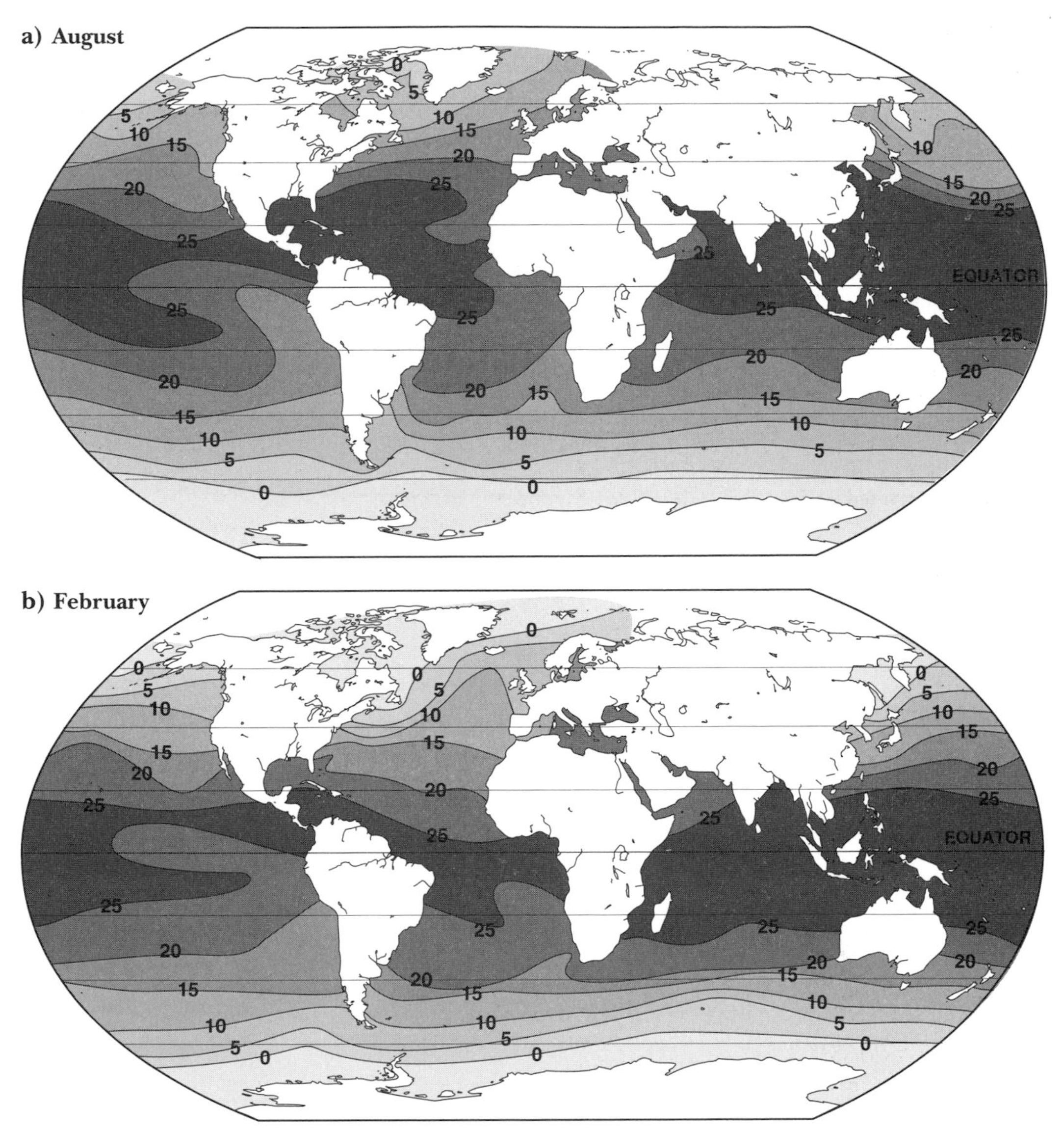

Figure 15. Sea surface temperatures (°C) for August and February.

are governed by fluctuations in temperature. Animals that cannot regulate their body temperature or life processes are known as cold-blooded (poikilothermic) and their body temperature follows the fluctuations of environmental temperatures. Warm-blooded (homeothermic) animals can regulate their body temperatures and maintain them at a relatively constant level. Warm-blooded marine mammals and birds are therefore less rigidly affected by sea or air temperatures and can migrate across the oceans, encountering large ranges of temperatures. To describe global animal and plant distribution, it is often useful to designate biogeographical zones based on average surface seawater temperatures:

Polar	-1.9°C to +2.0°C	(southern oceans)
	-1.9°C to +5.0°C	(Arctic waters)
Cold temperate	5.0°C to 10.0°C	(northern hemisphere)
	2.0°C to 10.0°C	(southern hemisphere)
Warm temperate	10.0°C to 15.0°C	(both hemispheres)
Subtropical	15.0°C to 25.0°C	(both hemispheres)
Tropical	25.0°C or greater	(both hemispheres)

In the tropical zones of both hemispheres the masses of water are characterised by a mixture of subtropical and subpolar waters which show seasonal variations in temperature. In temperate ocean environments the surface isotherms (lines of equal temperature) are therefore of little ecological significance due to frequent water mixing. The average daily temperature fluctuations of oceanic surface waters is generally less than 0.3°C (fig.15).

Even in coastal waters the fluctuation is unlikely to be more then 2.0°C over a 24 hour period, and it is therefore of little significance to nearshore marine communities. However, within the intertidal areas and rock pools of the seashores, large temperature changes can occur in a short period of time due to air exposure. Organisms found higher up the beach are exposed for longer periods of time and therefore have to endure more adverse conditions. Animals and plants that inhabit the intertidal zones have to be well adapted to cope with the enormous biological stresses placed upon them by the fluctuations of temperature, salinity and pH.

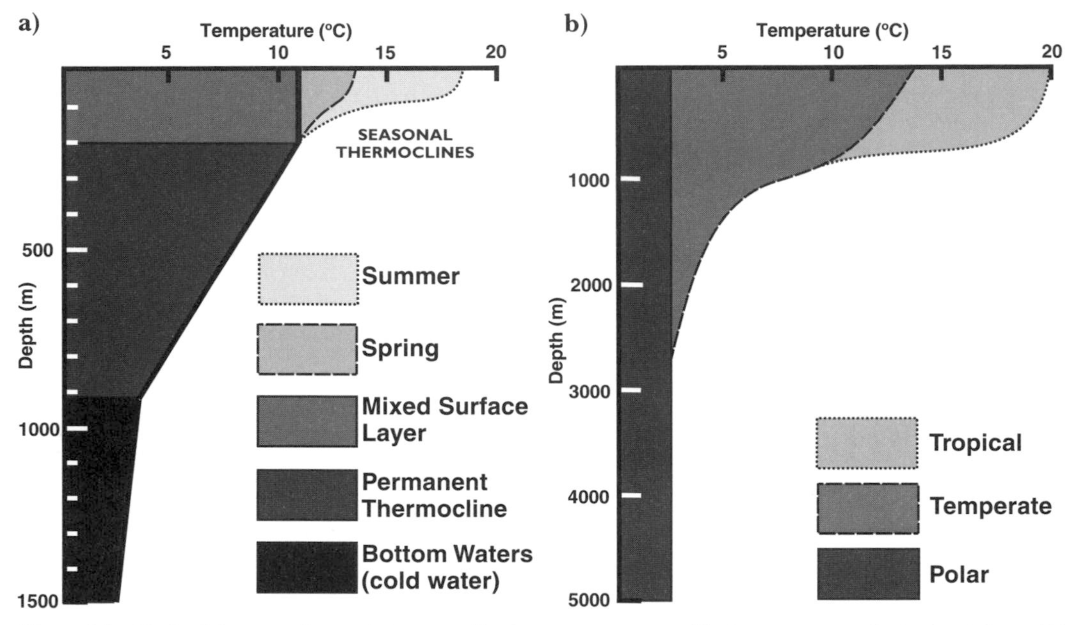

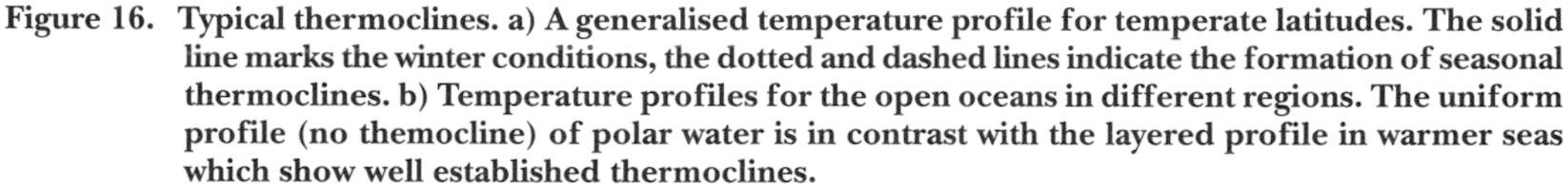

Figure 16. Typical thermoclines. a) A generalised temperature profile for temperate latitudes. The solid line marks the winter conditions, the dotted and dashed lines indicate the formation of seasonal thermoclines. b) Temperature profiles for the open oceans in different regions. The uniform profile (no themocline) of polar water is in contrast with the layered profile in warmer seas which show well established thermoclines.

Temperature gradients are also to be found with increasing depth, and temperature profiles reveal oceans to have a 'layered' thermal structure. In tropical and subtropical regions the warmer, lighter water floats above the cold, more dense water. This warm surface layer can be up to 500m in depth. The warm and cold layers of water are separated by a thin narrow band of stable water called a 'thermocline', which has a steep temperature gradient (fig.16a). In temperate and tropical oceans the thermocline is a permanent hydrological feature and occurs at depths from 200-1,000m. In the mid-latitudes a summer thermocline may be established in the first 100m of the surface water but it disappears due to mixing with the start of the autumn storms and gales. By their very nature, the cold waters and turbulent seas of the high polar regions do not develop a thermocline at any time, and there is little temperature variation between surface and deep water masses. Below the permanent thermocline in all other regions, the water temperature is remarkably uniform and stable with a temperature of about 4°C.

Thermoclines act as depth barriers to many plants and animals and they often mark the boundaries between hospitable and inhospitable water masses. A situation often arises where the nutrients from the surface layer are quickly depleted by the plants for photosynthesis, while the nutrients on the other side of the thermocline are abundant but unavailable to them. Therefore thermoclines control the availability of nutrients to the phytoplankton.

Water Movements and Oceanic Circulation Patterns

ATMOSPHERIC CIRCULATION AND SURFACE CURRENTS

The major driving force behind the movements of surface seawater is the wind. The patterns of atmospheric circulation reflect an uneven distribution of global temperature, which is the result of unequal solar heating. High levels of solar radiation at the equator produce hot air which rises as its density decreases and this creates a belt of 'equatorial low pressure' (fig.17). As the air rises, it cools, loses much of its moisture content and then descends as dry air in the subtropical regions. On its descent it will once more be warmed as it is compressed by the pressure of the overlying air. The relative high density of this air mass creates the subtropical belt of high pressure which is centred between latitudes of 30° north and south of the equator. This sinking column of air converges and then splits at the Earth's surface, with one part of the air stream moving towards the pole and the other part moving towards the equator. In both cases the air undergoes deflection because of the rotation of the planet (Coriolis effect) and this creates the familiar Trade winds and Westerlies of the northern and southern hemispheres. Eventually the Westerlies are pushed upwards along the sub-polar, low pressure belt and over the colder, more dense polar air. The Trade winds converge on the belt of equatorial low pressure, thus completing the circulation cycle. Therefore, the combined atmospheric actions result in a regular flow pattern of global winds which vary in strength according to their latitude.

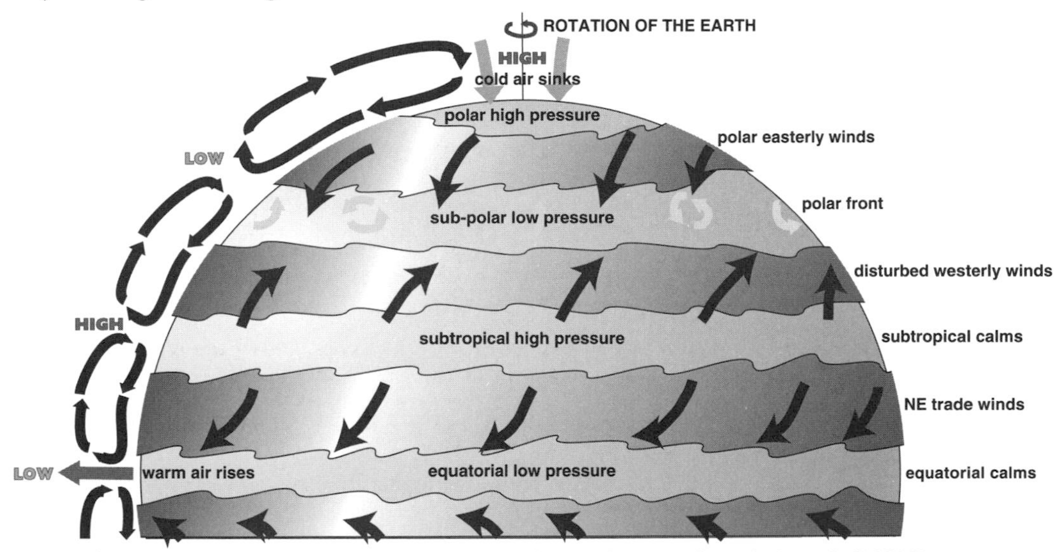

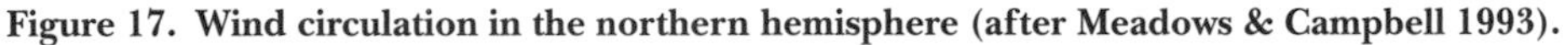

Figure 17. Wind circulation in the northern hemisphere (after Meadows & Campbell 1993).

In the northern hemisphere the prevailing winds lie between latitudes 10° and 30° north (NE Trade winds); between 60° and 90° north (Polar Easterlies); and between 30° and 60° north (Westerlies). The wind flow patterns in the southern hemisphere are a mirror image of those in the northern hemisphere. The prevailing winds set the surface layer of ocean waters moving and obstructions such as continental margins will deflect the water flow. The major surface currents of the oceans, which lie between the 60° lines of latitude, form large pairs of oval circulation patterns called 'gyres'. These ocean gyres with their associated water currents flow in a clockwise direction in the northern hemisphere. As a result of the prevailing winds and the Earth's rotation, surface currents flow at an angle of 45° of the angle of the wind. The speed of a water current decreases with increasing depth and direction of flow shifts in a spiralling fashion (Ekman's Spiral) until a certain depth when it travels in the opposite direction to the surface water. This spiral effect produces a net transport of water 90° to the right of the wind in the northern hemisphere and 90° to the left of the wind in the southern hemisphere. Therefore it is evident that the major prevailing wind systems produce the large oceanic gyres. In the southern oceans the circumpolar current circulating around the Antarctic continent is driven by westerly winds (The West Wind Drift). This forms a moving ring of water around the Earth between latitudes 45° and 60° south. There is no equivalent Circumpolar current in the northern hemisphere because land masses modify the flow patterns.

In all the major oceanic gyres of the world the water currents are narrower, deeper and faster along the western margins of the oceans (or eastern edges of land masses) compared with those along the western coastal margins. For example, the Gulf Stream currents have a flow rate ten times faster on the western boundary compared to its eastern boundary. Where adjacent gyre boundaries meet, large volumes of water will mix together with other bodies of water and their characteristics and composition will change. It is well known that the joining and mixing of cold, low salinity Labrador water with warm, high salinity Gulf Stream water produces the notorious fog banks off Newfoundland. However, it is not the difference between the water masses, but the temperature differences and moisture content of the air associated with the water masses that cause the fogs and mists. If the Gulf Stream current did not bring warm water from the Caribbean region, the seas and the climate of Western Europe would be much cooler. Within the centres of many of the ocean gyres are areas of relative calm where there is little or no circulation of water. A good example of this can be found in the North Atlantic gyre, where the stagnant central area is known as the Sargasso Sea, in which the familiar sargassum weed floats. These floating weed beds often have their own animal communities living in or under them. When ocean currents and wind directions are being described it should be remembered that ocean currents are described by the direction in which they are going or flowing to. In contrast, winds are described by the direction in which they are blowing from.

SUBSURFACE CURRENTS

The wind-driven surface currents influence only about 10% of the total volume of oceanic water. The remaining water is influenced by ocean currents forming well below the surface layer. They are driven by density differences between the water masses, which are dependent on temperature and salinity. The subsurface currents are therefore described as being driven by 'thermohaline' means. This type of circulation also produces a significant vertical component which allows thorough mixing within the deeper water masses.

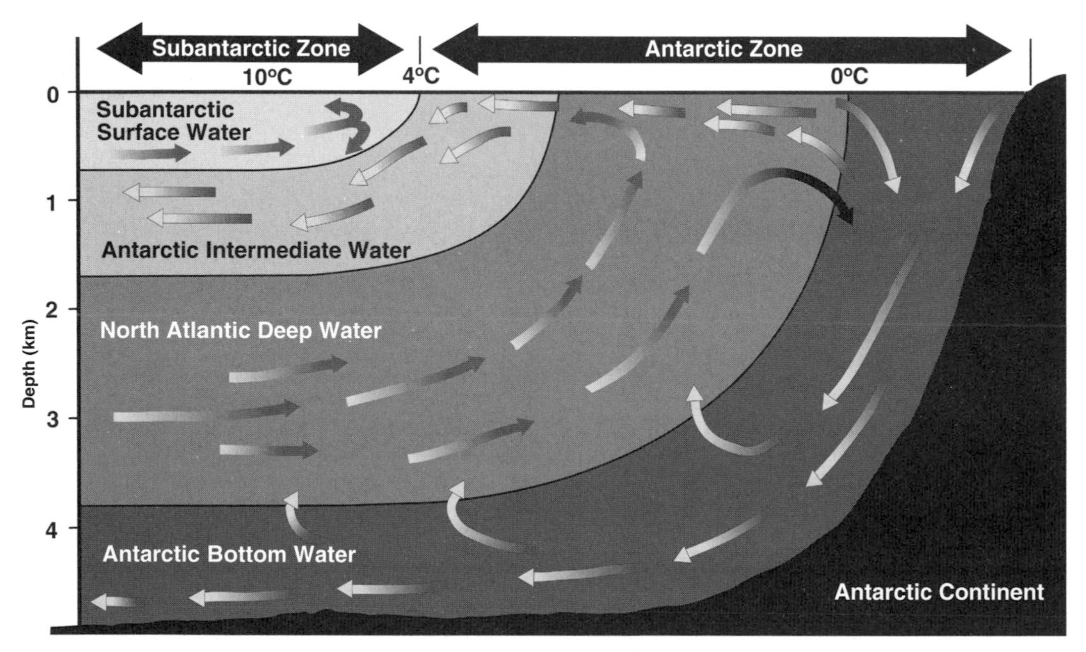

Figure 18. Antarctic subsurface water masses showing mixing patterns (based on Thurman 1994).

The deepest and most dense water masses originate mostly around Antarctica, Greenland and Iceland. During the winter months the seawater in these regions becomes dense and starts to sink. As it sinks, it flows towards the equator at a deeper level (fig.18). Antarctic Bottom Water can penetrate well into the northern parts of the Atlantic and Pacific oceans. From the Antarctic Convergence the low salinity Antarctic Intermediate Water sinks to a depth of about 900m. Sandwiched between the Antarctic Bottom Water and this intermediate layer is the North Atlantic Deep Water, which is rich in nutrients after hundreds of years in the deep ocean. The formation of deep-water masses is not pronounced in the Pacific Ocean and there seems to be no deep-sinking of surface waters. The Indian Ocean circulation system is heavily dependent upon atmospheric weather conditions. These include the 'monsoon' wind systems that blow from the northeast in the winter and from the southwest in the summer.

Deep water from the various oceans eventually returns to the surface to replace water that has moved away through current circulation. Vertical movements of deep water can be caused by the shape of the ocean floor or obstructions on the seabed. Slopes and rises of continental shelves will push the water upwards towards the surface or into other layers above it. This upwelling of deep ocean water will return nutrients such as nitrates and phosphates to the euphotic zone, were they can once more be utilised by plants to produce organic matter. Within the top surface layer of water (0-500m) there may also be an independent vertical circulation of water. Downwelling movements will transport oxygen-rich surface water to a lower depth in the surface layer. On its return journey it will bring nutrients towards the surface of the sea. This vertical circulation of water within the surface layer is probably caused by a combination of wind and proximity to coastlines. Just as upwellings can be caused by water masses moving apart from each other, downwellings can occur when water masses converge. In a similar way, where persistent winds blow water away from the coastline, water upwells to make good the loss at the coast (fig.19). The water may rise from depths of up to 500m in some cases, but if surface water is blown onto a coast it will downwell instead. The Trade Winds coupled with the Coriolis deflection will therefore cause areas of upwelling along western coasts of continents and downwellings along the eastern margins. Upwelling areas are important to primary production and the associated food chains, and they will be discussed further (see page 55).

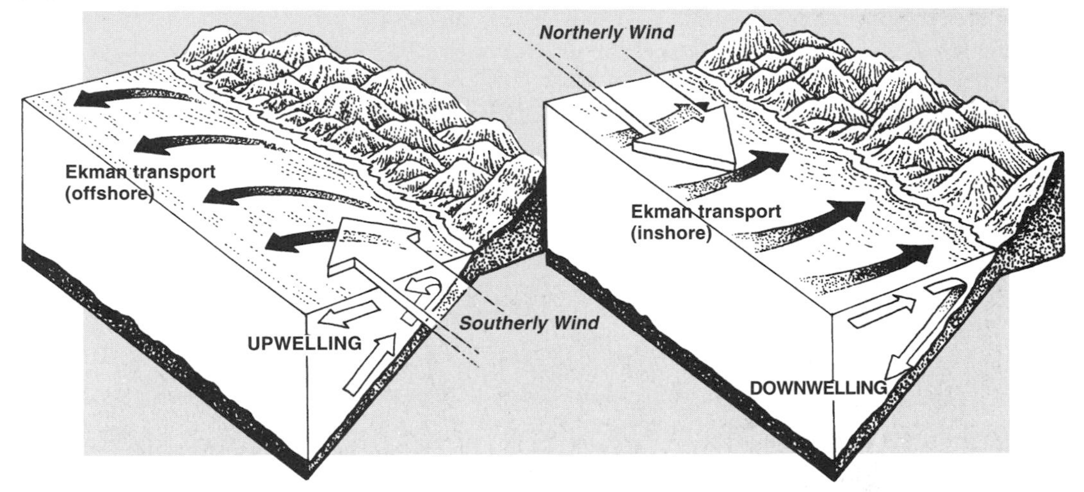

Figure 19. Coastal upwelling and downwelling in the southern hemisphere (after Thurman 1994).

TIDAL MOVEMENTS

Tidal effects are most easily seen at the coastal shorelines where the sea levels rise and fall in a regular pattern, usually twice daily (semidiurnal tides). However, because of certain physical conditions there may only be one daily tide (diurnal tide) in some areas of the world e.g. Gulf of Mexico. Tides can have a vertical range as little as one metre (e.g. Mediterranean and Jamaica) or as much as 15m (Bay of Fundy, Canada). Tides are caused by the interaction between the gravitational attraction of the Earth, the Sun and the Moon, plus the centrifugal forces resulting from the rotations of the Moon and Earth. The tide-generating force of the Sun is only about one half of that of the Moon due to relative size and distance from the Earth.

Tidal ranges are greatest during 'spring' tides. These occur twice each month when the Earth, Moon and Sun are in line and the gravitational fields of the sun and moon pull together. At the other extreme, tidal range is at its smallest during 'neap' tides; these occur when the gravitational force of the sun is acting at right angles to the moon (fig.20). In this alignment the two forces have a reduced effect on tidal variations. The 'high water mark' is the greatest height to which the tide rises on any day and the 'low water mark' refers to the lowest point to which the tide drops on any day. At any one point on a coastline the period between semidiurnal tides is 12 hours and 25 minutes and this is equivalent to half the time taken by the Earth to turn on its axis. This suggests that the time between successive high and low tides is about 12.4 hours. An incoming tide is referred to as a 'flooding' or 'flood' tide and one which is receding or going out is known as the 'ebbing' or 'ebb' tide. Spring and neap tides occur twice every lunar cycle, as it takes 28 days for the moon to revolve once around the Earth (fig.21). The influence of the sun's gravitational field is greatest at the time of the spring and autumn equinoxes when the sun is directly over the equator of the Earth. During these months the spring tides are at their greatest range.

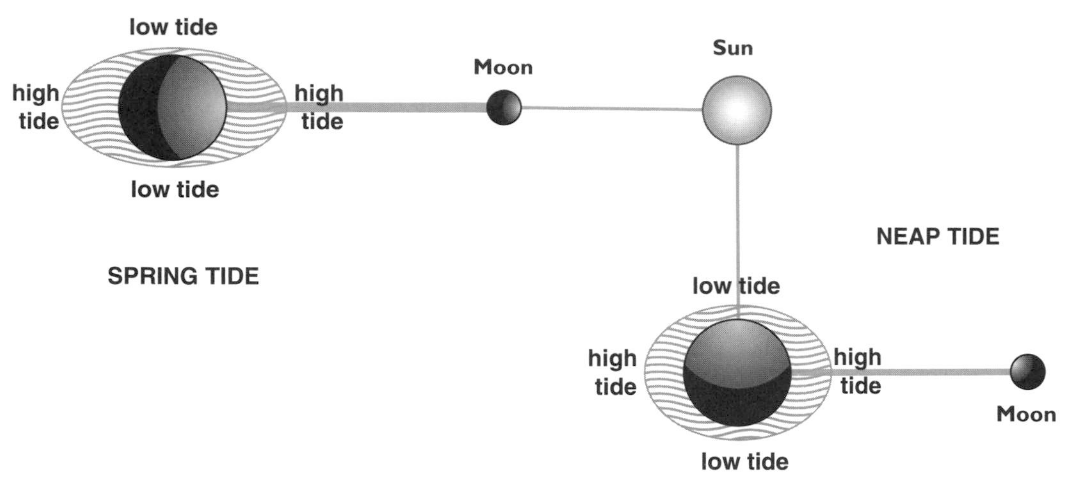

Figure 20. Positions of Sun, Moon and Earth in relation to the tides.

The timing of high and low water will vary from place to place and this variation is influenced by shapes of coastlines, and the presence of estuaries, channels and islands. Weather can also have an effect on the size and timing of a tide. Long periods of strong, prevailing winds can pile up the seas onto a coast. An onshore wind will therefore effectively push the tidal water higher up the shore, while an offshore wind will have the opposite effect. High and low pressure weather systems may also have an influence on the size of the tide. A low pressure system will try to suck up the seawater below it, while a high pressure system will push the water level down. A low pressure weather system can affect the height of a tide by as much as 5m. The most damaging time for a coastal area occurs when low pressure storms coincide with high spring and autumn equinox tides. The time of day at which high and low waters occur are fairly constant at any given location along a coastline. Therefore tidal predictions are fairly accurate and easily calculated for normal weather conditions. These are published as tidal prediction tables.

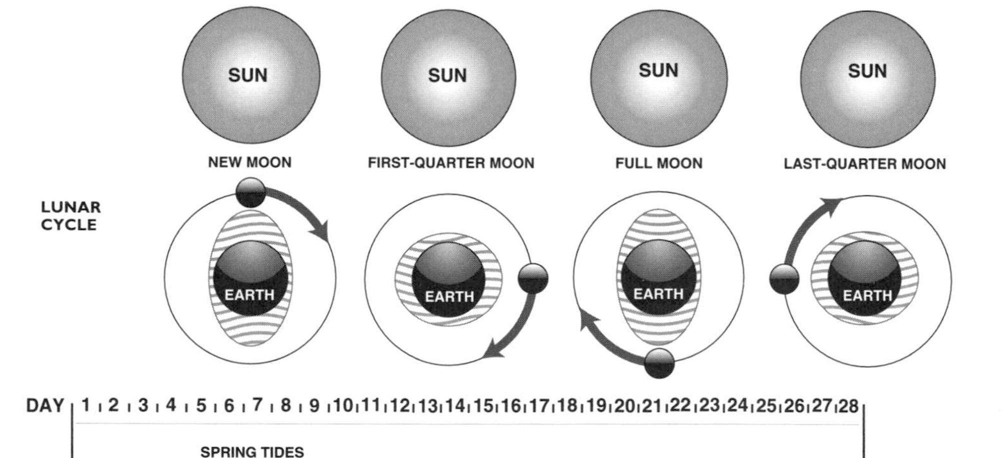

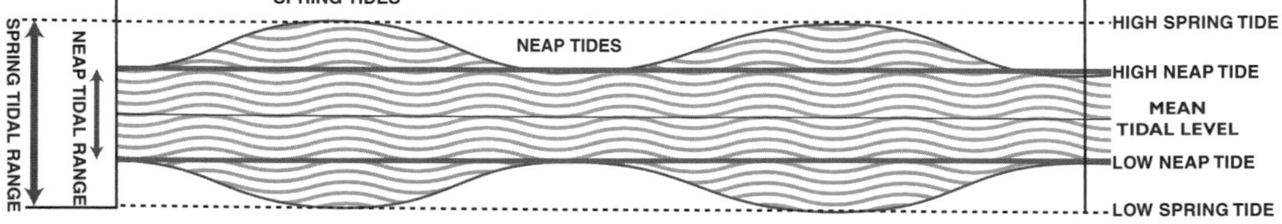

Figure 21. The lunar cycle and mean shore tidal ranges.

Tides have a profound influence on the distribution and types of animals and plants found on the shoreline. Many seashore organisms or visiting animals will synchronise their daily feeding activities or their reproductive cycles with the patterns of the tides. On some Californian (USA) beaches huge numbers of small fishes, California Grunion *Leuresthes tenuis*, arrive with the high spring tides of the spring and summer months. Just as the tides start to ebb, millions of eggs (about 1,000 per female) are laid, fertilised and then buried in the sand after the spawning fishes have been carried to the top of the beach by waves. The whole of this process takes less than 30 seconds. The damp sands and higher temperatures of the upper beach will incubate the eggs and the larvae hatch out at the next high spring tide. They are then washed away by the surf and currents. A similar phenomenon takes place on some east coast beaches of North America where the ancient-looking Horseshoe Crabs, *Limulus polyphemus*, arrive in large numbers on the high spring tides during May and June. Females will excavate a shallow scoop in the sand and lay up to 20,000 eggs each which are fertilised at the same time by the males that are clinging onto their backs. The wave action will cover up the eggs and at the next high spring tide the eggs will rupture and the tiny adolescents will be washed away with the surge of the water. Predation of Pacific Grunion and Horseshoe Crab young is high, but by laying eggs high up on the shore, predation from intertidal animals is greatly reduced. However, predation by terrestrial animals such as shore crabs and seabirds can be high. The vast numbers of eggs mean that some young will at least survive to maturity to carry on the life cycle. Synchronised coral spawning is also related to the lunar cycles and tides.

The intertidal or littoral (between tide marks) zones include all coastal areas which are periodically exposed to air by the falling tide and submerged by the rising water levels. The extent of the intertidal zone in any location is influenced by the slope of the beach, the shape of the coastline and by the local tidal range. There are a variety of coastal intertidal habitats and the type of habitat will determine the species of plants and animals which can survive in them. Seashore areas offer a rich diversity of habitats and life forms. Many of these areas are easily accessible for anyone to explore (see page 70).

WAVES

Waves on the sea are normally caused by wind blowing over the surface in a constant direction. After a period of time the waves will become more ordered to form 'swells' and these may move out of the windy areas. In a moving wave in the open ocean, the water does not travel with the wave but remains relatively stationary. The water particles within a wave will move in a near circular orbit, rising to meet each wave crest as it passes by (fig.22). The 'amplitude' of a wave is the height from wave crest to wave trough. The 'length' of a wave is the distance between successive crests or troughs.

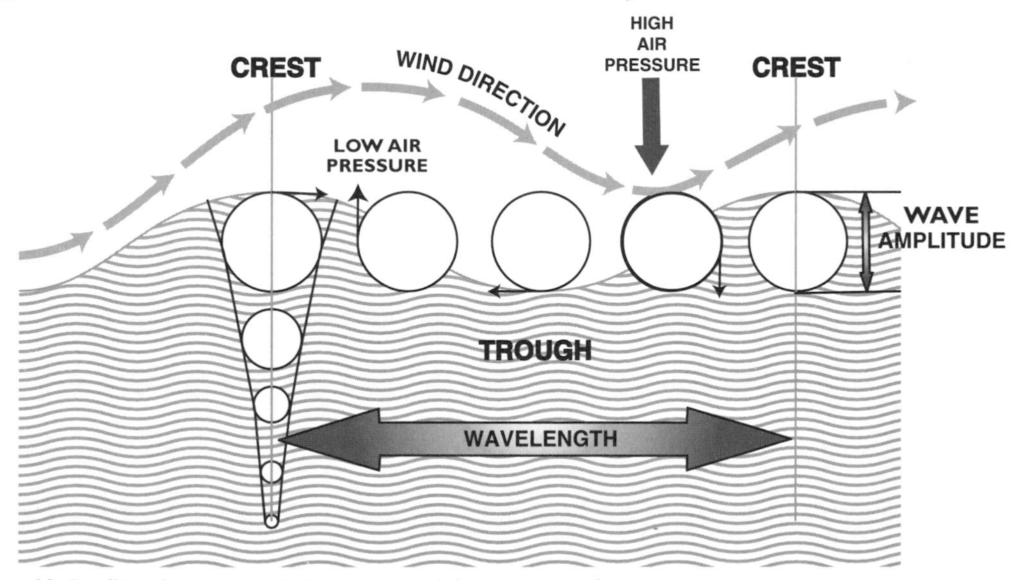

Figure 22. Profile of a wave, and the water particle rotation cycle.

Waves and the energy contained within them can travel thousands of miles across the oceans. Wave size will depend upon the wind's speed, duration and the distance over which it blows (the fetch). In theory the height of a wave cannot exceed 0.14 times the wavelength without collapsing. This results in white caps and breaking waves produced by strong winds. Wave motion quickly decreases with increasing depth of water and as every scuba diver knows, a raging sea on the surface may be of little consequence deeper down. As waves enter shallow coastal areas they slow down, shorten in length and increase in height as they come into contact with the seabed. The dragging effect of the seabed causes the wave crest to overtake the main body of the wave. The whole wave then becomes unstable until it finally collapses as 'surf'. The wave energy is turned into the energy of water movement as it surges up the beach. If waves impinge on a coastline at any angle other than at right angles, the waves will tend to break parallel with the beach. In areas that have bays and headlands the wave energy will be dispersed parallel with the beach of the bay and concentrated on headlands and promontories. During storm conditions these headlands and promontories can be spectacular areas as wave energy is released tens of metres into the air in a white explosion of spray. Destructive damage to these areas is inevitable and only the hardest of rock types can withstand the slow degradation from wave action. The exposed shoreline provides a demanding environment for the organisms living on it and, during stormy periods, great destruction of life can be the result. This is evident from combing the tideline after storms for the variety of sealife cast ashore. Wave action and water currents can be both destructive and constructive forces. Material that is removed from one area will be carried away and deposited elsewhere. Thus, nearshore habitats can be created and destroyed by waves and water movements and coastlines are constantly changing their shape. Waves stir up the shallow areas and this helps to clean the shores, oxygenate the water and circulate essential nutrients. The harsh conditions produced by mechanical wave action require all plants and animals living in areas of wave influence to be well adapted for survival (see Benthic Environments, page 66).

SURFACE STATE OF SEAS

Seafarers often need to describe the prevailing weather conditions at sea. For this purpose the Beaufort Wind Scale has been devised and is commonly used by all seafaring nations. It provides a means for seafarers to record the roughness of seas. The Beaufort scale, together with a description of the state of the sea, can be found in Appendix III.

BIOLOGY OF MARINE ENVIRONMENTS

by

Michael Burchett

INTRODUCTION

CLASSIFICATION OF ANIMALS (TAXONOMY)

It is easier to understand how plants and animals are identified if they are grouped together in a logical and systematic way. Frequent confusion arises when an animal or plant is given several different common names. Therefore each species of organism is assigned a scientific name according to the 'binomial' system (nomenclature). This was introduced in the eighteenth century by the Swedish naturalist Linnaeus. Under his system each species receives two scientific names. The first is the 'generic' name and is written with an initial capital letter. This is followed by the 'specific' name which is not capitalised e.g. *Solaster endeca* (Purple Sunstar).

Plants and animals of the same species are able to reproduce and generate fertile offspring, while organisms that are more distantly related cannot. Species with shared or common features are grouped together in a 'genus'. Related genera are further grouped into larger units called 'families' and several families together make an 'order'. Related orders are members of the same 'class' and several classes form the largest practical category of classification known as a 'phylum'.

At each level of classification a morphological description provides the characteristics for that level. It is important that a naturalist should be familiar with the distinguishing features of the major phyla of marine organisms for the purposes of correct identification. Examples of four important marine taxa are shown in Table 4, and an outline classification of marine plants and animals is shown in Table 5.

	Diatom	Copepod	Haddock	Sperm Whale
KINGDOM	Plantae	Animalia	Animalia	Animalia
PHYLUM	Chrysophyta	Arthropoda	Chordata	Chordata
CLASS	Bacillariophycene	Copepoda	Osteichthyes	Mammalia
ORDER	Centrales	Calanoida	Gadiformes	Cetacea
FAMILY	Chaetoceraceae	Amphascandria	Gadidae	Physeteridae
GENUS	*Chaetoceros*	*Calanus*	*Melanogrammus*	*Physeter*
SPECIES	*decipiens*	*finmarchicus*	*aeglefinus*	*catodon*

Table 4. Classification of four important marine organisms.

DIVERSITY OF MARINE ORGANISMS

The oceans cover about 70% of the Earth's surface but despite this the oceans have far fewer species of plants and animals than terrestrial and freshwater environments. It is thought there are about one million different species of animals on the planet but only 160,000 (16%) of these live in the marine environment. However, this lack of species diversity is compensated by the large population numbers of some groups such as the nematodes, pelagic copepods and benthic molluscs.Within the marine environment only 3,200 species (about 2%) live in the open oceans. The rest of the animals live on or near the seabed with a large proportion living on coral reefs. The community of sealife living on or close to the bottom of the abyss (abyssal benthos) is more remarkable in its diversity; estimates of undescribed species range from many thousands to millions.

Table 5. A simplified traditional classification of marine organisms based on the five kingdom hypothesis.

All phyla and divisions with true marine representatives are included. Subdivisions below phylum level are not comprehensive; only some of the more important subdivisions are given, in particular those well represented in the marine environment. A more detailed classification of marine invertebrates is given in table 7 on pp.116-122. The classification and taxonomy of vertebrate classes is discussed in the relevant chapters covering these groups.

‡ Groups not strictly represented in the marine environment other than as casual or peripheral occurrences.
* Phyla/divisions which straddle the plant and animal kingdoms, containing organisms which share characteristics of both kingdoms.

Kingdom	Phylum/Division	Subdivision	Description and/or common names included in group
Monera	**Schizophyta**		bacteria
Fungi	**Mycophyta**		fungi
	Lichenes		lichens
Plantae	Division: **Cyanophyta**		blue-green algae
	Division: **Rhodophyta**	Class: **Rhodophyceae**	red algae
	Division: **Phaeophyta**		brown algae
	Division: **Chlorophyta***		green algae, *Volvox*
	Division: **Chrysophyta***		diatoms and others
	Division: **Pyrrhophyta***		dinoflagellates
	Division: **Angiospermae**	Class: **Tracheophyta**	saltmarsh grasses, seagrasses, mangroves
Protozoa	Phylum: **Cryptophyta***		biflagellate phytoflagellates
	Phylum: **Haptophyta***		small phytoflagellates, includes coccoliths
	Phylum: **Choanoflagellida**		solitary and colonial zooflagellates
	Phylum: **Rhizopoda**		mobile amoebas (see also p.116)
	Phylum: **Actinopoda**		radiolarians and heliozoans (see also p.116)
	Phylum: **Ciliophora**		ciliated protozoans, including tintinnids (see p.116)
	Phylum: **Sporozoa**		parasitic protozoans (see also p.116)
	Phylum: **Microspora**		parasitic protozoans (see also p.116)
Animalia	Phylum: **Placozoa**		simple metazoan (see p.116)
	Phylum: **Porifera**		sponges (see p.116)
	Phylum: **Cnidaria**	Class: **Hydrozoa**	hydrozoans, siphonophores and hydrocorals (see p.116)
		Class: **Scyphozoa**	true jellyfish (see p.116)
		Class: **Cubozoa**	box jellyfish (see also p.116)
		Class: **Anthozoa**	sea anemones and corals (see pp. 116-117)
	Phylum: **Ctenophora**	Class: **Tentaculata**	comb jellies (see p.117)
		Class: **Nuda**	comb jellies (see p.117)
	Phylum: **Platyhelminthes**	Class: **Turbellaria**	flatworms or platyhelminths (see p.117)
		Class: **Monogenea**	flukes
		Class: **Trematoda**	flukes
		Class: **Cestoda**	tapeworms
	Phylum: **Orthonectida**		mesozoans (see also p.117)
	Phylum: **Rhombozoa**		mesozoans (see also p.117)
	Phylum: **Nemertea**		nemerteans or proboscis worms (see p.117)
	Phylum: **Gnathostomulida**		gnathostomulids (see also p.117)
	Phylum: **Gastrotricha**		gastrotrichs (see also p.117)
	Phylum: **Nematoda**		nematodes or roundworms (see p.117)

Kingdom	Phylum	Subphylum / Class	Common name
Animalia	Phylum: **Nematomorpha**		hairworms
	Phylum: **Kinorhyncha**		kinorhynchans (see p.117)
	Phylum: **Loricifera**		loriciferans (see p.117)
	Phylum: **Priapulida**		priapulids (see p.117)
	Phylum: **Sipuncula**		sipunculans or peanut worms (see p.117)
	Phylum: **Mollusca**	Class: **Monoplacophora**	archaic molluscs (see also p.118)
		Class: **Polyplacophora**	chitons (see also p.118)
		Class: **Aplacophora**	solenogasters (see also p.118)
		Class: **Gastropoda**	gastropods: snails and slugs (see p.118)
		Class: **Bivalvia**	bivalves (see p.119)
		Class: **Scaphopoda**	tusk shells or scaphopods (see also p.119)
		Class: **Cephalopoda**	cephalopods: cuttlefish, squid and octopuses (p.119)
	Phylum: **Echiura**		echiurans or spoon worms (see p.119)
	Phylum: **Annelida**	Class: **Polychaeta**	polychaete worms (see p.119)
		Class: **Hirundea**	leeches
	Phylum: **Pognophora**		pognophorans (see p.119)
	Phylum: **Arthropoda**	Subphylum: **Chelicerata**	horseshoe crabs and sea spiders (see p.120)
		Subphylum: **Crustacea**	crustaceans: 10 classes (see pp.120-121)
		‡ Subphylum: **Uniramia**	insects (c. 80% of known animal species), millipedes, centipedes and related groups (see also p.121)
	Phylum: **Phoronida**		phoronids (see p.121)
	Phylum: **Bryozoa**		bryozoans (see p.121)
	Phylum: **Entoprocta**		entoprocts (see p.121)
	Phylum: **Brachiopoda**		brachiopods or lamp shells (see also p.121)
	Phylum: **Chaetognatha**		arrow worms (see also p.121)
	Phylum: **Echinodermata**	Subphylum: **Crinozoa**	crinoids: sea lilies and feather stars (see p.121)
		Subphylum: **Asterozoa**	
		Class: **Asteroidea**	starfish (see p.122)
		Subclass: **Ophiuroidea**	brittle-stars (see p.122)
		Subphylum: **Echinozoa**	
		Class: **Echinodea**	sea urchins, sand dollars and heart urchins (see p.122)
		Class: **Holothuroidea**	sea cucumbers (see p.122)
	Phylum: **Hemichordata**	Class: **Enteropneusta**	acorn worms (see pp.122 and 290)
		Class: **Pterobranchia**	pterobranchs (see pp.122 and 290)
	Phylum: **Chordata**	Subphylum: **Urochordata**	ascidians or sea squirts and tunicates (see p.122)
		Subphylum: **Cephalochordata**	cephalochordates or lancelets
		Subphylum: **Vertebrata**	vertebrates
		Class: **Agnatha**	lampreys and hagfish (see p.290)
		Class: **Chondrichthyes**	sharks, rays and chimaeras (see p.290 and 345-347)
		Class: **Osteichthyes**	bony fishes (see p.290 and 348-350)
		‡Class: **Amphibia**	amphibians
		Class: **Reptilia**	reptiles (see p.351)
		Class: **Aves**	birds (see p.361)
		Class: **Mammalia**	mammals (see p.393)

Many of the plants and animals described in this book will be referred to according to their lifestyle and the naturalist should be familiar with the terms outlined below. Plants and animals that float or swim in the seas are called 'pelagic'. Only those creatures that live in the open oceans well away from coastlines, continental shelves and the seabed are said to be truly pelagic. Organisms which inhabit the sea floor are called 'benthic'. Animals that live within the bottom substrates are known as the 'infauna' and animals that crawl over the bottom are the 'epifauna'. Other groups of animals only inhabit the first few centimetres of the surface of the sea and these creatures are termed the 'neuston'. The neuston are generally microscopic creatures less than 0.2mm in size. All small organisms that drift or swim weakly are collectively known as 'plankton' or 'holoplankton'. If they are of plant origin they are grouped together as 'phytoplankton', while animal plankton is collectively called 'zooplankton'. The size limits of plankton are rather arbitrary, but 20mm is about the upper size limit for most planktonic organisms. Phytoplankton require sunlight to synthesise new organic material and are therefore in the 'euphotic' zone of the ocean. Herbivorous zooplankton are found accompanying the phytoplankton and feed upon it. Larger zooplankton which feed on the smaller zooplankton are carnivorous by nature and these creatures can inhabit much deeper waters. Animals larger than 20mm will generally have well developed swimming capabilities. These larger, free-swimming animals are collectively known as the 'nekton' and include fish, squid and whales. There are a few creatures that only live at the surface of the water and parts of their bodies protrude above the surface in a sail-like fashion. They are sometimes considered to be a special category of animals as they are transported passively by the wind instead of the ocean currents. These creatures are known as the 'pleuston' and typical examples include the By-the-wind Sailor *Velella velella,* and the Portuguese man-of-war *Physalia physalis.* Their gas-filled floats which protrude above the surface will catch the wind and stop the animal sinking below the surface. The long trailing tentacles of *Physalia* enable it to catch zooplankton and small fish as deep as 30m below the surface. *Velella* has only short tentacles and its food includes the much smaller zooplankton (e.g. copepods, larval crustaceans, shrimps and fish eggs). Both animals are cnidarians, with stinging cells or 'nematocysts' to paralyse prey during capture.

Another large group of planktonic animals are the 'mesoplankton'. These animals are not permanent drifting members of the plankton and are only free-swimming for part of their life cycle. They include the free-swimming larval stages of benthic organisms such as crab 'zoea', barnacle 'nauplii', snail larvae and worm larvae (trochophores). Many of the biological terms used to describe size are prefixed with 'micro' (20-200μm or 0.02-0.2mm), 'meso' (0.2-20mm), 'macro' (2-20cm) or 'mega' (20-200cm). (100μm = $^{1}/_{10}$mm).

ECOLOGICAL ZONES

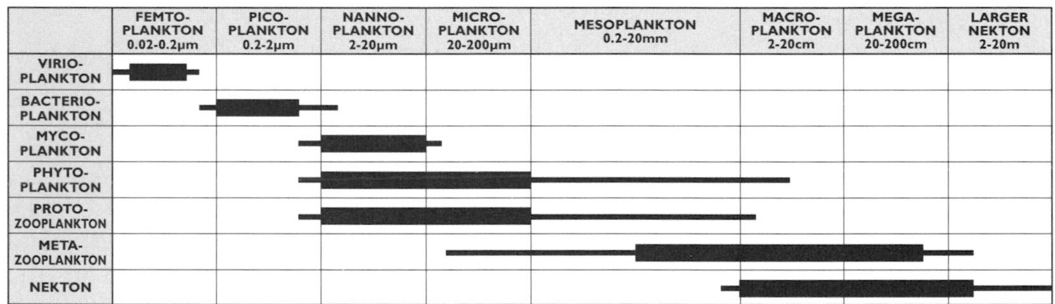

Table 6. Grade scale for size classification of pelagic organisms. (After Lalli and Parsons 1993).

The environments in which pelagic and benthic organisms live are grouped into four major ecological zones: the 'intertidal' or 'littoral' zone (between tide marks); the 'sublittoral' zone (low tide mark to edge of continental shelf); 'bathyal' zone and 'abyssal' zone. Marine animal distribution can therefore be assigned names according to depth and position e.g. bathypelagic (see fig.23).

Pelagic animals that live in the waters over a continental shelf live in the 'neritic province' and those creatures inhabiting open oceans come from the 'oceanic province'. Vertical environments can also be distinguished and the ocean province can be subdivided. The upper 'epipelagic' zone (0-100m) is characterised by strong sunlight and temperature gradients which will often fluctuate with the seasons. The 'mesopelagic' zone (100-1,000m) has dim light (twilight zone) or no light and is characterised by low, stable temperatures, lower levels of oxygen and higher levels of nutrients compared to the epipelagic zone. Below 1,000m, there is no light from the sun, only bioluminescent light produced by animals. Many fish and squid species that live adjacent to or on the seabed are known as 'demersal' in their distribution.

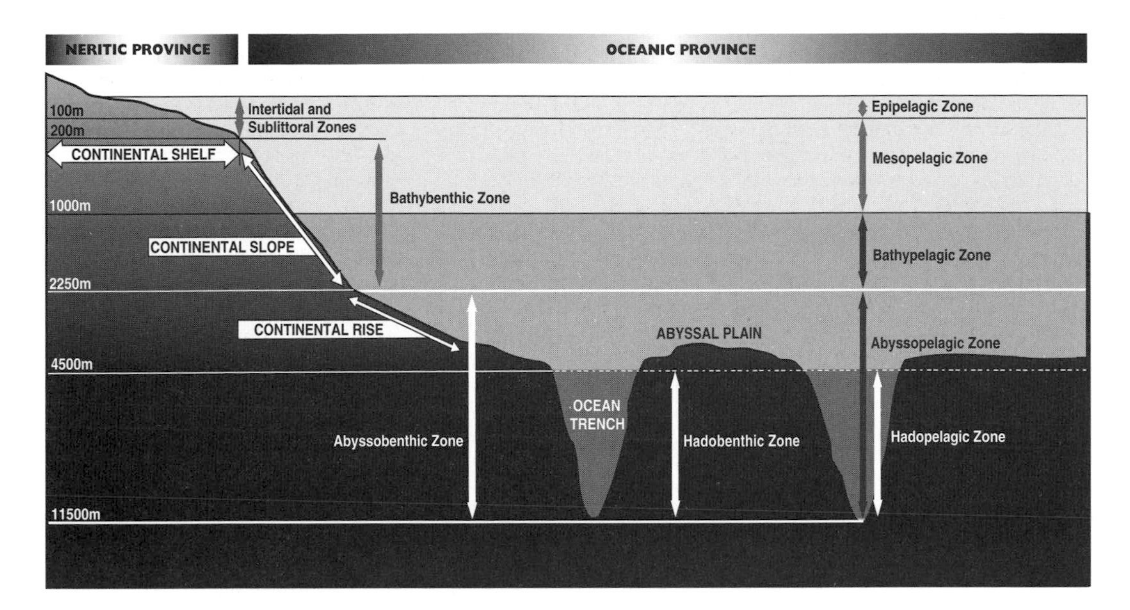

Figure 23. Cross-section showing major depth zones of an ocean. As cross-section profiles differ in the various oceans the depth limits of the different zones will also vary. (Adapted from Meadows & Campbell 1993). Epipelagic Zone = Euphotic Zone; Mesopelagic Zone = Disphotic Zone; Bathypelagic Zone and below to seabed = Aphotic Zone.

PROXIMITY TO COAST

All species of marine organisms inhabit the seas within preferred temperature and depth zones. Describing organisms by their position or proximity to a coast may be a useful indicator of habitat. Therefore species can be assigned to the following categories:

1. **Pelagic or oceanic species**: inhabit the open oceans and do not come near land, continental shelves or the seabed.
2. **Offshore species**: inhabit depths of water between 50m and 200m which are found over continental shelf areas and slope edges. These species seldom come near land but are not truly pelagic.
3. **Inshore or coastal species**: organisms found only near coastlines in depths of water less than 50m.
4. **Littoral species**: found close to the coastal fringe and therefore only occur in shallow depths. Many intertidal organisms fall into this category.

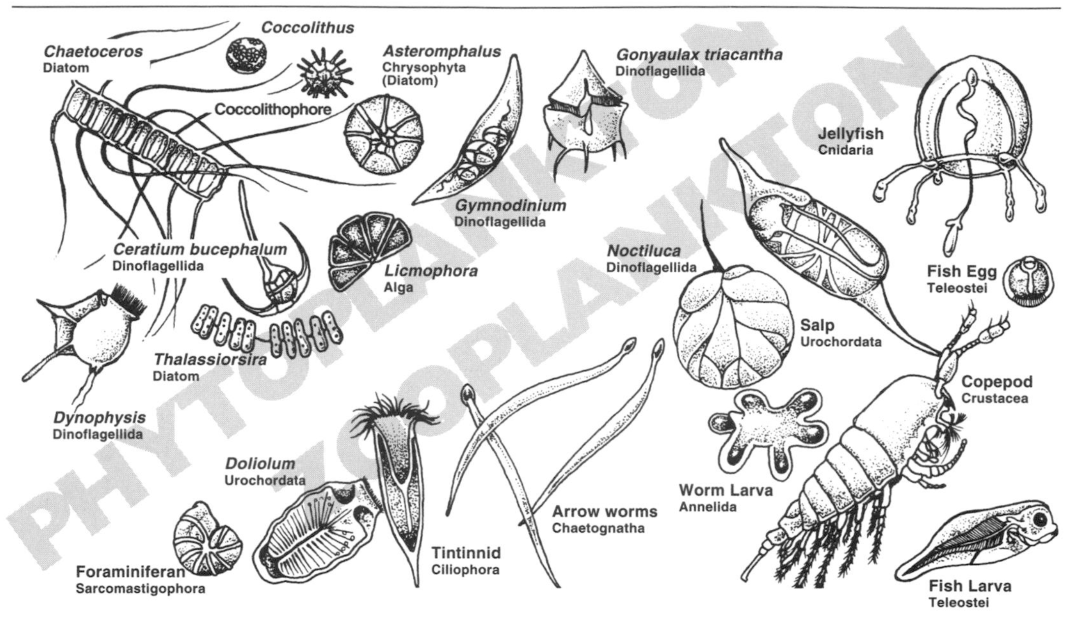

Figure 24. Representatives of phytoplankton and zooplankton. Many dinoflagellates are photosynthetic and some (such as *Noctiluca*) are carnivarous.

PLANKTONIC SYSTEMS OF SURFACE WATERS

TYPES OF PHYTOPLANKTON

Microscopic planktonic organisms of the oceans form the basis of marine food chains and the more complex food webs. A bucketful of seawater may contain several thousand microscopic plants and animals (fig.24). Microscopic plant life includes all the members of the phytoplankton that provide food for herbivores. The phytoplankton is composed of various types of 'unicellular' algae which range from a few microns to several hundred microns in size. The major groups of algae include the diatoms which are the dominant type of phytoplankton in the temperate and high latitudes. Planktonic diatoms have no swimming structures and are therefore not normally capable of independent movement. It is essential for diatoms to remain in the surface waters to carry out photosynthetic processes, and therefore these algae have developed a variety of adaptations to prevent sinking. Some diatoms have needle-like or feather-like projections which greatly increase their surface areas with little increase in body weight. A small body with a high drag factor will therefore sink at a very slow rate since drag is a surface-related effect. Other common types of phytoplankton include the dinoflagellate algae and these organisms have two whip-like 'flagella' for swimming. Another major group, the coccolithophores, lacks a flagellum but instead is characterised by an external shell composed of a number of plate-like structures called 'coccoliths'. The smallest plankton in the sea are the bacteria and the related prokaryotes (known as 'bacterioplankton'). Little is known about their biology but they are thought to play a significant role in the nutrient cycles as they are eaten by many other planktonic animals.

PRIMARY PRODUCTION

Sunlight is scattered and absorbed rapidly in surface waters and different wavelengths of light penetrate to different depths. In clear open oceans the light may penetrate much deeper than in the turbid waters of coastal regions. Seawater becomes progressively blue-green with depth because the longer wavelengths of light are rapidly lost through absorption. Plants need the sun's energy to combine carbon dioxide and water to produce organic carbohydrates. Oxygen is the by-product of this biochemical reaction. The carbohydrates then become available for use by animals, which can convert them into more complex compounds such as fats and proteins. The reverse process of photosynthesis is 'respiration'. During this process the high energy bonds of the carbohydrates are broken and energy is released for use in metabolic processes which keep an organism alive. Both plants and animals continually respire but photosynthesis, (which is carried out by plants) only takes place when sufficient light is available during the day. The simplified biochemical reactions for photosynthesis and respiration are shown in the following equation.

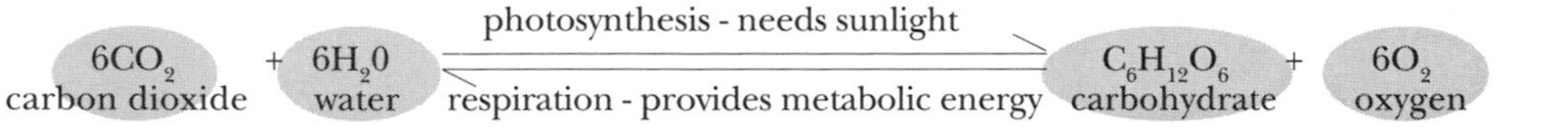

Primary production is limited by a number of environmental variables which include light, temperature and quantities of nutrients in the water. Temperature plays a secondary role as it determines the rate at which many chemical and biological reactions take place. The amount of light diminishes from the equator to the poles as the angle of the sun changes. In addition, seasonal variations in day length (and therefore light) increase north and south of the equator. This leads to seasonal variations of planktonic production in the high latitudes. There is little plankton production in tropical regions due to a lack of nutrients (oligotrophic conditions).

UPWELLING AREAS AND NUTRIENTS

Nutrient availability is often the key factor controlling phytoplankton production. Major nutrients required include phosphates, nitrates and in some cases, silicates for diatom shells. These nutrients allow the populations of microscopic plant life to multiply rapidly and produce the characteristic green, turbid waters which contrast so dramatically with the clear waters of less productive regions. Oceanic circulation and water movements play a vital role by bringing nutrient-rich deep water to the surface layer (euphotic zone) through a number of physical oceanographic processes. Off the coast of Peru and Chile (South America), Oregon (USA), Mauritania and Namibia (Africa) and Arabia, wind-driven upwellings (fig.25) occur between the coast and the main boundary currents (coastal upwellings). In a broad band stretching along the equator from the Americas almost to the western side of the Pacific, divergent water masses are formed by the interactions between the Trade winds and the Coriolis force (equatorial upwellings). This produces mid-ocean upwelling areas that are rich in nutrients and therefore regions of high primary productivity. In the high latitudes of both the northern and southern hemispheres, deep water may be brought to the surface by the intense mixing caused by winter storms (polar upwellings). The upwelling waters, combined with an increase of spring sunshine, form the basis for 'spring plankton blooms' in these regions. Local combinations of seabed topography and currents may also cause vertical mixing of water masses. Strong vertical circulations can occur in the wakes of offshore islands or headlands projecting into the paths of moving water. These processes work together and determine the levels of nutrients available for phytoplankton growth and production. Therefore the distribution of phytoplankton in the oceans of the world is largely reflected by the nutrient availability.

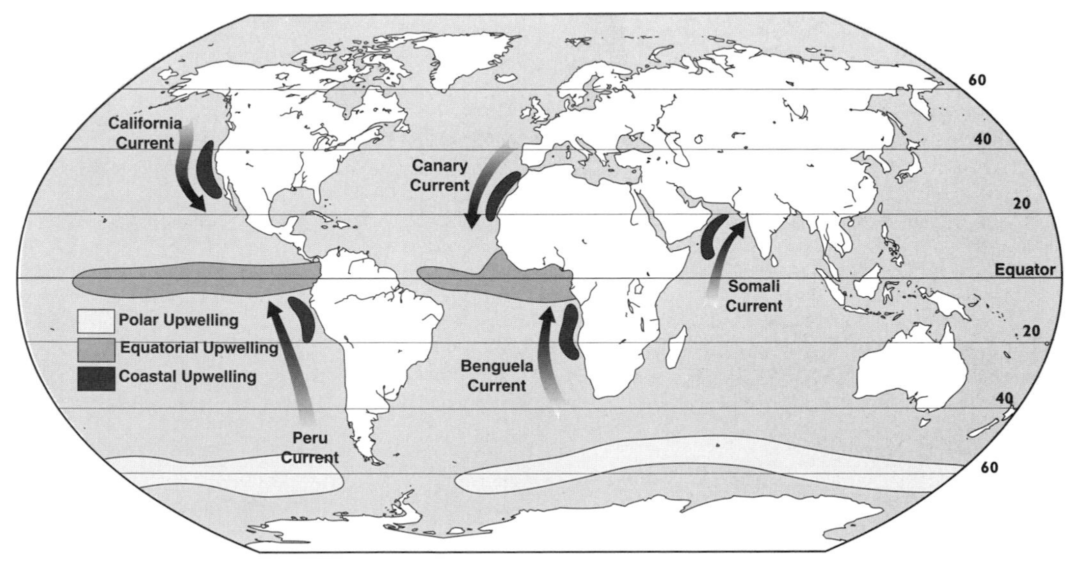

Figure 25. Upwelling areas.

The pattern of life among organisms that inhabit the rich upwelling areas of the oceans differs considerably from that in oceanic areas that are nutrient-poor and have little phytoplankton production. Within nutrient rich areas the species diversity of plant and animal plankton is poor. The living communities are dominated by large numbers of a few species, therefore the 'food chains' of these areas are relatively simple with few 'trophic' (feeding) levels. Here the herbivorous zooplankton species are far more numerous and diverse than the carnivorous species. Moreover, the diversity of carnivorous plankton species increases away from nutrient-rich regions of the world. Herbivorous crustaceans such as copepods *Calanus* spp. and krill (Euphausiacea) are the most abundant zooplankton of the ocean waters and are the main diet for large populations of fish, birds and sea mammals. The diversity of the higher predators is also lower in nutrient-rich areas. Major fish species include the clupeoids e.g. sardines, herrings and anchovies. These major stocks of fish provide important commercial catches for fishing fleets in the waters off California, Peru, Morocco and Namibia. Away from the rich upwelling areas, plankton communities and their associated predators become more species diverse, but with fewer individuals of each species. The fish species that live in areas with poor plankton production are quite different from the clupeoids of the nutrient-rich regions. They include large predatory species such as tunas and billfishes. These fishes remain independent of any localised food source as they routinely migrate across major oceans at appropriate times of year to find food. These fish have high metabolic rates and a capacity for sustained high-speed cruising.

VARIATIONS IN PLANKTON PRODUCTIVITY

Plankton production has been found to vary considerably with latitude (fig.26). In the various regions of the world the climatic and oceanographic conditions vary at different times of the year and there are subtle interactions taking place between temperature, light, nutrients and the mixing of seawater.

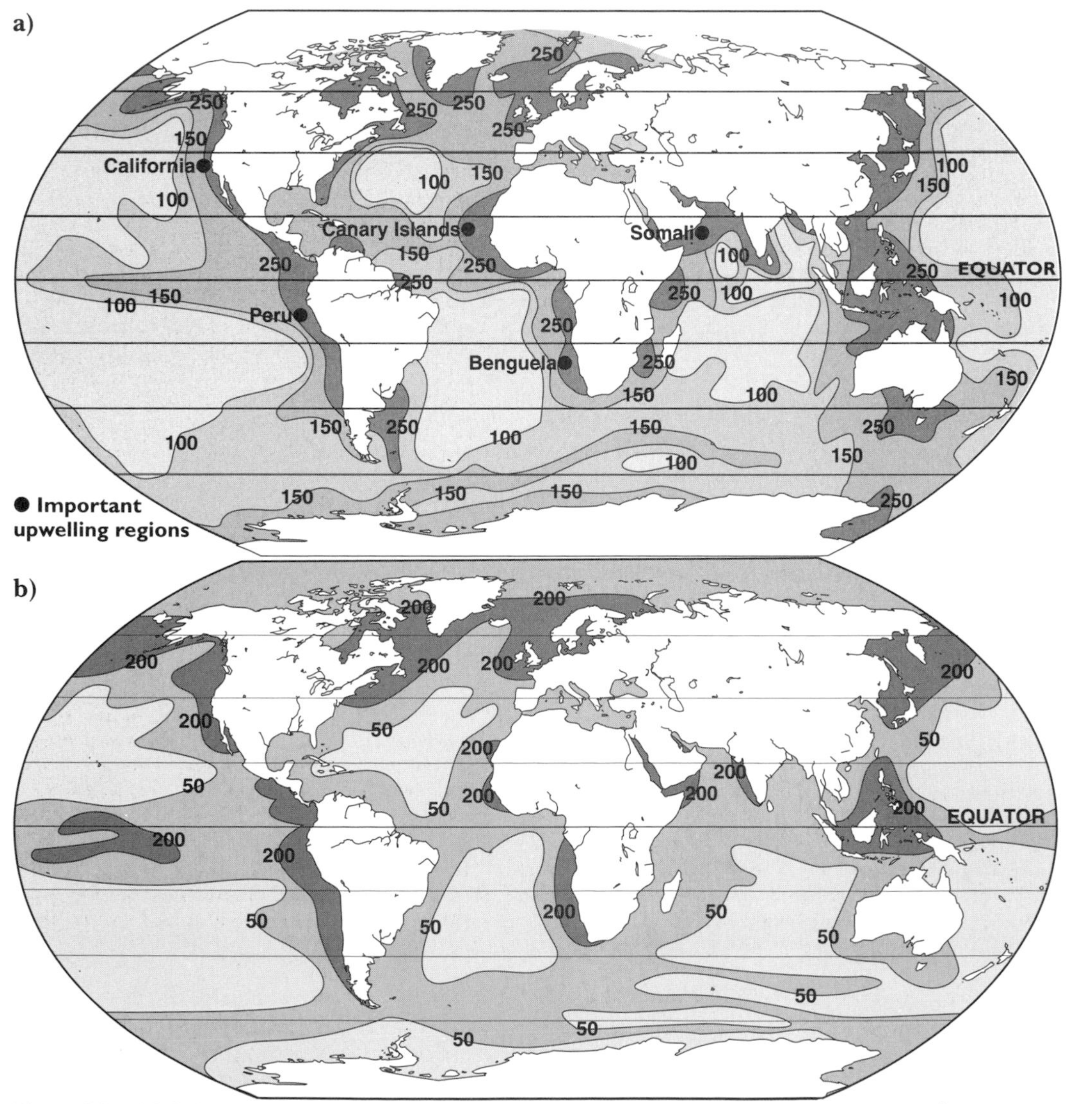

Figure 26. Global patterns of production. a) primary (phytoplankton) production (mgC/m²/day), b) secondary (zooplankton biomass) production (mg/m³).

Figure 27 clearly shows that there are seasonal variations in both the phytoplankton (primary production) and zooplankton (secondary production). There is always a delay period between maximum phytoplankton production and maximum zooplankton abundance.

Despite year-round high light intensities, the tropical regions of the world are generally low in productivity, except in areas of upwelling. In tropical waters the heat from the sun stabilises the water column and a 'thermocline barrier' forms. Between the surface of the sea and the thermocline, nutrients are quickly used up by the phytoplankton, after which the nutrients will remain in short supply. Conversely, in the polar waters of the Arctic and Antarctic regions, nutrients are in good supply and no thermocline is established to form a 'barrier'. However, these cold regions experience low levels of solar radiation except for the brief period during the summer months, when there is continuous 24 hour day length and brief but massive plankton blooms occur.

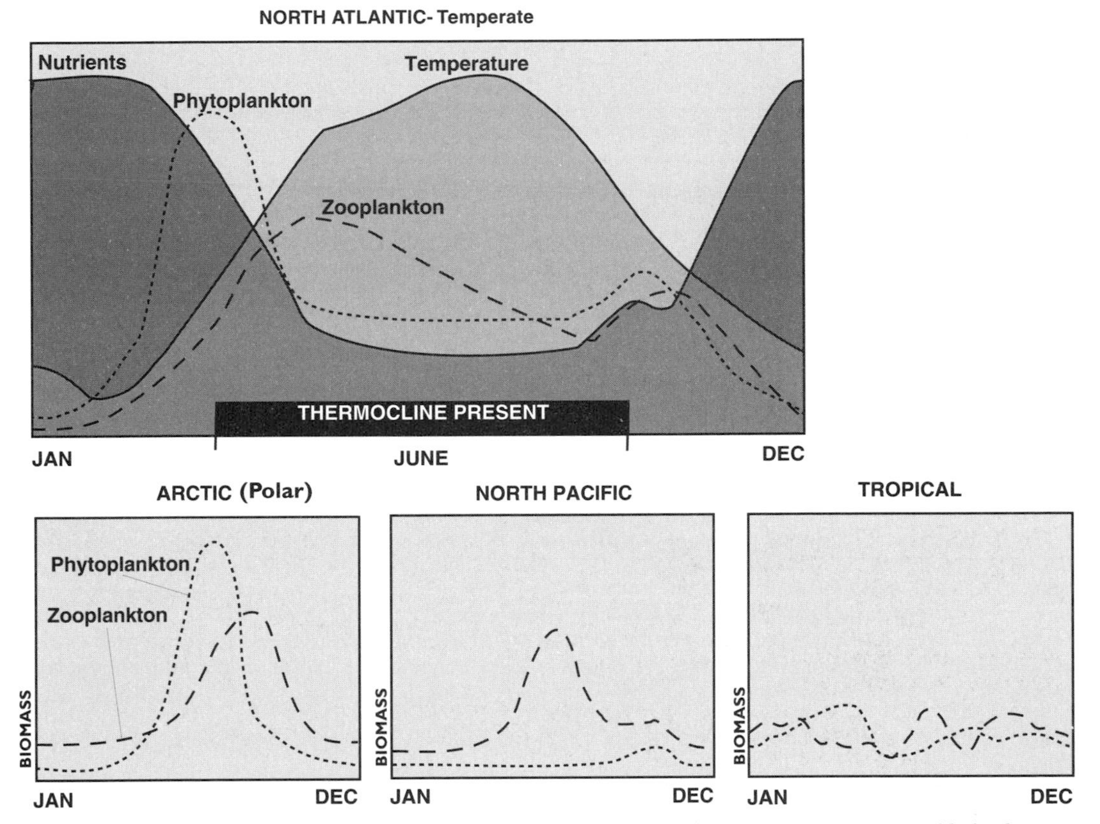

Figure 27. Seasonal variations of plankton production near the surface in different oceans and latitudes.

In temperate latitudes where both nutrient levels and light levels are sustained, total plankton productivity is higher than in any other region of the world. In temperate latitudes, distinct seasonal patterns take place with the changing seasons. During the summer months a thermocline will develop causing a reduction of nutrients and a decline in phytoplankton production. However, in spring and autumn rough seas break up the thermocline and nutrients from deeper waters once more become available which results in spring and autumn plankton production peaks. Although nutrient reduction will reduce phytoplankton production, grazing pressure from the herbivorous plankton will have the same effect. In the north Pacific the phytoplankton bloom never really becomes established as the zooplankton overwinter as a shallow layer of subadults and immediately consume the phytoplankton as soon as they increase in abundance. Therefore the phytoplankton do not reach high biomass levels.

GLOBAL PLANKTON DISTRIBUTION

It was once thought that plankton is evenly spread throughout the surface layers of the oceans. However, it is now known that this is not the case. Plankton is patchy in distribution both vertically and horizontally in the oceans. Open ocean plankton appears to be less patchy and also less abundant than the plankton of coastal and nearshore waters. Plankton patchiness is a complex phenomenon which is controlled by a number of biological and chemical interactions. The oceanic circulations and surface circulations play a major role in plankton distribution over wide areas. If plankton are transported to areas with the right environmental conditions for plankton survival and sufficient nutrients are available, then the population will flourish and form patches. However, if the ocean currents take the plankton into nutrient-poor, stagnant waters such as those found in the centres of many ocean gyres, there is no rapid plankton growth.

Studies on the plankton of the Pacific Ocean have revealed the presence of up to eight major types of plankton communities and that the distribution of plankton in the Pacific is controlled by current patterns. The water currents of the northern and southern hemispheres are virtual mirror images and this is reflected in the similar types of animals found within these plankton communities. The species composition in the various regions of the Pacific Ocean changes little over thousands of miles and there appears to be little exchange of plankton species between different regions. It is believed that the plankton communities of the Pacific are the oldest on Earth and that they have been in existence for the past 26 million years.

VERTICAL MIGRATION OF ZOOPLANKTON

One of the most interesting behavioural aspects of marine zooplankton is the daily vertical migrations carried out by many of the animals in the water column. Relative to the size and type of zooplankton, these vertical movements can vary from 400-1,000m over a 24 hour period. Many of these vertical migrants follow a similar daily pattern of movement. During the hours of daylight the zooplankton are relatively deep in the water column, but at the onset of darkness they ascend to the surface waters. At night they will disperse horizontally to feed, and just before dawn they congregate and once again descend to a certain depth in the water column. In polar regions the vertical migration does not take place on a daily basis, but on a seasonal basis. During the polar summer the zooplankton remain mostly at or near the surface and are only to be found at depth during the polar winter. In these high latitudes permanent sea ice covers the surface. The reason why many species of zooplankton undertake vertical migrations is not well understood. However, it may enable more effective feeding on phytoplankton, escape from surface predators during the day, decrease metabolic activity in the cool depths and provide better geographical dispersal through deep water movements.

The distribution of zooplankton has important effects on the distribution of higher predators which rely on zooplankton as a food source. Many fish, seabirds and whales need to find the patches or swarms of zooplankton in order to survive. Distance of zooplankton swarms from shore-based breeding colonies of many birds and seals may be one of the factors which contributes to the success or failure of breeding in a particular year. It is known that the penguins and seals of Antarctica will swim hundreds of miles to feed and some albatross species are known to fly thousands of miles in search of food. As many of the higher predators are dependent on the zooplankton and fish, exploitation of these same resources by humans should be approached with caution. Over-exploitation of a food resource near to a breeding colony could have severe consequences as the animals would need to cover larger distances to find sufficient food and these distances may prove too great. However, the patchy distribution of zooplankton may be the higher predators' salvation as it is a difficult resource for fishing fleets to find and exploit.

RED TIDES

Red tides occur in areas where large numbers of microorganisms become concentrated. The phenomenon can occur where currents meet to form a downwelling. The combination of the organisms swimming towards the surface and the water moving downwards results in a concentrated population which can be seen on the surface as long lines, streaks or windows in the water. Currents carry both nutrients and the organisms so that a red tide may persist for days or weeks. Red tides are so called because the water can appear rust coloured, but the colour varies according to the species responsible, with varying shades of red, pink, violet, orange, yellow, blue, green or brown being reported.

Sometimes red tides can be disastrous for other marine life. Firstly, when the phytoplankton has exhausted all the nutrients, or the temperature changes, or the amount of sunlight decreases, the entire population can die simultaneously. As the phytoplankton decomposes, the amount of oxygen in the water decreases as it is used up by bacteria to a level where many animals will suffocate. In 1977 a red tide off New Jersey, USA, caused an area of over 14,000 square kilometres to become deficient in oxygen. This resulted in the destruction of countless organisms, especially those living on the seabed (e.g. clams and mussels), which could not escape the anaerobic conditions.

The second way in which red tides can harm an environment is by producing poisonous by-products or 'toxins'. Some of these toxins cause death to fish and other cold blooded animals and can even affect marine mammals, seabirds and humans. Two planktonic dinoflagellate genera (*Gymnodinium* and *Gonyaulax*) include species which can produce these toxins. The west coast of Florida is known for frequent and extensive fish kills caused by red tides. The species responsible here is *Gymnodinium breve* which gives the water an oily appearance and produces a chemical that causes red blood cells to burst. The cell membranes rupture as they pass through the fish gills and simultaneously the toxin is absorbed by the blood. The ruptured blood cells are unable to carry oxygen and the fish die gasping on the surface.

Florida's tourist trade has sometimes suffered as large numbers of dead fish have been washed up along the shores and people exposed to windblown spray from red tide areas have suffered from respiratory irritations. In 1988, on the eastern seaboard of the North Sea, the coasts of Norway, Sweden and parts of Denmark were affected by a red tide bloom of plankton which killed thousands of fish including salmon and also shellfish. Cages of farmed salmon were pulled away from the path of the spreading plankton in a desperate attempt to save the them. In recent years there has been an alarming increase in the incidence of such plankton blooms in different parts of the world. Although the causes are not known, they could be associated with the increased run-off of fertilisers and sewage from the land which disrupts the natural nutrient cycles of coastal waters.

Another toxin produced by the *Gonyaulax* group is responsible for the disease known as 'paralytic shellfish poisoning' (PSP). Warm-blooded animals are particularly sensitive to this poison which affects the nervous system and leads to muscular paralysis and eventually death. Some cold-blooded animals and fish are affected, but shellfish such as clams and mussels are unharmed and the poison accumulates in their bodies. In extreme cases the outcome may be fatal for humans, even if just a few infected shellfish are eaten. Unfortunately, poisoned shellfish appear outwardly normal, so there are no obvious indications of the potential danger. Records of paralytic shellfish poisoning are worldwide. Most outbreaks have occurred along the west and north-east coasts of North America, the coasts of north-western Europe, around Japan, along the coast of British Columbia, in the southern Bay of Fundy and in the St. Lawrence river estuary. The annual bloom of *Gonyaulax* usually means that certain areas are closed for the fishing of shellfish, crabs and lobsters during that time of year. However, the increasing incidence of such plankton blooms in recent years has resulted in more fisheries being closed.

OTHER ASPECTS OF PELAGIC LIFE

FOOD CHAINS IN THE OCEAN

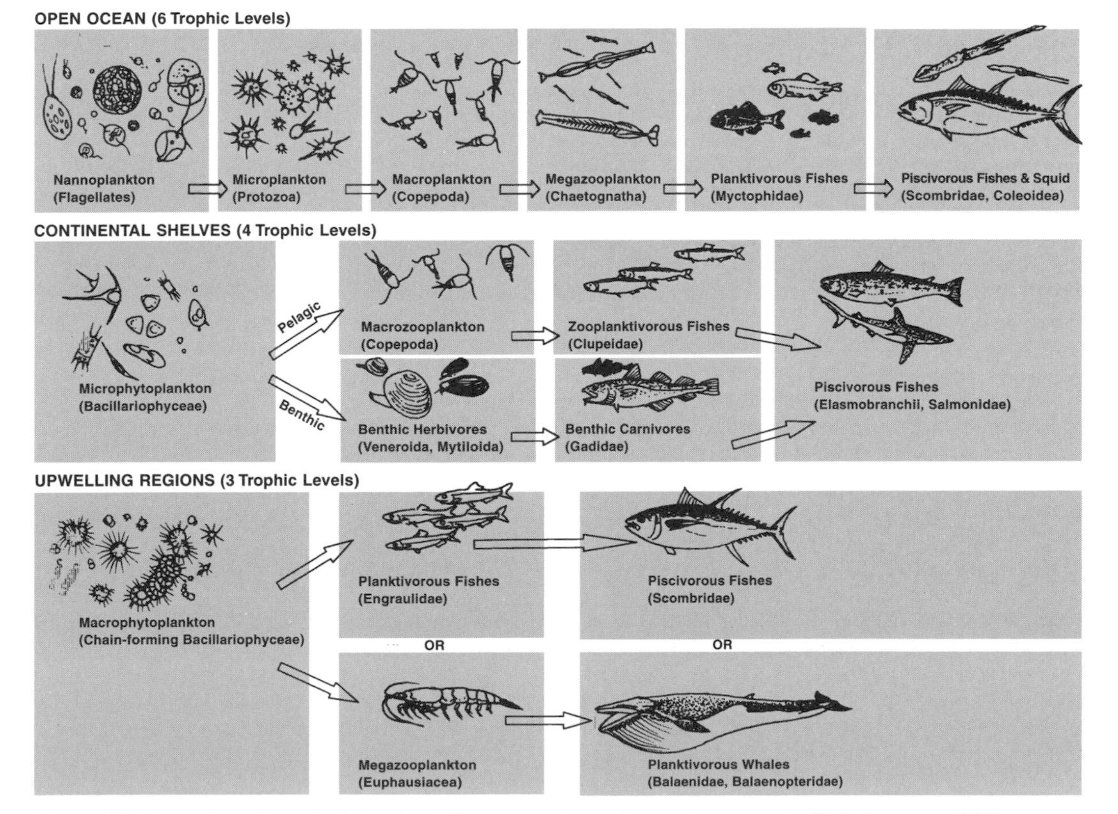

Figure 28. Examples of food chains in different marine habitats (based on Lalli & Parsons 1993).

In all ecosystems, animals rely on devouring each other to gain their energy (consumers). Plants and some bacteria are the only true primary producers of new organic material as they have the capacity to nourish themselves through chemosynthetic and photosynthetic processes. Virtually all animals therefore need to consume either plant material or other animals in order to survive. At each stage, when an animal eats a plant or another animal, a new trophic (consumer) level of feeding has been established.

A number of trophic levels where the energy of food is passed through several distinct stages is called a 'food chain' (fig. 28). There is always an energy loss with each step along the food chain as animals' metabolic demands require energy (fig.29). However, any waste products or dead organic material produced by animals can be broken down and recycled by bacteria. This process releases organic nutrients back into the water. The nutrients are then once more free to be taken up by phytoplankton for the purpose of primary production (fig.30).

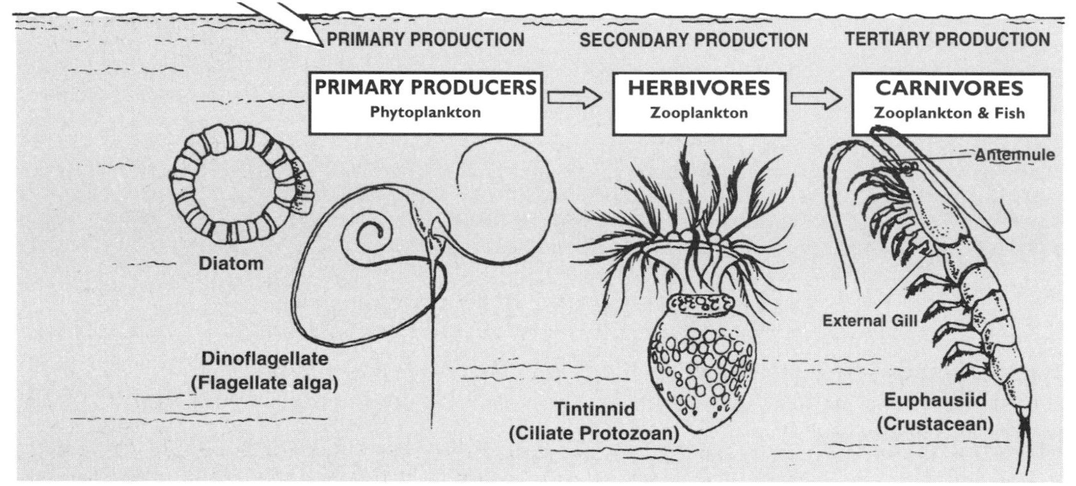

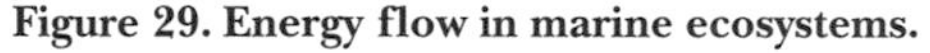

Figure 29. Energy flow in marine ecosystems.

There is a limit to the number of trophic levels that can exist in a food chain and generally there are fewer than six levels. By its very nature, a feeding population must be smaller in total population weight (biomass) than the one it is feeding on. In most food chains the individual members of a trophic level are either bigger in size or stronger than the animals on which they prey (fig.31). Food chains in nutrient-rich upwelling waters are characterised by having high primary productivity, few trophic levels and high biomass of fish or marine mammals. In the nutrient-poor open oceans, phytoplankton productivity is low due to the lack of nutrients and this leads to longer food chains with more trophic levels. With longer food chains more energy is lost, therefore there are smaller numbers of the top-level predators.

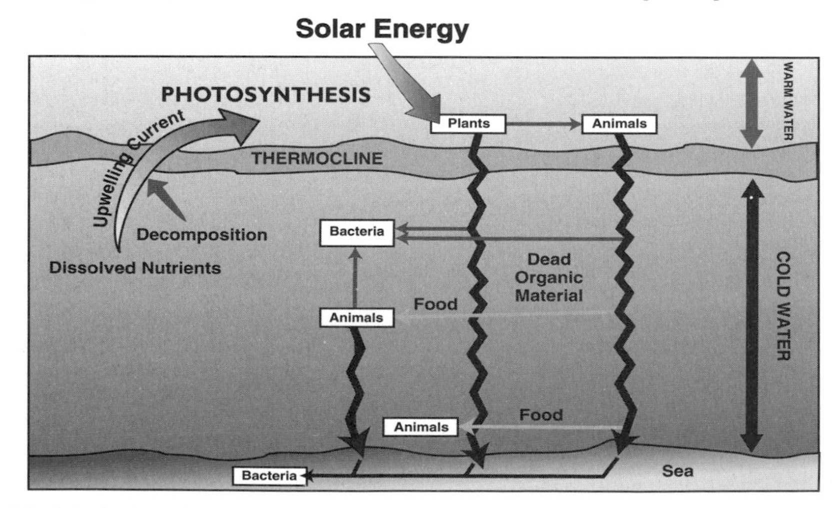

Figure 30. Biochemical recycling of nutrients from dead plant and animal matter (based on Pinet 1992).

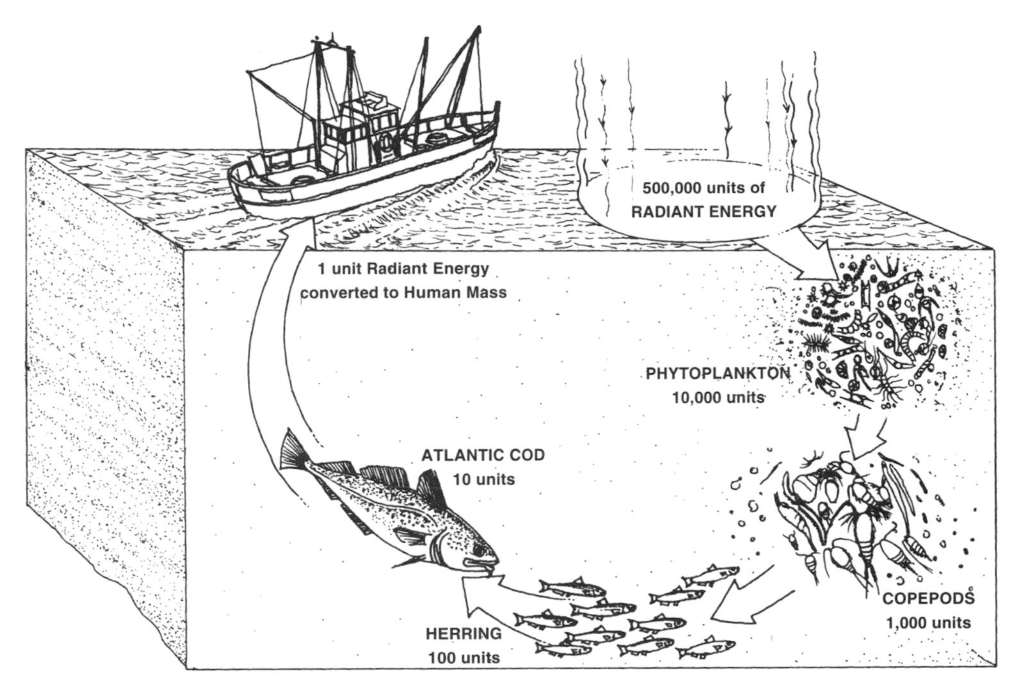

Figure 31. Energy flow in a North Atlantic food chain (based on Thurman 1994).

A food chain is a convenient and simple method of looking at trophic relationships (fig.32a), but in nature few simple food chains exist. Trophic relationships tend to be complex as predators often feed on more than one type of organism and the prey may come from several different trophic levels. Therefore, a 'food web' is a more accurate reflection of trophic relationships (fig.32b). Food webs are further complicated by several biological and behavioural factors, e.g. some animals compete with each other for the same food, others change their diet during their life cycles, some animals eat detritus (dead organic matter) and many animals exhibit cannibalistic tendencies.

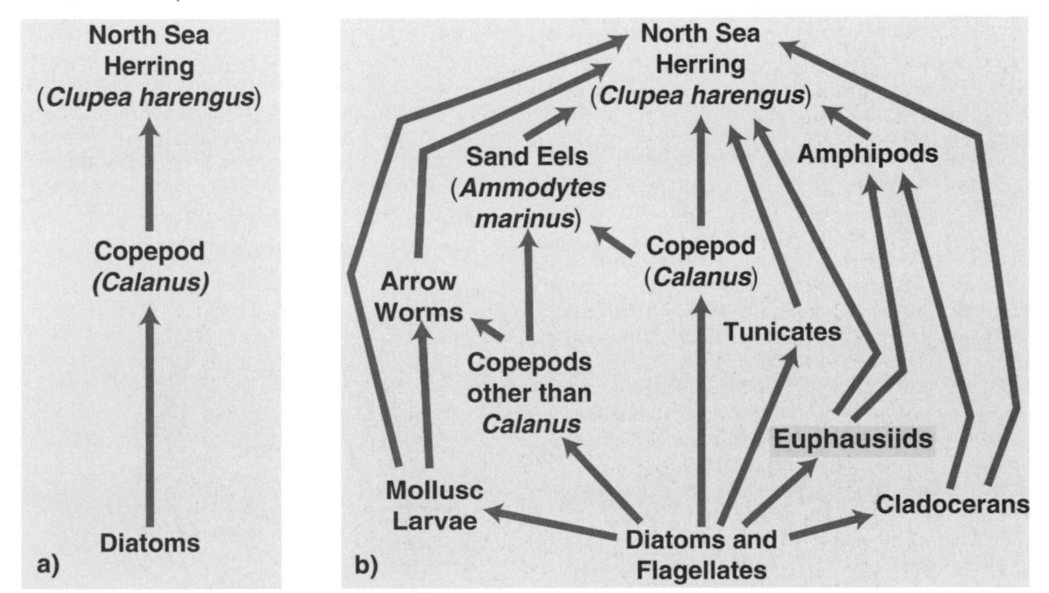

Figure 32. a) Food chain (three levels) and b) Food web of the North Sea Herring.

Among some of the best studied food webs of the world are those of the Antarctic region (fig.33). Much of the interest has been generated through the decline of the large whales and fisheries stocks. It was once thought that cold polar regions had low productivity with only simple food chains, but it is now known that some of the most world's productive waters are to be found there. The key organism in polar waters is the shrimp-like krill *Euphausia superba* which supports a wide variety of higher predators including fish, seabirds, penguins, squid, seals and whales. Over-exploitation of krill resources by man would have a disastrous effect on the polar ecosystem.

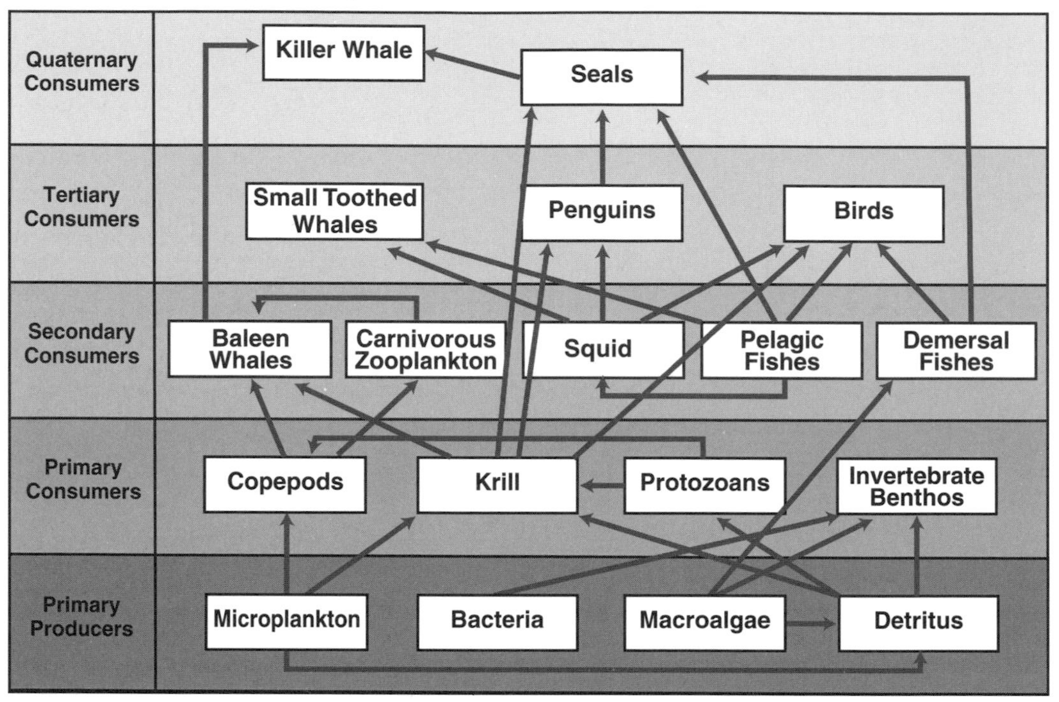

Figure 33. Food web of the southern oceans.

MIGRATION OF MARINE ANIMALS

Migrations are, by definition, active or passive movements of animals from one area to another and many migrations occur as seasonal movements. These regular seasonal movements enable certain species to make the best use of different areas at certain times of the year, or at different stages during their life cycle. The main functions of migration are for breeding and feeding purposes. Making the most efficient use of food resources will optimise survival of young and thus reproductive success. Therefore, migration can be seen as an adaptation to increase survival into adulthood. The fact that most pelagic and demersal fish populations are successful in terms of numbers is a clear indication that migration works well as a survival mechanism.

Migratory species of animals can be found in many of the larger and higher orders of marine animals such as squid, marine turtles and marine mammals including seals, whales and dolphins. Many true species of seabirds, including some albatrosses, petrels, frigate birds and the Arctic Tern *Sterna paradisea,* also undertake remarkable migrations. In all migratory marine animals there is a strong 'homing instinct' which guides them back to a particular area or place for the purpose of feeding or breeding. Among many fish species, finding suitable spawning grounds is essential if the young are to survive into adulthood. Often the spawning grounds are in areas of high planktonic productivity where the newly-hatched young can find a plentiful food supply of the right size.

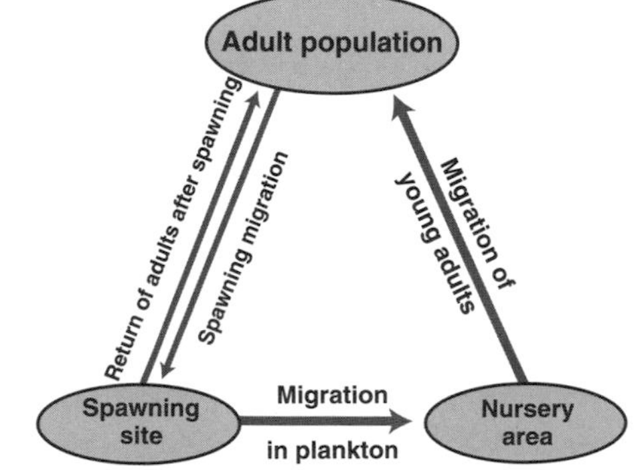

Figure 34. Marine migratory cycle (after Barnes & Hughes 1992).

A migratory 'cycle' or 'triangle' suitably describes the generalised migratory patterns of marine organisms. The circuit is represented by a spawning site, nursery area and adult feeding grounds (fig.34). Breeding in fish takes place on or over the spawning grounds and once the young hatch, they drift passively

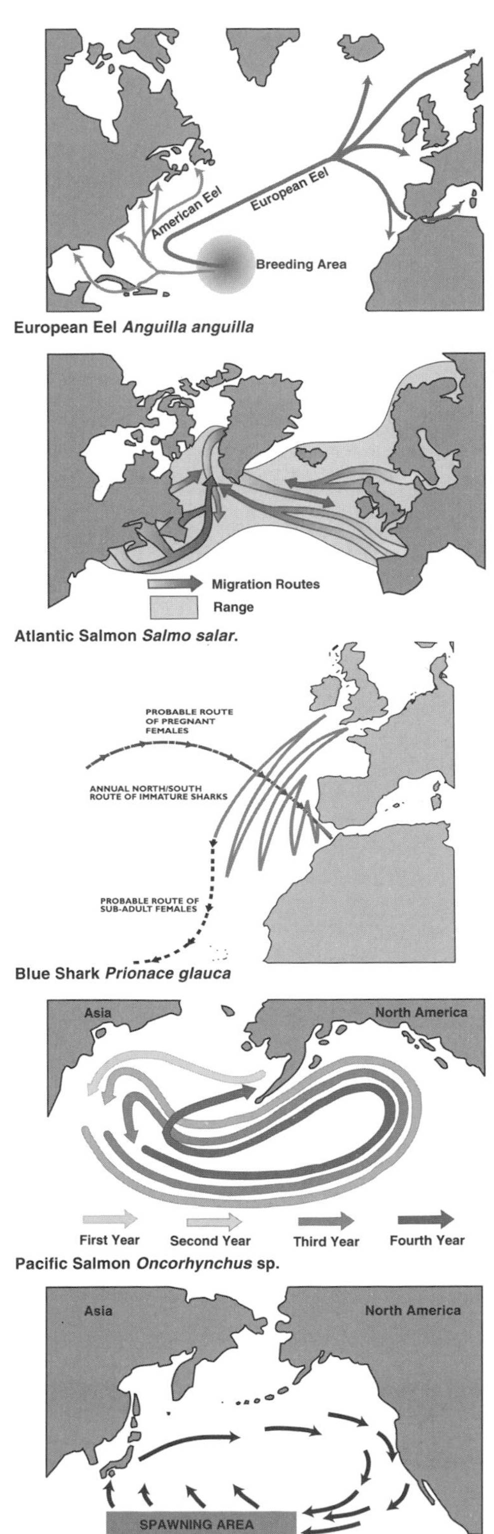

Figure 35. Migratory routes of some fishes.

in the prevailing water currents to a nursery area. The young then join the adults on the feeding grounds and once sexual maturity is reached, they return to the spawning grounds to complete the life cycle. In some deep-water fish species there is a vertical component to the migratory circuit. Buoyant eggs may be laid and these float to the surface zone where there is plenty of light and higher productivity levels. This provides the larvae with an abundance of suitable small food which is not available in the deeper waters. As the young fish mature to adulthood they migrate to the deeper levels to join the adult population.

Large distances can be covered during migrations: Herring *Clupea* sp. may travel 3,000km in a round trip, the European Eel *Anguilla anguilla* up to 3,000km and the Blue Shark *Prionace glauca* uses the whole of the North Atlantic basin during its migration (fig.35). Males are found seasonally off the eastern seaboard of the United States, where mating occurs; sexually mature females are found off European coasts and the Bay of Biscay, where they are thought to give birth.

The Gray Whale *Eschrichtius robustus* has an even longer migration of up to 18,000km (fig.36). Although migratory species are capable of swimming against prevailing currents, it does not make biological sense to use energy doing this. However, both migratory birds and whales use up large amounts of stored body fat as fuel on their migrations. Some migratory cycles take place within ocean gyres and marine animals use the currents to assist their passage, rather like an escalator. It is well known that many currents run in different directions or even in opposite directions at different depths. Many migratory animals will take advantage of this by regulating their swimming depth in the water column. The Pacific Hake *Merluccius productus* may migrate south to spawn by using the California Current, but returns north in the deeper counter-current system.

Many migrations take place along the same routes and the arrival at various areas may often coincide with the peak production of plankton or other food sources. Plankton production is often limited to short periods in temperate and polar regions, and therefore migrations need to take advantage of the food supply. The annual arrivals of migrants can be a regular event. Some animal species may have been arriving at the same locations every year for thousands of years. How an animal finds its way is not clearly understood. Rough guidance systems may include direction of current flow and familiarity with the water masses. Species like salmon *Salmo* spp. and Sea Trout *Salmo trutta* are able to locate the precise river and exact tributary were they were spawned. It is likely that some form of chemical imprinting takes place after hatching and the young fish never forget this 'water print'. The fish then 'home in' on this chemical beacon on the return journey as they approach the coastline or estuary. It is known that some 'straying' of salmon and trout does occur on the journey up the river systems.

However, this may be a beneficial mechanism to help colonise new areas and to find new spawning tributaries; these stray fishes are unlikely to make much difference to overall population survival if they do not breed. With all migrations, there is safety in numbers and there will always be some that will run the gauntlet of predation and return to spawn. Spawning in one area over a short period of time also makes it easier to find a mate (or many mates) and thus increases reproductive success.

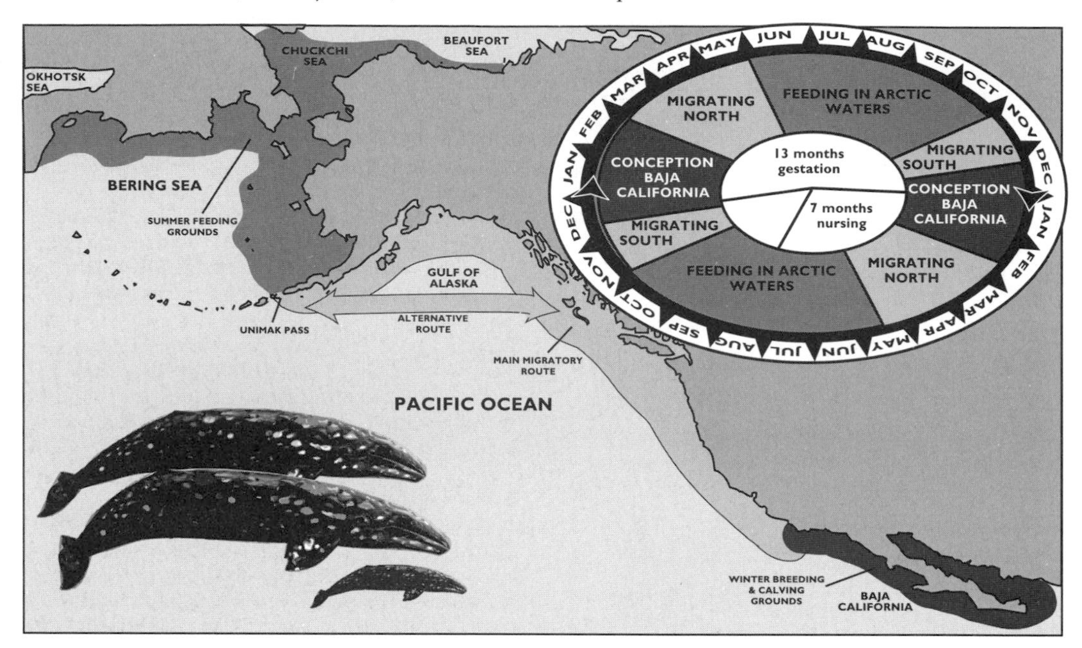

Figure 36. Migration and reproduction in the Gray Whale *Eschrichtius robustus*.

In tropical regions, migrations are not as common as in the higher latitudes as there is no peak production of planktonic food for the young migrants. However, many fish species and other higher predators will take advantage of upwelling areas, where nutrient-rich waters will yield a bonanza of plankton. Within the tropical belt, marine turtles such as the Green Turtle *Chelonia mydas* make precise migrations between their feeding grounds off the coast of Brazil and their nesting sites on the tiny island of Ascension in the mid-Atlantic. How the turtles manage to find the island is a mystery. Other species of animals in the seasonally productive waters of the higher latitudes may make lesser migrations to keep within the areas of high productivity. Quite often this will involve following the plankton blooms. Fishes such as cod *Gadus* sp. will take advantage of the Arctic summer bloom of plankton and then move further south to spawn in lower latitudes. They then follow the spring algal bloom as it spreads northwards the following year. Whales are well known for migrating to polar regions to feed on the rich spring plankton blooms.

Studying animals on their migrations is a difficult task. Information has been collected by tagging animals in one place, releasing them and hoping they will be found elsewhere, or back at the original tag site, at a later date. However, with microchip technology and satellite communications it is now possible to attach a relatively small electronic transmitter onto larger animals such as whales, turtles and seals. They can then be tracked electronically by satellite. A constant monitor of time, position, direction of travel and speed of travel can be recorded as they move around the world.

Seabirds exhibit some astounding feats of migration. Seabird migration is better understood as they are more easily seen on migration and many of their breeding locations are well known. Many bird species will use the prevailing winds to assist them. They will often follow the coastlines of major continents, only crossing open oceans when a favourable wind system is found. Many species of seabirds have evolved specialised anatomical and physiological adaptations to cope with the long journeys. True ocean birds can settle on the water to rest and can also drink seawater without ill effect. In the Wandering Albatross *Diomedea exulans* locking wing bones enable the wings to be kept rigid for long periods of gliding with little effort. These true ocean wanderers seek out the stormy areas of the southern oceans where large waves and swells provide updraughts to lift the birds above the water. They regularly perform circum-Antarctic navigations in search of food and nesting sites on remote subantarctic islands such as South Georgia.

Marine animal migration and its recording is a subject in which the seafarer can be of great value. Further data on sightings and migrations at sea are required to fill many of the gaps in our knowledge.

BIOLUMINESCENCE (ANIMAL LIGHT)

About, about in reel and rout
The death-fires danced at night,
The water, like a witch's oils,
Burnt green, and blue, and white.

Extract from *The Rime of the Ancient Mariner* by S.T. Coleridge

Bioluminescence refers to cold light emitted by organisms and it is known to be produced in many marine species of bacteria, dinoflagellates and invertebrates (both pelagic and benthic). Among vertebrates, fish are known to produce bioluminescence and it becomes increasingly important in the deep seas where it is the only form of light at depths greater than 1,000m, or at night.

The biochemical production of cold light is not completely understood, but it is thought to be the result of oxidation of organic compounds called 'luciferins' in the presence of the enzyme 'luciferase'. The biochemical reactions take place in special cells of the body called 'photocytes' or, in more complex animals, in organs known as 'photophores'.

Bioluminescence may have a communication function, but for many species the reasons for its use are not certain. In some planktonic species bioluminescent displays occur when the organism is disturbed in the wake of a passing boat or by a swimming animal. After disturbances some groups (e.g. siphonophores, ctenophores, jellyfish, ostracods and deep-sea squid) shed luminescent tentacles or release clouds of luminescent fluid. These chemicals act as an apparent decoy to distract potential predators. While the predator seeks out the source of light the prey swims off in a different direction. In some pelagic species bioluminescence may serve as a type of camouflage by breaking up the outline of the animal against surface light (similar to the camouflage clothing used by military personnel). Some invertebrates that have developed efficient light-sensing eyes (e.g. krill *Euphausia* spp.) may use bioluminescence to respond to, or to communicate with, others of its kind. It may be for the purpose of warning against an attacker or as a reproductive signal. Some deep-sea fishes and siphonophores use bioluminescent lures to attract prey within close range, as this saves energy which would otherwise be used to actively hunt prey. The lure may be dangling on top of the head or simply a patch of luminescent skin inside the mouth. Therefore, the ability to produce cold light has evolved independently in a variety of marine organisms and it has a number of uses ranging from 'searchlight', lure, warning signal, fright reaction, escape action, camouflage and communication for feeding and reproduction.

The types of luminescence observed at sea vary according to the type of organism producing it (although this is not known in all cases). Surface luminescence may take the form of:

1. **Sparkles**: pinpoints of light with 'comet-like' tails created as the organisms swim away. This is the most common form of bioluminescence produced by plankton after surface disturbance and it is found world-wide.
2. **Pulses and blobs**: large blobs of light which slowly fade after disturbance. The light may come from large salps and jellyfish and the whole animal may light up in the process of disturbance.
3. **Erupting luminescence**: dramatic balls of light that ascend from the depths, erupt at the surface and spread out as rings of light, with a diameter of 30cm to several metres. the effect is similar to a large gas bubble erupting at the surface causing ripples to radiate out from the centre. The source and mechanisms of these eruptions are not known.
4. **Luminescent bands**: large moving ribbons which are up to 400m in length and 0.5m in width. Bands may occur parallel to each other, 60m to 300m apart, or as individual ribbons.
5. **Luminescent wheels**: these have the appearance of a spiralling 'Catherine wheel' radiating out from a central hub or as 'spokes' of light moving away from the central hub. They may revolve in either direction and the wheels can be large.
6. **Milky seas**: the sea may have a ghostly appearance produced by diffuse bioluminescent light seen at or just below the surface.

Whatever type of luminescent light is encountered by the seafarer, a record of the type with the location, size, weather and sea conditions would make an interesting account in the ship's log or nature log. The bioluminescence of fishes is discussed in the Marine Fishes text (p.365).

Benthic Environments

SUBSTRATES AND ORGANISMS

The seabed offers a wide variety of diverse habitats, each with its own unique environmental and biological conditions. Types of bottom substrate may vary from hard rock, boulders and stones to the softer sediments of sand, mud and clay. Hard rocky areas provide a stable surface for attachment and many organisms e.g. algae, barnacles and mussels spend much of their lives attached to a hard substrate. These animals (epifauna) do not move around once they have settled and are known as 'sessile'. Crevices and rocky overhangs provide a different, more sheltered habitat which is colonised by organisms that cannot endure the full force of waves or currents. Crevices and cracks also provide shelter from predation by larger animals. By comparison, soft substrates provide a burrowing habitat for large numbers of animals that live in the surface layer (infauna). Soft substrates also provide a readily available food source and many burrowing creatures eat their way through the substrate (detritus feeders).

The species diversity of animals and their relative abundance in coastal areas will be determined by important environmental factors which include tidal levels, exposure to air, wave and water action, salinity and temperature fluctuations. At all depths in the sea the type of substrate will determine the relative proportions of epifauna and infauna. Biological factors that affect many of the animals include competition for food and living space, predation levels and the ways in which a community of plants and animals may develop.

Many benthic organisms have no free-swimming larval stage in their life cycles. Other organisms produce either 'planktonic' larvae or 'lecithotrophic' larvae. Planktonic larvae are produced in vast numbers and are widely dispersed by the water currents. These larvae are relatively small at hatching and may spend several months swimming and feeding on small planktonic matter. The majority of these larvae are consumed by other predators. However, some will survive and settle to a sedentary adult lifestyle if suitable substrates are found. Lecithotrophic larvae are produced in fewer numbers but the eggs are much bigger and the larvae hatch at a relatively larger size. During their time as free-swimming larvae they do not capture prey as the internal structures for food processing have not developed fully. They feed by consuming yolk originally laid down in the egg. Their relatively large size and shorter stay in the water column increase their chances of survival and therefore mortality rates are lower and recruitment to the adult population is less variable, compared with non-lecithotrophic planktonic larvae.

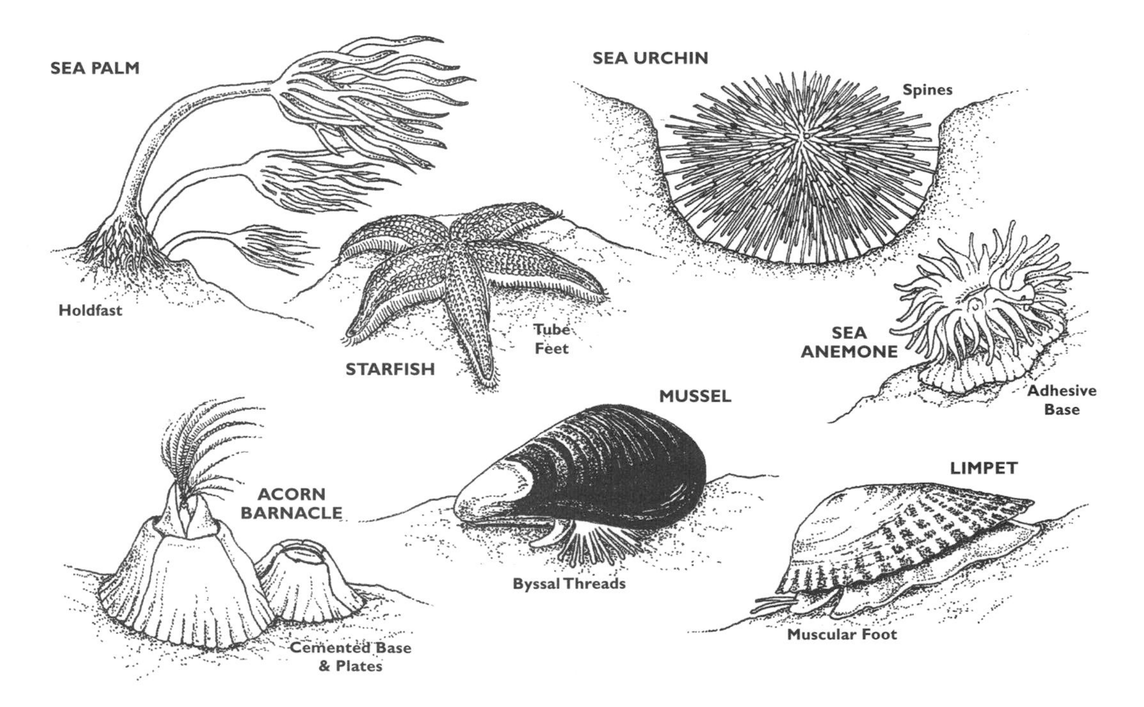

Figure 37. Attachment methods among various epifaunal organisms.

Benthic marine organisms exhibit a number of different lifestyles. Some benthic species live part of or all their lives in soft bottom substrates. These include many invertebrate species such as polychaete worms and bivalves. Some bottom-dwelling fish (e.g. mudskippers) and crustacea (e.g. shrimps and crabs) dig burrows for protection, but wander out onto the surface to feed. Many flatfish species and cuttlefish partially cover themselves in sandy substrate to avoid detection while waiting for a passing meal to come within striking distance. Many other benthic fish species also rest on the bottom and wander over it in search of food. Other fish hover above the seabed, living within the vicinity of the bottom and sometimes resting on it. These species are known as 'demersal'.

There are few infaunal organisms living on rocky surfaces as they are too hard to burrow into. However, a few bivalve molluscs bore into the softer rocks of limestone, sandstone and chalk. Other creatures such as limpets and some sea urchins can excavate shallow pits in the surface of soft rocks to give themselves some protection and better purchase. Epifaunal plants (algae) and animals live a sedentary life attached to a hard surface by a variety of special methods (fig.37). Although epifaunal organisms are found on both hard and soft substrates, they are far more diverse on hard substrates in shallow water reefs and intertidal areas. Common examples of epifaunal groups include mussels, many species of starfish, sponges, corals and barnacles.

The richness and diversity of benthic communities are closely linked to the primary productivity of phytoplankton which directly or indirectly provides the food supply for many benthic organisms. Therefore, shallow water areas that have high light levels and high nutrient levels provided by water mixing, upwelling and land run-off have high primary productivity and diverse benthic faunas. Deep-sea benthic faunas are characterised by a number of well represented invertebrate groups including the sponges (Porifera), brittle-stars (Echinodermata), snails (Mollusca), worms (Annelida) and crustaceans (Crustacea). Fish are also present.

Benthic organisms can be grouped according to their size and they include:

1. **Macrofauna** (or **macrobenthos**): these animals are larger than 1mm and include the large polychaete worms, corals, shellfish and starfish.
2. **Meiofauna** (or **meiobenthos**): these animals are between 0.1mm and 1mm and include small crustaceans such as copepods, small molluscs and worms.
3. **Microfauna** (or **microbenthos**): these include organisms smaller than 0.1mm such as protozoans and bacteria.

BENTHIC FEEDING

Benthic food webs are just as complex as any of those found in the pelagic environment. The majority of benthic organisms either rely on organic food which either slowly sinks to the bottom, or is brought to them on the water currents. Other organisms are carnivorous and eat other benthic animals for their food supply. Animals living in soft sediments consume the mud for its organic content. Many of the deep-sea organisms rely on food from the surface waters for their organic material. However, by the time much of the organic material reaches the bottom a high proportion of the nutritional contents may have been dissolved in the surrounding water or consumed and recycled by pelagic bacteria and fungi.

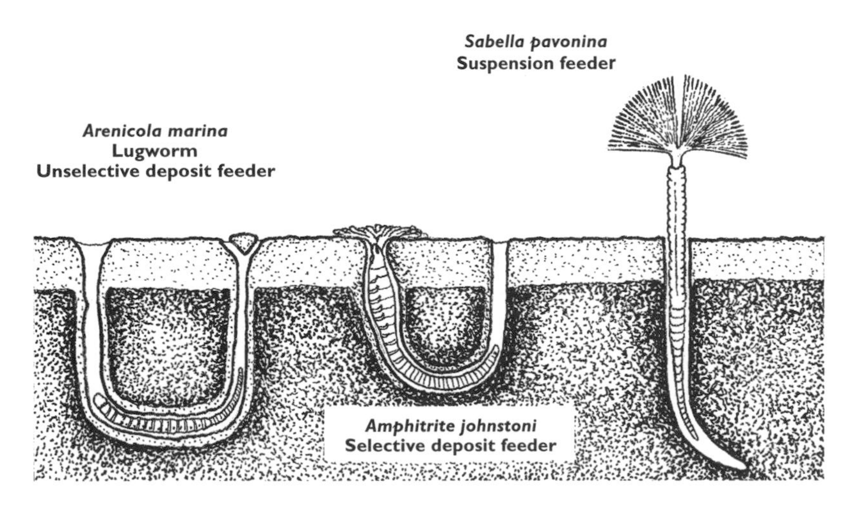

Figure 38. Feeding methods of some common muddy sediment animals (after Meadows & Cambell 1993).

Macrofauna

'Macrobenthic animals' have a variety of feeding mechanisms. Herbivorous 'grazers' include many molluscs (e.g. limpets, periwinkles, chitons etc.) and some sea urchins which actively graze on algae on both hard and soft substrates. Gastropods use their rasping teeth (radula) to scrape off the algae from the surface. Another group of organisms are the 'filter feeders' (or suspension feeders) and they filter the water for small organisms using a variety of fans, sieves and nets (fig.38). The tube-dwelling worms (sabellids and serpulids), bivalve molluscs and sea squirts (ascidians) all have ciliary methods of collecting the fine food particles. By creating their own current across moving strands of mucus the cilia can pass the food particles towards the mouth. Many of the suspension feeders will extend their feeding apparatus above the surface to avoid 'clogging' from the fine sediment (e.g. the bivalve *Angulus tenuis*). Feeding organs can be withdrawn into burrows or tubes quickly to avoid being eaten by predators. Sedimentary filter feeders consume their food inside a burrow or tube and the faeces are expelled into the surrounding sediment. Filter feeding is an efficient method of obtaining food and energy. It requires little effort as the food is brought to the sessile animals by the water currents. Therefore, many of these organisms are successful and this is reflected in their high numbers, fast growth rates and large biomass e.g. bivalves such as mussels, oysters, scallops and clams. On hard or rocky substrates many of the sedentary suspension feeders are firmly anchored to the bottom by cement (e.g. barnacles) or glued (e.g. 'byssal' threads of mussels) (fig.37). 'Deposit' (or detritus) feeders consume mud and sand as they burrow. In this way they consume organic material and microscopic organisms (e.g. bacteria and fungi) in the deposit. Much of the fine organic material has lost much of its nutritional value, therefore deposit feeders have to consume large amounts of sediment to obtain sufficient food energy. Most invertebrate animal groups are represented among the deposit feeders and include various crustaceans, polychaete lugworms (e.g. *Arenicola* spp.) and sea cucumbers (e.g. *Holothuria* spp.).

Meiofauna

'Meiofaunal animals' live between (interstitially) loose sand and mud particles and their numbers can be high. They can either be temporary residents of this habitat (e.g. young stages of larger macrofaunal adults such as the Edible Cockle *Cerastoderma edule*) or permanent residents. Important permanent groups include the harpacticoid copepods, ostracods, archiannelids, polychaetes, turbellarians and ciliates. The major environmental factors determining the abundance of these organisms include substrate particle size, salinity, type of sediment, depth of water, pH and oxygen levels within the sediment. The meiobenthic fauna in mud and clay is different to that in sandy sediments and this is a reflection of sediment particle size (mud particles are less than 100µm in size). Fine muddy substrates are dominated by the burrowing nematodes and harpacticoids while in fine sand (100-200µm) ciliates and harpacticoids are the dominant groups (fig.39). Meiobenthic animals show marked seasonal peaks of abundance in the littoral zone of the higher latitudes and many of the larger benthic animals (e.g. gobies, flatfishes, hydroids and polychaete worms) rely on the meiobenthic species as an important food source.

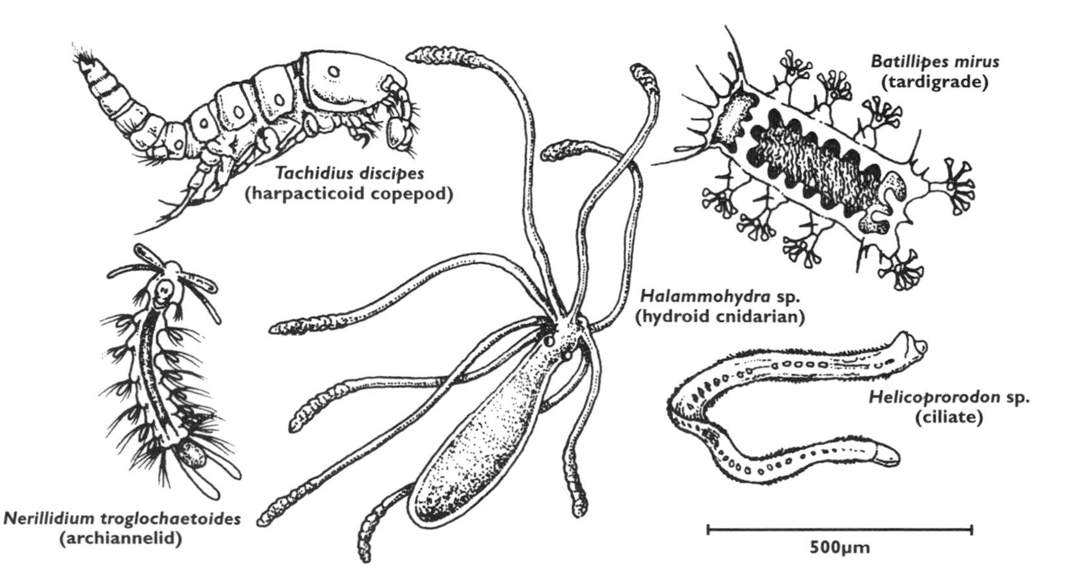

Figure 39. Representatives of the marine interstitial meiobenthos (after Meadows and Campbell 1988).

Microfauna

'Microfaunal animals' are the most diverse assemblage of organisms with large numbers of species living at the bottom. The best known groups include the foraminiferan protozoans of which there are several thousand species. These creatures vary greatly in size and can occur on or in the sediments. In shallow coastal waters they feed mainly on benthic diatoms and algal spores. In all other depth zones the protozoans feed on other protozoans, detritus, bacteria and fungi. The microfauna make up an important food component for the larger meiofaunal animals.

The larger predatory carnivores of benthic ecosystems include flatworms and ribbon worms (Turbellaria), various snails (Gastropoda) and echinoderms such as starfish. Larger molluscs including cuttlefish (*Sepia* spp.) and octopuses depend on bivalves and crustaceans as their major food source. Demersal fish feed on a variety of invertebrate organisms as well as other small fish and eggs. Many demersal fish have developed sensory 'barbels' to detect prey under the surface of sediments (e.g. rocklings *Ciliata* spp., and pouts *Trisopterus* spp.).

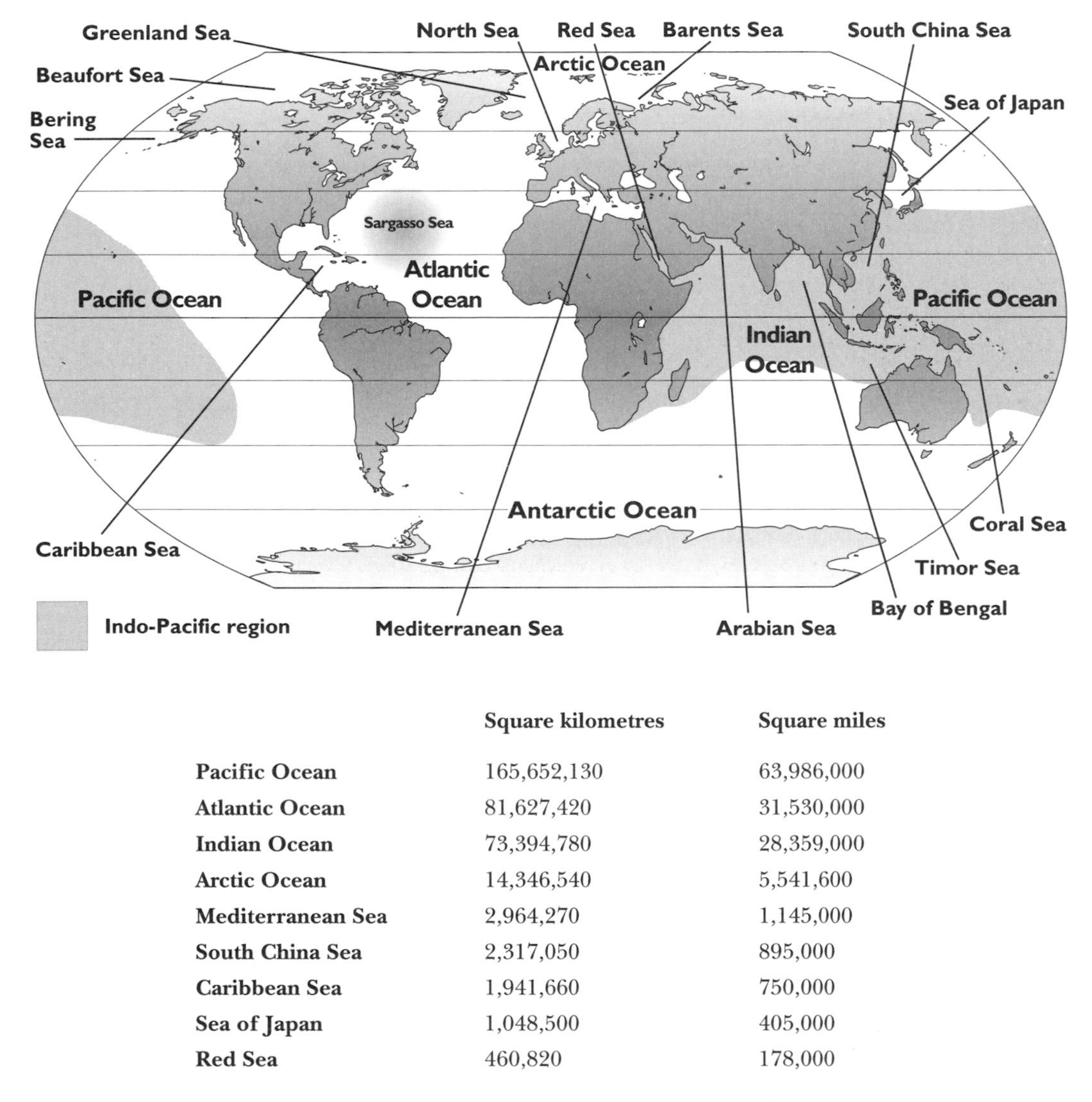

	Square kilometres	Square miles
Pacific Ocean	165,652,130	63,986,000
Atlantic Ocean	81,627,420	31,530,000
Indian Ocean	73,394,780	28,359,000
Arctic Ocean	14,346,540	5,541,600
Mediterranean Sea	2,964,270	1,145,000
South China Sea	2,317,050	895,000
Caribbean Sea	1,941,660	750,000
Sea of Japan	1,048,500	405,000
Red Sea	460,820	178,000

Figure 40. Major oceans and seas of the world, and their approximate surface area.

Coastal Sealife Ecology

Many of the major continents and smaller land masses are surrounded by a relatively shallow rim called the 'continental shelf' which extends down to a depth of about 200m. The shelf looks similar to the brim of a hat and the outer edge of the shelf drops away down the 'continental slope' to the much deeper ocean floor or 'abyssal plain' (fig.23). As land is approached from the seaward side, waters become shallower until a 'coastline' is reached. This marks the boundary between the marine and terrestrial environments. Along the coast is a 'shoreline' (or shore) which forms a narrow strip that lies between the tide marks and is therefore periodically covered by seawater. The 'coastal' region may extend away from the landward limit of the shore for several miles inland, as far as features that are related to the marine processes can be distinguished. The coastlines are marginal fringes of coasts that are formed by dynamic physical, chemical and climatic processes and where biological processes may also play a role. During the processes of coastal formation, materials are constantly being removed from one area and deposited in another. During this action material is slowly broken down into smaller particle sizes by mechanical and chemical processes. Large accumulations of terrestrial material can be deposited along coastlines, through the action of run-off, by fast flowing rivers. As river water nears the sea it slows down as the river bed becomes shallower and the suspended river sediment is deposited to form sand and mudbanks. If rivers are large enough or several rivers enter the sea at the same location, the deposited material may form wide shallow areas called 'deltas' (e.g. Mississippi and Nile deltas). In most coastal estuaries the deposited silt forms mudbanks and mudflats, and these may eventually become saltmarshes.

The sculpting of coastlines by various physical and climatic processes has created a variety of distinct habitats with their own diverse communities of organisms. The major types of coastal environments include:

1. **Shoreline environments**. These include those areas above the high tide mark and therefore they are part of the coastal region. The areas include:

   I. **The strandline** and beach areas above the high tide mark.

   II. **Sea cliffs**.

   III. **Sand-dunes**.

2. **Intertidal environments**. These include the areas between the high and low tide marks and are found mainly along seaward facing coastlines. In areas with large tidal ranges, clear habitat zones and their characteristic organisms can be distinguished (zonation). There are three distinct types of 'intertidal' habitats:

   I. **Rocky shores** (including rock pools, rock crevices and overhangs). These shores vary with a mixture of solid rock, boulders, cobbles, pebbles and gravel. Generally there will be no particles less that 2.0mm in size.

   II. **Sandy shores**. These commonly consist of mineral silica and particle sizes vary from 0.063mm to 2.0mm. There will be little organic material in the sand as it tends to be washed out.

   III. **Muddy shores**. These consist of varying amounts of organic material mixed with silt and clay particles to form what is known as mud. Particle sizes vary from 0.062mm down to 0.00195mm.

3. **Estuaries and Saltmarshes**. These are semi-enclosed, coastal areas of water in which the ocean water is significantly diluted by freshwater from land run-off. Many bays, inlets, sounds and gulfs may be considered as estuaries on the basis of their position and freshwater input. Saltmarshes are often found in estuaries adjacent to mudflats and occur in latitudes up to 65° north and south of the equator. Silt builds up at the top of tidal range and here plant colonisation takes place.

4. **Mangroves**. These areas replace saltmarshes in tropical and subtropical regions. They are common in a region 30° either side of the equator and comprise of trees and shrubs that are specially adapted to withstand water immersion and variable salinities.

5. **Macroalgal Kelp forests** (0-40m depth). These marine forests are found in cold temperate regions below 20°C. They start at the low water mark and extend down into subtidal areas. Kelp refers to the large brown algae up to 50m in length which have their own associated plant and animal communities. Kelp normally requires a hard substrate for attachment.

6. **Seagrass meadows** (0-50m depth). They occur in all latitudes and are the only true marine vascular plants. These monocotyledons prefer clear sheltered waters that are well lit, including inlets and coral lagoons. The dominant plant in temperate latitudes is Eelgrass *Zostera* which this is replaced by Turtlegrass *Thalassia* and other species in tropical regions.

7. **Coral reefs**. These shoreline environments are biologically constructed by the activities of living organisms of the phylum Cnidaria which secrete a supporting skeleton of hard calcareous material. Coral reefs are generally found in tropical regions of the world where the annual average minimum

sea temperature is above 18°C (64°F). Coral reefs require clear seawater as sediment will smother them and clog their feeding structures.

Many of these nearshore environments and their associated faunal and floral communities are often found close to human habitation. In many cases, human populations have established themselves and grown up in and around nearshore environments. Due to their importance and ease of access for study, each of the nearshore environments will be discussed in more detail. They are probably some of the best environments for marine naturalists to study.

SHORELINE ENVIRONMENTS

The influence of the sea extends well above the high tide mark and in many areas its effect is felt some considerable distance inland. The shoreline provides many diverse breeding sites for seabirds, shorebirds and terrestrial animals. Care must be taken during visits to shorelines and coastal areas during the breeding seasons. Many shoreline habitats now form parts of nature reserves where public access is controlled.

THE STRANDLINE

The strandline often looks like a wavy band of debris which snakes its way along the beach at the high tide mark. The position of the strandline will oscillate with the levels of high water and will be pushed furthest up the beach after storms and the spring equinox tides of spring and autumn. The debris which is referred to as 'flotsam and jetsam' is a mixture of dead plant material, dead animal remains and rubbish of human origin (fig.41).

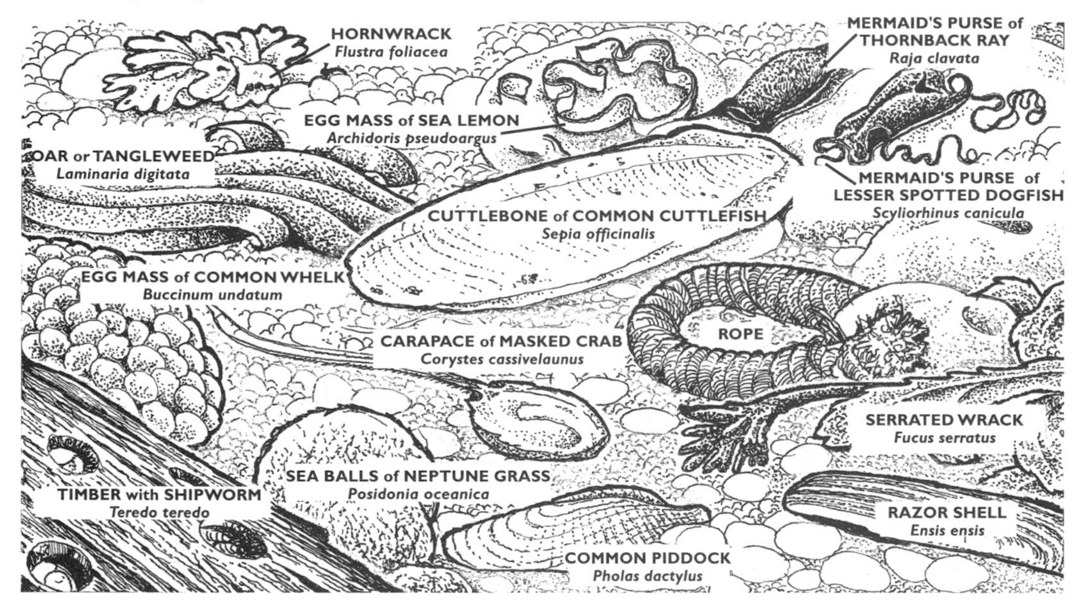

Figure 41. Some typical flotsam and jetsam washed up along the strandline.

The strandline is a rich area to explore and one that is often overlooked. One of the major delights of the strandline is trying to discover what the material is and where it originated from. Unusual pieces may include timber washed up from tropical regions with hole-boring animals (e.g. Shipworm *Teredo*) still visible inside. Shells from many mollusc species and egg cases from gastropod snails, skates and dogfishes (mermaid's purses) are also common finds. The strandline is an area where weak or dying animals may come to rest, especially after storms or during the breeding season. However, many of the weak animals are quickly killed by scavenging predators unless they are well hidden. Whales, dolphins and seals are the largest animals to be washed up or stranded by the ebbing tide. The cause of whale stranding is still not known and sometimes whole groups of animals (numbering many hundreds) can become beached. If smaller cetaceans are kept damp and cool they may be successfully refloated on the next high tide, without permanent damage. In many regions of the world the strandline is home to the scavenging Sandhopper *Talitrus saltator* and many land-dwelling insects including flies and beetles which feed on the dead organic debris. Although these animals provide an excellent 'clean-up service' they can be a nuisance to anyone sitting near the strandline during the summer months.

SEA CLIFFS

Cliffs provide suitable nesting sites for many species of seabirds such as fulmars, (Procellariidae), gulls (Laridae) and auks, (Alcidae). Cliff ledges give freedom from human interference and protection from predators and other scavenging animals. High cliff areas also provide an assisted take off from the updraughts of air passing up and over the cliff face. Other birds such as crows, rock doves and some birds of prey also use cliffs for breeding and hunting grounds. If severe weather conditions occur during the breeding season, a proportion of the young birds may fall victim to the elements and provide food for other birds. Seabirds returning from offshore feeding grounds may be robbed of their catch by other birds as they approach the colonies. Sea cliff areas provide habitats for other animal species including reptiles and a variety of plants which are tolerant to salt spray. Bird excretion results in high concentrations of nutrients which enrich the soils on the ledges. The nutrients are then available to the plants which flourish here and these plants help to bind the soil, thus reducing erosion.

Many seabirds nest on the flat tops of sea stacks or on isolated islands in preference to cliffs, but only when they are free from humans and scavenging animals. Because of their size many large seabirds have no choice but to nest in open areas. However, they prefer to choose a site which has a seaward slope that provides an updraught of air to aid takeoff. Many birds including some gulls and auks do not build a substantial nest, and eggs may even be laid directly onto bare rock or in a shallow scrape. Birds that do this often breed in dense colonies and in overcrowded conditions, probably as a defence mechanism against egg-stealing birds. Birds that choose cliff faces as nesting sites will seek out the harder types of rock which provide a safer platform. Chalk and sandstone cliffs are unstable environments prone to collapse and rock falls.

Puffins *Fratercula* spp., shearwaters and petrels (Procellariidae) nest in burrows and appear to fly almost straight into the burrow entrances as they return from sea. The breeding cycles of many seabirds are strictly seasonal and are often timed to coincide with the increased seasonal food availability. In the northern Holarctic region, the season for nesting seabirds is short and generally lasts from late April through to early August. During this period, the cliffs are alive with the noise of continually squabbling birds, and the frequent flights of the adults to and from their feeding grounds. In calm conditions 'rafts' of birds can often be seen resting on the waters below the nesting areas.

SAND-DUNES AND SANDBANKS

Sand-dunes are formed by sand blown up the beach by the wind. Large systems of dunes can form in river deltas to create a complex pattern of freshwater, brackish water and seawater marshes. The areas then develop into important habitats for shorebirds, wildfowl, seabirds, herons, cranes and even flamingos. Provided that the sand removed from the beach at low tide is being replaced by new supplies, the dunes can continue to grow in height. The highest coastal dunes are found on Moreton Island in Australia where they have reached a height of 250m. Dune systems may also be driven inland and this can be seen in the Coto Doñana dune system of Spain. Alternatively, dune systems can also grow out into the sea. The Carmargue region of the Rhone delta of France is moving out to sea because of the strong Mistral winds that blow down the Rhone Valley and offshore.

Sand-dune areas are inhospitable places for organisms. Strong winds, salty spray, hot sun, drifting sand, little water and poor nutrient retention make it a difficult place for plant and animal colonisation. However, certain plants including the beach grasses are able to capture and hold large quantities of sand in their extensive root systems. These hardy plants are fast growing and have the ability to grow up to the surface if they become buried by the sand. The grasses can withstand high winds, drought conditions and large temperature fluctuations. Marram grass *Ammophila* is especially good at stabilizing dune areas and as the vegetation becomes more extensive, the soil humus, nutrients and water retention slowly increase. Once dunes have been colonised by the 'pioneer' plants, other shrubs and low level plants take hold. Sea Spurge *Euphorbia,* Sea Sandwort *Honkenya,* Dune Lichen *Cladonia,* Sea Holly *Eryngium* and Sea Bindweed *Calystegia* are a few of the more common temperate plants. Eventually, evergreen shrubs will become established and this will be followed by coniferous plants. The last stage of dune colonization is the establishment of broad-leaved trees and their associated woodland communities (fig.42).

Lagoons may also form behind areas of sand-dunes and these areas become 'safehavens' for many of the rarer reptiles, amphibians, insects and the birds. Freshwater ponds that form in these areas become magnets for local wildlife and amphibians which require freshwater for reproductive purposes breed here. Sandbanks can form offshore on the seaward side of dune systems and provide important front line protection against severe weather. Coastal seawater lagoons develop between these banks and the main shoreline. Within their shallow protected lagoons, many young fish and shellfish take refuge from the harsher conditions of the shoreline.

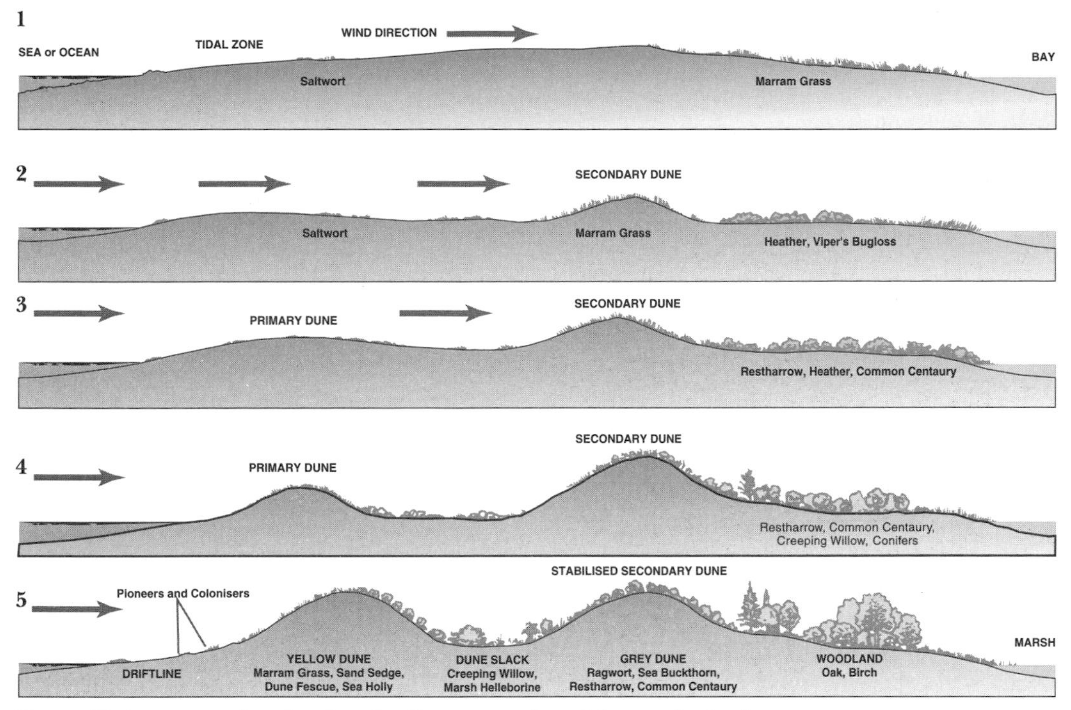

Figure 42. A cross section showing the five stages of sand-dune colonisation.

INTERTIDAL ENVIRONMENTS

INTERTIDAL CONDITIONS AND ADAPTATIONS

The 'intertidal' or 'littoral' environments include those that are periodically exposed to the air by the cyclic rise (flood) and fall (ebb) of the tides. The extent of the tidal range varies enormously with location, the seasons and time of the month. 'Spring' tides occur twice each month, two days after the 'full' and 'new' moons when the combined gravitational pull of the moon and sun are in a straight line. The 'neap' tides occur between spring tides after the first and third quarters of the moon, when the direction of the moon's pull is at right angles (figs.20/21). One tidal cycle (the time between successive high or low waters) may only occur once a day (diurnal) or, more commonly, twice a day (semi-diurnal). In some places the influence of islands and a large lagoon with narrow entrances together produce four daily tides (two major and two minor tidal cycles). This unusual variation can be seen in the lagoon complexes of Portsmouth and Langstone harbours on the south coast of England.

The average tidal range in North-West Europe is about 5.0m at springs and 3.8m at neaps. However, tidal ranges can be as great as 12-14m at some locations (e.g. St. Malo in France, Chepstow in UK). In many semi-enclosed seas that have narrow entrances, the water masses cannot escape quickly during the tidal cycle. Therefore, seas like the Black Sea, Red Sea, Baltic and Mediterranean have restricted tidal ranges of about two metres.

Tidal action and wave action have major influences on intertidal habitats. Tidal currents can transport large quantities of sediments and loose materials. The scouring action of water and loose material causes abrasion and mechanical battering of any protruding bottom substrate. Wave and tidal actions also stir the loose sediment into a suspension and the water becomes turbid and dark. The reduction of light through turbidity is not beneficial to plants but essential nutrients can be released through disturbance of the seabed. The actions of erosion and deposition by the interaction of waves, currents and climate sculpt the coastlines into their various formations. Therefore, these dynamic forces have the ability to create, alter or destroy nearshore and intertidal habitats. On exposed rocky coastlines wave action will be the key factor influencing organisms in this habitat. On more sheltered shores with less wave action, there may be little gap between the marine and terrestrial environments. Here the brown luxuriant algal growth may be only centimetres below saltmarsh grassland. By comparison, on the wave-swept shores algal growth will be stunted and the terrestrial zone separated from the marine habitat by a broad band of bare rock. The shore communities on exposed rocky areas will be characterised by the key species of mussels *Mytilus* spp.

and red algae (Rhodophyceae) which can withstand severe wave action. Where some shelter can be found from the pounding waves, barnacles *Balanus* spp. will be present in large numbers. Sand and shingle shores exposed to heavy waves and surf are generally unstable and support little life except perhaps at the microscopic level. On the more sheltered shores where mud and sand are stable, a rich diversity of animals will be found living in the soft substrate. The entrances to estuaries and sea lochs provide sediment habitats that have distinct zones of animals living within them. Large algae are generally missing from the soft intertidal sediments as there is little suitable surface for attachment. However, any solitary rocks protruding above the sediment may serve as a suitable surface for algal attachment.

Intertidal areas have the largest fluctuations of environmental conditions, when compared to any other marine habitat. Organisms living in intertidal areas have to endure periods of exposure to air, water currents, wave action, and fluctuations of temperature, salinity, oxygen and pH. In cold climates there may be the added problem of ice and ice scouring. To cope with these changes intertidal plants and animals have evolved a diverse variety of behavioural and physical adaptations in these harsh conditions. In soft sediments most of the organisms adapt to a burrowing lifestyle for protection against predation and exposure. Some of the burrowing creatures will only venture out at high tide when the water covers the habitat. On hard, rocky shores it is not possible to burrow for protection. Therefore, the organisms are biologically designed to a tougher specification to endure the harsh environmental conditions. One of the most successful adaptations is the development of a thick, hard and impervious shell which may be single (e.g. gastropod snails) or in two parts (e.g. bivalve clams and mussels). Barnacles have several calcareous plates which can be locked into position and, like all other shelled animals, the parts can be shut tight in adverse conditions. During periods of exposure to air or freshwater the shells need to be sealed tight to avoid desiccation (drying out) or physiological damage to the soft internal organs of the body. The creatures that have a single shell (e.g. winkles) withdraw into their shell and seal the opening with a hard 'operculum' attached to the foot. Bivalves remain closed by the use of strong 'adductor' muscles. Some of the larger gastropods, which include the abalone *Haliotis*, chitons (Polyplacophora) and limpets *Patella*, have a broad, flattened foot which provides a large suction area for attachment to hard surfaces. The shape of these shells is low with streamlined profiles and this helps to reduce the impact of the waves. Sea urchins (Echinoidea) have rigid exoskeletons (skeletons on the outside of the body) which help prevent damage in rough conditions. Like the rock-boring clams (Pholadidae) sea urchins can bore into the hard surface by mechanical and chemical means to create a deep pit for increased protection. They often leave these pits in search of food and return to them when conditions become adverse.

An animal surviving on exposed rocky areas must be able to attach itself firmly to the surface. To accomplish this, a number of different methods have evolved and these can be seen in figure 38. Some use a single large sucking foot (e.g. limpets and anemones), cemented plates (e.g. barnacles) and cemented byssal threads (e.g. mussels). There is also protection in large numbers and many mussels and barnacles become tightly packed to reduce the likelihood of being washed off by wave action. Close proximity to each other may also be beneficial for reproductive purposes. Many of the more mobile or delicate animals such as shrimps, crabs and isopods take refuge in rock pools, cracks, crevices and underneath rocks where wave action is reduced. Rock pools provide a survival capsule of water for many small creatures that could not survive exposure to air and enable organisms to survive higher up the beach.

Marine algae also exhibit adaptations to survive harsh conditions. Many of the large brown algae have thick flexible fronds which bend and flatten as waves pass by. Large algae often have strong stipes (equivalent to higher plant stems) which anchor the algae to a hard surface by means of an entangling 'holdfast' (figs.37/52). The holdfast will sometimes completely surround a boulder or stone which may be buried in a soft substrate. The surfaces of many algae are also covered in mucus which helps to reduce water loss during long periods of exposure. Algae such as *Fucus* and *Enteromorpha* can tolerate a remarkable 60% to 90% loss of water from their tissues without permanent damage.

INTERTIDAL FEEDING AND PREDATION

The main herbivores of the intertidal zone include limpets, many gastropod snails, chitons and sea urchins. These species effectively control the growth and colonisation of many algal species and therefore have an indirect effect on primary production of benthic algae. Many intertidal filter feeding organisms are dependent upon the plankton brought in on the tide. Major filter feeding groups include the tunicates, mussels, barnacles, clams, sponges, polychaete tube-dwelling worms (serpulids and sabellids) and corals. Major intertidal predators include some gastropod snails, starfish, anemones and fish. Starfish are voracious predators that eat a variety of shellfish including gastropod snails, mussels, barnacles and oysters. The predatory Common Dogwhelk *Nucella lapillus* also preys on a number of shellfish including clams, mussels and barnacles. Anemones are carnivorous by nature and will eat small worms, shrimps and fish caught by their tentacles. The scavengers of the rocky intertidal zone include crabs and isopods. Shorebirds and humans can have a predatory impact on shrimp, crab and winkle populations.

ZONATION

Why intertidal animals are found in 'zones' on the shore is not clearly understood. However, it is generally thought there are two major factors responsible for zonation: competition for suitable living space and competition for food resources. If plants and animals become better adapted to living under certain environmental and biological conditions there is a reduced chance of other organisms, which are less suited, encroaching. However, if biological and environmental conditions change, organisms that are best suited to the new conditions may move in. Within all intertidal habitats, stress to organisms may be biological or environmental in nature. Organisms living higher up on the shore will be subjected to higher levels of environmental stress (e.g. from desiccation, temperature, salinity and pH fluctuations) but to lower levels of biological stress (e.g. predation and competition for space and food resources). These stress factors have a profound influence on abundance and diversity of littoral organisms. At lower levels of the shore, species diversity is often high, but abundance of individuals within a species may be low. Higher up the shore, fewer species exist, therefore diversity is low, but within the populations that do exist, numbers of individuals may be high.

In intertidal areas there are distinct bands or 'zones' of animals and plants. Within any zone an individual species or groups of species may form associations. Zonation is most clearly seen on rocky intertidal coastlines where most of the marine inhabitants are on the surface. However some of the organisms will be underwater in rock pools or hiding in the cracks and crevices. Zonation on sandy or muddy shores is not as clear to see, as many of the organisms burrow, but underneath in the soft sediment, distinct zones will be present. Large algal species are not present in soft sediments, as there is little hard substrate onto which the weeds can anchor themselves.

The width and patterns of zones in intertidal regions are dependent on the gradient or slope of the shore, and the tidal range. On flattened shores the expanses of exposed areas may extend for many square kilometres when the tide is out. In locations with large tidal ranges even greater areas of shore will be uncovered at low water. On steep shores or those with cliffs that rise out of the sea, zonation may be reduced or compressed into narrow bands. These bands will be further compressed if the tidal range is small (fig.43).

Rocky Shores

Rocky shores usually have a steep cross-sectional profile. For this reason the wind and wave actions carry away much of the lighter sediments, leaving behind the heavier gravels, stones, boulders and rocks. Rock types may vary from the hard, impervious granitic type to the more porous soft rocks such as limestone, sandstone and chalk. The softer rocks degrade more easily to form cracks, crevices and rock pools. The rocky intertidal shores have three distinct zones comprising the 'supralittoral fringe' zone (splash or spray zone), 'littoral' (midlittoral) zone and 'sublittoral' zone:

SUPRALITTORAL FRINGE ZONE

These areas are only wetted at high spring tides or by breaking waves. This zone is best developed on shores with heavy wave action where spray may be thrown well above high tide levels. It is the transitional area between land and sea and forms an inhospitable habitat which only a few organisms have successfully colonised. They include a few terrestrial animals that can tolerate salt spray and occasional immersion, and some marine species that can resist exposure and desiccation. Organisms found in this zone include some gastropod snails, lichens, crabs, sea slaters, and occasionally flowering plants.

In temperate regions common examples of 'splash zone' lichens include the Encrusting Black Lichen *Verrucaria maura*, the crumb-like *Caloplaca marina*, the leaflike *Xanthoria parietina* and the upright *Ramalina silquosa*. Where a little soil has accumulated in the cracks and crevices, small salt-tolerant plants of Thrift (Sea Pink) *Armeria maritima* and Sea Campion *Silene maritima* may grow. A few small animals live in this zone such as the tiny gastropod snail *Littorina neritoides,* which is almost terrestrial in lifestyle but still depends on the sea for reproductive purposes. The scavenging Sea Slater *Ligia oceanica* is another occupant of this area, mostly venturing out at night and sheltering in damp cracks and crevices during the day. The occasional Acorn Barnacle *Semibalanus balanoides* may also be found in the splash zone but this species is not abundant away from areas regularly reached by spray.

LITTORAL (MIDLITTORAL) ZONE

This forms the largest zone on the shore and extends between the high and low tide marks. Within this zone barnacles are dominant in many regions of the world, but there are distinct local variations in abundance. A variety of organisms may successfully compete for resources especially in the lower parts of this zone. In temperate northern waters, key animal species include sedentary mussels, limpets, gastropod snails and polychaete worms. In tropical regions barnacles dominate but colonial soft corals may also become established alongside the barnacle populations. On rocky shores of the British Isles there is a well established succession of

Figure 43. Rocky shore intertidal zones and faunal limits (see fig.44 for key to zone abbreviations).

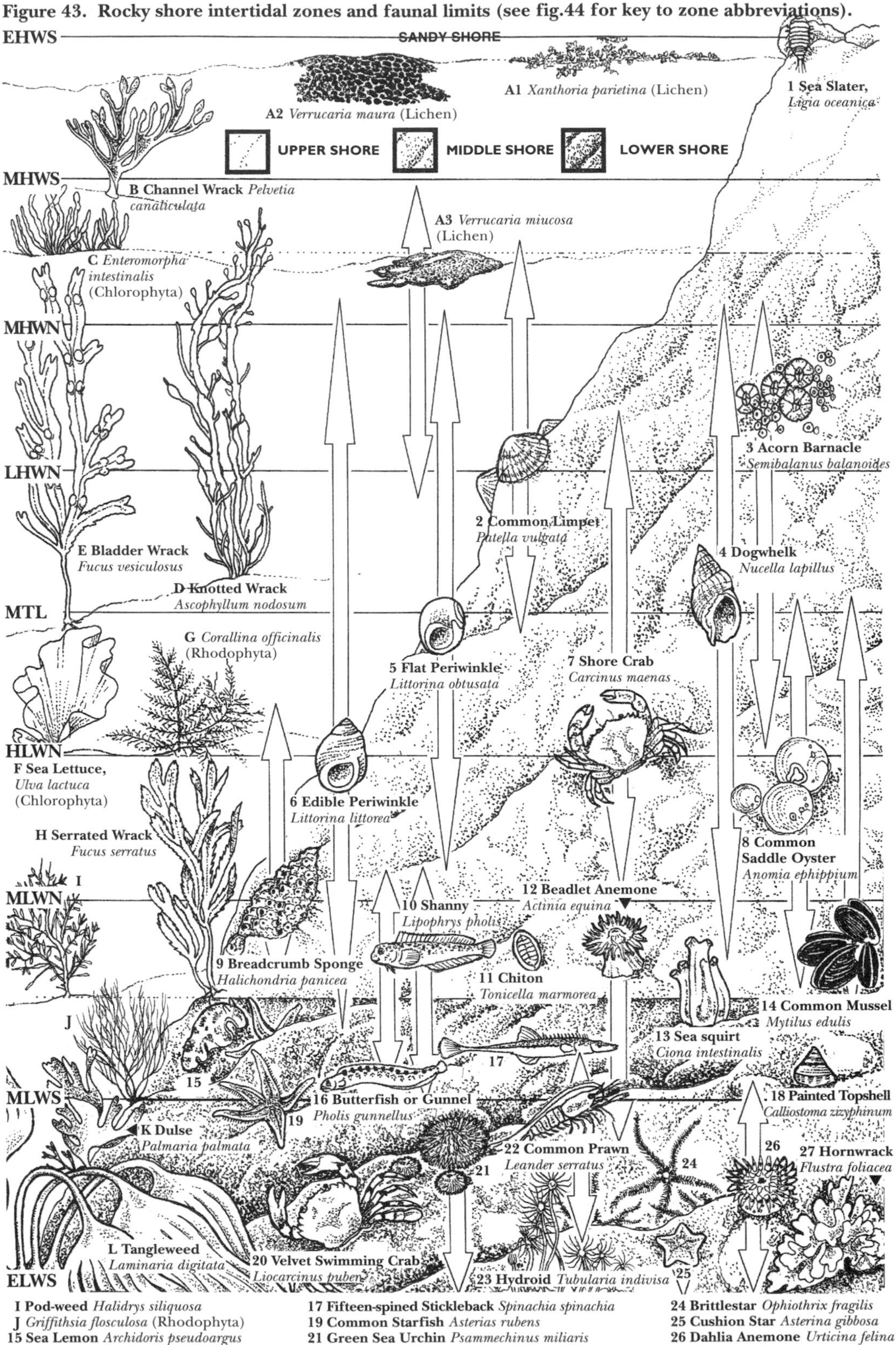

I Pod-weed *Halidrys siliquosa*
J *Griffithsia flosculosa* (Rhodophyta)
15 Sea Lemon *Archidoris pseudoargus*
17 Fifteen-spined Stickleback *Spinachia spinachia*
19 Common Starfish *Asterias rubens*
21 Green Sea Urchin *Psammechinus miliaris*
24 Brittlestar *Ophiothrix fragilis*
25 Cushion Star *Asterina gibbosa*
26 Dahlia Anemone *Urticina felina*

four different brown algae. From the top of the shore to the bottom, and in order of dominance they include Channel Wrack *Pelvetia canaliculata*, Spiral Wrack *Fucus spiralis*, Bladder Wrack *F. vesiculosus*, and Serrated or Saw Wrack *F. serratus*.

Many littoral animals such as bivalve molluscs, anemones and barnacles lead sessile lifestyles and have developed filter feeding methods to catch small prey or particles that drift by with the water currents. The anemones also prey on small fishes and shrimps by capturing and paralysing them with stinging tentacles; the food is then passed to the central mouth and ingested. Many of the more mobile animals can venture over large areas of the littoral zone in search of food when the tide is in. The winkles *Gibbula* spp., limpets *Patella* spp., chitons *Lepidochitona* spp., Shore Crab *Carcinus maenas*, hermit crabs *Eupagurus* spp., and Common Whelk *Buccinum undatum* move over the rocks and among the forests of swaying algae in search of food. Many herbivorous gastropod snails browse on young seaweeds that cover the rock surfaces, or feed on the fronds of larger algae. The browsing of the seaweeds controls the spread and growth of many algal species. Before the tide recedes to uncover the littoral shore, many of the mobile inhabitants seek the shelter of cracks, crevices and the undersides of stones. The more delicate animals may seek the refuge of rock pools, while gastropod snails withdraw into their shells and close the opening using the hard operculum attached to the foot. Wandering limpets will return to a shallow hollow which they have worn in the rock to exactly fit the rim of their shell. Here they will create suction pressure using their large foot to form a tight seal with the surface of the rock. (See figure 43 for common groups and species of organisms found in the intertidal zones of temperate rocky shores.)

Rock Pools, Cracks and Crevices

Athough rock pools are found at all levels on the shore, the largest pools generally are situated near the bottom of the shore. The depth of rock pools ranges from a few centimetres to a few metres, but all must contain some permanent water to be correctly designated a 'rock pool'. Rock pools provide a refuge from exposure and desiccation for many of the plants and animals which would not otherwise survive. Rock pools that provide the most stable conditions will be found near the bottom of the shore where they are exposed for shorter periods of time although large rock pools with substantial volumes of water also provide stable conditions even if higher up on the shore. Although rock pools may be stressful to organisms because of fluctuations of temperature, salinity, pH and freshwater run-off from the land, they allow animals and plants to survive higher up the beach and therefore they are able to colonise a much wider area of the littoral zone. For example, the limpet *Patella ulyssponensis* is usually found on the lower shore areas but can survive higher up the shore in rock pools. It may even become more numerous than the Common Limpet *Patella vulgata*. The same is true for some algal species. Red algae and fragile algae such as the Sea Lettuce *Ulva lactuca*, which normally live in the sublittoral zone, may also be found in intertidal rock pools. The zonation of brown fucoid species of algae may alter because of the protection offered by rock pools.

Common animal species found hiding under algae in rock pools includes winkles, crabs, small fish and the Glass Prawn *Palaemon elegans*. Sea squirts, anemones, starfish and urchins are also common inhabitants. Many large stones that lie exposed and surrounded by loose gravel or sand may also have accumulations of water underneath. These little subsurface pools of water are often good hiding places for smaller crustacea and polychaete worms. Another interesting inhabitant of the rock pools is the hermit crab *Eupagurus* spp.; this crustacean borrows the empty shells of other gastropods and uses them as a mobile home. As it grows, it must vacate the shell for larger premises and is most vulnerable to attack from predators during the move. The largest hermit crabs use the shells of whelks which are the only gastropod species that have suitably large shells.

Rocky overhangs provide a habitat for many of the more delicate organisms that inhabit the shore. The best examples are often seen at the bottom of the shore where they have been sculpted by long periods of wave action. The undercuts of large boulders form ready-made overhangs for colonisation by plants and animals. Rocky overhangs provide a cool, damp and sheltered environment which has no direct sunlight. At lower levels on the beach, trickling water and hanging weed improve the damp and humid atmosphere of the overhangs and evaporation will help to reduce temperature fluctuations. Typical animals of overhanging rock surfaces include sea anemones, sponges, sea squirts, soft bodied sea-slugs (nudibranchs) and other molluscs.

Cracks and crevices in and out of the water provide shelter for many plants and animals; the deeper cracks will have their own microclimate of dampness and cooler air. They also provide shelter from the direct forces of waves and currents during submersion. Many sedentary soft-bodied animals can be found in these cracks and crevices, including sponges, sea squirts, sea fans (Cnidaria) and sea mats (Bryozoa). Many of the smaller, more mobile gastropod snails and crabs will also seek refuge from shore predators by hiding in the cracks during exposed periods.

SUBLITTORAL ZONE

This region starts at the bottom of the littoral zone and the upper edge of it may only be exposed during the lowest spring tides. Rocky areas are characteristically covered by encrusting red algae *Lithothamnion* spp. In tropical regions small fucoids, marine grasses and colonial corals are dominant. In temperate latitudes and in the regions of 'upwelling', large brown macroalgae called kelp may become dominant. These large algal species often form kelp 'forests' that extend to sublittoral depths of 40m (fig.52). In many warmer latitudes the sublittoral zone may be dominated by large sea squirts and fine mats of red algae.

Sandy and Muddy Shores

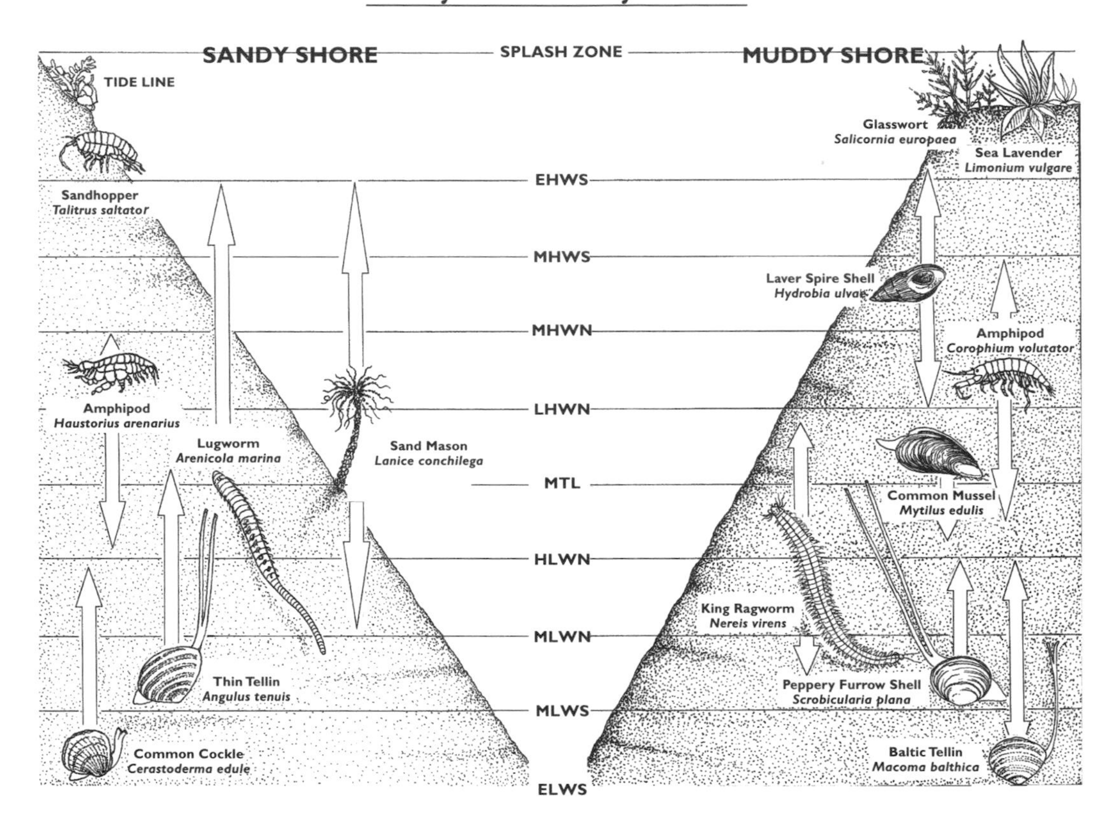

Figure 44. Muddy/sandy shore intertidal zones and faunal limits. EHWS - extreme high water springs; MHWS - mean high water springs; MHWN - mean high water neaps; LHWN - lowest high water neaps; MTL - mean tide level; HLWN - highest low water neaps; MLWN - mean low water neaps; MLWS - mean low water springs; ELWS - extreme low water springs.

At first glance the intertidal sand and mud shores are relatively barren, with few plants and animals to observe. However, below the surface there is a diverse community of burrowing animal life. Within the soft sediments, the animals are zoned according to their level on the shore. However, the zones of many of these burrowing animals may be quite wide over an extensive area of the littoral zone. A burrowing life-style will help the organisms to escape desiccation and predators such as birds. Burrowing also prevents animals being washed away by the surf on the more exposed sandy areas.

Intertidal sediment has a number of 'vertical layers' of different sediments that will vary in depth according to the location and level on the beach. A sandy surface layer may extend from the surface to a depth of 50cm on the higher parts of the shore. However, at lower levels on the beach, the mud layer may be close to the surface. This may cause the thin surface layer of sand to give way underfoot, leaving deep impressions in the surface. In areas where there is a good mixture of both mud and sand particles, burrows can be constructed without the problem of collapse. Many burrowing organisms will seek out these more stable areas and, as a result, the concentrations of organisms living in these areas can be high.

Animals living in soft sediments can be placed in a number of categories by reference to their lifestyle and behaviour. Some species live in the sediment and never venture onto the surface, while others will leave their burrows to scavenge and feed only when the tide is in. Some animals may only be temporary visitors, coming up with the incoming tide and retreating with it. Tidal visitors include adult flatfishes, the young of many demersal fish, shrimps and crabs. Along the edge of the waterline, seabirds and shorebirds can often be seen hunting for tideline animals including burrowing animals that are still near the surface.

SANDY SHORES

Sandy shores are normally found on gently shelving gradients where wave action and water movements are sufficient to remove any fine particles of mud. Sand composed of large grain sizes will not hold much water, detritus or nutrients, and many of the finer particles will be drained away or deposited in the deeper layers. On gently sloping shores, sand will take some time to dry out, therefore desiccation is not normally a problem for burrowing creatures. If conditions do get a little dry, the animals tend to burrow deeper into the sediment, where they will remain cool and damp. Some of the deeper layers will have reduced oxygen levels as this gas does not easily diffuse to these layers and the breakdown of organic material is instead carried out by anaerobic (not requiring oxygen) bacteria. Sandy and muddy shores have low diversities of animals compared to rocky shores and this is a reflection of the few types of habitats offered by soft substrates. Although the diversity may be low, the numerical abundance of burrowing animals can be high. In general, sandy environments have lower diversity and abundance than the muddy environments as they hold less organic matter, food and nutrients.

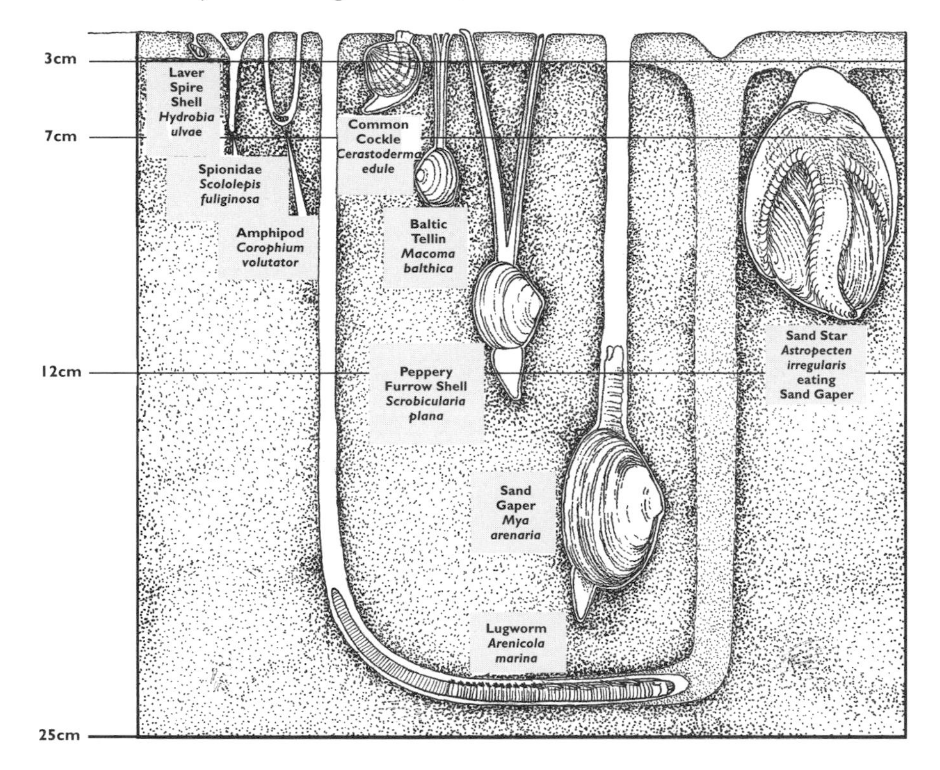

Figure 45. Common burrowing animals of sandy shores showing burrowing depths of adults.

In sandy environments there is a rich diversity of microscopic organisms. Some of these microorganisms live on the surface of the sand grains, while other animals live an 'interstitial' life between the particles of sand. Common microscopic groups include diatoms, dinoflagellates and blue-green algae. Larger fauna include molluscs (e.g. *Caecum glabrum*), sea cucumbers (e.g. *Leptosynapta minuta*), flatworms, polychaete worms, nematodes, hydroids and tunicates. The macrofaunal inhabitants of the sand include the burrowing polychaete Lugworm *Arenicola marina*, the bivalve cockle *Cerastoderma edule*, the razor shell *Ensis ensis* and many small crustaceans such as *Eurydice*, *Haustorius* and *Bathyporeia*. A few echinoderms such as sand stars *Astropecten* spp. and heart urchins *Echinocardium* spp. have adapted to a burrowing existence and are found in the lower intertidal regions.

Lugworms are probably the most common large invertebrates inhabiting sandy shores of NW Europe. In some areas densities can reach 70 individuals per square metre, but this will depend on the grain size of the sand. The lugworm is one of the most important prey items for many shorebirds and fish, and also provides bait for sea anglers. Below the tidelines of sandy shores and occupying the shallows, live the Sand Eel *Ammodytes lanceolatus*, Glass Shrimp *Palaemon elegans*, Butterfly Blenny *Blennius ocellaris* and the Shore Crab *Carcinus maenas*. These species are also found below the tideline of muddy shores, in pools around concrete and wooden structures, and in rock pools of rocky intertidal shores.

Along the highest reaches of the intertidal zones and the strandlines of sandy, muddy and rocky shores, a number of non-burrowing creatures will be found. In temperate latitudes the air-breathing, amphipod beach-hoppers or beach-fleas *Talitrus saltator* burrow into the sediments or hide under the rotting vegetation of the strandline. At night and sometimes during the day, they emerge to feed on the damp, decaying, organic matter washed up by the tide. Under ideal conditions, the numbers of beach-hoppers seem to reach plague proportions but this does not continue for long. In tropical regions of the world the main strandline inhabitant of sandy beaches is the scavenging ghost crab *Ocypode ceratophthalmus*.

Burrowing animals construct different types of burrows and live at different depths (fig.45). The Lugworm *Arenicola marina* constructs a 'U'-shaped burrow and at one end, fresh oxygenated water is drawn in. As the animal eats its way through the soft sediment it leaves a coiled faecal cast on the surface at the other end of the burrow. This cast is a clue to its presence for many predators. Bivalves burrow to different depths according to the species and siphon length. Most bivalves use a muscular foot which extends between the two halves of the shell for burrowing. The razor shell *Ensis* sp. is one of the fastest burrowers and quick digging with a spade is required to catch it. Cockles *Cerastoderma* spp. are particle feeders and they extend two short siphons to the surface. The fringed 'inhalant' siphon takes in water and particles, and the frills act as a coarse filter for the unwanted, larger pieces of material (fig.46). As water passes over the gills, oxygen is absorbed and food particles are filtered out. Water is then passed out by way of the 'exhalant' siphon. Another interesting bivalve is the Thin Tellin *Angulus tenuis* which has an inhalant siphon resembling an 'elephant's trunk'. When the sandy surface is covered by water it extends this trunk and moves it around the surface to 'vacuum' up small food particles.

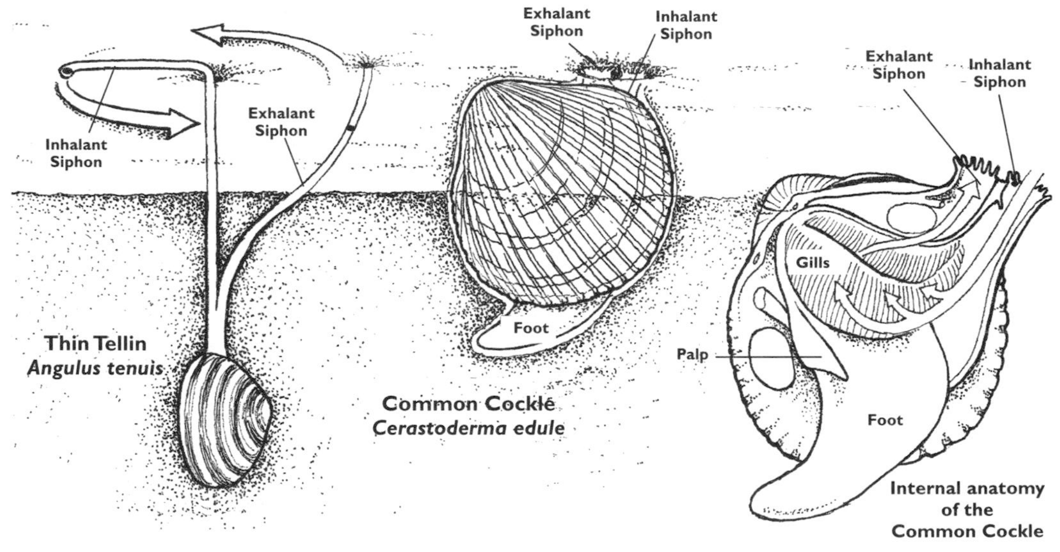

Figure 46. Filter feeding methods in two bivalves inhabiting sandy substrate.

MUDDY SHORES

On muddy shores the Lugworm *Arenicola marina* is present in large numbers along with other species of large polychaete worms, such as the ragworms *Nereis* spp. These species and the burrowing amphipod, *Corophium volutator* are found over large areas of the muddy intertidal habitat. At the lower level of the shore, key species include large numbers of the burrowing bivalves *Angulus, Cerastoderma* and *Macoma. Macoma* and *Cerastoderma* are also found in large numbers on sandy shores. Many of these key species have related countertparts in other regions of the world which occupy similar habitats. At lower shore levels, populations of mussels *Mytilus* spp. may become established and their shells provide a hard surface onto which other animals can attach themselves. Many of the soft sediment species are also found in estuary environments. However, the species and their abundance will vary according to their tolerances of brackish water or freshwater from land run-off. On estuarine mudflats *Hydrobia ulvae* is a good example of an animal which is well suited to a stressful environment. The densities of the *Hydrobia* populations can reach peaks of over 100,000 individuals per square metre on some mudflats. They are shallow burrowers and are found just below the surface of the mud.

ESTUARIES AND SALTMARSHES

Estuaries are semi-enclosed, coastal areas in which the seawater is significantly diluted by freshwater from streams and rivers feeding the estuaries. Many bays, inlets, sounds and gulfs can be considered estuaries on the basis of their physical structures. Freshwater from land run-off is nutrient rich, and when it mixes with salt water at the mouths of estuaries, plankton blooms often occur on the seaward side. Water flow decreases as the estuary widens and becomes shallower, and the sediment 'load' of the river settles out to be deposited as mud or sandbanks. Where large rivers or river systems converge, offshore deltas of mud and sand may form. In temperate latitudes, mudflats may subsequently rise above the normal levels of high tide through sediment and soil accumulations from river floodings. Colonisation of mudflats by marsh grasses will lead to permanent saltmarshes (fig.49). In subtropical and tropical regions, saltmarshes are often replaced by areas of 'mangroves' which occupy a similar habitat. However, in a few regions of the world, both saltmarshes and mangroves may be found growing in the same areas (e.g. in the Gulf States of USA and SE Australia).

ESTUARIES

The salinity of estuaries fluctuates and repeated changes in salinity increase stress levels in marine organisms. Species of some marine animal groups living in an estuarine environment can adapt physiologically to changing salinities (euryhaline animals). In some estuarine areas, the abundance of animals such as the oyster, crab, shrimp, ragworm, tubifix worm and *Hydrobia* snail can be high. Food webs within estuaries are complex and are not fully understood; however, a simplified food web is given in figure 47.

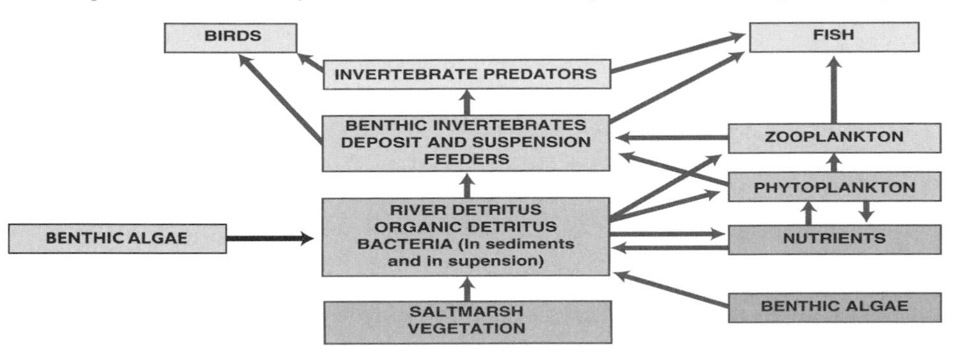

Figure 47. Estuarine food web.

The inhabitants of benthic estuarine communities can be divided into a number of groups according to their salinity tolerance. In European waters they include the following:

1. Marine species which penetrate estuaries to 30‰ (stenohaline animals). Key organisms include the Thin Tellin *Angulus tenuis* and the Common Cockle *Cerastoderma edule.*
2. Marine species that can tolerate salinities below 30‰, and can therefore penetrate further into estuaries. Key animal species include the Shore Crab *Carcinus maenas,* the Common Periwinkle *Littorina littorea,* the Common Mussel *Mytilus edulis* and the amphipod *Corophium volutator.* Algal species include the green alga *Enteromorpha.*
3. True brackish water species which only live in brackish waters but have marine relatives. They include the isopod *Sphaeroma rugicauda* and Horned Wrack *Fucus ceranoides.*
4. Freshwater species which penetrate into brackish water and include isopods such as *Asellus aquaticus,* some oligochaete worms, the coleopteran *Bembidion laterale* and other insect larvae.

Some benthic fish species such as the Flounder, *Platichthys flesus* and the young of a few other coastal species also inhabit the brackish waters of estuaries. Other species including the European Eel, *Anguilla anguilla,* the Atlantic Salmon *Salmo salar,* and the Sea Trout *Salmo trutta,* will acclimatise in estuaries during their migration between freshwater and seawater.

Mudflat areas are exposed at low tide and may extend to many square kilometres. Mudflats which are adjacent to saltmarsh areas provide important feeding grounds for many shorebirds and seabirds. The majority of invertebrate animals live within the mud to escape desiccation and predation, and some only emerge when they are covered by the rising waters.

Shorebirds have bills which are adapted for shoreline feeding, and different species of birds are able to take a variety of prey from different depths in the soft sediment (fig.48). Large flocks of shorebirds can often be seen probing the mud near the water's edge, foraging for prey and looking for the telltale signs of burrowing animals. Birds with long bills (e.g. oystercatchers, godwits, redshanks and curlews) can catch deep-burrowing prey including larger polychaete worms and bivalves. Many slender-billed birds have bills that are touch sensitive and flexible, and they are capable of lifting prey while keeping the mouth closed. The diversity and numbers of seabirds, shorebirds and wildfowl found on estuaries and their associated wetlands, reflect the enormous productivity of marshlands and mudflats. These areas afford the ornithologist with some of the best bird-watching sites. A boat moored in an estuary can make a comfortable observation platform for studying the wildlife of estuaries.

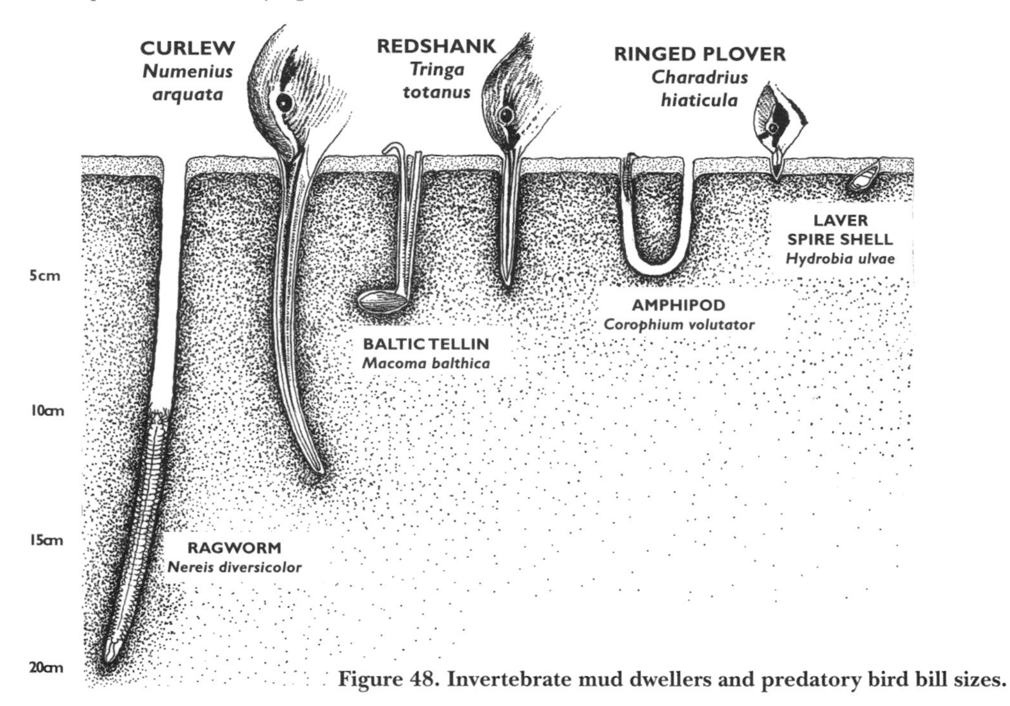

Figure 48. Invertebrate mud dwellers and predatory bird bill sizes.

SALTMARSHES

Saltmarshes are biologically important areas with high primary production. The marsh grasses hold sediment from the occasional river floodings which increases the richness of the developing soils. In North America, the major marsh grasses include *Spartina alterniflora,* which is found near the water's edge. Above this level, the rush *Juncus* sp. and other grasses (e.g. *Distichlis spicata, Spartina patens* and the glasswort *Salicornia* sp.) will occur. In Europe the most important pioneering colonisers include the green alga *Enteromorpha* at the water's edge, *Spartina* grasses and sometimes the flowering *Zostera nana.* On saltmarshes with higher levels of sand, the grass *Puccinellia maritima* may be the dominant species.

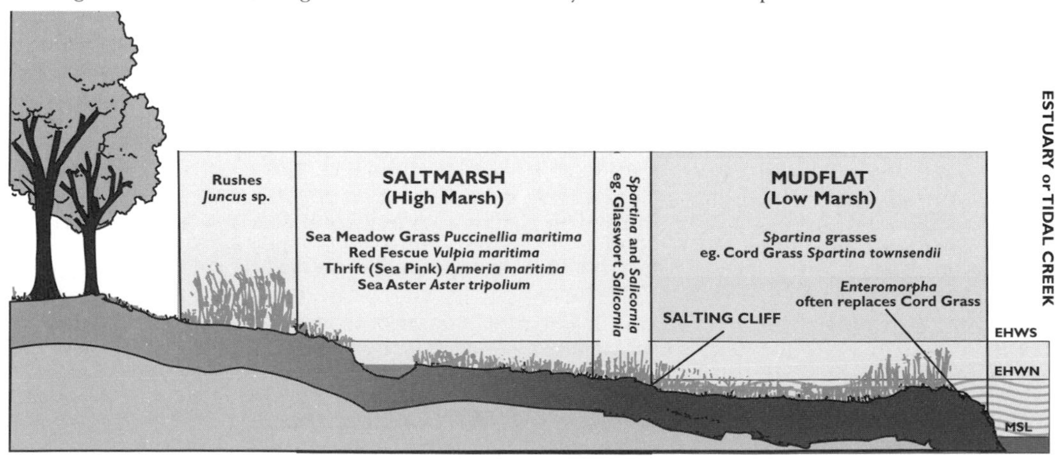

Figure 49. Saltmarsh in sectional view to show characteristic flora.

Saltmarshes form important habitats for many terrestrial animals including rats, snakes, insects and birds. Saltmarshes and their adjacent mudflats are important breeding and feeding grounds for many shorebirds, seabirds and wildfowl. The marshes may be used as 'staging posts' along migratory routes for some bird species, or overwintering grounds and summer feeding areas for others.

MANGROVES

The swampy intertidal areas of coastlines and estuaries are replaced by 'mangrove forests' in the subtropical and tropical regions of the world. Mangroves are a common sight along coastlines between latitudes of 30° either side of the equator and cover 60-70% of the shorelines (fig.50).

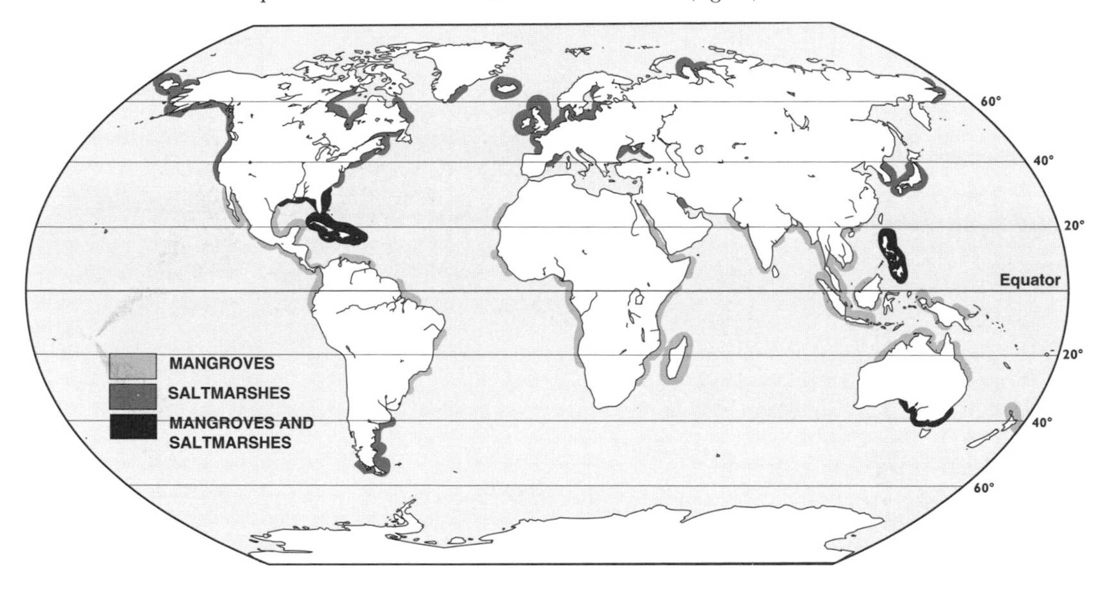

Figure 50. Mangrove and saltmarsh distribution.

Mangroves consist of a number of trees or tree-like shrubs that are tolerant to both saltwater and brackish water conditions. The trees and shrubs take about 20 years to mature and can reach a height of about 30m. However, they are generally lower than this and are kept in check by human activities and storm damage. They have extensive root systems which anchor them firmly in the soft substrate and some have specially adapted root systems to take in oxygen. There are about 12 genera of mangrove trees and shrubs with about 60 different species. The dominant genera are the red mangroves *Rhizophora*, and black mangroves *Avicennia* and *Bruguiera* (fig.51). In the Indo-Pacific mangrove forests there are about 30 different species, compared to 10 in the Atlantic mangrove communities. Therefore, as with many other ecosystems, the Indo-Pacific mangrove communities have a higher diversity of species.

Many of the salt-tolerant mangrove trees and shrubs can excrete salt and have physiological mechanisms for conserving water. This allows them to inhabit both saltwater coastlines and freshwater or brackish estuaries. All species must be able to tolerate a fine, soft-sediment substrate with low oxygen levels. To compensate for low oxygen levels, *Rhizophora* produces 'stilt-roots' which sprout from the sides of the lower trunk. These grow sideways and downwards into the mud to help prop-up and stabilise the tree. The black mangrove *Avicennia* produces pencil roots, which grow up and out of the mud to surround the tree like a forest of spikes, with a length of about 150-300mm. These roots and the aerial roots of *Rhizophora* take in oxygen when they are uncovered at low tide. Unusually, the seeds of mangrove trees germinate on the tree; the young plants then fall off into the water, drift away and take root elsewhere.

The size of mangrove forests depends largely on tidal ranges and the slope of the bottom. Broad, shallow areas with large tidal ranges have large expanses of mangroves and wide zonation bands of intertidal organisms on the mangrove tree trunks. The diversity of the intertidal communities varies with latitude and the salinity of the water. Animal communities in the brackish waters of mangroves are generally lower in diversity than those of the saltwater mangroves.

The intertidal mangrove forests provide a number of different habitats which include the muddy sediments, the surface of the mud, the surface of trunks and roots of the trees, and, above the waterline, the canopy of the forest. Sessile animals including barnacles, tunicates, oysters and sponges attach themselves to the tree roots which are free of clogging mud. Periwinkle snails are found crawling among the roots, grazing on the fine algae, and the *Littorina* species are zoned on the tree trunks in a similar way to that on rocky shores. Towards the landward fringe of the mangrove forests, several species of crabs are commonly found and include *Cardisoma* spp. and *Sesarma* spp. In the intertidal mud areas, the dominant large animal is the Fiddler Crab *Uca dussumieri* which comes out of its burrow at low tide to scavenge over the mud. Large-eyed Mudskippers *Periophthalmodon schlosseri* are also a common sight at low water and they use their modified fins and tail to crawl and hop across the surface of the mud. Some mudskipper species even crawl up the roots of the mangrove trees and out into the air. Other common organisms include burrowing polychaete worms, shrimps and amphipods.

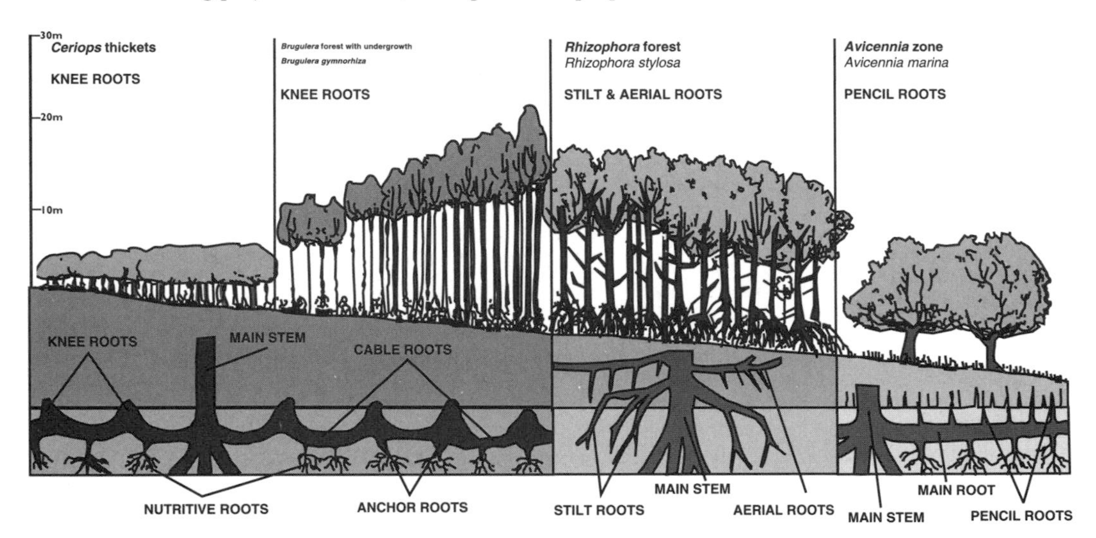

Figure 51. Mangrove zonation and associated flora with diagrams of stem and root structures.

In the subtidal areas of mangroves that are permanently covered by water, a high diversity of sealife is crowded together on the mangrove roots. Predominantly marine subtidal areas of mangroves also provide suitable nursery grounds for the juvenile stages of many important species of offshore fish. Rich populations of nearshore fish, shrimps, crabs, lobsters and shellfish are common in the saltwater shallows of mangrove forests. The forest canopy also provides a good habitat for many nesting birds (e.g. egrets) and bats, including the world's largest species, the fruit bats or flying foxes.

KELP FORESTS

In cold temperate and polar regions, the intertidal rocky shore communities merge subtidally into sublittoral macroalgal 'kelp forests'. The kelp forests prefer moderately exposed areas where there is reasonable water circulation. 'Kelp' is the name given to large brown algae that vary in size from 0.5 to 50 metres or more in length. Kelp that dominate the structure of kelp forests belong to the genera *Macrocystis* and *Nereocystis.* The kelp genera *Ecklonia* and *Laminaria* are smaller and do not form a surface canopy; they are termed kelp beds. In cool, clear waters the largest varieties such as the Giant Kelp *Macrocystis pyrifera* can grow to depths of 40m in the sublittoral zone, yet are still seen with their fronds trailing on the surface (fig.52). On gentle underwater gradients the kelp forests may extend for several kilometres offshore. All kelp plants need a hard surface or solid rock for attachment by means of the 'holdfast'. In some cases the holdfast may completely surround a boulder or small rock.

Kelp occurs in many cold areas of the world with rocky coastlines. It is found along the western coasts of North and South America and reaches into the subtropical latitudes where cooler 'upwelling' waters may be found. In the western Pacific, large areas of kelp are found off China and Japan (fig.53). In the Atlantic kelp is common off the seaboards of eastern Canada, southern Greenland and northern Europe. In the southern hemisphere, South Africa, the southern tip of Australia and New Zealand also have kelp beds and forests. Extending into the southern oceans, kelp forests are located around some of the subantarctic islands including South Georgia and the South Sandwich Islands. Kelp forests can also be seen around the east coast of South America and the Falkland Islands (Malvinas) where they have been exploited for many years. Alginates are extracted from kelp and the products are used extensively in the food, cosmetics and medical industries. They also provide a local source of fertilizer for many coastal farming communities. The global distribution of kelp is generally limited to the latitudes between the subpolar regions and the 20° summer isotherm. Above this temperature, corals begin to occupy similar coastal habitats and these increase towards the tropical regions. The extension of kelp into the subtropical latitudes is normally associated with colder currents produced by the upwelling waters which reduce temperatures and provide fresh supplies of nutrients. Examples of these upwelling areas include the Benguela, Peru and California currents which flow along the coasts towards the equator.

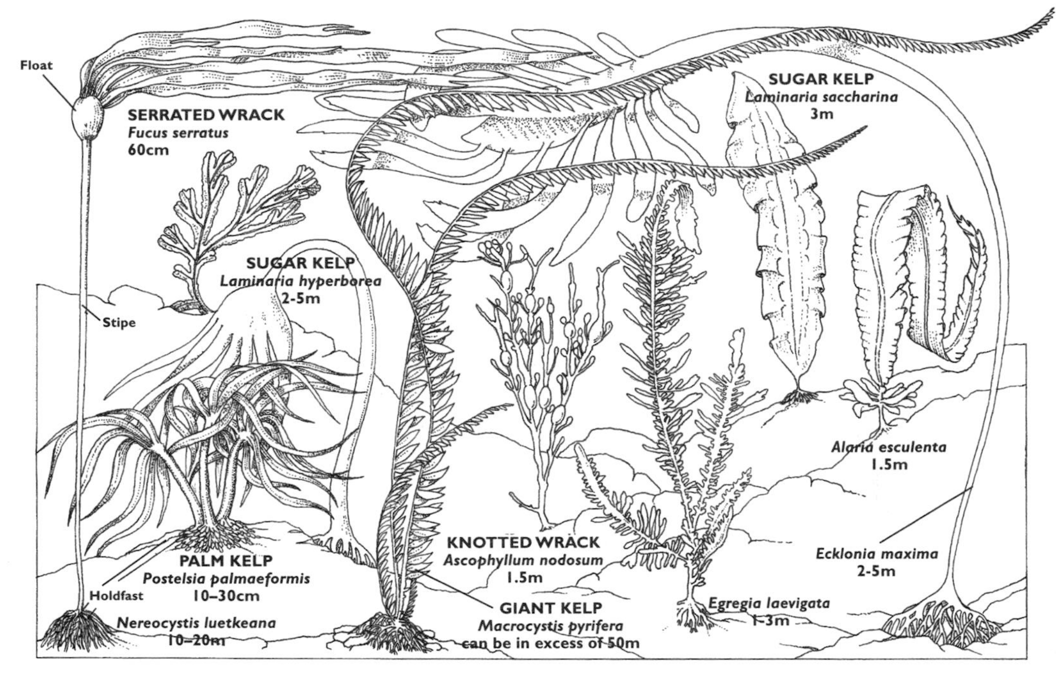

Figure 52. Representatives of major kelp species of northern temperate latitudes (not to scale).

Each kelp plant is attached by the entangling holdfast to the substrate but, unlike the true roots of higher plants, it does not take up nutrients or water. The large leaflike 'fronds' or 'blades' are attached to the holdfast by way of a flexible, stalk-like 'stipe'. Some macroalgal species such as *Nereocystis luetkeana* also have gas-filled floats called 'pneumatocysts' which buoy up the fronds and keep them nearer to the surface, where there is more light for photosynthesis. Macroalgae are highly productive plants and under ideal conditions some of the largest species can grow as much as 60cm per day. However, the normal rate of daily growth is between 5 and 25cm.

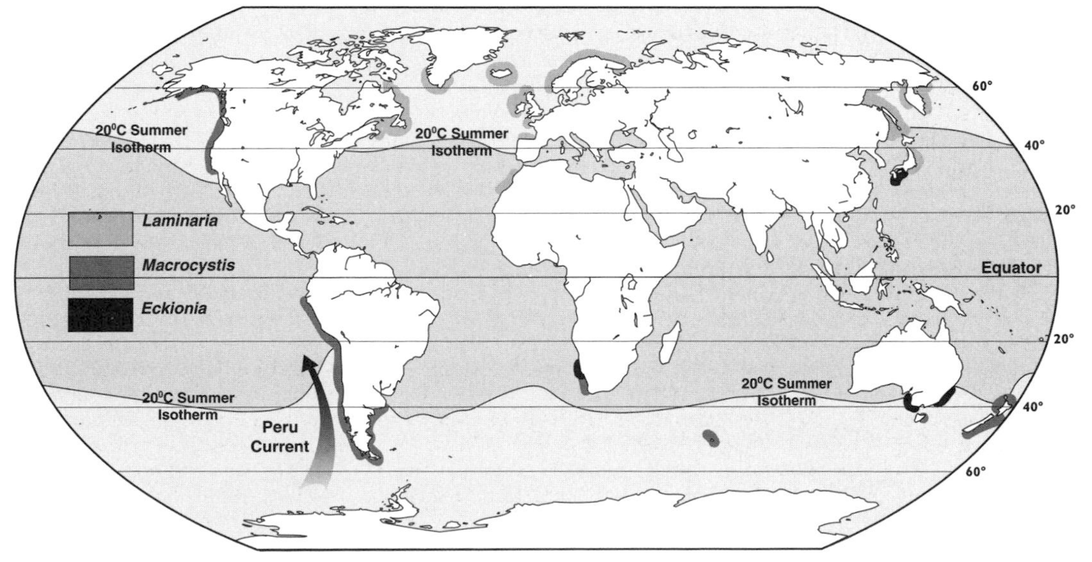

Figure 53. Distribution of kelp.

Kelp, like many other species of seaweeds, has a life cycle comprising two alternating generations. One part of the reproductive cycle is 'asexual' (no male or female germ cells are produced) and instead an adult 'sporophytic' form is produced (fig.54). The offspring from the sporophytic generation will reproduce sexually to produce male and female 'gametes' (equivalent to sperm and egg). Asexual methods of reproduction provide no mixing of male and female genes, therefore there is no genetic variation in the offspring. However asexual sporophytes are always present in the water and can quickly colonise bare areas of rock if adult plants have been uprooted and washed away by storms. A sporophyte simply has to settle on a rock and grow into an adult. The alternative strategy of sexual reproduction is a much more hit and miss method. Both male and female gametes have to find each other in the water and fuse to form a 'zygote'. However the long term evolutionary benefits of sexual reproduction make it worthwhile to a plant, even though it has to invest reproductive energy to produce gametes which may or may not develop into adults.

The *Laminaria* species produce a microscopic sexual stage compared to the large adult asexual, sporophytic plant. This strategy reduces the competition between the two generations for resources such as nutrients and living space. In other species of brown algae, the sexual generation has been lost altogether and only adult asexual, sporophytic plants are ever produced. Some species (e.g. the green Sea Lettuce *Ulva lactuca*) have both asexual and sexual adult plants which look similar. However, such species tend to be fairly short-lived. Among the kelp species, plants may be 'annual' or 'perennial' by nature. Perennial species will grow a new stipe and fronds from the holdfast region either each year or every few years.

The Palm Kelp *Postelsia palmaeformis* with its flexible, short and durable stipe thrives on wave-beaten intertidal rocks. The majority of shallow-living subtidal kelps, which include the laminarians, are not seen from the shore except at the low spring tides. In regions where *Macrocystis*, *Nereocystis* and *Ecklonia* beds flourish, the surface of the sea may look like a continuous layer of entangled fronds buoyed up by gas bladders.

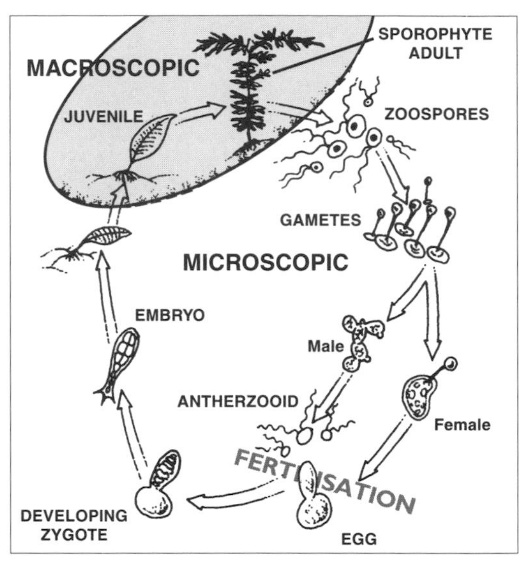

Figure 54. The life cycle of algae.

Kelp forests are similar to terrestrial forests in that they have diverse communities of animals associated with them (fig.55). The large fronds of many species provide a platform for sessile communities of smaller algae, diatoms, bryozoans and hydroids. Worms, crustaceans and molluscs are found wandering over the algal fronds. Some species such as the sea slug *Aplysia depilans* and other gastropod snails may feed on the kelp or on the smaller varieties of 'epiphytic' algae that grow on the frond surfaces. Many fish species that inhabit kelp beds have been found with pieces of kelp in their stomachs. However, this is probably due to accidental ingestion while hunting for prey on the algae.

The most voracious consumer of algae is the urchin *Strongylocentrotus* spp. In North America, the life cycle of the urchin and its association with the kelp is well known. The urchins can destroy whole areas of kelp forest, but once the algae are gone the urchin suffers starvation, disease and mass mortality. After the urchin population has declined the kelp forests regenerate and once more become established. Urchins of the kelp forests do provide a valuable food source for some of the larger predators. The activities of these predators reduce the numbers of urchins to a level where they do not destroy the kelp forests. In North America, Alaska and the Aleutian islands, the Sea Otter *Enhydra lutris* is the main predator of the urchin. The Sea Otter is known as a 'keystone' species because of the dramatic effects that its predation on urchins has on the regeneration of kelp. However, the numbers of these higher predators have declined through over-exploitation by humans. In many areas this has led to an increase in the urchin population and a decrease in the kelp and its associated communities. Another unusual animal which used to inhabit the kelp forests of the north Pacific was Steller's Sea Cow *Hydrodamalis gigas*, but this species has been hunted to extinction. This mammal required kelp for both food and shelter. Kelp forests are also home for many juvenile stages of commercially important offshore fish species that use the kelp as nursery areas.

Figure 55. Californian giant kelp forest with larger vertebrate inhabitants.

SEAGRASS MEADOWS

Seagrasses are the only true marine vascular plants and they have a distribution that covers a wide range of latitudes worldwide. Some species occupy the lower levels of the intertidal zone, but the most abundant areas of seagrass are found just below the tidal level, and they can occur down to a depth of about 50m. Seagrasses are found in clear, sheltered waters with a sandy bottom, but they also grow on fine sediments in turbid waters with low light intensities. Characteristically, they inhabit enclosed and sheltered bays, lagoons, lees of islands and offshore barriers. The dominant seagrass species in temperate latitudes are the eelgrasses *Zostera* spp., and they are replaced by turtlegrasses *Thalassia* spp., and other species in tropical regions.

Seagrass meadows are productive areas with quick growth and fast turnover of dead organic matter. This quick turnover releases nutrients for other users including bacteria, and other microbial organisms and detritus feeders; therefore, seagrass meadows are considered to be amongst the most productive areas of all shallow, sandy environments. The seagrasses provide shelter for diverse communities of invertebrate nematode worms, small epiphytic algae and sessile filter feeders, including hydrozoans, bryozoans and tunicates. Among the leaves snails, bivalves, polychaete worms and various large crustacea including lobsters and crayfish can be found. Many vertebrate fish species are attracted to the seagrass areas to feed and spawn. The seagrasses are used as nursery grounds for many young fish which will later leave for adult feeding grounds elsewhere. Large animals that consume seagrass include some herbivorous fish, some marine turtles and the sirenians (Dugong and manatees).

Pg 70 ~~100~~

CORAL REEFS

Coral reefs are well known for their underwater splendour, vivid colours and rich species diversity. The brilliant colouring of many coral species, combined with their 'radial symmetry', often creates a beauty that is unsurpassed (fig.61). Most of the true coral reef builders belong to only one phylum, the Cnidaria which includes among others, the hydras, jellyfish and sea anemones. Reef-building corals are colonies of tiny individual animals called 'polyps'. Each polyp secretes a calcium carbonate exoskeleton (calyx or corallite) which protects the soft, sack-like body inside (fig.56).

Coral reef areas at low latitudes exhibit a high rate of growth, even though the surrounding waters may be low in food and nutrients. This is possible as much of the food for many coral species is provided by microscopic green algae called zooxanthellae. These microscopic plants live inside the cells of the polyp's tissues and are also found in the tissues of other species including some molluscs (e.g. giant clams *Tridacna*), various sea squirts (ascidians), and colonial anemones. The zooxanthellae are immobile dinoflagellates and numbers of individuals within corals can reach 30,000 per cubic millimetre. They do not harm the animal but live with it in partnership (mutualism). The coral polyp provides protection, living space and nutrients (ammonia and carbon dioxide waste products) for the zooxanthellae. The zooxanthellae provide food and oxygen for the living coral by recycling waste products and carrying out photosynthesis. It is thought that zooxanthellae provide up to 90% of the polyp's nutrient requirements and a large proportion of its biomass. Zooxanthellae are also involved in calcium fixation and skeleton building. Coral polyps occasionally lose their zooxanthellae partners and whole areas of the reef may die, leaving large white expanses. The reason for this is not known. As zooxanthellae require clear waters and sunlight for photosynthesis, most of the new coral reef growth takes place near the surface. Corals can be divided into those species that have zooxanthellae (hermatypic) and those that do not (ahermatypic). The ahermatypic species are not confined to the shallow surface waters as they do not require sunlight. Therefore they are found in all areas of the world from the tropics to the polar regions and at a wide range of seabed depths.

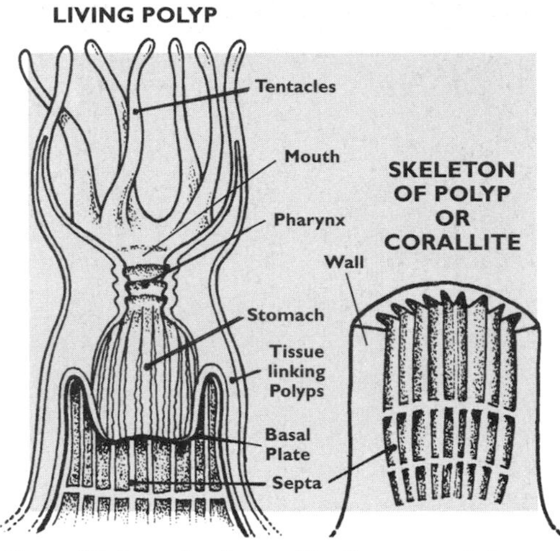

Figure 56. A coral polyp and its limestone skeleton.

Hermatypic coral regions occur around the world in warm subtropical and tropical regions, to 35° north and south of the equator. The temperature range for coral reef development is between 18°C and 40°C. However, for maximum reef development, a temperature range of 25°C to 30°C with clear waters and strong sunlight is necessary. A 5 to 20m depth zone is the most appropriate and productive zone for maximum coral growth. For general geographical purposes the distribution associated with coral reef growth is taken to be the 20°C temperature isotherm either side of the equator (fig.57).

Key areas where coral reefs may be found are Australia, Indonesia, the Philippines, the Pacific islands, the Caribbean, Bahamas, Brazil, the Gulf of Guinea and the eastern seaboard of Africa including the Red Sea. Average rates of coral growth and corallite accumulation are about 5mm per year. However, under ideal conditions some foliaceous coral species (*Acropora*) can grow as much as 27cm in length per year. In coastal areas where turbid waters may be encountered, coral growth rates are substantially reduced. Turbid waters cut down the amount of sunlight available for zooxanthellae photosynthesis and large amounts of sediment suspended in the turbid water can 'blanket' and choke a reef. Coral reefs are not present in coastal areas with substantial river estuaries or deltas, as many corals have low tolerance to freshwater.

The diversity of coral species is much lower in tropical Atlantic regions (about 60 species) compared to the Indo-Pacific region where there are about 700 species, although the total number of species is still not fully known. Within the reef building colonies of the Atlantic, calcareous red algae (mostly *Porolithon*) play a major role in reef building and help bind the reef together as a whole. The species diversity of the associated reef communities is probably the highest of all biological habitats in the sea. A single reef may have as many as 3,000 animal species living on or in it, and this number does not include all the smaller animal species that are not easily seen. In the Indo-Pacific region there are over 5,000 species of molluscs and 2,000 species of fishes. The Atlantic coral reef areas, by contrast, have only 1,200 species of molluscs and 600 species of fishes. This remarkable difference in species diversity provides further evidence of much older Pacific Ocean ecosystems compared to other regions or oceans of the world. As coral reefs have such large diversities of organisms, it is not surprising that the food webs are also some of the most complex in the animal world.

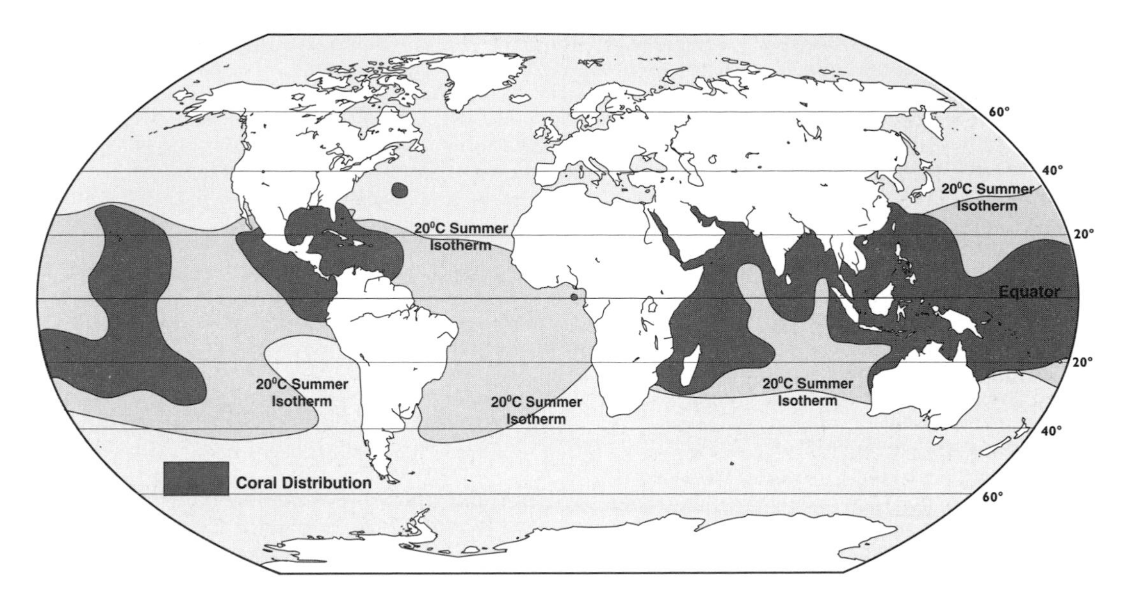

Figure 57. Coral reef distribution. Islands and coastlines located within the range shown are capable of supporting coral reef growth.

It has been found that diversity of coral reefs and their associated communities declines with decreasing temperatures, either latitudinally or with increasing depth of water. On exposed reefs which are open to the prevailing winds and wave action, only the more robust coral species and associated organisms are found. Although diversity may be lower here, abundance of a particularly well suited species may be high because of reduced competition for resources and living space. With increasing depth of water, the sponges, seawhips, gorgonians and ahermatypic corals become more common. At the bottom of the coral reef 'slope' a fine sediment will have accumulated and here a typical benthic infauna and epifauna for that depth becomes apparent.

Among the animals associated with coral reefs are many of the herbivorous feeders and suspension feeders, similar to those filling the same niches on rocky bottoms of sublittoral temperate coastal areas. Limestone reefs are habitats with cracks, crevices, galleries and tunnels for animals to occupy. Many of the suspension feeders such as the bivalve molluscs and serpulids become embedded into the reef structure. Major groups of invertebrate organisms associated with reef communities include sea squirts (ascidians), sponges (poriferans), starfish (echinoderms), snails, limpets and clams (molluscs), worms (polychaetes), sea slugs (nudibranchs), shrimps and lobsters (crustaceans) and bryozoans. Carnivorous invertebrates

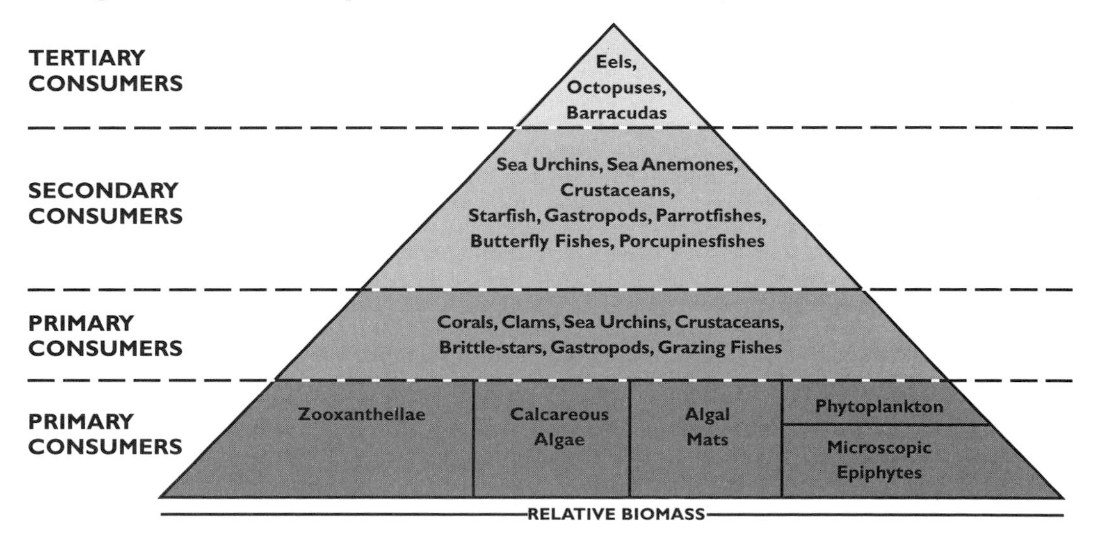

Figure 58. Trophic levels of coral reefs (after Pinet 1992).

include certain polyclad worms, sea slugs, gastropod snails and starfish such as the voracious Crown-of-thorns Starfish *Acanthaster planci.* This starfish appears to have cyclic population explosions with high numbers destroying whole areas of reef, leaving them white. After this, the starfish population declines through lack of food and disease, allowing the reef to slowly recover. There are a number of fish species that will also eat corals. Some butterfly fishes (Chaetodontidae) and file-fishes (Monocanthidae) have cutting teeth or long snouts to prise out or nip off polyps. Parrotfishes (Scaridae) have powerful jaws that bite off chunks of coral which they break up, consuming the soft polyps and ejecting the hard exoskeletons.

Reef fishes are the dominant vertebrates among the coral reef communities. The fishes comprise about 70% carnivorous, 20% herbivorous and 10% omnivorous species. The larger fish predators include sharks (Carcharhinidae), barracudas (Sphyraenidae), groupers (Serranidae), and moray eels (Muraenidae). Surgeon-fishes (Acanthuridae) are the dominant herbivores of the coral reefs, but other species including blennies (Blenniidae), damselfishes (Pomacentridae), and triggerfishes (Balistidae) will also gnaw algae from the surface of corals. The large numbers of animal species leads to intense competition for food and living space (fig.58). Therefore behavioural, reproductive, physiological and feeding strategies are wide and varied, and high levels of predation are found at all levels in the complex food webs.

In corals there are both dioeceous species (males and females are separate individuals) or 'hermaphrodite' species (an individual with both male and female organs). A polyp which is produced by sexual reproduction, often through mass synchronised coral spawning, grows into an adult and then reproduces 'asexually' to produce the rest of the coral colony. Asexual reproduction enables coral reefs to grow quickly. The adult will 'bud' a new individual on its side. Using this method of reproduction, all individuals in a colony will be identical and carry the same genetic material. New generations of coral polyps form on top of the skeletons of the previous generation, causing the reef to grow upwards and outward with time. There are three main types of coral reefs which are recognised by their shape and formation:

1. **Fringing reefs** e.g. north coast of Jamaica (Caribbean).
2. **Platform reefs and Barrier reefs** e.g. Great Barrier Reef (Australia) and Great Sea Reef (NW Fiji).
3. **Atolls** e.g. Maldives (Indian Ocean) and many Pacific islands.

Reef formation often begins with a coastal reef developing along a rocky shoreline where larval polyps can attach themselves to a hard substrate. A 'fringing' reef is formed as a band along a coastline (e.g. Red Sea), or around an island (e.g. Caribbean islands) where it gradually extends seawards. The width of the reef 'terrace' that develops will vary according to the slope, depth of water, seabed substrate and clarity of water (fig.59). Gently sloping shores with a hard substrate and clear water develop wide coral terraces. If the fringing reef is attached to a coastal landmass or island which is slowly sinking, the coral will continue to grow upwards towards the surface, as long as the upward growth rate can keep pace with the rate of sinking. If this continues over a long period of time, offshore 'barrier' reefs are formed and these are separated from the coastline by a deeper 'lagoon' of open water (fig.60). The lagoon waters are often connected to the open sea by deep 'surge channels' through which the water drains. The last stage of island reef development occurs when the island slowly sinks below the water to leave a ringed reef or 'fringing' reef around a central lagoon. This is known as an 'atoll'. The Indian Ocean and south Pacific have many such examples where underwater volcanoes (seamounts) are topped with coral reefs.

Coral reefs, as with many rocky shores, exhibit sublittoral zonation patterns with depth down the 'face' of the coral reef. Zonation is affected by bottom topography, wave action, exposure to air, water clarity, temperature and light levels. The shores of many reef areas are composed of low sandy areas which may be fringed with coconut palms above the high water levels. These trees are highly tolerant of salt and a sandy substrate, and can withstand tropical storms. Where lagoons have developed, the bottom may become filled with loose coral rubble and sand. If the lagoon is fairly shallow, towering coral columns (coral knolls or patch reefs) may grow from the lagoon floor and reach the surface. In shallow waters beds of seagrasses, such as turtlegrass *Thalassia,* may develop on the sandy bottoms where the water is clear and there are high levels of light. Among the seagrasses, animal communities are present with a rich diversity of organisms, including polychaete worms, molluscs, decapod crustacea and fish. Lagoon areas are often visited by marine turtles. Dolphins and Dugongs also appear to use lagoons as safe recreational and breeding areas as they are relatively free from predators such as sharks.

The outermost 'fringe' of a coral reef often consists of a coral reef 'crest' (algal ridge) and a seaward 'face' (known as a 'drop-off' to scuba divers, fig.59). The reef crest which lies on the seaward side of a reef, varies in width from 10-50m and is often marked by a line of breaking surf. The heavy, breaking surf renders the surface inhospitable to many coral organisms, but red encrusting algae often dominate the crest. At the low tide level the coralline algae may give way to large brown algae. The seaward face of the outer reef varies in depth, and the depth band between 5 and 50m provides the most suitable zone for coral growth and establishment of the associated reef communities. The 'windward' and 'lee-side' of many islands will have different zonation patterns. The more sheltered lee areas have a richer species diversity of plants and animals compared to the more exposed windward coral areas.

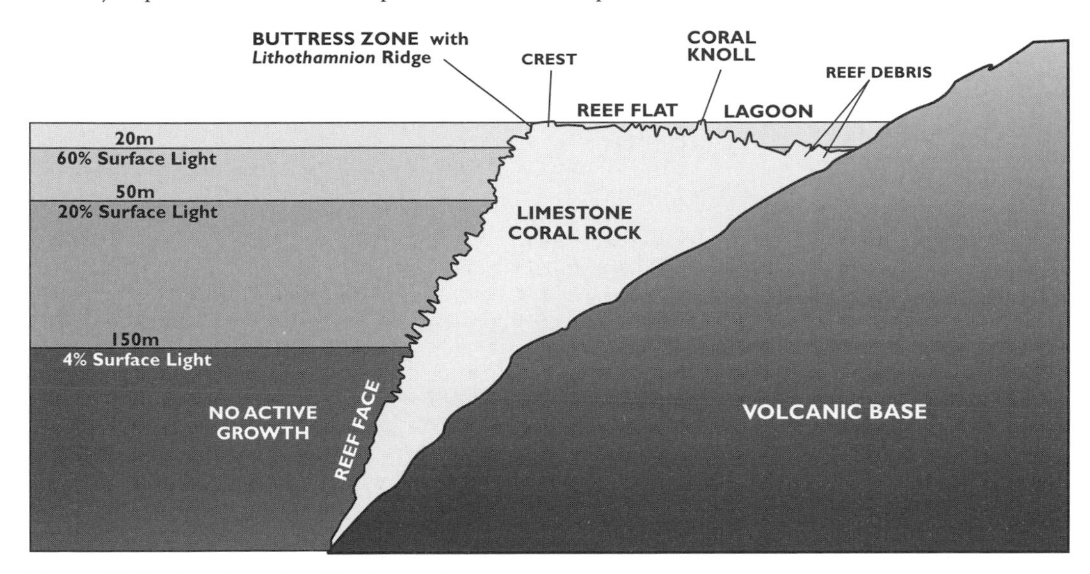

Figure 59. Section through a fringing reef.

Coral areas attract many visitors for snorkelling and scuba diving. The tourist trade also brings in its wake other requirements for accommodation, local food (including seafood), sewage treatment and other associated human waste disposal needs. Great care should accompany the growth of tourism in these sensitive marine areas. Visitors should be shown the art of watching without causing disturbance or damage; souvenir collecting must be discouraged. Controversy still remains as to the best approach to coral reef conservation. Should tourists be taken to only a few sites and leave other areas unspoilt, or should their impact be reduced by fewer numbers visiting many areas?

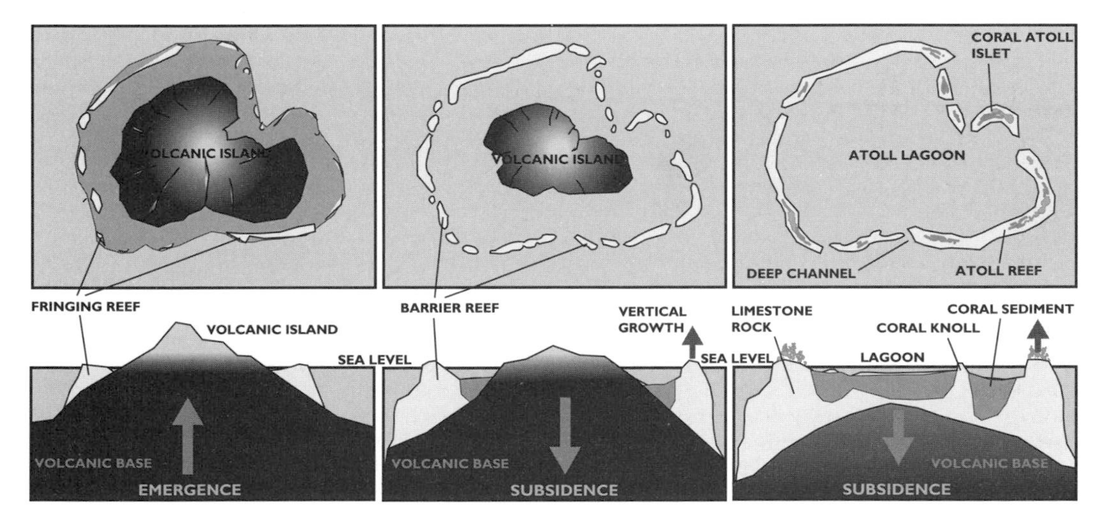

Figure 60. Atoll formation in aerial view and side profile.

In most cases, market forces intervene as tourists are taken to the more convenient areas. Only the intrepid seekers of wilder areas will venture further. Let us hope that governments, scientists and the tourist industry can work together to preserve the coral areas that have taken thousands of years to establish.

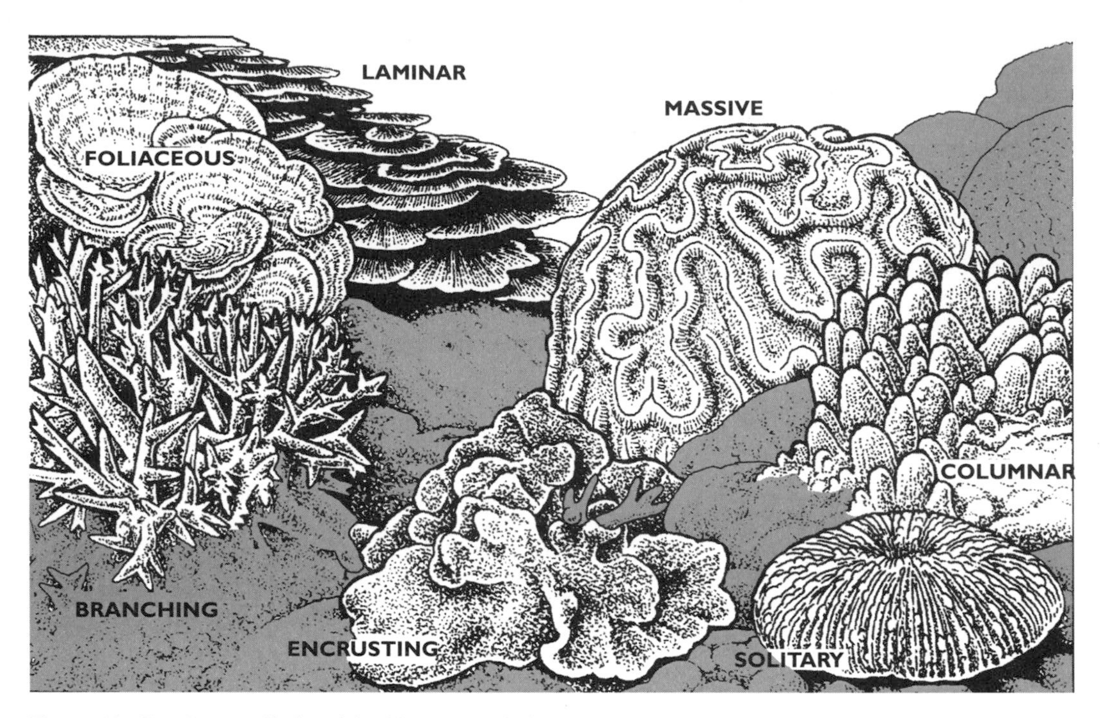

Figure 61. Coral types distinguished by external shape.

Summary: The Marine Ecosytem

The previous sections have described the development and relationships of marine ecosystems. This section brings together the various facets which make the marine environment a 'dynamic working whole' and summarises the common features of marine ecosystems.

The majority of organic production in seawater is confined to the shallow, photic zone where light levels are sufficiently high. In these surface waters, microscopic plant organisms produce new organic material through the biological process of photosynthesis. This biochemical process requires sunlight, carbon dioxide and nutrients in the presence of the unique biological pigment chlorophyll. Primary production through photosynthesis is carried out by algae. The microscopic floating algae (phytoplankton) provide food for many herbivorous animals and the majority of these animals are small to microscopic in size. Nutrients that leak out from the plants into the surrounding water are free to be recycled by the planktonic marine bacteria, which are in turn eaten by other small zooplankton. As dead plant and animal matter sinks slowly through the water column towards the seabed, it becomes the food source for scavenging zooplankton. However, some dead organic matter does eventually reach the sea floor where it becomes incorporated into the soft sediment and enriches it.

On the seabed, and often below the photic zone, communities of benthic sediment dwellers feed on the detritus. The small sediment dwellers and herbivorous zooplankton are a food source for many of the larger zooplankton. The larger zooplankton become the food source for many of the larger marine animals which may be sedentary (sessile) or mobile in lifestyle. Because of the nature of primary production in shallow waters, the most productive areas of the oceans in terms of biomass are the shallow coastal seas and continental shelf areas, where water varies in depth to about 200m.

Secondary production (animal tissue production) declines with increasing distance from shallow areas and with increasing depth. Food and nutrient availability controls the efficiency of food utilisation in marine systems. Where food and nutrients are abundant (e.g. upwelling areas, coastal and estuarine areas) utilisation is least efficient. It is as its most efficient in deep water communities, in pelagic open ocean areas, and in coastal coral reef communities. Nutrient supply is the main controller of primary production and therefore has an indirect effect on the rest of the food webs. Waste products are released by all organisms when they break down food to release the stored energy, which is required for body metabolic processes. Waste products from respiratory processes and excretion include carbon dioxide, water, phosphates and ammonia. These products are used by plants during photosynthesis and oxygen is a by-product of this biological reaction. Therefore a balanced dependency between plants and animals has been established.

In shallow waters, nutrients are often brought to the surface by the mixing of water through tides, currents, wind and wave action. However, nutrients are lost to the deeper oceans by sinking water masses in the cold, polar regions or in downwelling areas. These are eventually returned to the surface by deep water circulations being forced upwards by underwater obstructions or continental slopes. Nutrients lost to deep ocean water masses may not reappear at the surface for several hundred years. When the cold, nutrient-rich waters do appear they are known as 'upwelling' areas. In shallow water communities, high productivity with limited availability of nutrients can only take place if there is rapid recycling of nutrients (high turnover). This can take place through high levels of predation, short organism life cycles or by mutualism (beneficial arrangements between two organisms).

In many marine communities, if organism diversity is high (e.g. coral reefs), the numbers of individuals belonging to each group or species will often be low; the reverse is also true. In areas where diversity is low, the number of organisms of certain species may be very high (e.g. estuaries). Energy flowing through marine ecosystems takes place through complex food webs and there may be several trophic (consumer) levels. At each level there are fewer individuals and a smaller biomass than the one below it. Therefore a trophic pyramid has a base of primary production, followed by planktonic herbivores, planktonic carnivores, larger carnivorous animals and finally the higher order predators including fish, cetaceans, squid, seabirds, sea mammals and humans. At each trophic level animals will often be bigger to overcome the individuals they prey upon in the trophic levels below them.

The species diversity, biomass and numbers of individuals in a particular population will vary according to the biological and environmental stress. Apart from nutrients, other factors which influence marine plant and animal populations include temperature, light, pH, salinity, exposure to air and the amount of living space. In open oceans, salinity and temperature are relatively stable in comparison to coastal intertidal areas and in estuaries. Sea surface temperature plays a key role in the geographical distribution of marine organisms and controls the rate at which metabolic reactions take place.

Most animals reproduce sexually to pass on genetic material (DNA) from one generation to the next. DNA which is passed on through sexual reproduction from parent to offspring is never exactly the same. Alterations or mutations in DNA passed on to the next generation are a feature of sexual reproduction . Mutations are likely to be retained if they have adaptive value. It has been suggested that the only explanation of life is the survival of genes from one generation to the next. A population of plants or animals is maintained when production of offspring balances the numbers of individuals that die. Organisms have developed a wide range of reproductive methods and strategies to maintain the survival of offspring, so that sufficient individuals will reach maturity to produce the next generation. In reproduction that takes place asexually (e.g. corals) a new individual is 'budded' off the adult and therefore the offspring are genetically identical to the parent. However, this method of reproduction does not allow genetic changes to occur; many marine organisms show alteration between asexual and sexual generations.

The proportion of shallow water benthic organisms that produce planktonic larval stages varies with latitude and therefore temperature. In higher latitudes towards the polar regions, there are fewer species of marine organisms that produce planktonic larvae. Here, growing seasons are short and rates of tissue development are reduced in cold water. In the colder regions many of the larvae are hatched at a larger and more developed stage and this will reduce the time required to reach maturity. It also increases the chances of survival through reduced levels of predation. However, if egg size needs to be large then the numbers of eggs produced is often much lower. Therefore, fewer eggs can be produced by adult females at any one time, but there may be successive broods. Production of fewer, larger eggs is a strategy that is also used by many of the larger or higher predators such as sharks and rays.

With an overall understanding of marine ecosystems and the links between environmental, biological and evolutionary systems, the seafaring naturalist has the fundamental knowledge with which to explore the natural world. Observing and understanding plants and animals as part of a whole, dynamic ecosystem is as important as looking at one aspect, one organism or just one part of its life cycle.

BIBLIOGRAPHY

Barnes, R. S. K. 1994. *The Brackish-water Fauna of Northwestern Europe.* Cambridge University Press, Cambridge.

Barnes, R. S. K. & Hughes, R. N. 1992. *An Introduction to Marine Ecology.* Blackwell Scientific Publications, Oxford.

Bramwell, M. (ed.) 1977. *Atlas of the Oceans.* Colour Library Books, Guildford.

Hayward, P. J. 1994. *Animals of Sandy Shores.* Naturalist Handbooks 21. Richmond Publishing, Slough.

Hayward, P. J. & Ryland, J. S. 1995. *Handbook of the Marine Fauna of North-west Europe.* Oxford University Press, Oxford.

Lalli, C. M. & Parsons T. R. 1993. *Biological Oceanography: An Introduction.* Pergamon Press, Oxford.

Meadows, P. S. & Campbell, J. I. 1993. *An Introduction to Marine Science.* Blackie Academic and Professional, London.

Pinet, P. R. 1992. *Oceanography: An Introduction to the Planet Oceanus.* West Publishing, New York.

Tait, R. V. 1975. *Elements of Marine Ecology.* Butterworth, London.

Tchernia, P. 1980. *Descriptive Regional Oceanography.* Pergamon Marine Series Vol. 3. Pergamon Press, Oxford.

Thurman, H. V. 1994. (7th ed.) *Introductory Oceanography.* Macmillan, New York.

Vaughan, R. 1994. *The Arctic:. A History.* Alan Sutton Publishing, Stroud.

Wilson, D. 1989. *The Circumnavigators.* Constable, London.

Identifying Marine Life

Part One: Invertebrates

PLATE 1: OPEN OCEAN INVERTEBRATES

Some of the larger invertebrates that might be encountered in pelagic environments around the world are represented here. The species on this plate are mainly surface dwellers (the Portuguese Man-of-war occurs only at the surface), the exception being the Chambered Nautilus (11) which is normally found at a depth of 200-300m.

CNIDARIANS
Phylum: Cnidaria

CLASS: SCYPHOZOA (Jellyfish)

True jellyfish that exist as a pelagic medusoid stage for most of their lives.

1. Lion's Mane Jellyfish *Cyanea capillata*
World's largest jellyfish reaching 2m+ in diameter in Arctic waters. Average bell diameter 1m. Flat, transparent bell which contracts into large dome when swimming, opaque in large older individuals. 'Body' seen through bell is reddish-brown to red. Surrounding the mouth are four thick, undulate oral arms. Numerous long marginal tentacles, up to five times longer than bell diameter form fine net for capture of prey (small fish and plankton). Occurs worldwide. *C. lamarckii* is similar but smaller, less than 30cm in diameter, and has deep blue to purple 'body'. Can both deliver painful sting.

2. Rhizostome Jellyfish *Rhizostoma pulmo*
Long thin marginal tentacles absent. Eight oral tentacles partially fuse in mature specimens forming floret-like structures, above free arms. Bell transparent to opaque with yellow, reddish-purple to blue margins. Temperate seas, common in warm shallow waters. Bell diameter 1m.

3. Jellyfish *Mastigias papua*
Translucent bell with scattered reflective light spots which increase in number towards margins. Yellow-brown 'body' with dappled oral tentacles, which are fringed white. Occurs IP. Bell diameter 10cm.

4. Compass Jellyfish *Chrysaora hysoscella*
Translucent white to opaque yellow. Distinctive brown lines radiating from centre, hence its common name. 24 marginal tentacles up to 60cm long. Four thick, frilled, central oral tentacles longer than marginals. Occurs in most temperate seas. Bell diameter 25cm.

5. Luminescent Jellyfish *Pelagia noctiluca*
Pink-spotted, mushroom-shaped jellyfish. Four oral tentacles same length as eight marginal tentacles. Spots on bell are groups of stinging cells (warts). Produces luminescent mucus which looks like luminous paint when on skin. Sting not powerful, like series of pin pricks lasting 30 seconds. Usually found in surface waters of nAO, nPO (+MS). Bell diameter 10cm.

6. Moon Jelly *Aurelia aurita*
Common saucer-shaped jellyfish. Transparent when small, becoming pinkish or purplish when larger. Four obvious overlapping 'figure of eight' reproductive structures. Many fine short tentacles around margins. Four oral tentacles. Worldwide in coastal waters. Bell diameter 50cm.

7. Jellyfish *Cotolorhiza tuberculata*
Central dome surrounded by deep groove and long thin tentacles extending from 'body'. Green-brown colour caused by commensal algae living in tissues. Pelagic, sometimes in shallow water. Bell diameter 20cm. Often accompanied by the young of Horse Mackerel *Trachurus trachurus.*

CLASS: HYDROZOA (Hydroids)
Order: Siphonophora (Siphonophores)

These are not true jellyfish and consist of a colony of individual polyps which have different roles.

8. Portuguese Man-of-war *Physalia physalis*
Common siphonophore occurring at sea surface. Gas-filled, clear blue float with pink crest unmistakable. Long, blue, coiled stinging tentacles up to 10m+, hanging below blue to purple-coloured colony. Often known as 'bluebottles'. Float up to 30cm in length, has crest which acts as sail. Blown along by wind and carried by currents. Shows 'tumbling' behaviour when the animal dips each side alternately in the water, so keeping the float moist. Can deliver painful sting. Temperate waters worldwide. The **Bluebottlefish** *Nomeus gronovii* is often found in association with *Physalia.*

MOLLUSCS
Phylum: Mollusca

CLASS: CEPHALOPODA (Octopuses and Squid)

9. Squid *Todarodes pacificus*
Large eyes with three pairs of fleshy crests near 'neck'. Triangular fin sharply tapering to rear tip. As with most squid, body coloration can change rapidly. Red-brown with paler underside and tentacles, with wide black stripe along mid-dorsal line. Occurs wPO, Kuril Islands to Hong Kong. Mantle 30cm+ with longest tentacles 40% of mantle length.

10. Brown Paper Nautilus *Argonauta nodosa*
Pelagic relative of octopus. White to pale buff mantle and tentacles. Female secretes shell throughout life and lays her eggs in the shell. Eggs hatch when females move inshore to 'spawn' and die. Egg cases often washed ashore on beaches. Southern temperate seas. Male shell 10cm in diameter, female 50cm.

11. Chambered Nautilus *Nautilus pompilius*
Tentacles in two concentric rows. Eyes primitive pit-type. Shell white to orange with white stripes and black central whorl. Shell filled with gas and liquid which it adjusts to control buoyancy. Lives in deep water moving into shallows when 'spawning'. Occurs wIP. 20cm.

TUNICATES
Subphylum: Urochordata

CLASS: THALIACEA (Salps)

12. *Pyrosoma* colony
Transparent, cylindrical and colonial; tropical species luminescent. Individuals arranged in the form of an open cylinder, atrial siphons face inwards. Pharyngeal cilia produce water current for locomotion and feeding. Worldwide, more common in warmer seas. *Pyrosoma atlanticum* (AO) 3cm in length, *P. spinosum* (tropical seas) 3m+ length.

1
8
7
3
2
4
12
6
11
5
9
10

PLATE 2: KELP FOREST (Monterey Bay, California)

At Monterey Bay two current systems meet. The temperate California Current is cool and the tropical Davidson Current is warm. The area therefore teems with life since diverse species from normally colder and warmer habitats coexist. Kelp forests (*Macrocystis*) are common around the coast and an abundant invertebrate fauna that resides on or between them, on exposed rock faces, in crevices and on soft substrates. Only a fraction of the total species known can be shown here.

SPONGES
Phylum: Porifera

1. Sponge *Mycale adhaerens*
Pink to orange. Normally found encrusting the shells of scallops (**7**). Exhalant siphons under scallop's shell. 8 cm.

CNIDARIANS
Phylum: Cnidaria

2. Strawberry Sea Anemone *Corynactis californica*
Bright red to deep red. Thousands cover vast areas of exposed rock (free from kelp holdfasts). 2cm high.

3. Soft coral *Gersemia rubiformis*
Colour varies, white to pink. Resembles most cold water soft corals with limited finger-like branching. 15cm high.

4. Sea pen *Ptilosarcus guneyi*
Filter-feeding polyps. If disturbed deflates by expulsion of water, disappearing into substrate. Green luminescence when stroked. 60cm high.

SEGMENTED WORMS
Phylum: Annelida

5. Red Tubeworm *Serpula vermicularis*
Distinctive, 40 pairs of red to light pink filaments. Filaments up to 10cm across. Subtidal to 100m depth.

MOLLUSCS
Phylum: Mollusca

CLASS: POLYPLACOPHORA (Chitons)
6. Lineate Chiton *Tonicella lineata*
Distinctive patterning. Feeds at night on coralline algae. Length 5cm. Intertidal, sometimes subtidal.

CLASS: BIVALVIA (Bivalves)
7. Scallop *Chlamys hastata*
Brightly coloured orange and white mantle folds. Commonly covered in encrusting sponges (**1**). Filter feeder. Width 8cm.

8. Northern Horse Mussel *Modiolus modiolus*
Rough exterior to shell, dark with purplish tinge. Bluish-white inside. Darker orange flesh than *Mytilus edulis*. Abundant, often associated with kelp. Filter feeder. Width 13cm. Low tide to 150m depth.

CLASS: GASTROPODA (Gastropods)
9. Green Abalone *Haliotis fulgens*
20cm diameter.

10. Blue Topshell *Calliostoma ligatum*
Yellow shell with numerous concentric fine black bands. Feeds on bryozoans and hydroids, also kelp. 5cm high.

11. Purple-ringed Topshell *Calliostoma annulatum*
Brightly coloured orange and cerise shell with chequered concentric banding. Same diet as *C. ligtum*. 5cm high.

12. Spanish Shawl *Flabellina iodinea*
Brightly coloured for warning. Produces toxins in mucus, and has undischarged nematocysts in orange cerata. Feeds on hydroids. Length 8cm.

13. Nudibranch *Melibe leonina*.
Transparent. Swims using gills as keels; 3 animals shown filter-feeding with bell-shaped oral hoods. Length 25cm.

CLASS: CEPHALOPODA (Squid and Octopuses)
14. Giant Octopus *Octopus dofleini*
Red-brown, but can quickly change colour. Contrasting white suckers in double rows. Active hunter. Length 150cm; weight 68kg.

CRUSTACEANS
Subphylum: Crustacea

ORDER: DECAPODA (Decapods)
15. Red Rock Crab *Cancer productus*
A pair of distinctive deep red rock crabs shown courting. Mating may take up to 7 days. Carnivorous and a scavenger. Carapace width 10cm.

16. Butterfly Crab *Cryptolithodes typicus*
Camouflaged like a small, grey-brown pebble. Feeds on detritus and plankton. Carapace width 2.5cm.

ECHINODERMS
Phylum: Echinodermata

CLASS: STELLEROIDEA (Starfish and Brittle-stars)
17. Sunflower Star *Pycnopodia helianthoides*
Colour variable, turquoise to red. Among fastest moving and largest of starfish. Territorial. Adults have 24 arms, juveniles five. Feeds on crustaceans and fish. Width 90cm.

18. Bat Star *Patiria miniata*
Coloration variable, dark red to orange to mottled white. Short triangular arms. May be cannibalistic. Width 20cm. Subtidal to 300m depth. Common in ePO.

19. Giant Sea Star *Pisaster giganteus*
Rich orange-brown upperparts, creamy-yellow underside. Feeds on molluscs. Width 30cm.

20. Spiculate Brittle-star *Ophiothrix spiculata*
Coloration variable. Arm length 5-8 times diameter of disc. Prominent spines on arms and disc. Catches plankton with sticky mucus. Width 14cm. Subtidal to 2,000m depth.

CLASS: ECHINOIDEA (Sea Urchins)

21. Sea urchin *Strongylocentrotus franciscanus*
Long, deep red spines. Primary kelp grazer. Diameter 20cm.

BRYOZOANS
Phylum: Bryozoa

22. Sea Mat *Membranipora membranacea*
Off-white colonies found encrusting kelp fronds and stipes. Individuals rectangular. No two colonies show same zooid patterning. Colony up to 15cm width. Intertidal to 30m depth.

TUNICATES
Subphylum: Urochordata

CLASS: ASCIDIACEA (Sea Squirts/Ascidians)
23. Light Bulb Tunicate *Clavelina huntsmani*
Distinctive bright yellow 'filament'. Smaller in winter months with regression of the tips of the stolons. 5cm high, colonies up to 50cm across. Subtidal to 40m depth.

14
3
18
13
17
21
19
8
20
2
9
15
21
5
4
12
7
1
22
23
11
10
6
16

PLATE 3: ROCKY TEMPERATE SHORE (NW Europe) I

Plates 3 & 4 illustrate typical flora and fauna of NW European rocky shores, from the tidal zone down to 20m depth. Only the invertebrate fauna are shown (most algae omitted for clarity). The species shown are unlikely to be found at the same site together. Cephalopod lengths are given as total length.

SPONGES
Phylum: Porifera

1. Purse Sponge *Scypha ciliata*
Creamy yellow. Surface hairy, ring of longer hairs at free end. 3cm high. Lower shore to 100m. Widespread.

2. White/Lace Sponge *Clathrina coriacea*
White calcareous sponge. Associated with *D. grossularia* (**25**).

CNIDARIANS
Phylum: Cnidaria

3. Sea Pen *Pennatula phosphorea*
Feather-like. Red with white polyps. Luminesces brightly when touched at night. 40cm high. 20m+ depth.

4. Jewel Anemone *Corynactis viridis*
Varied brilliant colours. Characteristic white knobs on tips of tentacles. Often found in large numbers. 5mm diameter. Extreme lower shore to 100m depth.

5. Plumose Anemone *Metridium senile*
Creamy-white to brown, column smooth. Numerous slender plumose tentacles. Can be in dense aggregations. 8cm+ high. In PO, up to 60cm high. Intertidal to 3m depth.

6. 'Parasitic' Sea Anemone *Calliactis parasitica*
Pale yellow column with red-brown vertical lines of spots. Up to 200 creamy tentacles. Commensal with *Dardunus* and *Pagurus* hermit crabs. 8cm high. Subtidal to 60m depth.

SEGMENTED WORMS
Phylum: Annelida

7. Sea Mouse *Aphrodite aculeata*
Oval shaped. Mat of grey-brown hairs cover scales below. Golden-brown iridescent chaetae. Length 10cm. 10-30m+ depth.

MOLLUSCS
Phylum: Mollusca

CLASS: BIVALVIA (Bivalves)

8. Queen Scallop *Aequipecten opercularis*
Shell with c20 ribs, colour varies, often encrusted by sponges. 'Ears' either side of hinge equal in length. Active swimmer. Width 9cm. Extreme lower shore to 200m depth.

CLASS: GASTROPODA (Gastropods)

9. Common/Painted Topshell *Calliostoma zizyphinum*
Conical with up to 9 shallow whorls. Yellow-pink with dark red stripes. 2.5cm high. Extreme lower shore to 200m depth.

10. Common Whelk/Buckie *Buccinum undatum*
Distinctive large heavy shell, often encrusted. Voracious predator on polychaete worms. 16cm. Intertidal to 80m depth.

11. Green Ormer *Haliotis tuberculata*
Green to brown to red shell. Thick layer of mother-of-pearl inside shell. 8cm. Lower shore to shallow subtidal.

12. Common Periwinkle *Littorina littorea*
Grey, black or brown shell with contrasting striations. Feeds on algae. Abundant. 2.5cm shell height. Intertidal.

13. Slipper Limpet *Crepidula fornicata*
Distinctive white-cream oval shell. Found in chains. Diameter 2.5cm. Shallow water, especially in oyster beds.

14. Sea Hare *Aplysia punctata*
Brown with red or green hues. 4 tentacles. Ejects purple dye when disturbed. Can swim. Length 15cm. Subtidal.

15. Nudibranch *Onchidoris bilamellata*
White with red-brown blotches. Gill cluster at rear. Secretes acid when disturbed. Length 4cm. 5 - 20m depth.

CRUSTACEANS
Subphylum: Crustacea

ORDER: DECAPODA (Decapods)

16. Spiny Squat Lobster *Galathea strigosa*
Orange-red with blue lines. Abdomen tucked under thorax. Aggressive. Body length 12cm. Lower shore to 35m depth.

17. Common Lobster *Homarus gammarus*
Dark blue-grey with orange markings. Nocturnal. Feeds on molluscs. Body length 50cm+. 20-100m depth.

18. Common Hermit Crab *Pagurus bernhardus*
Adults found in whelk shells. Red-brown claws. Up to 8 *C. parasitica* (**6**) anemones on shell. Length 10cm. Subtidal.

19. Velvet Swimming Crab *Liocarcinus puber*
Carapace red-brown with numerous fine hairs giving muddy-brown appearance. Red eyes, joints blue. Aggressive, rears up with claws outstretched. Width 8cm. Subtidal.

ECHINODERMS
Phylum: Echinodermata

CLASS: STELLEROIDEA (Starfish and Brittle-stars)

20. Common Brittle-star *Ophiothrix fragilis*
Disc colourful. Slender arms banded. Deeper species with more intense colours. Feeds on algae, carrion, or captures small prey in mucus 'net'. Width 20cm. Up to 475m depth.

CLASS: ECHINOIDEA (Sea Urchins)

21. Purple Heart Urchin *Spatangus purpureus*
Red to violet. Covered in short 'hairy ' spines with occasional longer spines. Burrower. Diameter 12cm.

CLASS: HOLOTHUROIDEA (Sea Cucumbers)

22. Pudding *Cucumaria frondosa*
Dark leathery body with pale bushy tentacles. Filter feeder attached to rocks. Large (length 50cm). 5 - 30m depth.

BRYOZOANS
Phylum: Bryozoa

23. Ross/Rose Coral *Pentapora foliacea*
Red-brown flattened fused branches. Dome-shaped colony up to 1m across, 20cm high. Strong current areas. 10 - 50m.

24. Hornwrack *Flustra foliacea*
Brown to yellow to grey hues, fades when dead and washed ashore. Leaf-like branching. 20cm high. Subtidal to 100m.

TUNICATES
Subphylum: Urochordata

CLASS: ASCIDIACEA (Sea Squirts/Ascidians)

25. Gooseberry Sea Squirt *Dendrodoa grossularia*
Red-brown colour. Solitary or large congregations in sheltered positions. 2.5cm high. Lower shore to 100m+.

1
22
17
4
2
25
24
3
14
16
20
10
7
11
23
5
21
19
6
18
8
9
15
12
13

PLATE 4: ROCKY TEMPERATE SHORE (NW Europe) II

RED ALGAE
Division: Rhodophyta

1. Red Coralline Alga *Crustose corallines*
Pink to red. Encrusts hard substrates. Photosynthetic. Patches can be 1m+ across. 5m to 30m+ depth.

SPONGES
Phylum: Porifera

2. Sponge *Amphilectus fucorum*
Red to pale orange, soft texture, encrusting. Tasselled form illustrated.Diameter 15cm. 5 to 20m depth.

3. Breadcrumb Sponge *Halichondria panicea*
Normally green, sheet-like form, can be white to orange-brown. Diameter 20cm. Middle shore to deeper water.

CNIDARIANS
Phylum: Cnidaria

CLASS: HYDROZOA (Hydroids)

4. Squirrel's Tail *Sertularia argentea*
Brownish-white, often dense furry mats. Polyp tentacles feed on plankton. 20cm. High energy sites, 10 - 30m+.

5. Hydroid *Tubularia indivisa*
Yellowish straight stems with pink polyps. Network encrusts substrate. 18cm high. Lower shore and below.

CLASS: ANTHOZOA (Sea Anemones and Corals)

6. Red Sea-fingers *Alcyonium glomeratum*
Pink to red soft coral. Similar to Dead Man's Fingers *A. digitatum.* 15cm high. 10m to 30m+ depth.

7. Sea anemone *Sagartia elegans*
Four variations of colour. When disturbed ejects white threads (acontia). 6cm high. Lower shore to 50m.

8. Devonshire Cup Coral *Caryophyllia smithii*
Solitary coral. Buff skeleton with conspicuous ridges. Polyps vary in colour. 1.5cm high. Lower shore to 50m.

FLATWORMS
Phylum: Platyhelminthes

9. Candy Stripe Flatworm *Prosthecereus vittatus*
Easily recognised with 2 conspicuous tentacles. Carnivorous. 3cm. Found under stones, 5 to 30m+.

SEGMENTED WORMS
Phylum: Annelida

10. Fanworm *Bispira volutacornis*
Two whitish multifilamentous gill clusters. Membranous tube with fine mud particles. Gills 3cm across. Filter feeder. Often in colonies. Length 15cm. 5 to 30m+ depth.

MOLLUSCS
Phylum: Mollusca

CLASS: GASTROPODA (Gastropods)

11. European Cowrie *Trivia monacha*
Pink to brown shell with three dark spots. Feeds on sea squirts. Length 1.5cm. Lower shore to shallow subtidal.

12. Nudibranch *Thecacera pennigera*
Dappled yellow and black. Feeds mainly on bryozoan *Bugula.* Length 5cm. Lower shore to deeper water.

13. Common Grey Sea Slug *Aeolidia papillosa*
Grey to brown. Dorsal cerata on either side of mid-line. Feeds on anthozoans. Length 12cm. 5 to 30m depth.

CLASS: CEPHALOPODA (Octopuses and Squid)

14. Curled/Lesser Octopus *Eledone cirrhosa*
Usually red-brown above, white below. Single row of suckers. Nocturnal hunter. Length 50cm. 5 to 30m+ depth.

15. Common Cuttlefish *Sepia officinalis*
Flattened body, 8 small arms and 2 long retractable tentacles. Nocturnal hunter. Length 30cm. Coastal waters.

16. Common Atlantic Squid *Loligo vulgaris*
Normally translucent pink to red-brown. Relatively large elongate squid with diamond-shaped fins. Shield projects over head. Length 60cm. Mainly neritic.

CRUSTACEANS
Subphylum: Crustacea

ORDER: DECAPODA (Decapods)

17. Crawfish/Spiny Lobster *Palinurus elephas*
Spiny red-brown carapace. Claws on 5th appendage only in females. Hunter and scavenger. Length 50cm.

18. Edible Crab *Cancer pagurus*
Pink to brown oval carapace. Large claws tipped black. Hunter and scavenger. Width 20cm+; weight 6kg. Subtidal.

19. Spiny Spider Crab *Maja squinado*
Pink to white triangular carapace, often encrusted. Largest spider crab in area. Scavenger. 20cm. Lower shore to 50m.

ECHINODERMS
Phylum: Echinodermata

CLASS: CRINOIDEA (Feather Stars)

20. Feather Star *Antendon bifida*
Purple to orange, 5 paired, feathered arms. Sessile. Filter feeds with tube feet. 20cm diameter. To 200m depth.

CLASS: STELLEROIDEA (Starfish)

21. Common Sunstar *Crossaster papposus*
Red-brown, white underside. 8-13 blunt arms. Eats other starfish, also cannibalistic. 25cm diameter.

22. Bloody Henry *Henricia oculata*
Blood red to purple. Stiff arms, circular cross-section. Carnivorous, hunting by smell. 10cm+. 5 to 30m+ depth.

CLASS: ECHINOIDEA (Sea Urchins)

23. Common/Edible Sea Urchin *Echinus esculentus*
Red to purple test, spines whitish. Omnivorous, important subtidal grazer. 17cm diameter. Subtidal to 50m depth.

CLASS: HOLOTHUROIDEA (Sea Cucumbers)

24. Cottonspinner *Holothuria forskali*
Black and warty, buff belly with 3 rows of tube feet. Ejects white sticky threads if disturbed. Sucks up organic material from substrate surface. Length 20cm. Lower shore to 70m.

BRYOZOANS
Phylum: Bryozoa

25. Bryozoan *Bugula neritina*
Brownish and bushy. Zooecium large and rectangular. Filter feeds with lophophore. Lower shore to deeper water.

TUNICATES
Subphylum: Urochordata

CLASS: ASCIDIACEA (Sea Squirts/Ascidians)

26. Sea squirt *Ascidia mentula*
Solitary and in groups. Red or green thick tunic. Markings round siphons. Filter feeder. 10cm high. To 200m depth.

20
16
22
15
23
11
22
24
7
8
25
19
17
4
13
18
4
5
3
14
9
6
6
10
21
12
2
22
26

PLATE 5: REEF CORALS

A coral reef is a complex and dynamic environment. Corals live in low nutrient, low primary productivity areas in tropical seas and show a remarkable symbiosis between coral polyp and microscopic algal zooxanthellae. This plate illustrates some of the forms that coral colonies can take. Most coral polyps feed only at night, or at low light intensity, when the coral reef becomes a palette of colours. However, it is in the daytime, when the zooxanthellae photosynthesise, that the food for the polyps is produced.

RED ALGAE
Division: Rhodophyta

Encrusting or crustose algae are important on coral reefs. They are especially abundant in high energy sites, e.g. outer margins of reefs, and areas of intense herbivorous grazing. The algal mats protect the reef's more eroded surfaces, cementing the reef substrate together.

1. Encrusting Alga *Porolithon onkodes*
Common pink encrusting alga, often covering the bases of corals.

2. Leafy Coralline Alga *Lobophora variegata*
Leafy red-brown encrustations found in the same sites as *P. onkodes* (**1**).

GREEN ALGAE
Division: Chlorophyta

Green algae are represented on the reef by, for example, Turtle Weed *Chlorodesmis fastigiata*. There are also many encrusting green algae, such as *Halimeda opuntia*.

3. Grape Weed *Caulerpa racemosa*
Easily identified, often found amongst coral boulders, or lying over reef substrate.

SPONGES
Phylum: Porifera

4. Boring sponge *Cliona* sp.
Coloration varies, commonly purple. Envelops living coral and dissolves, with an organic acid, the limestone skeleton. Quick growing. Length 30cm.

CNIDARIANS
Phylum: Cnidaria

CORALS
Classes: Anthozoa and Hydrozoa

ORDER: ALCYONACEA

5. Soft coral *Sarcophyton trocheliophorum*
This soft coral retracts its polyps when disturbed or at low tide, taking on a distinctive folded, leathery appearance; further irritation gradually causes colony to deflate. Colony up to 1m height.

6. Bladed Soft Coral *Dendronepthya mucronata*
This coral lacks zooxanthellae; with its polyps fully extended the skeletal spicules (sclerites) can clearly be seen. Colony up to 60cm height.

ORDER: GORGONACEA

7. Gorgonian *Acabaria* sp.
These gorgonian fans grow at right angles to the current and with the polyps fully extended are efficient plankton feeders. Commonly found under ledges. Colony up to 1m width.

8. Whip Coral *Juncella fragilis*
White, grey or yellowish. Length 1m.

ORDER: HYDROCORALLINA

9. Hydroid coral *Stylaster elegans*
Common under overhangs. Hard colony skeletons retain coloration when dead. Height 23cm.

10. Fire coral *Millepora tenera*
Found on sides or top of reef. Polymorphic, some polyps for feeding, others for defence. Stings can penetrate skin, causing continuous itching. Height 1m.

ORDER: MADREPORARIA

11. Coral *Montipora efflorescens* (Close-up)
Green, blue, pink or cream. Upper reef slope. Massive.

12. Brain coral *Oulophyllia bennettae*
Greenish-grey, (brown form on W coast of Australia). Massive.

13. Whorled Coral *Montipora capricornis*
Distinctive shape, blue, purple or brown. Mainly in lagoons. Massive.

14. Staghorn coral *Acropora nobilis*
Arborescent. Cream to blue and green. Upper reef slope or lagoons.

15. Table Coral *Acropora cytherea*
Wide, flat table-shape. Pale cream to brown and blue. Upper reef slope.

16. Solitary Mushroom Coral *Diaseris distorta*
Circular, capable of lifting itself off substrate.

17. Coral *Acropora sarmentosa* (close-up)
Dull greenish-grey or brown with pale tips. Bottlebrush branching. Upper reef slope.

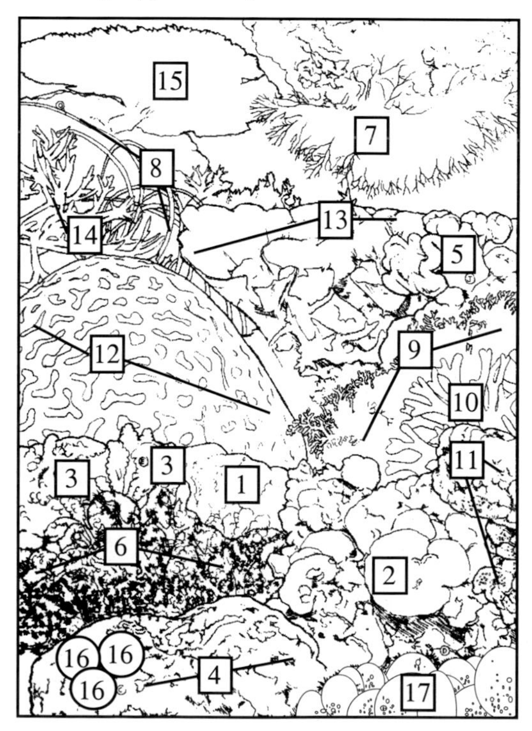

PLATE 6: CORAL REEF I

Coral reefs contain many invertebrate species and a large number are undescribed. Plates 6 and 7 show a representative selection of invertebrates found in and around Indo-Pacific coral reefs.

SPONGES
Phylum: Porifera

1. Staircase Sponge *Caulospongia perfoliata*
Colour orange to red. Strong skeleton. Filter feeder on plankton. Deeper water species, up to 1m high. 10 to 25m depth.

FLATWORMS
Phylum: Platyhelminthes

Often found searching for food as they glide along the reef; many species can also swim.

2. Flatworm *Pseudoceros* sp.
Wide, carnivorous diet. Length 2cm.

3. Flatworm *Pseudoceros* sp.
Wide, carnivorous diet. Length 3cm.

SEGMENTED WORMS
Phylum: Annelida

4. Christmas-tree Worm *Spirobranchus giganteus*
Coloration varied, from blue to yellow or red. Delicate twin spiralling radioles. Up to 50 worms per m^2. Filter and detritus feeder. 15cm across radioles. To 30m depth.

5. Southern Fan Worm *Sabellastarte indica*
Wide variety of colours, from red to white. Distinctive striped radioles, distinguishing it from *S. sanctijosephi*. Filter and detritus feeder. 15cm tube length. To 30m depth.

MOLLUSCS
Phylum: Mollusca

CLASS: GASTROPODA (Gastropods)

6. Giant Triton *Charonia tritonis*
Large distinctive shell. Feeds on sea urchins and starfish. Shown eating a Crown-of-Thorns Starfish *Acanthaster planci* (**16**). Length 45cm. Exploited by shell collectors.

7. Spanish Dancer *Hexabranchus sanguineus*
Large strikingly coloured nudibranch. Usually crawls over sea bottom, but also swims, as shown, by vigorous undulations of body. Feeds on sponges. Length 30cm.

8. Textile Cone Shell *Conus textile*
Obvious triangular shell markings. Nocturnal. Feeds on gastropods, normally species without opercula. Stuns or kills prey with venomous dart shot out from proboscis. In daytime, burrows in substratum. Length 12cm. Toxic to humans.

9. Geographer Cone Shell *Conus geographus*
Similar to above but with less distinct markings. Feeds on small reef-dwelling fish. Geographer Cones' toxin known to be fatal to humans. Length 15cm. Intertidal.

10. Green-spotted Nudibranch *Nembrotha nigerrima*
Distinctive velvet black and green-spotted coloration. Feeds on colonial ascidians. Produces toxins to deter predators. Distinct markings used as a warning. Widespread. Length 10cm. 3 - 8m depth.

CLASS: CEPHALOPODA (Octopuses and Squid)

11. Blue-ringed Octopus *Hapalochlaena lunulata*
Easily identified by bright blue rings. Paralyses prey with venomous bite from beak, which can be lethal to humans. Length 15cm. Feeds mainly on crabs and molluscs. N Australia to Japan.

CRUSTACEANS
Subphylum: Crustacea

ORDER: DECAPODA (Decapods)

12. Banded Coral Shrimp *Stenopus hispidus*
Distinctive long white antennae, a well-known cleaner shrimp. Found in pairs or groups at 'cleaner stations' around reefs. Females carry green eggs beneath abdomen. Scavenger. Length 6cm. To 30m depth.

13. Painted Dancing Shrimp *Hymenocera picta*
Distinctive shape and coloration. Almost always found in pairs, female larger. Flared claws used for display. Feeds at night on starfish. Length 6cm. Low tide to 10m depth.

14. Xanthid crab *Trapezia* sp.
Bright red with silver-grey markings over carapace and appendages. Lives on corals; larger corals have numerous pairs of these crabs (especially branching corals). Scavenger, can defend coral territory from *A. planci* (**16**). 3cm wide.

ECHINODERMS
Phylum: Echinodermata

CLASS: CRINOIDEA (Feather Stars)

15. Noble Feather Star *Comanthina nobilis*
Two main colour forms, bright yellow or black with numerous other colours. Filter feeds on plankton from high coral outcrops. Host to many commensal species. 40cm diameter. 5 - 90m depth.

CLASS: STELLEROIDEA (Starfish and Brittle-stars)

16. Crown-of-thorns Starfish *Acanthaster planci*
Distinctive with long dorsal spines; may cause severe injury if handled. Feeds on coral polyps, and has caused destruction of large areas of coral reefs. 40cm diameter.

CLASS: ECHINOIDEA (Sea Urchins)

17. Savigny's Sea Urchin *Diadema savignyi*
Test normally black but some, as shown, may have lighter tests. Long black venomous spines cause painful skin punctures in humans. Herbivorous grazer on algae. 23cm diameter. Widespread in IP. To 25m depth.

18. Mathae's Sea Urchin *Echinometra mathaei*
Brown, pink or purple spines with white circle at base. Rock-borer. Nocturnal herbivorous grazer on algae. 9cm diameter. Widespread. To 5m depth.

CLASS: HOLOTHUROIDEA (Sea Cucumbers)

19. Pineapple Sea Cucumber *Thelenota ananas*
Orange to pink, red or brown. Rigid dorsal papillae. Host to commensal scale worms. Suction feeder on detritus. Length 75cm, 5 - 40m depth.

TUNICATES
Subphylum: Urochordata

CLASS: ASCIDIACEA (Ascidians/Sea Squirts)

20. Blue-pod Ascidian *Pycnoclavella diminuta*
White or blue. Small-stalked ascidian. Filter feeds on plankton. Colonies up to 25cm width.

1
15
11
19
17
6
16
9
7
12
8
13
2
18
16
20
10
3
5
4
14

PLATE 7: CORAL REEF II

SPONGES
Phylum: Porifera

1. Thin Yellow Fan Sponge *Ianthella basta*
Thick, stiff fibres hold sponge erect; no spicules. Inhalant side faces prevalent water current direction. 30cm high.

CNIDARIANS
Phylum: Cnidaria

2. Box-jelly/Sea Wasp *Chironex fleckeri*
Translucent jelly bell. Pendalium at each of the 4 corners with up to 16 tentacles. Largest of the Cubozoa. Feeds on plankton and fish. Virulent sting. Tentacles to 1m length.

3. Ritter's Sea Anemone *Radianthus ritteri*
Column dark orange brown, tentacles pale green to dark green. One of the largest single-structured tropical anemones. Host to many commensals, especially the anemone fish *Amphiprion akindynos*. Varied diet from plankton to fish. Up to 1m across. In areas with strong water currents, 5 - 15m depth.

SEGMENTED WORMS
Phylum: Annelida

4. Fireworm *Eurythoe complanata*
Distinctive red and pink polychaete worm. Numerous chaetae are long, thin and toxic. They break off in skin when touched, causing severe irritation. Omnivorous, feeding on invertebrates and algae. Length 10cm.

MOLLUSCS
Phylum: Mollusca

CLASS: POLYPLACOPHORA (Chitons)
5. Gem Chiton *Acanthopleura gemmata*
One of the largest tropical chitons. Normally found in 'homing' hollow created over generations. Nocturnal herbivorous grazer. Length 15cm. Intertidal to deeper water.

CLASS: BIVALVIA (Bivalves)
6. File shell *Lima* sp.
Mantle bright red. Long sticky tentacles around mantle edge often break off when touched, allowing fast escape by swimming. Width 4cm. Found under coral rock.

CLASS: GASTROPODA (Gastropods)
7. Tun Shell *Tonna perdix*
Foot and mantle dark brown with pale blue mottling. Carnivorous; feeds on sea cucumbers by extending body almost completely out of shell, as shown in illustration. Length 14cm.

8. Red-mouthed Frog Shell *Bursa lissostoma*
Distinctive shell often inhabited by hermit crabs. Rarely seen alive. Females nearly twice size of males. Egg capsules are transparent tubes 25mm high. Carnivorous on ascidians. Length 20cm. Wide depth range (to 120m).

9. Ocellate Nudibranch *Phyllidia ocellata*
Unique markings. White skin patches are glands that secrete toxins for defence. Nocturnal predator on sponges. Found under ledges and in caves. Length 5cm. Common in IP. 5 - 35m depth.

10. Red-lined Flabellina *Flabellina rubrolineata*
Dark red or purple line along centre of back diagnostic. Whole animal pale pink to pale blue. Feeds on 3 species of hydroid. Length 4cm. To 30m depth.

CLASS: CEPHALOPODA (Octopuses and Squid)
11. Cuttlefish *Sepia latimanus*
Uses coloration and mantle projections to break up outline. A large species common in reef waters. Carnivorous. Length 50cm+.

CRUSTACEANS
Subphylum: Crustacea

ORDER: DECAPODA (Decapods)
12. Slipper Lobster *Parribacus antarcticus*
Pale yellow 'shaggy' appearance. Carnivorous, diet includes giant clams. Large, up to 25cm in length.

13. Spotted Porcelain Crab *Neopetrolisthes maculatus*
Distinctive pale coloration with red spotting. Lives within several species of sea anemones. Filter feeds with hairy maxillipeds. Width 3.5cm. Low tide to 25m depth.

14. Decorator Crab *Camposcia retusa*
Hides itself by attaching debris to hook-like chaetae over carapace and appendages. Width 5cm. Carnivore and scavenger.

ECHINODERMS
Phylum: Echinodermata

CLASS: CRINOIDEA (Feather Stars)
15. Basket Star *Astroboa nuda.*
Large web-like arms. Nocturnal feeder. Traps prey (crustacean and fish larvae) up to 30mm in length. These are impaled on tiny hooks on branched arms. Diameter 1m+.

CLASS: ECHINOIDEA (Sea Urchins)
16. Slate Pencil Urchin *Heterocentrotus mammillatus*
Easily recognised sea urchin with long, thick, banded spines, probably used to wedge animal in crevice. Herbivorous, grazing on algae. Not common. Diameter 25cm.

17. Flower Urchin *Toxopneustes pileolus*
Distinctive white, flower-like pedicellariae. Seaweed and shell valves carried on back. Larger specimens are presumed to be toxic to touch, possibly fatal to humans. Herbivorous, grazer on algae. Diameter 12cm. Subtidal to 30m depth.

CLASS: HOLOTHUROIDEA (Sea Cucumbers)
18. Leopard Sea Cucumber *Bohadschia argus*
Easily identified 'eyed' sea cucumber. Background colour varies but markings fairly constant. Mucus from animal reported to cause damage to eyes. Ejects sticky white threads from cloaca (Cuvierian tubules) in defence, as illustrated. Suction feeder. Length 25cm.

BRYOZOANS
Phylum: Bryozoa

19. Little Fan Bryozoan *Lanceopora obliqua*
Distinctive colony shape. Anchors itself to substratum with a stalk. Filter feeds on plankton using lophophores. 15 - 30m depth in areas of strong currents.

TUNICATES
Subphylum: Urochordata

CLASS: ASCIDIACEA (Ascidians/Sea Squirts)
20. Giant Jelly Ascidian *Polycitor giganteum*
Distinctive colonial ascidians. Two colour forms, zooids either white or orange. Zooids interconnected. Filter feeder. Colonies up to 30cm width. 10 - 40m depth.

15
2
11
20
8
12
7
16
18
1
9
5
17
4
6
19
14
13
3
10

PLATE 8: CORAL LAGOON

Behind the protection of the reef lies the lagoon. The waters are calmer, with less erosion and wave damage. Temperatures are higher, the waters less deep and larger areas are exposed to the air. This lagoon habitat is different from the outer reef slope and has its own, sometimes unique, inhabitants.

CNIDARIANS
Phylum: Cnidaria

CLASS: ANTHOZOA (Sea Anemones and Corals)

1. Coral *Porites cylindrica*
Branching coral sometimes encrusting at base.

2. Stony coral *Montipora venosa*
Pale brown to yellow polyps; coral has gnarled appearance.

CLASS: SCYPHOZOA (Jellyfish)

3. Upside-down Jellyfish *Cassiopea andromeda*
Bell flat above. Arms with bladders filled with zooxanthellae. Found usually lying upside-down on sandy bottoms of bays.

MOLLUSCS
Phylum: Mollusca

CLASS: BIVALVIA (Bivalves)

4. Giant Clam *Tridacna maxima*
Iridescent blue or green mottling on mantle. World's largest bivalve. Mantle contains photosynthetic tissue pockets with symbiotic zooxanthellae present. Also filter feeds. Width 1m+.

CLASS: GASTROPODA (Gastropods)

5. Baler Shell *Melo amphora*
Grey-brown smooth shell. Large siphon. Feeds on other molluscs. Largest volute; length 56cm. Found only in N Australian waters.

6. Spotted Volute *Amoria maculata*
Easily recognised by orange stripes on pale or white foot. Varied shell patterning. Mainly nocturnal. Feeds on other molluscs. Length 8cm. Endemic Australian species, common around the Great Barrier Reef. Low tide to 30m depth.

7. Eyed Cowry *Cypraea argus*
Distinctive eyed pattern to shell. Not much known about habits, possibly feeds on sponges. Length 12cm.

8. Ass's Ear Abalone *Haliotis asinina*
Frilled mantle with varying shades of green can cover the entire shell, unlike any other abalone. Shell therefore not encrusted like other species. Shell with row of distinctive holes for exhalant passage of water. Largest tropical abalone, up to 18cm. Herbivorous on algae.

9. Bubble Shell *Haminoea cymbalum*
Delicate green coloration with tiny black eyes. Transparent fragile shell. Occurs intertidally. Herbivorous on algae. Length 1cm.

10. Nudibranch *Phyllodesmium longicirra*
Beige mottled coloration. Easily recognised by large paddle-like cerata. Feeds on soft coral *Sarcophyton* with razor-like radula, slicing open the polyps and removing zooxanthellae. These are stored in the cerata. Length 10cm.

11. Elizabeth's Chromodoris *Chromodoris elisabethina*
Blue and yellow pattern characteristic. Diurnal, found on reef walls. Feeds on encrusting sponges. IP. Length 5cm.

12. Spectrum Sap Sucker *Elysia* sp.
Brightly coloured sacoglossan nudibranch. Slices open algal cells and sucks out protoplasm. Length 5cm.

CRUSTACEANS
Subphylum: Crustacea

ORDER: DECAPODA (Decapods)

13. Painted Rock Lobster *Panulirus versicolor*
Largest tropical lobster. Young lobsters often in groups, but adults solitary. Carnivorous on molluscs and a scavenger. Length 46cm.

14. Red Hermit Crab *Dardanus megistos*
Red colour and large size help identification; largest reef hermit crab. Usually solitary. Predatory but can scavenge and is able to feed on detritus. Length 25cm.

15. Red-and-White Painted Crab *Lophozozymus picto*
Thick body, characteristic deep red colour with white spots. Shelters below coral slabs or in holes. Common on reefs. May be poisonous. Diet includes algae and plankton. 20cm diameter.

16. Mantis shrimp *Odontodactylus scyllarus*
Large, colourful mantis shrimp. Powerful claws catch prey. Chelipeds carried folded up beneath thorax. Feeds on crustaceans and other mobile invertebrates. Length 25cm.

ECHINODERMS
Phylum: Echinodermata

CLASS: STELLEROIDEA (Starfish and Brittle-stars)

17. Blue Sea Star *Linckia laevigata*
Commonly blue, but can be grey to pink or yellow. Deeper living form generally khaki. Lagoon form bright blue. Omnivorous. 40cm diameter. Common throughout IP.

18. Brittle-star *Ophiarachnella gorgonia*
Grey-white with green bands. Arms with small spines. Strong, active crawler. Loops round its prey. Omnivorous diet includes detritus. 20cm diameter.

CLASS: ECHINOIDEA (Sea Urchins)

19. Cake Urchin *Tripneustes gratilla* (spawning)
Coloration varied, pattern of spaces between spines is a distinctive feature. Large tube feet may be purple to red. Feeds on algae and seagrass. 15cm diameter.

CLASS: HOLOTHUROIDEA (Sea Cucumbers)

20. Sea cucumber *Pseudocolochirus axiologus*
Blue body wall and orange tube feet. Suspension feeder, tentacle shown being 'cleaned'. Length 25cm. Normally at base of reef mounds.

TUNICATES
Subphylum: Urochordata

CLASS: ASCIDIACEA (Ascidians/Sea Squirts)

21. Common Solitary Ascidian *Polycarpa aurata*
Distinctive white tunic with pink stripes. 2cm width.

22. MacDonald's Ascidian *Diplosoma macdonaldi*
Tunic transparent, pale zooids. Colony 20cm width.

23. Soft Didemnid Ascidian *Didemnum molle*
Green colour due to blue-green algae. Colony 25cm width.

1
19
14
13
3
15
5
20
4
6
17
21
11
8
7
23
10
18
16
12
9
22
2

Marine Invertebrates

by

Marc Dando

INTRODUCTION

Figure 62. Common Lobster *Homarus gammarus* losing a battle with a Common Octopus *Octopus vulgaris.*

The invertebrates dominate the animal kingdom and one major group, the arthropods, which includes insects, chelicerates and crustaceans, is by far the most numerous of the invertebrates. The arthropods consist of over a million species (85% of all animal species) and are found, sometimes in large congregations, throughout the world. By comparison, the vertebrates comprise only about 40,000 species , a number roughly equivalent to all the known marine crustacean species. This chapter presents an overview of the marine invertebrates, a fascinating and sometimes overlooked group.

Zoologists now agree that there are about 30 phyla in the animal kingdom, all but one of which are invertebrates. These invertebrate phyla have over 70 classes and 220 orders. To help understand the complexities of invertebrate classification table 7, on pages 116-122, illustrates the most important subdivisions of the invertebrate phyla, concentrating only on marine groups. There is a single phylum of non-invertebrate animals called Chordata. This comprises vertebrates and related groups, and its marine representatives are discussed from page 289 onwards.

Due to the plethora of species, correctly identifying and classifying invertebrates has always been a major problem. Even distinguishing between two unrelated phyla may be difficult to an untrained eye.

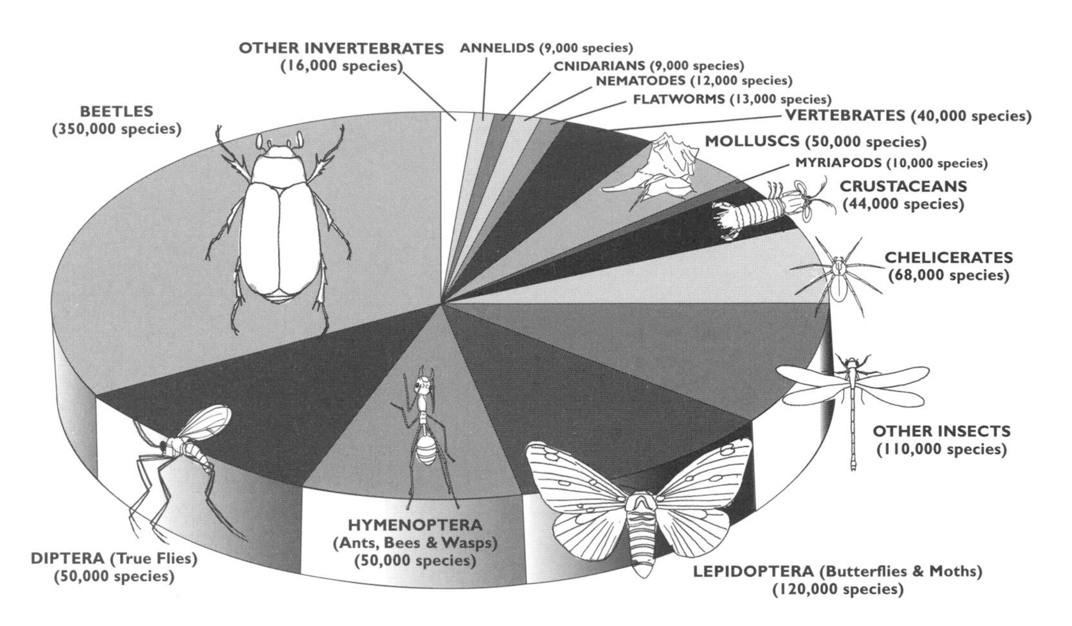

Figure 63. Pie chart showing the major animal groups according to approximate numbers of species.

For example, the soft didemnid ascidian *Didemnum molle* was at one time thought to be a sponge (phylum Porifera), but is now correctly recognised as a sea squirt (phylum Chordata, subphylum Urochordata). By contrast two very dissimilar looking invertebrates may be closely related to each other. Anyone looking at the large, gelatinous, free-swimming Lion's Mane Jellyfish *Cyanea capillata* (plate 1) would probably not immediately associate it with the rock-like, reef-building corals; but both belong to the same phylum, Cnidaria. This can be confusing but invertebrate anatomy gives a clue to their taxonomy. Figure 64 below illustrates the basic body plans of eight of the more commonly encountered marine invertebrate groups, and the distinguishing features of the phylum to which they belong.

Swarms of bees and armies of ants show just how abundant invertebrates are on land, but in the marine environment invertebrate numbers are even greater. A study undertaken off the coast of the Netherlands has shown that one square metre of mud contains some 4.5 million marine nematode worms. As a group, the 12,000 named free-living and parasitic nematode species are thought to be so abundant that they could form a living layer around the entire planet.

Figure 64. Basic body plans of eight common marine invertebrate groups (continued opposite).

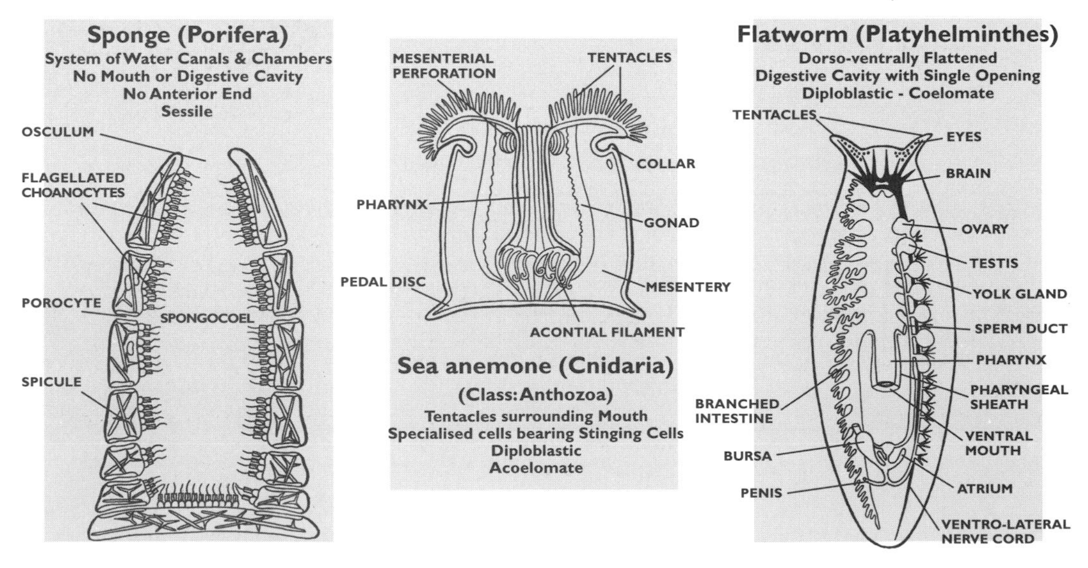

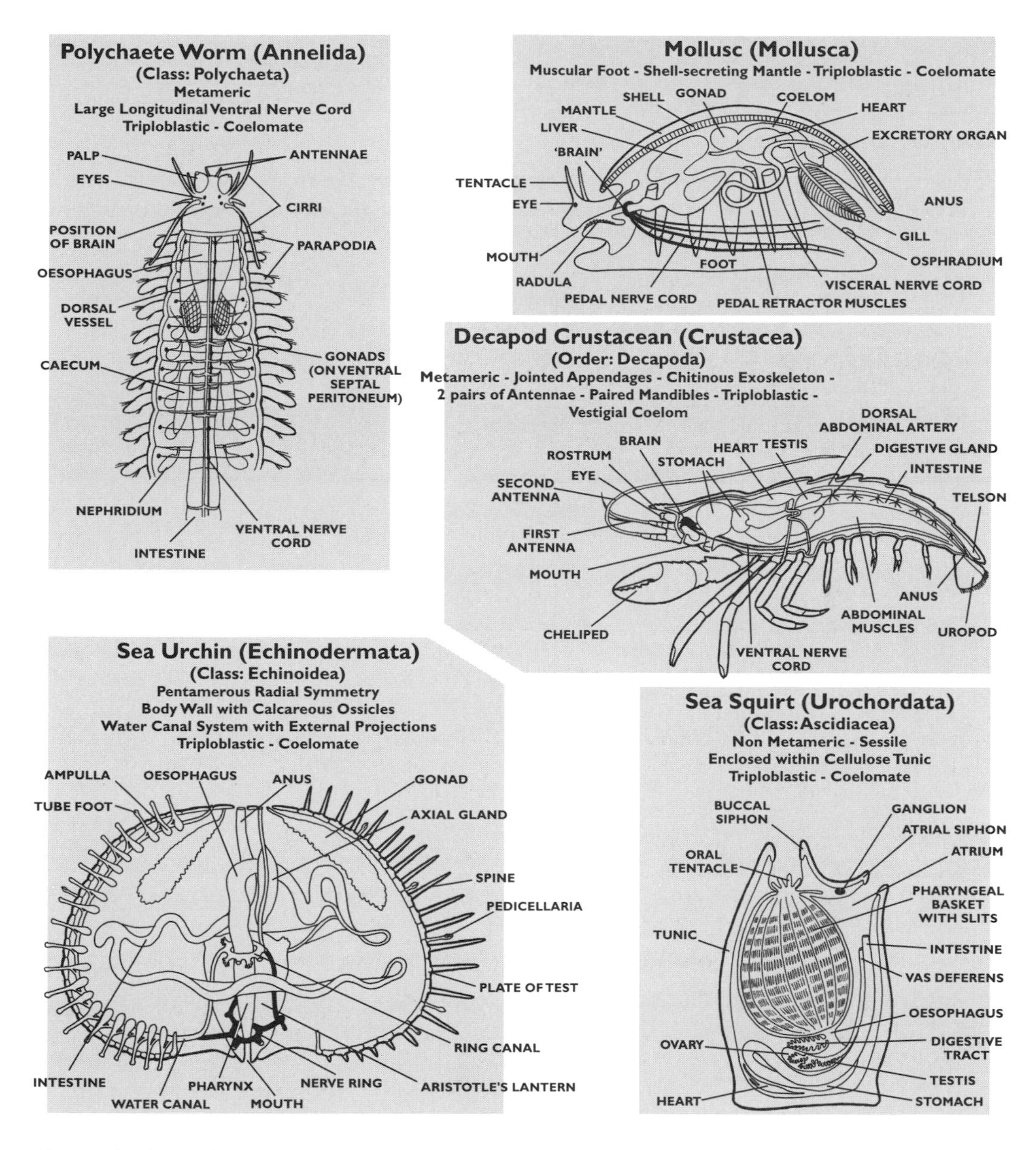

The marine invertebrates comprise over 160,000 known species. Two groups dominate, the molluscs and the crustaceans, which together contain over 90,000 species.

The marine environment has many advantages over terrestrial and freshwater environments. Its relative stability facilitates homoeostasis, that state of equilibrium that assists respiration and osmoregulation. Its buoyancy also gives structural support and enables aquatic invertebrates (and vertebrates) to grow to substantial sizes compared to their terrestrial counterparts. For instance, the largest known living invertebrate, the Giant Squid *Architeuthis* sp. has been reported to grow up to 19.5m long. By contrast one of the longest known terrestrial invertebrates, the giant earthworm *Megascolides australis*, has only been recorded to 3.6m.

The overwhelming abundance of marine invertebrates is rarely fully appreciated, even though they are a key factor in oceanic food webs. The following sections provide a glimpse into the diversity, complexity and importance of the invertebrates within the marine environment.

Table 7. An abridged classification of the animal kingdom comprising the important marine invertebrate groups. The taxonomy given here is a provisional scheme which has recently been subject to radical revision. A comprehensive scheme including all the invertebrate phyla has not yet been agreed.

KEY

The genera listed are discussed in the text.

SbP- Subphylum SpC- Superclass SbC- Subclass SpO- Superorder SbO- Suborder IO- Infraorder SF- Superfamily F- Family

Numbers in brackets show the approximate number of species in certain groups. An asterisk denotes groups that straddle the animal and plant kingdoms and are included as plants in other classifications.

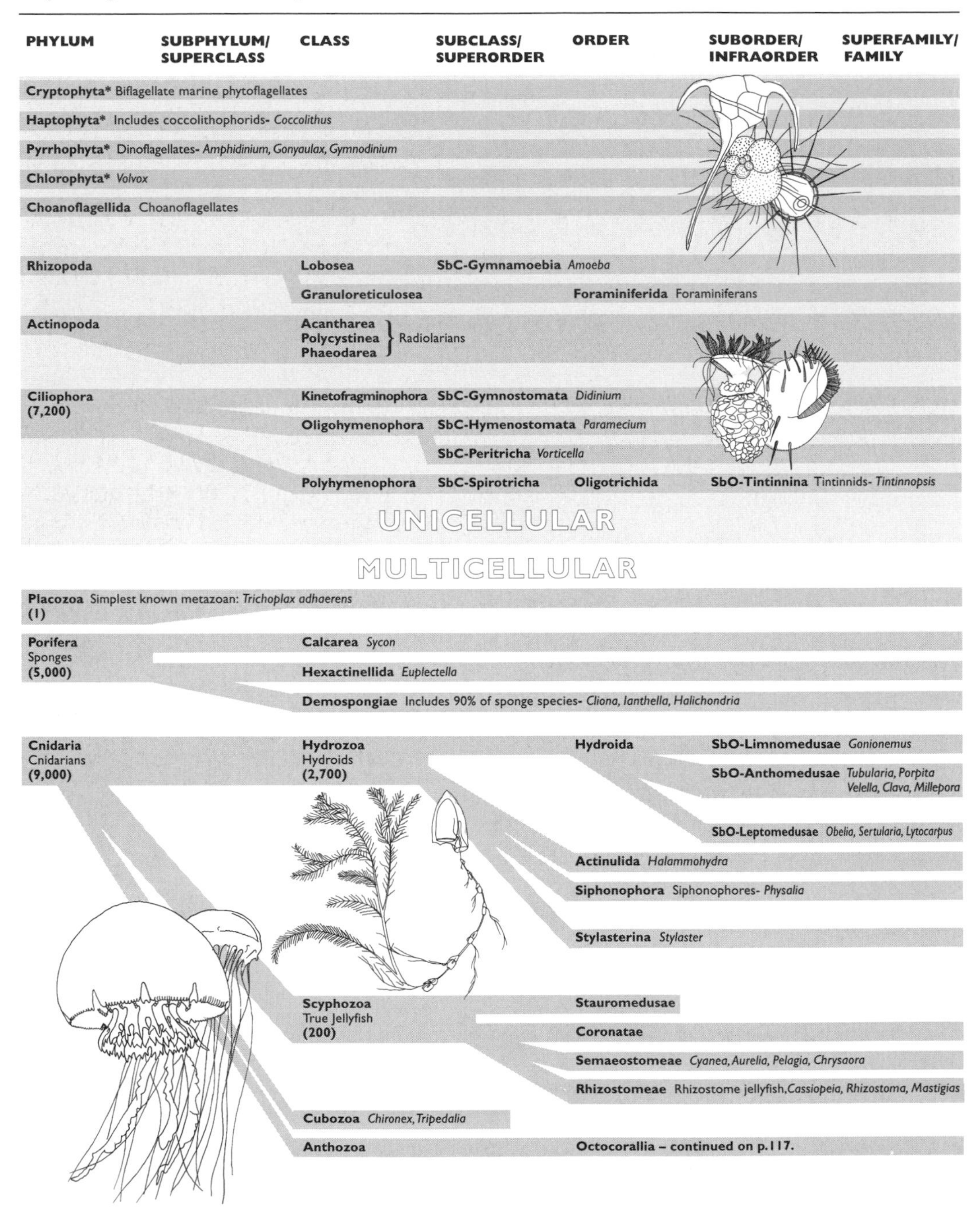

PHYLUM	SUBPHYLUM/ SUPERCLASS	CLASS	SUBCLASS/ SUPERORDER	ORDER	SUBORDER/ INFRAORDER	SUPERFAMILY/ FAMILY
Cryptophyta* Biflagellate marine phytoflagellates						
Haptophyta* Includes coccolithophorids- *Coccolithus*						
Pyrrhophyta* Dinoflagellates- *Amphidinium, Gonyaulax, Gymnodinium*						
Chlorophyta* *Volvox*						
Choanoflagellida Choanoflagellates						
Rhizopoda		**Lobosea**	**SbC-Gymnamoebia** *Amoeba*			
		Granuloreticulosea		**Foraminiferida** Foraminiferans		
Actinopoda		**Acantharea**, **Polycystinea**, **Phaeodarea** } Radiolarians				
Ciliophora (7,200)		**Kinetofragminophora**	**SbC-Gymnostomata** *Didinium*			
		Oligohymenophora	**SbC-Hymenostomata** *Paramecium*			
			SbC-Peritricha *Vorticella*			
		Polyhymenophora	**SbC-Spirotricha**	**Oligotrichida**	**SbO-Tintinnina** Tintinnids- *Tintinnopsis*	
UNICELLULAR						
MULTICELLULAR						
Placozoa Simplest known metazoan: *Trichoplax adhaerens* **(1)**						
Porifera Sponges **(5,000)**		**Calcarea** *Sycon*				
		Hexactinellida *Euplectella*				
		Demospongiae Includes 90% of sponge species- *Cliona, Ianthella, Halichondria*				
Cnidaria Cnidarians **(9,000)**		**Hydrozoa** Hydroids **(2,700)**		**Hydroida**	**SbO-Limnomedusae** *Gonionemus*	
					SbO-Anthomedusae *Tubularia, Porpita Velella, Clava, Millepora*	
					SbO-Leptomedusae *Obelia, Sertularia, Lytocarpus*	
				Actinulida *Halammohydra*		
				Siphonophora Siphonophores- *Physalia*		
				Stylasterina *Stylaster*		
		Scyphozoa True Jellyfish **(200)**		**Stauromedusae**		
				Coronatae		
				Semaeostomeae *Cyanea, Aurelia, Pelagia, Chrysaora*		
				Rhizostomeae Rhizostome jellyfish, *Cassiopeia, Rhizostoma, Mastigias*		
		Cubozoa *Chironex, Tripedalia*				
		Anthozoa		**Octocorallia** – continued on p.117.		

PHYLUM	SUBPHYLUM/ SUPERCLASS	CLASS	SUBCLASS/ SUPERORDER	ORDER	SUBORDER/ INFRAORDER	SUPERFAMILY/ FAMILY
Cnidaria		**Anthozoa** Sea anemones & Corals **(6,000)**	**SbC-Octocorallia**	**Stolonifera** Organ-pipe corals, *Tubipora*		
				Telestacea		
				Alcyonacea Soft corals, *Sarcophyton*		
				Helioporacea Blue coral, *Heliopora*		
				Gorgonacea Gorgonian corals, *Gorgonia, Juncella*		
				Pennatulacea Sea pens, *Pennatula, Ptilosarcus*		
			SbC-Hexacorallia	**Actiniaria** Sea anemones, *Metridium, Actinodendron*		
				Madreporaria Stony corals, *Acropora, Porites, Fungia*		
				Corallimorpharia *Corynactis*		
				Ceriantharia *Cerianthus*		
				Antipatharia Black Corals		
Ctenophora Comb jellies **(50)**		**Tentaculata**		**Cydippida** *Pleurobrachia*		
				Lobata *Bolinopsis*		
				Cestida *Cestum*		
				Platyctenida		
		Nuda		**Beroida** *Beroe*		
Platyhelminthes (12,700)		**Turbellaria** Flatworms **(3000)**		**Acoela** *Convoluta, Archaphanostoma*		
				Nemertodermatida Small marine flatworms		
				Macrostomida Small marine & freshwater flatworms		
				Haplopharyngida Small marine flatworms		
				Polycladida *ProsthecerOs*		
		Monogenea } Flukes		**Rhabdocoela** Large group of marine & freshwater flatworms		
		Trematoda } Flukes		**Proseriata** Small marine flatworms		
		Cestoda Tapeworms		**Tricladida** *Procerodes, Dugesia*		
Orthonectida Rhombozoa - Minute parasitic marine invertebrates (mesozoans) **(50)**						
Nemertea Nemerteans **(900)**		**Anopla**- *Lineus*				
		Enopla				
Gnathostomulida Small marine interstitial worms **(80)**						
Gastrotricha Gastrotrichs- Microscopic flattened marine worms **(460)**						
Nematoda Nematode worms **(12,000)**						
Nematomorpha Hairworms- only one genus is marine; *Nectonema* adult is pelagic **(320)**						
Kinorhyncha Small marine worm-like invertebrates **(100)**						
Loricifera Spiny microscopic interstitial species in shelly gravel- *Nanaloricus mysticus* **(1)**						
Priapulida Cucumber and worm-like marine burrowing invertebrates **(13)**						
Sipuncula Peanut worms- Cylindrical marine worms with frilly tentacled mouth **(320)**						

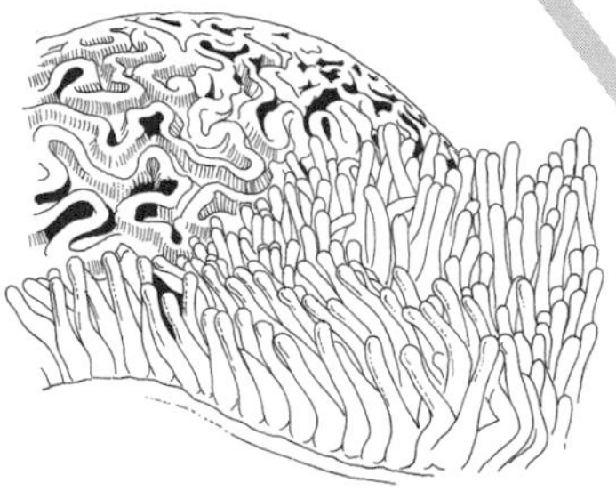

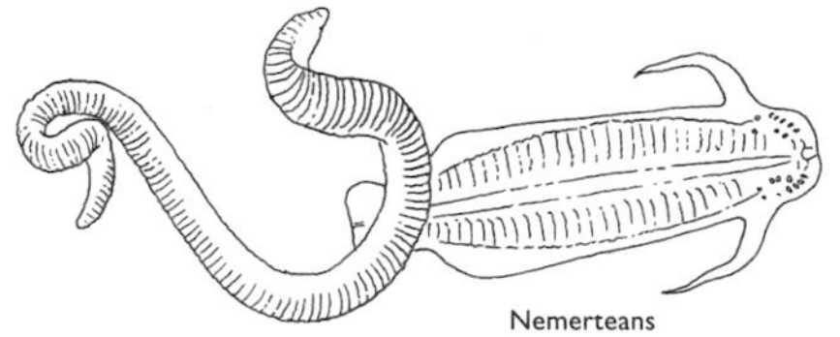

Nemerteans

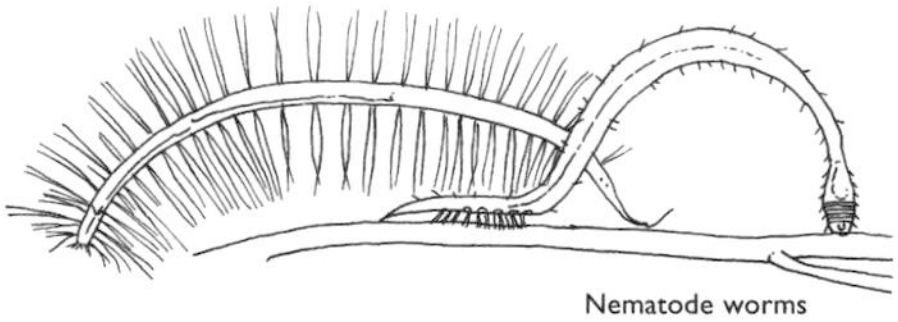

Nematode worms

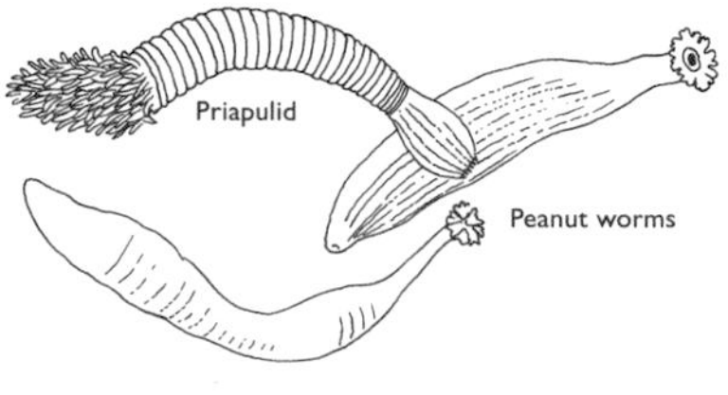

Priapulid

Peanut worms

PHYLUM | **SUBPHYLUM/ SUPERCLASS** | **CLASS** | **SUBCLASS/ SUPERORDER** | **ORDER** | **SUBORDER/ INFRAORDER** | **SUPERFAMILY/ FAMILY**

- **Mollusca** Molluscs **(50,000)**
  - **Monoplacophora** Deep-water archaic molluscs- *Neopilina* **(8)**
  - **Polyplacophora** Chitons- *Tonicella, Acanthopleura, Boreochiton* **(500)**
  - **Aplacophora** Solenogasters- Worm-shaped molluscs- *Lepidomenia* **(180)**
  - **Gastropoda** Gastropods **(35,000)**
    - **SbC- Prosobranchia (18,000)**
      - **Archaeogastropoda**
        - Superfamilies
        - **Pleurotomariacea** Slit shells/Abalones *Haliotis*
        - **Fissurellacea** Keyhole Limpets *Diodora*
        - **Patellacea** Limpets- *Patella*
        - **Trochacea** incl. Top shells *Calliostoma, Gibbula*
        - **Neritacea** *Nerita*
      - **Mesogastropoda**
        - **Littorinacea** Winkles- *Littorina*
        - **Rissoacea** *Hydrobia*
        - **Cerithiacea** incl. Turret & worm shells
        - **Epitoniacea** Pelagic sea snails- *Janthina*
        - **Calyptraeacea** Limpet-like shells, Slipper shells- *Crepidula*
        - **Strombacea** Carrier shells, Conch shells- *Strombus*
        - **Cypraeacea** Cowries- *Cypraea*
        - **Heteropoda** Pelagic snails- *Carinaria*
        - **Naticacea** Moon shells- *Natica*
        - **Tonnacea** Helmet shells, Bonnet shells, Tritons- *Charonia*, Tuns- *Tonna*
      - **Neogastropoda**
        - **Muricacea** Drills- *Murex*
        - **Buccinacea** Whelks- *Buccinum, Nucella*, Tulip shells, Mud snails
        - **Volutacea** Olives, Mitre shells- *Mitra*, Harp shells Volutes- *Voluta, Amoria*
        - **Conacea** Cone shells- *Conus*, Terebrid shells
    - **SbC-Opisthobranchia**
      - **Cephalaspidea** Bubble shells- *Haminoea*
      - **Anaspidea** Sea hares- *Aplysia*
      - **Notaspidea** *Pleurobranchus*
      - **Sacoglossa** Sacoglossan sea slugs- *Elysia*
      - **Thecosomata** Shelled pteropods (Sea butterflies)- *Cavolina*
      - **Gymnosomata** Naked sea butterflies- *Cliopsis*
      - **Nudibranchia** Nudibranchs (Sea slugs)
        - Doridaceans- *Chromodoris, Onchidoris, Archidoris*
        - Aeolidaceans- *Aeolidia, Glaucus, Flabellina, Phyllodesmium*
        - Dendronotaceans- *Tritonia*
        - Arminaceans

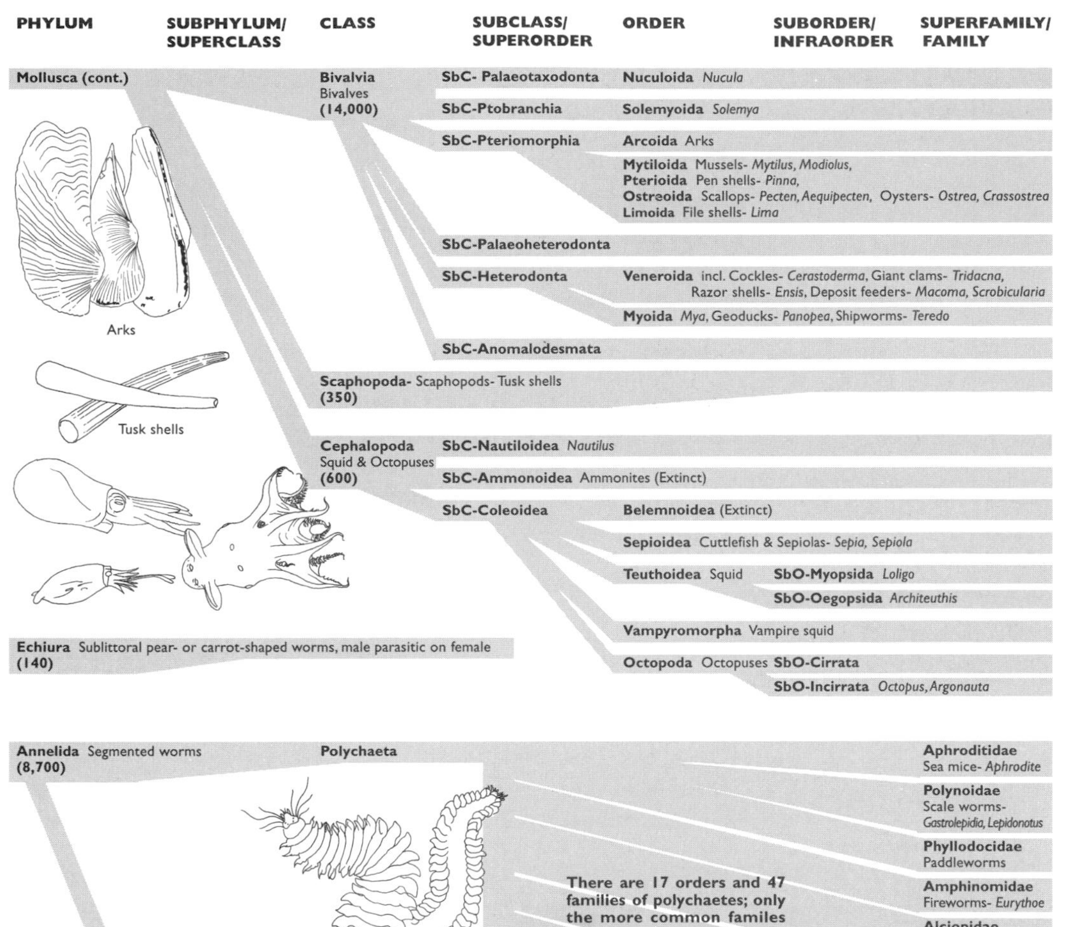

Annelida Segmented worms
(8,700)

- **Polychaeta**
  - **Aphroditidae** Sea mice- *Aphrodite*
  - **Polynoidae** Scale worms- *Gastrolepidia, Lepidonotus*
  - **Phyllodocidae** Paddleworms
  - **Amphinomidae** Fireworms- *Eurythoe*
  - **Alciopidae** Planktonic worms with large eyes
  - **Nereididae** Ragworms- *Nereis*
  - **SF-Eunicea** *Eunice*
  - **Spionidae**
  - **Cirratulidae**
  - **Capitellidae**
  - **Arenicolidae** Lugworms- *Arenicola*
  - **Maldanidae** Bamboo worms
  - **Sabellariidae**
  - **Terebellidae** Terebellid worms- *Lomia*
  - **Sabellidae** Sabellid fan worms- *Sabella, Sabellastarte, Bispira*
  - **Serpulidae** Serpulid fan worms- *Serpula, Spirorbis Spirobranchus*
  - **Pectinariidae** Sand Masons
- **Oligochaeta** Terrestrial worms
- **Hirudinea** Leeches

There are 17 orders and 47 families of polychaetes; only the more common familes are given here and those families or species that have been mentioned in the text.

Pogonophora Giant deep-water tube-dwelling worms, one Galapagos Rift species, *Riftia pachyptila*, recorded at 1.5m body length
(80)

PHYLUM	SUBPHYLUM/ SUPERCLASS	CLASS	SUBCLASS/ SUPERORDER	ORDER	SUBORDER/ INFRAORDER	SUPERFAMILY/ FAMILY
Arthropoda Arthropods **(1,000,000+)**	**SbP-Trilobita** Trilobites (Extinct)					
	SbP-Chelicerata	**Merostomata**	**SbC-Eurypterida** Extinct eurypterids			
			SbC-Xiphosura Horseshoe crabs - *Limulus* **(4)**			
		Arachnida Scorpions, spiders & mites **(67,400+)**				
		Pycnogonida Sea spiders **(1,000)**				
	SbP-Crustacea Crustaceans **(44,300+)**	**Cephalocarida** Most primitive living crustaceans **(9)**				
		Branchiopoda Large class of mostly freshwater forms, some marine Cladocera				
		Ostracoda Ostracods (Mussel shrimps) **(7,000+)**		**Myodocopida** *Cypridina, Conchoecia*		
				Halocyprida		
			SC- Podocopa- *Cypris*			
		Copepoda Copepods **(11,500)**		**Calanoida** Planktonic- *Calanus*		
				Misophrioida Benthic- *Misophria*		
				Harpacticoida Benthic, *Harpacticus, Tisbe, Canuella*		
				Monstrilloida Parasitic larvae, planktonic adults		
				Siphonostomatoida Parasitic adults- *Penella, Lepeophtheirus, Caligus*		
				Cyclopoida *Halicyclops, Oithona, Sapphirina, Oncaea*		
				Poecilostomatoida Parasitic adults		
		Mystacocarida Small interstitial forms **(9)**				
		Branchiura Ectoparasites of marine and freshwater fishes **(130)**				
		Tantulocarida Recently discovered deep-water ectoparasites **(4)**				
		Remipedia Another recent discovery, polychaete-like, primitive crustaceans **(8)**				
		Cirripedia Barnacles **(900)**		**Thoracica**	**SbO-Lepadomorpha**	*Lepas, Pollicipes, Conchoderma*
					SbO-Verrucomorpha	*Verruca*
					SbO-Balanomorpha	*Balanus, Xenobalanus Coronula, Semibalanus*
				Acrothoracica Naked boring barnacles mainly in shells & corals		
				Ascothoracica Parasites on echinoderms & corals		
				Rhizocephala Parasites primarily on decapods- *Sacculina*		
		Malocostraca Malocostracans **(28,070+)**	**SbC-Phyllocarida**	**Leptostraca** Most primitive living malocostracans		
			SbC-Hoplocarida	**Stomatopoda** Mantis shrimps- *Ondontodactylus* **(300)**		
			SbO-Peracarida	**Mysidacea** Opossum shrimps- *Mysis* **(780)**		
				Cumacea Burrowing peracaridans **(800)**		
				Tanaidacea Have features of both cumaceans & isopods; benthic **(550)**		
				Isopoda Isopods- nearly all marine- *Idotea, Ligia, Aega* **(10,000)**		
				Amphipoda Amphipods- mostly marine- *Corophium, Paraleucothoe Phronima,* Whale lice- *Cyamus, Gammarus* Sandhoppers- *Talitrus* Skeleton shrimps- *Caprella* **(5,500)**		

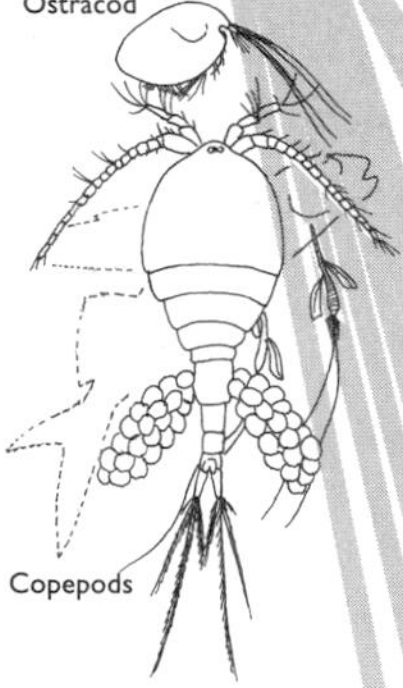

Ostracod

Copepods

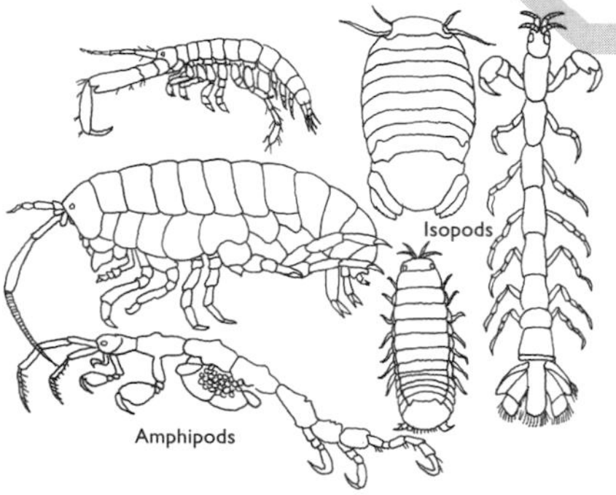

Isopods

Amphipods

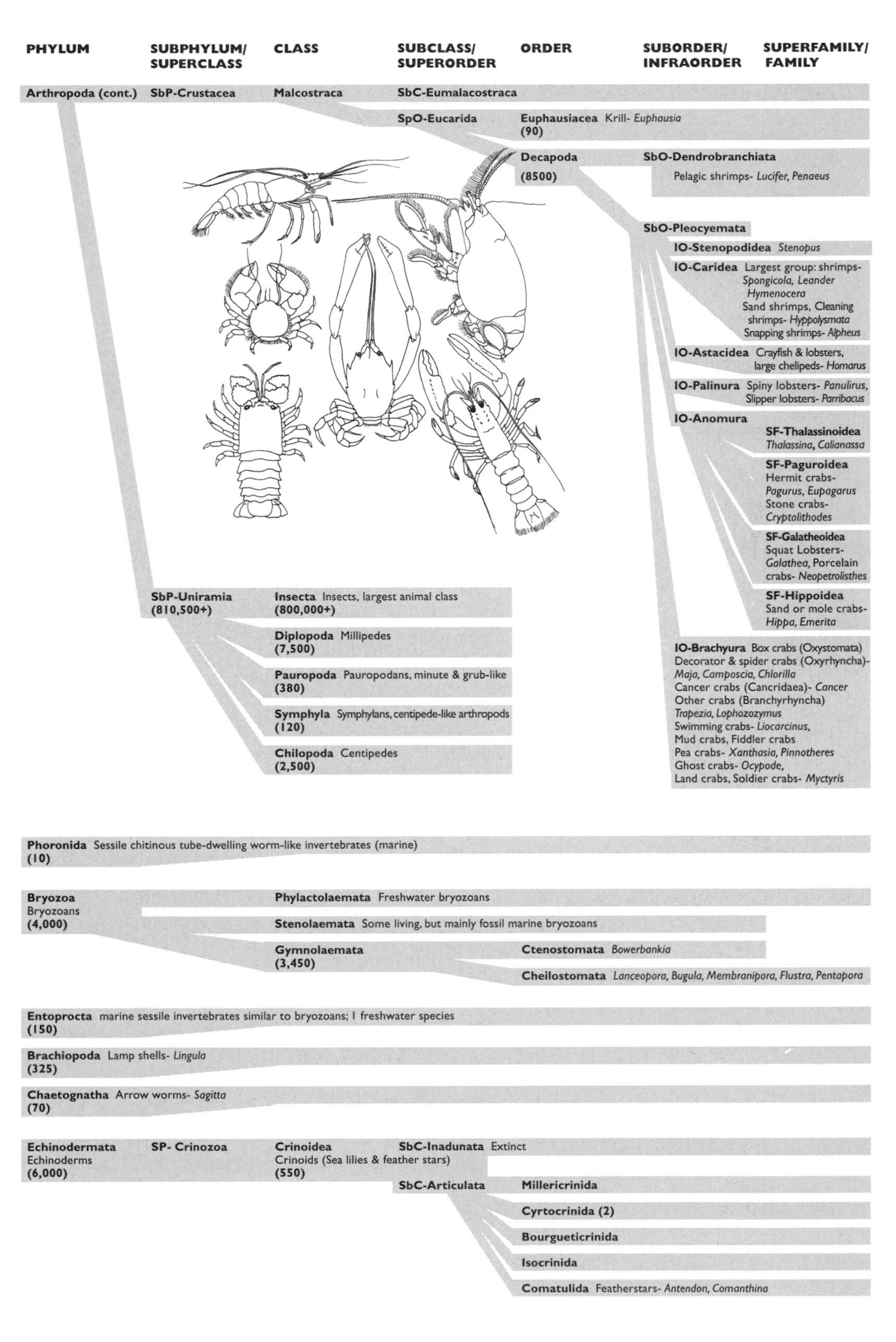

PHYLUM
SUBPHYLUM/ SUPERCLASS
CLASS
SUBCLASS/ SUPERORDER
ORDER
SUBORDER/ INFRAORDER
SUPERFAMILY/ FAMILY
Arthropoda (cont.)
SbP-Crustacea
Malcostraca
SbC-Eumalacostraca
SpO-Eucarida
Euphausiacea Krill- Euphausia (90)
Decapoda (8500)
SbO-Dendrobranchiata
Pelagic shrimps- Lucifer, Penaeus
SbO-Pleocyemata
IO-Stenopodidea Stenopus
IO-Caridea Largest group: shrimps- Spongicola, Leander Hymenocera Sand shrimps, Cleaning shrimps- Hyppolysmata Snapping shrimps- Alpheus
IO-Astacidea Crayfish & lobsters, large chelipeds- Homarus
IO-Palinura Spiny lobsters- Panulirus, Slipper lobsters- Parribacus
IO-Anomura
SF-Thalassinoidea Thalassina, Calianassa
SF-Paguroidea Hermit crabs- Pagurus, Eupagarus Stone crabs- Cryptolithodes
SF-Galatheoidea Squat Lobsters- Galathea, Porcelain crabs- Neopetrolisthes
SF-Hippoidea Sand or mole crabs- Hippa, Emerita
IO-Brachyura Box crabs (Oxystomata) Decorator & spider crabs (Oxyrhyncha)- Maja, Camposcia, Chlorilla Cancer crabs (Cancridaea)- Cancer Other crabs (Branchyrhyncha) Trapezia, Lophozozymus Swimming crabs- Liocarcinus, Mud crabs, Fiddler crabs Pea crabs- Xanthasia, Pinnotheres Ghost crabs- Ocypode, Land crabs, Soldier crabs- Myctyris
SbP-Uniramia (810,500+)
Insecta Insects, largest animal class (800,000+)
Diplopoda Millipedes (7,500)
Pauropoda Pauropodans, minute & grub-like (380)
Symphyla Symphylans, centipede-like arthropods (120)
Chilopoda Centipedes (2,500)
Phoronida Sessile chitinous tube-dwelling worm-like invertebrates (marine) (10)
Bryozoa Bryozoans (4,000)
Phylactolaemata Freshwater bryozoans
Stenolaemata Some living, but mainly fossil marine bryozoans
Gymnolaemata (3,450)
Ctenostomata Bowerbankia
Cheilostomata Lanceopora, Bugula, Membranipora, Flustra, Pentapora
Entoprocta marine sessile invertebrates similar to bryozoans; 1 freshwater species (150)
Brachiopoda Lamp shells- Lingula (325)
Chaetognatha Arrow worms- Sagitta (70)
Echinodermata Echinoderms (6,000)
SP- Crinozoa
Crinoidea Crinoids (Sea lilies & feather stars) (550)
SbC-Inadunata Extinct
SbC-Articulata
Millericrinida
Cyrtocrinida (2)
Bourgueticrinida
Isocrinida
Comatulida Featherstars- Antendon, Comanthina

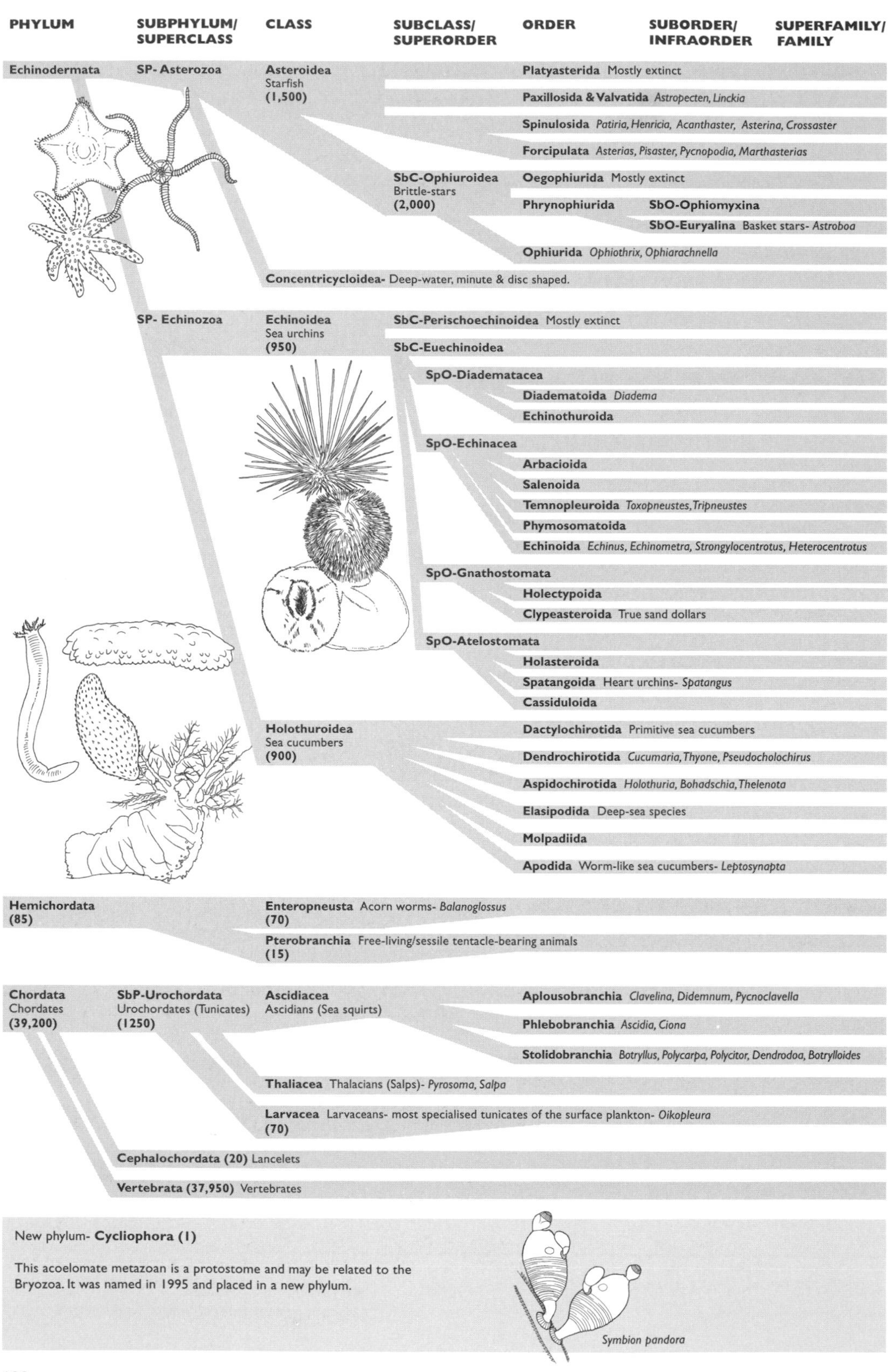
PHYLUM
SUBPHYLUM/ SUPERCLASS
CLASS
SUBCLASS/ SUPERORDER
ORDER
SUBORDER/ INFRAORDER
SUPERFAMILY/ FAMILY
Echinodermata
SP- Asterozoa
Asteroidea
Starfish
(1,500)
Platyasterida Mostly extinct
Paxillosida & Valvatida Astropecten, Linckia
Spinulosida Patiria, Henricia, Acanthaster, Asterina, Crossaster
Forcipulata Asterias, Pisaster, Pycnopodia, Marthasterias
SbC-Ophiuroidea
Brittle-stars
(2,000)
Oegophiurida Mostly extinct
Phrynophiurida
SbO-Ophiomyxina
SbO-Euryalina Basket stars- Astroboa
Ophiurida Ophiothrix, Ophiarachnella
Concentricycloidea- Deep-water, minute & disc shaped.
SP- Echinozoa
Echinoidea
Sea urchins
(950)
SbC-Perischoechinoidea Mostly extinct
SbC-Euechinoidea
SpO-Diadematacea
Diadematoida Diadema
Echinothuroida
SpO-Echinacea
Arbacioida
Salenoida
Temnopleuroida Toxopneustes, Tripneustes
Phymosomatoida
Echinoida Echinus, Echinometra, Strongylocentrotus, Heterocentrotus
SpO-Gnathostomata
Holectypoida
Clypeasteroida True sand dollars
SpO-Atelostomata
Holasteroida
Spatangoida Heart urchins- Spatangus
Cassiduloida
Holothuroidea
Sea cucumbers
(900)
Dactylochirotida Primitive sea cucumbers
Dendrochirotida Cucumaria, Thyone, Pseudocholochirus
Aspidochirotida Holothuria, Bohadschia, Thelenota
Elasipodida Deep-sea species
Molpadiida
Apodida Worm-like sea cucumbers- Leptosynapta
Hemichordata
(85)
Enteropneusta Acorn worms- Balanoglossus
(70)
Pterobranchia Free-living/sessile tentacle-bearing animals
(15)
Chordata
Chordates
(39,200)
SbP-Urochordata
Urochordates (Tunicates)
(1250)
Ascidiacea
Ascidians (Sea squirts)
Aplousobranchia Clavelina, Didemnum, Pycnoclavella
Phlebobranchia Ascidia, Ciona
Stolidobranchia Botryllus, Polycarpa, Polycitor, Dendrodoa, Botrylloides
Thaliacea Thalacians (Salps)- Pyrosoma, Salpa
Larvacea Larvaceans- most specialised tunicates of the surface plankton- Oikopleura
(70)
Cephalochordata (20) Lancelets
Vertebrata (37,950) Vertebrates
New phylum- Cycliophora (1)
This acoelomate metazoan is a protostome and may be related to the
Bryozoa. It was named in 1995 and placed in a new phylum.
Symbion pandora

STRUCTURAL SUPPORT

After death the siliceous or calcareous skeletons of many marine invertebrates are left behind. The most familiar of these are the shells of molluscs that are washed up along the shoreline. However, other more striking structures are those left by tiny ancient marine invertebrates. Over millions of years their shell deposits (mainly from protozoan foraminiferans) have been compressed within the earth's crust to form the rocks we know as chalk and limestone. Prehistoric siliceous shells of radiolarians, another protozoan group, have produced flint and chert deposits around the world and these are among the oldest known fossils. Today, the sedimentary rocks of the future are being produced by the remains of these marine invertebrates settling on ocean floors around the world. Living corals also show thousands of years of skeletal (calcium carbonate) deposition, forming gigantic reefs in tropical waters. Some, like the Great Barrier Reef, can be seen from satellites orbiting the earth.

In contrast, many other marine invertebrate species have no hard shell or skeletal structure and their soft body tissues have left little fossil trace. Generally there is little fossil evidence of any early marine invertebrates, but in beds such as those of the Flinders range of Australia, a combination of factors occurred millions of years ago that allowed soft-bodied animals to be preserved. These fossils only provide a glimpse of the fauna in the seas at the time, but they help our understanding of these marine invertebrates and their possible evolution (see p.179).

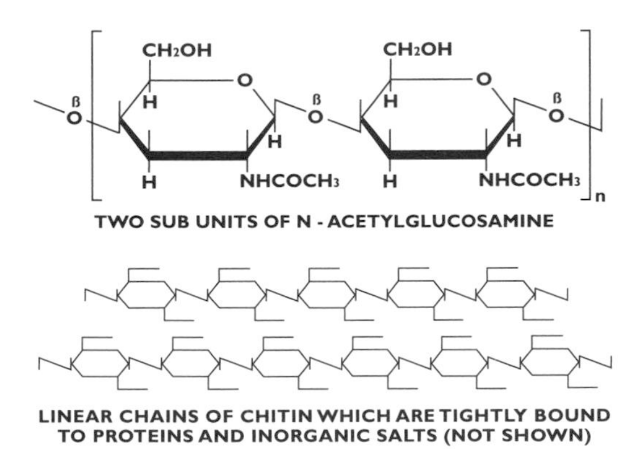

Figure 65. The polymeric structure of chitin.

Similar molecular structures found in various marine invertebrate skeletons can indicate a common ancestry. The hard exoskeleton of crustaceans is composed of the polysaccharide chitin, a homologous polymer of glucosamine (fig.65) bound with proteins and inorganic salts, mainly calcium carbonate. This differs from the skeleton of other arthropods which have a high proportion of the protein sclerotin for their exoskeleton's molecular framework. This implies the possible early divergence of the crustaceans from all other arthropods (e.g. insects, millipedes and centipedes).

Chitin also occurs in the internal skeleton (endoskeleton) of the squid. The skeleton consists of a long, flattened plate which is often called a 'sea pen' due to its resemblance to a quill. However, the squid's early ancestors had a much bulkier skeleton that has been modified in various ways during evolution as indicated by the divergence of modern cephalopod groups (fig.173, p.186).

Soft-tissued invertebrates have their own form of structural support. The marine worms, which include nematodes, polychaetes and flatworms, have a toughened exterior skin, the cuticle, and a soft interior that, with its muscles, gives the animal a structure like an inflated bag. Cuticular strength and flexibility is derived from collagen, a fibrous protein that is also present in higher animals. Collagen is often found with other structural proteins such as sclerotin and keratin which are present in various combinations to strengthen or harden the worm's cuticle. The cuticle for the most part is flexible, but in regions of the body where abrasion or stress is high (e.g. polychaete jaws) sclerotin and keratin are found in higher proportions.

Many 'soft-bodied' invertebrates do not utilise a thickened cuticle for their protection. For example, the tubiculous polychaete worms (Sabellariidae (fig.38), Pectinariidae, Terebellidae and Maldanidae) collect fine sand, silt and other small particles as building materials. These particles are combined with a cement secreted from glands on the ventral surface of their bodies to manufacture external protective tubes (fig.66).

Use of an external protective layer is not confined to the tubiculous polychaete worms. Hermit crabs utilise the empty shells of gastropod snails to protect their delicate abdomens (plates 3 & 8), but when a crab becomes too large for its shell it has to search for a suitable larger replacement. Once found, the potential new shell needs to be tried out before the old one is discarded. Vacant shells are crucial to the hermit crab. Scarcity of suitable empty gastropod shells results in a decrease in growth and reproduction of the local hermit crab population.

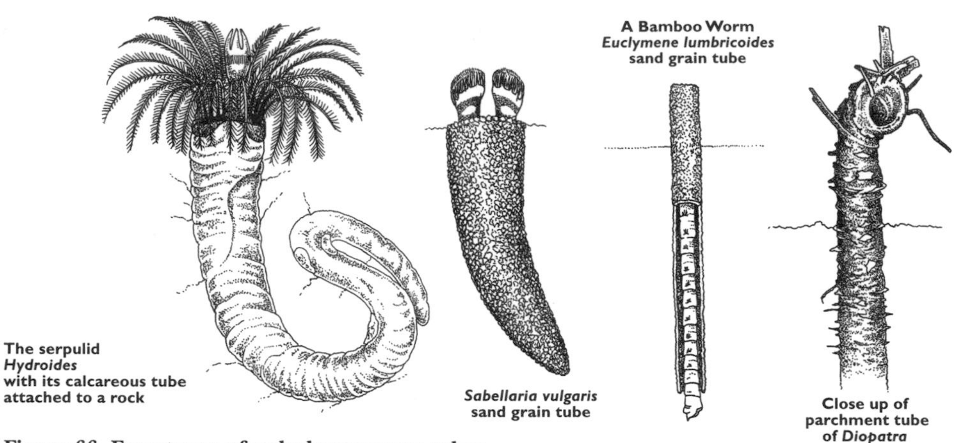

Figure 66. Four types of polychaete worm tubes.

Gastropod molluscs, the original inhabitants of the hermit crab's protective shell, often produce elaborate protective structures. However, the archetype mollusc's dorsal surface may originally have been covered by an oval, convex shell that protected the underlying visceral mass. The calcified shell was secreted by the underlying epidermis (the mantle). This secretion was most active around the edges of the mantle, with new shell being added to the older parts. As the mollusc grew, the shell grew in diameter and thickness. In today's molluscs each order has its own particular shell structure, whether it is single, bivalved or even, in some species, not externally visible. A shell is absent altogther in octopuses.

Bivalved molluscs typically have two ovoid convex halves (the valves) which are hinged along their dorsal edge and a protuberance (the umbo) which is the oldest part of the shell (fig.67a). The valves are shut together by two large adductor muscles sealing the bivalve from predatory attack and desiccation. Bivalve shells have an outer layer (the periostracum) overlaying two to four calcareous layers. This proteinaceous layer is of variable thickness. The clams (Myoida) have a particularly thick layer that gives the shell a dull, rough appearance. The periostracum protects the growth margins of the shell and can form a watertight seal on closure of the valves. Shell form is a function of the rate of shell deposition from the underlying mantle. The mantle-secreted calcareous shell is made from one or more forms of microscopic crystalline structure:

Nacreous (mother-of-pearl)	Horizontal layers of crystals similar to tiny crazy-paving slabs
Foliate	Horizontal lathes of crystals
Prismatic	Vertical prisms

The crystals' matrix is made up of various proteins, mucopolysaccharides and inorganic secretions, and is produced in the extrapallial space, which lies between the shell and the mantle. Foreign bodies, such as sand grains, may become lodged in this space. A sand grain forms a nucleus around which a nacreous shell layer is deposited and may eventually result in a pearl.

Bivalve shells show a wide range of modifications to their surface sculpturing. One common form is corrugation or ridging, which increases the strength of the valves (e.g. Common Cockle *Cerastoderma edule* (fig.46) and the Giant Clam *Tridacna maxima* (plate 8). In other bivalve molluscs the shell shape often

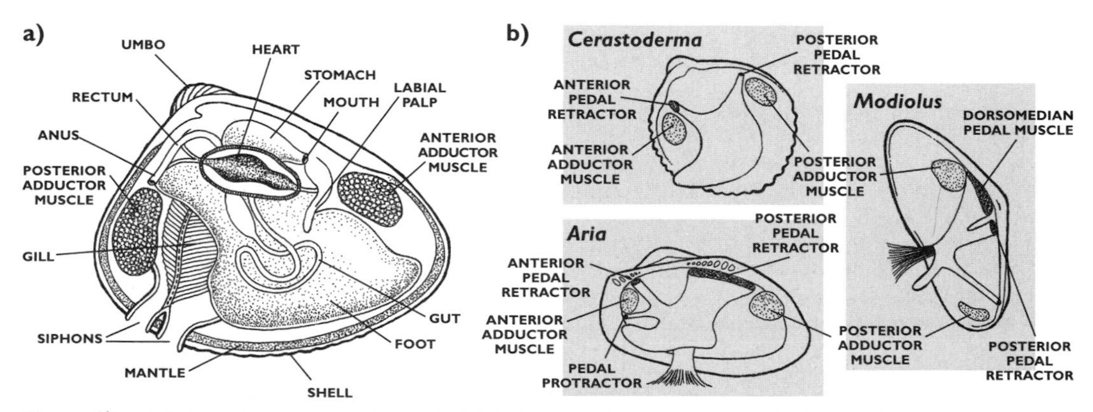

Figure 67. a) Internal structure of a typical bivalve and b) the foot and shell muscle arrangements of three common bivalve genera.

suggests their habitat and lifestyle. For example, razor shells are smooth, long and slender and this enables them to slice through the sand and burrow rapidly. On the other hand wood-boring bivalves (e.g. the Shipworm *Teredo navalis*, fig.107, p.148), use their strong, serrated valves to grate into wood with a rocking motion, ingesting the debris as they burrow. Some scallops, Pectinidae (plates 2 & 3), can even use their valves for escape (see p.148). Growth bands on bivalve shells, formed by the laying down of new shell by the expanding mantle, can be used to provide information about the age of the animal and the environmental conditions during its life.

Another group of molluscs, the gastropod snails, have shells that are normally conical and spiral (see plates 2, 3 & 7). The spires are formed by tubular whorls that contain the animals' visceral mass. The apex of the cone has the smallest and oldest whorls and these gradually increase in size around a central axis (the columella) to form the largest whorl (body whorl). When the animal is viewed from above and the spire is uppermost, gastropod shells display either a right-handed (dextral) spiral where the aperture opens to the right, or a left-handed (sinistral) spiral opening to the left (fig.68).

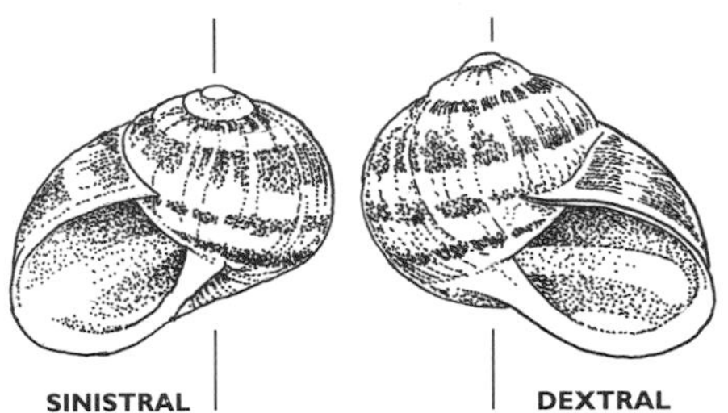

Figure 68. Dextral and sinistral spiralling in a gastropod shell.

Gastropod shells, unlike bivalve shells, have four layers. The first is the outer periostracum (as in bivalves) consisting of a conchin protein layer that is normally quite thin. This layer may be absent, as in the cowries, or thickened, like the whelks with their distinctive roughened surface texture. Below the periostracum is the outermost calcium carbonate layer, the crystals of which are normally laid down vertically in a thin protein matrix. Inside this is the inner calcium carbonate layer with two or more sheets (lamellae) of crystals and these themselves overlie a thin organic matrix that completes the shell. Coloration of the shell results from pigments in the periostracum. The many combinations of colours and patterns, which can be quite stunning, are produced by variations in the pigments' densities. The form and sculpturing of the gastropod shells are more complex than bivalve molluscs. However, there are two distinct shell variations that commonly occur.

The first variation is typified by the abalones (see plates 2, 3 & 8) where the shell of the adult has a single, large, symmetrical, expanded body whorl. This condition is also found in the slipper shells (see plate 3), but in these gastropod snails the asymmetry of the juvenile is retained (see p.173 and evolution section). The limpets, however, do become secondarily symmetrical in the adult and this results in the typical pyramidal cone shape. Cowries (see plates 4 & 8) on the other hand are a 'halfway house' between the last two gastropod groups for they have a superficial bilateral symmetry. In the adult the last and largest whorl has completely overgrown the previous whorls, reducing the shell's aperture and hiding all traces of the earlier juvenile shell.

The second major variation is the reduction or loss of the shell as seen in the nudibranchs (see plates 2, 3, 4, 6, 7 & 8) which have lost all but a small part of their ancestral shell. In the bubble shells (plate 8) the remnant of the shell remains within the mantle and is more obvious to the naked eye than in nudibranchs. In both groups structural support comes from the gastropod's muscles and dermis.

The last group of molluscs, the cephalopods, includes the Chambered Nautilus *Nautilus pompilius* (plate 1) which has retained a primitive type of cephalopod shell. Inside the shell are partitions (septa) that are perforated by a cord of body tissue (the siphuncle) forming up to 30 chambers in a mature adult. The siphuncle link extends from the visceral mass to all but the smallest chambers at the centre of the coil and the tissue can secrete gas into the saline-filled chambers to achieve neutral buoyancy. The secreted gas is a mixture of nitrogen and argon and is normally, when within the chambers, at pressures just below one atmosphere. The nautilus can migrate up and down the water column to depths below 200m where it usually hunts and scavenges, using the gas/liquid ratio within the chambers to alter its buoyancy. The siphuncle tissue actively transports sodium chloride from the saline solution in the chambers causing the liquid to become hypo-osmotic to the surrounding water. Water is therefore withdrawn from the chamber and the gas mixture diffuses in to replace the fluid loss. The advantage to the nautilus is that it is able to move within the water column with little energy expenditure.

Other cephalopods such as cuttlefish (Sepioidea) also use their shells for buoyancy. The chalky honeycombed centre of the shell is filled with liquid and gas (mainly nitrogen at a pressure of 0.8 atmospheres). A membrane beneath the shell pumps liquid in and out of the matrix thus varying its density in a similar method to that of the Chambered Nautilus. During daylight hours the shell is dense containing more liquid and the cuttlefish resides on the bottom, often burying itself in the sand. At night the shell becomes more buoyant; liquid is replaced by gas, and the cuttlefish goes hunting. The buoyancy of the animal is a photic response controlled by ambient light intensity.

Some cnidarians also have buoyant skeletons. By-the-wind Sailors *Velella velella* are colonial hydroid polyps that hang from a flat, horn-like, gas-filled float with a 'sail' projecting above the surface (fig.69). The float and the sail enable the By-the-wind Sailors to catch the wind and currents at the sea surface. However, because the animals are at the mercy of the elements they are often washed up on coasts in large numbers.

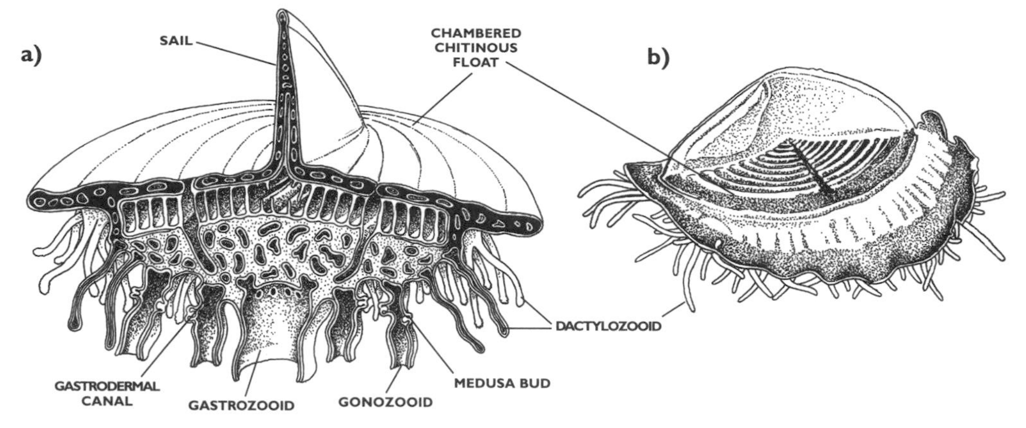

Figure 69. a) By-the-wind Sailor *Velella velella*, float shown in section and b) a colonial hydroid with sail.

The planktonic larvae of crustaceans (zoea) also use their exoskeletons as a means of dispersal within the surface waters. These tiny larval crustaceans often have flattened bodies with long feathery spines on their legs and tail that substantially increase their surface area. This larger surface area increases their drag and so decreases their rate of descent within the water column (fig.70). The longer they can remain in the surface waters the further they may be carried. However, in tropical seas the warmer water is less viscous and so the drag on the animal is less. To compensate for this, tropical zoea have larger spines and endoskeletal projections (processes) than their temperate counterparts.

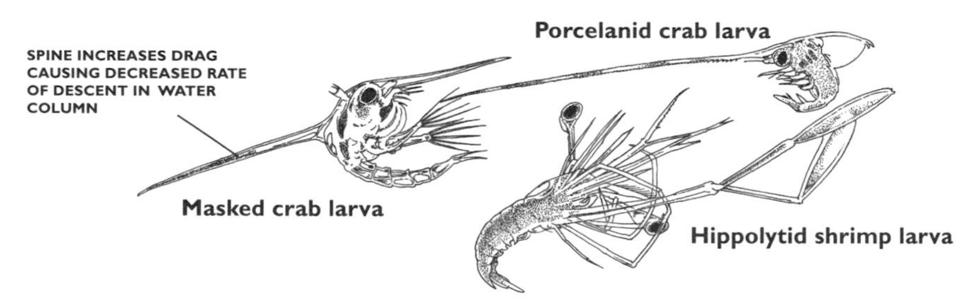

Figure 70. Zoea larvae of three crustaceans found in the zooplankton.

Animals with exoskeletons or shells incur problems when growing, as the outer casing can limit the animal's size. Invertebrates solve the problem of growth within a fixed shell in ingenious and sometimes complex ways.

Chambered invertebrates, like the radiolarians and nautiloids, are able to add on more growing room by extruding new, larger chambers to the older shell. Gastropods and bivalves produce new whorls or larger shells to accommodate their increased size.

Cuticles also limit growth and the segmented worms increase their size by adding body segments. However, they can, as in other collagen-cuticled worms, stretch their skin to its limit and then add further growth tissue interstitially, between the cells and not at segment margins, to the taut outer layer.

Nematode worms and crustaceans (and all terrestrial insects), have to moult before they can increase their body size. Nematode worms usually moult four times during their lives. Their moulting starts at the anterior end of the body, and as the old cuticle separates the new cuticle is secreted. It is a hazardous but simple procedure.

Crustaceans moult many times during their lives, a procedure complicated by their body form and their many appendages. Firstly, as the time for moulting approaches the epidermis below the cuticle, or shell, secretes a pro-enzyme. Soon after, in a procedure called apolysis, the epidermis (often referred to as the hypodermis in arthropods as it lies below another outer covering) separates from the cuticle. The hypodermis can then secrete a new thin outer cuticle below the old one. The now activated pro-enzyme acts as a catalyst for the digestion of the detached endocuticle (the old inner cuticle). The products of this breakdown, mainly important salts, are absorbed into surface capillaries. When this phase is completed, the new wrinkled cuticle, or procuticle, is secreted. Finally the endocuticle splits open along predetermined lines and the crustacean emerges from the old translucent husk. The animal's body then swells by absorbing water, stretching the wrinkled procuticle into its fully enlarged shape. This 'bloating out' allows the crustacean future growing room. Soon after exposure to the water and the stretching of the new cuticle, the tanning process hardens the shell. This moulting process (ecdysis) is not often seen as the animal needs to protect itself by hiding before ecdysis begins. Ecydysis is controlled by the release of the hormone ecdysone that acts directly on the cells of the hypodermis. Ecdysone's release is controlled by neuronal stimulus, but its release has many contributing factors such as day length, limb loss (some crabs have even been seen pulling off their own claws to stimulate ecdysis) or a combination of other factors. The exact control mechanism is often not known, but without the stimulus to moult the animal cannot grow.

One particular group of marine invertebrates has evolved an elaborate and permanent hard protective exoskeleton, the stony corals of the phylum Cnidaria. Corals are colonial animals made up from hundreds if not thousands of polyps. Each polyp has a ring of tentacles and a hard outer calcareous skeleton. This skeleton is typically white when the coral is dead, but when living there is a thin translucent tissue layer over it which gives the coral reefs their sometimes brilliant daytime colours. Solitary corals (found throughout temperate and tropical seas) are similar to small sea anemones but live in a cup of hard calcareous minerals. The colonial stony corals of the reef-building (hermatypic) type only occur in shallow tropical waters. Each stony coral colony consists of a thin sheet of living tissue overlying the calcareous skeleton below. Above this sheet protrude the individual polyps, their mouths fringed with numerous tentacles (figs.134 & 135). The structural support for each polyp is derived from its skeleton which is folded into mesenteries. These mesenteries 'dovetail' with the calcareous septo-costal plates to form the polyps' tube skeleton, which is at right angles to the tube's corallite wall (fig.56, p.88). The coenosteum comprises horizontal plates which link the corallite walls and their associated plates together, forming a solid structure. These horizontal plates can be a continuous sheet, a matrix of tiny plates, or sometimes an elaborate structure that is more prominent than the corallites themselves (e.g. brain corals, plate 5).

The coral's skeleton is defensive and supportive, but many marine invertebrates use their skeletal structures for locomotion. Often in these species there is a balance between practical structural support and ease of movement. Other marine invertebrates display a surprising array of locomotive structures which are nearly always attached in some way to their structural framework.

LOCOMOTION

One of the simple forms of locomotion is shown by the amoebae. Their cytoplasmic streaming enables the organism to produce outgrowths (the pseudopodia) of varying shapes. The more common forms are either the wide, blunt, lobose pseudopodia or the thin, more pointed filopodia of the smaller amoebae. Foraminiferans (marine amoeboid protozoans) generally have thread-like branched pseudopodia (reticulopodia, fig.71). These foraminiferans live in calcium carbonate shells where one or more chambers are used as floats.

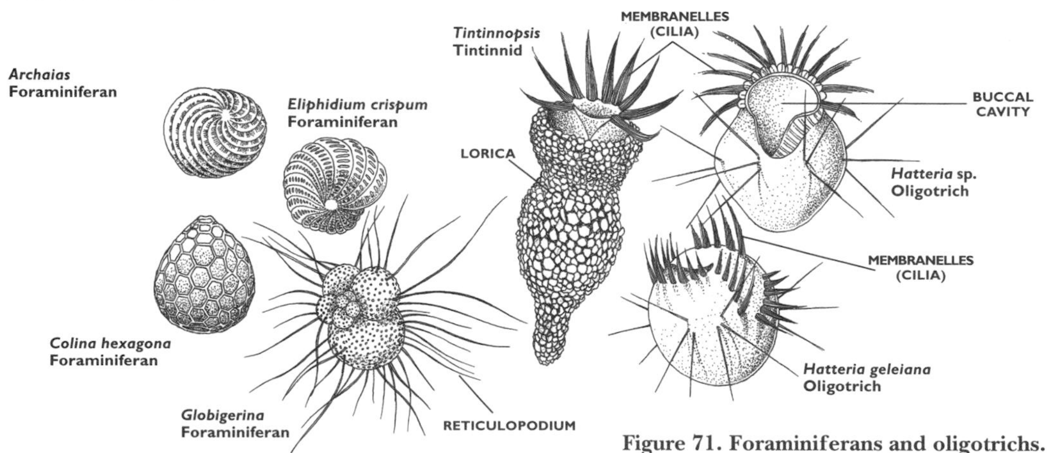

Figure 71. Foraminiferans and oligotrichs.

Both amoebae and foraminiferans use cytoplasmic streaming for locomotion. This process uses the property of some protein chains to undergo extension and contraction. Protoplasm containing extended protein chains, like uncoiled springs, is fluid and is known as endoplasm. When the chains contract, like a fully coiled spring, the protoplasm is gelatinous and resides outside the endoplasm layer as the outermost boundary of the cell, the ectoplasm. In the change from endoplasm to ectoplasm, the protein chain's contraction, at the pseudopodium tip, extends the outgrowth and pulls the preceding extended protein chains along in a conveyor belt-like flow of endoplasm inside the cell. At the posterior end the ectoplasm is converted to endoplasm which then flows through the cell to start the process again. The ectoplasm is therefore pulled along towards the posterior of the organism and cell membrane flow glides the amoeboid protozoan through, or over, its habitat.

The more common bottom-dwelling foraminiferans anchor their reticulopodia in the sand and then move their bodies along by contraction of these outgrowths. Radiolarians and heliozoans, which are also amoeboid protozoans, do not use their pseudopodia for locomotion, but utilise them for feeding. Locomotion in these species is achieved by buoyancy.

Another protozoan group, the ciliates, have a specialised locomotive organelle. One group of ciliates, the oligotrichs (fig.71), are common zooplankton throughout the oceans, but they only represent a small proportion of the free-swimming marine ciliates. Like all ciliates they have small hair-like projections called cilia (fig.72a) that are embedded in the plasma membrane of the cell wall. Accompanying the organelles are the adjacent alveoli that contribute to the stability of the outer cell membrane. Cilia have the same structure as flagella, but are generally more numerous and shorter. At the base of each cilium is a basal body, the kinetosome, which is connected via fine fibrils (kinetodesma) to other kinetosomes on the same longitudinal row. The whole structure is termed a kinety.

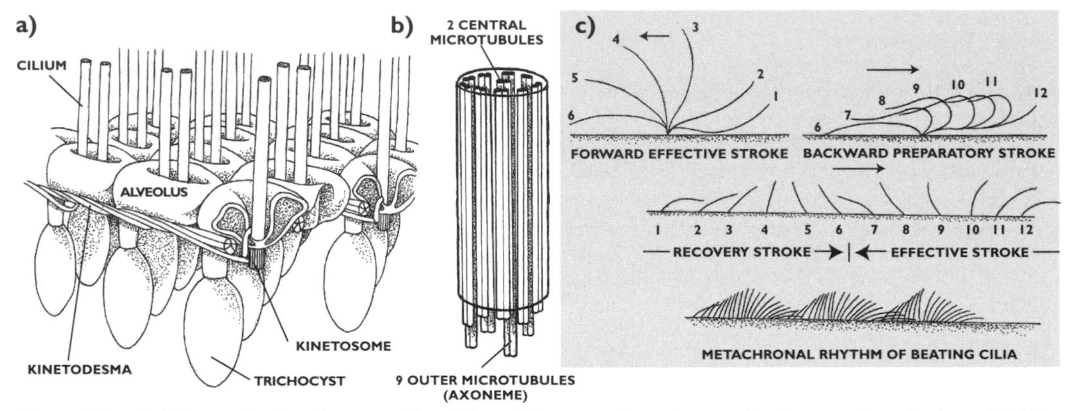

Figure 72. a) Cilia and related organelles. The trichocysts function as food capturing devices. b) The ultrastructure of a cilium. c) Stroke sequence and metachronal rhythm of cilia.

When cilia beat they produce an effective stroke and a recovery stroke (fig.72c). The recovery stroke offers less water resistance, similar to feathering an oar, and the effective stroke propels the animal. If looked at closely, a metachronal wave action can be seen to ripple across the surface of many ciliated protozoans. This synchronised ciliary beat can be reversed, giving ciliate protozoans the ability to move in different directions. When a ciliated protozoan encounters an object it wishes to avoid, it reverses, moves round a few degrees and tries again until it has avoided the object.

This form of movement is not confined to the ciliate protozoans as it occurs in many marine invertebrate groups. One such group is the ctenophores or comb jellies (plate 1) which are abundant in many coastal waters and also occur in oceanic waters. Their gelatinous bodies have eight ciliated bands, called comb rows, which beat (as in the ciliated protozoans) with a metachronal wave and so propel the ctenophore through the water. These cilia can often be seen as a spectrum of shimmering colour on the surface of their transparent bodies.

Flatworms (turbellarians, plates 4 & 6) also use cilia to provide the force for swimming (in smaller species) and creeping along the substratum (in larger species). The creeping flatworms depend on an adhesive mucus blanket for traction. This mucus is produced by rhabdites and other secretory glands around the anterior end of the flatworm. The cilia are able to gain traction in the mucus-covered surface during the effective stroke. There is also some muscular contraction along the body, but most of the time flatworms seem to glide, with ease, across the substratum of their habitats.

Although some flatworms can grow relatively large (the Mexican Flatworm *Pseudoceros mexicanus* can reach up to ten centimetres in length) ciliary locomotion becomes increasingly ineffective as the animal's size increases. For larger invertebrates more powerful locomotive methods are necessary.

Round worms use different means of locomotion from the flatworms. For example the polychaete worms have numerous hair-like structures projecting from their metamerically segmented bodies. These hair-like bristles (chaetae) are fine filaments secreted from the parapodia (fig.73). The parapodia are paired

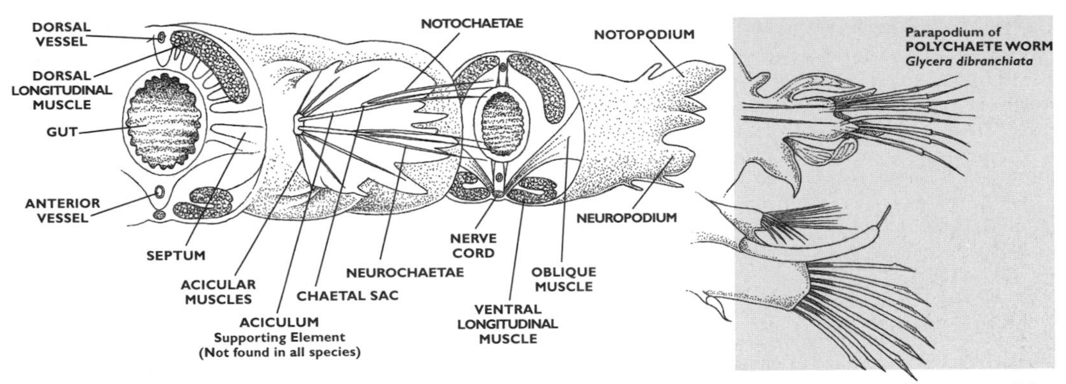

Figure 73. A polychaete worm illustrating parapodia and chaetae. Each parapodium comprises an upper lobe (notopodium) and a lower lobe (neuropodium). Chaetae are moved by the acicular muscles.

lateral appendages generally extending from each body segment except the head and the first and last segments. Crawling polychaetes use the parapodia and chaetae to the same effect as flatworms use cilia; the chaetae push against the substratum to move the worm along. A wavelike action travels down each side of the body, but in this case mucus is not normally needed for traction. Instead a sinusoidal muscular wave action helps to gain grip on slippery surfaces. This movement is typical of ragworms (*Nereis*) as they travel across sand (fig.74), and of planktonic swimming polychaetes.

Tube-dwelling polychaetes have smaller parapodia which are used for gripping the inside walls of their tubes. The Lugworm *Arenicola marina* (fig.45) uses its hydrostatic skeleton for locomotion inside its tube. It has lost the septa that form 'bulkheads' between each segment (except for the head and tail regions where the septa still exist) and has a more or less open fluid system. During movement the worm pushes all its body fluid to the anterior end. Then by alternate contractions of the body wall muscles the Lugworm thrusts its head deeper into the sand or mud and the rest of the body is hauled behind it into the substrate. This peristaltic action also occurs to a greater or lesser degree in most polychaete worms.

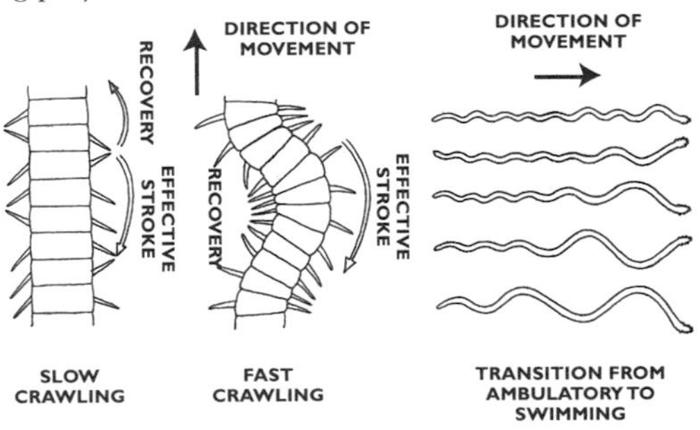

Figure 74. Locomotion in a polychaete worm.

Hydrostatic muscular action is found in many other burrowers, especially bivalve molluscs such as the Common Cockle *Cerastoderma edule* (fig.46) and the Peppery Furrow Shell *Scrobicularia plana* (fig.45). Burrowing bivalve molluscs have a long slender muscular foot that can be protruded, extended and then inserted into the substratum (fig.75). Once dug in, the foot is then swollen by blood pumped in from the mantle and viscera. Acting now as an anchor, the shell retractor muscles contract to pull the body and shell down onto the foot. The foot is deflated by blood flowing back into the mantle and viscera. This action is repeated until the bivalve is buried deep enough to escape danger or to prevent desiccation as the tide ebbs.

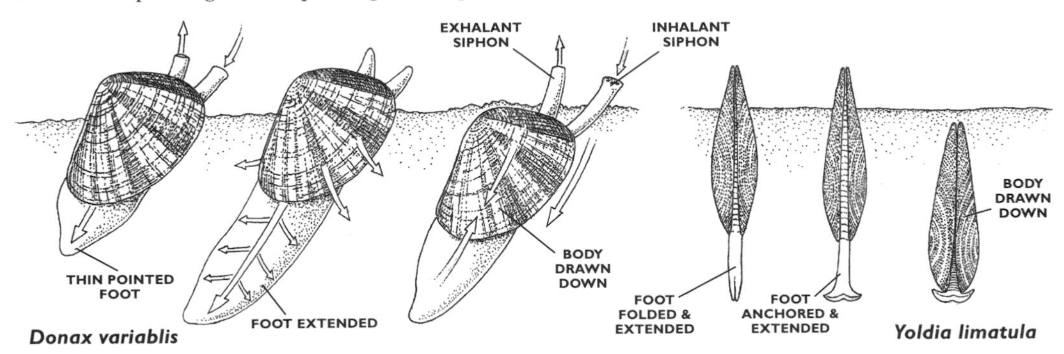

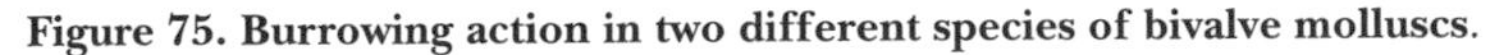

Figure 75. Burrowing action in two different species of bivalve molluscs.

Gastropod molluscs use a different mechanism of movement associated with the muscular foot, similar to that of flatworms, which involves keeping the ciliated, flat, muscular surface of the foot in contact with the

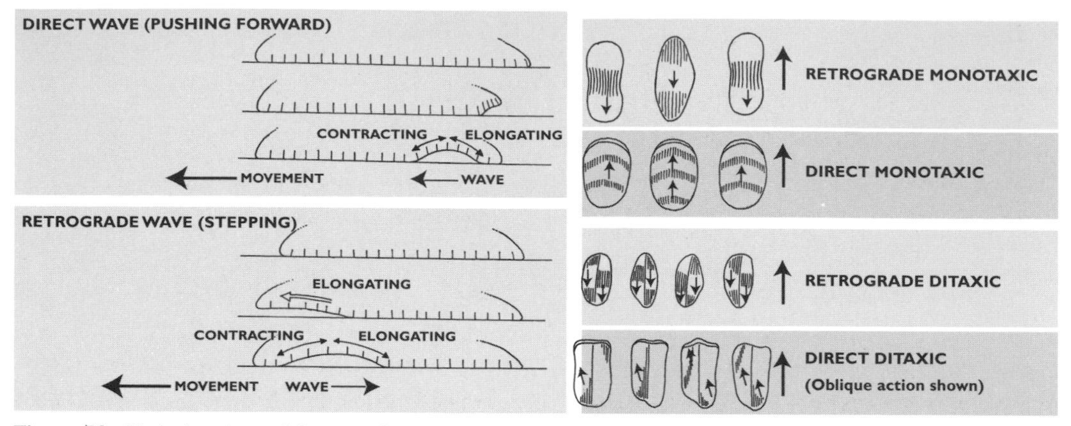

Figure 76. Variation in pedal waves in gastropods. Small arrows show direction of waves. Large arrows show direction of movement of whole animal.

substratum. Thick mucus is produced which can be seen in the trails left by terrestrial slugs and snails. Gastropod mucus has two states, gelatinous and liquid. Where the foot needs traction on the substratum the mucus is gelatinous but where the foot is moving to gain another traction point the mucus is liquid. This change in physical state is brought about by local shearing action of the muscles on the mucus layer beneath. Even though the gastropod foot is ciliated it does not rely on ciliary action for propulsion. Powerful peristaltic muscular waves that ripple along the foot surface perform this function. In some species each muscular, or pedal, wave travels along the foot from back to front, in the same direction as the animal's movement; this is called a direct wave (fig.76). Each individual ripple along the foot lifts the gastropod up and forward a small amount. The faster the waves, the faster the animal moves. Other species exhibit retrograde waves where the pedal wave action is in the opposite direction to the movement of the animal (fig.76). The wave action starts at the anterior end and the gastropod steps forward a small amount at every ripple along the foot. In both forms the whole foot is involved; this is monotaxic locomotion (fig.76). However, many species only use one side of the foot at a time and this is called ditaxic locomotion (fig.76). Many marine prosobranch gastropods use retrograde ditaxic locomotion which involves an oblique wave action across the foot (fig.76).

Not all molluscs use this form of locomotion. Some species like the moon shells, *Natica* and *Polinices*, plough through the sand with a protective shield made of a flap from the foot. Others such as the conch shells, *Strombus*, do not even use the foot for pedal wave locomotion but simply claw their way along with a 'piton' formed from the operculum.

The echinoderms have a specialised hydraulic locomotive system. This water-vascular system consists of canals (derived from the coelom) lined with cilia and associated appendages called tube feet, which are

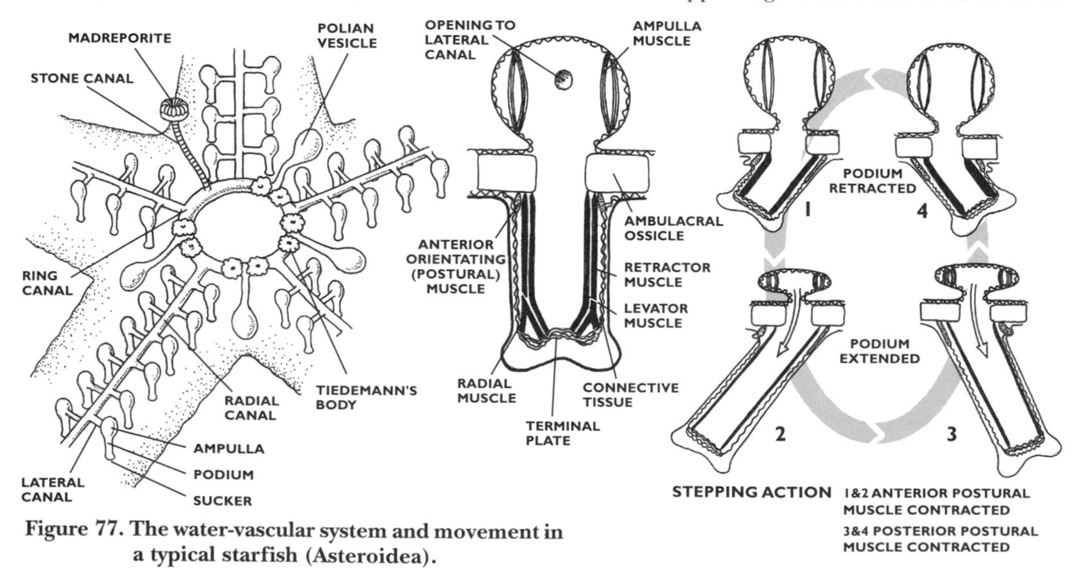

Figure 77. The water-vascular system and movement in a typical starfish (Asteroidea).

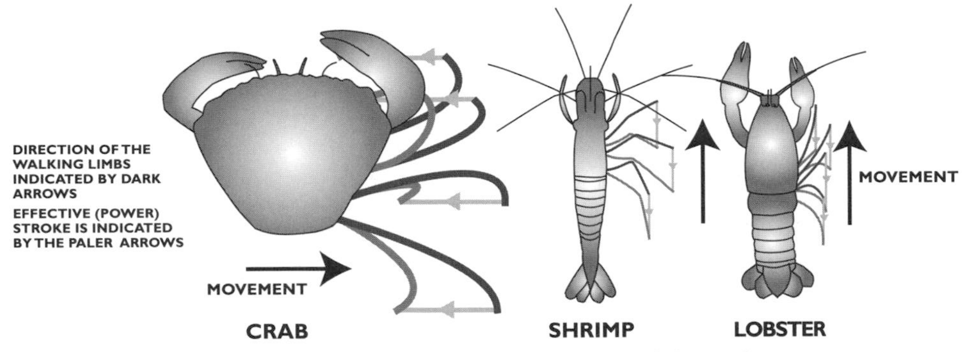

Figure 78. Stepping motion and limb movement in three types of decapod crustaceans.

outgrowths of the body wall. The internal water-vascular canals are connected to the outside via a button-shaped madreporite (fig.77). This has furrows of ciliated epithelium with pores opening internally into the pore canals. In turn these pore canals lead into a single stone canal that subsequently joins the circular ring canal (fig.77). Radial canals extend from the ring canal, along the oral side of the animal to the tips of the arms. The final ramifications of this system are the lateral canals that branch alternately along the entire length of the radial canals (fig.77). Each lateral canal possesses a valve, a terminal bulb (the ampulla) and a tube foot (the podium). Attached to the podium are muscles that enable the appendage to bend, and each tube terminates in a sucker. The whole system is filled with fluid similar in composition to that of seawater but with a small amount of protein and a higher potassium content.

Movement of the tube foot has four distinct phases (fig.77), which commence after the ampulla contracts and the canal valve closes forcing water into the podium which elongates it. The sucker adheres to the surface with a secretion from its terminal region which is similar to the locomotive mucus of flatworms and gastropods. When adhesion needs to be broken so that the foot can be lifted, a chemical is released from the terminal region that breaks down the sticky secretion. The four phases of movement consist of i) the podium swinging forward, ii) gripping the substratum, iii) articulating backwards and iv) finally releasing itself. Coordination of movement is sectional. Each independent section contributes its own phasing to the stepping action. Therefore an adjacent section will always be slightly out of phase with its neighbour. Those sections further apart will accordingly be further out of phase with each other. Overall locomotive action, however, is highly coordinated; the sectional movements appear as a wave of podia movement along the underside of the walking surface. Many echinoderms are highly manoeuvrable animals. They can negotiate nearly all surfaces and are able to climb up vertical rock faces.

Brittle-stars also use their muscular arms when they move, often in a rowing motion; each arm is able to curl around solid objects increasing the animal's hold on the substratum. Only the crinoids, sea lilies and feather stars have a different mode of locomotion from the other echinoderms. Sea lilies are sessile and fixed to the substrate by stalks, but feather stars move either by swimming (achieved by raising and lowering alternate sets of arms) or by crawling along using their arms and cirri to grasp the substratum. Most of the time feather stars remain fixed to the substratum by their rootlike cirri and swimming only takes place when they are threatened.

The crustaceans also use a stepping action for locomotion. Their many-jointed tubular limb system gives them excellent manoeuvrability . Their muscle blocks are attached, by specialised cells, to the inner surfaces of their exoskeleton. Flexion and extension is controlled by the contraction and relaxation of the muscle blocks acting in opposition. When the flexor muscles contract the extensor muscles relax and the limb is flexed. Extension of the limb is effected by contraction of the extensor muscles and relaxation of the flexor muscles.

Crustacean exoskeletons do not possess rotational joints. To overcome this they have evolved multi-jointed limbs with vertical and horizontal flexion points which enables them to move a limb in almost any direction. For easier limb articulation all crustaceans have their limbs mounted ventrally. However, even with the support of seawater, larger crustacean species (e.g. large crabs and lobsters) sag under their own body weight between opposing limbs, and this limits their body size. These ventrally located walking limbs have movement arcs that are distinctive to particular crustacean groups (fig.78). Whatever the species each limb's movement is divided into two separate stepping phases. The effective step occurs when the tip of the limb is in contact with the substratum. The following phase is the recovery step; the limb is lifted away from the body and swung forward in a wider arc than the effective step.

Many crustaceans find walking at speed difficult to coordinate and so have developed shortened paddle-like limbs, called pleopods. Mantis shrimps, copepods and other planktonic crustaceans use their pleopod limbs for swimming (fig.79). For shrimps and lobsters, swimming and crawling are interchangeable, depending on the substratum. Should they be threatened they can use their tail for rapid escape. The tail tip is made from a central plate called the telson and smaller side plates (uropods). One violent jerk of the posterior region can propel the animal backwards, away from immediate danger. This is often seen when shrimps leap clear out of the water to avoid a hand-held shrimping net.

One group of crustaceans, the ostracods, uses their elongated antennae for locomotion (fig.79). These specialised antennae are modified in various ways according to the species and its habitat.

Not only crustaceans can swim. Gastropod nudibranchs, like the Spanish Dancer *Hexabranchus sanguineus* and the sea hares *Aplysia*, swim with undulations of their mantles (plate 6, fig.80). Other gastropod molluscs, such as pteropods, can also swim. The pteropods consist of pelagic, shelled and non-shelled species. Both the shelled pteropods (sea butterflies) and the non-shelled pteropods (naked sea butterflies) use their modified foot as a fin-like organ (fig.80).

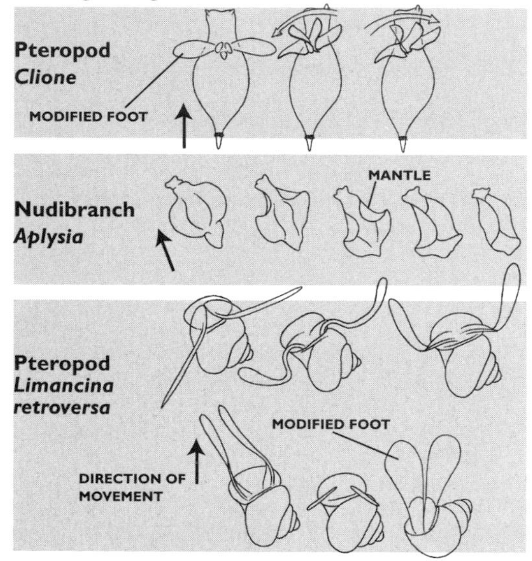

Figure 80. Swimming movements in two pteropods and a nudibranch.

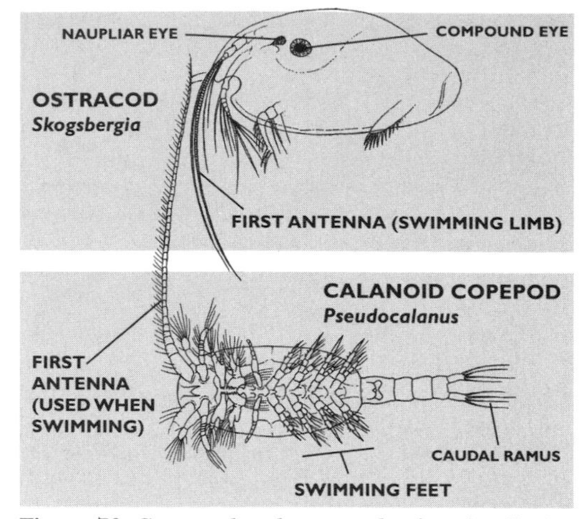

Figure 79. Copepod and ostracod swimming limbs.

Most pteropods swim upside down and their locomotion is relatively slow. However, one group of molluscs has mastered the art of swimming at speed in a most ingenious way.

Squid (plates 1 & 3) and cuttlefish (plates 4 & 7), in a similar way to flatfishes and rays, send a gentle sinusoidal wave-motion undulating along their translucent fins. The cuttlefish's fin fringes around the mantle margin. The squid's fin is confined to the posterior mantle margin. But the fins are not the main source of propulsion and are used mainly for stability. It is through jet propulsion produced by rapidly expelling water from the mantle cavity (fig.81) that these animals move at speed through the water. This process has an inhalant phase and an exhalant phase. During the inhalant phase, the circular muscles of the mantle relax and the radial muscles contract, thus increasing the volume of the mantle cavity. Water rushes in, through a vent between the head and the edge of the mantle, and fills the vacuum created by the mantle cavity's expansion. In the exhalant phase the circular muscles contract and the radial muscles relax. This decreases the volume

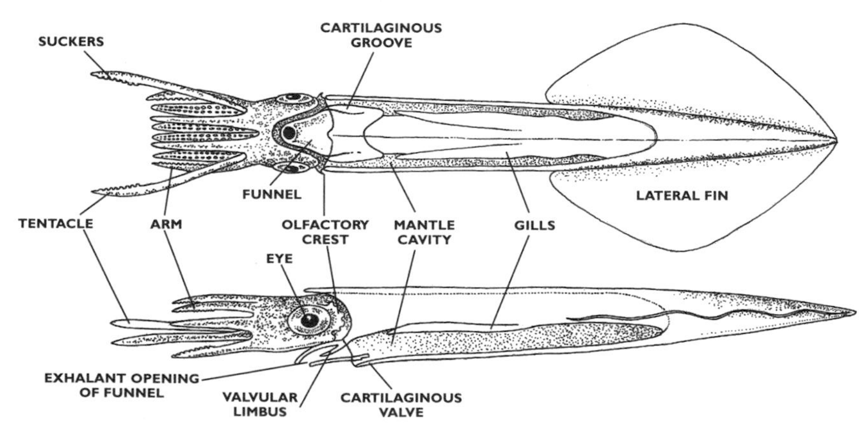

Figure 81. Structure of the squid (*Loligo*) to show position of mantle cavity and funnel.

of the mantle cavity, thereby increasing the water pressure inside. The hydrostatic pressure is increased by the mantle cavity being sealed off to the outside by cartilaginous valves around the head. The only way the pressurised water can escape is through the tubular funnel below the head, as the ventral mantle cavity is also sealed off by valve flaps. A strong water jet is therefore emitted through the funnel and propels the cephalopod in the opposite direction to its flow. Flying squids (Onycoteuthidae) can launch themselves up to six metres into the air above the water's surface and over 15m horizontally before they re-enter the water. The jet-propulsion of cephalopods can also be very subtle. The funnel can be swivelled to the front or rear allowing the animal to hover and manoeuvre in water with great agility. The greatest propulsive power is delivered during escape and if threatened some species of squid can attain burst speeds reaching 22 knots. These are the fastest of all aquatic invertebrates and rival many of the fastest swimming vertebrates.

A less sophisticated version of water propulsion is common to many marine cnidarians. The scyphozoans and hydroid medusae can swim by pulsing movements of their gelatinous bells (fig.82). This pulsation is brought about by rhythmical contractions of a band of powerful muscle fibres that pull against the gelatinous bell and its water volume. When these muscles relax the elastic mesogloea (a gelatinous internal layer) springs back. Jellyfish and hydroid medusae thereby pulse, the bell rhythmically expanding and contracting to propel the medusa along. Some cubozoans (box jellyfish, plate 7) and rhizostome jellyfish (plate 1) can control the direction of their swimming. However, most medusae drift with the current and only swim to vary their posture or position within the water column.

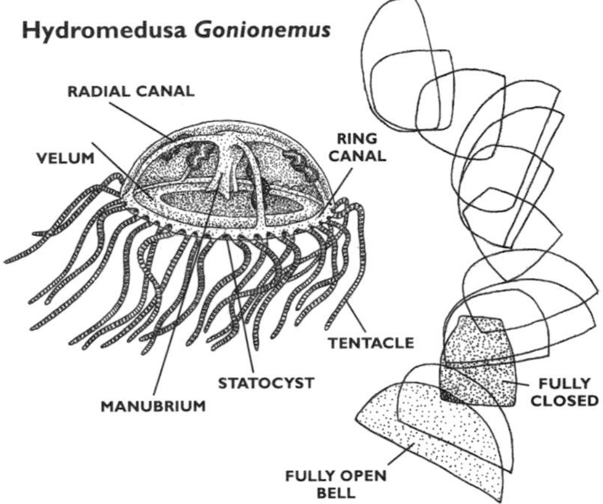

Figure 82. Locomotion in the hydromedusa *Gonionemus*.

FEEDING

Competition for food resources has led to the evolution of many feeding strategies and mechanisms. Specialisation in a particular type of diet or mode of feeding can enable an organism to compete more successfully for a food resource. Some marine invertebrates have found unusual sources of nutrition. For example, the marine shipworm *Teredo navalis* and the gribbles *Limnoria* spp. bore into submerged wooden structures, feeding on the wood fragments and fungi that rot the wood. Other invertebrates are less specific in their diet. Sea urchins are notorious for eating almost anything encountered as they browse the substratum.

Herbivorous marine invertebrates use a variety of feeding mechanisms to consume plant matter. The simplest of these herbivores, the protozoans, use phagocytosis to ingest their food (normally single-celled algae or microalgae). They engulf food usually through a mouth, the cytostome. After ingestion a membrane-bound sac, called a food vacuole, is produced (fig.83) which undergoes changes in acidity, chemical content and size as the plant material is slowly digested. Pinocytosis is then used to absorb products of this chemical breakdown as soluble matter.

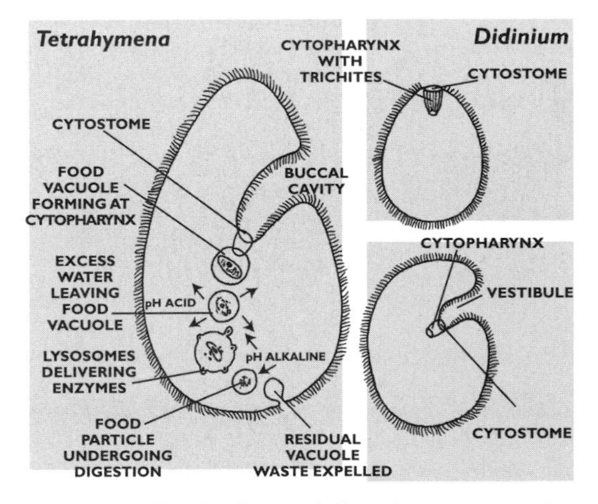

Figure 83. The feeding and digestion process of ciliate protozoans.

The cytostome of ciliate protozoans is generally located towards their posterior end (fig.83). It opens out into an inner tube, the cytopharynx, which is where food vacuoles are pinched off; in higher ciliate protozoans it opens out into a buccal cavity. In these species food is wafted in by the action of compound ciliary organelles. These cilia are so densely packed that connections between the kineties (see p.128) enable them to beat together. Because ciliate protozoans are of a larger size than many protista, they may be

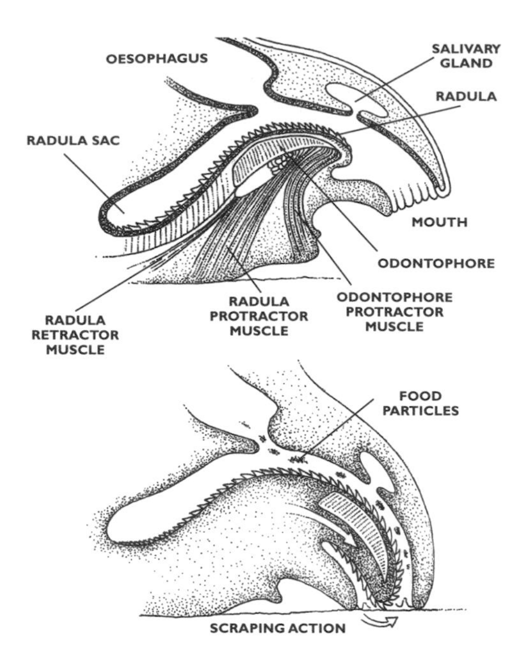

Figure 84. The head of a snail in longitudinal section showing the typical gastropod radula and its action.

carnivorous and feed on other invertebrates such as rotifers and gastrotrichs. The tintinnids (fig.71) are typical ciliate carnivorous suspension feeders and use their large buccal cavities and specialised cilia to drive suspended matter into their cytostomes.

Gastropods are successful grazers. Most of them feed on fine algae rasped from rock surfaces and other substrata. Some even feed on larger algae such as kelp.

The commonest group of herbivorous gastropods found on rocky shores, mangroves and marine marshes are periwinkles (Littorinidae, plate 3). At low tide these animals withdraw into their shells and close their opercula (a thin hard protective plate on the foot). The lip of the shell is sealed to the solid substratum with mucus to prevent desiccation at low tide. So at high tide, when their habitat is flooded, they re-emerge to graze on the covering of algae including endolithic algae just below the rock surface.

To graze on such fine material the periwinkles use an organ unique to all molluscs, except bivalves, called the radula (fig.84). The gastropod mouth is normally located below the anterior part of the animal and opens into a pocket-like buccal cavity containing the radula sac. The radula apparatus has two parts, the cartilaginous base (the odontophore) and the radula itself with its longitudinal rows of chitinous teeth. When feeding the odontophore is pushed forward into the mouth opening which flattens the radula over the odontophore tip and erects the radula teeth (fig.84). As the odontophore tip 'licks' the surface with a forward and upward action, the teeth cut and scoop the surface. The loosened food particles are then swallowed. The radula is continually being worn down by this process so new teeth are formed at the posterior of the radula and, like a slow moving conveyor belt, are brought forward to the working surface to eventually replace the worn teeth at the tip. Tooth production is rapid and some species may secrete as many as five rows per day. As in most cutting and grinding actions, lubrication is important to the process. The radula is lubricated by mucus provided by at least one pair of salivary glands. This sticky mucus strings the loosened food particles together, facilitating the passage of food into the oesophagus. Each species has its own specialised radula teeth arrangement which is suited to the plants or animals it feeds on (fig.85).

Some gastropods are specialised herbivores such as the sacoglossan opisthobranchs. These diminutive and often highly coloured nudibranchs have a single row of teeth which they use to cut open individual algal cells. One European species, *Elysia viridis* (about three centimetres long), can incorporate chloroplasts into its own digestive gland cells by phagocytosis and use them for photosynthesis. Similarly, the Australian aeolid nudibranch, *Phyllodesmium longicirra* (plate 8), ingests the symbiotic zooxanthellae from the soft coral *Sarcophyton* and stores them in its body. Its large, numerous, paddle-like mantle outgrowths (cerata) are distinctive, unlike the normal thin tubular processes of aeolid nudibranchs. These specialised cerata are where the captured microscopic cells are stored and the photosynthetic products of these zooxanthellae are taken up and used. Therefore, this nudi-

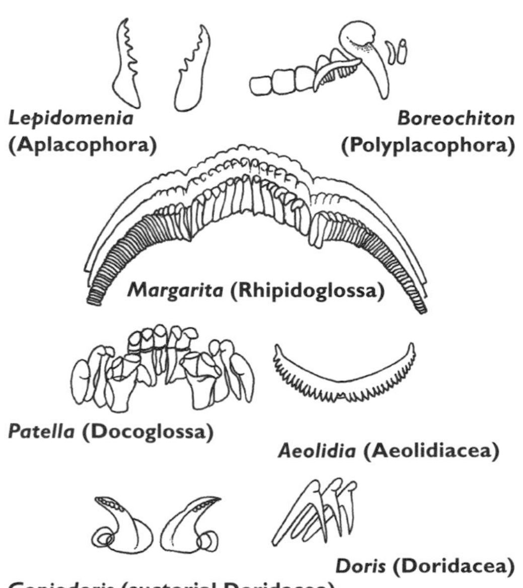

Figure 85. Radula teeth of various gastropods.

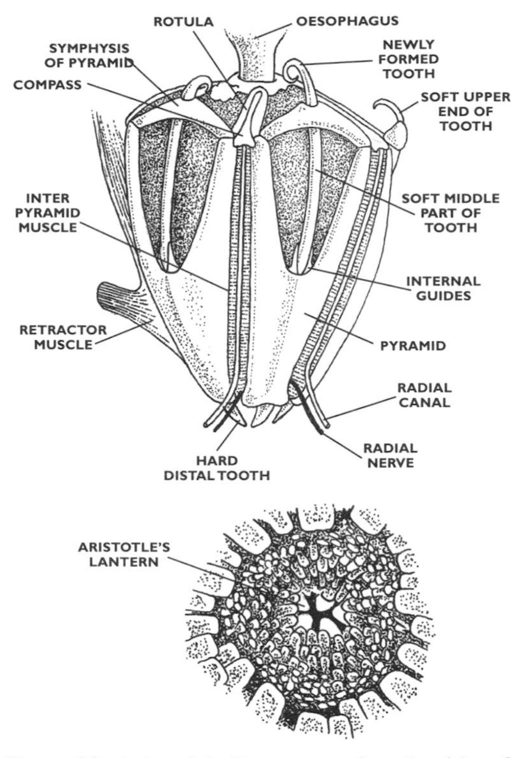

Figure 86. Aristotle's Lantern and underside of Slate-pencil Sea Urchin *Heterocentrotus mammillatus*.

branch has to feed only infrequently to replace dead zooxanthellae.

Sea urchins (echinoids) are widespread and generally omnivorous. Sea urchins of the Euechinoidea use an organ called Aristotle's lantern for grazing (fig.86). Like the radula of the gastropods it is a highly developed scraping tool composed of five calcareous plates called pyramids. These radially arranged pyramids are connected to each other via transverse muscles. Each of the connected plates has a calcareous band running down its centreline. At the upper end of this band there is a region where new tooth material is secreted, normally about one to two millimetres per week (fig.86). The terminal end is formed into a hard pointed tip (the distal tooth). Aristotle's lantern is used to scrape away at the surface on which the sea urchin is resting. The lantern can also be raised, lowered and swung laterally, and is therefore able to feed in a wider arc than if it were static.

Most sea urchins are dependent upon the substrates over which they move for their diet. They often feed in vast numbers, grazing large areas clean of algae or any other sessile surface dwellers. Normally the algae quickly recover after the sea urchins have left but when the urchins are not held in check by predators the algae may be destroyed. In the late 1960s the sea urchin *Strongylocentrotus franciscanus* (plate 2) was responsible for the devastation of the Californian kelp beds. One documented sea urchin 'plague front' was recorded advancing at 30 feet per month, leaving nothing but barren rock in its wake. The major cause of this plague was the near extermination of the Sea Otter *Enhydra lutris* (fig.312, p.431), the sea urchin's main predator.

Echinoderms include the holothuroideans (sea cucumbers). These differ from the sea urchins in that they are, with few exceptions, deposit and filter feeders. They either stretch out their branched tentacles into the water column, or sweep them over the bottom. In both cases the particles are collected on sticky papillae on the surface of the tentacles. The structure of the animal's tentacles is indicative of its feeding method. Fine bushy tentacles are used by fine suspension feeders and robust, thickened tentacles are used by those that feed on larger, heavier particles on the bottom. Once the tentacle is loaded with food particles it is bent orally and inserted into the pharynx (plate 8). The particles are ingested and the tentacle is returned to its feeding position. The adjacent tentacle undergoes the same routine until the animal finishes removing food from all its tentacles and the cycle begins again. Mobile sea cucumbers, such as the Leopard Sea Cucumber *Bohadschia argus* (plate 7) and the Pineapple Sea Cucumber *Thelenota ananas* (plate 6), are suctorial feeders. These are aspidochirote holothuroideans with flattened round tentacles designed to gather sediment particles. They are mainly non-selective feeders, ingesting everything within the sand or mud and leaving distinctive casts in their wake. Sedentary sea cucumbers are all dendrochirote holothuroideans as typified by the Pudding *Cucumaria frondosa* (plate 3), which has greatly expanded, finely branched pinnate tentacles for filter feeding. When they find an area with a suitable current these animals will attach themselves to the substratum with their tube feet, often remaining there for long periods.

Deposit and suspension feeding methods are common among the marine invertebrates. Sponges exhibit some of the simplest mechanisms. These creatures are constructed around a system of water canals with a central atrium that can vary in complexity (fig.87). The organism requires a constant flow of water throughout its water canal system. This water current is produced and regulated by the beating of the choanoflagellate cells, similar in structure to elongated ciliated cells. Unlike the cilia of ciliate protozoans and ctenophores there seems to be no overall co-ordination, each flagellum beating separately in a spiral motion. Water is drawn into the chambers through small pores called porocytes (fig.88) and then driven into the atrium. Finally it passes through the osculum to the water outside.

Fine particles are brought in by this water current, especially minute organic particles which are uti-

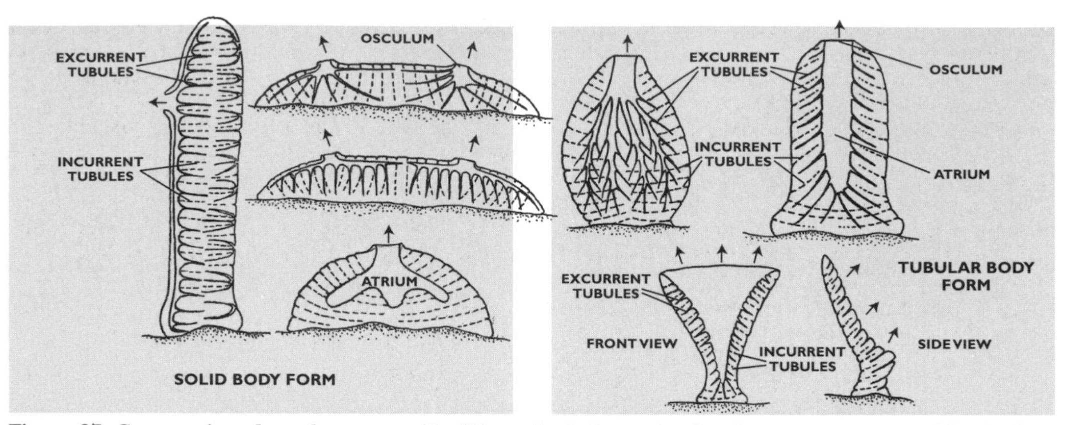

Figure 87. Cross sections through sponges with different body forms showing the water currents used for feeding.

lised as food by the sponge. To make sure that only small particles enter the system there are screens across the incurrent canals and external dermal pores (porocytes). Sponge cells absorb microparticles by phagocytosis; large microparticles (5-50μ) are absorbed by pinacocyte cells, and smaller microparticles (<5μ) by choanocytes. These particles are then transferred to another cell type, the amoebocytes, where digestion takes place. Additional nutrition is also obtained from blue-green algae embedded in the wall matrices of some sponges.

Sea squirts (ascidians) are another widespread group of sedentary filter feeders. Water is drawn in through the buccal siphon by the beating of lateral cilia in the margins of the stigmata, slits in the pharyngeal basket (fig.89). A large quantity of water is strained through the stigmata to remove the plankton that is necessary for the animals' survival. To bind the collected plankton together, a mucus sheet is secreted by the endostyle, which is a single deep groove along the length of the pharyngeal wall. Two continuous mucus sheets flow out across the inner sides of the pharyngeal basket, which meet at the dorsal lamina. The plankton laden mucus then passes down a 'gutter' to the oesophageal opening. To stop feeding the sea squirt simply closes its buccal siphon and halts the ciliary beat. Some species of ascidians (e.g. the Soft Didemnid Ascidian *Didemnum molle* plate 8), like some sponges, contain symbiotic algae and the nutrients produced by photosynthesis are utilised by the host as a food source.

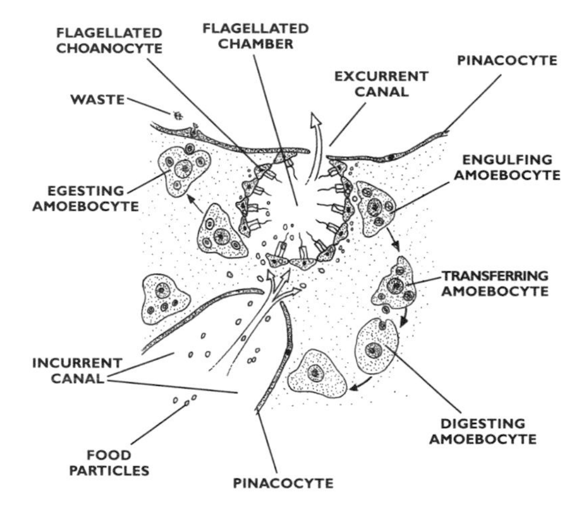

Figure 88. Cellular digestion in a sponge.

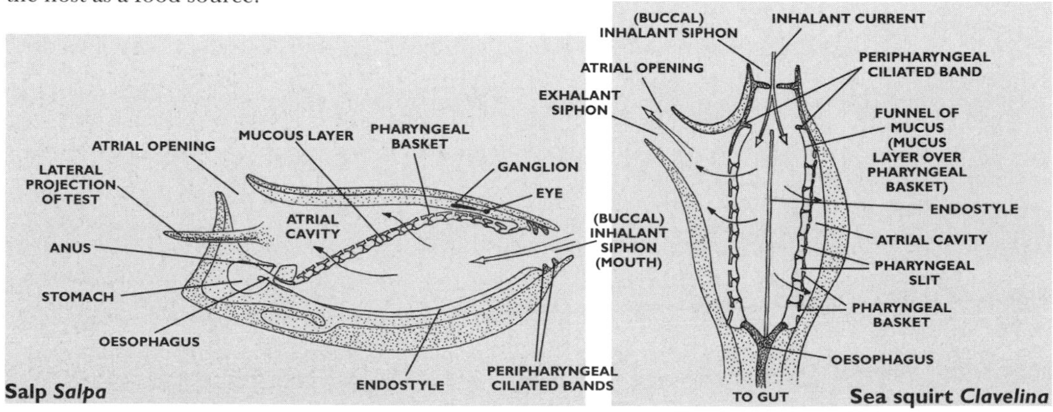

Figure 89. Cross sections of typical ascidians.

Although the filter feeding bryozoans often look like fungal growths, sometimes even plant-like (plates 3, 4 & 7), they are colonial zooids. In most species each zooid is encased in a protective covering, either boxlike or tubular with a single opening (fig.90). The filter feeding tentacles (lophophore) are supported on a calyx and can be withdrawn for protection. When the tentacles are expanded they form a funnel. The lateral ciliated tracts along the tentacles create a current that flows into the funnel and then out between the fine processes (fig.90). When a food item, such as small plankton, is drawn in between the tentacles it touches the lateral cilia which locally reverse their beat and it is thrown back into the inside of the funnel's tentacles. It is drawn again through the tentacles, but this time at a lower point down the funnel. Again this causes a local ciliary reversal and it is thrown once more deeper into the funnel. This process carries on until it is eventually swallowed by the mouth. This method of filter feeding is called an 'upstream ciliary collecting system'.

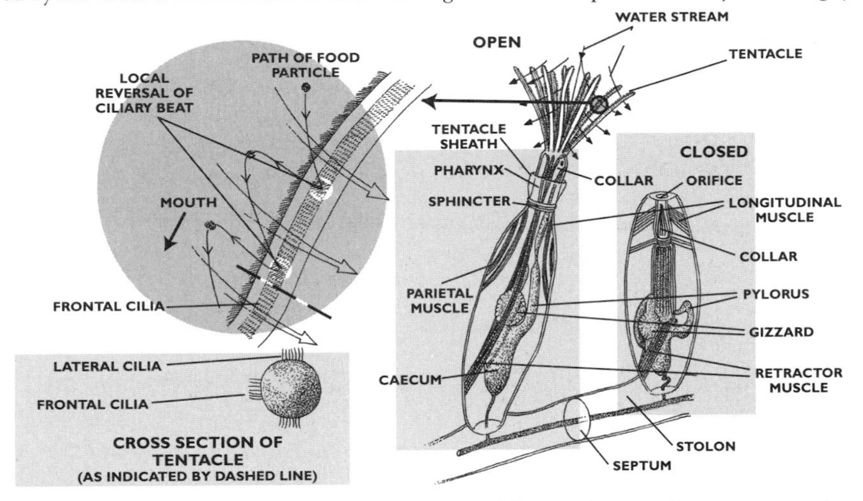

Figure 90. The bryozoan *Bowerbankia* filter feeding, with an enlarged view of a tentacle and its cilia.

Viewed through a microscope some bryozoans may appear as minute rows and columns of barnacle-like creatures. The true barnacles (class Cirripedia) also filter feed, but their fine tentacle-like processes are derived from the legs (fig.91). These feeding organs (cirri) are adaptations of the locomotive appendages of the larval stage which have many branching chaetae. During feeding the paired protective scuta and terga open and the cirri unroll. Once unrolled the cirri form a fine basket that scoops downwards in a clenching action. The first and smallest pairs of cirri scrape the captured particles off the other cirri and transfer them to the mouth parts. Many barnacles like the Acorn Barnacle *Semibalanus balanoides* feed on small plankton but larger species such as the stalked or goose barnacles, *Lepas* and *Pollicipes*, capture larger copepods, isopods and amphipods. Similar feeding mechanisms also occur in many other invertebrate groups. Barnacles have a wide distribution and can survive on the most unusual surfaces, including whale skin, turtle shells, boat hulls, cooling ducts and many other marine substrates.

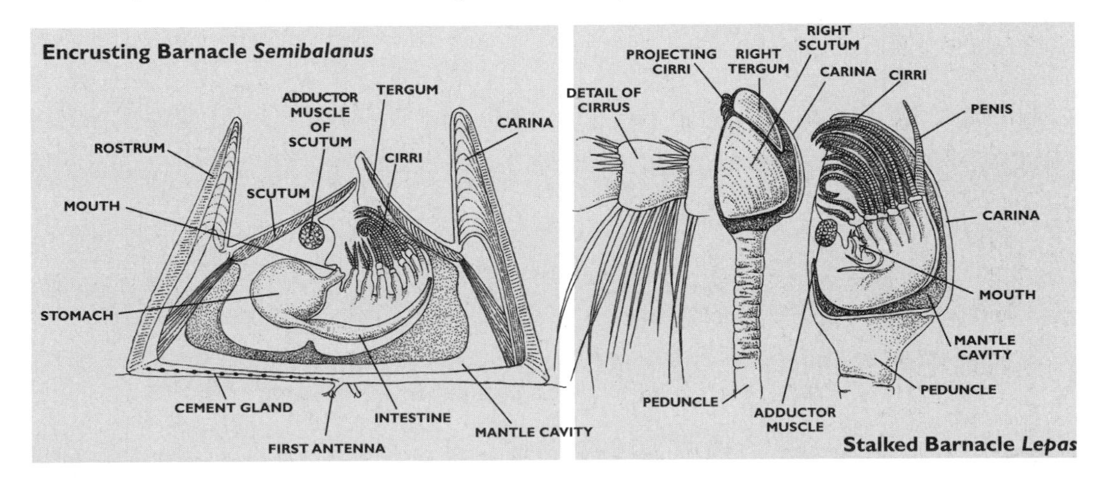

Figure 91. Two barnacles, *Semibalanus* and *Lepas*, showing their feeding organelles.

Another widely distributed filter feeding group, the tube dwelling polychaete worms, are mainly surface deposit feeders which are capable of filter feeding if conditions allow. The most striking of these are the fanworms, such as the serpulid (order Serpulidae) Christmas-tree Worm *Spirobranchus giganteus* (plate 6) which lives on the surface of live corals and which secretes calcareous tubes. The feeding appendages on the head of most polychaetes (prostomial palps) have been modified to form spiral crowns consisting of many pinnate processes, called radioles. The similar sabellids (order Sabellidae), such as the Southern Fan Worm *Sabellastarte indica* (plate 6) and *Bispira volutacornis* (plate 4), build a tough membraneous sand-grain tube and also bear radioles. These filter feeding segmented worms use cilia to generate a water current that flows through the radioles into and out of the funnel (fig.92).

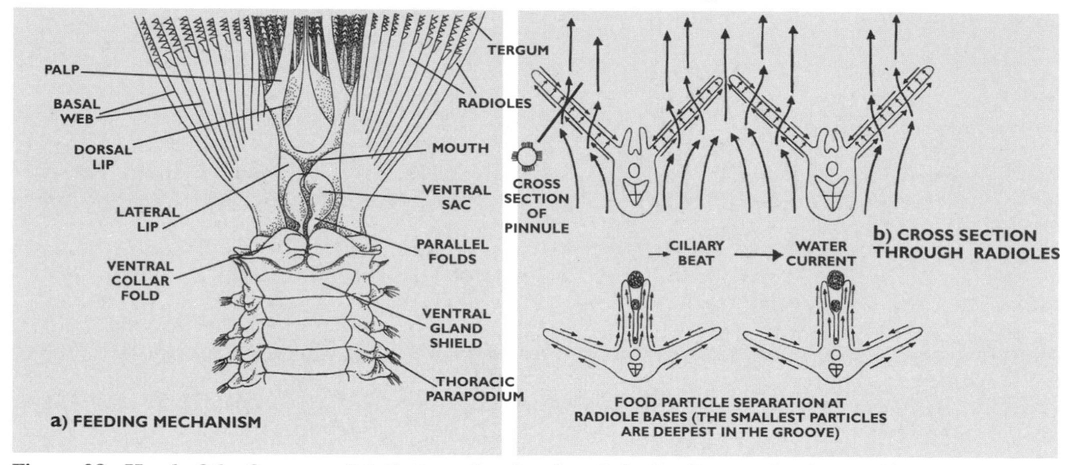

Figure 92. Head of the fanworm *Sabella pavonina* showing a) the feeding mechanism and b) a cross section through two radioles.

Trapped particles on the fine processes of the radioles, the pinnules, are transferred to a groove along each radiole by way of the cilia. In this groove collected particles are carried along to the base of the radiole where the largest particles are rejected. The fine material is then carried to the mouth by a ciliated tract.

With the exception of sabellid and serpulid polychaetes most worms have an extendable pharynx which is used for feeding. However, some sedentary polychaete worms have a specialised buccal structure that can extend over a large area of the substratum. These animals are selective deposit feeders that use a method similar to that of the serpulids and sabellids to convey collected food particles, using a mucus secretion, to the mouth. An example of a selective feeder is the terebellid worm *Terebella lapidaria* (fig.93). Its banded contractile feeding tentacles stretch over the sand. These tube-like tentacles (fig.93) creep out over the surface by ciliary action and the collected particles travel back to its tube dwelling to be sorted into three sizes. The larger particles are rejected, the medium-sized ones are used to repair or build the tube dwelling and the smallest ones are ingested. After a time there is a build up of the larger rejected particles which can hamper the worm's feeding. The worm subsequently 'coughs', by a violent contrac-

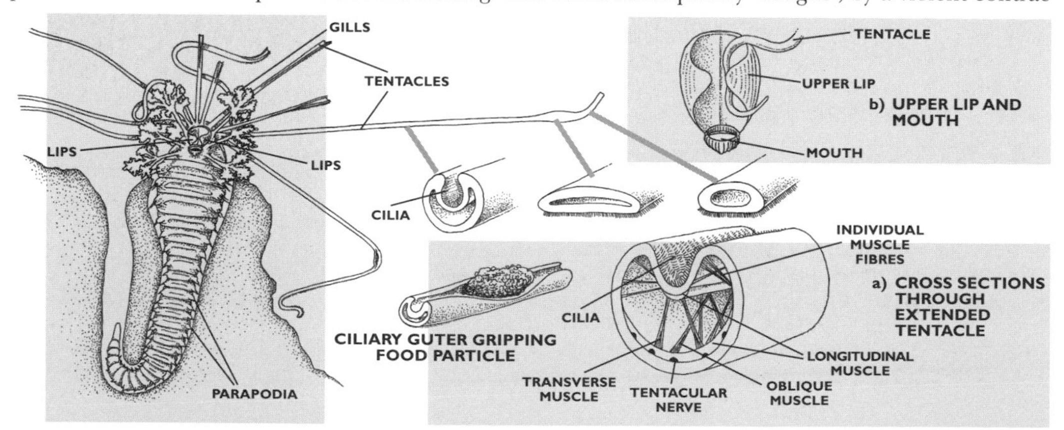

Figure 93. The terebellid worm *Terebella lapidaria* with a) cross sections through one extended tentacle showing the feeding mechanism and b) detail of the upper lip and mouth, showing tentacle being wiped by an upper lip.

tion of its body, to remove unwanted sediment away from the furthest reaches of its contractile tentacles.

Another tube-dwelling polychaete, the Lugworm *Arenicola marina* (fig.45), is a non-selective deposit feeder that consumes sand and mud directly by ingestion. Its gut processes the deposits and periodically the worm reverses up its tube to defecate the unwanted silt onto the surface of the sand, forming characteristic spiral casts.

Many bivalve molluscs also leave casts on beaches and in this group selective deposit and suspension feeding is the commonest feeding method. The largest of all invertebrate filter feeders is a large bivalve mollusc, the Giant Clam *Tridacna maxima* (plate 8), although it is mainly dependent on symbiotic algae for its food (see p.154).

Bivalve molluscs generally feed by straining the surrounding water or sifting fine deposits on the sea bed. To do this they have modified their gills, evolving huge respiratory organs far greater in area than is necessary for breathing. In its simplest form the apparatus has a 'W' shape (fig.94a), with four broad filtering surfaces (lamellae). A food groove extends along the entire underside of each gill demibranch and the frontal cilia on the demibranchs carry food particles vertically to the food groove. The lateral cilia also found on the demibranchs produce a water current through the gills and the laterofrontal cilia (a bundle of fused cilia not present on ancestral bivalves) act as a screen to filter particles from the water (fig.94b, c). The mucus and collected particle chain passes down the food groove to the labial palps and finally to the mouth.

As with polychaete suspension feeders the collected particles are processed according to size. In some species, such as oysters, scallops and razor clams, the cilia carrying this food chain are so deep inside the lamellae that they may become clogged. Therefore, when the water they are filtering is loaded with sediment the gills contract and the outer sides engage large cilia to drive out much of the obstructing sediment to the ventral groove. Here the particles are dropped into the bottom of the mantle cavity to be ejected, often by convulsive movements of the valve muscles. All bivalves have to eject such unwanted material, pseudofaeces, and normally this is accomplished through the exhalant siphon. The Giant Clam judders visibly when ejecting its pseudofaeces and discharges a plume many metres into the surrounding water.

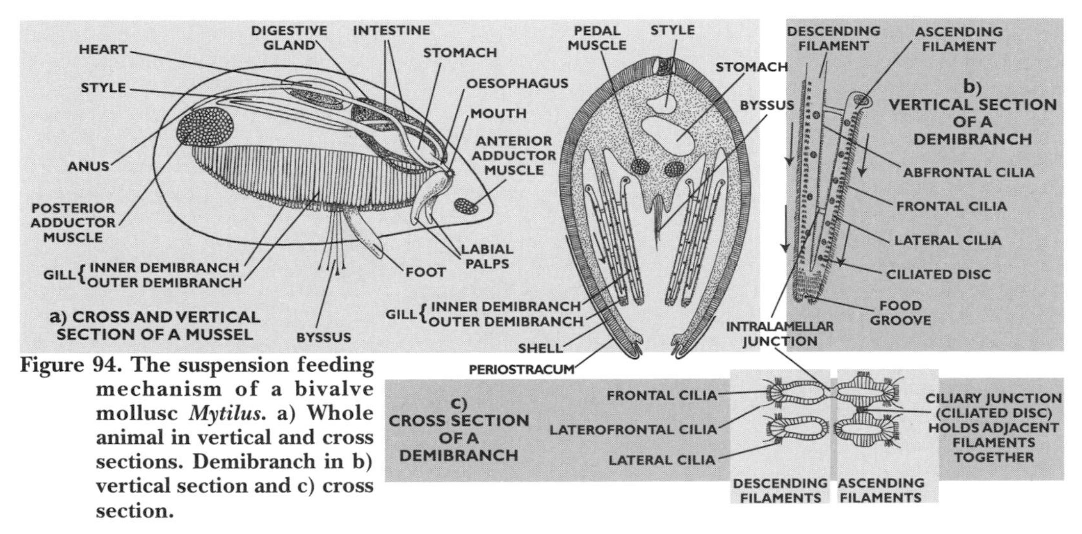

Figure 94. The suspension feeding mechanism of a bivalve mollusc *Mytilus*. a) Whole animal in vertical and cross sections. Demibranch in b) vertical section and c) cross section.

The razor clams, cockles, tellins, trough shells and many other molluscs spend most of their time buried in sand and silt (see p.78). However, even there they can filter feed using their extended exhalant and inhalant siphons. The largest of these burying bivalves is the Geoduck *Panopea generosa*, a Californian bivalve whose valves can reach the size of a coconut. When fully extended its siphons stretch to nearly a metre in length so that they can no longer be totally accommodated within the valves.

Bivalve molluscs use a variety of methods to collect the nutrient-rich sediment (see p.79 on intertidal feeding). Some bivalves even attach themselves permanently to the substrate for example, Northern Horse Mussel *Modiolus modiolus* (plate 2), and oysters such as the European Oyster *Ostrea edulis*. The mussels use byssal threads to anchor themselves, whereas the oysters attach their shells to the substratum by secreting a cement from the left valve mantle.

Permanent attachment to the substratum may be a risky strategy as it limits the animal's ability to escape from predatory attack, but it affords better protection from powerful currents and waves. Permanent attachment usually occurs in habitats where there is a constant movement of water replenishing nutrients and enabling the young to be easily dispersed.

Stony corals thrive using this strategy and two methods of nutrient intake are used. The first feeding method utilises the stinging tentacles of the polyps. At night when the polyp tentacles extend into the current flow, suitable prey is caught and brought to the mouth where it is ingested. As with most cnidarians these tentacles have thousands of stinging cells, called nematocysts or cnidae (fig.116), embedded in the epidermis. These are found all over the body but are most abundant on the tentacles. Among hydrozoans and scyphozoans the outside of the nematocyst has a short, stiff bristle (the cnidocil), while the anthozoans have a ciliary cone complex. Both these structures are trigger mechanisms. Inside the nematocyst is a coiled thread bearing many barbs covered by an operculum lid. When the cell is triggered, by a combination of mechanical and chemical stimuli, the operculum opens and the thread is shot out. As the thread is coiled it rotates rapidly counter clockwise, driving it deep into any tissue that it encounters. These penetrating nematocysts are the most important type of cnidae used for prey capture and many deliver a lethal or paralysing dose of toxin to the prey. The toxic effect is not normally perceptible to humans, although in some cases it may cause irritation (cnidae that are used as defence mechanisms are discussed on p.145). Some stony corals and sea-anemones possess an extra capturing cell (the spirocyst) that shoots out long adhesive threads into their prey. The sea anemones that have these spirocysts are known to use them for attachment purposes. Various anthozoans, including soft corals like Red Sea-fingers *Alcyonium glomeratum* (plate 4), sea anemones (e.g. *Sagartia elegans,* plate 4) and the sedentary adult hydrozoans (e.g. *Sertularia argentea* and *Tubularia indivisa,* plate 4) also feed on small animals which brush against their outstretched tentacles.

The stony corals and a few sea anemones, such as *Anemonia viridis,* have a second method of nutritional intake through a symbiotic relationship with zooxanthellae which allows the corals to grow in areas where polyp feeding alone would not provide sufficient nutrients (see page 154).

Floating hydrozoan colonies, such as the Portuguese Man-of-war *Physalia physalis* (plate 1), consist of a colony of individual polyps hanging below a gas-filled float whose stinging tentacles come from one type of polyp group (the fishing zooids). When brushed against, they release nematocysts which contract bringing the prey into contact with the tubular feeding polyps immediately below the float. Once the prey is digested by the feeding polyps, the nutrients diffuse into the communal body cavity for distribution to all polyps in the colony.

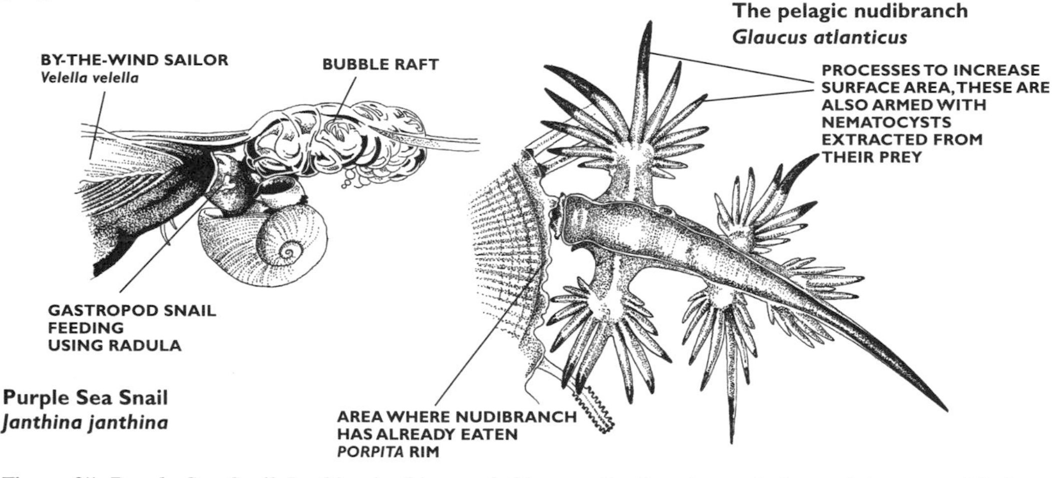

Figure 95. Purple Sea Snail *Janthina janthina,* and *Glaucus atlanticus,* two pelagic predators on cnidarians.

Even though scyphozoans and siphonophores have barrages of stinging cells they are not immune to predation and at least two groups of gastropods voraciously attack smaller pelagic cnidarians (the predators of larger jellyfish include the turtles, especially the Leatherback Turtle *Dermochelys coriacea,* plate 24). One group are the violet sea snails, Janthinidae, which create a bubble raft of mucus (fig.95) to which they attach themselves. Then, at the mercy of the currents, they await their prey, normally the By-the-wind Sailor *Velella velella* or *Porpita* spp. Having found its prey the sea snail attaches its foot to the raft of the hydrozoan and extends its mobile mouth. The radula proceeds to gnaw away at the tentacles ignoring, or maybe immune to, the nematocysts. Similarly, a group of floating nudibranchs, e.g. *Glaucus atlanticus,* which gulp in air to keep themselves afloat, feed on the same cnidarians (fig.95). The fine processes all over its body not only help it to move around in the surface film of the sea, but also increase its surface area which helps the animal to remain afloat. *Glaucus* is immune to nematocysts and leaves behind only the inedible remains of the float once it has fed.

Benthic aeolid nudibranchs also devour cnidarians. The Grey Sea Slug *Aeolidia papillosa* (plate 4) feeds on the Beadlet Anemone *Actinia equina*, and the large Plumose Anemone *Metridium senile*. This animal can look similar to its prey, with its long dorsal tentacle-like outgrowths. These outgrowths contain undischarged nematocysts obtained from feeding on sea anemones. Interestingly the sea anemone does not discharge the most toxic nematocysts when it is ingested by the Grey Sea Slug. Somehow, the gut of the gastropod can distinguish these most toxic nematocysts from the rest of the matter and transport them whole and undischarged to the cerata. The similar flabellid nudibranchs (e.g. the Red-lined Flabellina *Flabellina rubrolineata,* plate 7) even advertise their toxic 'cargo' with their bright colours and obvious white-tipped cerata.

Certain gastropods are among the most voracious of all carnivorous marine invertebrates and use their long proboscis and file-like radula teeth to hunt other gastropods and bivalve molluscs. The Dogwhelk *Nucella lapillus* uses its proboscis to rasp a round hole in the shell of a mussel or to prise apart the calcareous plates of a barnacle. When it reaches the animal it uses its radula to scrape away the soft parts for ingestion. The Necklace Shell *Natica* sp. produces sulphuric acid secreted from the proboscis glands, found at the proboscis tip, to accelerate the process of boring through the shell of its prey. The American (Oyster) Drill *Urosalpinx* sp. causes considerable damage to commercial oyster beds. When it encounters a bivalve it pushes out the front part of its foot to apply a special gland to the shell surface. The gland secretes an acid which demineralises the shell matrix; the animal then scrapes with the radula for about a minute and applies the gland for a further half-an-hour until the shell is pierced and the soft parts can be reached. The penetration of a shell two millimetres thick can take up to eight hours and experiments have shown that shells up to five millimetres thick can be breached. Drills may also inject a toxin into their prey, either by using the radula or a toothed salivary papilla.

Cone shells are more direct and swift in their predation and have advanced the radula and chemical attack one step further. All cone shells have a highly manoeuvrable proboscis and when fully projected a single barbed radula tooth is released from the radula sac into the buccal cavity (fig.96). The radula tooth can then be thrust like a harpoon into its prey and the immobilising neurotoxin within the tooth is released. The neurotoxin of some southern Pacific species of cone shell may be highly toxic to humans. The most dangerous species are the Geographer Cone *Conus geographus* (plate 6) and the Tulip Cone *Conus tulipa.* The former species has caused several deaths; in one case death occurred within four hours.

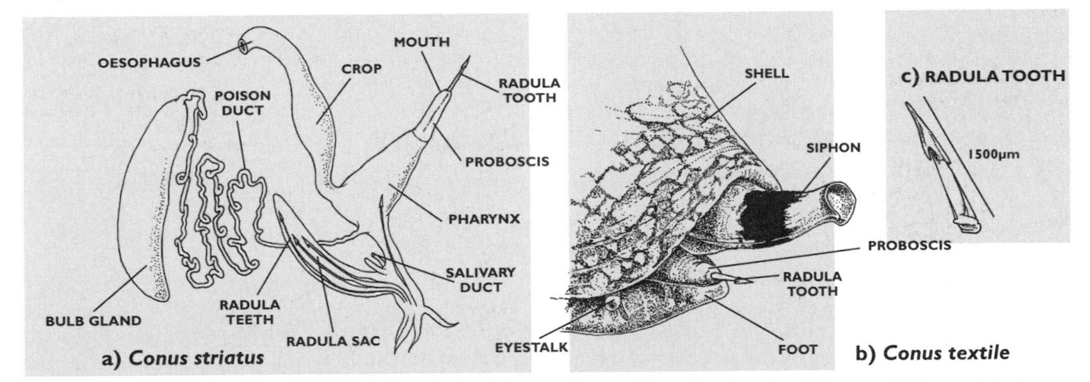

Figure 96. Cone shell feeding apparatus. a) anatomy of the poison gland and radula sac of *Conus striatus*. b) external view of a cone shell *Conus textile* with everted proboscis. c) radula tooth.

The Giant Triton *Charonia tritonis* (plate 6) is one of the few animals known to feed on the Crown-of-thorns Starfish *Acanthaster planci* (plate 6). It was once thought that the collecting of the Giant Triton kept the Crown-of-thorns in check thus preventing extensive damage to the coral reefs. Recent research indicates this is a false assumption; tritons feed on several echinoderm species and have little influence on the total starfish population.

The Tun Shell *Tonna perdix* (plate 7) is a large gastropod found on coral reefs which engulfs prey as large as sea cucumbers. To devour its prey the Tun emerges almost completely out of its shell and only when the mollusc has gradually broken down its victim into small pieces can it retreat back into it. This may take many hours and therefore, if the animal is threatened during feeding, it can regurgitate its prey and withdraw into its shell.

During their evolution, cephalopods have partly or entirely lost their ancestral molluscan shell, and in doing so they have become highly mobile predatory invertebrates. It is sometimes overlooked that they belong to the same group as gastropods and bivalves, as they have a different body form, a higher 'intelligence' and greater speed of movement. Their prey, unusually for molluscs, is located through the use of their highly developed eyes. The mobile tentacles are used to capture the prey with the aid of cup-shaped horny adhesive discs (the suckers). Squid and cuttlefish have ten appendages arranged in five pairs. Eight are shorter and heavier and are normally referred to as arms, while the longer, more mobile fifth pair are called tentacles and have suckers only on their spatulate ends. A few cephalopods do not possess suckers; some squid species have curved tentacular hooks and the nautilus has transverse ridges on the inner sides of its 38 arms. The octopuses, of course, have only eight arms with stalkless suckers.

Squid feed mostly on bony fishes (teleosts) and accelerate towards their prey using water jet propulsion. Two long tentacles are shot out to trap the fish. Once captured the prey is brought within reach of the buccal cavity and its powerful beak-like jaw (fig.97). Large pieces of tissue are torn off and the radula, which acts like a tongue, pulls the flesh into the buccal cavity where it is swallowed. The 'beaks' are often found in the stomachs of squid-eating cetaceans, including the Sperm Whale *Physeter catodon,* Risso's Dolphin *Grampus griseus* and the beaked whales. The pelagic squid *Loligo* spp. (plate 1) has been seen to dart into schools of young mackerel, grab a fish and then quickly bite a piece off behind the head or remove the head completely. It can then devour the fish at its leisure, leaving only the gut and tail which it discards.

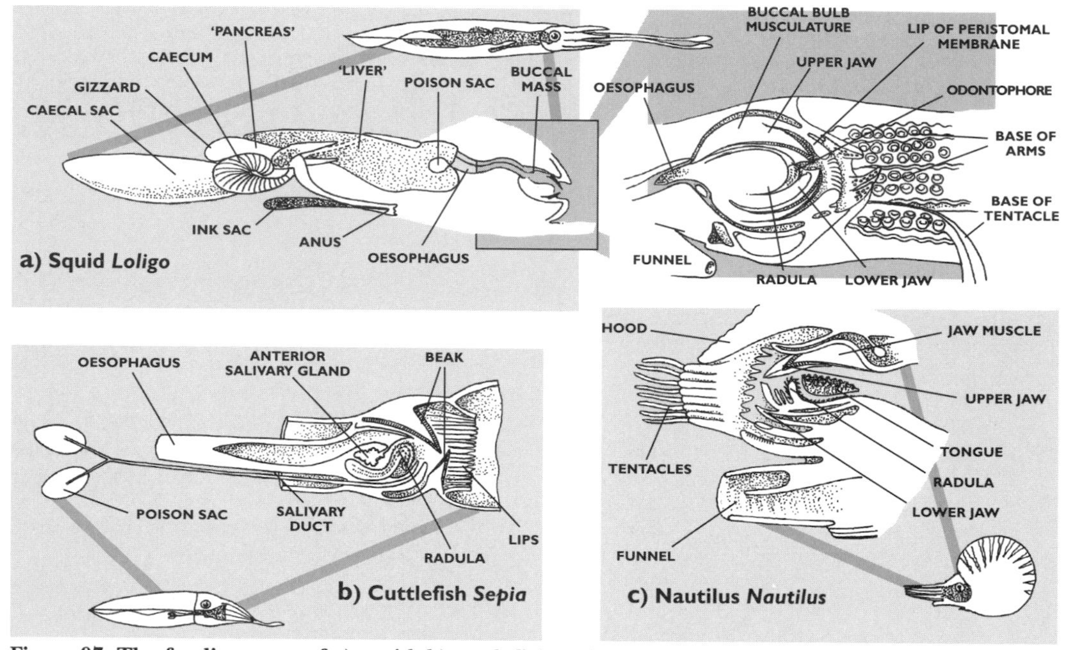

Figure 97. The feeding parts of a) squid, b) cuttlefish and c) nautilus.

Cuttlefish (Sepioidea, plates 4 & 7) use stealth and often wait half buried in the sand to catch their prey. Like squid they can rapidly change colour and do so to camouflage themselves. Most of their prey are bottom dwelling invertebrates, such as shrimps and crabs. The cuttlefish uses its two uppermost arms to lure its prey. When within easy reach, the two long mobile tentacles dart out quickly to engulf it. With the 'excitement' of capture the cuttlefish may completely change colour to a monochrome banding pattern. It tears its victim apart with its horny beak and swallows the smaller pieces of flesh.

Octopuses (plates 2, 4 & 6) live either in cairns made of rubble or in caves. Often they leave these dens to search for food on the sea bed, but at other times they wait in their lairs to seize prey as it passes by. Clams, gastropods and fish are all seized by the tentacles. Crabs are their favoured prey. Unlike cuttlefish, octopuses can also feed by day though this depends on the species. They usually leap upon less mobile prey engulfing it in their arm webs (a membrane between the basal ends of the arms). In contrast to other cephalopods, octopuses inject the prey with a toxin administered by the jaws, or sometimes by the salivary glands which flood the engulfing arm web with toxin. Once immobilised, the prey is then flushed with a variety of proteolytic and other enzymes. The partly digested flesh is then ingested and the remains are left discarded on the bottom.

The Chambered Nautilus *Nautilus pompilius* (plate 1) is a scavenger and predator which roams the bottom in search of decapod crustaceans using its sensitive extended tentacles. When the prey is caught it uses similar mouthparts to that of the squid to eat its prey.

Decapod crustaceans, the common prey of benthic cephalopods, are themselves predators and scavengers. All have ornate mouths with six pairs of biramous feeding appendages (fig.98). Food is caught or picked up with the chelipeds (large, manipulative thoracic appendages) and passed forward to the maxillipeds. These then push the food forwards between the other mouthparts to the mouth. Simply put, these mouthparts consist of the mandible that holds the portion of food while the maxillae and maxillipeds

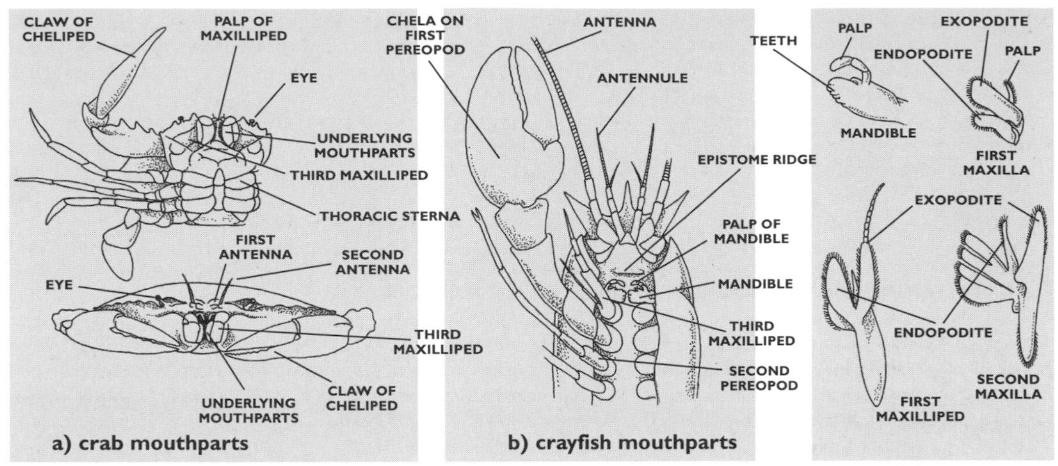

Figure 98. The external mouthparts of two decapod crustaceans, a) a crab and b) a crayfish.

tear it apart before swallowing. Most decapod crustaceans remain well hidden, venturing out only to seize prey or to scavenge food nearby. Only larger crabs and lobsters expose themselves to view and even then only at night.

The size and shape of a decapod crustacean's cheliped is often indicative of its feeding habit. Those that have dimorphic chelipeds (normally a heavier and larger right claw) are more likely to prey on gastropods and bivalves. The larger cheliped is used to crush the shell and the more slender one for cutting up the exposed animal and passing it to the mouth. Less dimorphic chelipeds may indicate a more scavenging lifestyle.

Chelipeds are not always used in feeding and the decapods may use slightly modified maxillipeds for filter feeding. The porcelain crabs are predominantly filter feeders and the Spotted Porcelain Crab *Neopetrolisthes maculatus* (plate 7) can be found on reefs using the coral polyps and sea anemones for protection. Another small crab, the mole crab *Emerita* spp., also filter feeds by using its setose antennae to filter the water. The chelipeds wipe them clean when they are 'full' and then pass the food to the mouthparts.

As a group, the crustaceans are avid scavengers and will scavenge maimed or dead of their own kind. They can track down potential prey through their long sensitive antennae.

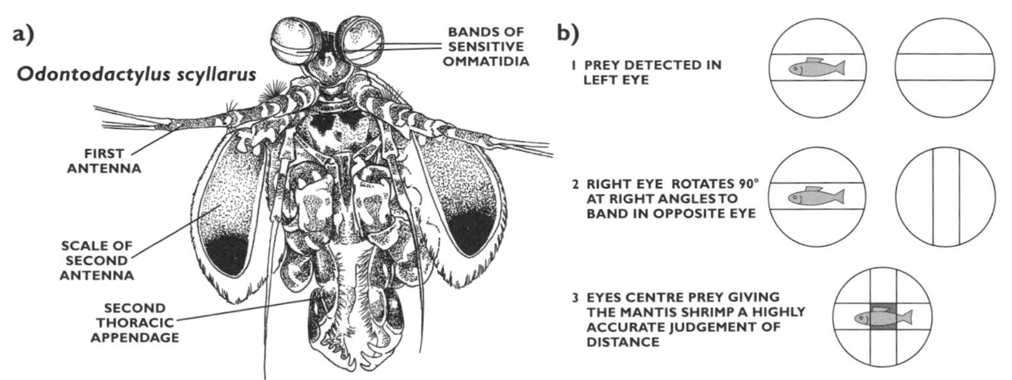

Figure 99. a) A mantis shrimp's eyes and b) the mechanism of targeting its prey.

The mantis shrimps, Stomatopoda, are specialised predators of fish, decapods and molluscs. The mantis shrimp *Odontodactylus scyllarus* (plate 8), is a typical example. Its well developed, stalked compound eyes (fig.99a) are the most highly evolved eyes among the crustaceans and rival the vision of many insects. They are able to detect the smallest of movements and are highly accurate in judging distance. Each eye can independently rotate and has a linear band of sensitive colour sensing ommatidia, light sensing organs that make up part of the compound eye (see p.158). When prey is located by one of the roving eyes, it is fixed in the linear band of that eye. The other eye is then centred on the prey and its linear band is rotated at right angles to the band position in the opposite eye (fig.99b). Through this mechanism the animal is able to calculate the distance and position of its prey more accurately than most marine invertebrates. It strikes at the prey, either crushing it with its large second thoracic limb, or spearing softer-bodied prey with sharp raptorial appendages (fig.100).

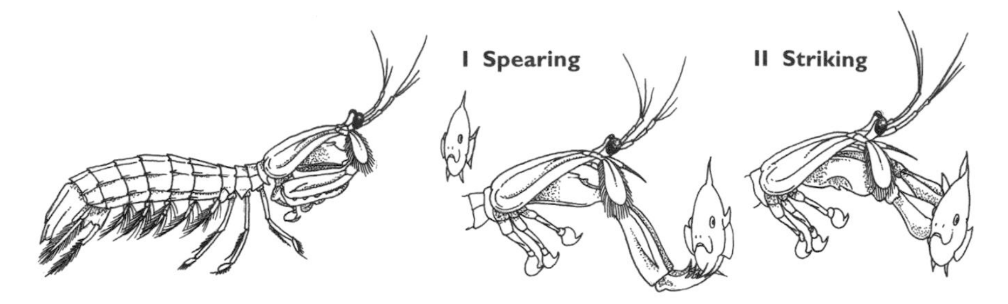

Figure 100. Feeding method in a mantis shrimp capturing soft-bodied prey.

Other crustaceans use their front appendages in different ways to capture their prey. For instance, the snapping (pistol) shrimps, Alphaeidae, can use the force of sound waves to capture their prey. They live in naturally occurring holes and crevices in rocks and coral rubble or in burrows. All have greatly enlarged right chelipeds the base of which possesses a large tuberculate process that fits into a socket (fig.101). The movable finger is 'cocked' by two special adhesive discs that engage when the finger is fully raised. Only when the adductor muscle has overcome this adhesive force will the finger snap shut which it does with such force when the approaching prey is within a certain distance that a popping sound is generated. At close quarters this can be sufficient to stun its prey.

Another shrimp, the Painted Dancing Shrimp *Hymenocera picta* (plate 6), also has unusual feeding habits. These are secretive shrimps that live in coral reefs, usually in pairs. They feed at night on starfish, especially *Nardoa galathea* and *Fromia elegans.* The starfish are found either on the reef itself or in the rubble and sand of the lagoons. Once the starfish has been captured it is taken back to the shrimp's lair and turned on its back. Here the shrimp uses its strong, sharp, feeding chelipeds to cut up the echinoderm arm by arm, leaving the disc until last. This gruesome way of feeding keeps the starfish alive for several days and the shrimp maintains a fresh larder of food to eat. Gnathophyllid shrimps are also known to prey on the Crown-of-thorns Starfish but, like the Giant Triton, they are too sparsely represented on the reef to make an impact on starfish numbers. This starfish is also attacked by another decapod. The xanthid crabs *Trapezia* sp.(plate 6), are found on most Indo-Pacific coral reefs living between the coral polyps. Many reports suggest that these crabs, in defending their territory against intruders, nip off the tube feet of the starfish thereby causing it to withdraw.

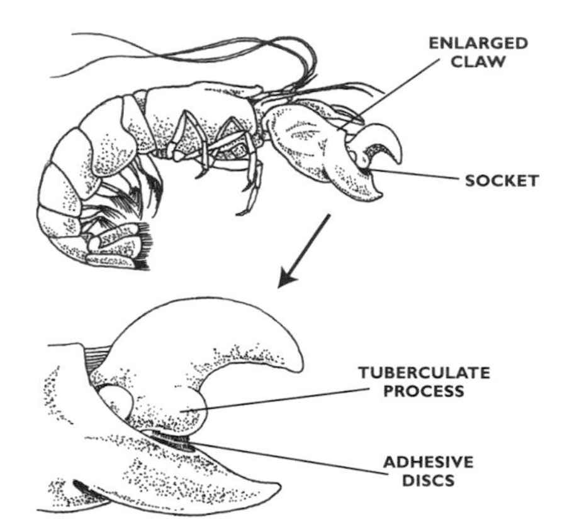

Figure 101. The cheliped of a snapping shrimp.

Many starfish and brittle-stars, Asteroidea and Ophiuroidea, are voracious carnivores and generally feed in one of two ways: engulfment (many of the brittle-stars) or extra-oral digestion. The Crown-of-thorns Starfish is a typical extra-oral feeder. On finding a suitable piece of coral it spreads its everted stomach over the prey. Some hours later, when the digestive enzymes have dissolved the coral tissue and the resultant 'soup' has been absorbed, the stomach is withdrawn leaving the white coral skeleton. In this way the starfish can consume an area as great as its own disc (up to 30cm in diameter) in one day. In one year an individual can destroy five square metres of coral reef and in a plague year they may number many thousands of individuals in a single locality alone.

Primitive starfish such as *Astropecten,* which cannot evert their stomachs and have suckerless feet, swallow their prey whole and digestion takes place inside their stomachs. Starfish are notorious predators on oyster beds. Species such as the Spiny Starfish *Marthasterias glacialis* and the Giant Sea Star *Pisaster giganteus* (plate 2) are voracious gastropod and bivalve hunters, and will also take other marine invertebrates too slow to avoid confrontation. Those starfish that feed exclusively on bivalves pull the valves apart, the strong suckers adhering to the valves to give a firm grip. Eventually the valves of the shell are pulled wide enough apart for the starfish to evert its stomach through the gap (sometimes as small as 0.1mm). The gap increases as digestion of the adductor muscles occurs and eventually the contents of the shell are absorbed. The stomach is then withdrawn inside the starfish. The time taken to digest the mollusc varies; for example, one species of Japanese starfish *Asterias* sp. takes between two-and-a-half and eight hours to tackle its prey.

The Bat Star *Patiria miniata* (plate 2) is an omnivore which spreads its stomach over the seabed, digesting all types of organic matter it finds. Other starfish are cannibalistic like the European Common Sunstar

Crossaster papposus (plate 4); it feeds mainly on other starfish including smaller Common Sunstars.

In this section only some of the feeding techniques of invertebrates have been examined. Each species has developed its own variation of a particular feeding method. Due to the multitude of predators (including vertebrates) marine invertebrate species have developed highly successful means of escape.

DEFENCE AND CAMOUFLAGE

In coral seas, because many invertebrates are toxic, their brilliance of colours and exquisite patterning often serve as warning devices. The flabellid nudibranchs (plate 7) carry undischarged nematocysts in their brightly coloured cerata. Nematocysts are also used by their original owners as a means of defence and some species of cnidarians can cause fatalities, even to large animals. Such species are normally widely respected and one common example is the Portuguese Man-of-war *Physalia physalis* (plate 1). This siphonophore is circumglobal in its distribution and is easily recognised by its iridescent purple float. However, some of the most toxic cnidarians are not so brightly coloured and one group, the box jellyfish (Cubozoa), typically have transparent cube-shaped bells less than 15cm across. Each corner of the jellyfish possesses a cluster of long stinging tentacles. One, *Chironex fleckeri* (plate 7), the largest of its class, has tentacles up to three metres in length. This species has been responsible for the deaths of many people and it uses its feeding mechanism for defence. Sessile cnidarians may also have nematocysts capable of wounding larger animals and humans. Divers testify to the pain of brushing against stinging corals like the White Stinging Hydroid *Lytocarpus philippinus* of the Indo-Pacific. This hydroid should be avoided by divers as should the fire corals, such as *Millepora tenera* (plate 5), and Fire Anemone *Actinodendron plumosum*, which looks like a miniature fir tree.

Cold temperate regions have far fewer toxin-producing invertebrates although the flabellid nudibranchs are common. One typical example is the Common Grey Sea Slug *Aeolidia papillosa* (plate 4) which is widely distributed throughout colder northern seas. Toxic jellyfish such as the Lion's Mane Jellyfish *Cyanea capillata* (plate 1) are seen by many northern seafarers, as is the smaller jellyfish, *Pelagia noctiluca* (plate 1). Both of these scyphozoans can inflict painful stings to unprotected skin and should be avoided when seen in the water (they can even inflict stings when stranded). In the European waters of the North Sea a particular bryozoan sometimes causes injury to fishermen. *Alcyonidium* forms jelly-like colonies up to 20cm long and these are often caught up in the nets of the trawlers. Contact with these colonies, or even ropes that have touched them, can produce painful blistering, leading in some cases to dermatitis. In the North Sea area this is known as Dogger Bank Itch. It is produced by the bryozoan's defence mechanism and deters browsing predators such as sea urchins, gastropod molluscs and decapod crustaceans.

Irritations and rashes may also be brought on by handling some marine polychaete worms, such as the larger ragworms *Nereis* spp. The tropical fireworms are notorious for causing skin irritations. One species, *Eurythoe complanata* (plate 7), has spines (chaetae) that are so sharp that they can penetrate through the gloves of divers. Each of these chaetae has its own venom gland that injects the toxin directly into the pierced skin. The embedded spine may also break off causing further local irritation.

The examples discussed so far have been of passive defensive mechanisms. Echinoderms are able to defend themselves with greater vigour. Sea urchins bear minute pincer-like pedicellariae on their dorsal surface, some of which secrete toxins. They are normally used to kill and remove organisms that may clog up the skin and spines of the urchin. In some cases the toxic substances delivered by the pedicellariae may also defend it against larger predators. Flower Urchins *Toxopneustes pileolus*

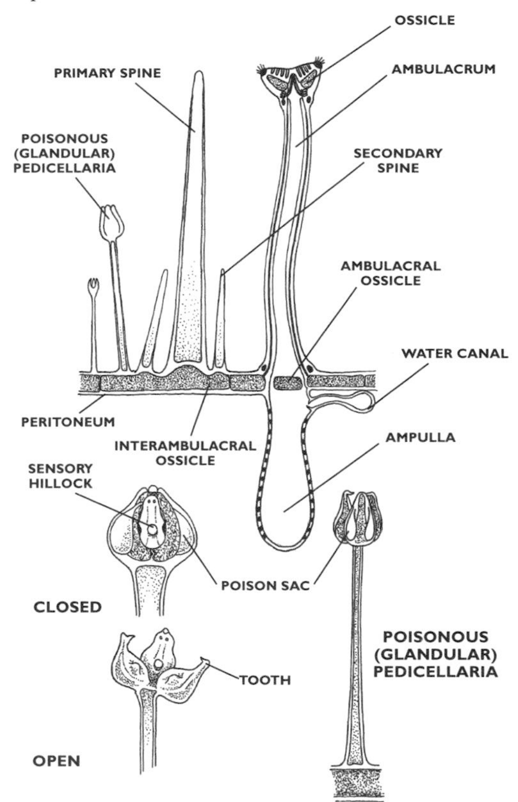

Figure 102. Venomous pedicellariae and spines of a sea urchin.

(plate 7) have been reported to have caused the deaths of three divers in Japan. However these are isolated cases and careful handling of smaller species has no adverse effects. Larger specimens from the SW Pacific are known to contain large venomous pedicellariae (fig.102) that may be harmful to touch. The pedicellariae are not the only means of defence in sea urchins as the spines that cover the dorsal surfaces not only provide a very effective physical barrier but are in some cases poisonous. Savigny's Sea Urchin *Diadema savignyi* (plate 6), and related species, have hollow spines filled with a toxin produced at the articulated base. These toxic spines can then be directed towards a potential threat which is detected by a shadow passing over light sensitive cells on the surface of its shell.

Molluscs also have their share of toxic defence mechanisms. Two well known examples are the cone shells and the octopuses. The toxin of the Geographer Cone Shell *Conus geographus* (plate 6), a fish feeder, often causes some form of paralysis in its prey. It is believed that the neurotoxin used for stunning prey affects humans, not because there is an excessive production of toxin, but because humans are particularly susceptible to cone shell toxins. Highly toxic cone shells have caused death in 20% of those cases where the victim had been 'stabbed', a figure that rivals fatalities from some of the deadlier snakes.

All octopuses inject their prey with poison, and this poison can be used for defence purposes as well as for attack. The most toxic octopod is the Blue-ringed Octopus *Hapalochlaena lunulata* (plate 6) of Australia but, as is usual with defensive strategies, the octopus much prefers to escape rather than defend itself.

Many marine invertebrates use decoys or diversions as part of their escape tactics. Some sea cucumbers have lavish, and physiologically costly, mechanisms. When sufficiently disturbed the sea cucumber can eject a mass of sticky white threads (Cuvierian tubules), from the anus (see plate 7, Leopard Sea Cucumber *Bohadschia argus*). Some Cuvierian tubules are adhesive while others liberate a toxin called holothurin. Extracted holothurin is used by Pacific islanders as a means of narcotising fish in enclosed tide pools. Holothurin is present in many sea cucumbers, not only those that eject tubules, and in all cases the organic toxin is unstable when released and only remains effective for a short time. The Cuvierian tubules lie between the branches of the respiratory tree usually attached to the base of the left-hand one (fig.103). When threatened the anus is aimed at the attacker and the body wall rapidly contracts, rupturing the cloaca and shooting out the tubules. During expulsion the tubules are greatly lengthened by water forced into their hollow central region (the lumen). When these thin elongate tubules finally break free they can render small crabs and lobsters immobile, and the attackers are often left to die while the sea cucumber crawls away. Some genera can expel one or both respiratory trees, the digestive tract and gonads all at once. This is called evisceration and is a common sea cucumber defence mechanism. One genus, *Thyone*, even ejects its tentacles and pharynx at the anterior end. However, this may be a seasonal occurrence and is not normally associated with defence. All three processes are followed by a regeneration period when new organs grow from the remnants of the eviscerated tissues.

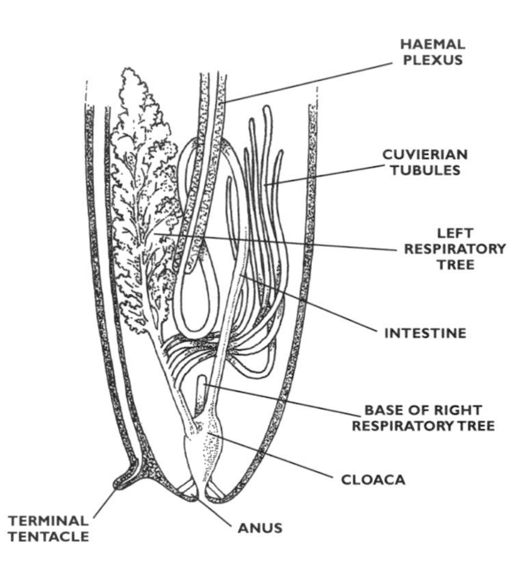

Figure 103. Cuvierian tubules in a sea cucumber.

The loss of organs or appendages for escape purposes occurs frequently among marine invertebrates. Starfish may be found washed up on the beach with arms missing, or with small stubs where regeneration has started at the base of severed arms. Crabs even have a weak point on their chelipeds where a natural break can occur if the need arises. This process, although reasonably costly in terms of regeneration, is less costly in terms of the invertebrate's life. In higher animals the cost of limb replacement is much greater and therefore few vertebrates exhibit this method of defence.

If an animal cannot be seen by a predator no elaborate or physiologically costly defence mechanisms are needed. Cryptic coloration or camouflage is one of the most important defensive strategies. Colours that blend with the background, disruptive patterning and countershading that breaks up the animal's outline all help to make a motionless animal 'disappear'. Many invertebrates can change colour, and those that cannot 'decorate' themselves to blend with the environment.

The latter behaviour is displayed by the decorator crabs (e.g. *Camposcia retusa,* plate 7). These small crabs cover themselves with fragments of animals, such as bryozoans and sponges, or debris to mask their outline. If the decoration becomes inappropriate the crab will change it to blend into the background. The masking material is held in place by hundreds of tiny hooks (chaetae) that cover the surface of the exoskeleton.

Many spider crabs (which include the decorator crabs) exhibit bizarre colours and shapes to disguise themselves. Should the animal need to move away from its matching background it may expose itself to attack. However, some invertebrates can react quickly to such a change and instantaneously alter their cryptic coloration.

Cephalopods are masters of colour, pattern and texture changes. To facilitate fast changes of colour the pigment cells (chromatophores) are under nervous control. The colour variation is controlled by the size of the cell. Expansion of the chromatophores is caused by the contraction of the small radial muscles, drawing the cell into a large flat plate and making it more visible. To reduce the size of the cell these muscles are relaxed and the elastic cell wall returns to its smaller 'original' size. These chromatophores can be of different colours with similar coloured cells often grouped together. Coloration of the chromatophores is enhanced by reflector cells located beneath the pigment cells and chromatophore filters between the layers of cells. The Common Cuttlefish *Sepia officinalis* (plate 4) can exhibit complex patterning, sometimes mirroring its background of sand, gravel or rock, or bold patterns to reflect its behavioural 'moods'. Atlantic Squid *Loligo vulgaris* are generally pale in colour but darken when disturbed (see plate 4).

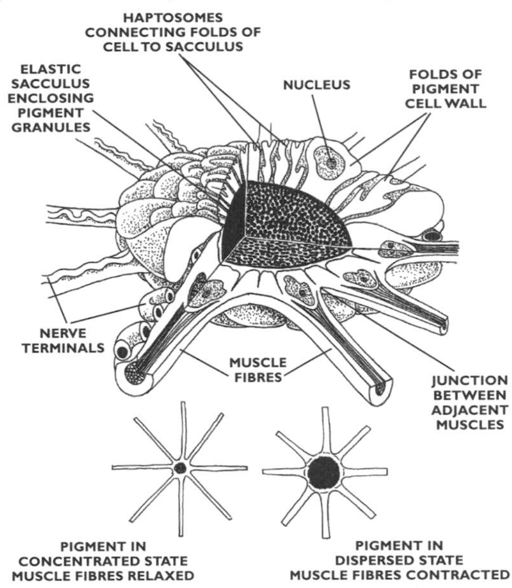

Figure 104. The chromatophore of a squid.

Cephalopods have two other methods of evading attack, bioluminescence and ink ejection. Luminescent photophores are arranged in various patterns, according to the species, to break up the outline of the squid. Some squid use symbiotic bacteria, while others produce glandular bioluminescent secretions, but in both cases muscular control opens and closes an 'iris' in the photophore to vary the light emitted.

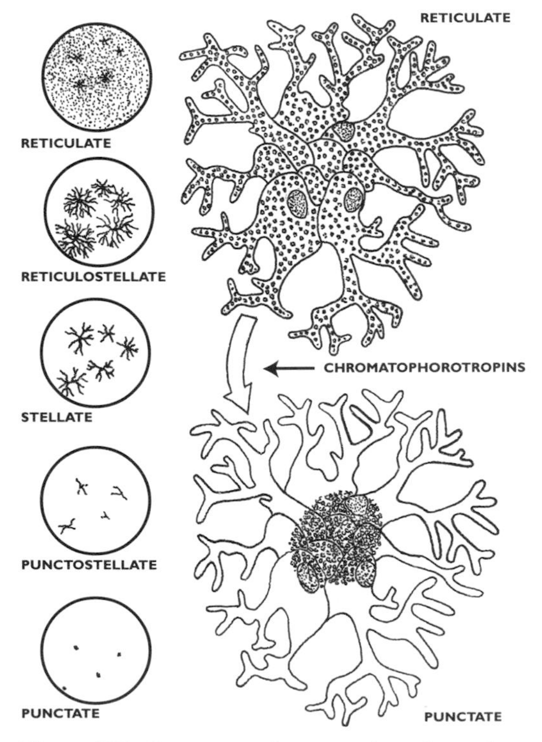

Figure 105. Crustacean chromatophore 'organ'.

Cephalopod ink (once used as ink in pens) is a brown or black fluid which contains a high concentration of melanin. The ink is stored in a large ink sac that opens into the rectum, behind the anus. A cloud of ink is released through the anus creating a diversion which may also act as a repellent to the predator.

Crustaceans are also known to change colour but as this is under hormonal control the change is often slow. The chromatophore in this case is a multibranched non-contractile pigment cell that is in close association with other chromatophores forming a multicellular 'organ' (fig.105). When the animal is in its pale phase (punctate state) granules of pigment are concentrated in the centre of the cell. If a hormonal change is induced the granules disperse within the chromatophore (stellate state) and the animal darkens. Red, yellow and blue pigments are carotene derivatives and the red coloration in boiled crabs, lobsters and shrimps comes from one of these carotene pigments called astaxanthin. In crustaceans, especially shrimps, the pigment cell of the chromatophore may contain up to four pigments (a polychromatic cell). The hormones that control this colour change are released from glands below the eye stalk and other parts of the central nervous system. It has been discovered that each pigment has its own antagonistic set of hormones, one to produce the punctate state and the other to bring about the more vivid stellate state. Many crabs blanch (lighten) or darken their coloration, but some shrimps, such as *Palaemonetes*, are also known to vary their coloration.

Gastropods change colour at an even slower rate than crustaceans. The Sea Hare *Aplysia punctata* (plate 3) changes its colour according to its diet. While feeding on green algae this nudibranch becomes green in hue; red algae turn it ruddy. This allows the animal to hide inconspicuously within the algae that it grazes upon. However, most gastropods have a predetermined genetic shell colour pattern and the cryptic coloration of buffs, beiges and browns in many species is used to blend into the background. One type of gastropod pigment, Tyrian purple, is produced by Mediterranean whelks *Murex* sp. and the Dogwhelk *Nucella lapillus*. The pigment is produced by a mucus gland in the mantle cavity. This pigment is a photochemical and changes from its original cream colour through green and finally to purple. It is thought to have an unpleasant taste, and the secretion does deter predators (in the past the Egyptians, Phoenicians and Romans used the pigment to dye cloth).

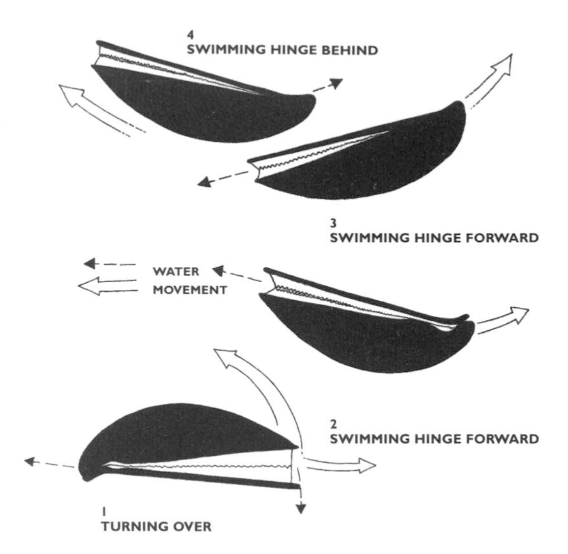

Figure 106. Swimming in a scallop.

Most molluscs, if other methods of camouflage and defence fail, can retreat into their shells. However, this defence does not always work and many molluscs have another mechanism.

Common Whelks can, when they are attacked by starfish, throw their whole visceral mass and mantle into a violent rocking motion, shaking off all but the most persistent predators. Abalones, such as the Ass's-ear Abalone *Haliotis asinina* (plate 8), use a similar tactic when touched by starfish. The gastropod raises its shell on its muscular foot and turns vigorously from side to side in a twisting motion. While this method can protect the abalone from its natural predators, Maori fishermen use it to prise abalones from the rocks. When diving the Maori fisherman carries a starfish with him which he places on the abalone shell. This makes it raise the shell, thus allowing the diver to lift it from the rock more easily. The abalone may also jeopardise others in the vicinity, for the release of a chemical messenger (an alarm substance) triggers other abalones to carry out the same defence behaviour. The Rough Keyhole Limpet *Diodora aspera* uses another technique. By extending its smooth mantle over the outside of its shell it effectively becomes too slippery for the starfish to grip.

Many scallops can actively swim away from danger (fig.106). If a sudden shadow passes over the animal, and is detected by a certain number of eyes at once, it can rapidly open and close its valves to propel itself

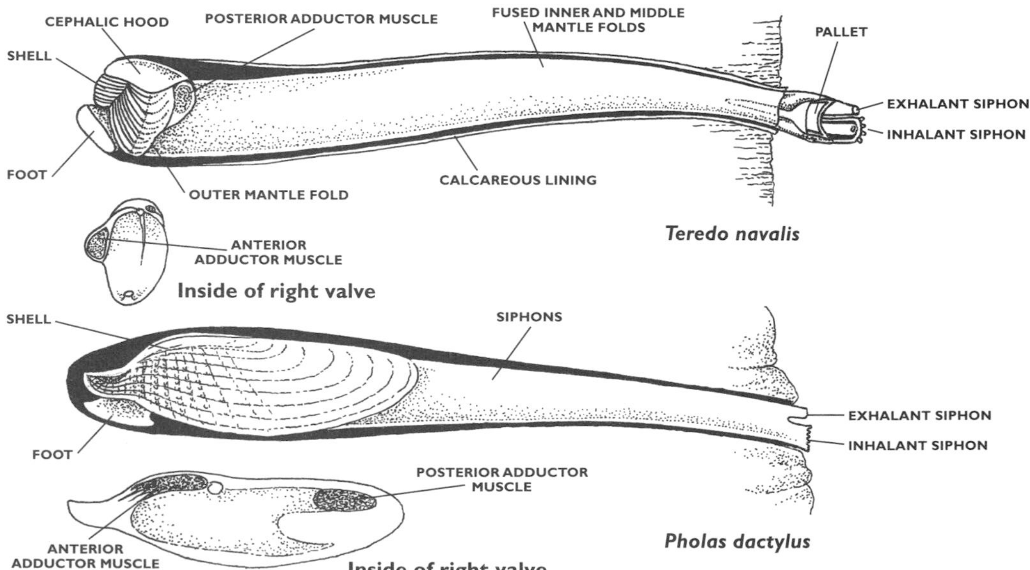

Figure 107. Boring bivalves, a shipworm (above) and a piddock (below) with adductor muscle arrangement.

through the water. The direction of movement is controlled by mantle flaps deflecting the water flow out of the shell.

One of the most effective ways of preventing predation is to disperse into an area that predators have difficulty in reaching. Bivalve molluscs are adept at this (figs.44 & 45). Cockles, razor shells, venus shells, tellins and many others bury themselves in soft substrates such as sand, silt and mud. Shipworms and piddocks go one step further and grind themselves into hard surfaces such as wood and rocks (fig.107). In these hard substrates they find near perfect shelter. These animals cause serious problems to marine structures, especially those made of wood; but they may have ecological importance by helping to decompose seaborne wood which could otherwise choke marine habitats.

Crustaceans such as crabs and amphipods also make use of burrowing tactics. When the tide ebbs crabs like the Soldier Crab *Mictyris longicarpus* (fig.108) which inhabit muddy and sandy bays in the tropical Pacific, burrow into the sand with a strange corkscrew action. Fiddler crabs do likewise, and only come out to feed at low tide, when the beach seethes with these diminutive crustaceans. Tropical species only feed at low tide during the daytime, but temperate species in the summer also feed at low tide during the night and during the cold winter months remain dormant.

Tube worms are some of the most obvious burrowers. Their tubes are often permanent and the creature's time is taken up with feeding and maintaining the tube, which can be very elaborate. Whatever coastline may be visited in the world, the casts, tubes, holes or signs of these filter and deposit feeders are always present.

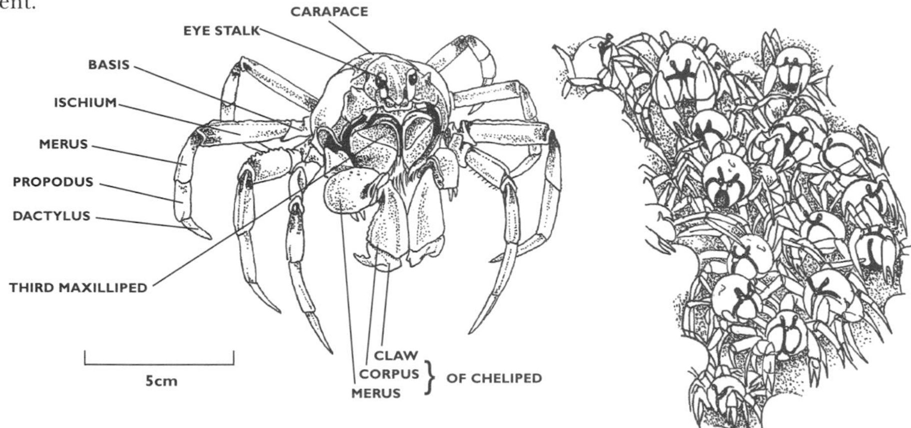

Figure 108. The Soldier Crab *Mictyris longicarpus*.

PARASITISM, COMMENSALISM AND SYMBIOSIS

Parasitism, commensalism and symbiosis are general terms that describe animal and plant associations. This book uses the term 'parasitism' to describe animals that benefit at the expense of the host, an association that often leads to severe injury or the eventual death of the host. Commensalism, as described by van Beneden in 1876, is interpreted as one organism, the commensal, that benefits nutritionally from another, the host, without harming it. The term symbiosis (or mutualism) can be described as an association between different species that is mutually beneficial to both the host and the symbiont. In many cases neither species can survive without the other.

Marine invertebrate parasites are common and often bizarre in lifestyle and form. These parasites do not incur many of the problems associated with terrestrial invertebrate parasites. The problems of protection against desiccation before they reach their host are minimal, as they are surrounded by a near isotonic medium. The larval parasite can be carried long distances in the plankton enabling it to infect a new host and achieve maximum dispersal. The difference between these two environments can be demonstrated by comparing two similar parasitic groups. The flukes have two classes, the Monogenea and the Trematoda. The parasitic Monogenea are mainly marine and have one generation per life cycle, the egg developing directly into an adult. However, the freshwater Trematoda (the largest group of parasitic flatworms) have complex life cycles involving two to four hosts and as many larval morphs with indirect development. Many of these flukes use an aquatic host to distribute the intermediate stages to the main host or further intermediate hosts. Many marine invertebrates and vertebrates are commonly infected with protozoan parasites, but because of their microscopic size they are only briefly mentioned here.

Larger parasitic invertebrates are well represented by one class of crustaceans. Barnacles (Cirripedia) have numerous parasitic species found mainly in the orders Ascothoracica and Rhizocephala (the latter group is wholly parasitic). The former group are parasitic mostly on corals and echinoderms, while rhizocephalans generally parasitise decapod crustaceans.

These parasitic barnacles have lost all external resemblance to an adult free-living form and can only be identified as cirripeds from the horned nauplius larvae. *Sacculina carcini* is a typical example and is usually parasitic on the Shore Crab *Carcinus maenas.* It occurs as a growth on the underside of the crab (fig.109) and can be readily distinguished from the crab's own eggs by its smooth appearance. The adult parasite consists of a branching rootlet system (much like a fungal growth) which invades the tissues of the crab, eventually forming a reproductive sac, the external growth. It remains in the crab for three to four years and takes up so much of the crab's nutrients that the host is not able to moult or grow. Only when *Sacculina* dies and drops off will normal growth resume.

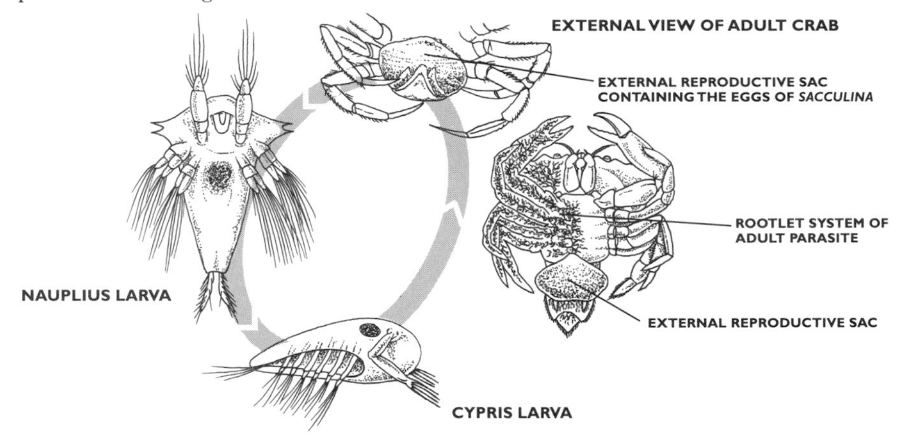

Figure 109. The parasitic barnacle *Sacculina carcini* on the Shore Crab *Carcinus maenas.*

A closely related group to the barnacles, the copepods, contains over 1,000 parasitic species. Most of the ectoparasitic copepods are found on fish hosts and have highly adapted mouthparts able to pierce or suck the host's tissue. Some species have a frontal gland that produces a button (or bulla) that attaches the parasite to the gill filaments of a teleost fish. One such parasite, *Lernaea branchialis,* lives on the gills of the Atlantic Cod *Gadus morhua.* The modified head and thorax consist of branched processes in and around the gill filaments of the fish and only the abdomen and egg sacs of the parasite protrude outside the operculum like a small worm. Other species are endoparasites, that is they live entirely within the body of the host. These forms are mostly parasitic on polychaete worms, echinoderms and bivalve molluscs. One genus of cyclopoid copepods, *Sapphirina,* occurs in areas where there are large drifts of salps (planktonic tunicates). The juvenile copepods invade the body cavity of the salp and then proceed to eat it from within (fig.110). Juveniles that develop into mature males leave the salp soon after the final moult. Females remain inside the salp husk until they are ready to mate and then only briefly leave the host to copulate with a male. After mating the female returns to brood the eggs in the husk. After hatching the minute larvae leave the salp to find and invade a new host.

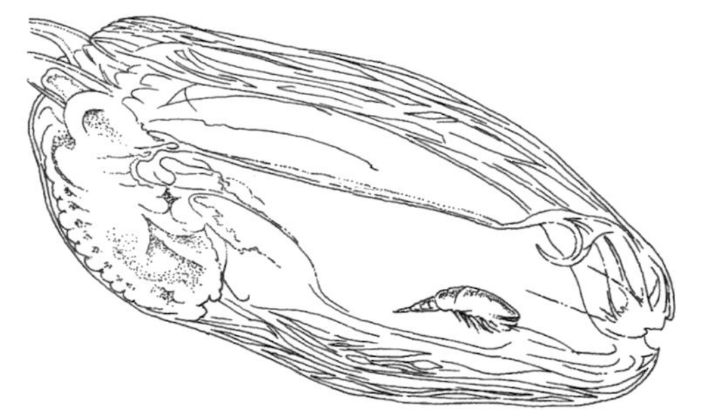

Figure 110. The copepod parasite *Sapphirina angusta* within its salp host. The copepod is about 5mm in length.

The sheltering of one animal within or on the body of another can be quite benign. This is called phoresis and the invader a phoretic organism. However, there are occasions when the transporting or sheltering host may be considered as being parasitised. For example, the pearlfish *Carapus acus* (similar to the Caribbean Pearlfish *Carapus bermudensis,* on plate 14), will break through the cloacal wall of its sea cucumber host to browse on the gonads. Most species of pearlfish do not harm their host, indeed the fish cannot mature until it finds a phoretic partner. The fish has become modified for this unusual lifestyle by losing its scales and pelvic fins so that it can swim backwards into the holothuroidean anus.

Other fish also benefit from marine invertebrate phoretic hosts. The Bluebottlefish *Nomeus gronovii* shelters in the Portuguese Man-of-war's tentacles (plate 1). The fish has some immunity to the toxic nematocysts, whereas its predators do not. Other driftfish have looser relationships with jellyfish. The

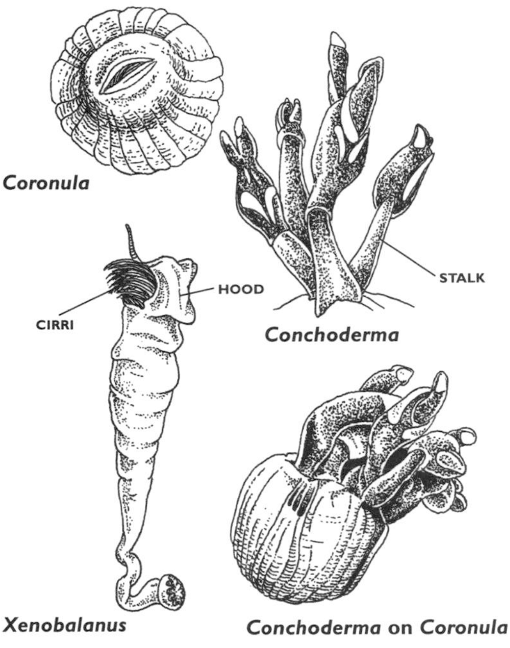

Figure 111. ***Coronula, Conchoderma*** **and** ***Xenobalanus,*** **three types of phoretic barnacles on whales.**

Lion's Mane Jellyfish *Cyanea capillata* (plate 1) is often seen with silvery fish darting amongst its potentially lethal tentacles. These are normally juveniles of pelagic fish species, but at least one species of shrimp and an unknown species of anglerfish also live in association with these pelagic cnidarians.

Barnacles not only exhibit parasitic behaviour. They also attach themselves to whales, providing another example of phoresis which may even border on commensalism. Three types of barnacles exhibit such lifestyles; the acorn barnacles, stalked barnacles and pseudostalked barnacles. Acorn barnacles are represented by the genera *Coronula* and *Cryptolepas,* the former often forming a close association with another barnacle, *Conchoderma,* which is a stalked barnacle. For most of the time *Conchoderma* fasten themselves onto *Coronula* (fig.111), although they can attach themselves directly onto the skin and teeth of whales. Barnacles found on whale skin appear not to harm the host and are often very visible. However, the barnacles themselves are often grazed on by whale lice. Pseudostalked barnacles such as *Xenobalanus* penetrate deeper into the whale's skin, without causing apparent injury.

Amongst commensal marine invertebrates there are numerous examples of smaller organisms sheltering on or within larger animals. Marine polychaetes often associate with hermit crabs, shrimps, clams and echinoderms. The ragworm *Nereis fucata,* lives in the upper whorls of whelk shells occupied by hermit crabs. When the hermit crab is feeding, the worm edges out under the shell lip and snatches pieces of food from the host.

Scale worms (Polynoidae) often live commensally in the burrows of other animals and even on the surface of the host. The scale worm, *Hesperonoë,* lives in the long deep burrows of the American burrowing shrimp *Calianassa californiensis.* Once a burrow has been found by the scale worm it will be defended against other scale worm intruders. The Club-spined Scale Worm, *Gastrolepidia clavigera* (fig.112), lives on the undersides of tropical sea cucumbers' mouths. As free-living specimens of this species have never been found it is assumed that these are host dependent, arriving as larvae and settling directly onto the host. One commensal echinoderm scale worm has even been found with up to one third of its body length inside the stomach of the European coastal starfish, *Astropecten irregularis.*

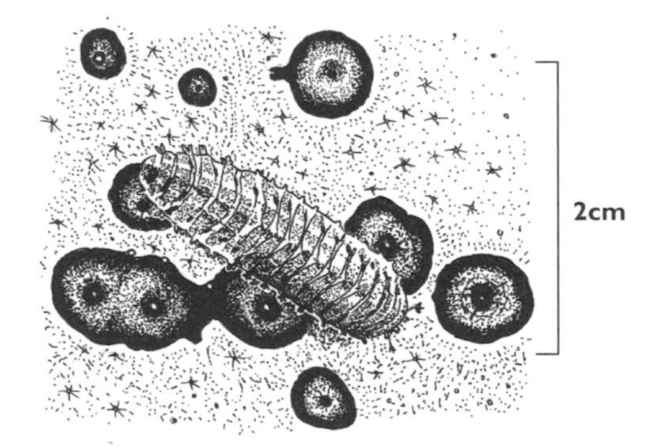

Figure 112. Club-spined Scale Worm ***Gastrolepidia clavigera,*** **a commensal on tropical sea cucumbers.**

Sponges, tunicates and bivalve molluscs also seem to attract their share of commensal creatures. Many species of shrimps and worms occur inside these sedentary animals. Sponges in particular play host to a wide variety of commensals. An interesting example is Venus's Flower-basket *Euplectella* sp. (fig.113). This beautiful siliceous hexactinellid sponge contains a young female and male shrimp, *Spongicola,* that enter through the osculum of *Euplectella.* As the shrimps mature and grow they eventually become too large to escape and therefore their entire lives are spent imprisoned within the confines of the sponge. Inside the sponge plankton is brought in by the water current produced by the sponge. The excess plankton not taken up by the host feeds the shrimps. The Japanese collect the skeletons of both the sponge and the trapped crustaceans, and give them away as wedding gifts to symbolise the eternal bond between the married couple. Other examples include the spider crab, *Chorilla,* and an isopod, *Aega,* which are also found in species of *Euplectella.* These commensals are probably a considerable nuisance as they can block the sponge's canal system. Sponges remove commensals either by the use of toxic chemicals or by actually eating the smaller larval stages.

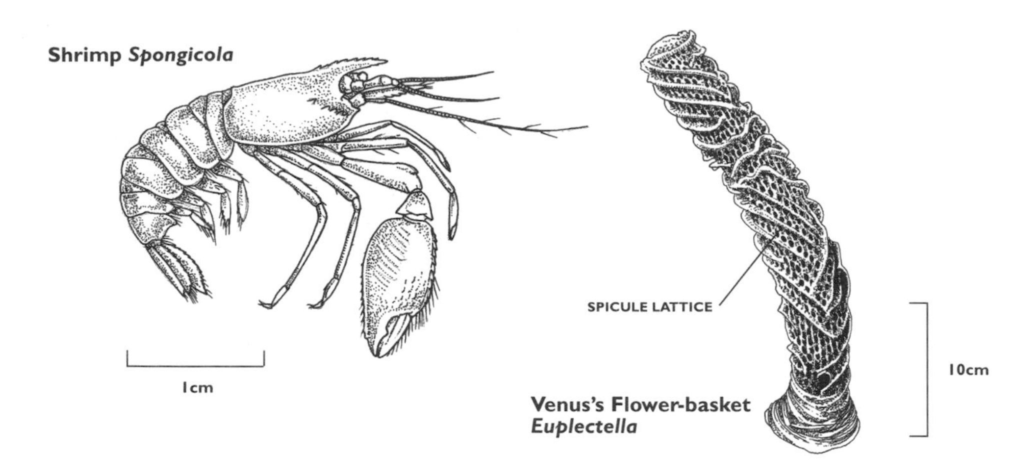

Figure 113. The shrimp *Spongicola* and its host sponge the Venus's Flower-basket *Euplectella*.

Tunicates have a similar range of commensals, including amphipods, copepods, nemertean worms and small shrimps. These commensals normally live in the pharynx or the atrial cavity of the host. A small, bright red commensal amphipod *Paraleucothoe novaehollandiae* lives in the pharynx of the Giant Sea Squirt *Pyura spinifera*, a large ascidian found on the coast and offshore reefs of Australia. The sea squirt may also carry a covering of pink commensal encrusting sponges over its body and stalk. Pearlfishes are known to inhabit the cavities of tunicates. One example of a pearlfish host is the Red-throated Ascidian *Herdmania momus*, which is found in most tropical and southern temperate coastal waters. In one marine survey a solitary tunicate of the genus *Ascidia* was found to contain two different species of bivalves, a pea crab, an amphipod and two species of copepod, all within its seven centimetre high body.

Pea crabs (Pinnotheridae) are more commonly found within the mantle cavities of bivalve molluscs. One such species is *Xanthasia murigera* which lives in the Giant Clam *Tridacna maxima* (plate 8). It is a particularly large commensal species which grows up to two centimetres across the carapace. Its size probably reflects the large size of the host. Only one male and one female are found inside each clam and canabalism of post-larvae prevents other pea crabs settling in. *Pinnotheres pisum*, a smaller pea crab, is associated with mussels (e.g. Northern Horse Mussel *Modiolus modiolus*, plate 2). The pea crab *Pinnotheres ostreum* which lives in the American Oyster *Crassostrea ostreum* can damage the oyster's gills and therefore could also be considered a parasite. Like all Pinnotheridae, this crab is considerably modified for its commensal existence. The female has a soft exoskeleton, whereas the male, who leaves the host in search of a mate, retains the chitinised cuticle. Pea crabs find a potential host by detecting substances in the water that are released by the host. Crabs then follow this 'scent trail' to their host and establish themselves within its body.

Hermit crabs carry their own 'homes' on their backs and some species live in symbiotic association with sea anemones. One such anemone is the 'Parasitic' Anemone *Calliactis parasitica* that forms a relationship with the Common Hermit Crab *Pagurus bernhardus* (plate 3). The crab perhaps acquires protection from potential predators which may avoid the stinging tentacles of the anemone. The anemone takes advantage of the crab's mobility and obtains scraps of food from the host while it is feeding. When the crab moves to a larger shell it will encourage the anemone to leave the old shell for the new one. To induce the anemone to move the hermit crab drums on the old shell and the anemone's base with its chelipeds. This causes the animal to relax its hold on the old shell so that the crab can carefully pick it up and transfer it to the new shell. *P. striatus* and *P. arrosar*, two other closely related hermit crabs, will forcibly place a 'Parasitic' Anemone onto their shells if they do not have one. The strangest of the hermit crab and sea anemone relationships occurs between the crab *Pagurus*

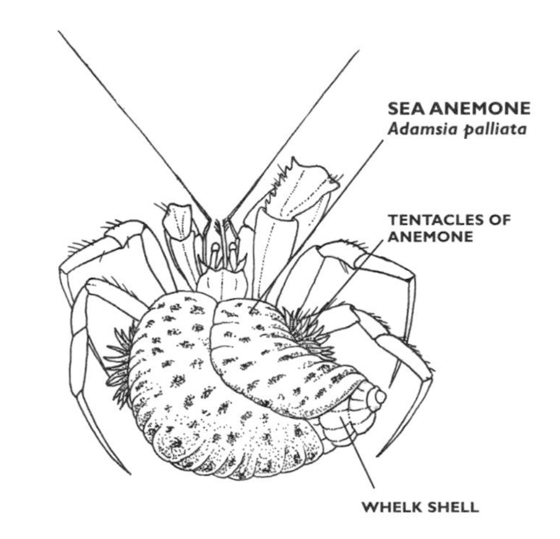

Figure 114. The symbiotic relationship between the hermit crab *Pagurus prideauxi* and the sea anemone *Adamsia palliata*.

prideauxi and the sea anemone *Adamsia palliata* (fig.114). This anemone envelopes the front parts of the crab and its shell which protects the exposed and vulnerable cuticle. The tentacles of the anemone lie under the shell and the crab where it can easily pick up any stray morsels of food. This relationship, as with many of the parasitic, commensal and symbiotic associations, is highly specific. *A. palliata* is never found on any other species of hermit crab. This association is so close that once beyond the juvenile stage the animals cannot exist without each other, a feature of all mutualistic relationships.

One group of teleosts, the anemone fishes or clownfishes of the tropics, are wholly reliant upon sea anemones. Some species of clownfishes choose only one species of anemone while others may colonise several different anemone species. Each pair of fish that occupies an anemone can become highly territorial, 'owning' a particular large anthozoan which they defend vigorously. One such clownfish, *Amphiprion akindynos*, lives symbiotically among the tentacles of large Indo-Pacific anemones, such as Ritter's Sea Anemone *Radianthus ritteri*, which it 'prefers' to all others. This giant sea anemone measures up to one metre across and is found in areas where there are strong currents. The clownfish that live within its barrage of nematocysts need to acclimatise to the anemone. They do this by swimming slowly in close proximity to it with a characteristic motion. At the same time the fish alters its mucous coating, raising the threshold of the nematocysts' discharge. This makes it possible for the fish to live among the mass of tentacles, affording it greater protection. However, it is not a quick process and may take anything up to an hour to acclimatise. The advantages of this relationship to the fish are obvious and the sea anemone also seems to benefit. The clownfish may protect the anthozoan from smaller predators like polychaete worms.

A specialised group of symbionts spend their entire lives cleaning other animals. Of this group at least six species of shrimp clean larger animals. The Californian Cleaner Shrimp *Hippolysmata californica* works in teams. The fish become docile at the approach of the crustaceans and their specific tactile behaviour. The shrimps remove encrusting animals, parasites and dead tissue which they selectively eat. More specialised cleaner shrimps advertise themselves through bold coloration. Two examples are the Pederson Shrimp *Periclimenes pedersoni* of the West Indies with its conspicuous white stripes and violet spots and Grabham's Cleaner Shrimp *Hippolysmata grabhami* with its brilliant white antennae and striped back (fig.115). These colours are accentuated by the waving motions of the antennae and body. When a 'customer' arrives at a cleaning station, an area established by the cleaning shrimps where fish come to be cleaned, the shrimp leaves its lair, normally a large sea anemone; unlike the Californian Cleaner Shrimp, these species are solitary. One of the best known Pacific cleaner shrimps is the Banded Coral Shrimp *Stenopus hispidus* (plate 6). These normally live in pairs, the male being much smaller, often hitching a ride on the female's back. They normally work at night and can be seen walking over sleeping fish, removing external parasites and eating them as they go. Like other cleaner shrimps they too have bold colours and long triple-branched antennae. This is an effective means of attracting potential 'customers' to their cleaning station.

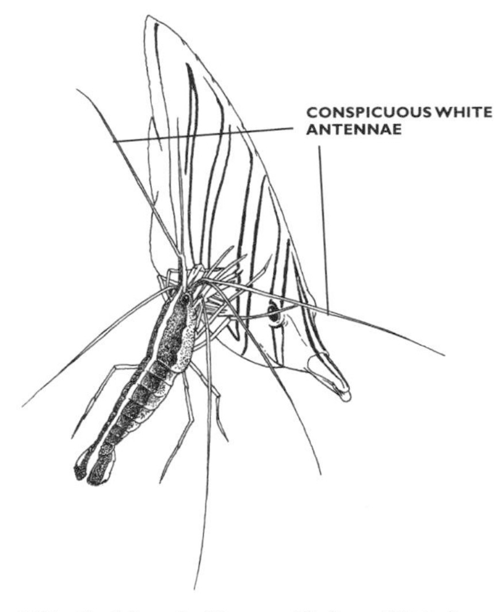

Figure 115. Grabham's Cleaner Shrimp *Hippolysmata grabhami* cleaning parasites from a fish's scales.

Some marine invertebrate species exhibit such close bonding that in some cases the host would die if the association did not exist. The symbiont, in return for the production of useful nutrients for its host, receives shelter and the removal of its waste products.

Algae are present in many symbiotic marine invertebrates. On the beaches of Brittany, France, a small green flatworm, *Convoluta roscoffensis,* has within its tissue the dinoflagellate alga *Amphidinium carterae.* At the juvenile stage the flatworm is entered by the alga and as the animal grows the gut degenerates until it can no longer feed itself by ingestion of food. The flatworm then changes its metabolism to digest its algal symbiont. After egg laying, the animal digests the algae at such a rate that they cannot regenerate quickly enough. The flatworm dies when all the algae have been eaten.

Sessile invertebrates also show varying degrees of dependence on algal symbionts. The Giant Clam has dinoflagellate algae (zooxanthellae) inside the tissue of its mantle lips and siphon. In these large bivalve molluscs the algae are 'farmed' in an unusual way. The single-celled algae are clustered in specialised cells of the clam's tissues that lie below organic lenses which focus light onto the algal cells. When the light is intense, around midday, a pigment can be moved across the lenses to protect the zooxanthellae. After a time the older zooxanthellae are digested by the clam's blood phagocytes, its kidney being adapted to rid the blood of the waste by-products. The clam is not wholly dependent on the algae but it does have a substantially reduced filter feeding mechanism and would, without the zooxanthellae, have difficulty in nourishing itself.

Some cnidarians also possess zooxanthellae within their tissues. They include mobile species like the Upside-down Jellyfish *Cassiopea andromeda* (plate 8) which is found in shallow lagoons, intertidal sand-banks or mudflats and mangroves around the Indo-Pacific. This species houses certain dinoflagellate algae in its mesogloea. With adequate light the animal has no need to rely on its normal scyphozoan feeding techniques and can live on the products of algal photosynthesis passing to the jellyfish's tissue. Other species of *Cassiopea* are common in the mangrove swamps of Florida, 'basking' upside-down on the bottom of the shallow waters. One species lives in the brackish lagoons of Truk, in the Caroline Islands of the Pacific, and has become so specialised that it no longer needs nematocysts to catch prey. Instead it follows the sun as it tracks across the lagoon to gain maximum benefit for its zooxanthellae.

Anthozoan corals are the largest group of symbiotic hosts. Over 60 genera of corals contain zooxanthellae within their gastrodermal cells and this includes nearly all reef building (hermatypic) corals. In some coral species the concentration of the algae can be so high that half of the nitrogenous proteins of the colony are produced from zooxanthellae proteins. The coral's nutritient needs are in part supplied by tentacular feeding on the plankton. However, the major part of its diet is supplied by the activity of its algal symbionts (food caught by the polyps is important in replenishing minerals, especially nitrates, necessary for algal growth). Some 94 - 98% of all the organic carbon these symbionts produce passes into the host cells, mostly in the form of glycerol with smaller amounts of glucose and alanine (an amino acid). The symbionts also facilitate the deposition of skeletal calcium carbonate. Hermatypic corals are able to deposit the limestone framework two to three times faster during the day than at night. As a result, hermatypic corals grow mainly during daylight hours. At night the deposition can be so slow that it barely keeps pace with the destruction caused by the erosion of the waves. This association between host and dinoflagellate algae is so intimate that the larvae of hermatypic corals also contain zooxanthellae.

On occasions environmental stress causes loss of zooxanthellae. Coral stress may be induced in several different ways, for example too much or too little light, low salinity, high temperatures or even disease. In these circumstances, the coral host dies.

SENSORY PERCEPTION

Cnidarians have a simple nervous system. In its basic form a characteristic 'nerve net' is found in the sedentary stage of hydrozoans. These nerve nets have bipolar neurons (two-process cells), or tripolar neurons (three-process cells) neurons in the sub-epidermal layer of the skin (fig.116).

Nerve nets can take the form of a syncytium (animal tissue with multiple nuclei inside a single cell membrane) or have synapses where nerve impulses pass across junctions between nerve cells on the release of a chemical transmitter (similar to higher animals). Synaptic junctions in the cnidarians are different from most others as nerve impulses can pass across the gap (synaptic cleft) in both directions. More advanced cnidarians such as the scyphozoans have a higher level of nervous organisation. The Moon Jelly *Aurelia aurita* (plate 1) possesses two nerve nets each with a different structure and function (fig.117).

The first is a diffuse nerve net of slow conducting multipolar neurons, which are nerve cells with many fine filament-like processes. The second is known as a through-conducting system and is considerably faster at carrying nerve impulses than the first system. This system comprises unidirectional bipolar neurons. The interrelationship between the two networks is shown in figure 117. The slower nerve net is well dispersed whereas the fast through-conducting system has fewer, more simple connections and pathways. Swimming cnidarians also have

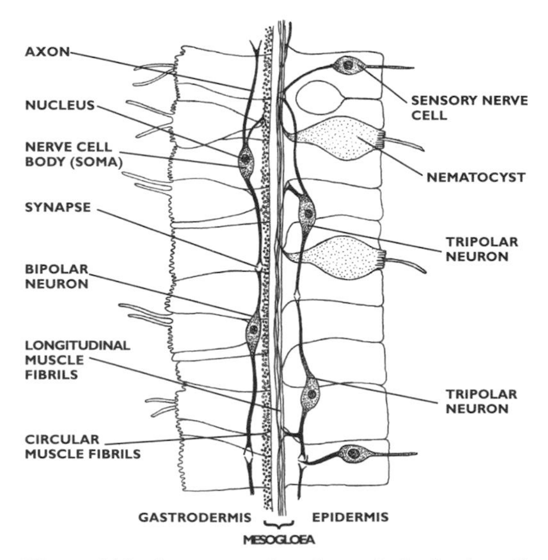

Figure 116. A cross section through the body wall of a hydrozoan showing the simple nerve net.

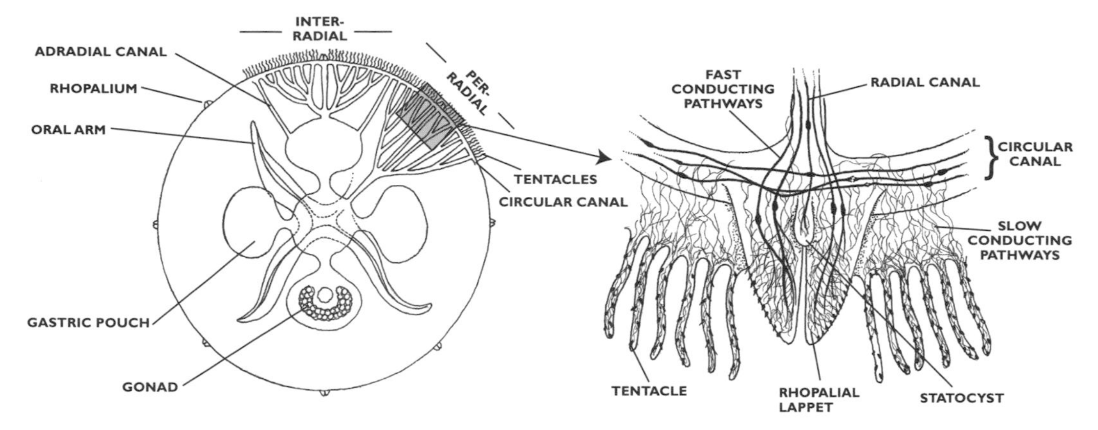

Figure 117. The nervous system of the Moon Jelly *Aurelia aurita*, seen from below.

concentrations of nerve cells (ganglia) and two nerve fibre tracts (the inner and outer nerve rings). The Moon Jelly has eight ganglia around the bell margin. The rhopalia are sense organs, with both mechano- and photoreceptor centres. The former are known as statocysts and the latter ocelli, primitive 'eyes' which are little more than a collection of light receptive cells (fig.118).

In the Moon Jelly these organs are contained together in the rhopalium which sends information to the two nerve rings. The rings ultimately communicate with the nerve net and the musculature. Experiments have demonstrated that stimulation of the statocyst inhibits muscular contraction on the stimulated side of the bell. This promotes the opposite side to push more water out from under the bell which becomes tilted higher. Stimulation of the statocyst arises through movement of a small solid body, normally calcium sulphate, which by knocking or touching a small hair triggers a nerve impulse. This system constantly corrects the level of the animal, as the lifting of one side of the bell is rectified by the raising of the other side. The ocelli have different effects depending on the species and its feeding habits. Day feeders have ocelli that are positively phototactic, night feeders have negatively phototactic ocelli.

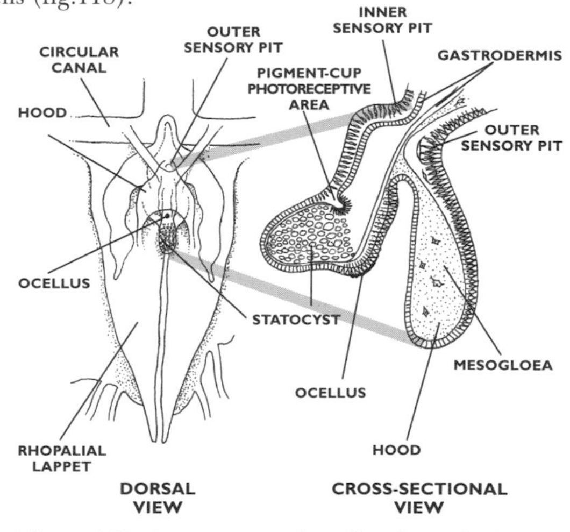

Figure 118. A statocyst and ocellus from the Moon Jelly *Aurelia aurita*.

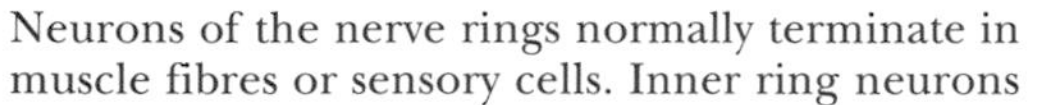

Neurons of the nerve rings normally terminate in muscle fibres or sensory cells. Inner ring neurons of jellyfish medusae are large and control the swimming muscles. It is this inner ring that is responsible for coordination of the rhythmic bell pulsations. Understanding the nervous system of cnidarians poses a problem since the bi-directional nerve cells may have two or more processes going to the muscles and sensory cells. Another type of nerve cell to be found in the cnidarian nervous system is the interneuron. These nerve cells are more important in higher animals (especially vertebrates) as they communicate between other neurons. It is from these interneurons that the central nervous system of vertebrates may originally have evolved.

Echinoderms also have a nervous system organised on a radial plan. The starfish's nerve centre surrounds its mouth in a pentagonal arrangement called the oral ring (fig.119). From each angle of the pentagon a large radial nerve communicates with the peripheral nerve plexus that innervates the muscles and sense organs of the arm. Two different nerve networks are present; one innervates the muscles and the other carries sensory information, and both run in parallel. This one-directional information transfer along specific nerve pathways is quite well developed, but is not as clearly defined as the more complex nerve networks of higher animals.

There are five nerve centres in the oral ring, one for each arm. When the leading arm moves, it causes a temporary dominance over the other nerve centres and in most species of starfish any arm may be a leading arm. However, in some species there is a permanent dominant arm which always leads when the animal moves. Unlike the cnidarians, starfish do not have specialised sense organs, but only small, simple

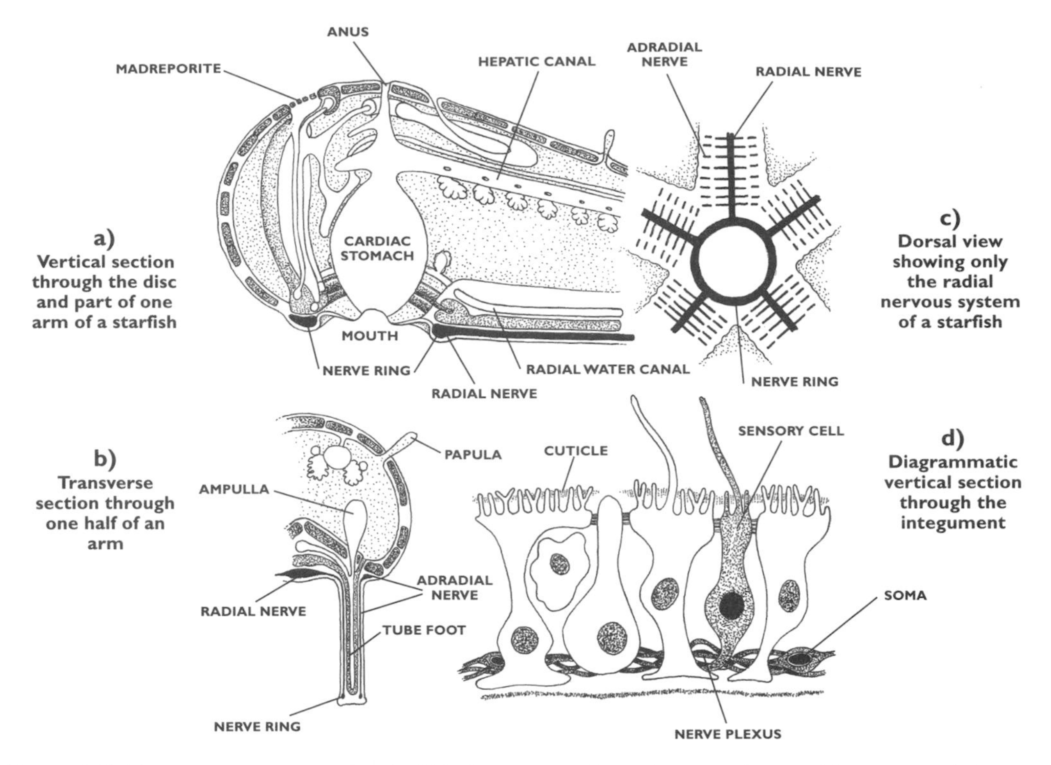

Figure 119. The nervous system of a starfish showing position of nerve ring, radial nerves and structure of sensory cells.

eye spots at the tips of the arms. These eye spots lie beneath the integument and comprise 80 to 200 pigment-cup ocelli forming an optic cushion. Most starfish are attracted towards light (positively phototactic), although the intensity of the reaction varies between different species.

All echinoderms possess dispersed sensory cells (fig.119d) that are contained within the epidermis. Starfish have up to 70,000 of these cells on the ends of their tube feet and the groove under their arms. Sea urchins also have sensory cells on extendible ampullae and on the spines. All echinoderms have statocysts (spheridia) that are used to orientate the animal when righting itself. The nervous system of sea cucumbers is similar to that of the starfish except the radial nerves run longitudinally along the length of the body. Experiments have shown that starfish are capable of learning. When the animal is taught to turn over in a certain way, the experience acquired is retained for up to five days without any reconditioning or reinforcement. Nervous coordination of sea urchins is demonstrated by their response to shadows. When a shadow passes over them they rapidly move their spines as if towards a potential threat.

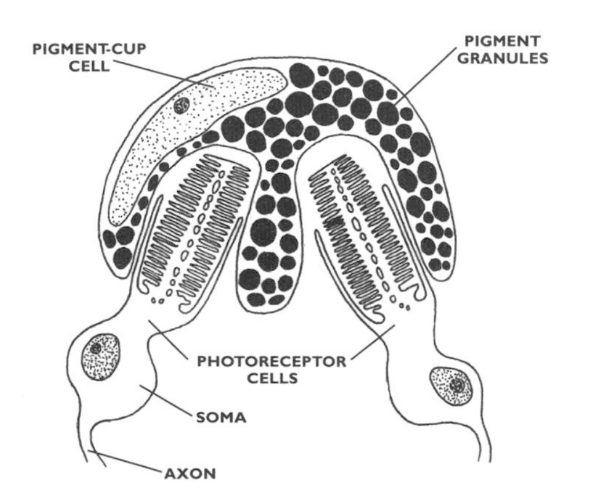

Figure 120. Pigment-cup eye of a platyhelminth.

A simple central nervous system occurs in the Platyhelminthes. The development of a head to tail orientation, the head being first to encounter a new stimulus, logically implies that this region should have the major concentration of sensory structures. It is this localisation of sensory structures and associated neurons that is associated with a ganglionic mass. Throughout the platyhelminth nervous system there are no other ganglia except those in the head. This is believed to be a primitive feature. Most marine flatworms have pigment-cup eyes (fig.120) and it is not uncommon for them to have two or three pairs. These eyes cannot focus light but function to react to the presence of light. The majority of marine flatworms move away from light (negatively phototactic).

Ciliary receptors are concentrated around the tentacles and body margins of the platyhelminth. On the head region chemoreceptors are located in ciliated sunken pits or grooves. The cilia maintain a water flow across the sensory cells that detect changes in the chemical nature of the surrounding water. These are effective if they are paired on either side of the head. The animal can then detect a possible direction of the stimulus; as soon as it swings its body too far one way from the water flow, it loses the stimulus on that side of the head. The flatworm then swings its body round in the opposite direction to regain the stimulus. This process carries on as the flatworm moves along towards the source of the stimulus (fig.121).

The platyhelminth ganglionic mass is connected to the rest of the body via one or more longitudinal nerve cords and these make contact with a peripheral nerve plexus which is associated with the muscles. The central neurons of the ganglion and nerve cord are usually monopolar, nerve impulses only travelling in one direction. This arrangement of impulse direction and its associated synaptic connections is also characteristic of the higher invertebrates and all vertebrates.

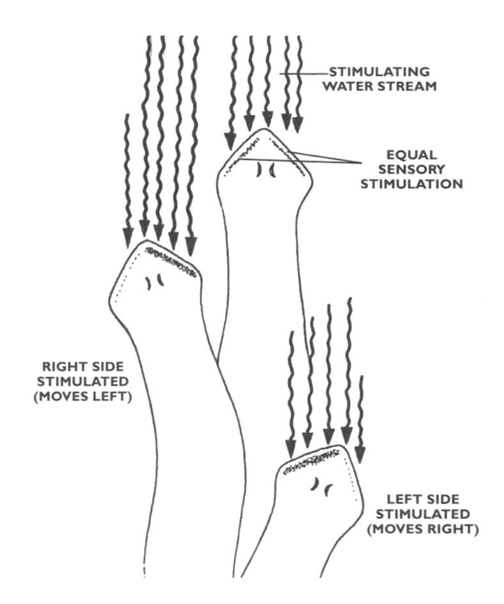

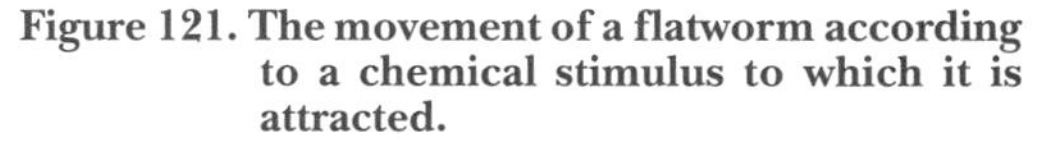
Figure 121. The movement of a flatworm according to a chemical stimulus to which it is attracted.

Polychaete worms, (including ragworms, lugworms, fanworms and scale worms) show a clear division of the nervous system into central and peripheral regions. The brain is usually bilobed and lies just below the dorsal epithelium. It is connected to the rest of the body via a primitive ventral nerve cord. The nerve cord consists of paired, segmentally arranged ganglia joined by paired connectives. This arrangement is most distinct in the fanworms, but in other polychaetes the paired connectives are fused to varying degrees (fig.122a). Nerve fibres in the ventral nerve cord are thicker in diameter. They mediate the fast contraction reflex of tube-dwelling polychaetes. This thickening can be likened to that of wire cabling where the diameter of the wire is inversely proportional to its resistance, so the thicker the wire the less its resistance. In nerve cords this enlarged diameter allows more rapid impulse transmission. Thickened nerve fibres, called giant axons, are found in many invertebrates where they function in speeding up impulse rates along important nerve routes (e.g. startle response). Experiments have shown that ordinary longitudinal nerve tracts conduct impulses at about half a metre a second and giant axon fibres at twelve metres per second. Therefore in a twelve centimetre long polychaete worm it takes only 1/100th of a second for a nerve impulse in a giant axon to pass from the head to the tail (fig.122b).

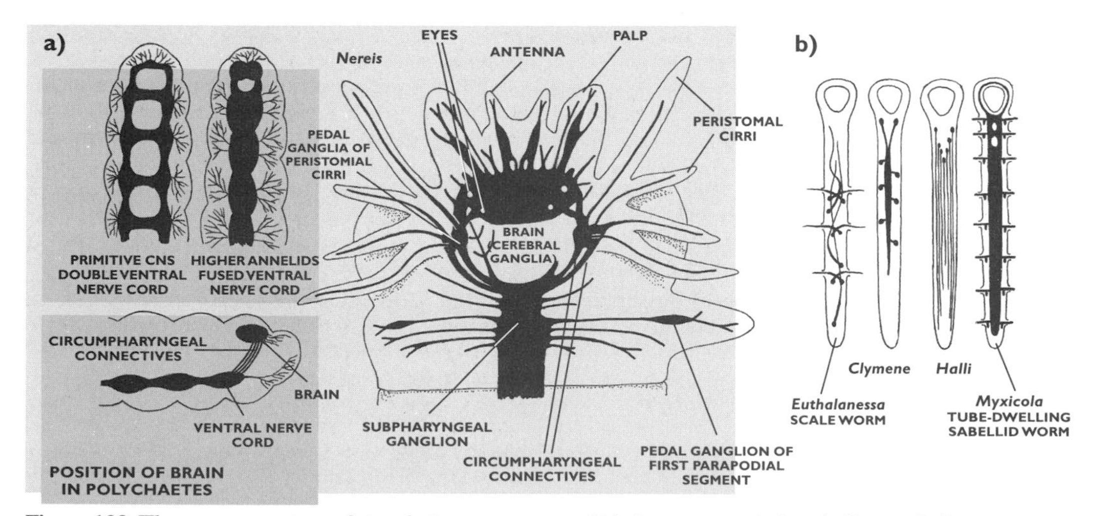

Figure 122. The nervous system of a) polychaete worms and b) giant axon variations in four polychaete worms.

Sense organs are more advanced in the annelids than the platyhelminths and the eyes are similar to higher invertebrate types as they have a cup-like form (fig.123a). Eyes of most polychaete worms are only able to detect light intensity and the direction of its source. But the pelagic carnivorous Alciopidae (transparent polychaetes with two large eyes) can form images on a retina with the aid of a crystalline lens and its associated structures (fig.123b).

Like some platyhelminths, the polychaetes have a pair of ciliated sensory pits or slits located on the head (nuchal organs). These are the main sensory organs for detecting food and they are enlarged in the predatory species. Statocysts are also present in the tube-dwelling polychaetes, for example in the Lugworm *Arenicola marina* where they are located in the body wall of the head. They are used to orientate the animal during burrowing and if they are removed then the animal loses the ability to burrow at right angles to the surface.

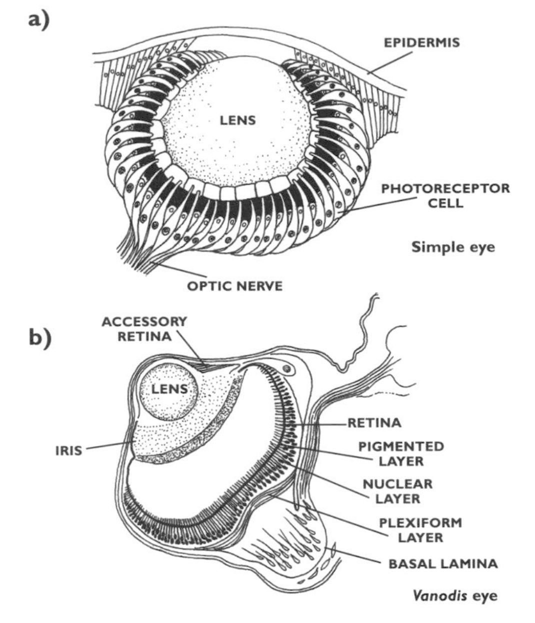

Figure 123. Two types of light-sensing organs in polychaetes shown in section: a) ventral cup type and b) crystalline lens type.

The marine arthropods (which include crustaceans, horseshoe crabs and sea spiders) have a well developed dorsal anterior brain with circumoesophageal connectives and a ventral nerve cord connected to segmented ganglia (not unlike the polychaetes). Increase in brain size is probably related to the increase in number and complexity of sense organs. The segmental ganglia are double structures which can fuse together and in some species there is only one highly fused ganglion. Unlike the polychaete worms, nerve nets and nerve plexuses are virtually absent in arthropods.

Crustacean sense organs include the compound eye. This structure is composed of many long, tubular units (ommatidia) capable of light reception (fig.124). Each ommatidium consists of an outer translucent cuticle or cornea, a crystalline cone and the receptor or retinular cells. The cornea and crystalline cone function as primary and secondary lenses, respectively. The external facets of the corneal cells form a hexagonal pattern (square in lobsters). Each retinula has a central translucent cylinder (rhabdome) with seven or eight photoreceptive (retinular) cells arranged around it. Surrounding this core lie the special pigment cells that are used in combination to vary the light intensity within the ommatidium (fig.124). In arthropods that live in high light intensity habitats these pigments are fixed and act as opaque filters to stop any stray light from entering other adjacent ommatidia. This is called an apposition eye (fig.124d, f). Lobsters living in low light intensity habitats have superposition eyes (fig.124c, e) where the pigment can migrate up and down the sides of the ommatidium. This protects the retinular cells from high light intensities, and increases sensitivity in low light levels very much like an iris in a vertebrate or cephalopod eye. Each retinular cell has an axon that connects to an optic ganglion. In some cases the impulse is collected by an eccentric cell that then passes on the signal. This ganglion is either found in the brain or, in crustaceans, within the eye stalk. The image is therefore created from a mosaic of light spot intensities. It is believed that each of the retinular cells within a single ommatidium is sensitive to a particular wavelength of light, produced by pigment cells absorbing the other frequencies. This crude, coloured mosaic is interpreted by the brain and the smaller and more tightly packed the ommatidia, the finer the image resolution will be. Experiments have shown that the compound eyes of crustaceans can focus clearly up to 20cm, but beyond this distance their focus is poor. However, the compound eye has advantages; some particularly large-eyed crustaceans with eye surfaces that arc over 270° have nearly all-round vision.

Arthropods are able to sense the environment through the impervious layer of the cuticle. Sensory modifications of the chitinous exoskeleton called sensilla are present. These are often in the form of hairs or chaetae, but others may be peg-like structures, pits or slits. Sensilla consist of one or more sensory neurons with a number of housing cells. Some may respond to one type of stimulus, while others may be multi-functional responding to many chemo- and mechanostimulants. Perforated sensilla appear to be chemoreceptive while chaetae are normally mechanoreceptive. Some mechanoreceptors may take the form of a slit and these detect tension across the exoskeleton. New sensory structures are laid down horizontally below the old cuticle before moulting occurs. When the dendrite (nerve cell process) is severed from the old sensory structure another dendrite is already in contact with the new sensillum. These chaetae and other larger sensilla can clearly be seen across the carapace and limbs of larger deca-

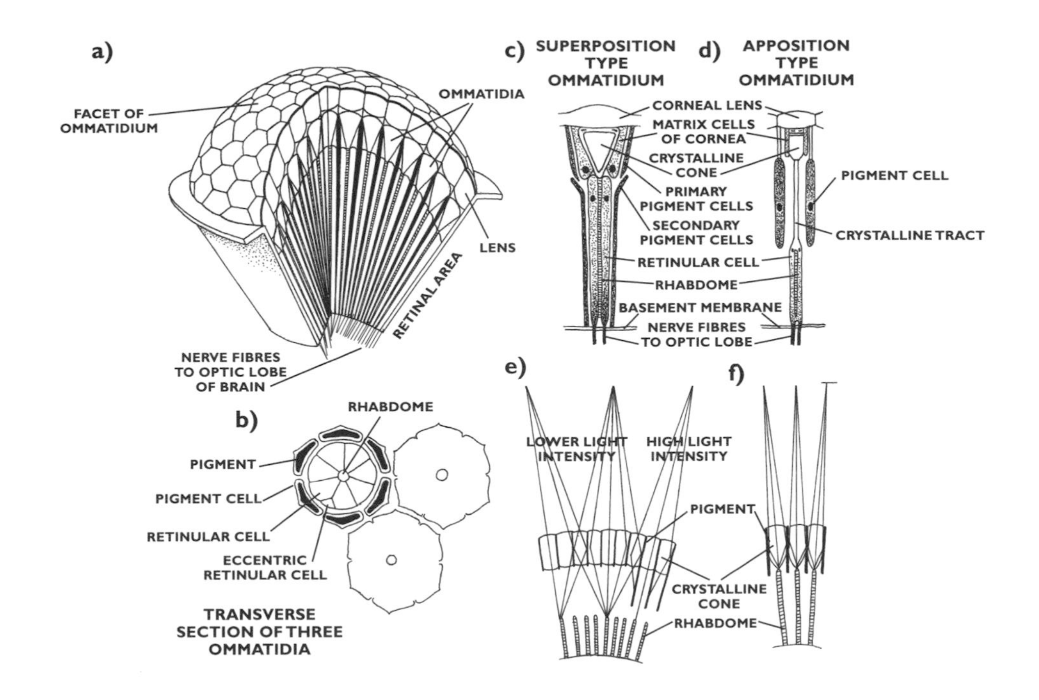

Figure 124. A simple compound eye and superposition and apposition ommatidia.
a) Internal structure of a compound eye showing its main regions.
b) Transverse section of three ommatidia showing how pigment cells are arranged around the retinal cells and rhabdome.
c) Superposition-type ommatidium with primary and secondary pigment cells.
d) Apposition-type ommatidium with short pigment cells.
e) Position of pigment under different light intensities in a superposition ommatidium.
f) Position of pigment (fixed) under different light intensities in an apposition ommatidium.

pod crustaceans as stiff hairs and pimples. Chemoreceptors and chaetae are particularly numerous on specialised head structures (antennae). These structures can be long (for example the antennae of shrimps, lobsters and some crabs may be longer than the body length) and are used as sensory structures. Arthropods have specialised sensory organs and enlarged or complex brains. In the marine environment one molluscan group, the cephalopods, has a particularly well developed central nervous system.

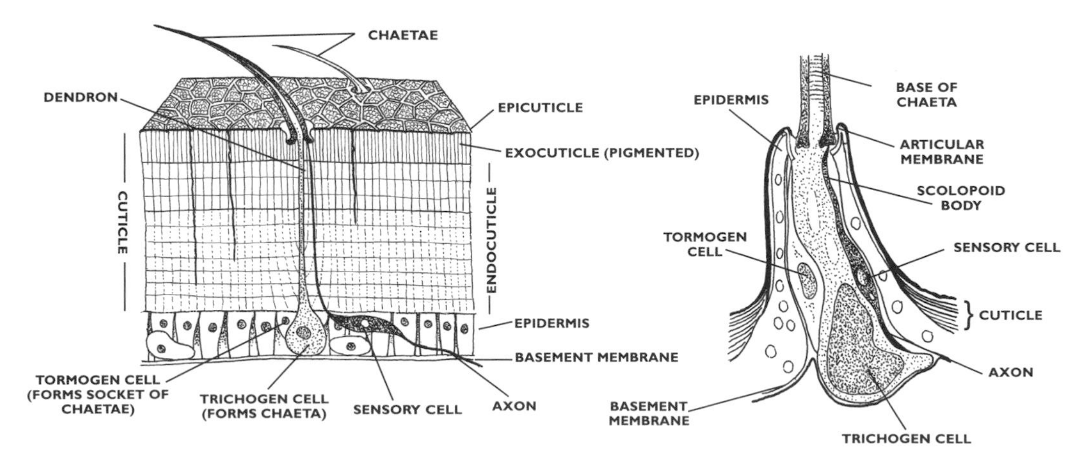

Figure 125. The sensillum of crustaceans.

The basic plan of the nervous system in molluscs consists of a series of three or more paired ganglia that are connected by paired nerve tracts to form a complex brain mass. Two paired nerve tracts extend posteriorly; one is the ventral (pedal) nerve tract that innervates the foot and the other is the dorsal (visceral) tract that controls the mantle and visceral organs (fig.126). Molluscs possess tentacles, a pair of eyes, a pair of statocysts and a chemoreceptive organ (osphradium).

Bivalves have the simplest nervous system which retains the basic plan of three pairs of ganglia (cerebroganglia) and two pairs of large nerve tracts. One pair of ganglia lies each side of the oesophagus. These are connected dorsally to each other and each pair gives rise to a pair of nerve tracts. Theupper tract connects to a pair of visceral ganglia which lie on the surface of the posterior adductor muscle. These ganglia innervate the posterior adductor muscles as well as the siphon. The foot and the anterior adductor muscle are innervated by the pedal ganglia that are connected to the main cerebroganglia by the ventral pair of nerve tracts. Coordination of the foot and valve movements is therefore controlled by the cerebroganglia sending impulses to the two pedal ganglia.

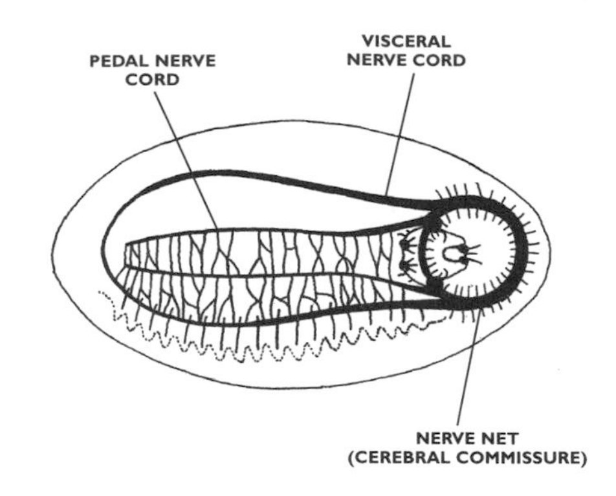

Figure 126. The nervous system of an archetype mollusc.

Bivalve sense organs are usually located on the margins of the mantle. In the scallops and file shells (plates 2, 3 & 7) the entire margins bear pallial tentacles which contain tactile and chemoreceptor cells. However, the tentacles are more often restricted to the apertures of the siphons, as shown by the fringed siphons of the cockle.

Statocysts are also present within the foot and these are innervated by the cerebroganglia and, like the tube-dwelling polychaete worms, are used to orientate the animal while burrowing. Sedentary forms like the oysters and clams have substantially reduced statocysts.

Ocelli are interesting sense organs in bivalves which are present in many species. For example, large clams such as the Giant Clam *Tridacna maxima* (plate 8) have many thousands of ocelli in their mantles. These are able to detect sudden changes in light intensity and warn of approaching danger. Some of the mussels (Mytilidae) retain their larval cephalic eye which is located, in the adult, in front of the gill axis. This eye is similar to the gastropod eye with a photoreceptive pit and a simple lens which is only capable of detecting changes in light intensity. Scallops possess two rows of numerous metallic blue eyes (see plate 3, the Queen Scallop *Aequipecten opercularis*). One row is of the rhabdomeric type similar to the eyes found in the flatworms, while the other row is of the ciliary kind found in higher animals.

Rhabdomeric photoreceptor cells (fig.127b), predominate in the protostome animals and terminate in microvilli. Ciliary retinal cells (fig.127a) are mainly found vertebrates which have cilia instead of the microvilli.

The blue sheen of the scallop's eyes is caused by the reflective layer or tapetum behind each eye and is similar to the reflective layer found in cats' eyes. These eyes are not able to detect shapes, but they can detect movement in the immediate vicinity. Cockles such as the Common Cockle *Cerastoderma edule* possess only small eyes around the top of the siphon, which is normally the only part exposedwhen feeding. If movement is detected the siphon is withdrawn leaving little trace of the animal except for a shallow impression in the sand.

Bivalves also have another molluscan sensory trait, the osphradium. This organ is located immediately below the posterior adductor muscle in the exhalant chamber. It is responsible for chemoreception

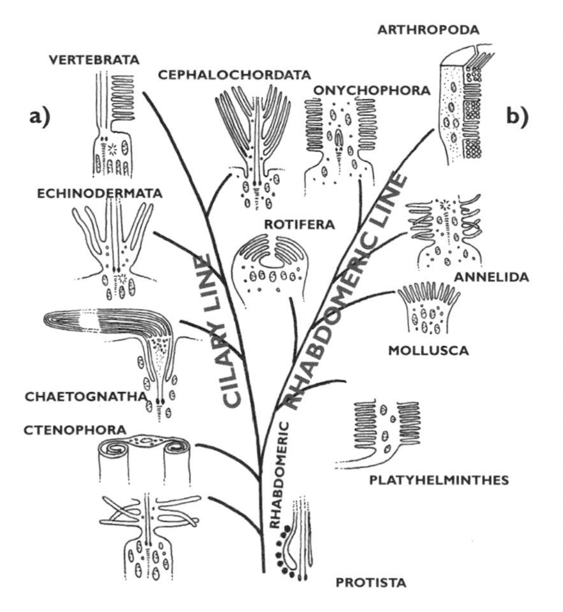

Figure 127. Photoreceptor cells, a) ciliary and b) rhabdomeric.

and monitors the water flowing through the mantle cavity. Although this is called an osphradium and is responsible for chemoreception there is some doubt whether it is derived from the same embryonic tissue as the gastropod's osphradium.

The gastropod osphradium in archetypal forms is present on each gill, but the more advanced, single-gilled molluscs have only one osphradium. The organ is normally filamentous or convoluted to increase the surface area available for chemoreceptive cells. In the carnivorous and scavenging prosobranchs, where the organ is large and highly developed, prey can be detected from at least two metres away. To help find its food, the prosobranch's siphon is waved about to monitor the water current from various directions.

Some gastropods have primitive eyes. Limpets have a simple photoreceptive pit. Other gastropod eyes are more complex, such as those of the murexes which have a lens and a closed-over vesicular retina. Nudibranchs possess few eyes, while the conch shells have paired stalked eyes (similar to land snails), and many gastropods have numerous simple eyes. As in the bivalves, only general light intensity is detected, although studies on periwinkles have shown they may be able to recognise vertical bands.

As is common in bivalves, statocysts are present in many active gastropods and are located near the pedal ganglia in the foot. However, the more sessile members of the group have none.

The nervous system of gastropods is complicated by torsion, the 180° twisting of the visceral mass and mantle at the post-larval stage (fig.128). The visceral nerve tracts are twisted into a figure of eight and one parietal ganglion is higher than the other (the parietal ganglia innervate the gills and the sensory osphradium). Eyes, tentacles, statocysts and buccal ganglia (that control the muscles of the radula) are all innervated by the cerebral ganglia. The adjacent pleural ganglia are connected to the pedal ganglia which regulate local reflexes of the locomotive foot muscles that in turn control the muscular wave action. Advanced gastropods have fused ganglia that create a primitive brain structure around the oesophagus. But it is in another group of molluscs, the cephalopods, where this centralisation of the ganglia has become most advanced.

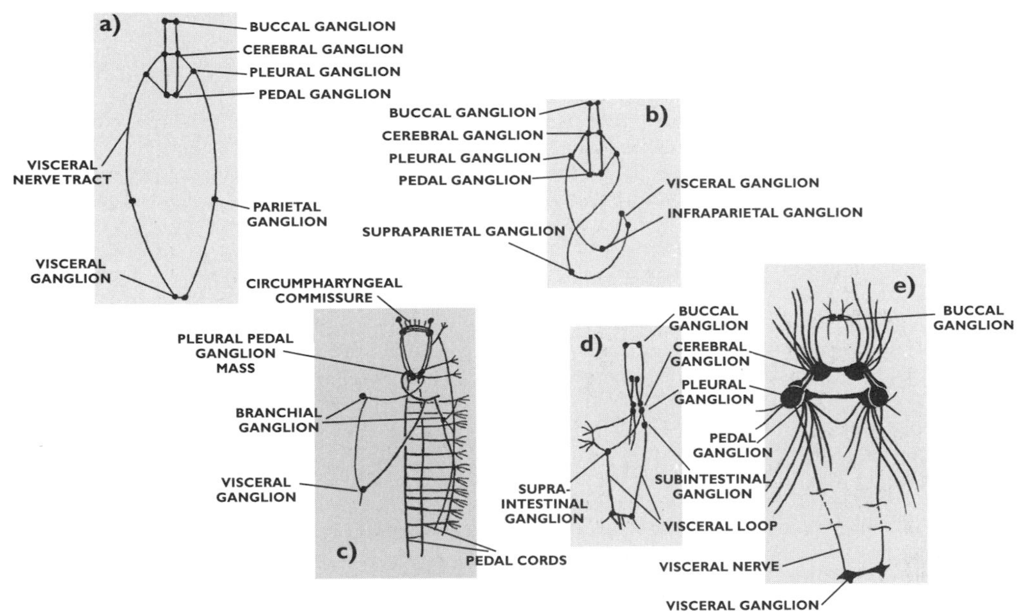

Figure 128. Torsion in the gastropod nervous system. Pre-torsion (a) and post-torsion (b) appearance of visceral nerve tracts. The CNS of three gastropods is shown in (c) an abalone *Haliotis*, (d) a conch shell *Triton* and (e) a Sea Hare *Aplysia*.

Cephalopods are among the more advanced marine invertebrates known and possess a sophisticated CNS. They are active predators with well developed nervous systems based around the molluscan plan of central ganglia and peripheral plexuses. However, the nervous system is highly concentrated in the head region (cephalisation). During evolution this has resulted in the molluscan ganglia, covered by a cartilaginous 'cranium', concentrating and fusing around the oesophagus.

The innervation of the cephalopod's tentacles and siphon by the pedal and branchial ganglia indicate that these tissues are homologous to the foot of the gastropods and bivalves. The mantle has a large pair of nerves originating from the visceral ganglia that control normal swimming actions (fig.129). Rapid

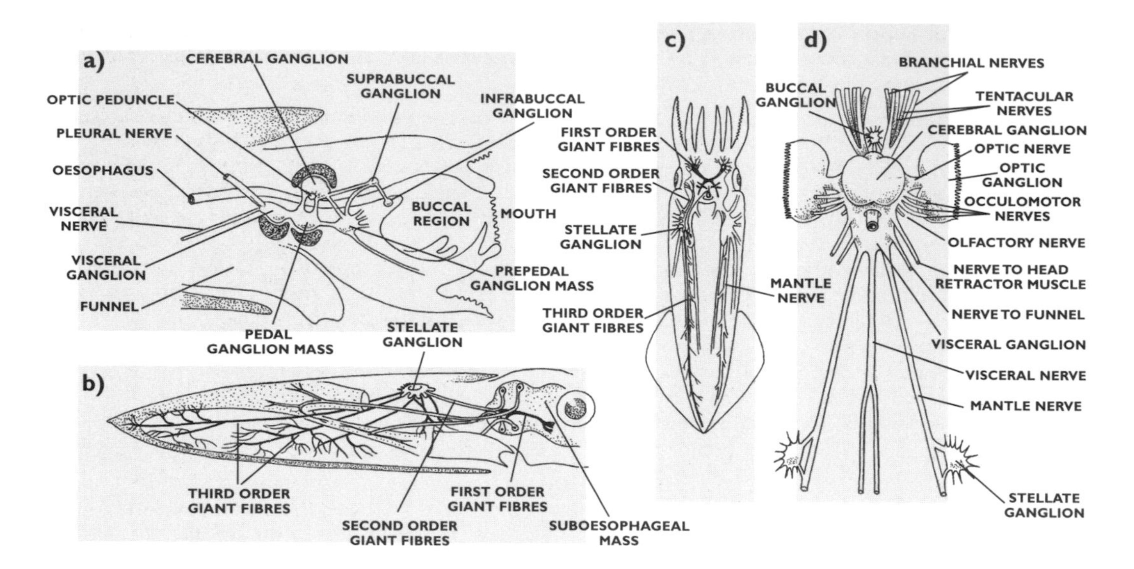

Figure 129. Central nervous system of a cephalopod *Loligo*. Transverse section of a) head region and b) body. Top view of c) whole animal and d) brain region with stellate ganglia.

swimming used during escape or attack is initiated by three sets of giant axons (fig.129) that are in turn controlled by connections in the stellate ganglia. When enough sensory cells in the head region are stimulated they cause the primary axons (first order giant fibres) to fire. These axons stimulate the secondary axons (second order giant fibres) which send impulses to the stellate organ. Here the secondary axons innervate the tertiary giant axons (third order giant fibres) that innervate the mantle muscles. The tertiary giant axons are inversely tapered; the further the axon fibre is away from the nerve cell body the larger is its diameter. This creates a near perfect simultaneous contraction of the mantle muscle, giving the squid the sudden contraction necessary for rapid movement.

One of the most striking features of the cephalopods is their well developed eyes, which are structurally similar to the eyes of vertebrates (fig.130). A spherical cartilaginous capsule surrounds the eye with a variable focus lens, iris and extra-ocular muscles - all are highly advanced features. Focusing is facilitated by a forward and backward movement of the lens, similar in action to a camera lens. Light entering the eye is limited by the slit-like iris and the migration of pigments in the retina. *Octopus* can discriminate objects as small as five millimetres at a distance of one metre. The Nautilus, however, lacks a lens and cornea. Instead its eye has a small aperture that is open to the external water and this functions in much the same way as the aperture of a pinhole camera with its fixed depth of focus.

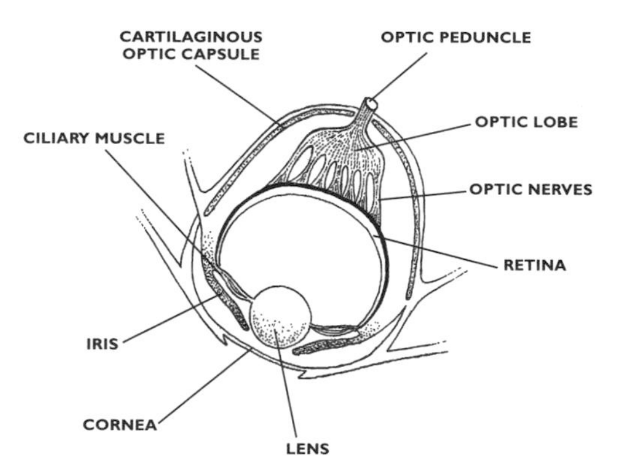

Figure 130. The complex eye of a cephalopod.

One of the cephalopod group, a squid, possesses the record for the largest known eyes. A Giant Squid *Architeuthis* sp. (plate 45) found washed up on the shores of New Zealand in 1933 was recorded to have an eye 40cm across.

The cephalopod statocyst is also advanced and is comparable to the analagous structures located in the inner ear in vertebrates. It is embedded in the cartilage either side of the brain and has two functions. Firstly, it provides the animal with static spatial orientation and allows it to position itself in relation to the earth's gravity. This type of statocyst function is found generally throughout the invertebrates. In addition, it informs the animal of its inertial position or, more simply, its changes in motion (i.e. the cephalopod's acceleration and deceleration).

The cephalopod's arms and tentacles are well endowed with chemo- and mechanoreceptors. This is particularly striking in the benthic feeders such as octopuses which may be 100 to 1,000 times more sensitive to certain stimuli, such as chemicals in the water, than humans.

The range of behavioural patterns in cephalopods is also complex. Most of this behaviour is innate, such as the squirting of ink or preferences for certain prey and hunting methods. Cuttlefish and octopuses appear to be capable of learning and inventing new actions to suit different situations. Octopuses may also recognise polarised light and shape.

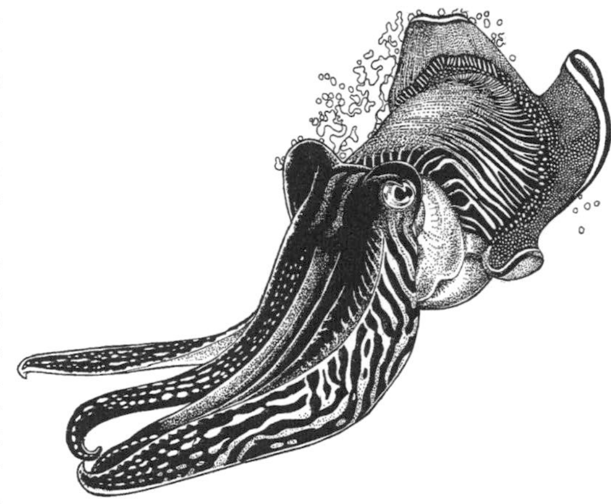

Figure 131. The threat display of a male Common Cuttlefish *Sepia officinalis*.

Cephalopod behaviour is associated with the patterns created by the skin chromatophores that are under nervous control. A large part of the brain is concerned with colour control. For instance, the cuttlefish's defence responses are highly variable. They can send waves of monochrome stripes rippling across the body, or produce two large black menacing eye spots and other striking defence patterns. Stripes are also used by males to threaten potential adversaries (fig.131). Complex behaviour patterns are also associated with vertebrates where they are found only in animals that have a high degree of cephalisation with its accompanying expansion of the central nervous system and metameric spinal nerve cord.

Tunicates have taken a step further towards an advanced CNS with the evolution of the nerve cord and its accompanying rudimentary notochord. The tunicates (subphylum Urochordata), which include the sedentary ascidians and the planktonic larvaceans and thaliaceans, have a larval form that possesses a hollow dorsal nerve cord. The nerve cord in its simplest form carries impulses to the tiny muscle blocks. In the ascidian larvae there is a dilation of this tube at the anterior end which later develops into the cerebral ganglion and a pigmented cup with a statocyst (fig.154,p.176). In most species the larval stage only lasts up to 36 hours and the rudimentary nerve cord is never fully developed. However enigmatic the formation of the notochord and the neural tube may be in this group, it does appear to be a precursor of the vertebrate condition.

REPRODUCTION

Invertebrates have two different reproductive methods. In the first method each copulating parent transfers half its chromosome number to form a new individual (sexual reproduction). Secondly, invertebrates use asexual reproduction in which an individual can reproduce by budding off a small replica of itself, with the identical chromosome complement to its parent.

Asexual reproduction occurs mainly in the lower invertebrates, but there are exceptions. Two such groups are the ascidians and the echinoderms. In ascidians, especially the colonial tunicates, this ability to regenerate asexually is highly developed. Budding may occur in different regions of the body, whereas most other invertebrates can bud from only one fixed body part. In tunicates the bud is called a blastozooid and its germinating tissue is formed in various parts of the parent according to the species. In primitive species of the Clavelinidae (e.g. Blue-throated Ascidian *Clavelina australis* and Light Bulb Tunicate *Clavelina huntsmani*, plate 2), the blastozooid originates from the rootlike extensions of the parent's base (the stolon). Some ascidians show budding in the larval stage, but buds generally appear from the abdominal region in the adult.

This type of subdivision can result in large tightknit colonies, and colonies of the Giant Jelly Ascidian *Polycitor giganteum* (plate 7) can reach over 30cm in diameter. An alternative to budding is increasing size by growth. Some small solitary ascidians (e.g. Common Solitary Ascidian *Polycarpa aurata*, plate 8, and Gooseberry Sea Squirt *Dendrodoa grossularia*, plate 3), barely reach two and a half centimetres high. However, there are exceptions and some individual ascidians are much larger than the norm; for example *Ascidia mentula* (plate 4) can reach ten centimetres high and the southern temperate Club Ascidian *Polycarpa clavata* is over 20cm high. Ascidians enlarge themselves by budding and this can be seen by the abundant star-shaped patterns of the botryllid ascidians which are found throughout the world's oceans (fig. 132).

Echinoderms cannot use their asexual regeneration methods to colonise areas in the way that ascidians do. In many starfish asexual reproduction is normal and usually involves a division of the central disc, the animal breaking into approximately two halves. In the Multipore Starfish *Linckia multifora* of the Indo-Pacific, one arm moves away from the others which are gripping the substratum in order to facilitate the tearing at a preset point. The ends of the break reseal and regeneration occurs. This type of regeneration may take up to a year, but in the case of *Linckia* it takes no more than a month. These partially regenerated

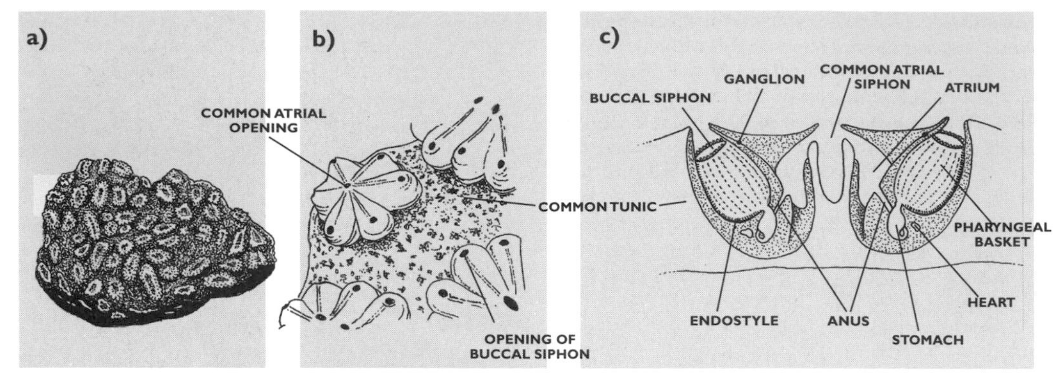

Figure 132. The colonial sea squirt *Botryllus schlosseri*, showing a) overall view of colony b) detail of colony and c) vertical section of two individuals with their common atrial opening.

'comet' forms can be seen on reefs or washed up on beaches (fig.133). Sea cucumbers may also reproduce in this way. For example, *Holothuria atra*, a common black sea cucumber of the Indo-Pacific reefs, is often found either regrowing a head or a tail. This method of reproduction is only used by starfish (including brittle-stars) and sea cucumbers; crinoids and sea urchins are not known to reproduce in this way.

Asexual reproduction is common among the cnidarians, and in the simpler hydrozoans budding takes place in the warmer months of the year. Similarly, the budding of sea anemones commonly occurs when conditions are at an optimum, the animals using this rapid method of reproduction when growth rate is high. Several methods are employed. One is similar to the starfish where the animal moves away from the substratum leaving behind part of the pedal disc (pedal laceration). Another method involves the action of pinching off a piece of the disc wall; the detached portion then regenerates into a small sea anemone. A third method is by splitting or asexual fission reproduction. In this the animal either separates longitudinally (halving the column down the vertical axis) or transversely (breaking the sea anemone horizontally across the column).

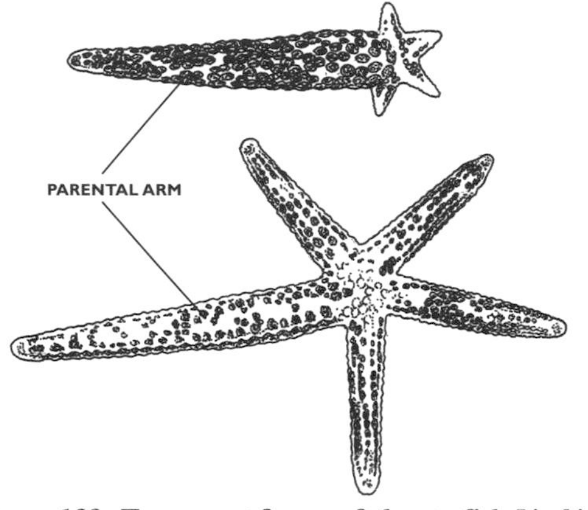

Figure 133. Two comet forms of the starfish *Linckia multifora*.

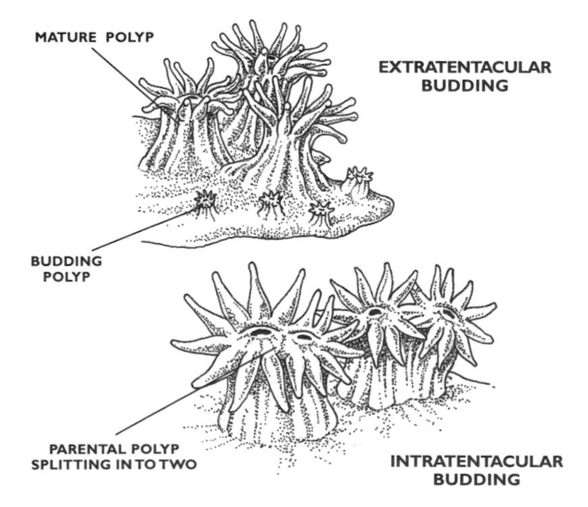

Figure 134. Budding methods in corals. Extratentacular budding (above) and intratentacular budding (below).

Corals also use asexual reproduction. Some coral colonies break up to produce several daughter colonies, while others employ budding to expand a colony. Budding can take place at the base of old polyps (fig.134) or from the discs of the parent polyp by asexual longitudinal fission. In the majority of hermatypic corals this budding is of the complete type and the polyps are fully separated forming the typical externally pitted coral surface. However, the brain corals (plate 5) show an unusual type of intratentacular budding which occurs without the bud completely breaking away after the new mouth is formed. This partial fission produces the unusual surface structure of these corals with rows of polyps sharing a common oral disc with numerous mouths (fig.135). The more common type of extratentacular budding produces deposition of the basal skeleton with new corallite walls and septocostae. Brain corals secrete ridges of shared corallite walls and associated structures resulting in the characteristic globular shape. Marine hydroids such as the Squirrel's Tail *Sertularia argentea* and *Tubularia indivisa* (plate 4) use both asexual and sexual reproduction. In most species the branching hydroid colony is anchored by a stolon to the substratum. Each side branch ends with a terminal bud. These terminal buds are not all feeding gastrozooid polyps, but are interspersed with reproductive polyps (gonozooids) (fig.136). Using asexual budding, gonozooids produce the sexual or medusoid stage. Medusoids are found in a variety of forms and sizes, usually from 5 - 60mm in diameter and resemble small jellyfish. Some hydromedusae (the term used to distinguish hydroid medusae from the 'true' jellyfish medusae of the Scyphozoa) in the genus *Gonionemus* can crawl over the bottom, but *Tubularia* actually keeps its hydromedusae attached to itself (fig.136). The hydromedusae are normally dioecious, producing eggs and sperm from different individuals. Fertilisation occurs either externally on the surface of the tube-like mouth (manubrium) or internally where the eggs start their development in the gonads. Whatever method of fertilisation occurs, a free-swimming, minute flatworm-like larva (planula) is produced. After several hours, up to a day or more, the planula attaches itself to the substratum and develops into a new hydroid colony.

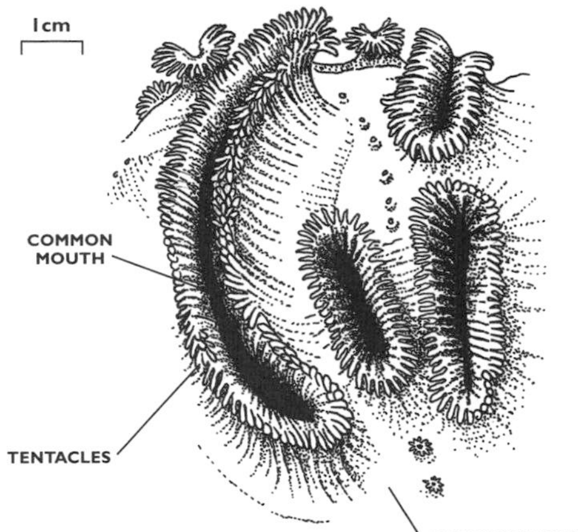

Figure 135. Polyp rows in a brain coral.

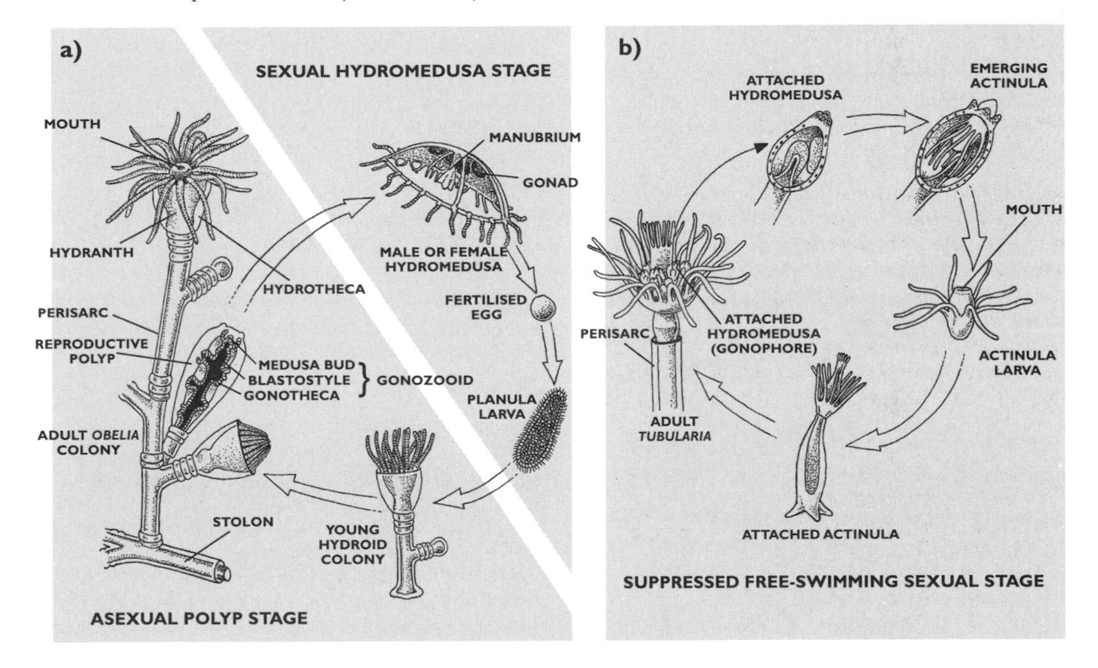

Figure 136. A typical hydroid colony and its reproductive cycle shown in *Obelia* (a) and *Tubularia* (b).

In the floating hydroids *Porpita* and *Velella* the planula remains free-swimming, developing into a floating pelagic hydroid colony suspended from a chambered chitinous float (fig.69). Siphonophores are similar to *Velella* and *Porpita* in that they float and exist as large pelagic colonies. The Portuguese Man-of-war, *Physalia physalis* (fig.137) has gonozooids clustered together to form a gonophore. As in other hydrozoans, *Velella* produces hydromedusae from the gonophores which are the sexual stage in the life cycle.

Among scyphozoans or true jellyfish, the stage that is equivalent to the hydrozoan polyp is the scyphistoma stage. This stage feeds in the same way as most polyps and it asexually buds to produce young medusae (ephyrae) that are released at certain times of the year. Some species, like the cubozoan *Tripedalia cystophora*, do not show this budding sequence and the scyphistoma metamorphoses into a young medusa. In the Lion's Mane Jellyfish *Cyanea capillata* (plate 1) and the Compass Jellyfish *Chrysaora hysoscella* (plate 1) the larvae are released at the gastrula stage. In some species the resultant ephyra may take up to two years to reach sexual maturity, while other species like the Moon Jelly *Aurelia aurita* (plate 1) take only about three months. Like the hydromedusae, the medusae of the scyphozoans are dioecious, but unlike the hydroids, the gonads are formed in the gastrodermis. When the eggs are released they often lodge in pits on the oral arm, as seen in the Moon Jelly. The eggs are temporarily held in these pits for fertilisation by the sperm and for their early development. Once released from the oral arms, the early planula larvae briefly have a free-swimming existence before they settle and develop into the scyphistoma.

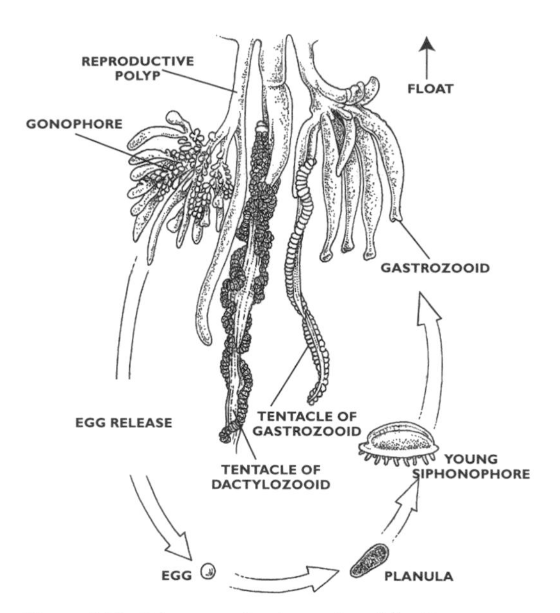

Figure 137. The reproductive cycle of the Portuguese Man-of-war *Physalia physalis*.

Corals and sea anemones also release sperm and eggs, but corals release their gametocytes only on certain nights of the year. Both the hard and the soft corals synchronise the release of their eggs and sperm in this way, turning the clear reef waters milky white. During the preceding months the gonads, which are located in the gastrodermis as bands down the mesenteries, slowly ripen. Corals can be dioecious or hermaphrodite (where sperm and eggs come from the same individual). In both types of corals egg development begins first and thesperm production starts later, requiring less time to ripen. The rate of gamete production (gametogenesis) increases as the water warms in the spring season. Once eggs and sperm are fully developed the corals wait for the full moon when gamete release takes place. This peaks at the fifth night after first release. The over-abundance of gametes ensures fertilisation and production of planula larvae even after heavy predation. However, not all corals adopt this pattern and there are species which brood.

Here fertilisation occurs within the polyps themselves and the planula larvae partly develop inside before emerging. The free-swimming larvae, produced by both methods of fertilisation, attach themselves to a suitable substrate and a new coral colony begins.

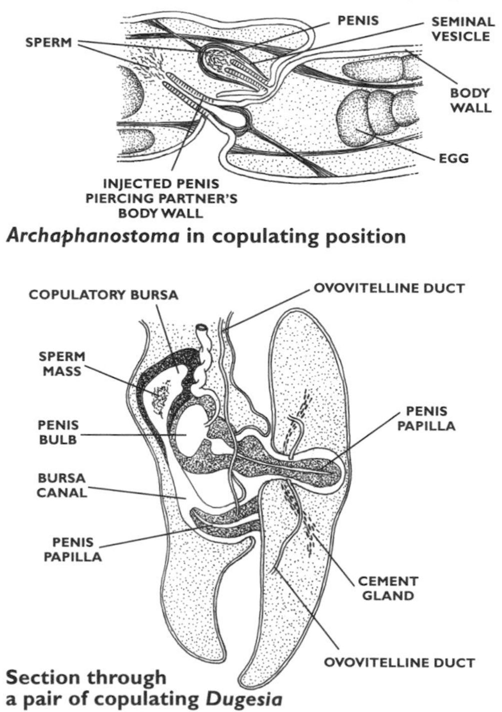

***Archaphanostoma* in copulating position**

Section through a pair of copulating *Dugesia*

Figure 138. Copulation in two species of flatworm.

Many marine invertebrates are viviparous. The European Oyster *Ostrea edulis* broods its larvae in its gills. The term 'white sick' describes the creamy egg accumulation in the oyster and 'black sick' when pigmentation has been laid down in the developing larvae. The viviparous *Ostrea* produces only 1.8 million eggs; however, it is estimated that oviparous oysters (e.g. *Crassostrea*) need to produce 100 million eggs to achieve an equivalent rate of larval survival.

To ensure fertilisation many marine invertebrates, other than the cnidarians, adopt internal fertilisation. Flatworms use this method and each animal has both male and female sexual organs, including a penis (fig.138). The flatworms' hermaphroditic condition and their internal fertilisation increase their chances of successful reproduction. Successful matings have to occur in all encounters with the same species, hence many small marine invertebrates are hermaphroditic. The penises of flatworms are muscular, sometimes with hollow stylets, although some species may have multiple penis bulbs and stylets. These organs are sometimes used as a means of defence. The male gametes are stored for use in the bursa or the seminal vesicle (or both in some cases), while the female parts may temporarily store the eggs in the uterus. Copulation is usually reciprocal, the stylet being rammed through the body wall of the partner to inject the sperm into the interstitial cells (parenchymal cells). After fertilisation the eggs are released and capsules may be attached to stalks on the bottom. This method of reproduction is common to most polyclad flatworms whose eggs do not produce larvae but develop directly into young flatworms.

Polychaete worms are thought only to reproduce sexually. In ancestral worms most of the body segments probably produced gametes and this is normal for the majority of present-day polychaetes. Those polychaetes with more distinct segmentation (e.g. ragworms) normally have gonads confined to the anterior segments. The hermaphrodite fanworms produce eggs in the anterior segments and sperm in the posterior segments. Once the gametes have been produced by the gonads they reach the outside through external openings. Some polychaete worms, such as the capitellids (e.g. *Capitella capitata*), have separate gonoducts, one per segment. Others use the nephridiopores,which are the openings for the excretory organs, the nephridia. The small, delicate free-living syllid polychaetes have swollen nephridia as their gonoducts. This is also the case for the frontal-horned burrowing spionids, but ragworms use the anus to release their sperm. A few syllids and the large free-living eunicids, become pelagic at sexual maturation. When they reach the sea surface the body wall ruptures to release the sperm and eggs, after which the adult dies.

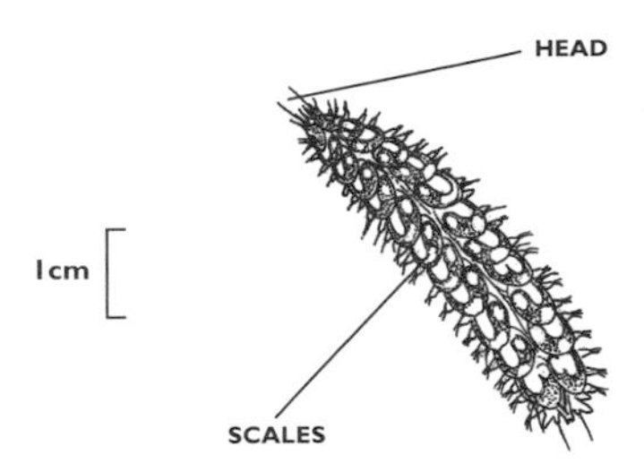

Figure 139. Dark-marked Scale Worm *Lepidonotus melanogrammus.*

Scale worms (Polynoidae and Sigalionidae) such as the Dark-marked Scale Worm *Lepidonotus melanogrammus* (fig.139), may brood their developing young after fertilisation. The plate-like scales cover the eggs throughout early development before the larvae are released. *Serpula vermicularis,* a serpulidian similar in form to a fanworm, has a calcareous white tube that encrusts empty shells and within this the early stages of the young are brooded.

Some polychaetes produce an epitoke, which is a pelagic reproductive form. This occurs in the nereids, syllids and eunicids, and one of the best known examples is the Samoan Palolo Worm *Eunice viridis* (fig.140). When mature, the posterior epitokal region consists of a chain of egg-filled segments which separate from the atoke region at a specialised joint between the segments. This type of reproductive behaviour occurs at the beginning of the last lunar quarter in October or November around the shores of Samoa. When the time is right there is a mass release of the white writhing epitokes, which are harvested by the natives as a great delicacy. Similarly the West Indian Palolo Worm, *Eunice schemacephala,* swarms in the last quarter of July at about 0300. By dawn the epitokes in both species that have survived predators are lying in the surface water. At first light, they rupture releasing their gametes. Fertilisation ensues soon afterwards. The resultant fertilised eggs attain the ciliated larval stage only a day later and these eventually sink to the bottom to begin a new generation of palolo worms.

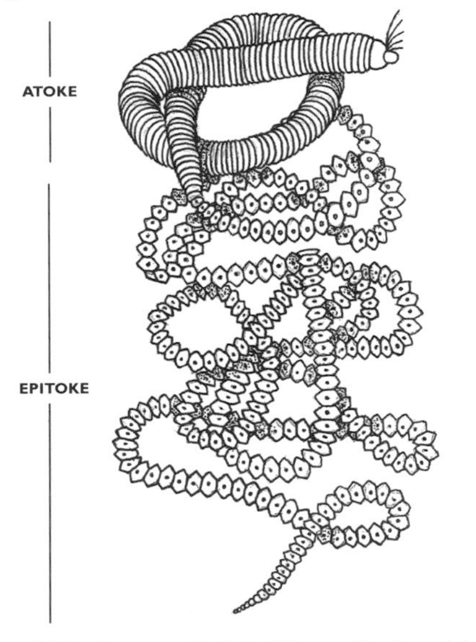

Figure 140. Samoan Palolo Worm *Eunice viridis.*

Crustacean copepods and decapods transmit their sperm in neat packages called spermatophores. These are delivered to the female by a pair of copulatory pleopods on the male during mating. In most of the marine decapods this occurs shortly after moulting. The sexes are attracted to each other by pheromones produced before or just after ecdysis. These more complex marine invertebrates often have a courtship ritual. The male Hermit Crab *Pagurus bernhardus* (plate 3) taps and strokes the female and even rocks her to coerce her into mating.

To be first to mate with a receptive female often dominates the behaviour of mature male Cancridae, (for example the Edible Crab *Cancer pagurus* (plate 4) and the Red Rock Crab *Cancer productus* (plate 2), and male Portunidae, such as the Common Shore Crab *Carcinus maenas.* The male crab even attends the female before she moults, often carrying her beneath his sternum. Other crabs attract mates using an enlarged cheliped. Tropical fiddler crabs are seen signalling near their sand burrows, each species having its own particular signalling ritual (fig.141). Acoustics may also be used in association with signalling, either by hitting the elbow of the cheliped on the sand or by rapid flexing of the walking legs. Male ghost crabs use this technique, while the Stalk-eyed Ghost Crab *Ocypode ceratophthalmus* makes a platform for its rapping. In most cases shrimps, lobsters and crabs take up positions opposite each other for the mating ritual and the hermit crab partially emerges from its shell. Fertilisation occurs externally at the same time as egg laying.

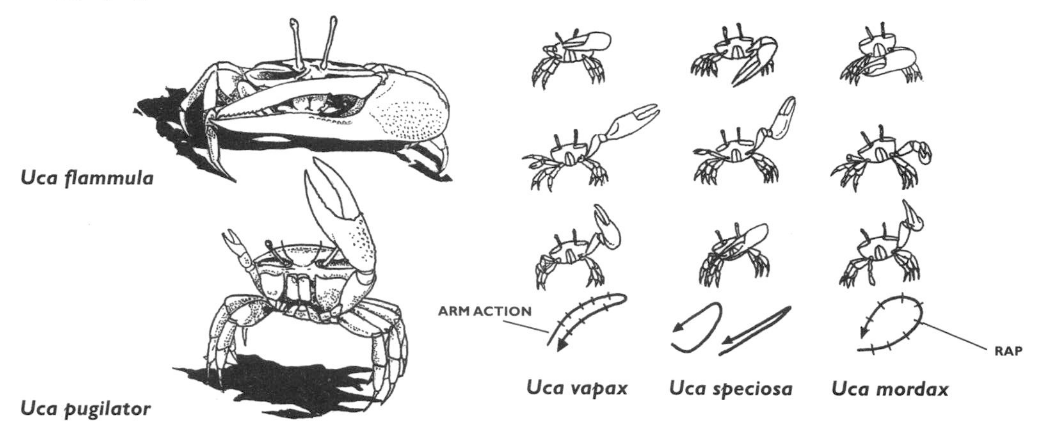

Figure 141. Two tropical fiddler crabs and the mating 'semaphore' signals of three species of fiddler crab.

Once fertilisation has taken place some species release their eggs into the sea while others retain them on their body surfaces. The majority of shrimps, lobsters and a few crabs develop special egg-carrying chaetae on their limbs where the eggs are kept until hatching. Crabs carry their eggs under their abdomens attached to fine chaetae and are said to be 'in berry'.

Molluscs can also be attentive parents, for example the European Oyster *Ostrea edulis*. However, female octopuses are probably among the most dedicated of invertebrate parents. Female octopuses hang their fertilised eggs in strings from rocks in caves or crevices. Here they remain in attendance, aerating and cleaning the eggs with their siphons, without eating until hatching occurs. The female of the Common Octopus *Octopus vulgaris* is so weakened by constantly brooding her eggs that she subsequently dies.

Most remarkable of all octopuses is the pelagic Brown Paper Nautilus (Argonaut) *Argonauta nodosa* (plate 1). From enlarged membranes on the female's two dorsal arms a paper-thin shell is secreted. This is not used as a home by the female but as a delicate floating brood chamber into which she eventually lays her eggs. The shell takes time to secrete, therefore these opalescent calcareous brood chambers are slowly constructed from birth. Towards the end of the female's life, a suitable male mates with her and after mating he dies. The female then lays her eggs in the shell and positions herself with her posterior just inside the egg filled case. The female Argonaut then drifts and swims to spawning grounds of sea grass meadows, shallow reefs or sandy flats. Here she dies and thousands of neatly packed tiny Argonauts hatch which eventually leave the shallows and head out to sea.

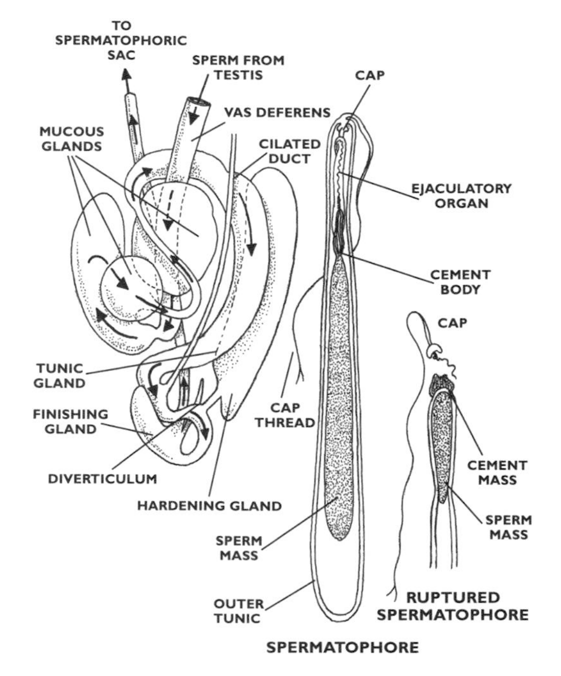

Figure 142. The spermatophore and male genitals of the Atlantic Squid *Loligo vulgaris*.

The male Chambered Nautilus *Nautilus pompilius* (plate 1) has a copulatory organ which it uses when the partners mate head on. The eggs produced after fertilisation have complex capsules that are individually attached to the substratum. Octopuses have a spoon-like depression on the tip of a specialised arm used in mating to deposit the adhesive spermatophores in the female, and the female Argonaut also has a cavity to store the sperm.

A male cuttlefish (e.g. the Common Cuttlefish *Sepia officinalis*, plate 4, and *Sepia latimanus*, plate 7) performs various displays to identify itself to a potential mate. He normally assumes the striped pattern (fig.131) which temporarily bonds the two partners and deters other potential suitors.

The male can then present his spermatophore via the hectocotylus (the modified arm similar to that found in the nautilus and octopuses) to the buccal membrane of the female. Like most other cephalopods, each spermatophore consists of a tapered tube filled with the sperm mass, a small cement body and a coiled ejaculatory organ with a cap (fig.142). Once the cap is removed by tactile friction during the placing of the spermatophore, the ejaculatory organ uncoils, launching the sperm mass out of the case. The cement body which accompanies the sperm adheres to the mantle of the female, or her buccal wall, and there the sperm is slowly released. Once the 100 or so eggs are fertilised, the female attaches each group onto seaweed or rocks where they look like small bunches of black grapes. She will remain with her eggs until they hatch. Both sexes die soon after mating and brooding.

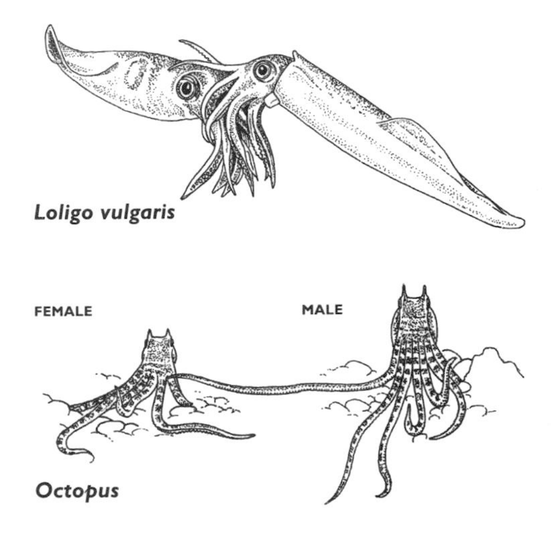

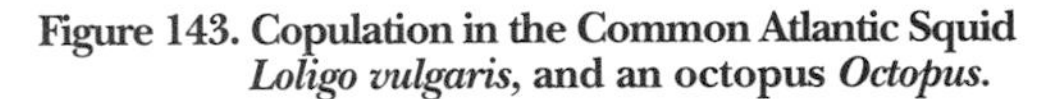

Figure 143. Copulation in the Common Atlantic Squid *Loligo vulgaris*, and an octopus *Octopus*.

Squid show similar patterns of behaviour but because of their pelagic existence there are some variations. For example, the Common Atlantic Squid *Loligo vulgaris* congregates in large numbers to copulate and spawn. Copulation takes place while swimming head to head (fig.143), the male inserting the spermatophore into the female's mantle cavity under her mouth. Once the eggs are fertilised by the sperm the female then attaches them to the sea bed in clutches of 10 - 50 strings, each with up to 100 eggs. Mass spawnings may result in a large mound of strings on the bottom. The capsules harden and swell to at least twice their original size. The adults die and a new generation is ready to hatch.

Other molluscs such as the Dogwhelk, common along Atlantic temperate coasts, lay egg capsules on the seabed which are sometimes found washed up on the beach. They usually lay their eggs in large groups as is generally the case in neogastropods (olives, drills, mitres, volutes and cone shells). Mesogastropods and opisthobranchs (worm shells, slipper shells, cowries, moonshells, tritons, bubble shells, sea hares, pteropods and other nudibranchs) lay a gelatinous mass, whereas in the neogastropods the gelatinous mass has a leathery covering (fig.144a). The Common Cerith *Cerithium vulgatum* and the Common Periwinkle *Littorina littorea* produce this gelatinous mass from a gland surrounding the oviduct. In higher prosobranchs capsules are produced by secretions from a modified jelly gland called the capsule gland.

Mating among these animals is similar to that of flatworms. Males have a penis and internal fertilisation takes place. The gonads of both sexes are associated with the excretory organs, the nephridia. In more advanced species the right nephridium has degenerated, the top half being modified to form a genital duct that reaches the outside near the opening of the mantle cavity via the ciliated pallial duct. Sperm storage enhances the reproductive success of the neogastropods (fig.144 d, e); this is reflected in their diversity and huge numbers.

Two other groups of gastropods, the patellacean limpets and the slipper limpets, exhibit not only hermaphroditic reproduction but also protandric hermaphroditism. One common example is the Slipper Limpet *Crepidula fornicata* (plate 3) that lives as chains of individuals, with the foot of one limpet anchored to the top of the shell of the next. The youngest and the smallest member is at the top and is always a male. Individuals immediately below and in the middle of the pile are intersexes. The largest members at the base are mature egg-producing females which use the sperm stored from earlier insemination to fertilise their eggs. The sex of each individual is influenced by the overall sex ratio of the group, probably mediated by pheromones. Older males will remain male longer if attached to a female, but once removed, they will change sex. Once an individual has become female it will remain female for life. Sex changes may occur in the Common Limpet *Patella vulgata* as the population of younger individuals has a high proportion of males, indicating that maturity may bring about feminisation.

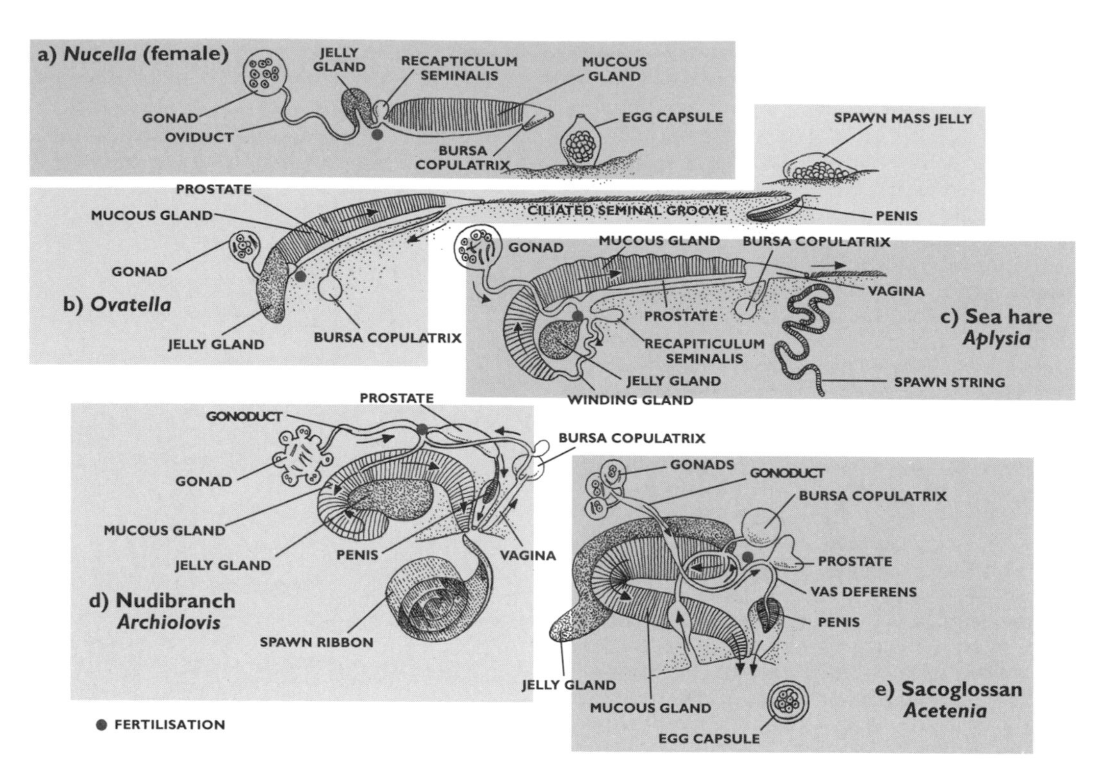

Figure 144. Gastropod genital duct system and jelly gland. The form of the egg cluster is shown for each of the five species. Arrows show the movement of gametes and dots (●) indicate the region where fertilisation is believed to occur.

A small number of bivalves show protandric behaviour; the European Oyster *Ostrea edulis* may change sex many times during its life cycle. All the young start off as males, but once they have shed their sperm they become female. The females are then fertilised by sperm shed into the surrounding water by younger individuals. Once the fertilised female has shed her eggs she reverts to being a male. The frequency of this sex change is dependent upon water temperature and food availability. In colder waters this happens biannually but in warmer temperate waters it may occur several times a year. Each individual will be in or out of phase with a neighbour, allowing the population to have both sperm and eggs present at any one time. As with all bivalves, the sperm is released through the mantle cavity and is sucked in by the inhalant current of the female.

Sea urchins have a similar method of reproduction to that of most bivalves, shedding their sperm and eggs into the sea water (see Cake Urchin *Tripneustes gratilla*, plate 8). The eggs are normally fertilised in the surrounding water, but some cold water species and heart urchins brood the eggs around the peristome (the mouth and adjacent spines). Irregular sea urchins, like the sand-dollars, also brood eggs in deep concavities on their petaloids (aboral grooves on their upperside).

Starfish likewise release their sperm and eggs into the surrounding water and fertilisation takes place externally. This normally happens once a year, with some females releasing up to two and a half million eggs. In many polar species, however, the yolky eggs are held on the underside at the base of the tentacles. The eggs develop directly into small starfish with no planktonic larval stage. Along the cold temperate Atlantic shores lives the Cushion Star *Asterina gibbosa* which, while it does not brood as such, exhibits some brood care by attaching its eggs to stones, seaweed and other substrates.

By contrast sea cucumbers only possess one gonad and unlike other echinoderms are dioecious. The majority of the 900 species of sea cucumbers are external fertilisers, with only 30 brooding species. Most of these species, as with starfish, are polar in distribution, the majority being found in Antarctica. Two species of sea cucumbers, *Thyone rubra* (found in Californian waters) and the Worm Cucumber *Leptosynapta inhaerens* of the North Sea, are known to nurture their young within the coelom of the female. The young eventually leave by rupturing the anal region.

Crinoids such as the Noble Feather Star *Comanthina nobilis* (plate 6) have no distinct gonads: the gametes develop from germinal epithelium within the expanded coelom of the pinnules. Rupturing of the pinnule wall releases the eggs or sperm. In some species eggs are stuck onto the outer surface of the pinnules. The eggs then hatch to produce larvae. Like all other echinoderms, cold water species brood their young. These crinoids have saclike tucks of the pinnule near the genital canal, the eggs entering there by the rupture of the canal wall.

In the mostly hermaphroditic sponges, the eggs and sperm are released at different times. The sperm are produced from the flagellated choanocyte cells and the eggs develop from either the choanocyte or archaeocyte cells. The egg cells engulf adjacent nurse cells to increase their food reserves. When ripe the sperm plumes out of the exhalant siphon, sometimes in large milky clouds up to two to three metres high. This may trigger other adjacent sponges to 'ejaculate', causing a synchronised release of sperm into the surrounding water. The sperm may then be sucked into the inhalant siphon of another sponge of the same species (fig.145). Once inside the flagellated chamber they enter the cell cytoplasm of a choanocyte. Once the sperm has been transferred to the egg cell by the choanocyte, fertilisation occurs *in situ*. Development of the larvae is normally viviparous in sponges, but some Demospongiae are known to be oviparous, the eggs developing in the surrounding sea water.

Like the sponges, bryozoans are mainly hermaphrodite and produce eggs and sperm from the same colony. Sperm and eggs are normally released at the same time. Male zooids may have many testes, the developing sperm bulging into the basal coelom of the animal. When ripe the one or two ovaries

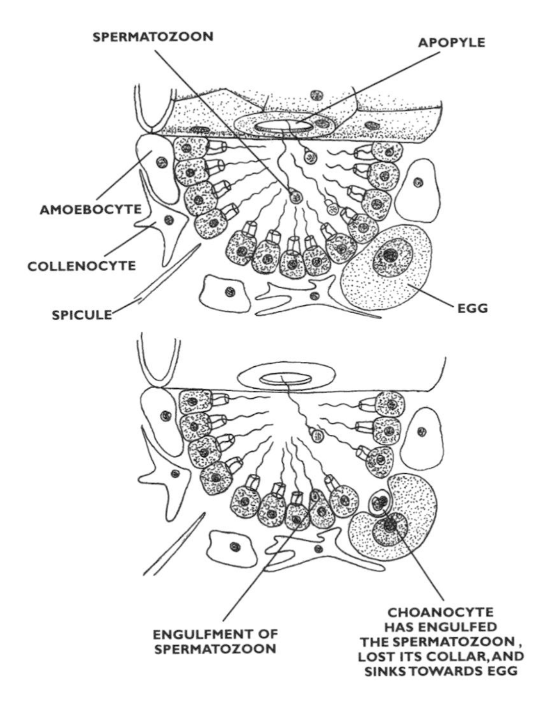

Figure 145. Fertilisation in a typical sponge. Top, entrance of sperm into flagellated chamber. Bottom, choanocyte engulfs sperm and sinks towards egg.

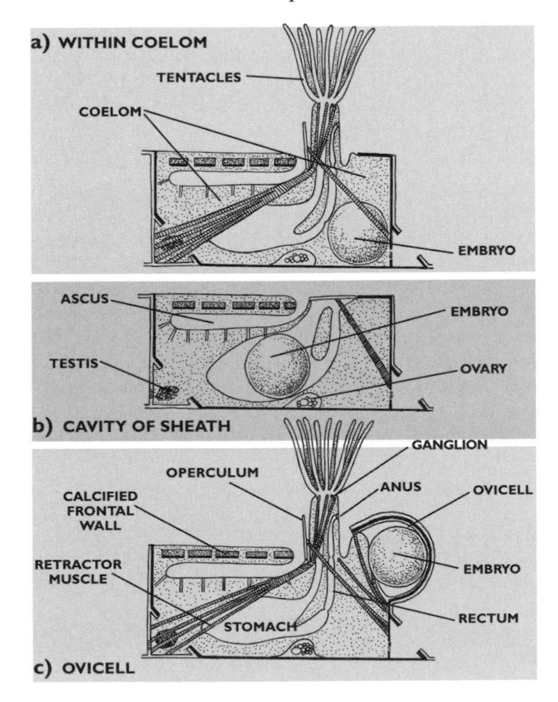

Figure 146. Bryozoan brood chamber a) coelomic, b) degeneration of lophophore and gut c) ovicell type.

of the females constrict the distal end of the coelomic cavity. *Electra* and *Membranipora* (see *Membranipora membranacea* on plate 2) shed their eggs into the surrounding water, but most species brood their fertilised eggs. A small number of species retain the eggs within their coelom (fig.146a), others resorb the digestive tract and lophophore forming an external area for the eggs (fig.146b). Some may use the cavity of the tentacular sheath or tucks of the atrial wall for brooding. Other similar species such as *Bugula neritina* (plate 4) have an external chamber, the ovicell, where a single egg is brooded. The egg's nutrition may come from a yolk or, as in *B. neritina,* a placenta-like connection (fig.146c) from the maternal zooid. Once the egg has developed the larva is released via a special opening in the lophophore which is sometimes mounted on a tube-like projection.

Solitary ascidians are similar to the sponges as they shed their sperm via the exhalant (atrial) siphon. The young of viviparous species are also shed from this opening. Oviparous ascidian eggs normally have a floatation membrane or, in the case of colonial species, a richer yolk. These are normally brooded in the atrium or sometimes inside a special pocket formed from the atrial wall. Hatching occurs at the larval stage, and some larvae may reside in the atrium throughout their development to miniature adults.

Adult marine invertebrates are usually either sessile or slow moving which makes their dispersal difficult. The vast majority of species have a larval stage which is the dispersive phase in the life cycle.

LARVAE AND SETTLEMENT

The plankton consists of myriads of pulsating, swimming and drifting, transparent and translucent larvae. Many planktonic organisms are invertebrates, their form often bearing no resemblance to their adult appearance. Larvae are generally transparent except for certain internal organs. This makes them nearly invisible in the surrounding waters. Their often bizarre long processes and lobes increase their surface area and help to slow down their rate of descent from the surface waters. Oil droplets or ammonium chloride are incorporated in the body cells of larvae and form a buoyancy mechanism.

The change (metamorphosis) from a planktonic to adult morph is dramatic in some species. Often much of the larval body form may be discarded and the adult then develops from just a small part of the original larva. The advantages of this drastic process are not always clear, especially in view of the high predation levels on planktonic larvae.

Dispersal is the most obvious adaptive feature of planktonic larvae but there are others. Feeding and growth in an environment where food is abundant is important to species the adult habitat of which is either competitive or does not provide ample food resources. In addition there can be no competition for food between larva and adult during the period when the larvae are dispersed.

However, the balance between larval success and failure through predation is a fine one. It has been speculated that an overproduction of larvae is a waste of adult resources or, worse still, if too many new adults survive the food supply is exhausted. By contrast too few successful larvae would lead to a steady or even dramatic decline in a species' population.

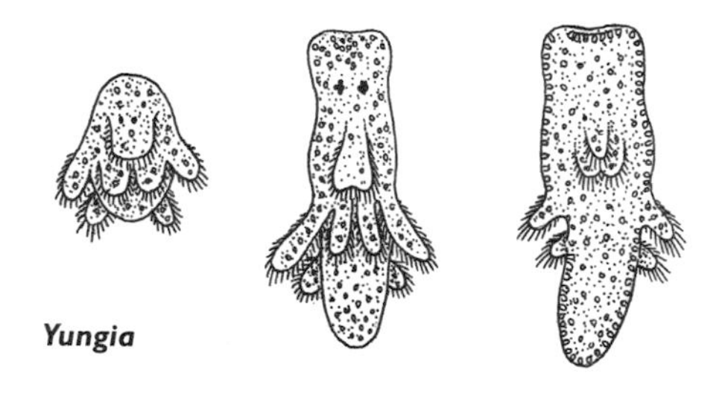

Figure 147. Metamorphosis of a Müller's larva.

Not all invertebrate larvae are planktonic. Some cnidarian planula larvae use ciliary action to creep about over the bottom, after being dispersed by their pelagic parents, the medusae. Other planulae may utilise cilia to swim, but few of these are found in the plankton; it is the sexual stage (medusa) that lives in the plankton.

Surprisingly few marine flatworms produce larvae and only a small number have a free-swimming stage (fig.147). Those that do, produce a small Müller's larva that lives only for a few days in the plankton before settling to the bottom to become a young flatworm.

Trochophore larvae, the larvae of the polychaete worms, are common in plankton. All are similar in form and have long flagella-like cilia at the apex (the apical tuft), and a girdle of cilia around the widest part of the body (the prototroch) (fig.148). These larvae often have two pigmented eyespots and a nerve net under the apical tuft. The mouth is positioned under the prototroch which is conveniently placed to collect the food that is captured and transported by the cilia. Polychaete eggs have little yolk and the food stores are rapidly exhausted during trochophore development. Consequently the larvae start to feed early in life. As the larva grows more girdles of swimming cilia are formed. When the trochophore reaches the stage of metamorphosis, the larva sinks out of its planktonic habitat to become an adult.

Many species which live only one or two years and spawn only once produce large numbers of small eggs whose trochophores are well developed, feeding in the plankton for over a week. Perennial species, breeding more than once a year, produce a small number of large yolky eggs. These trochophores are benthic and do not feed, developing quickly instead into small polychaete worms. In polychaetes that have a short life span with several generations per year, similar larvae to the latter group are produced with yolky eggs and a short benthic trochophore phase.

The planktonic larvae start their metamorphosis by budding off segments from the hind end, becoming increasingly worm-like. At this stage the sabellids and other similar polychaetes have chaetae on the last segment that become elongate, thus helping the larva to maintain its position in the water column and to deter some smaller predators. Budding continues as the larva grows until the body becomes too large and heavy to remain in the planktonic environment and gradually sinks to the bottom. When the larva reaches a suitable substratum it casts off its cilia and begins a sedentary existence.

The evolutionary link between the annelids and the molluscs can be seen in their larval stages. Chitons, limpets and some bivalves have a trochophore as their first larval stage (fig.149). It is short lived, rapidly transforming into a ciliated diaphanous larva (the veliger). It has two round-headed lobes with a fringe of cilia surrounding the lip, a foot region and a gut covered by a horny larval shell (fig.149). Those molluscs

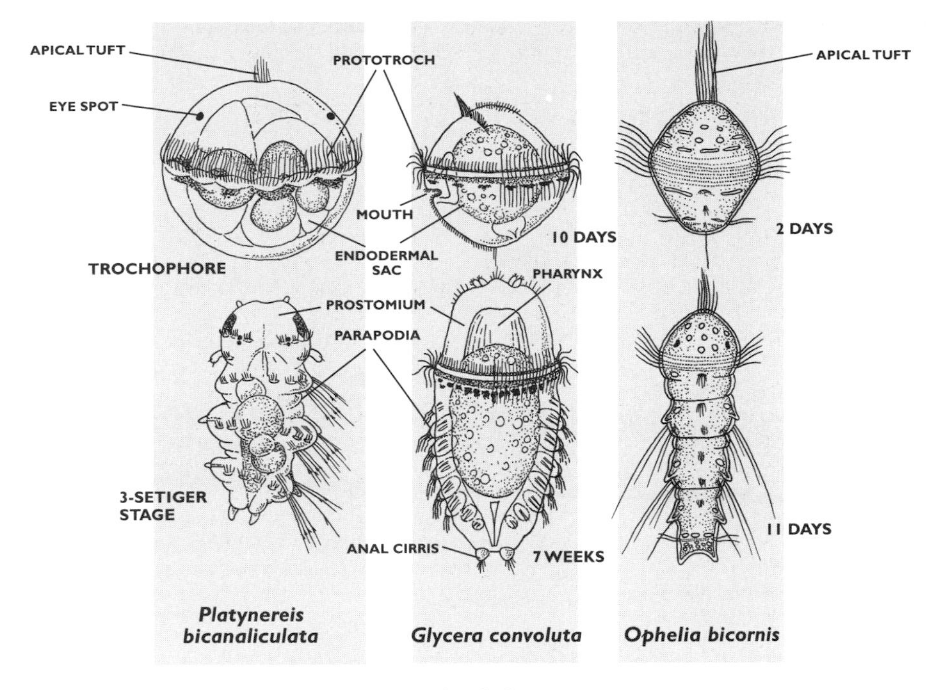

Figure 148. Trochophore larvae of three species of polychaete worms.

that do not disperse their larvae, retain the veliger in the egg, from which a small gastropod emerges (e.g. some marine prosobranchs). Other molluscs omit this stage and hatch directly into young adults (e.g. cephalopods). A few gastropods have planktotrophic veligers that live for almost three months feeding in the upper layers of the sea before metamorphosing into adults. Others may have a brief yolk-laden lecithotrophic veliger stage that lives only a few hours in the plankton. These are normally produced in much smaller numbers than the longer-lived planktonic types.

The veliger stage of all marine gastropods undergoes torsion, whereby the whole top of the body twists round 180° on the foot. This torsion is not only an advantage to the adult, permitting it to withdraw a larger and more developed head (as postulated by Stasek 1972), but also gives the veliger the ability to withdraw its ciliated velum, allowing it to plummet out of the reach of a predator's attack.

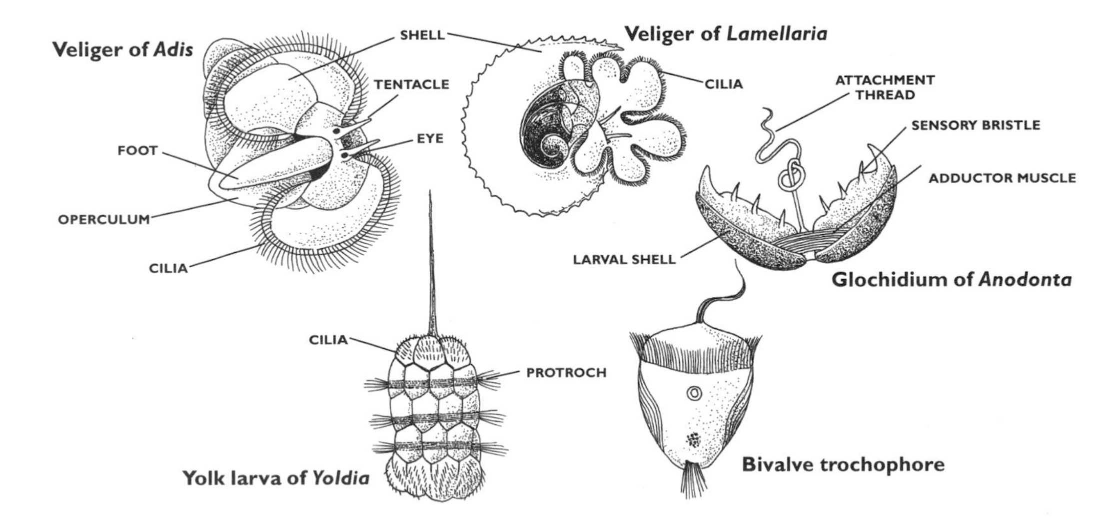

Figure 149. Trochophore and veliger larvae of molluscs.

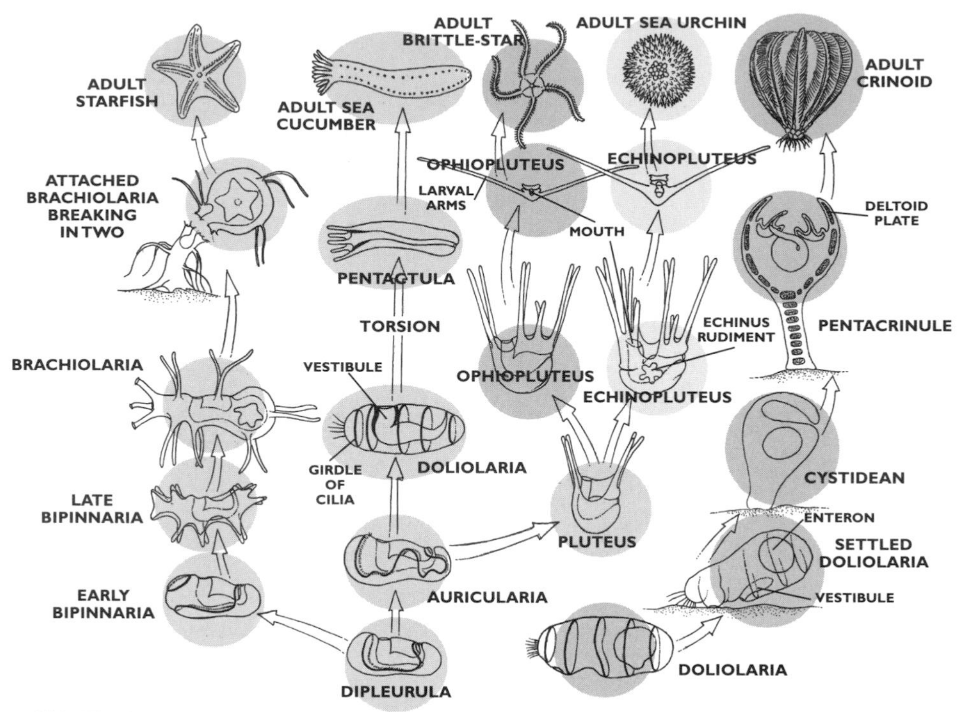

Figure 150. The larval types of echinoderms and metamorphosis to adult stages.

Echinoderms have different types of larvae from those of the molluscs and polychaetes (this indicates a possible differing evolutionary lineage). The basic form of the larva (dipleurula) (fig.150), because of the convoluted ciliated band that circumscribes the body and the four lobes, has a heavily ciliated appearance. Each of the classes of Echinodermata have their own larval variations. The sea cucumbers have early auricularian larvae with ear-like ciliated flaps. As the larva progresses in its growth it develops into a barrel-like form (doliolaria) that has three to five ciliated girdles (fig.150). Some species of sea cucumber produce a non-feeding barrel-shaped vitellaria. This type of larva may also be found among the crinoids and a few of the brittlestars. From either the vitellaria or doliolaria larval stage there is a gradual metamorphosis, with little loss of larval features, and the gradual appearance of the buccal tentacles. The result is a larval stage called the pentactula. Finally the functional podia appear and the young sea cucumber settles on the substratum to begin adult life.

Sea urchins have echinopluteus larvae shaped like an upside down umbrella with six fine long arms supported by upward pointing calcified rods edged with a ciliated band (fig.150). Development of the larvae may take several months while the adult exoskeleton is slowly forming. After a period in the plankton the larvae sink to the bottom where a metamorphosis takes place in less than an hour.

Starfish development is even more dramatic. The ciliated bipinnaria larva gradually develops a heavily ciliated band above the mouth that fuses into a closed loop. These ciliary extensions are far more flexible than those of the echinopluteus larva and just before metamorphosis three additional arms appear at the anterior end producing the branchiolaria larva. These arms have sticky tips that attach to the substratum and once fixed to the bottom metamorphosis takes place. The young starfish develops rapidly at the posterior end of the larva, the anterior end degenerating to form only an attachment stalk (fig.150 & 151). The one millimetre-wide starfish then separates from the stalk

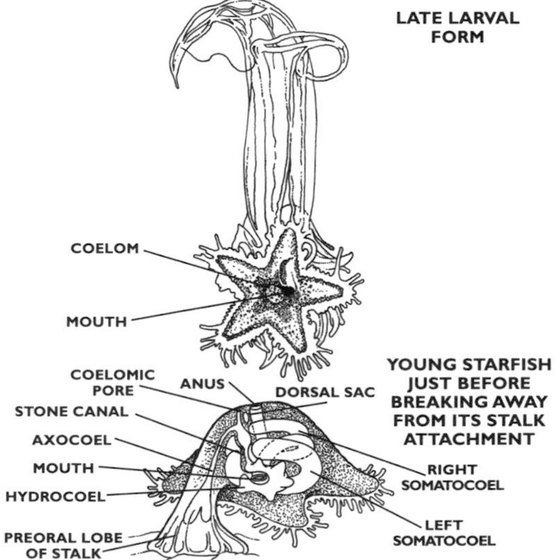

Figure 151. Metamorphosis of a starfish larva.

leaving the remains of its larval body behind.

The ophiopluteus brittle-star larvae have four pairs of distinctive elongate arms (fig.150). Unlike the starfish, metamorphosis occurs while the larva is still free-swimming, with no attachment stage. The development from egg to young brittle-star takes between 14 and 40 days. In single-brooded and oviparous brittle-stars the larval stage may last up to seven months.

The larvae of crustaceans are more akin to the adults than many other invertebrate larvae. Free-swimming, planktonic larvae are characteristic of most marine species. The basic form is the nauplius, a pear-shaped, pellucid, thin-cuticled larva with three pairs of limbs including the antennules (fig.152). The nauplius larva moults several times, adding successive trunk segments and additional appendages during a series of post-naupliar stages before the adult stage is reached (fig.152).

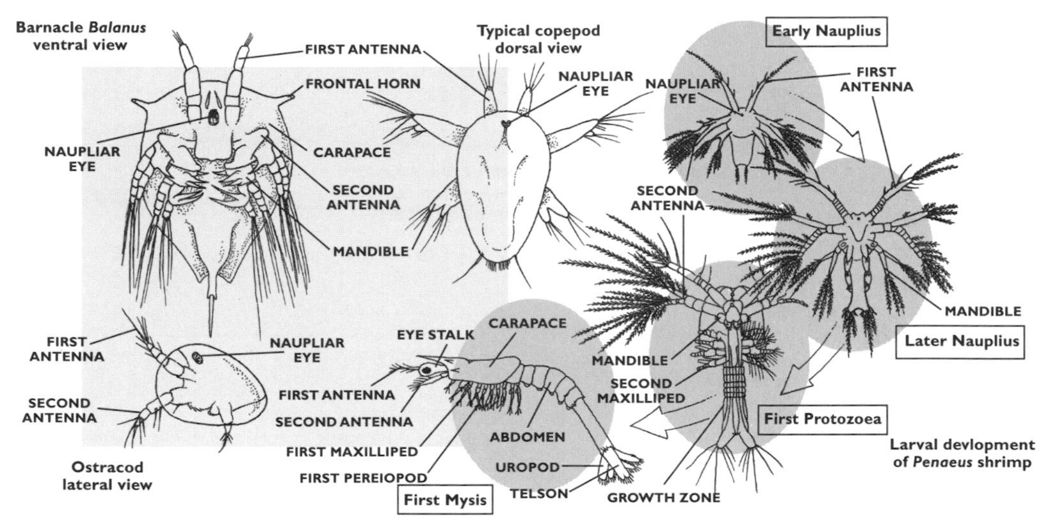

Figure 152. Nauplius larvae in the crustaceans and development of the shrimp *Penaeus*.

Larvae of large crustaceans, such as the crabs and many shrimps, have a zoea stage. These larvae have a jointed abdomen with a spined carapace behind the head and large eyes (figs.70 & 153). The glassy zoea of a porcelain crab is quite bizarre; the animal seemingly hangs from a thin horizontal strut (fig.70). This is the long thin head spine, many times longer than its body length which helps to maintain the creature's position in the water column and may even prevent the larva from being swallowed by some predators. In many species the stage succeeding the zoea is the post-larval form and is similar to the adult form. In others there may be marked differences between the post-zoea and adult appearance. In the crabs, where this stage is called a megalops larva, the cephalothorax (the carapace section) is similar to that of the adult but the abdomen is still large and extended (fig.153). However, after further additional moults all the adult features are present. The diversity of this subphylum can be judged from the variety of these larvae and their various stages of development. Among the shrimps and crabs the nauplius stage is omitted and the zoea hatch straight from the egg. In other groups, such as the peracarids, the post-larva is the only pre-adult stage and all other larval stages may be suppressed. Cold and deep water crustaceans, as in many other invertebrate groups, tend to have shortened larval stages.

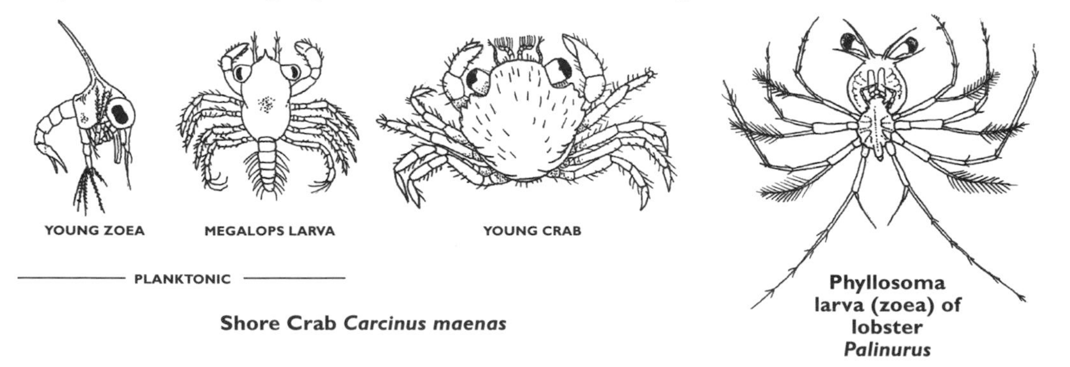

Figure 153. Zoea and megalops larvae of the Common Shore Crab *Carcinus maenas*, and a zoea of a lobster.

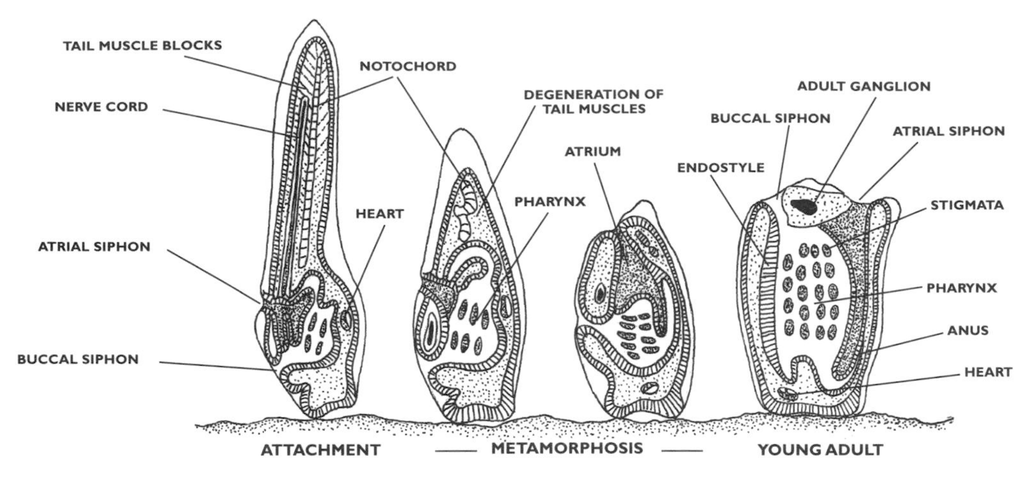

Figure 154. The metamorphosis of a sea squirt's tadpole-like larva to a young adult sea squirt.

The tadpole-like larvae of the tunicates have rounded heads with sharply demarcated tails (fig.154). The tail, which they use for swimming, has an elastic strengthening rod, the notochord, to which the muscle blocks are attached. These tadpole larvae have a free-swimming period of up to 36 hours but in the genus *Botryllus* it can be as little as a few minutes. When the ascidian larva metamorphoses into the adult stage, it cements its head to the bottom and casts off its tail. The adult is formed from only the head region with its associated pharyngeal basket (fig.154).

In one group of tunicates the larvae have become sexually mature, producing a different type of adult. This group, the larvaceans, has about 70 known species present in surface plankton throughout the oceans. They have retained some larval characteristics and are said to display neoteny. All larvaceans secrete a gelatinous 'house', such as *Oikopleura* (fig.155) with its walnut sized 'glasshouse' that it sheds and replaces every four hours.

Thaliaceans are urochordates that include the planktonic salps and doliolids. These alternate their existence between a solitary larval sexual stage and a colonial asexual stage that is similar to some cnidarians.

Pelagic larvae are able to detect those sites that would be advantageous to the sedentary adult stage. In some cases they are able to postpone metamorphosis for long periods, until they find a suitable site. The larvae appear able to recognise environmental clues as to a substrates' suitability.

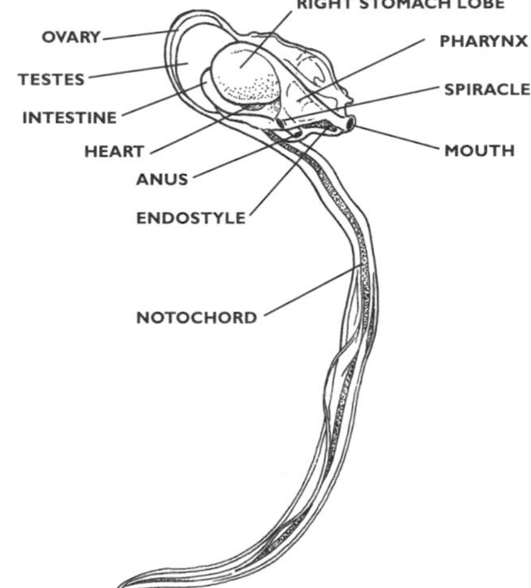

Figure 155. The larvacean *Oikopleura* ('house' omitted).

Due to problems caused by attachment of the Acorn Barnacle *Semibalanus balanoides* to the hulls of ships and their cooling conduits, its mechanism of settlement has been extensively studied. The larvae that hatch from the barnacle's eggs are nauplius larvae (fig.156), which is a dispersive stage. *Balanus glandula* may produce two to six broods of up to 30,000 nauplii every spring and winter, which allows a good chance for at least some of these larvae to survive. It is the next larval stage that finds an appropriate substratum for settlement. After six moults the nauplius changes into a bivalved cypris larva (fig.156) with its characteristic pair of compound eyes (the median single nauplius eye is also present). This is a non-feeding settling stage, crawling around on the surface of rocks or suitable attachment areas. It is sensitive to water currents (weaker currents encouraging attachment) illumination, depth and most importantly chemical stimuli.

Contact with quinone-tanned proteins that form the cuticles of settled larvae of adults or dead barnacles, encourages further settlement. This stimulus is not species specific and the Acorn Barnacle is induced to settle through cuticular proteins used by any one of three different species. Some settlement areas can become over-dense, so much so that special 'tussocks' are formed of long thin individuals. Gregariousness is an advantage to all species that are immobile, as they have to be in close proximity to copulate. Barnacles are mostly hermaphrodite and in their role as a male they protrude a penis, up to twelve millimetres long, into the shell plate of a neighbour.

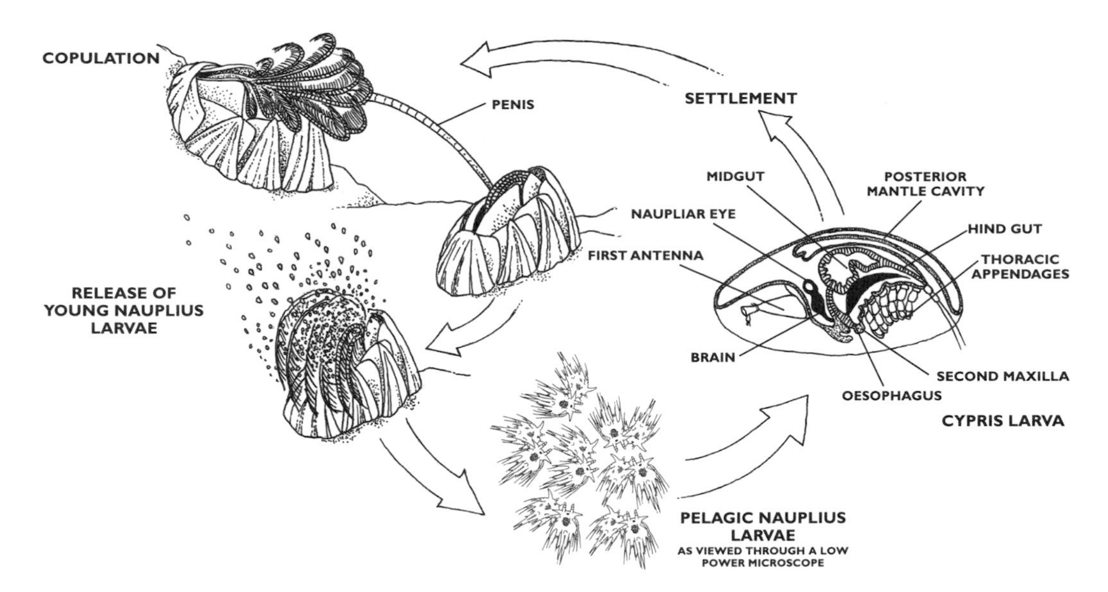

Figure 156. The life cycle of the Acorn Barnacle *Semibalanus balanoides.*

If a cypris larva does not find a suitable settlement site then it may become incapable of metamorphosis and eventually die. In some cases when the biological time period has just passed, a faulty metamorphosis may occur.

The right chemical stimulus does not always have to be present, as barnacles will settle on freshly painted hulls. These pioneering barnacles will encourage others to settle which induces further settlement, creating a chain reaction of barnacle infestation. If these pioneers could be deterred in the first place, then hulls and cooling ducts might be kept clean. Antifouling paint is the normal method of prevention, but it is too broad a toxin and destroys many other benign forms of animal life within close proximity. Further research is needed to understand the mechanism involved in settlement and only then may a specific antifouling agent with no harmful environmental side effects be produced.

Understanding larval settlement can have positive advantages. The European Oyster *Ostrea edulis* has been cultivated for over two millennia. Ropes with dead shells or piles of stacked curved tiles, called 'parcs', are placed in the water as suitable settlement sites for the larval oysters. Like gastropods, the European Oyster has a long-lived planktonic veliger which may feed for up to two weeks before settling (fig.157). The first stage of settlement involves the veliger coming into contact with a suitable potential substratum. It then crawls over the surface, anchored by a secreted thread, to explore it. Should it come into contact with traces of secretions from previous individuals it is induced to metamorphose and settle. However, its nutritional state and the length of time the veliger has been swimming also play a part in the final stimulation to shed the vellum. The young adult then secretes a powerful cement attaching itself firmly to the substrate.

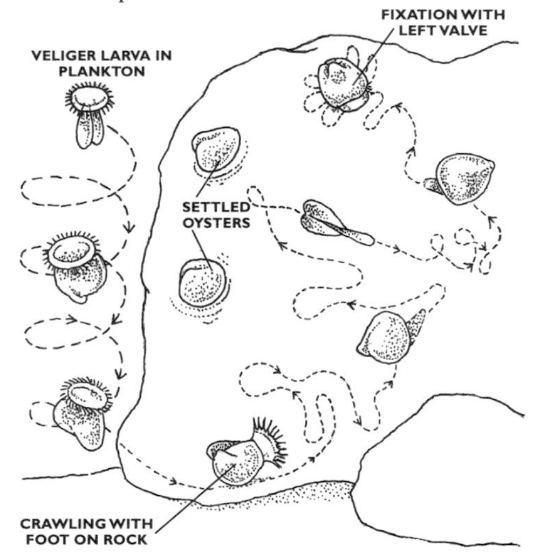

Figure 157. Larval settlement in the European Oyster *Ostrea edulis.*

Polychaetes are also gregarious settlers. The sabellid *Sabellaria alveolata* builds extensive sand-based tube reefs that are common along sandy beaches (fig.158). The chemical secretions in the tubes of already settled individuals attract more trochophores to metamorphose adjacent to these areas.

The trochophore of *Spirorbis borealis* is induced to settle out of the plankton by the chemical stimulus of algal fronds on which these coiled, chalky, tube-dwelling polychaetes live. In experiments, extracts of algae have induced the larvae to settle in seawater samples.

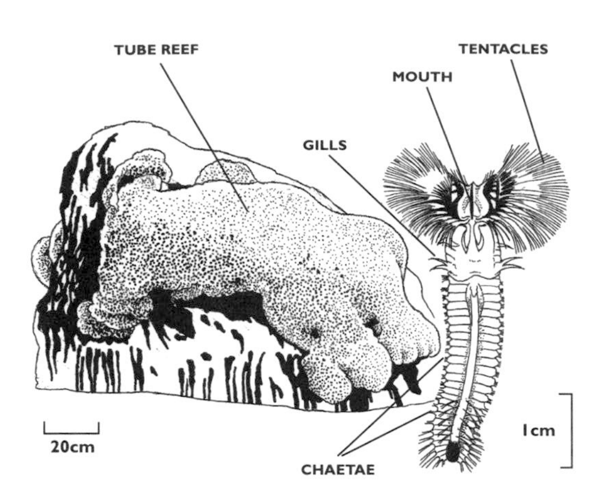

Figure 158. The polychaete worm *Sabellaria alveolata* and its tube reef.

Among the echinoderms, all but the sea cucumbers are vulnerable to this type of fouling. To prevent it sea urchins and starfish have stalked pincer-like projections (pedicellariae) scattered over their dorsal area. In starfish there are two types of pedicellariae for keeping the body surface clear of potential encrusting organisms. Firstly there is the stalked kind, such as that found covering the Common Starfish *Asterias rubens.* There are three small movable ossicles, similar to forceps, that can pinch or cut off the settling larvae (fig.159b). Normally these form a circlet around the base of a spine although they may be found on the spines themselves. The second variety is found on some species of starfish, mainly the Valvatida (e.g. *Astropecten, Linckia*). These have sessile pedicellariae that consist of two or more movable spines which can be more valvular and act like clamps (fig.159a). The sea urchin pedicellariae have long stalks with very obvious jaws comprising three parts. These stalked pedicellariae can be raised, lowered and angled, and with the pincer movements of the jaws are able to protect the animal from larval settlement (fig.159c). These are the same pedicellariae that carry a toxin, in some sea urchins, as a means of defence (see page 145). Once the pedicellariae of both the starfish and the sea urchins have broken up the larvae and the debris, the remains are removed from the surface of the animal by ciliary action.

Bryozoans also need to deter larval settlers and remove detritus, and to do this they have two types of organelles. The simplest type, the vibraculum (fig.160), is a modified operculum that has been adapted into a long bristle which is moved in one plane, sweeping away detritus and settling larvae.

The more complex avicularium is quite different in form and works like a minute transparent bird beak (fig.160). This organ too is a modification of the operculum. It has movable jaws that 'peck' at larvae, and even at larger predatory organisms, kill-

Larvae of nudibranchs which feed almost exclusively on one type of animal, are induced to settle by the secretions of their prey. Contact with these hydroids, bryozoans or sea squirts has to occur before metamorphosis is triggered. The larva of the nudibranch *Tritonia hombergi* settles only on the whitish soft coral, Dead Man's Fingers *Alcyonium digitatum,* and not on the closely related species Red Sea Fingers *Alcyonium glomeratum* (plate 4). Other nudibranchs are not so specific and feed on a wider range of related species. Since the settlement of these animals is related to feeding behaviour, dead colonies have no attraction to the veliger, and in this respect the settlement mechanism differs from that seen in barnacle and oyster larvae.

Sedentary or sessile marine invertebrates may be clogged by the larval 'rain' of potential invertebrate encrustations. For example, scallops, oysters and clams can become thickly encrusted with polychaete tubes and barnacles.

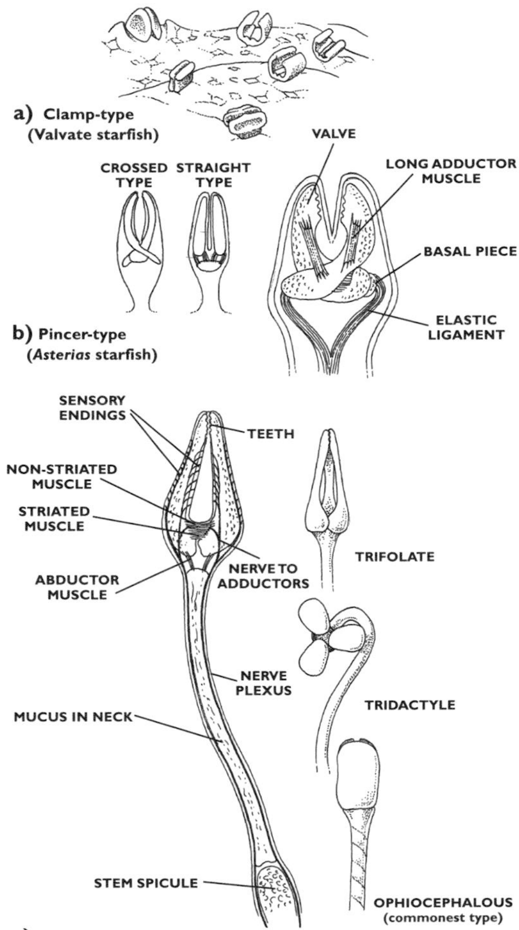

Figure 159. Cleaning organs of starfish and sea urchins.

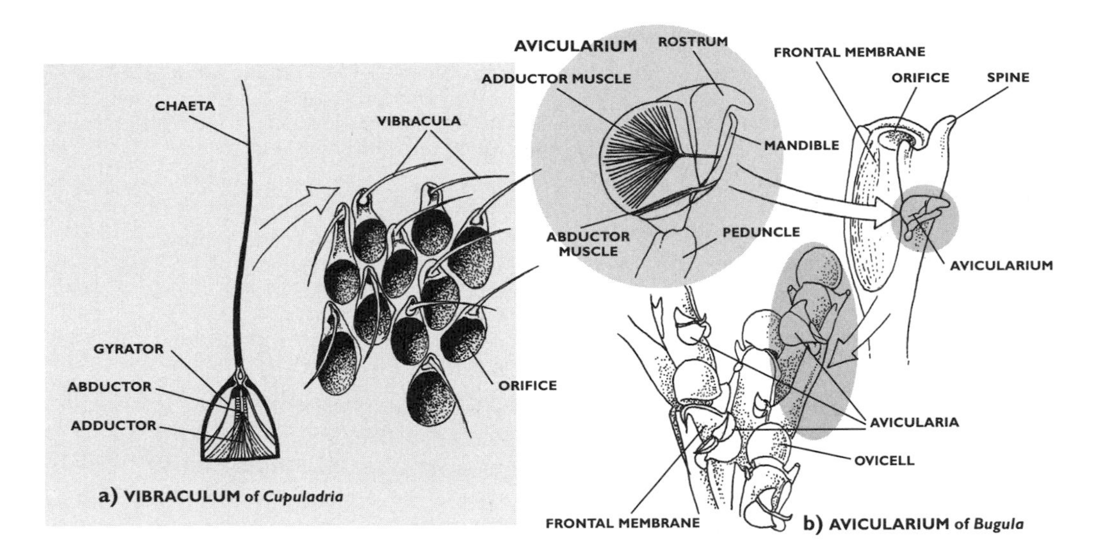

Figure 160. The cleaning organs of bryozoans: a) vibraculum and b) avicularium.

ing them or antagonising them enough so that they depart. Some of these avicularia are sessile while others are stalked (e.g. *Bugula*), but both types are generally used on larger crawling animals that try to graze on the colonies.

Once the free-swimming larvae settle a new generation can begin. In the struggle amongst the marine invertebrates between predator and prey for food sources, a wide variety of competitive strategies has evolved.

EVOLUTION

Fossil evidence of invertebrates is difficult to find. Fossilisation of soft-bodied animals occurs only in rare conditions, although in the well known Burgess Shales fossil invertebrates are found in abundance. From the study of the structure of living and fossil invertebrates, an invertebrate evolutionary history has been constructed. Recent advances in cell biology and genetic studies have helped this process along.

EARLY EVOLUTION OF INVERTEBRATES

The earliest forms of life may have evolved some 3,000 to 4,000 million years ago (mya), in the form of bacteria-like organisms which thrived in the seas of that period. It is hypothesised that from these simple organisms evolved the blue-green algae (cyanophytes) which for the first time were able to produce new organic material by the process of photosynthesis. These algae are present today in some freshwater lakes around the world (e.g. Lake Superior, USA) and some saltwater shores such as Hamelin Pool on the northwestern coast of Australia. Hamelin Pool provides a glimpse into the distant past when lime-producing cyanophytes secreted countless pillars (stromatolites) that covered the shallow waters of the ancient oceans. Over the eons the stromatolites were compressed to form the hard, transversely-ringed rock known as Gunflint Chert. These are the earliest known fossils of life on the Earth.

During a period spanning 1,500 million years, blue-green algae may have dominated life in the seas. Microfossil data indicate that by around 1,200mya the first primitive protozoans were present. The evolutionary origin of these protozoans is the subject of speculation and some authorities have hypothesised that certain species of bacteria present at that time may have ingested cyanophytes. These were incorporated within the cytoplasm of the bacteria instead of being digested. Protozoans flourished and started to diversify into the phyla that we recognise today, which include the amoeboid and flagellated protozoa, and the ciliates. The parasitic sporozoans may have evolved later when suitable host animals had appeared. At about this time an evolutionary split may have occurred between animal and plant life with many of the Ciliophora possessing structural traits of both kingdoms.

THE EVOLUTION OF MULTICELLULARITY

The evolution of long flagella or the more numerous, shorter cilia enabled protozoans to move and capture prey. The Protozoa were, however, limited by their size for as a cell grows, the internal structure and chemical processes become less efficient. One group of flagellate protozoans solved the problem of size by evolving into a loosely knit colony. Today this trend can be seen in the colonial *Volvox*, whose cells are independent of each other and yet remain together working as a unit (fig.161).

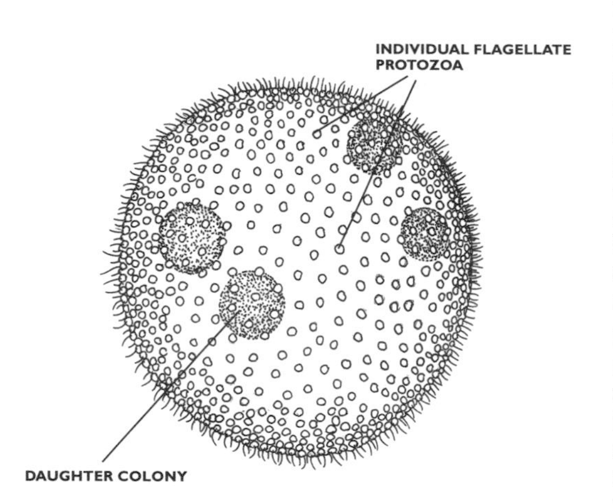

Figure 161. The loosely colonial protozoan *Volvox*.

The Colonial Theory, put forward by Haeckel in 1874, later modified by Metschnikoff in 1887 and then revived by Hyman in 1940, suggests that the ancestral metazoan was a ball of cells similar to *Volvox*, but was of a choanoflagellate nature. This ancestor became a double-walled organism (an ovoid gastraea, fig.162) either by a' glove-finger intucking' of the wall or alternatively by proliferation of the cells within the blastula in the hollow centre of the ball of choanocytes. This free-swimming, radially symmetrical organism (a planula) may therefore represent the ancestral metazoan form.

In 1883 a small multicellular organism, *Trichoplax adhaerens*, was discovered in a European seawater aquarium. This showed a number of features thought to have been present in the planuloid metazoan ancestor. However *Trichoplax adhaerens* is not radially symmetrical, a feature present in the proposed ancestor. Whether the flattened asymmetrical body of *Trichoplax adhaerens* is an ancient (primary) or newly evolved (secondary) feature is not known. It is the simplest known metazoan and is placed in its own phylum, the Placozoa.

From a basic planula-like form, two lineages may have evolved. One was represented by the sponges which first appeared between 800 to 1,000mya, in the Proterozoic period, in the seas. The sponges are a zoological enigma because they demonstrate primitive, low level differentiation of cells but also show a specialised body structure with anterior and posterior ends. However, most zoologists agree that sponges are an early branch from the main metazoan line (fig.164). This is inferred by the sponges' flagellated collar cells which are comparable to those of the ancestral choanoflagellate protozoans. The evolution of sponges from a planula ancestor could have been similar to the post-larval development of an amphiblastula larva (fig. 163) resulting in the adult post-larval rhagon stage. Sponges are often placed in a sub-kingdom, the Parazoa, because there are no other species in the animal kingdom with which they can be closely allied. They have no obvious way of communicating between cells, and no integrated muscles or other coordinating tissues.

No evidence of the second lineage in invertebrate evolution was found until the 1940s when some unusual shapes were seen by geologists in the now famous Ediacara Sandstones of the Flinders Range in southern Australia. These strange impressions in 650 million-year-old rocks were probably the marks left by a stranded jellyfish on a beach eons ago.

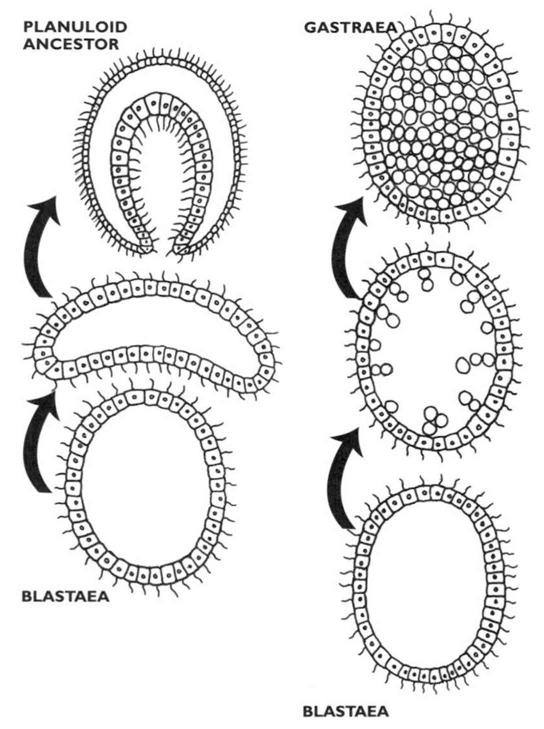

Figure 162. The possible evolution of the first metazoan according to the theory of Metschnikoff (left) and Haeckel (right).

It is thought that cnidarians evolved from a planktonic planula-type larva which had acquired in its evolution certain features of the present-day cnidarians (i.e. a mesogloea layer, some cellular differentiation and the medusoid sexually reproductive stage). From this type of free-swimming cnidarian, an actinula may have evolved, a form that was adapted to a more benthic way of life, eventually developing into a polyp having typical cnidarian features. This sedentary polyp-type cnidarian form was successful for a number of reasons; it was able to exploit a new food supply, it had an extended larval life and it was able to reproduce rapidly and asexually. It may have produced, in further development, a hydromedusa, formerly by direct transformation and later by asexual budding. This asexual budding gave this polyp-like form the great advantage of multiple medusa development from a single polyp. In this way the hydroid life cycle seen today may have evolved.

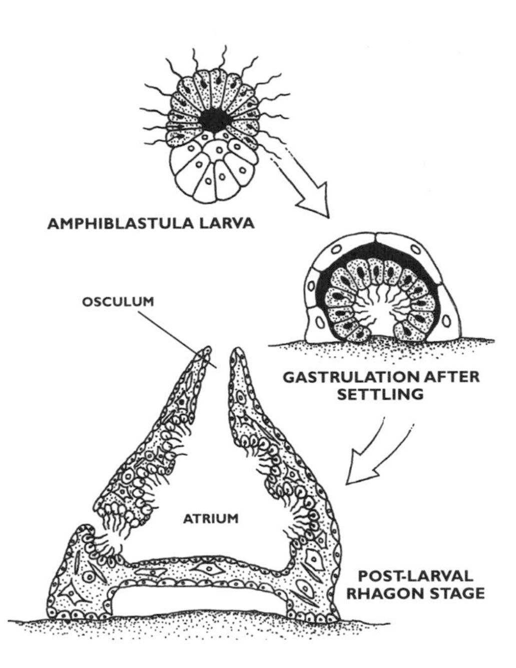

Figure 163. Sponge post-larval development.

Further hydroid development led to the suppression and, in some species, the loss of the medusoid stage. Some primitive species still retain the hydroid life cycle (hydrozoans). One group, however, shows development and enlargement of the medusoid stage and increased longevity. This group comprises the modern jellyfish (scyphozoans). It is probable that another early group of cnidarians evolved in a different way by thickening the mesogloea; these are the planktonic comb-jellies (ctenophores).

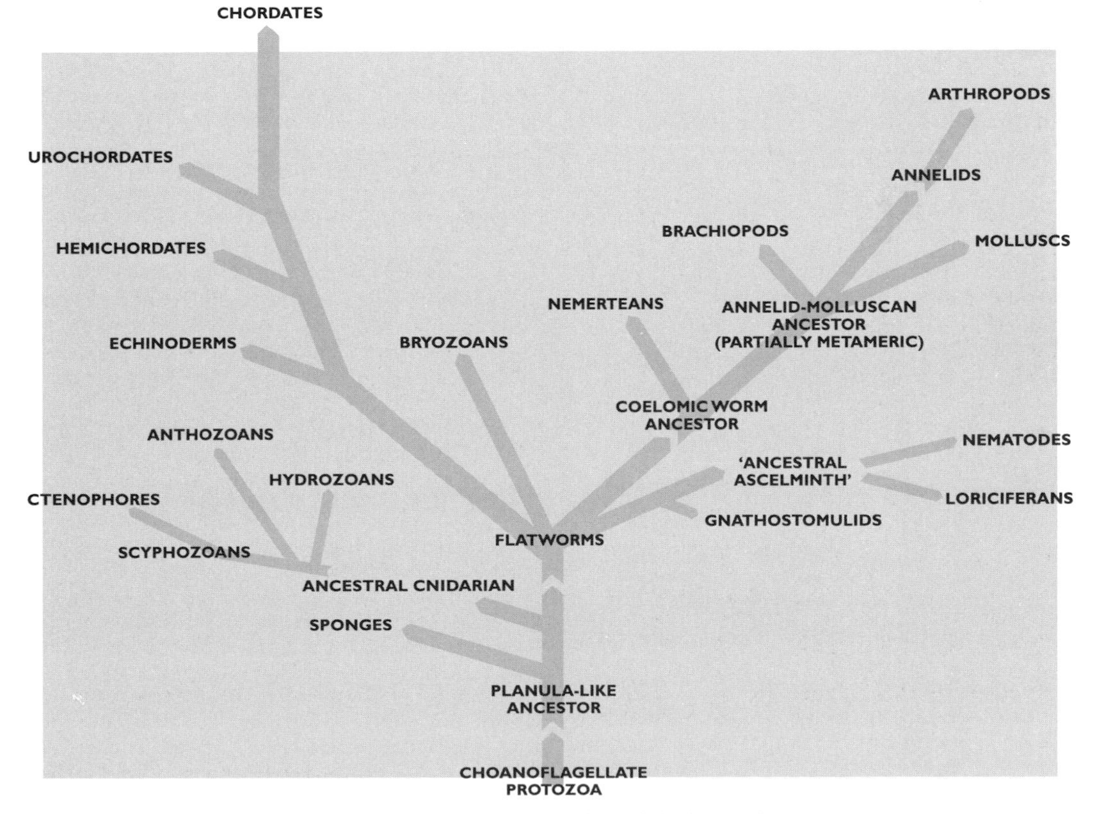

Figure 164. A possible simplified scheme for the evolution of the invertebrates.

SYMMETRY AND COELOMS

An important landmark in invertebrate evolution was the change from radial symmetry to bilateral symmetry. A widely held theory proposes that the flatworms (platyhelminths - particularly the free-living turbellarians) are closely related to ancestral bilateral organisms. Like the cnidarians, which are radially symmetrical, they only have one opening to the gut with ingestion and egestion occurring through the same orifice. They have no gills and the cilia are used for locomotion as in the planula larvae of cnidarians.

The Schizocoel Theory proposes that the coelom evolved from the acoelomate flatworm condition, through the hollowing out of the packing or parenchymal cells. This is similar to the process of coelom development that occurs in larval annelids and molluscs. Therefore, if the flatworms are the ancestral group of bilaterally symmetrical organisms, this would place them earlier in the invertebrate evolutionary line than the coelomic worms. The more recent Enterocoel Theory proposes that the ancestral coelomic worms were the ancestors of bilaterally symmetrical organisms. Its supporters contend that the coelom developed from the gastric pouches of a cnidarian ancestor. This is similar to the process of mesoderm and coelom formation in the echinoderms, hemichordates and chordates. Controversially this suggests that the flatworm acoelomate condition is secondary and that flatworms have the same coelomate ancestor as nemerteans, rotifers, annelids and echinoderms. For the purposes of this chapter, the traditional Schizocoel Theory has been followed.

Recent support for the Enterocoel Theory comes from the study of the worm *Lobatocerebrum*. It has been discovered (Rieger 1980) that this worm is structurally intermediate between the coelomate worms and acoelomate flatworms. *Lobatocerebrum* has a cuticle, ventral nerve cord and is metameric like the annelids. However, like the flatworms, it is ciliated. The parenchyma is composed of fluid-filled cells with little space between them (a typical flatworm feature). However, *Lobatocerebrum* has more features in common with the oligochaetes, the group in which it has been classified. This worm may be a halfway stage between annelid and flatworm as predicted by the Enterocoel Theory.

The ribbon worms (nemerteans) may have appeared at about the time that the split between radial and bilateral symmetry occurred, possibly from an ancestral annelid stock. The nemerteans are structurally similar to flatworms, but are generally cylindrical with a distinct mouth and anus, and are much longer. The Atlantic and North Sea ribbon worm *Lineus longissimus* which hides under rocks in the littoral zone can grow to five metres in length.

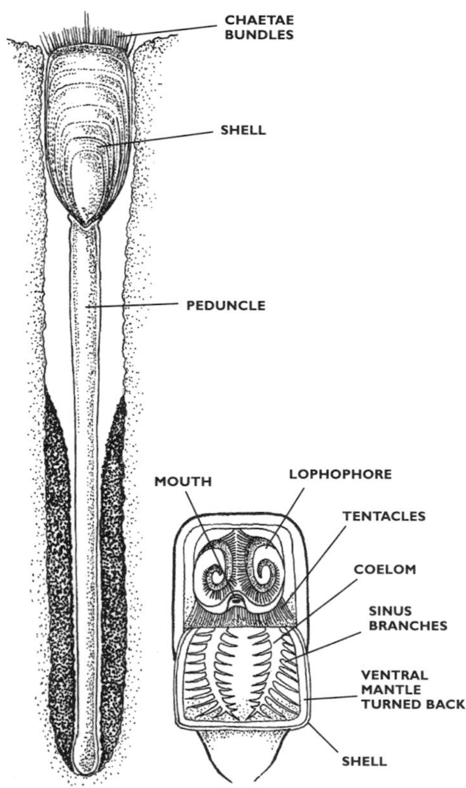

Figure 165. The 'living fossil' brachiopod *Lingula*.

According to the Schizocoel Theory, the annelids evolved from a flatworm ancestor. These early, cylindrical, burrowing worms were more successful in the bottom sediment than the flatworms; the sediment provided a perfect habitat for food and protection from predators. Development of a primitive coelom would have helped in the build-up of hydrostatic pressure enabling the animal to burrow more efficiently. With metamerisation, the burrowing condition was further facilitated with pairs of locomotive organs on each body section. These longer, multi-segmented worms thrived on the sea bottom during the Cambrian period. It is likely that the marine polychaetes were the earliest descendants of the ancestral annelid stock, and that the oligochaetes and leeches evolved later from a polychaete ancestor.

Another descendant of these annelid-like ancestors probably buried itself in the sediment with just its mouthparts showing (rather like the later adaptations of some polychaetes, e.g. fanworms). Cilia were probably used to create a current of water to filter food from the surrounding environment. The worms secreted a protective tube and, over time, the tubes may have evolved into a collar with a slit-like opening to facilitate the flow of water while the naked tail remained buried in the protective mud and silts. Later this collar may have become modified into two flat protective shells as seen in the brachiopods, or lampshells, an important fossil phylum that still has living representatives today.

The brachiopod *Lingula* is said to be a living fossil (fig.165) and is one of only 325 modern brachiopod species, compared to over 12,000 fossil species that were abundant in the seas of the Palaeozoic and Mesozoic eras.

EVOLUTION OF THE MOLLUSCS

After the discovery of the primitive monoplacophoran *Neopilina* in 1952, new theories have been put forward on the evolutionary origins of the molluscs. The gastropods, bivalves, scaphopods and cephalopods may all have evolved from an ancestral form similar to the limpet-like *Neopilina*. *Neopilina* (fig.166) is less than three centimetres long and has five to six pairs of gills alternating with six pairs of kidneys. This segmental arrangement of body organs suggests that the common ancestor of molluscs and annelids may have been partially metameric. If annelids evolved by an increase in metamerism, then molluscs become progressively less segmented in evolution. The worm-like aplacophorans (often called solenogasters) and the polyplacophorans (chitons) probably diverged early on in the molluscan lineage, before the single-shelled *Neopilina* stage evolved. The absence of a shell in the aplacophorans would therefore be a primitive feature and not a newly evolved (secondary) adaptation (fig.168).

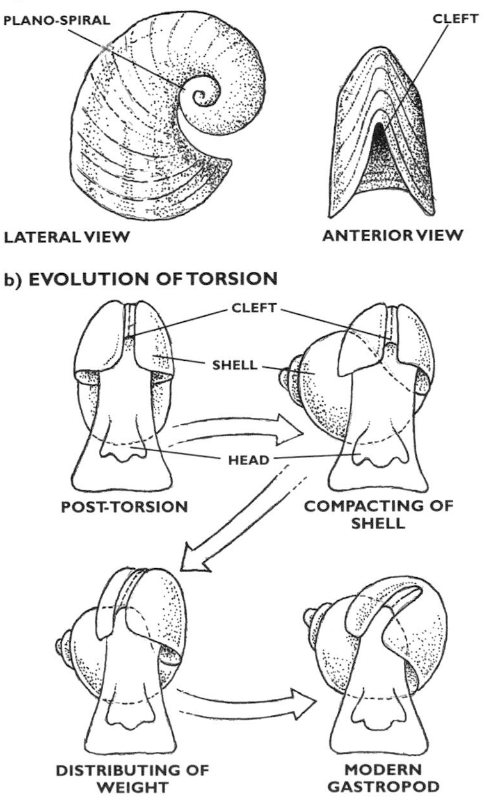

Figure 167. a) the fossil genus *Strepsodiscus* and b) the evolution of torsion.

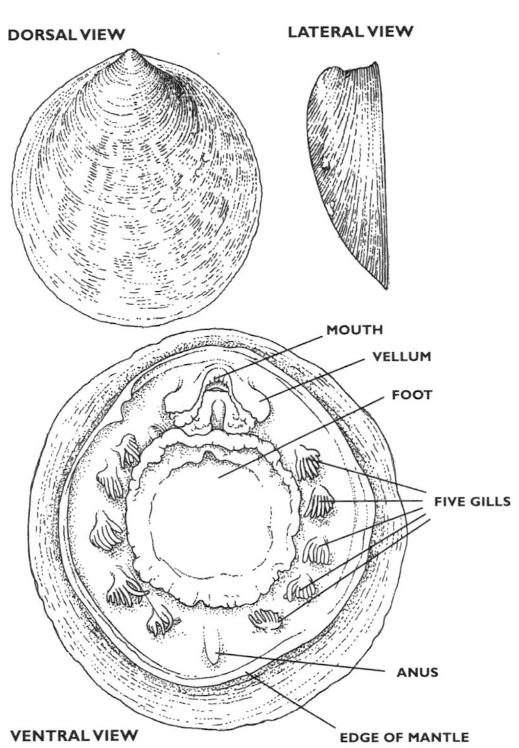

Figure 166. The primitive mollusc *Neopilina galathea*.

The ancestral gastropods probably produced a shell similar to *Neopilina* which shows equal growth around the circumference of the mantle, so that a simple, cone-like shell is produced. These primitive forms appeared in the early Cambrian period; but cone-like shells were unwieldy, especially for larger individuals, and these molluscs were unable to exploit confined, narrow habitats. The evolution of spiralling solved this problem. Here the front mantle secretes shell material faster than the rear to form a shell with a flat spiral (plano-spiral). This is seen in the fossil genus *Strepsodiscus* (fig.167), which does not exhibit the normal torsion of modern gastropods.

A counterclockwise 180° rotation of the visceral mass, mantle and its cavity occurs in the larva, mirroring the phylogeny of the gastropods (i.e. ontogeny recapitulating phylogeny). The gills, anus, nephridiopores and mantle cavity were then located at the anterior end of the mollusc. The early plano-spirally torsioned gastropods had a cleft in the shell to allow a constant flow of water through the mantle cavity, with the inhalant and exhalant currents well separated. Water flowed in through the shell low over the mantle and then made a U-turn passing up the back of the cavity, then along the top and out through the widened shell aperture.

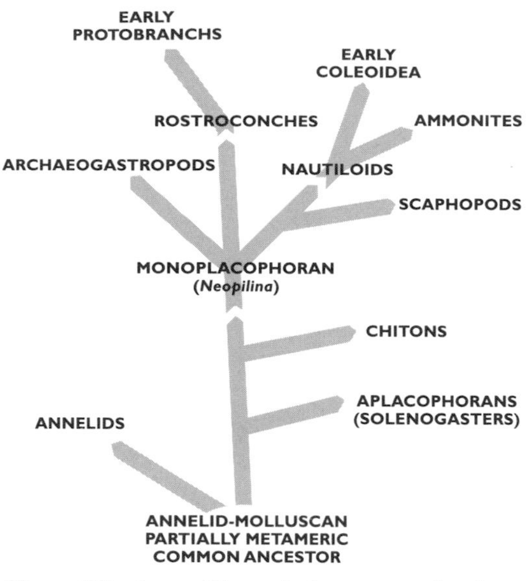

Figure 168. A possible evolutionary tree for the evolution of the Mollusca.

Tusk shells (scaphopods) with their characteristic shape (fig.169) are probably an evolutionary side-branch that diverged at an early stage in bivalve evolution. From this it can be deduced that early bivalves had only one shell, or valve. These molluscs evolved as shallow mud burrowers, similar to the now extinct rostroconch molluscs (fig.168). This group is included in the subclass Protobranchia that contains the oldest known fossil bivalves and some living species like the Common Nut-shell *Nucula nucleus* of Atlantic and Mediterranean shallow sandy waters.

The nut-shells have a single pair of gills with the lateral gill cilia creating a respiratory current and the frontal cilia removing trapped sediment. These living protobranchs are selective deposit feeders as were, it is assumed, the archetypal bivalves. They have lost the radula but have a pair of tentacles to maintain contact with the substratum because the mouth is raised from its lower archaic position. The tentacles are associated with two large flaps, the labial palps, that are sited on either side of the mouth. In some of the early prosobranchs filter feeding evolved. This may have been the trigger for the massive evolutionary radiation of the bivalves that followed. The evolution of the primitive lamellibranch-type heralded the dominance of this form of bivalve over all others (fig.170). As bivalves evolved, detritus particles in the ventilatory current may have been used more as a food source than those sucked up from the bottom by the mouth, with the gills developing as food filters.

An important evolutionary change of the gastropod shell was the asymmetrical spiralling of the shell with the coils laid down around a central axis (the columella). This produced a much more compact shell with a longer length fitting into a smaller space. However, this would have been top heavy if it had remained in the same position on the gastropod's body, so as the shell evolved it became repositioned giving a better weight distribution. This resulted in the modern, obliquely positioned shell. Additionally, occlusion of the right hand side of the mantle cavity occurred, resulting in the degeneration of some organs. This development led to the modern gastropods with their lack of the right-side gills, auricles and nephridia.

Nudibranchs evolved from carnivorous gastropods which lost their shells in evolution. Some present-day nudibranchs such as the Spanish Dancer *Hexabranchus sanguineus* (plate 6) are able to swim for short distances and other nudibranchs have become pelagic like the pteropods (sea butterflies) and the naked pteropods (naked sea butterflies).

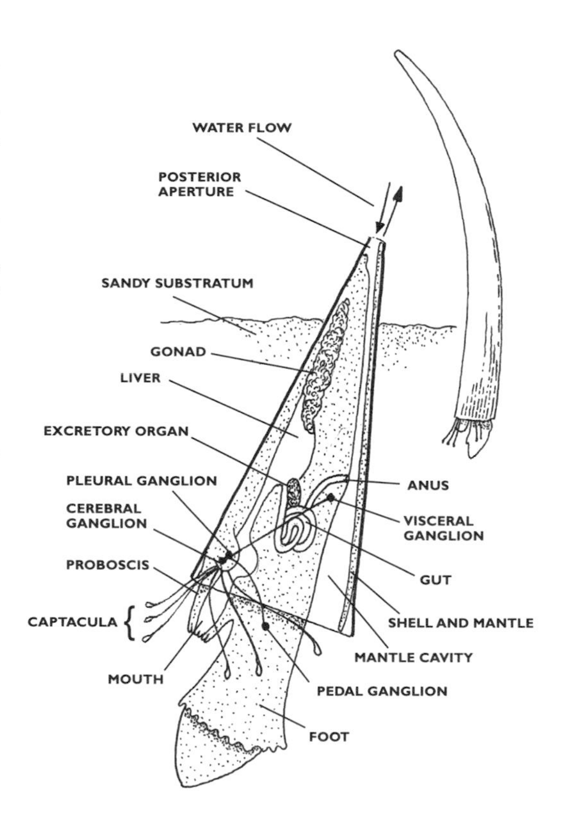

Figure 169. Structure of a scaphopod, *Dentalium*.

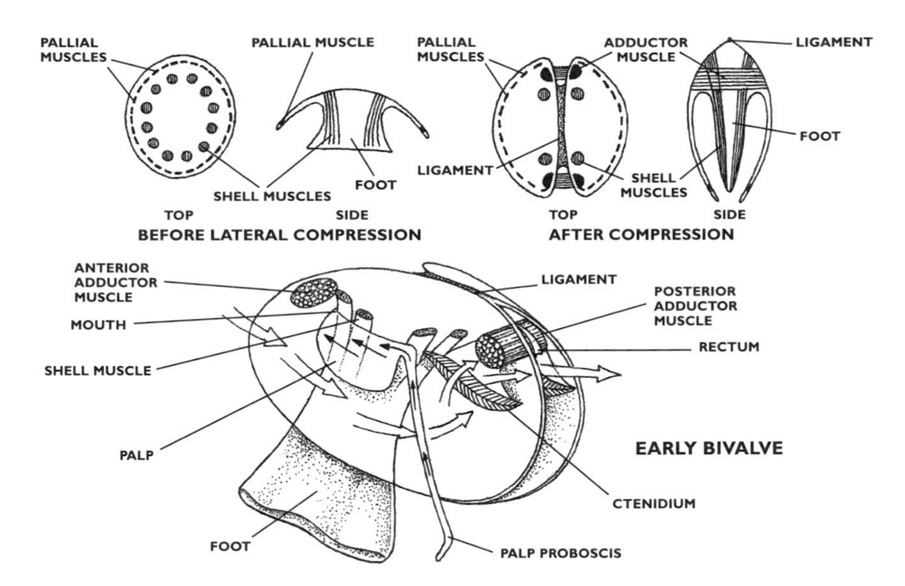

Figure 170. Derivation of bivalve mollusc organisation.

As bivalves evolved, new forms developed from the filter feeders. Carnivorous bivalves appeared (the Septibranchia), as did those that could partly photosynthesise with the help of zooxanthellae symbionts (see p.153). Others became rock and wood borers, utilising entirely new sources of nutrition.

Early in their evolutionary history, one branch of the molluscs evolved in such a way as to become more mobile, yet retain the protection of a shell. These were the early cephalopods which are thought to have evolved from high-coned monoplacophorans (fig.171) and appeared about 550mya in the Cambrian period. Their apical chambers were first filled with fluid but when the siphuncle evolved, gas took the place of fluid to form a buoyancy organ. These early shelled cephalopods had curved cones that later developed into straight and then the more familiar coiled shells. The straight shells of some giant nautiloids in the Ordovician period were over five metres long. The progressive coiling and uncoiling of their shells occurred throughout their evolution.

One of these early species still survives today as a 'living fossil', the Chambered Nautilus *Nautilus pompilius* (plate 1). The genus *Nautilus* with its five living species is the only extant remnant of this dominant prehistoric group of marine cephalopods.

The nautiloids gave rise to a variant group, the ammonites, which ruled the seas until about 100mya in the Cretaceous period. The ammonites had elaborate sutures between the chambers, compared to the simpler sutures in nautiloids (fig.172). The greater strength of this type of construction allowed for a thinner and lighter shell. Evidence for the ammonites' abundance in the oceans is shown by their extensive fossil remains. In some rock deposits their remains lie so thickly that they form visible solid bands. They were abundant in the Mesozoic era but then became extinct.

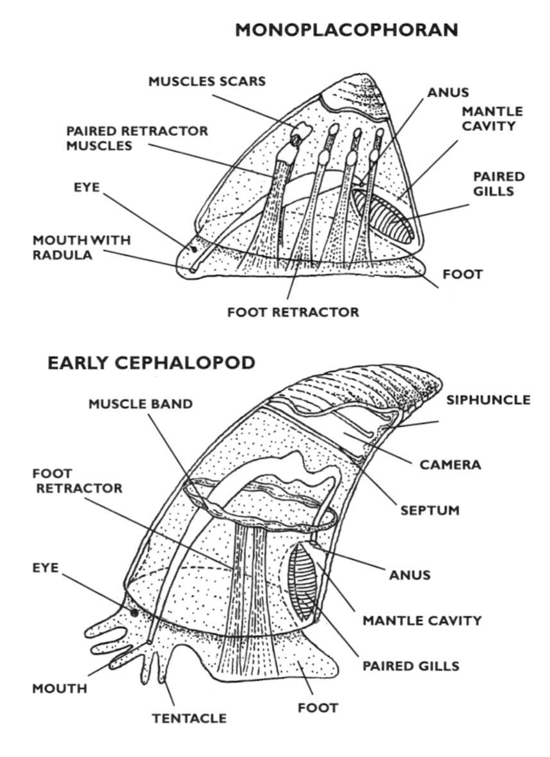

Figure 171. A monoplacophoran and an early cephalopod showing the similarities in structure.

One group that evolved from the early straight-shelled nautiloid line remains today, sometimes occurring in great numbers. These are the Coleoidea whose mantles overgrew their shells during evolution to produce the first true endoskeleton. From primitive coleoidian ancestors there evolved four distinct living groups (fig.173); the spirulans, cuttlefish, squid and octopuses. All evolved elaborate eyes and enlarged brains.

The ancestral octopuses probably diverged from the main lineage quite early and lost the primitive shell. The *Spirulirostra* forms had a thickened wall around a coiled inner shell which was lost in later evolution. The inner shell underwent further coiling resulting in the modern forms typified by *Spirula*. The third lineage, the squid group, initially had a reduced shell (e.g. *Belemnoteuthis*) which then became flattened (e.g. *Conoteuthis*). It was further reduced in size as in modern squid (e.g. *Loligo*). This uncalcified shell or gladius is sometimes referred to as a sea pen because of its resemblance to a quill. The final group, the cuttlefish, were the only cephalopods to retain the shell but the shelf and thickened wall (fig.173) were all but lost resulting in the typical cuttlefish 'bone' (e.g. *Sepia*) that is found washed up on shores around the world.

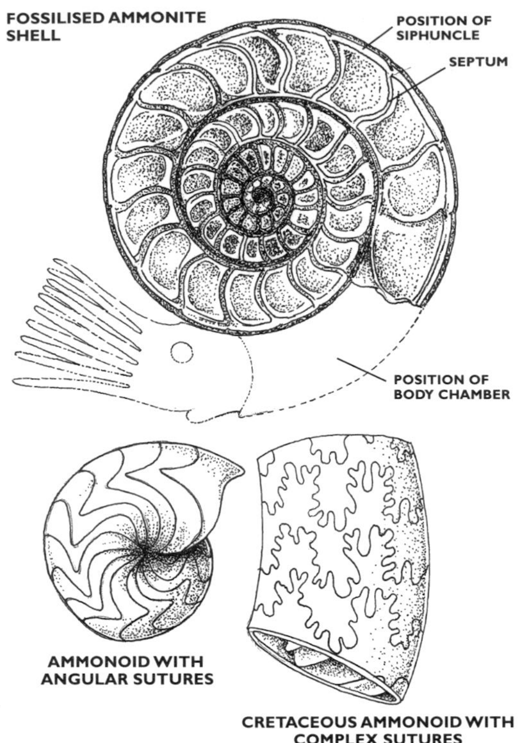

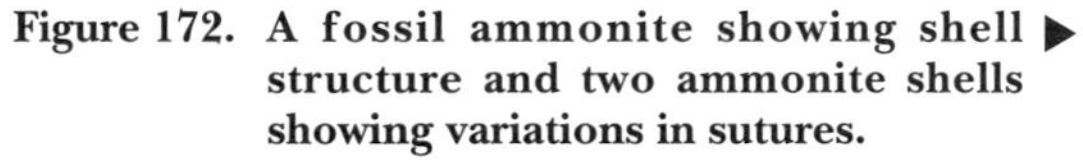

Figure 172. A fossil ammonite showing shell ▶ structure and two ammonite shells showing variations in sutures.

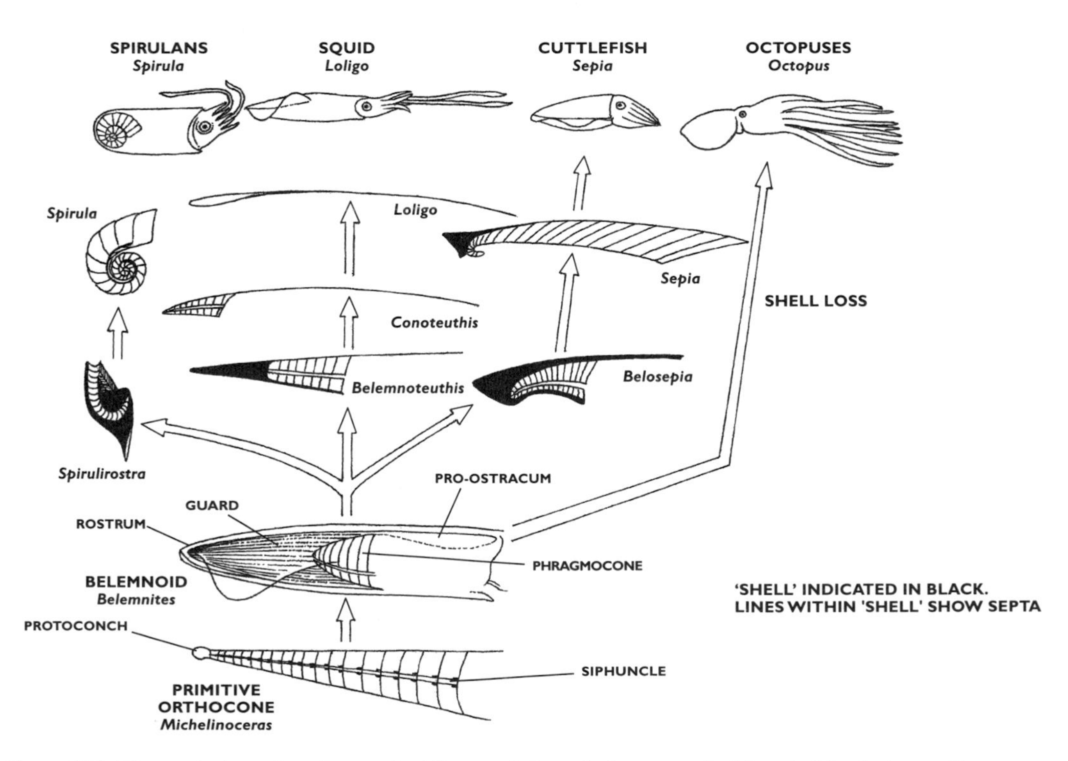

Figure 173. The evolution of modern coleoidian cephalopods from a probable primitive belemnoid ancestor and variations in shell reducton in four orders.

EVOLUTION OF MARINE ARTHROPODS

At around the same time as the appearance of the early nautiloids (550mya in the Cambrian period) and found in the same fossil rocks, are impressions of the most primitive arthropod group, the trilobites. These were the first animals to evolve a high-definition eye; the cephalopod camera-type eye developed much later in the coleoidians. Trilobites were an exclusively marine group (fig.174) that was once abundant and spread throughout the seas of the world, diversifying into a large number of species. They reached their evolutionary peak during the Cambrian and Ordovician periods. Then, like the ammonites after them, they became extinct and left no closely related living species.

Trilobites were between half a millimetre to one metre in length (averaging about three to ten centimetres) and most of them lived on the sea floor. They generally moved on four pairs of appendages that had two parts, one for walking (the telopodite), and the other probably for digging or swimming (the pre-epipodite). The posteriorly directed mouth had two lip-like flaps, the labrum. The trilobites also had long sensory antennae which are homologous with crustacean antennae. As they evolved, pelagic forms developed which scanned the sea bottom below with specialised eyes as they paddled on their backs. Deep water species were also present and some species could roll themselves up like a pill bug when threatened. The smallest of these early arthropods were probably planktonic.

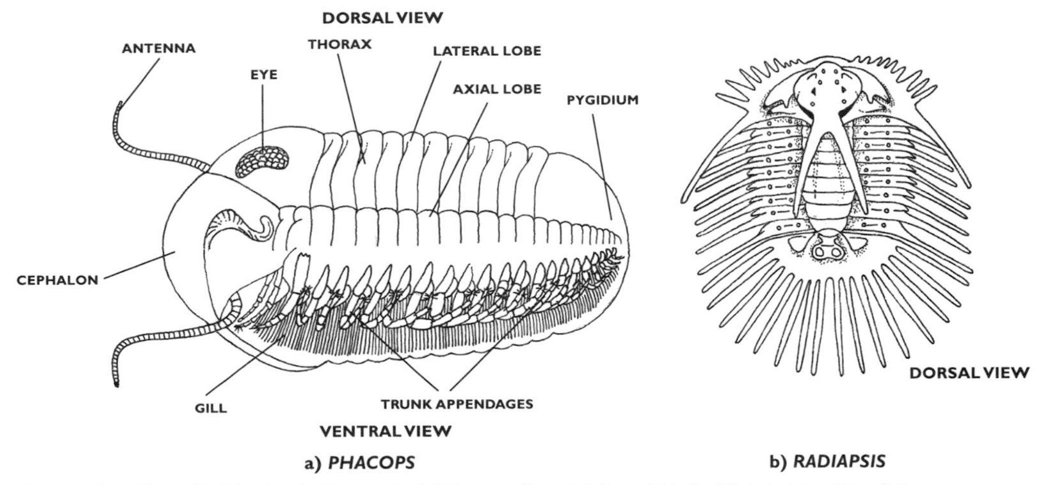

Figure 174. The trilobite body form of a) *Phacops* (benthic) and b) *Radiapsis* (planktonic).

The only living invertebrates similar to the trilobites are the horseshoe crabs (xiphosurans) which have a lineage dating back to the Cambrian period along with the extinct eurypterids (both groups constituting the Merostomata). The merostomes may have evolved from an ancestor close to the stem trilobites but not directly from them. The eurypterids were the largest arthropods that ever existed; the genus *Pterygotus* reached about three metres in length. They were similar in their general body plan to the living horseshoe crabs (*Limulus*). The Arachnida (spiders and mites) were ancestors of the marine merostomes. Another offshoot from the merostomes may have been the pycnogonids (sea spiders). These are early arachnids that never became terrestrial. Modern deep-sea species of pycnogonids can be relatively large, reaching leg spans of up to 75cm, a greater span than any living terrestrial arachnid.

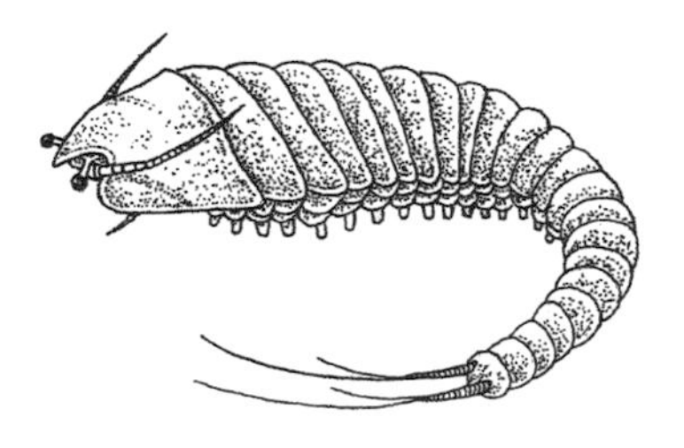

Figure 175. A possible ancestral crustacean.

Another diverse marine arthropod group, the crustaceans, may also have evolved from the same ancestor that gave rise to trilobites. It is known that ancestral crustaceans, which appeared at about the same time as the trilobites, were small epibenthic invertebrates with a head and many trunk segments. The head had two antennae, two stalked compound eyes, a nauplius eye, a backward pointing mouth with a pair of mandibles and two pairs of maxillae (fig.175).

These simple crustaceans survived during the millions of years of trilobite dominance of the seas and subsequently evolved into the 42,000 species known today. During the 550 million years of crustacean evolution, three main groups have appeared. The first are the simple branchiopods, the second the barnacles and copepods and the third, the most numerous, the malacostracans (fig.176). The malacostracans are subdivided into two major subgroups, the peracaridans and the eucaridans.

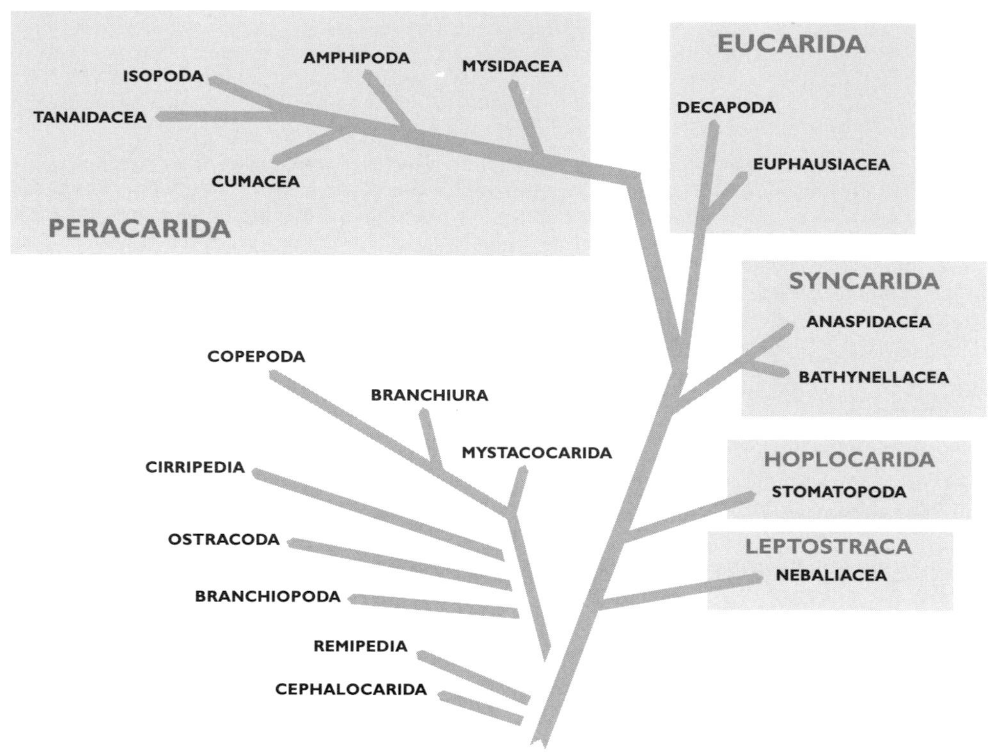

Figure 176. A possible scheme for the evolution of the crustaceans.

EVOLUTION OF THE ECHINODERMS AND THEIR DESCENDANTS

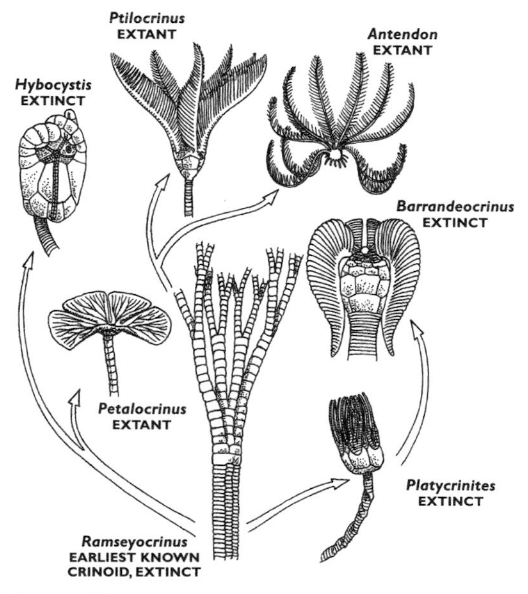

Figure 177. Crinoid evolution.

The protostome lineage ends with the arthropods, which include the most numerous of all the animal groups, the insects. The second major line of invertebrate evolution comprises the deuterostomes. A change in the method of coelom formation may have occurred at about the same time as the controversial divergence of new invertebrate lineages (including the deuterostomes) from early protostomes and the emergence of the first brachiopods and bryozoans. The ancestral, crinoid-like echinoderm lineage may have begun with a bilaterally symmetrical, tricoelomate, free-living animal.

The crinoids (sea lilies) are the only living echinoderms that are permanently attached to the substrate, as were the Palaeozoic crinoids. In these prehistoric seas, the crinoids were suspension feeders using their outstretched arms with their fine pinnules (branchioles) to catch and direct the food down the ambulacral groove into the upward facing mouth (fig.177). Some of the early fossil forms were gigantic with stems 20m high. Unlike the trilobites, their contemporaries of the Ordovician period, the sea lilies dwindled but did not die out entirely.

Soon after the appearance of the first crinoids, other early echinoderms evolved. Thus began the successful evolutionary radiation of the starfish, sea cucumbers and other echinoderms which we see today.

Some of these early tricoelomates assumed a sessile way of life and took on a radially symmetrical form that is often found in sedentary animals. One group may have attached itself by the anterior end, the left side becoming the upper or oral side and the right side the aboral. Then the mouth moved to the oral side and the two left side coelomic sacs became the water vascular system which at that time probably filtered food from the sea water.

The primitive sessile deuterostomes evolved a plated exoskeleton for protection, the five-part symmetry of form evolving in the echinoderms where it is such a characteristic feature. This pentamerous symmetry is thought to be an adaptation to prevent weak points occurring in the radial skeleton (fig.178).

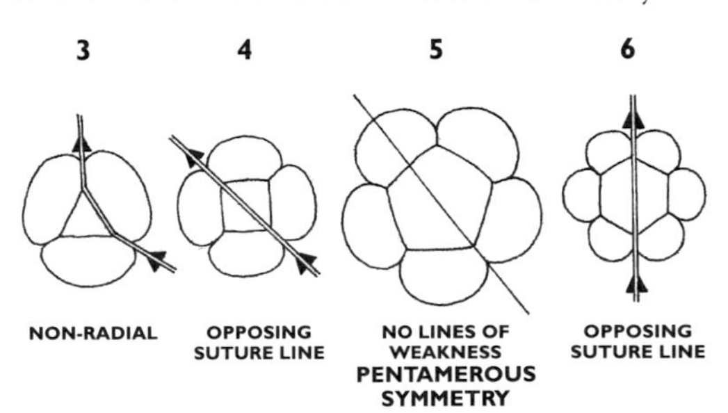

Figure 178. A possible explanation for the pentamerous symmetry of echinoderms showing lines of weakness (arrows)

Some early sessile echinoderms may have become detached from the substratum and resumed a free-living existence. However their mouths were on the upperside so when the later species evolved, they turned over, the oral side facing down to the substrate where food was to be found. An important phase in the development of the modern echinoderm form was the development of the water vascular system for locomotion. From an early 'starfish-like' form it is clear to see how the other groups could have evolved. The sea cucumbers (holothuroideans) probably split from the main stock early on, their arms doubling back to form a tube with one arm still retaining podia for locomotion. The sea urchins (echinoids) followed a similar pattern, but formed a squat ovoid shape, developing the spines into an effective defence mechanism.

Echinoderms are highly diverse and are found throughout the seas of the world, but their hydrostatic skeletal and locomotive system has its structural limitations. The echinoderm form did not develop further, although the deuterostome lineage continued to evolve.

Early in the evolution of the deuterostomes, the ancestor of the modern pterobranchs probably evolved. The close similarity between the pterobranchs and the echinoderms was discussed by Hyman in 1959 and Nielsen in 1985.

The next evolutionary step, the hemichordates, show one new particular feature, pharyngeal clefts, which are present for the first time in the invertebrate evolutionary line. Paired pharyngeal clefts occur only in one other group, the chordates, which includes the vertebrates. Some hemichordates also have a hollow dorsal nerve cord, a feature possibly homologous with the hollow dorsal nerve cord of the chordates. All members of the phylum Chordata have a rod-like supporting structure or notochord, absent from the hemichordates. This implies that hemichordates diverged just before the deuterostome line split into those animals with a notochord and those without. The latter may have developed, by that time, a rudimentary, dorsal, hollow nerve cord.

It is possible that the notochord first evolved in invertebrate larval forms as it is present in the larval tunicates. The tunicates represent the penultimate branch in the deuterostome line, the main evolutionary line progressing further to the cephalochordates and finally to the first vertebrates. The tunicates evolved into the great variety of forms seen today including sedentary forms (e.g. the ascidians), planktonic forms (colonies like the salps and doliolids that can number thousands of individuals), or solitary forms such as the larvaceans.

When looking back through time, it took about 3,000 million years for life to evolve from its first fossil traces to the stage of the first organised metazoan (the cnidarians), but from then on, in a relatively short period of about 100 million years, all the known forms of invertebrate life as well as the first true vertebrates evolved.

BIBLIOGRAPHY

Barnes R. D. & Rupert E. E. 1995 (6th ed.). *Invertebrate Zoology*. Saunders College, Florida.

Barrington, E. J. W. 1972. *Invertebrate Structure and Function.* Thomas Nelson, London and Nairobi.

Bullough, W. S. 1968 (2nd ed.). *Invertebrate Anatomy*. Macmillan, London.

Campbell A. C. 1993. *Guide to the Seashore and Shallow Seas of Britain and Europe.* Hamlyn, London and New York.

Coleman, N. 1991. *Encyclopedia of Marine Animals.* Blandford, London.

Erwin, D. & Picton, B. 1990. *Guide to Inshore Marine Life.* Immel, London.

Gore, R. 1990. Between Monterey Tides. *National Geographic* 177(2): 2-43.

Green, J. 1961. *A Biology of Crustacea.* H. F. & G. Witherby, London.

Haekel, E. 1874. The gastaea-theory, the phylogenetic classification of the Animal Kingdom and the homology of the germ lamellae. *Q. J. Micr. Sci.* 14: 142-165, 223-247.

Hyman, L. H. 1940. *The Invertebrates: Protozoa through Ctenophora.* Vol.I. McGraw-Hill, New York.

— 1959. *The Invertebrates: The Smaller Coelomate Groups.* Vol.5. McGraw-Hill, New York.

Kohl, L. 1980. British Columbia's Cold Emerald Sea. *National Geographic* 157 (4): 526-551.

Marsh K. 1977. *How Invertebrates Live.* Gallery Press, Leicester.

Metschnikoff, E. 1877. *Embryologische Studien an Medusen, mit Atlas.* A. Holder, Vienna.

Morton, J. E. 1971. *Molluscs.* Hutchinson, London.

Nichols, D. 1969 (4th ed.). *Echinoderms.* Hutchinson, London.

Nielsen, C. 1985. Animal phylogeny in the light of the trochaea theory. *Biol. Jour. Linn. Soc.* 25: 243-299.

Phillips Dales, R. *Annelids.* Hutchinson, London.

— (ed.) 1971. *Practical Invertebrate Zoology.* Sidgwick & Jackson, London.

Readers Digest. 1986. *Great Barrier Reef.* Readers Digest, Sydney.

Rieger, R. M. 1980. A new group of interstitial worms: Lobatocerebridae nov. fam. (Annelida) and its significance for metazoan phylogeny. *Zoomorphologie* 95: 41-48.

Sabelli, B. 1982. *Shells.* MacDonald, London.

Stasek, C. R. 1972. The Molluscan Framework. *Chemical Zoology* 3: 41-48.

Usherwood, P. 1981. *Nervous Systems.* Edward Arnold, London.

Vernon J. E. N. 1993. *Corals of Australia and the Indo-Pacific.* University of Hawaii, Honolulu.

Whitfield P. J. 1979. *The Biology of Parasitism.* Edward Arnold, London.

Wilson J. A. 1979 (2nd ed.). *Principles of Animal Physiology.* Macmillan, New York.

Young A. 1994. *Marine Wildlife of Atlantic Europe.* Immel, London.

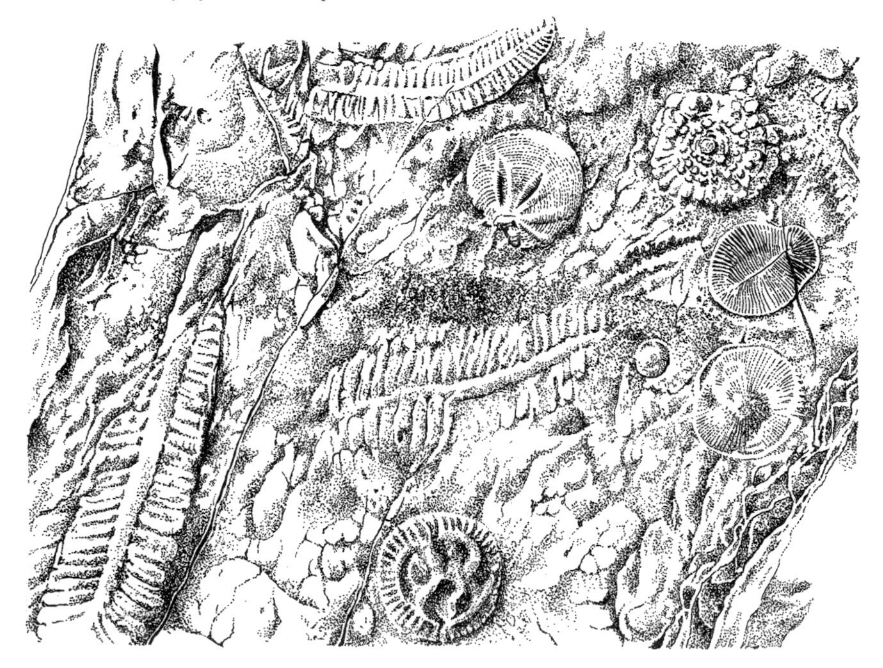

Identifying Marine Life

Part Two: Vertebrates

PLATE 9: COASTAL SHARKS

This plate shows a variety of inshore sharks including commonly encountered species in cold temperate waters (1,2,3and13). The requiem sharks illustrated are all active predators on coral reef fishes. Zebra Sharks (10) use their heavy crushing jaws to tackle hard-shelled invertebrates such as gastropods and crabs.

CATSHARKS Family: Scyliorhinidae

1. Lesser Spotted Dogfish *Scyliorhinus canicula*
Sandy colour with small spots. Dorsal fins well back. Origin of second dorsal fin level with back end of anal fin. eAO, Norway to Senegal (+MS),1m. **Nursehound/Greater Spotted Dogfish** *S. stellaris.* Similar in form and distribution to *S.canicula;* sandy or greyish with larger spots, second dorsal origin more forward over anal fin. 1.62m.

SMOOTH HOUNDS Family: Triakidae

2. Starry Smooth Hound *Mustelus asterias* 471
Sleek shark with large fins, tapering head, roundish snout and white spots on greyish dorsal area. Flat, crushing teeth. eAO, North Sea to Mauritania(+MS). 1.4m. **Smooth hound** sp., *M. mustelus.* Similar in form but lacks spots. eAO, North Sea to Cape (+MS). 1.64m. Other smooth hound species in the genus *Mustelus* have similar body shape and occur worldwide.

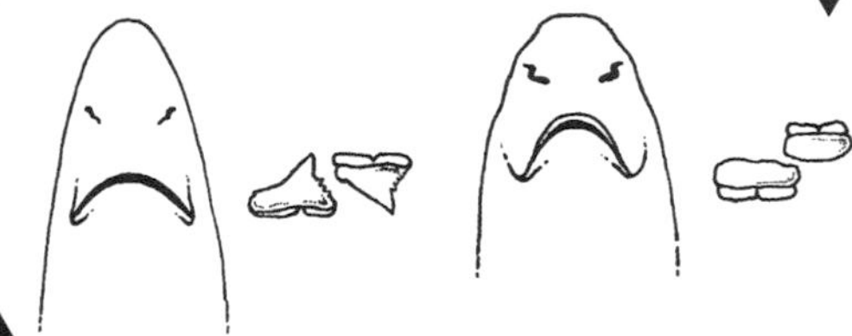

3. Tope *Galeorhinus galeus*
Slender-bodied shark with pointed snout, tail distinctive with deep notch about half way along upper lobe. A strong open water swimmer, to 400m+. Mainly cold and warm temperate distribution, coastal and offshore seas except nwPO, nwAO and IO. 1.95m. Also known as Soupfin Shark or School Shark.

REQUIEM SHARKS Family: Carcharhinidae

4. Grey Reef Shark *Carcharhinus amblyrhynchos*
Large, sleek with broadly rounded snout, black trailing edge to caudal fin, black tips to second dorsal, anal and pelvic fins, and underside tips of pectorals. Tip of first dorsal may be entirely grey or have a white edge (compare to Silvertip, plate 11). Red Sea east to wPO. 2.33m.

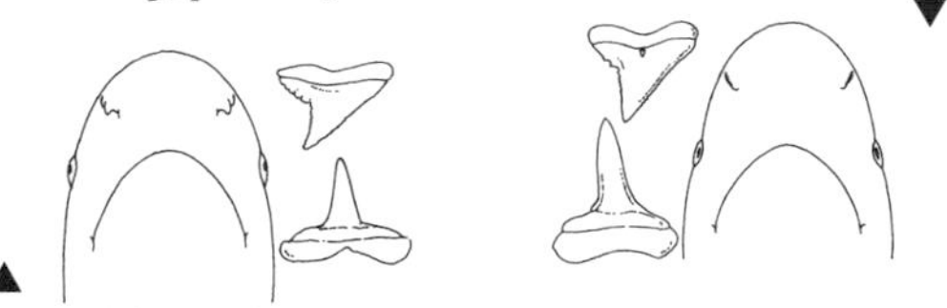

5. Blacktip Reef Shark *Carcharhinus melanopterus*
Very striking with white flash on the flank, black tip to the first dorsal fin (with white band below), rest of fins edged or tipped black and broad black margin to the pectoral fin. Overall profile similar to Grey Reef Shark (4) (snout shorter in *C. melanopterus*). Tropical and warm temperate IO and wPO including many island groups of western central PO; has entered MS from Suez Canal. 1.8m.

6. Blackspot Shark *Carcharhinus sealei* 471
Small reef shark, large black spot on second dorsal (short rear tip). Short rounded snout. Tropical shallow seas, IO and wPO. 95cm. **Whitecheek Shark** *C. dussumieri* is similar but first dorsal more triangular in adults and wider mouth than *sealei.* Occurs in same area as *C. sealei.* 1m.

7. Spottail Shark *Carcharhinus sorrah*
Distinct black tip to pectorals, second dorsal fin and lower lobe of tail. First dorsal may have black edge. Tropical shallow seas, IO and wPO. 1.6m. **Pondicherry Shark** *C. hemiodon* similar; second dorsal fin higher and more teeth distinguish it from *C. sorrah.* Indus coast east to wPO, may enter rivers. 1.5m+.

8. Whitetip Reef Shark *Triaenodon obesus*
Combination of white tips to first or both dorsal fins and upper caudal lobe, together with blunt snout with obvious nostril flaps is unmistakable. Upper body grey, underside white, sometimes with dark spots on sides. A common reef shark. Tropical IO and PO. 2.1m.

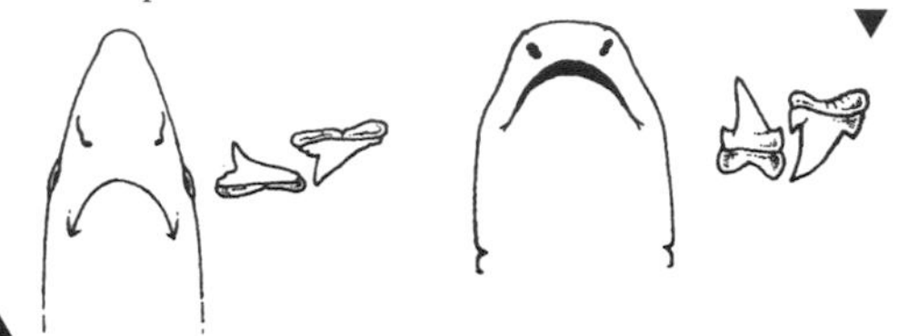

9. Milk Shark *Rhizoprionodon acutus*
Sleek, with long narrowly rounded snout, small pectoral and first dorsal fins, anal fin originating in front of second dorsal. Tropical seas, eAO, IO and wPO. 1m.

ZEBRA SHARK Family: Stegostomatidae

10. Zebra Shark *Stegostoma varium*
Caudal fin almost as long as rest of the body. Broad rounded snout, two dorsal fins close together. Juveniles darker with vertical yellow stripes (10a). Adults spotted. Tropical and warm temperate IO and wPO. 3.5m.

WOBBEGONGS Family: Orectolobidae

11. Spotted Wobbegong *Orectolobus maculatus*
Both dorsal fins well to rear, flattened body adapted to bottom and crevice dwelling lifestyle. Colour patterns on body and shape and number of mouth flaps distinguish the different species in this genus. S Australia. 3.2m.

BULLHEAD/HORN SHARKS Family: Heterodontidae

12. Port Jackson Shark *Heterodontus portusjacksoni*
Note the steep 'forehead', fin spines and body pattern. Markings, head profile and teeth important to distinguish between similar species in this genus; some species with humped backs. Feeds on bottom invertebrates, especially echinoderms. Australia and NZ. 1.65m.

DOGFISH SHARKS Family: Squalidae

13. Spurdog *Squalus acanthias*
Bottom-living, often in schools, voracious predator. Cosmopolitan distribution but patchy, not in tropical waters. 1.6m.

1
2
3
4
6
7
8
9
5
13
10
10a
12
11

PLATE 10: GIANT SHARKS AND RAYS

This plate shows the remarkable variety of body forms of some of the largest sharks and rays. The plankton feeders are the largest of these although fossil relatives (*Megalodon*) of the present day Great White Shark were also large, possibly up to12m in length; they have a variety of internal modifications of their gill chambers that serve to trap and collect food suspended in the sea water. They also have relatively tiny teeth of unknown function in their jaws.

PLANKTON-FEEDING SHARKS AND RAYS

The movements of these sharks and rays are related to the availability of their food. They often congregate in plankton-rich currents deflected by islands, headlands and underwater ridges, upwelling regions and current boundaries. Poorly known seasonal factors also play a part in distribution. Whale Sharks and Basking Sharks sometimes come to the surface when they appear to be relatively inactive ('basking'). It is important to keep detailed records of any observations made at sea; little is known about these elusive species.

1. Megamouth Shark *Megachasma pelagios*
An unmistakable species (Family Megachasmidae), only recently discovered (1976), despite its size (5.1m). Small gill slits, long caudal fin with notched, triangular upper lobe about half the length of rest of body and short lower lobe, no tail keel. Pectoral fin elongate, straight edges with white tips and underside. Head of living specimen **(1a)** shows bulging cheeks and wide gums with hundreds of tiny hook-like teeth in each jaw. Mouth lining silvery, may be luminescent. PO and seIO (so far known only from a few records).

2. Basking Shark *Cetorhinus maximus* **472**
Skin colour grey to black with white or grey band along underside. Large gills that almost encircle the throat are distinctive. This species is placed in its own group (Family Cetorhinidae). Hundreds of tiny hook-like teeth in each jaw. Small eyes seemingly out of place and set well forward on the tubular snout above the hoop-like mouth **(2a)**. Tail with side keel and crescentric, notched caudal fin. When feeding, the mouth is often pulled open to such an extent that the gill covers fan right out in the slipstream showing the gill cavities between them. The raised tip of the snout often shows above the water's surface, together with the large first dorsal fin and the tail tip. Can jump completely out of the water. Cooler waters in AO and PO, also seIO (Australia). Much to learn about movements (especially in winter when these sharks may become inactive in deep water).World's second largest fish; 9.8m and more than 4 tonnes.

3. Whale Shark *Rhincodon typus* **472**
World's biggest fish and largest cold-blooded vertebrate (Family Rhincodontidae). Blue-grey above, white below. Complex pattern of vertical white bars and spots. Markings behind last gill slit are individually specific. Enormous head, wedge-shaped in side profile, flat-fronted and squarish from above; small barbels, cavernous mouth opening 1.5m wide. Hundreds of tiny, hook-like teeth in each jaw. Eyes inconspicuous, closed by retracting and rotating eyeball backwards in socket. Dorsal fins behind mid-body, ridge running from first dorsal to head. Flank ridges prominent, lowest forms massive, blade-like keel on each side of tail. Tail fin huge, span about one third of total body length, upper lobe triangular (no notch) and longer than lower lobe. Filter feeds on plankton, also takes fish and squid. Often accompanied by young Golden Trevally *Gnathanodon speciosus*. Migratory, frequents tropical reefs, circumglobal. Inquisitive and harmless, does not jump. 12m, unconfirmed reports to 18m.

4. Manta Ray *Manta birostris*
A giant ray (Family Myliobatidae) which has a low dorsal fin at the base of whip-like tail (lacks sting) and large mouth flaps. These are moveable and guide food to the slot-like mouth. Hundreds of tiny pillar-shaped teeth in lower jaw only. Black above, the lower surface white with some grey speckling and edging to the wings. Often has remoras (*Echeneis naucrates*) attached. As a plankton feeder, it is both harmless and graceful when seen gliding through the water or leaping clear of the surface **(4a)**. Circumglobal in tropical seas. Reaches a span of 6.7m, reported to 9.1m.

PREDATORY SHARKS

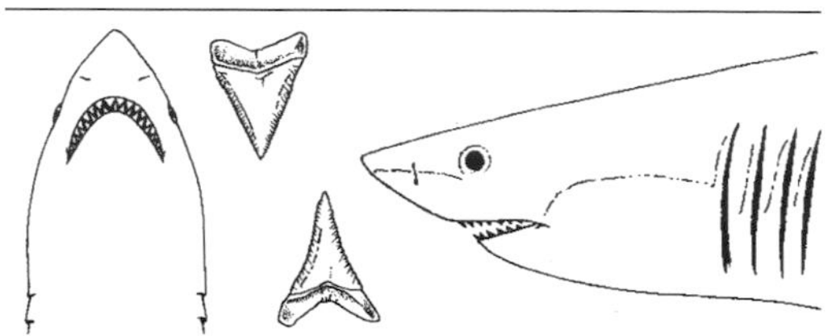

5. Great White Shark *Carcharodon carcharias* **473**
A mackerel shark (Family Lamnidae) similar to Mako Shark (plate 12) but with a blunter snout; both species often confused when young (<2m in length). Crescentric caudal fin. White belly and mouth, sharp dividing line with dark grey back. Large serrated triangular teeth. Takes fish and seals. A known man-eater, but attacks are rare and some probably occur when Great White Sharks mistake swimmers for large fish or seals. Found worldwide around coasts and shallow seas, including oceanic islands, often near seal colonies in temperate seas (+MS) (7°C+). 6.0m. A well fed 5m female Great White Shark weighed 1.8 tonnes. Now gaining protection.

6. Greenland Shark *Somniosus microcephalus*
A giant dogfish shark (Family Dalatiidae) living under arctic ice, the largest shark of polar seas. Lacks fin spines. Dark brown to grey in colour. May occur seasonally in shallow waters, but usually deeper (200-600m). Feeds on a wide range of prey, including fish, seals, birds and carrion. Polar and cold temperate nAO, also known from widely scattered localities in S hemisphere. 6.5m.

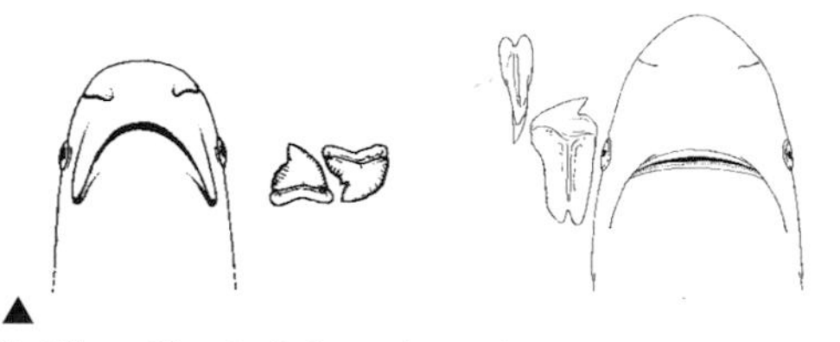

7. Tiger Shark *Galeocerdo cuvier*
Requiem shark (Family Carcharhinidae) with massive, broad head. Banding on back and comb-like teeth make the Tiger Shark unmistakable. Takes fish, turtles, sea snakes, birds and flotsam. Nocturnal feeder, enters shallows on bathing beaches. A known man-eater, the most dangerous shark in tropical coastal waters. Worldwide in tropical and warm temperate seas (also reported in colder seas, even off Iceland, probably following warm currents). Usually less than 5m, reported to reach 7.4m and 3.1 tonnes.

2a
1
2
4
1a
3
7
6
4a
5

PLATE 11: COASTAL AND OCEANIC SHARKS I

On this plate, the requiem sharks are divided into two arbitrary groupings on the basis of the presence or absence of fin markings to assist recognition. It should be noted that, when present, fin markings vary between species as well as between young and old individuals of the same species.

REQUIEM SHARKS
Family: Carcharhinidae

I. WITHOUT DISTINCTIVE FIN MARKINGS

1. Copper Shark/Bronze Whaler *Carcharhinus brachyurus*
Metallic grey upper body, white belly. Underside of pectoral fin tips dusky, other fins (especially dorsals and upper caudal) variably black-tipped or dusky. Coastal and offshore in tropical and warm temperate seas (+MS). Migratory in N of range, not north of MS in nAO. 2.9m.

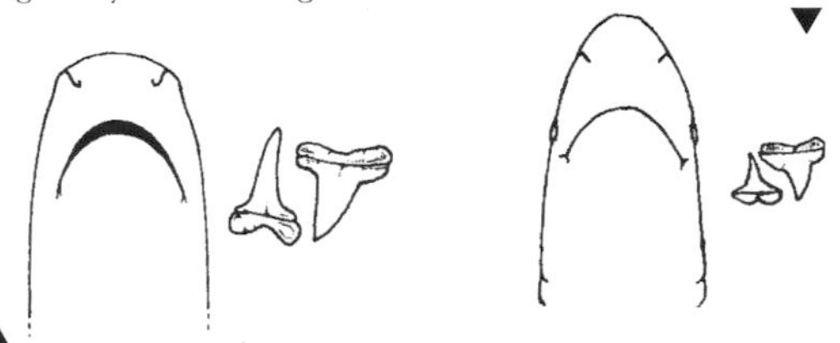

2. Indo-Pacific Lemon Shark *Negaprion acutidens*
Blunt snout, yellowish colour, small eyes, large rear fins (second dorsal fin almost as large as first) and broad pectoral fins, similar to Sand Tiger Sharks (12), but first dorsal further forward. Tropical seas, IO and wPO as far east as Tahiti. 3.1m. **Lemon Shark** *N. brevirostris*, has less pointed dorsal fins, occurs in ePO and AO (rare in eAO). 3.4m. Lemon sharks are said to become bad tempered when disturbed, and potentially dangerous to man.

3. Blue Shark *Prionace glauca* **472**
Sleek, lightly built with vivid blue coloration on back, underside white, long pectoral fins and long snout, first dorsal fin well back. Small tail keel. Surface of oceanic waters, circumglobal except polar seas, 7°C to 16°C, often deeper in tropical seas. 3.8m. Dangerous to man.

4. Bull Shark *Carcharhinus leucas* **472**
Heavy-bodied shark. Blunt, broadly rounded snout, with conical side profile. Grey body colour, fins can have dusky margins. Robust, triangular teeth in upper jaw. Worldwide in tropics and warm temperate seas, often in shallow waters, rivers and even lakes; found well upstream in the Amazon, also Mississippi, Ganges and Tigris rivers. 3.4m. Known man-eater. In Ganges and Indus, the poorly known **Ganges Shark** *Glyphis gangeticus* probably occurs with the Bull Shark. It has a short snout and dagger-like lower teeth. 2m.

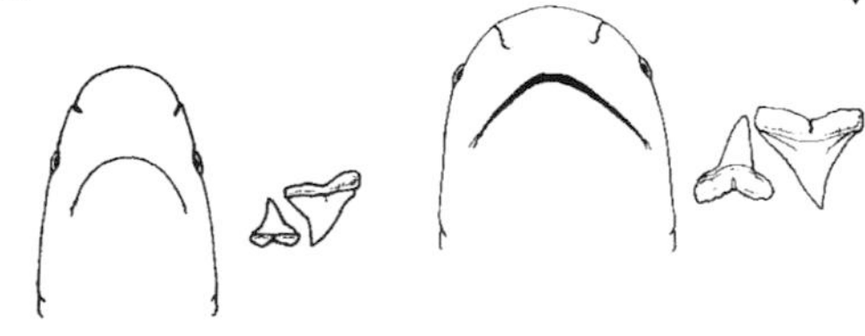

5. Sandbar Shark *Carcharhinus plumbeus*
Stocky, first dorsal fin well forward, high and erect. Short rounded snout, colour grey-brown. Circumglobal distribution in warm temperate and tropical seas, coastal (+MS). 2.4m.

6. Bignose Shark *Carcharhinus altimus*
Similar to Sandbar Shark **(5)**, in form and habit. Narrower upper teeth distinguish this species. Circumglobal. 2.8m.

7. Silky Shark *Carcharhinus falciformis* **472**
Sleek, lightly built, bluish to grey, long rounded snout, low first dorsal fin behind inner pectoral fin corner. Long free rear tip to second dorsal and anal fins. Common. Circumglobal in warmer seas (23°C+). 3.3m.

II. WITH DISTINCTIVE FIN MARKINGS

8. Silvertip Shark *Carcharhinus albimarginatus*
Striking white edges to the fins, enhanced by dark 'metallic' grey coloration of upper body. Common, often on outer reef edge. Tropical PO and IO 3.0m.

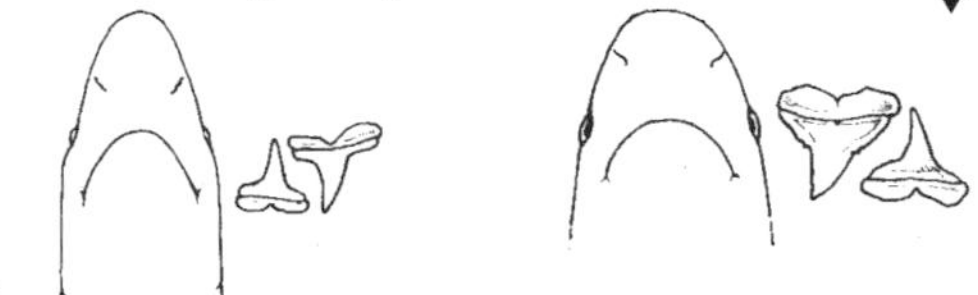

9. Spinner Shark *Carcharhinus brevipinna*
Slender shark with a long, pointed snout and small eyes. Small first dorsal fin. Fins tipped black (except in young). Active swimmer and leaper, spinning on its axis as it makes feeding passes through schools of fish. Coastal, worldwide in warmer seas (+MS). 2.78m.

10. Blacktip Shark *Carcharhinus limbatus*
Similar to (9), but generally lacks spot on anal fin and stouter build. Coastal, worldwide in warmer seas (+MS). 2.55m.

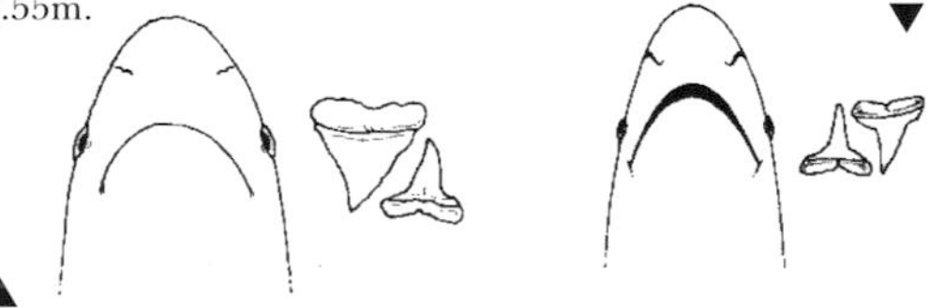

11. Oceanic Whitetip Shark *Carcharhinus longimanus*
Large rounded first dorsal and pectoral fins, short blunt snout, fins broadly tipped white, often with diffuse or speckled edge. Grey to brownish. Found in warmer waters than Blue Shark (20°C+). Circumglobal, common in tropical and warm temperate seas (?MS). 3.5m. (Shown with pilot fish *Naucrates ductor*). Known man-eater.

SAND TIGER SHARKS
Family: Odontaspididae

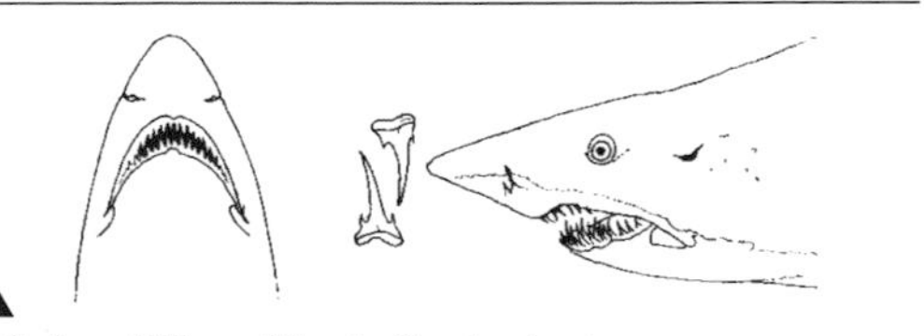

12. Sand Tiger Shark *Carcharias taurus*
Body colour brown with numerous brown spots. Long pointed snout. Large, curving needle-like teeth in mouth give a ferocious appearance, but usually harmless. Surf zone to a depth of 200m+. In warmer waters of AO, IO, wPO and MS. 3.1m.

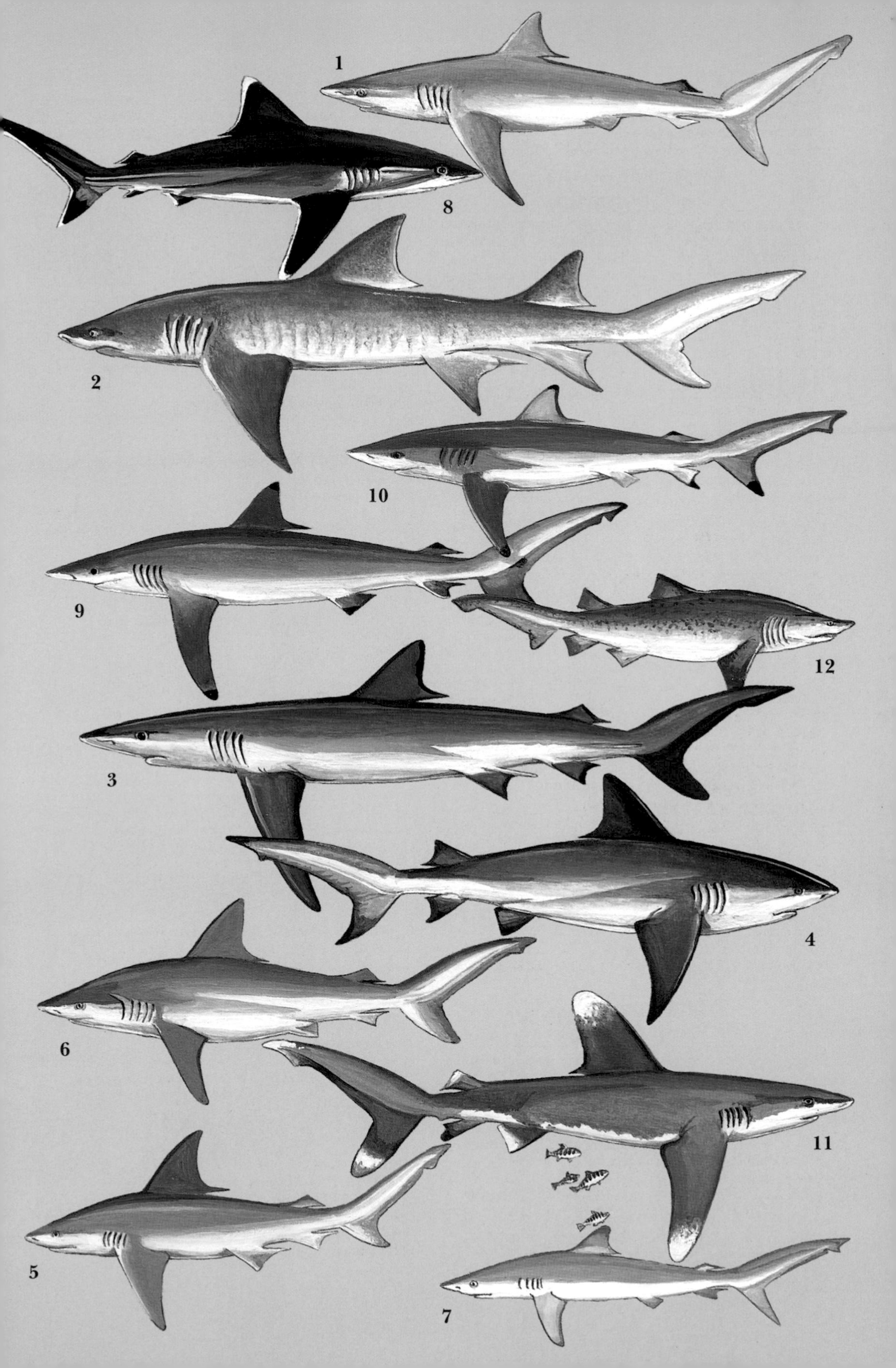
1
8
2
10
9
12
3
4
6
11
5
7

PLATE 12: COASTAL AND OCEANIC SHARKS II

Mackerel and Thresher Sharks are open ocean species that prey on other sharks, bony fish and squid. The Shortfin Mako is one of the few known shark species swift enough to prey on the Swordfish (11); sailfishes (plate 16), which are close relatives of the Swordfish, have been recorded to swim at speeds of 59 knots.

MACKEREL SHARKS
Family: Lamnidae

Conical snout, long gill slits, high first dorsal fin, long and falcate pectoral fins, small second dorsal and anal fins, obvious tail keel(s) and crescentric tail fin. Warm-blooded. Makos can be distinguished most easily from Porbeagles by tooth shape.

I. MAKOS *(Isurus)* First dorsal fin origin behind pectoral fin, second dorsal origin in front of anal fin, single keel, gill slits extending partially onto upper surface of head and narrow triangular teeth without cusplets.

1. Shortfin Mako *Isurus oxyrinchus* **473**
Sharply pointed snout, short pectoral fin (less than head length) and elongated curving triangular teeth at front of jaws. Vivid blue back. Mouth, throat and belly white. Circumglobal, inshore and oceanic, except colder waters (16°C+). 3.94m. Good game fish, spectacular leaper when hooked. Dangerous, non-fatal attacks recorded.

2. Longfin Mako *Isurus paucus*
Dark blue, white restricted to belly, long pectoral fins about equal to the head length, tips of front teeth straighter, mouth and throat dusky. Circumtropical. 4.17m.

II. PORBEAGLE AND SALMON SHARK (*Lamna*) Deeper-bodied than Makos, origin of first dorsal above a shorter pectoral fin, the second dorsal fin over anal fin, a second keel present, shorter gill slits and narrow triangular teeth usually with cusplets. Not so swift as the Shortfin Mako. Often feed on schooling fishes such as cod and mackerel.

3. Salmon Shark *Lamna ditropis*
Stubby snout shorter than Porbeagle, free rear tip of dorsal fin dark, dusky blotches on belly. Found in the colder nPO (as far north as Bering Sea); feeds on salmon. 3.05m.

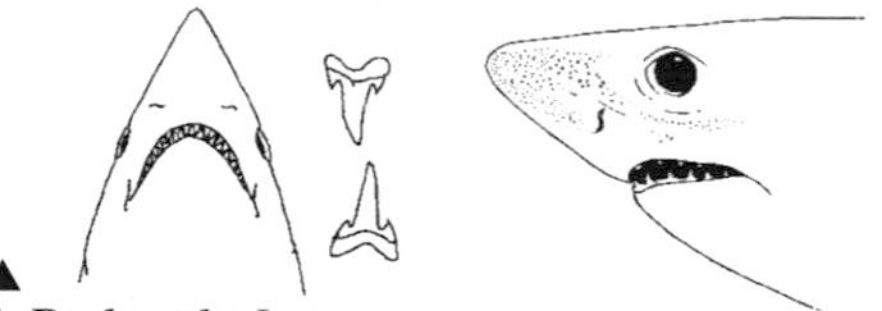

4. Porbeagle *Lamna nasus*
Snout pointed, free rear tip of first dorsal fin whitish. Belly lacks blotches. Coast to open sea, in colder waters of nAO as far north as Spitsbergen and circumpolar southern seas including off S Georgia and Kerguelen Is. Not in nPO. 3.7m.

THRESHER SHARKS
Family: Alopiidae

Three widespread species with a unique scythe-like upper tail lobe which equals the length of the rest of the body, conical snout, long pectoral fins, small second dorsal and anal fins. Feed on schooling fish and squid. Tail fin can be used to herd and stun prey; can been seen protruding out of the water with the first dorsal fin. Teeth low and triangular, with smooth cusp (except Pelagic Thresher where teeth are notched). Heavily fished in some areas, and much confusion between species in the past.

5. Pelagic Thresher *Alopias pelagicus* **473**
'Metallic' blue back clearly defined from white on belly which does not extend above pectoral fin base. Pectoral fins with straight leading edge and broadly rounded tips. Often mistaken for Thresher Shark (7). Tropical IO and PO, not often inshore. 3.3m.

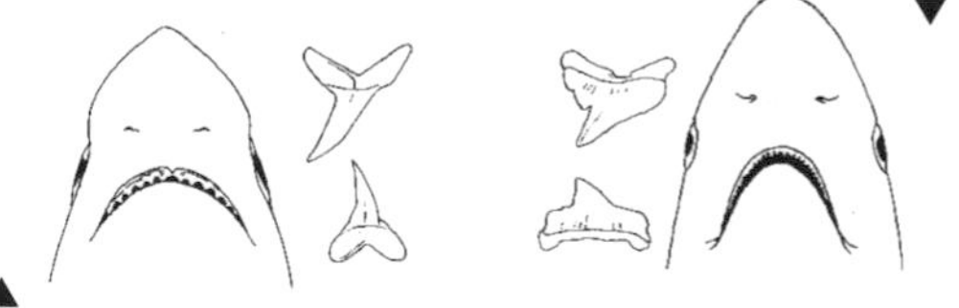

6. Bigeye Thresher *Alopias superciliosus*
Head crease behind huge oval eyes, unmistakable. Grey above, pale underside. Pectoral fins falcate with broadly rounded tips. Probably circumglobal distribution in tropical and warm temperate waters from coasts to open sea and surface to 500m+. 4.6m.

7. Thresher Shark *Alopias vulpinus* **473**
Grey above, white on belly extending well above bases of pectoral fins. Mottling between dark back and white underside. Falcate pectoral fins with pointed tips. Can attack and eat seabirds. Probably circumglobal distribution, largest and most widespread thresher, ranges into cooler waters than Bigeye. 6.1m.

REQUIEM SHARKS
Family: Carcharhinidae

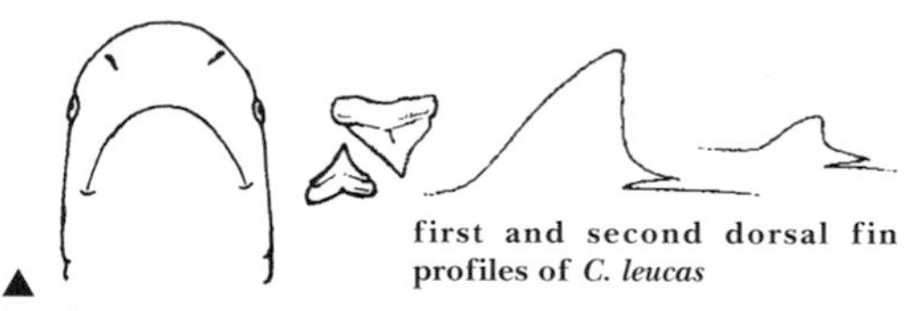

first and second dorsal fin profiles of *C. leucas*

8. Pigeye/Java Shark *Carcharhinus amboinensis* **473**
Body shape like Bull Shark (plate 11), also with blunt snout but differs in taller first dorsal fin (over 3.1 times height of second dorsal, 3.1 or less in Bull Shark) and less rows of teeth in lower jaw. Warmer coasts of eAO, IO and wPO. 2.8m.

HAMMERHEADS Head expanded as a wide hydrofoil. These two widespread species can be distinguished by differences in shape of the head and pelvic fins.

9. Great Hammerhead *Sphyrna mokarran*
Tall first dorsal fin and large rear fins, pelvic fin with concave rear edge. Head has fairly straight leading edge with notch in the middle. Circumglobal in tropical and warm temperate seas (+MS). 6.1m.

10. Scalloped Hammerhead *Sphyrna lewini*
Has an arched leading edge to the hammer, pelvic fin rear edge straight. Social, sometimes in spectacular schools. Circumglobal in tropical and warm temperate seas (wMS only); most common hammerhead species. 4.2m.

SWORDFISH
Family: Xiphiidae

11. Swordfish *Xiphias gladius*
Sleek tuna-like body, large eyes and bill-like jaws. Pelvic fins absent, tail keel single, body scaleless in adults. Bill flat in cross section. Cosmopolitan distribution except polar waters. 4.5m, 0.5 tonnes.

1
2
11
3
4
5
6
8
7
9
9
10
10

PLATE 13: RAYS AND SKATES

This plate illustrates species from some of the commonly occurring ray families. All have five gill slits on the underside behind mouth except *Hexatrygon* (4), a rarely seen deep-water ray. Eagle rays are neritic or pelagic; skates are bottom-dwellers, adapted to life in coastal, offshore and deep-sea habitats.

WHIPTAIL STINGRAYS
Family: Dasyatidae

Tail long with sting or spine but lacking caudal fins. Tail spine can inflict painful wound on bathers. Benthic, also in freshwater in tropics.

1. Violet Stingray *Dasyatis violacea*
Body smooth with no thorns. Uniformly round bow-shaped front edge to body with pointed wing tips, a lower caudal fin fold and a long sting. Eyes flush with head surface. Uniformly coloured, purplish-violet to bluish-grey above and greyish below. Unusual pelagic lifestyle. Circumglobal in warmer seas. 80cm disc width.

2. Blue Stingray *Dasyatis marmorata*
Body smooth, wing tips rounder and snout more pointed compared to Violet Stingray. Blue blotching on a golden-brown background above, underside white. Benthic and coastal, including surf zone. eAO, Senegal around Cape to Natal. 75cm disc width. **European Stingray** *D. pastinaca*, similar, but disc olive or plain grey-brown, sometimes blotched above. eAO (+MS). 60cm disc width, 2.5m total length.

BUTTERFLY RAYS
Family: Gymnuridae

Disc broader than long with a triangular front profile. Short whip-like tail with a small spine (in some species only). Body smooth, some species uniformly coloured, others with spots. Underside white. Tropical and warm temperate seas worldwide. Disc width to 4m. About 12 species.

3. Butterfly Ray *Gymnura japonica*
Disc rhomboid, spotted, short banded tail. No caudal or dorsal fins. S. Japan to China Seas. 1.8m total length.

LONG-NOSED RAYS
Family: Hexatrygonidae

4. Sixgill Stingray *Hexatrygon bickelli*
Unique elongate snout is more robust than in long-nosed skates. Six gill slits. Body smooth. Colour dark violet to brownish above, white below. Continental slope below 200m. Little known, wPO and off S. Africa. 1.1m+ total length.

SKATES
Family: Rajidae

Compared to stingrays, skates lack a sting and some species have two reduced dorsal fins at the end of the tail. Many species have long snouts. Upper surface often heavily spined, extending onto tail in most species. Some species reach a total body length of 2m. Benthic.

5. Twineye Skate *Raja miraletus*
A typical short-nosed skate with two blue ocelli above, underside white. AO and swIO (+MS). 60cm total length.

EAGLE RAYS
Family: Myliobatidae

Graceful rays that 'fly' through the water. This family is divided into three groups, each with distinctive type of head shape. Harmless to man.

I. MANTA RAYS, DEVIL RAYS

Devil rays (*Mobula*) have a subterminal mouth with teeth in both jaws (mouth terminal in *Manta*). Separation between the Devil Ray species is difficult; colour, curvature of wings, head profile and presence or absence of denticles and sting useful. *Mobula* species usually less than 2m disc width; European *M. mobular* can reach about 5m disc width. Warmer seas worldwide.

6. Devil Ray *Mobula diabolus*
Upper surface bluish-black, underside whitish. No spine on tail. Short mouth flaps. Plankton feeder. Occurs in IO and wPO. 1.8m disc width.

II. COWNOSE RAYS

Similar to *Mobula* but with different head shape and obvious sting. Front of head concave, distinct 'groove' between head and wing when the head is viewed from the side. Tropical coasts worldwide. Disc width to 2.1m.

▲

7. Cownose Ray *Rhinoptera javanica*
Upper surface brown, underside whitish. Large spiracle behind eye. Occurs in large schools, feeds on bivalves. Occurs in IO and wPO. 1.5m disc width.

III. EAGLE RAYS

Similar to cownose rays, but the jaws are massive and the rounded or conical snout is extendible, used to root out benthic prey. Head shape, body pattern, position of dorsal fin and teeth useful to separate species. Upto 3m disc width.

8. Common Eagle Ray/Bullray *Myliobatis aquila*
Upper surface dark with bronzy tone, white underside. Warmer coasts of eAO (+MS) and wIO (Natal). 1.5m disc width. The Pacific **Bat Ray** *M. californica* of warmer waters of coastal N. America is similar.

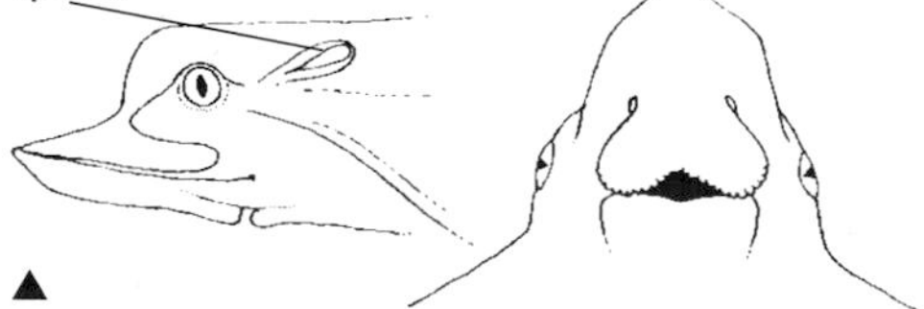

▲

9. Spotted Eagle Ray *Aetobatus narinari*
Bulbous snout and blackish body covered in pale spots and rings. Large spiracle behind eye. Circumglobal in warmer seas (+MS). 2.4m disc width. **Duckbill Ray** *Pteromylaeus bovinus* has a similar pointed snout and is often banded brown or blue-grey in juveniles, plain brown in adults with brownish-red tips to wings. Large tail spine. eAO and coast of S. Africa (+MS). 1.8m disc width.

1
2
3
4
5
6
7
8
9

PLATE 14: BIZARRE FISHES

This plate illustrates some of the world's most unusual fishes. These species live on coral reefs, in the open ocean and in the deep sea; the Pearlfish (13) has the peculiar habit of living inside the bodies of other reef animals.

SPADEFISHES
Family: Ephippidae

Compressed and circular tropical fishes that stray into warm temperate waters. Large dorsal and anal fins give distinctive profile. Can form large schools, often on the edge of reefs and over sandy areas.

1. Atlantic Spadefish *Chaetodipterus faber*
Distinctive dark grey or brown stripes on a silvery to brownish background. Young are brown. Dorsal fins have elongate tips. Similar to large angelfishes (plate 22), but body more circular. wAO. 91cm.

2. Longfin Batfish *Platax teira*
Unusual profile of this deep-bodied fish is enhanced by long, rounded fins. Adults silvery-brown with dark bars. Graceful swimmers. Young are yellow-orange. IO and wPO. 60cm. **Circular Spadefish** *P. orbicularis* similar but lacks blotch in front of anal fin. IO and wPO. 57cm. **Pinnate Batfish** *P. pinnatus* also similar but indented forehead and protruding mouth. Young almost black **(2a)**, mimic poisonous flatworm. IO(?) and wPO. 45cm.

OPAHS AND ALLIES
Families: Lampridae, Regalecidae

Curious oceanic and deep-sea fishes, silvery with red fins and protrusible mouths. Most are slender with eel-like bodies. Both the elongate **Oar-fish** (Family Regalecidae) and the rounded **Opah** or **Moonfish** (Family Lampridae) sometimes occur close to the surface.

3. Opah/Moonfish *Lampris guttatus*
Oval and compressed body, bluish with white spots, shading to pink on the belly and elongate red fins. Jaw is protrusible, toothless. Worldwide in tropical and temperate waters, mainly at depths of 100-400m. Reaches 2m+, 1 tonne+.

4. Oar-fish *Regalecus glesne*
Elongate, toothless fish with a dorsal crest and thread-like pelvic fins. World's longest bony fish. Similar range to the Opah but deeper depths (200-1000m). Known to reach 8m.

PORCUPINEFISHES
Family: Diodontidae

Similar to pufferfishes (plate 23) except with fleshy spines and solid beak without a central groove. Worldwide in warmer seas.

5. Spotted Porcupinefish *Diodon hystrix*
A typical porcupinefish. Body can be enormously inflated **(5a)** with sea water (or air) and is almost impossible for predators to swallow. Adults on reefs, nocturnal. Young are pelagic. Worldwide in warmer seas may reach 91cm, usually smaller.

OCEAN SUNFISHES
Family: Molidae

Fishes with disc-like bodies lacking a tail.

6. Ocean Sunfish *Mola mola*
Silvery to brownish-grey with grooves on the body. Swims by sculling motion of large vertical fins. Feeds on jellyfish. Worldwide except polar seas. 4m+ and 2 tonnes+. The related S hemisphere species *M. ramsayi* is equally large.

FROGFISHES
Family: Antennariidae

Both illustrated species of frogfishes use a 'lure' on the snout to catch fishes (not present in all species). Some frogfishes are very colourful and blend in with coral backgrounds by altering skin colour. Warmer seas worldwide.

7. Sargassum Fish *Histrio histrio*
This species has a unique pelagic habitat in sargassum seaweed. Uses camouflage and lure when feeding on small fish. Body can be inflated with sea water to ward off predators. Worldwide, except ePO. 14cm.

8. Roughjaw Frogfish *Antennarius avalonis*
Colour can vary greatly, orange and black ocellus above tail base. Tropical ePO, in shallow coastal water to 300m depth. 34cm.

MORAY EELS
Family: Muraenidae

Eels that lack pectoral and pelvic fins. Large adults can inflict serious bites if provoked.

9. Green Moray *Gymnothorax funebris*
Large, adults greenish, young blackish with a white chin. Tropical to warm temperate wAO. 2.5m.

10. Spotted Moray *Gymnothorax moringa*
Large moray with leopard-like markings. Tropical and warm temperate wAO east to St. Helena, Ascension Is. 3m. There are other moray species with similar markings.

11. Dragon Moray *Enchelycore pardalis*
Boldly marked with white, black, yellow and orange plus tube-like nostrils, posterior pair just in front of the eyes. IO and wPO. 80cm. Some morays have leaf-like nostrils (e.g Ribbon Eel *Rhinomuraena quaesita,* from Indo-Australian reefs). There are other boldly speckled, blotched and striped species, including the uniquely striped **Zebra Moray** *Gymnomuraena zebra* from the IO and PO. 1.5m.

SCORPIONFISHES
Family: Scorpaenidae

Large-headed, generally benthic fishes with toxic fin spines.

12. Turkeyfish/Lionfish *Pterois volitans*
Boldly striped pattern and feather-like dorsal and pectoral fin filaments. Ranges from Malaysia east to Austral Ridge in PO. 38cm. **Devil Firefish** *P. miles* very similar but smaller spots on dorsal fin filaments. Feeds by 'corralling' small fish and crustaceans with pectoral fins. Red Sea and IO, east to Sumatra. 30cm. Both species with poisonous spines, should be avoided by swimmers.

PEARLFISHES
Family: Carapidae

Fishes with transparent larvae, adults generally living in association with benthic invertebrates.

13. Pearlfish *Carapus bermudensis*
Scaleless, eel-like fish living inside the body of sea cucumbers during the day, leaving its host only at night to feed out on the reef. Tropical wAO. 30cm.

1
5
9
5a
3
4
6
2
7
8
10
12
11
2a

PLATE 15: OCEANIC FISHES

This plate includes species found in surface waters of the open ocean. These fishes have a streamlined shape for fast swimming and keels on their tails for manoeuvrability. Tail fins are crescentric providing high thrust and low drag. Tunas are amongst the swiftest fishes, able to out-distance all but the fastest predators; Wahoo (8) and Yellowfin (2) have burst speeds of about 40 knots (measured for fishes about 1m long). Some tuna species are warm-blooded; body temperature is maintained above that of sea water. Most species are heavily fished.

TUNAS AND MACKERELS
Family: Scombridae

Tunas are large, heavily muscled, powerful fishes occurring worldwide from cold temperate to tropical seas. They prey on small fish and squid. Their main predators are other tunas, billfishes, mackerel sharks and toothed whales. The following fishes are divided into two arbitrary groups based on the extent of scale development to assist in recognition. Body length is measured from snout tip to the end of the tail (central outer margin).

I. LARGE SCALES BEHIND HEAD AND AROUND PECTORAL FINS ONLY

1. Longfin Tuna/Albacore *Thunnus alalunga*
Distinctive long, dark pectoral fins. Thefirst dorsal fin is yellow (often with pale tip), dorsal finlets yellowish, caudal fin margin white. Worldwide. 1.3m.

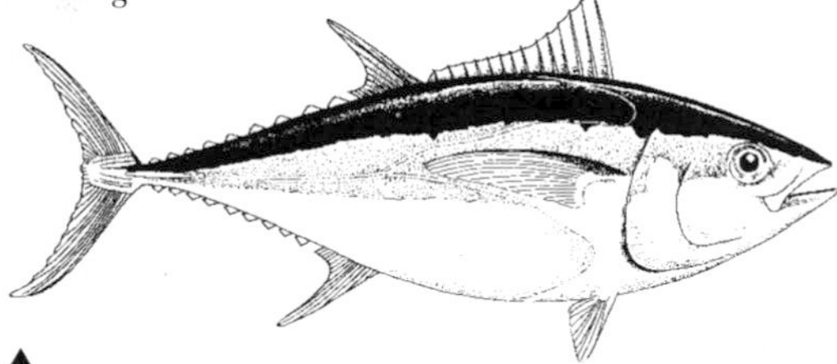

▲
Can be confused with **Bigeye Tuna** *T. obesus* which also has enlarged pectoral fins (pronounced in fish <1m in length) but lacks white margin on tail. Worldwide, absent in MS. 2m.

2. Yellowfin Tuna *Thunnus albacares*
The second dorsal and anal fins are yellow (like the finlets) and can be very long (more than 20% of body length). Pectoral fin ends about level with origin of the anal fin. Often has golden stripe on side. Mainly tropical, worldwide, absent from MS. 3m.

3. Southern Bluefin Tuna *Thunnus maccoyii* **474**
Short pectoral fins never reaching as far back as notch between dorsal fins. First dorsal fin blue or yellow, anal fin and all finlets yellowish-edged with black. Central tail keel yellow in adults. S hemisphere, south of 30°S in colder waters. 2.25m, 150kg.

4. Mackerel Tuna/Little Tunny *Euthynnus alletteratus*
There are four to six spots below the pectoral fins. Complex wavy barring on rear portion of upper body. Warmer AO (+MS). 1m. Two other very similar species: **Kawakawa** *E. affinis* has a different geographical distribution, warmer IO and PO 91cm; **Black Skipjack** *E. lineatus* has a different colour pattern (barring on body is horizontal). ePO. 99cm.

5. Skipjack Tuna *Katsuwonus pelamis* **474**
Body with long distinctive belly stripes, barring on rear of back. Trunk and tail scaleless, iridescent purple-blue above, silvery-white belly. Worldwide. 1.2m.

6. Frigate Mackerel *Auxis thazard*
Barring on upper body, well developed scales around pectoral fins extending down lateral line. 51cm. **Bullet Mackerel** *A. rochei* similar, differs in detail of lateral line scales 51cm. Both worldwide in warmer seas.

II. SCALES EVENLY DEVELOPED COVERING WHOLE BODY

7. Atlantic Mackerel *Scomber scombrus*
Wavy bars on upper body, belly unmarked. nAO (+MS). 56cm. **Chub Mackerel** *S. japonicus* is similar, also nAO (and cosmopolitan) but belly marked with spots or lines. 50cm.

8. Wahoo *Acanthocybium solandri*
Streamlined with a long snout. Body bright silvery metallic colour with blue stripes on the flanks (fade soon after death). Mainly tropical, inc. MS. Popular game fish. 2.1m.

BARRACUDAS
Family: Sphyraenidae

Swift and sleek predators which have an infamous and misleading reputation. They are ferocious predators but are unlikely to attack people; all known instances were in turbid water or resulting from spear-fishing. Can form large schools, bigger fish tend to be solitary. Powerful jaws, protruding below, formidable fang-like teeth, well separated dorsal fins, forked tail and obvious lateral line. Silvery fishes; the Great Barracuda is the largest species in the Sphyraenidae.

9. Great Barracuda *Sphyraena barracuda*
Greenish cast, blotches on the flanks, dark bars on the upper body (not always obvious). Jaws unequal length. Worldwide in warmer seas, except ePO. 1.5m+.

TARPONS
Family: Megalopidae

Large-mouthed fishes with mirror-like sides. large scales. Last ray of the dorsal fin stretches over back. Adults breed at sea and young live in nurseries in mangrove swamps where they gulp in air with the distinctive upturned mouth (air bladder acts as a lung). Occur in warmer seas, also in freshwater. Popular game fishes.

10. Tarpon *Megalops atlanticus*
Compressed body. Note angle of lower jaw. AO. 2.4m,159kg. **Pacific Tarpon** *M. cyprinoides* is smaller, 90cm.

DOLPHIN FISHES
Family: Coryphaenidae

Pelagic predators with an unmistakable profile, highly coloured, long dorsal fin, blunt head and forked tail. Head much steeper in males. Occur worldwide in warmer seas, fast swimmers, shooting out of the surface in pursuit of flying fishes. Popular game fishes.

11. Dolphin/Dorado *Coryphaena hippurus*
A beautiful range of brilliant colours on the dorsal surface; red, blue, green and yellow through to silvery white on the belly. Larger fish in schools show zebra-like stripes when feeding. 2m. **Pompano Dolphin** *C. equiselis* is smaller, chubbier with a less abrupt forehead (as female Dorado), 56 or less rays in the dorsal fin (Dorado has up to 65). Possibly more tropical and oceanic than Dorado. 90cm.

1
2
3
8
9
10
11
7
6
4
5

PLATE 16: LARGE OPEN OCEAN FISHES

Sailfishes, Spearfishes and Marlins together form a group known as billfishes that share distinctive jaws shaped like a beak or bill. The Swordfish (plate 12) is also included with the billfishes and has the distinction of being the only truly cosmopolitan species in the family. The billfishes illustrated on this plate are confined in their distribution to the Atlantic Ocean and/or Mediterranean Sea. The curious Louvar is similar to billfishes in some of its external features and its oceanic habit; it is included here for this reason. Billfishes are fine blue-water sport fish for anglers, also excellent food fish usually caught by longlining. Body length in billfishes is measured from the lower jaw tip to the end of the tail (central outer margin).

BILLFISHES
Family: Xiphiidae

Scales, teeth, pelvic fins and two tail keels on each side of tail base present in the species below. Bil round in cross section.

1. Atlantic Sailfish *Istiophorus albicans*
Dorsal fin sail-like with elongate central rays and distinctive spotted, vivid purple-blue membrane. Sides of body flat, bill long and thin, lateral line simple, pelvic fins long and filamentous, forehead not humped. Colour blue-black above, white ventrally. About 20 full-length, vertical stripes along body sides, each stripe formed by numerous iridescent blue spots.

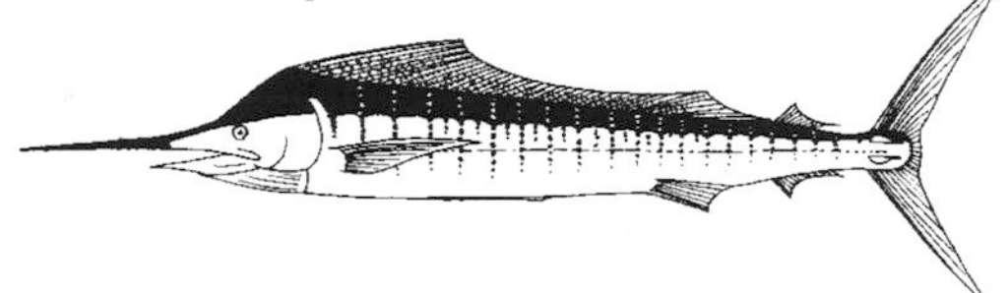

Sailfishes 'fold away' their large dorsal fin into a groove in the back during fast swimming. There is also a belly groove for the long pelvic fins. The dorsal fin is opened when feeding on schooling fish, perhaps helping to 'bunch' a dispersed school. Prey is stunned and slashed by rapid strokes of its bill. Spectacular changes in body coloration occur during feeding (in sailfishes and marlins); gill covers, pectoral fins and flanks turn bright blue. Flank stripes also appear to 'glow'. Warm temperate and tropical AO (+MS), usually oceanic, also migratory inshore. **Indo-Pacific Sailfish** *I. platypterus* is said to differ in fin proportions and larger adult size. 3.6m, 101kg.

2. Mediterranean Spearfish *Tetrapturus belone* **474**
Body colour blue-black above, white below. Metallic blue stripe from eye to tail. Body sides flat, lateral line simple, forehead not humped, bill short. Small, pointed pectoral fins 10-13% of body length. First dorsal fin relatively low at front, less than body depth, up to 46 rays, unspotted. MS only. 2.4m, 70kg. Potential confusion with **White Marlin** (4) which also occurs in MS, but can be separated on size of first dorsal fin and shape of pectoral fins.

3. Longbill Spearfish *Tetrapturus pfluegeri*
Body colour blue-black above, white below, similar to Mediterranean Spearfish (2) but with larger bill, longer pectoral fins (18% of body length or more), and higher first dorsal fin (up to 53 rays).Warm temperate and tropical AO (40°N to 35°S) in offshore waters (not MS). 2.0m+, 45kg+.

4. Atlantic White Marlin *Tetrapturus albidus* **474**
Body blue-black above and white below, sometimes 12-20 light blue to lavender vertical stripes on flanks. Sides of body flat, lateral line simple, forehead not humped. First dorsal fin high and rounded at front edge, deep blue membrane with rows of large spots. Pectoral fins blackish-brown, wide with rounded tips. First anal fin tip rounded. Warm temperate and tropical AO (+MS), 22°C+, oceanic. 2.8m+, 82kg+.

5. Atlantic Blue Marlin *Makaira nigricans*
Body blue-black above, white below, with about 15 pale blue stripes, each formed by numerous spots and/or bars. Body sides more rounded than flat, lateral line a complex meshwork, forehead distinctly humped (**5a**). First dorsal fin low, sometimes with black spots on posterior membrane. Tagging studies show this species makes trans-Atlantic migrations. Warm temperate and tropical AO, not MS, oceanic, 22°C+. 3.75m+, 580kg. **Indo-Pacific Blue Marlin** *M. mazara* similar, but differs in detail of lateral line and larger adult size. 4.4m+, 900kg+.

LOUVAR
Family : Luvaridae

6. Louvar *Luvarus imperialis*
Sides of body flat similar to sailfishes rather than tunas, blunt-headed with bright pink fins. Young with black spots. Toothless, said to feed on plankton, usually in deeper water off continental shelf. Temperate and tropical waters of N and S hemispheres. 1.9m.

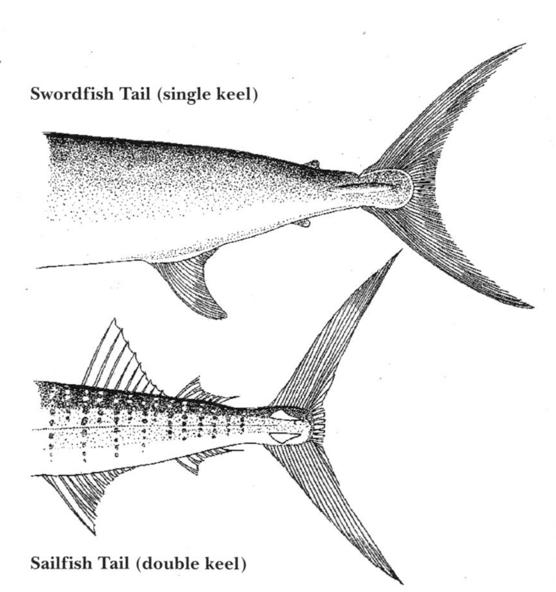

Swordfish Tail (single keel)

Sailfish Tail (double keel)

5a
1
2
3
4
5

PLATE 17: MULLETS, BLUEFISHES, JACKS AND ROOSTERFISH

This plate shows a variety of inshore and pelagic fishes. Jacks (Carangidae) often occur in 'pelagic schools' at the edge of tropical reefs.

MULLETS
Family: Mugilidae

1. Thick-lipped Grey Mullet *Chelon labrosus*
Chunky, striped body with two separate dorsal fins, the first spiny. Mouth small, neAO (+MS). 75cm.

BLUEFISHES
Family: Pomatomidae

2. Bluefish *Pomatomus saltator*
Renowned as a voracious fish and highly prized by fishermen. Often in schools, will chase prey into the shallows and even known to bite bathers. Head large, keeled above, low first dorsal fin and mouth extends behind eye. Darker above with silvery sides and dark blotch at base of pectoral fins. Greenish to bluish dorsal area becomes more obvious after death. Worldwide (but patchy) in temperate and tropical waters, pelagic along the continental shelf. 1.1m (usually 40-60cm). Known as Tailor in Australia and Elf in S.Africa.

JACKS
Family: Carangidae

Fast swimming predatory fishes of tropical reefs and open seas. Includes large deep-bodied fishes, such as Giant Kingfish (6), a voracious predator. See examples of the genus *Caranx* below. Other species are tuna-like with elongate body form, such as Giant Queenfish (9). Certain species are round-bodied such as Permit (10).

For convenience here, species are divided into three groups according to the form of their plate-like lateral line scales (scutes).

I. JACKS WITH LARGE SCUTES COVERING WHOLE LATERAL LINE

3. Horse Mackerel/Scad *Trachurus trachurus*
Bluish to greenish above, silvery below, black mark on gill, high first dorsal fin. neAO (+MS), Iceland to Cape Verde Is., on bottom or pelagic on continental shelf. 70cm. Other similar species in the genus worldwide; three species overlap range of *T. trachurus* (*T. picturatus, T. mediterraneus* and *T. trecae*). These differ in details of their lateral line scales.

II. JACKS WITH LARGE SCUTES ON TAIL LATERAL LINE ONLY

4. Bluefin Trevally *Caranx melampygus*
This species has beautiful blue fins and blue-speckled flanks. IO and PO. 1m.

5. Creville Jack *Caranx hippos*
Steep-headed, fins often yellowish. Black area on edge of gill and along base of pectoral fin. Warmer wAO, Ascension, St Helena and eAO. 1.5m (usually smaller). **Horse-eye Jack** *C. latus* similar but lacks spot on pectoral fin. May form mixed schools with Creville Jacks. Scutes usually blackish. Similar range, as far east as St Paul's Rocks. 75cm.

6. Giant Kingfish *Caranx ignoblis*
Deep-bodied, large and robust. Dorsal surfaces dark olive-green fading to white on belly, sometimes yellow tinge to flanks. Small dark spots cover dorsal and upper sides. ventral fin with white edge. IO & wPO, average 60-100cm, can attain 1.7m and 68kg.

III . JACKS WITH NO LATERAL LINE SCUTES

7. Rainbow Runner *Elagatis bipinnulatus*
Sleek & slender with a double blue stripe divided by a yellow stripe. Has small anal finlets (compare to Amberjack) & a low first dorsal fin. Worldwide & pelagic in warmer seas over 20°C (not MS), commoner in Pacific, 1.1m. Young can mimic pilotfish.

8. Greater Amberjack *Seriola dumerili*
Dusky head stripe that runs back to the low 1st dorsal fin & yellow band along flank. Can be bluish above & silvery below or with bronze cast. Fleshy tail keel. Worldwide in warmer seas, 1.8m & 80Kg. **Yellowtail**, *S. lalandi* is very similar but lacks dark head stripe & has a yellow tail. Worldwide in warmer seas, 1.5m. **Highfin Amberjack** *S.rivoliana* also similar but lacks yellow colouration. Circumtropical, 1.1m.

9. Giant Queenfish *Scomberoides commersonianus*
Large mouthed, blunt snouted, elongate fish. Dusky geen above, silvery flanks often with golden yellow cast & white belly. Five to eight dark blotches along the sides above lateral line. Pectorals short, 1st dorsal fin with six to seven spines, anal with two detached spines, finlets present, the large caudal is deeply forked. Dorsal & anal fin spines poisonous. IO & wPO, 1.2m.

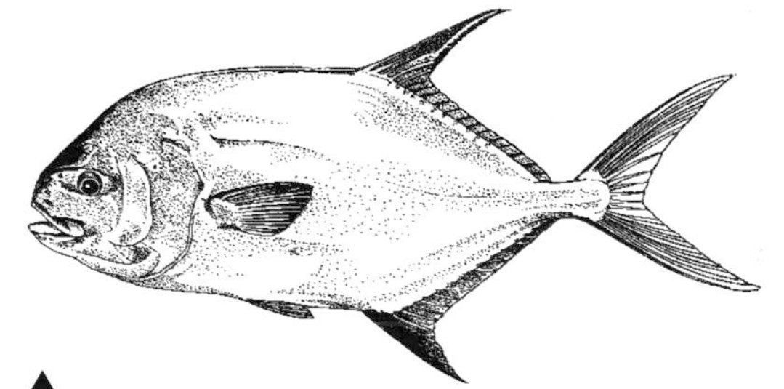

▲

10. Permit *Trachinotus falcatus*
A beautiful, deep-bodied compressed silvery fish, with black edging on elongate fins. May have circular black area behind base of pectoral fin. Orange mark at anal fin base. Inshore, feeds on molluscs. Good sport fish. Warmer wAO & e to Africa, 1.15m. **Palometa**, *T. goodei*. Similar but smaller with body bars (pale to blackish), black colour on the fins often more extensive. Around reefs, warmer AO, 50cm.

11. Lookdown *Selene vomer*
Long sloping forehead, elongate fins, compressed body. Swims over bottom tilted forward appearing to 'look down', inshore habit. Often has a brassy cast. Tropical to warm temperate AO, 30cm.

ROOSTERFISHES
Family: Neumatistiidae

12. Roosterfish *Nematistius pectoralis*
Large jack-like fish easily recognised by 'comb' of elongate spines of 1st dorsal fin & two dark curved stripes on body. ePO, S. California to Peru inc. Galapagos Is. in shallow inshore areas. Good sport fish. 1.2m.

1
2
3
4
5
6
7
8
9
10
11
12

PLATE 18: TEMPERATE SEA FISHES

This plate shows common coastal fishes from cold temperate waters. Most species are from the northeastern Atlantic Ocean. The Common Seahorse (9) and Giant Goby (14) are two species that are not frequently encountered in British waters.

CODS Family: Gadidae

1. Atlantic Cod *Gadus morhua*
Stout-bodied with long barbels, small eye, three dorsal fins and two anal fins. Demersal and schooling. 2m, weight in historical times up to 90kg, 45 kg today.

LUMPSUCKERS /LUMPFISHES Family: Cyclopteridae

2. Lumpsucker *Cyclopterus lumpus*
Body scaleless, rows of bony plates on sides. Sucker disc on belly. Moves inshore in summer to spawn, male guards eggs. Benthic on hard substrates. 61cm.

DORIES Family: Zeidae

3. John Dory *Zeus faber*
Body tall and compressed from side-to-side with protrusible mouth. Body has dappled markings, golden or brown hue. Large, dark ocellus on flank. Solitary. 70cm.

SALMONS Family: Salmonidae

4. Sockeye Salmon *Oncorhynchus nerka*
An anadromous species. Sexual dimorphism distinct during breeding; males (4a) become hook-jawed ('kype') and bright red on back and sides with prominent fleshy hump; white below. nPO. 65cm, 3kg.

WOLF-FISHES Family: Anarhichadidae

5. Wolf-fish *Anarhichas lupus*
Head large, strong crushing jaws with canine-like teeth. Pelvic fins absent. Bottom-dwelling on hard substrates. Diet includes crabs, sea urchins and molluscs. 1.25m, 12kg.

LEFT-EYED FLATFISHES Family: Scophthalmidae

6. Turbot *Psetta maxima*
Broad in comparison to length. Scaleless, with numerous bony tubercules in the skin though these may be absent in the upper (eyed) side. Females larger than males. 1m.

RIGHT-EYED FLATFISHES Family: Pleuronectidae

7. Plaice *Pleuronectes platessa*
Head and jaws relatively small, teeth larger in jaws of blind side. A line of 4-7 bony knobs behind eyes. 1.1m.

PIPEFISHES AND SEAHORSES Family: Syngnathidae

8. Greater Pipefish *Syngnathus acus*
Body slender with pronounced rings, tail squarish. Fins reduced except for dorsal fin which is main means of propulsion. Male has a brood pouch formed of a double fold of skin under tail. Benthic amongst algae. 46cm.

9. Common Seahorse *Hippocampus ramulosus*
Long-snouted with prehensile tail, no caudal fin. Male has brood pouch. Inshore amongst algae and sea grass, rare visitor to northern European waters. Feeds on small crustaceans. 15cm (crown of head to tip of tail).

EELS Family: Anguillidae

10. Common (European) Eel *Anguilla anguilla*
A catadromous species. Bottom-dwelling and nocturnal in freshwater. Migrates to Sargasso Sea to spawn. 1m+ (females), males to 50cm.

HERRINGS Family: Clupeidae

11. Atlantic Herring *Clupea harengus*
Silvery fish, no lateral line or second dorsal fin. Coastal pelagic and migratory, feeds on zooplankton. 40cm.

GURNARDS Family: Triglidae

12. Grey Gurnard *Eutrigla gurnardus*
Body grey with white spots above, white below. Dark blotch on first dorsal fin. Lower pectoral fin rays separated, used for support and searching for food. Benthic. 50cm.

GOATFISHES Family: Mullidae

13. Red Mullet *Mullus surmuletus*
Deep red body colour with dark stripe from eye to tail and yellow stripes on flanks. Stripes form an indistinct marbled pattern at night, then reform during the daytime. Two long barbels probe soft substrate, used in finding food (benthic invertebrates). 45cm.

GOBIES Family: Gobiidae

14. Giant Goby *Gobius cobitis*
Littoral, in the U.K. restricted to pools high on rocky shores. Feeds on crustaceans and green filamentous algae. Breeding males have white-edged dorsal and anal fins. 27cm. The Gobiidae are represented worldwide by about 1900 species. The many European species are difficult to separate and regional guides should be used.

STARGAZERS Family: Uranoscopidae

15. Northern Stargazer *Uranoscopus scaber*
Hides partially buried in mud, exposing eyes and mouth. Attracts fish prey with lure on lower jaw. This is the only family of electric marine teleosts; the electric organ is formed from modified eye muscles. 38cm.

BLENNIES Family: Blenniidae

16. Butterfly Blenny *Blennius ocellaris*
A large-eyed blenny with a high first dorsal fin ray and an ocellus with a white or blue border on the dorsal fin membrane. Colour brownish with dark bars. 20cm.

DRAGONETS Family: Callionymidae

17. Common Dragonet *Callionymus lyra*
Body is flattened, scaleless, spines in front of gill cover. Adult males have long first dorsal fin and blue markings on head, body and both dorsal fins (17a). Females and immature males with series of six brown blotches on side and three saddles across back (17b). Two similar species also occur in cold temperate eAO (*C. maculatus* and *C. reticulatus*). Males larger than females, 30cm.

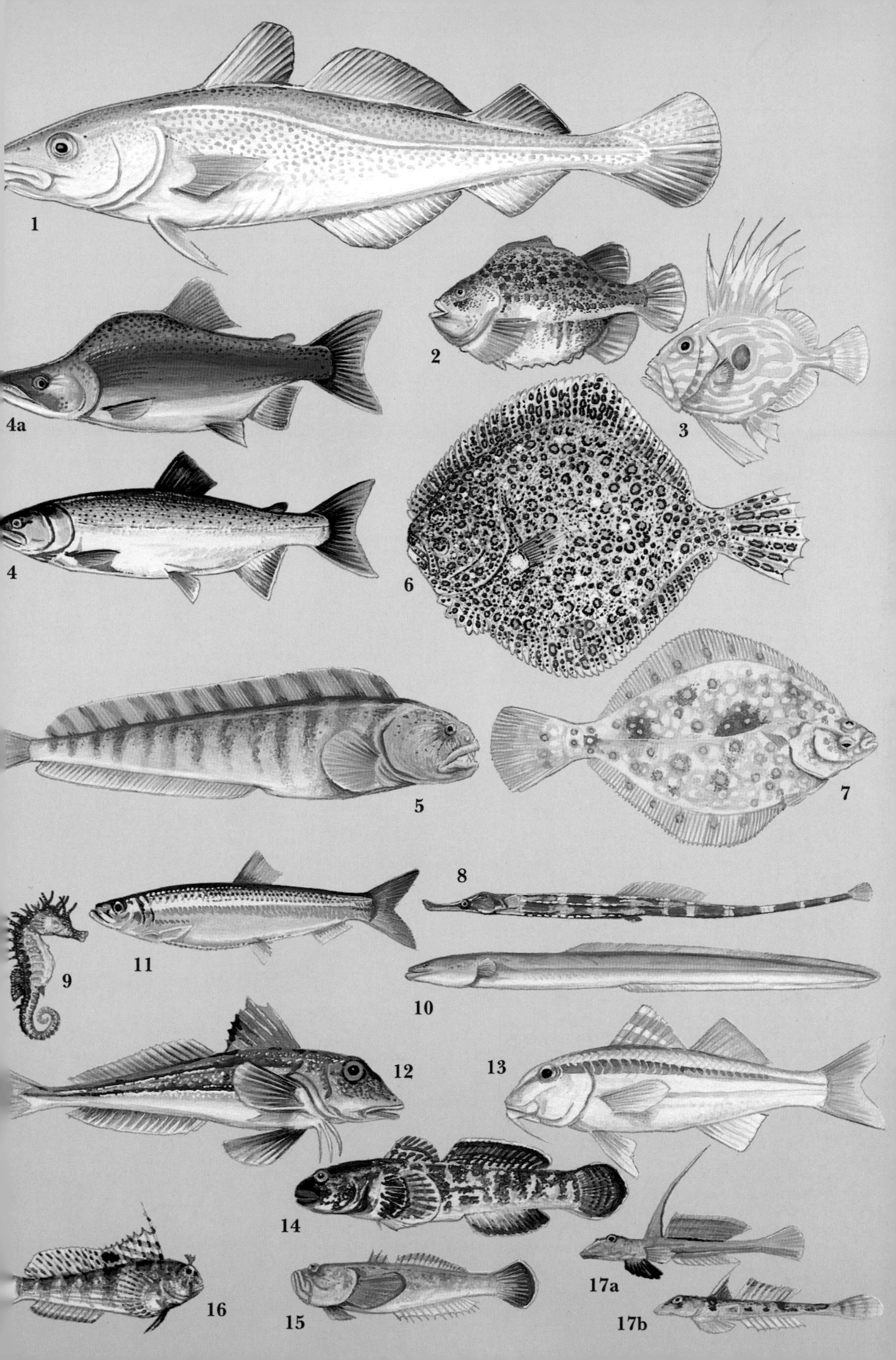
1
2
3
4a
4
6
5
7
8
9
11
10
12
13
14
16
15
17a
17b

PLATE 19: GROUPERS AND OTHER BASSES

Basses (Serranidae) are very diverse and include many of the largest fishes that occur on tropical reefs. Dorsal fin usually continuous and the jaw is usually large with an exposed maxilla and three spines on the gill cover. In Sea Basses (Moronidae) the dorsal fins are separate and two spines are present on each gill cover.

SEA BASSES
Family: Moronidae

1. Striped Bass *Morone saxatilis*
Silvery fish, bluish or greenish on back. Seven to eight black stripes on side, central stripes longest. Dorsal fins separate. nwAO and Pacific Coast of USA. 1.8m and 57kg.

BASSES
Family: Serranidae

I GROUPERS

Medium to large fishes that are typically bottom-dwelling predators feeding on fish, crustaceans and cephalopods. Ambush prey on reefs. Some species are inquisitive and become friendly when fed by divers.

2. Barramundi Rockcod *Cromileptes altivelis*
A distinctive humpbacked fish with small head. Mottled brown with dark brown to black polka dot patterning. Darker pectoral, anal and caudal fins. Tropical IO and wPO. 66cm.

3. Coney *Cephalopholis fulva*
A small grouper very variable in colour from uniform red **(3a)** and yellow **(3b)** to bicoloured form **(3c)**. Two black spots on caudal peduncle always distinctive, with variable blue spotting on body. Tropical wAO. Up to 41cm (usually less than 30cm).

4. Tomato Rockcod *Cephalopholis sonnerati*
A striking tomato red with concave head margin and darker fin margins in older fish. May have fine white spots and blue lines on the head. IO and wPO. 60cm.

5. Yellowtail Rockcod *Epinephelus flavocaeruleus*
A robust grouper, glossy black with yellow-orange fins (partially tipped black), lips and forehead. IO and wPO. 90cm.

6. Coral Hind *Cephalopholis miniata*
A striking bold orange to red colour with numerous blue spots. The fins can be edged with blue or brown. IO and wPO. 50cm.

7. Peacock Rockcod *Cephalopholis argus*
Body dark purplish-brown with darker bands, covered by fine blue dots. IO and wPO, uncommon in Red Sea, introduced to Hawaii. 50cm.

8. Nassau Grouper *Epinephelus striatus*
Dark 'tuning fork' mark on forehead with broader second band passing through eyes and a dark spot on the peduncle. This broadly banded grouper varies in a range of colours from very pale (8a) to dark forms (8b), or tiger-striped patterns or even bicoloured. Usually has small spots around eye. Warmer wAO. 120cm. **Red Grouper** *E. morio* is similar to Nassau Grouper but lacks dark blotch above the tail. Warmer wAO. 1.1m.

9. Rock Hind *Epinephelus adscensionis*
Background skin colour varies from buff to individuals with dark-banded body covered with reddish spots. Black blotches at base of dorsal fin, some with pale margins to fins. Dark saddle on tail base. Warmer wAO. 60cm. There are other similar species in this area. **Red Hind** *E. guttatus* lacks the dorsal blotches and saddle on the caudal fin, posterior fins have broad dark margins. 67cm. The **Graysby** *Cephalopholis cruentatus* has four black or white spots (not blotches) at base of dorsal fin (not on caudal fin) 30cm.

10. Black Grouper *Mycteroperca bonaci*
Variable blackish to grey with dark rectangular blotches. Warmer wAO 1.3m and 82kg. **Yellowfin Grouper** *M. venenosa*, can resemble the Black Grouper but blotches more oval, and lacks dark markings on upper part of tail base. Tropical wAO. 90cm.

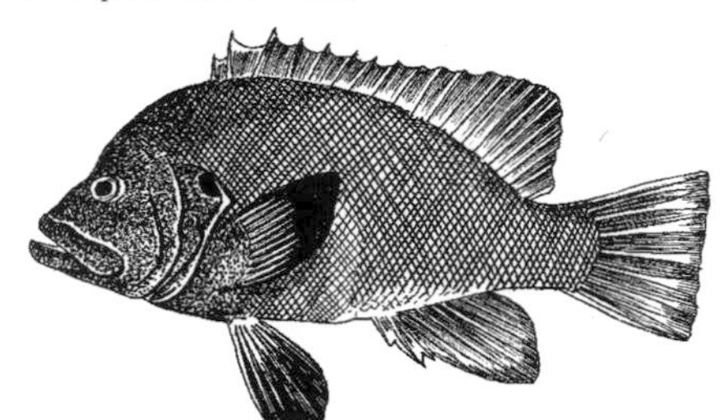

Red-mouthed Grouper ***Aethaloperca rogaa*. IO and wPO. 60cm.**

II HAMLETS

Small fish with great diversity in body coloration. Three of the ten colour types are illustrated here; all belong to the same species *Hypoplectrus unicolor*. Tropical wAO. 13cm.

11a. Barred Hamlet *Hypoplectrus unicolor*
Buff-coloured with blue lines.The most common form in the West Indies.

11b. Golden Hamlet *Hypoplectrus unicolor*
Yellow-coloured with face mask (blue-edged spot below eye).

11c. Shy Hamlet *Hypoplectrus unicolor*
'Two-tone' coloration with yellow tail.

III SEA PERCHES

Complex combination of reds, pinks and yellows with elaborate fin extensions and deeply forked tails (especially males). Tropical seas worldwide, small schooling fishes common on coral reefs.

14. Orange Sea Perch *Pseudanthias squamipinnis*
Common 'goldfish' of IP coral reefs in large schools. Feeds on tiny planktonic crustaceans and fish eggs, sheltering from predators amongst coral heads. 15cm.

2
3a
3b
3c
1
4
5
6
7
8b
9
10
8a
11a
11b
11c
12

PLATE 20: WRASSES, PARROTFISHES AND TRIGGERFISHES

The three families illustrated here are brilliantly coloured fishes with numerous species; only a few species are illustrated to demonstrate their diversity. All have terminal mouths, with canine-like teeth in the wrassee, bird-like beaks in the parrotfishes and fused tube-like jaws in the triggerfishes. Colours are important for identification but can be confusing due to changes with age, sex and 'mood'. In general, wrasses and parrotfishes swim with their large pectoral fins; triggerfishes swim with their fan-like dorsal and anal fins. Wrasses, with about 500 species, are the second largest family of marine fishes.

WRASSES
Family: Labridae

Terminal mouth, thickened lips and one or more pairs of protruding canine-like teeth.

1. Cuckoo Wrasse *Labrus mixtus*
A slender wrasse with a pointed snout. Female and immature males are yellowish to pinkish with three blotches on the rear of the back, breeding males are spectacularly coloured (illustrated) and could be mistaken for a different species. Makes a seaweed nest and displays to females. Coastal eAO, Norway to Senegal. (+MS). 35cm.

2. Yellowtail Wrasse *Coris gaimard*
Adult is brown-violet with blue spots and yellow tail. Young (illustrated) are orange with white stripes on the upper body. Indonesia east to Hawaiian Is. 38cm. Replaced by the similar **African Wrasse** *Coris africana* in the IO which lacks a yellow tail in the adult.

3. Dragon Wrasse *Novaculichthys taeniourus*
Ragged and blotched appearance in juveniles which resembles seaweed; its swimming imitates the drifting motion of seaweed in the current. Adult with blue head and red belly. IO and PO. 27cm.

4. Hogfish *Lachnolaimus maximus.*
Deep body usually reddish but may vary; can be yellowish to silvery, with black spot at rear base of dorsal fin. Steep forehead (with a darker brown band in males, and a more pronounced snout), lyre-shaped tail and first 3 dorsal fin spines elongate. Warmer wAO. 91cm.

5. Common Cleaner Wrasse *Labroides dimidiatus*
Dark central stripe widening towards tail. Adults are greyish-white over the front half and blue behind. This wrasse cleans parasites from other fish. IO and wPO. 10cm. There are other species of cleaner wrasses in the genus *Labroides.* The blenny *Aspidontus taeniatus* mimics this species, but takes a bite out of the fish, rather than removing its parasites!

6. Pudding Wife *Halichoeres radiatus*
Appears greenish-blue from a distance, more complex close up with blue lines on head and blue-spotted body. Juveniles with yellow stripes on whitish body, five white blotches along the back and three black blotches behind. Illustration shows an intermediate stage with fine blue lines which are less evident in adults. Adults have two yellowish lines from the pectoral fins to the belly and mid-body bar. Warmer wAO. 51cm.

7. Blue-headed Wrasse *Thalassoma bifasciatum*
The adult male is very distinctive with blue head and lyre-shaped tail. Juveniles yellow **(7a)**. Tropical wAO. 18cm.

8. Ladder Wrasse *Thalassoma trilobatum*
Adult males have ladder-like rows of blue flank stripes on a coppery background. Fins edged bluish. IO and wPO. 30cm.

9. Rainbow Wrasse *Thalassoma purpureum*
Adult males are bluish-green with orange stripes along the body, on the head, and the dorsal and anal fins. IO and wPO as far east as Easter Is., 43cm. Juvenile of this species is almost indistinguishable from Ladder Wrasse (8), but differs in minor details of head markings.

10. Blackfin Wrasse/Hogfish *Bodianus bilunulatus*
An elongate wrasse with a black 'saddle' behind the dorsal fin. Body becoming reddish in large males. May have a yellow belly and other dark markings additional to saddle. Teeth usually obvious. IO and wPO. 60cm.

PARROTFISHES
Family: Scaridae

Teeth fused into beak-like plates.

11. Queen Parrotfish *Scarus vetula*
Adult male is greenish with blue and yellow outer bands to dorsal and anal fins, lyre-shaped tail and yellow stripes on face. Female **(11a)** dark brown with a pale flank patch (hidden in illustration). Warmer wAO. 61cm.

12. Blue Parrotfish *Scarus coeruleus*
Large males with humped forehead (illustrated). Young and subadults of both sexes blue with bright yellow forehead. Warmer wAO. 60cm.

13. Stoplight Parrotfish *Sparisoma viride*
Males (illustrated) have a greenish body with orange edges to scales. Yellow blotch present on base of tail. Dorsal and anal fins, and facial lines yellow, orange or pink. Females almost gaudy with a purplish-brown upper body (with scattered paler scales), bright red belly and tail. Warmer wAO. 51cm.

TRIGGERFISHES
Family: Balistidae

Fused tube-like jaws and large spine on first dorsal fin.

14. Rectangular Triggerfish *Rhinecanthus rectangulus*
Brownish-yellow body with blackish-brown band across body from 'crown' to anal fin. Blue lines above mouth and near eye, orange line at base of pectoral fins, whitish rear fins. IO and wPO. 30cm.

15. Queen Triggerfish *Balistes vetula*
A colourful triggerfish with elongate tips to the fins. Note the fine colour detail on the body and fins. Some individuals are more pinkish or orange on the belly, young dull brown. Tropical AO. 60cm.

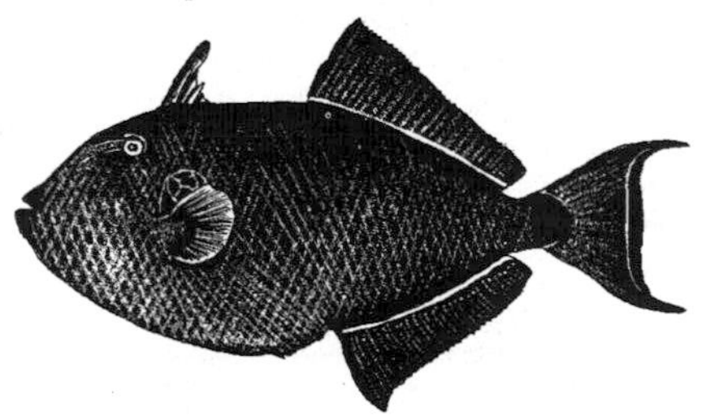

Black Durgon *Melichthys niger.* Circumtropical. 35cm.

2
1
3
5
4
6
7
8
7a
9
10
11
11a
14
12
13
15

PLATE 21: REEF FISHES I

These fishes are common around tropical reefs. Grunts and snappers feed at night. Squirrelfishes and soldierfishes also forage at night using their well developed eyes. During the day they hide in caves and under ledges on the reef.

SQUIRRELFISHES AND SOLDIERFISHES
Family: Holocentridae

Soldierfishes lack gill spine which are present in squirrelfishes.

1. Squirrelfish *Holocentrus ascensionis*
Dull red to pinkish, sometimes blotched. Dorsal fin transparent with yellow tips to spines, other fins whitish. Cold temperate to tropical wAO, east to eAO. 30cm.

2. Longspine Squirrelfish *Holocentrus rufus*
Reddish, stripes can be blotched. Tips of dorsal fin spines white, other fins yellowish. Tropical wAO. 28cm.

3. Blackbar Soldierfish *Myripristis jacobus*
Rounded head, black bar behind gill, body reddish, can be silvery below, white edges to fins. Tropical AO. 20cm.

4. Red Soldierfish *Myripristis murdjan*
Red with a black bar behind the gills and black margins to second dorsal, anal and caudal fins. Large dark eye. Red tips to dorsal fin spines. Changes body colour from dark red hue (illustrated) to silvery hue at night. Warmer IO and wPO. 27cm.

SNAPPERS
Family: Lutjanidae

Continuous dorsal fin, large canine-like teeth in both jaws.

5. Schoolmaster *Lutjanus apodus*
Long sloping forehead, brownish body with 8 pale bars and yellow fins. A dark blue line runs through eye and a broken blue line below it (solid in young). Warmer AO, most abundant snapper in West Indies. 60cm.

6. Emperor Snapper *Lutjanus sebae*
Steep-headed. Adults uniformly red with silvery wash due to pale centre to many scales. Blackish body bands and fin margins. IO and wPO, absent from oceanic islands, 80cm, 27kg.

7. Scribbled Snapper *Lutjanus rivulatus*
Variable in colour; boldly marked individuals have blue lines on the head, dark bars on a reddish brown body and yellowish fins. Small white dots in centre of scales. IO and wPO. 75cm.

8. Yellowtail Snapper *Ocyurus chrysurus*
A 'two-tone' snapper with a mid-body yellow band that extends to the deeply forked tail. Note the subtle bluish greens and yellows on upper body and striped white belly. Caudal fin with transparent margins. Warmer wAO, east to Cape Verde Islands. 75cm.

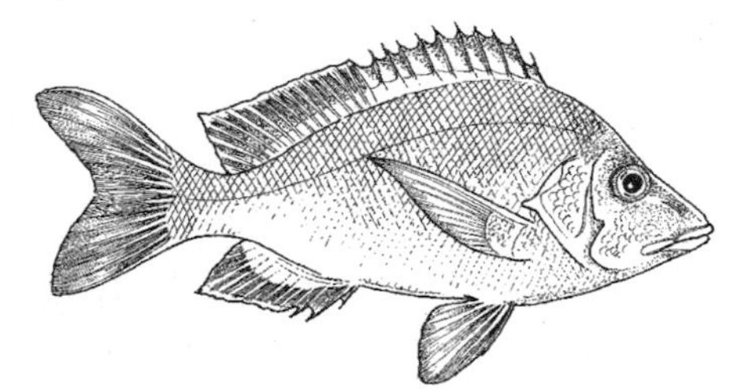

Humpback Snapper ***Lutjanus gibbus*****. IO and wPO. 60cm.**

MOJARRAS
Family: Gerreidae

Single dorsal fin and deeply forked tail.

9. Yellowfin Mojarra *Gerres cinereus*
Body often looks silvery; at close range it shows brownish bars on the body and yellowish fins. Note the slightly concave lower jaw. Tropical wAO. 41cm.

GRUNTS
Family: Haemulidae

Continuous dorsal fin and small conical teeth in jaws. Make grinding noises with swim-bladder and teeth. Examples illustrated are from tropical wAO.

10. Porkfish *Anisotremus virginicus*
Distinctive black-edged white band across gill region with blue and yellow stripes on trunk. Fins yellow. Young with black spot on tail base. 40cm.

11. French Grunt *Haemulon flavolineatum*
Whitish to slightly bluish body with fine yellow stripes, those below lateral line curve upwards. Fins yellow. 30cm. Many similar species worldwide, often with horizontal bands or heavily spotted.

12. Margate *Haemulon album*
Obvious sloping head, body silvery to pearly grey, often with distinct black band from dorsal fin along rear of back onto tail. Scales have dark spot at base on upper body. Dark caudal fin. 60cm. **Sailor's Choice** *H. parra* is similar but has a blunter snout and lacks dark caudal fin (although it may be dusky).

13. Spanish Grunt *Haemulon macrostomum*
Dark stripes across whitish flanks. Yellow area below dorsal fin. Fins are dark (except pectoral) with pale yellow edges. Forehead slightly concave. 43cm.

14. Blue-striped Grunt *Haemulon sciurus*
Alternating blue and yellow stripes. Dark dorsal and caudal fin edged yellow. 45cm.

15. Black Margate *Anisotremus surinamensis*
Deeper-bodied than the Margate, with a broad dark band behind the pectoral fins and prominent black fins in silvery specimens. Scales on upper body with darker base. Young lack the dark band but have a dark spot at the base of caudal fin. 65cm.

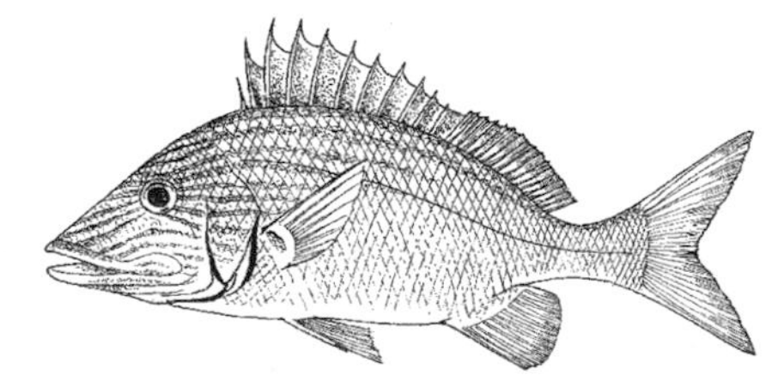

White Grunt ***Haemulon plumieri*****. wAO. 40cm.**

1
2
3
4
5
6
7
8
9
10
11
12
13
14
15

PLATE 22: REEF FISHES II

This plate includes examples of some of the most colourful of all tropical reef fishes. All the families are diverse. Caution is required with identifications since juveniles can look quite different from adults (see examples of angelfishes on this plate).

BUTTERFLY FISHES
Family: Chaetodontide

Fishes lacking gill spines and tail spines.

1. Foureye Butterfly Fish *Chaetodon capistratus*
Small species with a white-ringed, black ocellus below the rear of the dorsal fin (also found in young of Banded Butterflyfish). Body is covered with fine, dark, dashed diagonal lines that converge towards the gill area and merge at a distance to create a silvery-grey to bluish hue. Note the subtle banding on the fins and distinctive eye band. Tropical wAO, 15cm.

2. Banded Butterfly Fish *Chaetodon striatus*
Caudal fin has obvious white rear margin. Well marked adults have fine dark horizontal lines that converge towards the mid-line of the body. Tropical wAO. (?eAO). 15cm.

3. Golden Butterfly Fish *Chaetodon semilarvatus*
Striking yellow body with black 'teardrop' marking around the eye. Groups of these fish are a spectacular sight. Common in its range; Red Sea and Gulf of Aden. 23cm.

4. Longnose Butterfly Fish *Forcipiger flavissimus*
Distinctive forceps-like snout used for feeding on soft parts of benthic invertebrates. Widespread in IP, outer reef near ledges and caves. 22cm.

MOORISH IDOL
Family: Zanclidae

This family with a single highly coloured species is related to surgeonfishes.

5. Moorish Idol *Zanclus cornutus*
Long snout and trailing first dorsal fin, lacks tail spine. In schools, feeds on sponges. Widespread in IP. 16cm.

SURGEONFISHES
Family: Acanthuridae

Brilliantly coloured fishes with varied fin and body shapes. One or more pairs of blade-like tail spines.

6. Sohal Surgeonfish *Acanthurus sohal*
Bright yellow or orange spot behind pectorals and at tail base. Common on outer reef; Red Sea and Arabian Gulf. 40cm.

7. Dussumier's Surgeonfish *Acanthurus dussumieri*
Adults distinctly marked with black gill-cover membrane and white spine on each side of tail which can be tucked into a groove beneath it. IO and wPO. 54cm.

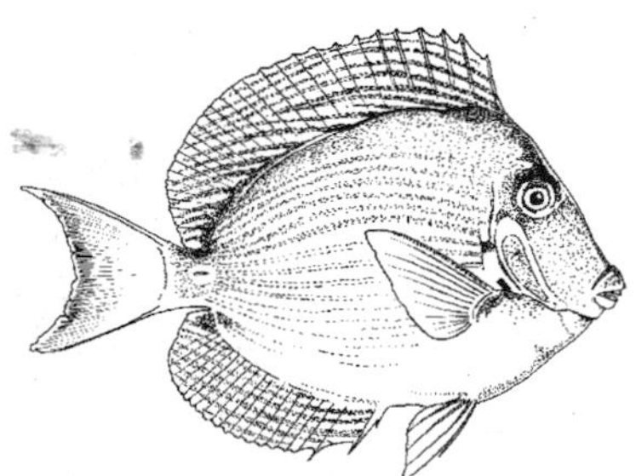

Blue Tang ***Acanthurus coeruleus*****. wAO. 36.5cm.**

DAMSELFISHES
Family: Pomacentridae

Small, plump-bodied fishes. Many species boldly protect territories, aggressive to divers. Some are found only with certain species of reef anemones. Damselfishes species vary from all brown to bicoloured or banded, with or without distinct spots.

8. Sergeant Major *Abudefduf saxatilis*
Silvery-grey belly, yellow above, with 5-6 dark stripes (more difficult to see in bluish colour phase of nesting males). Tolerates high temperatures and is common in rock pools. Forms loose schools on coral reefs and sea grass beds. Distribution from cold temperate nwAO south to tropical and warm temperate swAO and east to Africa. 15cm. *A. vaigiensis* from IP is similar.

9. False Clown Anemonefish *Amphiprion ocellaris*
Orange-bodied fish with three broad white bands (third on tail) that is associated with three species of anemones. eIO and wPO. 11cm. Numerous similar species which vary in the number and position of bands on the body and body colour, e.g. the yellow to brown **Twobar Anemonefish** *Amphiprion bicinctus* of the Red Sea and Gulf of Aden. 14cm.

10. Yellowtail Damselfish *Microspathodon chrysurus*
Brownish-black body with blue spots in young adults which decrease in number with age. Tail yellow. Juvenile bluish-black with many light blue spots and white tail; found amongst fire coral (*Millepora*) where they act as cleaners for other fish. Tropical wAO. 20cm.

ANGELFISHES
Family: Pomacanthidae

Superficially similar to butterflyfishes but prominent spines on lower gill cover.

11. Rock Beauty *Holacanthus tricolor*
A distinctive angelfish with a yellow tail. Juvenile yellow with blue-ringed, black ocellus. Tropical wAO. 20cm.

12. Queen Angelfish *Holacanthus ciliaris*
Vivid blue with a yellow spot on each scale. Note the colour of fins, blue crown spot and yellow cheek. Juvenile striped with a yellow tail and pectoral fins. Tropical wAO. 45cm.

13. French Angelfish *Pomacanthus paru*
Body and fins bluish-black with yellow-edged scales, vivid blue head. Yellow base to pectoral fin and tip of the dorsal fin. Juvenile stripes can be seen in young adults as ghostly bars. Striped young **(13a)** have a large black spot on caudal fin. Tropical wAO and Ascension Is. 30cm.

14. Emperor Angelfish *Pomacanthus imperator*
Yellow lines extend over the brownish to bluish body and onto the dorsal fin. Note the rounded fins (like many butterflyfishes), orange yellow tail and blue bands on the black face mask. Juveniles **(14a)** bluish-black with concentric blue and white lines. IO and wPO, 40cm.

15. Blue Angelfish *Pomacanthus semicirculatus*
Adult less colourful than young with bluish-black spots. Note the blue markings on the head. The common name is more apt to the juvenile **(15a)** - navy blue to black with curving white and light blue lines. Tail square-cut with transparent margin. IO and wPO. 40cm.

1
2
3
4
5
7
8
11
6
12
13
15a
14a
14
13a
9
15
10

PLATE 23: POISONOUS AND STINGING FISHES

The fishes illustrated here include some of the most toxic marine animals known. All these live in coastal areas that are also frequented by swimmers and seafarers; since these fishes are all dangerous to humans, there is a special need to be aware of them. Pufferfishes (Tetraodontidae), trunkfishes (Ostraciidae) and scorpionfishes (Scorpaenidae) are widespread in warmer waters and all species in these families are considered toxic to humans, though in varying degrees.

PUFFERFISHES
Family: Tetraodontidae

The skin, viscera and gonads are toxic in these fishes and are dangerous if eaten. The toxin (tetrodotoxin) causes fall in blood pressure, respiratory failure and death. Hospital treatment should be sought in suspected cases, artificial respiration will be necessary for unconscious patients (after adequate care has been taken to remove any traces of toxin from around the mouth and lips) until medical help can be obtained. The toxin can induce an almost complete cessation of outward signs of life (origin of 'zombie' stories). Captain Cook almost succumbed after eating pufferfish (*Lagocephalus scleratus*) in 1774.

1. White-spotted Puffer *Arothron hispidus*
Distinctive white banding on body. Widespread in tropical IP. 48cm.

2. Black-saddled Toby *Canthigaster valentini*
Two distinctive dark bands on back. IO and wPO. 10cm. The non-toxic filefish, *Paraluteres prionurus* **(2a)**, resembles *C. valentini* which it mimics. IO and wPO. 10cm.

3. 'Fugu' *Takifugu rubripes*
Body covered by prickles, a large round black spot (ocellus) edged with white just behind pectoral fins. Sea of Japan and Yellow Sea. 60cm. In Japan, the flesh (muscle) of fishes of this genus is considered a delicacy and specially skilled chefs remove the toxic organs before eating. However, there are still fatalities each year, often when less skilled people prepare Fugu for their own consumption. Fatalities from tetrodotoxin have also been reported from Australia.

TRUNKFISHES
Family: Ostraciidae

These box-like fishes are enclosed in a bony 'carapace' formed of poygonal plates which act as an effective anti-predator device. They are further protected by toxin secreted from the skin (ostracitoxin).

4. Cube Trunkfish *Ostracion cubicus*
Yellowish-brown body colour with distinctive blue spots. IO and wPO. 45cm.

5. Ornate Cowfish *Aracana ornata*
Body with hook-like projections on back and flanks. S. Australia, in sea grass beds. 15cm.

SOLES
Family: Soleidae

6. Moses Sole *Pardachirus marmoratus*
Upper surface with grey spots. Secretes bitter toxic substance from base of dorsal and anal fins which is effective in deterring predators, including sharks. wIO. 26cm.

ROUND RAYS
Family: Urolophidae

7. Round Ray *Urolophus halleri*
Disc nearly round, upper surface blue-black with yellow spots, lower surface white or yellow. Short, stout tail with obvious sting, causes painful injuries (non-fatal) to bathers who step on this ray in shallow water. Seasonally common off beaches in summer months when they move inshore to breed. N. California to Panama. 56cm total length.

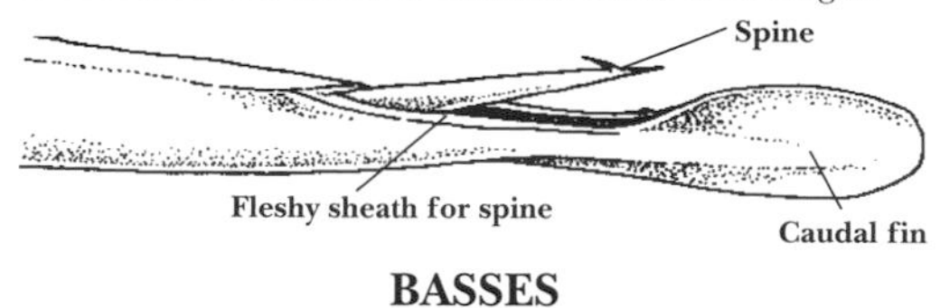

BASSES
Family: Serranidae

8. Greater Soapfish *Rypticus saponaceus*
Upper surface of head concave in side profile. Secretes a bitter skin toxin (grammistin) which can kill other fish when confined with it. In ledges on coral reefs. Tropical AO. 33cm.

WEEVER FISHES
Family: Trachinidae

9. Lesser Weever Fish *Echiichthys vipera*
The first dorsal fin is black with poisonous spines and is often exposed (with the eyes) when this fish lies buried on the bottom. Moves inshore during the summer to gravelly, sandy or muddy littoral areas and frequently occurs near bathing beaches. Causes painful stings (non-fatal). eAO (+MS). 15 cm. Considered the most dangerous of the 4 species of European weevers, both for its toxin and inshore habitat.

SCORPIONFISHES
Family: Scorpaenidae

10. Estuarine Stonefish *Synanceia horrida*
Similar to Stonefish (11), but head lacks flat side profile. India to Papua New Guinea and NE Australia, north to China. 30cm. Venomous spines in the dorsal fin have grooves containing poison glands which rupture on puncturing the victim's flesh. Sting fatal within 1-2 hours, immerse wound in hot water to destroy venom and seek immediate medical advice.

11. Stonefish *Synanceia verrucosa*
This fish is well camouflagued by its dull, blotched coloration and lumpy skin when hidden amongst the coral rubble and sand of shallow coral lagoons. The eyes are placed on top of the head **(11a)** and are the only visible sign of its presence when buried in sand. Sting fatal. IO and wPO. 35cm.

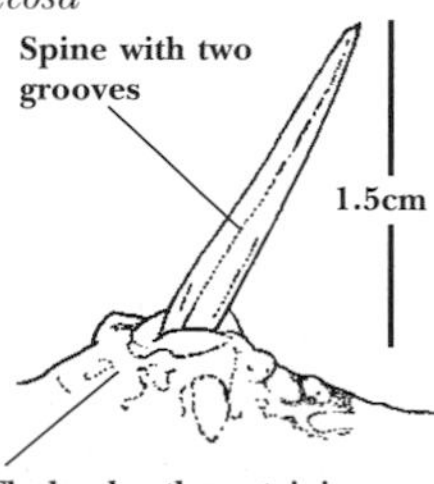

5
1
2
2a
4
6
3
7
8
9
10
11a
11

PLATE 24: MARINE TURTLES AND SEA SNAKES

Marine turtles have hard carapaces and vary in size from the small ridley turtles to the giant Leatherback Turtle, which lacks external horny plates (comparable to the scales of other reptiles). Appearance and shape of carapace varies with age. Young turtles have more uniform scutes to the carapace, often with a more obvious central ridge. Old turtles have thick necks and humped backs; males have longer and bulkier tails. The identification of these species is difficult if seaweed or barnacles cover the carapace. Refer to the factsheets for additional details; unless stated, distribution is circumtropical/warm temperate, including MS. Stragglers drift into colder seas. Lengths given as maximum carapace length.

MARINE TURTLES

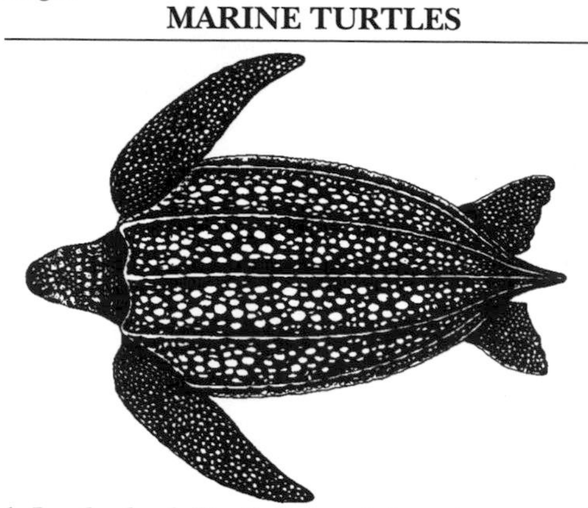

1. Leatherback Turtle *Dermochelys coriacea* **475**
Rubbery carapace, blackish with pale spotting covered by a leathery skin with 7 longitudinal ridges. Upper jaw has tooth-like projection. Only sea turtle lacking head scales. Long front flippers lack claws. Wide diet from crustaceans to seaweeds, prefers jellyfish. Deep diver (1500m+). World's largest turtle, up to 1.88m and 0.5 tonnes+. Reported to be common off S Ireland during summer months.

2. Hawksbill Turtle *Eretmochelys imbricata*

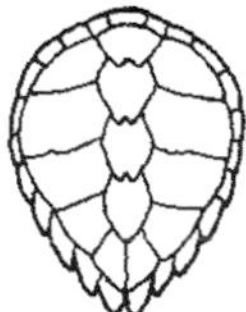

Overlapping carapace scutes diagnostic, but can be difficult to see (scutes lie side-by-side in old turtles). Carapace often a rich flecked tortoiseshell-brown with a slight keel and saw-toothed rear edge. Has a long beak and two pairs of prefrontals. Two claws on each flipper. Varied diet, able to digest sponges, often found on coral reefs. A medium-sized turtle, up to 90cm. More tropical than temperate, usually in seas over 22°C. Unlikely in MS.

3. Green Turtle *Chelonia mydas*

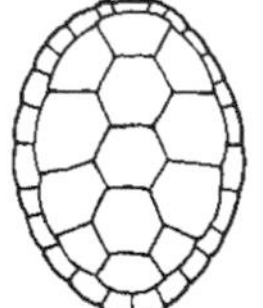

Oval to rounded carapace lacks a keel and is highly variable in colour from olive-brown to the 'tortoiseshell' of a Hawksbill. Has one pair of prefrontals, one claw on each flipper and beak is only slightly hooked, if at all. Young are carnivorous, adults prefer to feed on seaweeds and in sea grass meadows. Named after the colour of its fat. Large turtle, up to 1.24m. Breeds in MS.

4. Black Turtle *Chelonia mydas agassizi*

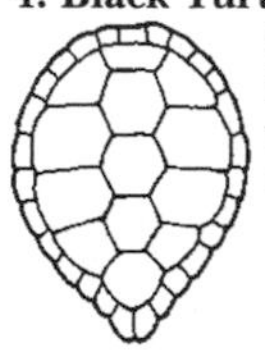

Olive-brown to blackish carapace; shape is higher and narrower than Green Turtle, with obvious sloping sides, a slight keel, tapering towards rear. ePO.

5. Flatback Turtle *Natator depressa*

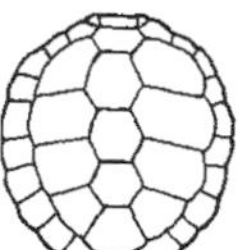

Like Green Turtle with a flattish, smoother, well rounded grey to brown rather pliable carapace. One pair of prefrontals. Carnivorous diet is retained in adults. Restricted to N/NE Australia, vagrants to Papua New Guinea. 1.0m.

6. Loggerhead Turtle *Caretta caretta*

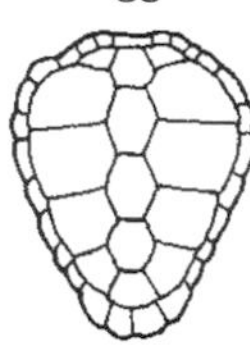

Red-brown carapace broad and shield-like with 5 costals. Carapace often encrusted with barnacles (6a). Beak robust, head very broad with two pairs of prefrontals (often with additional scales wedged between them). Neck often thick. Two claws on the flippers. Three inframarginals (without pores) on the bridge. Diet includes echinoderms,molluscs and crustaceans (e.g. crabs) which are crushed with its enormous beak. Large turtle, up to 1.2m. Most likely hard-shelled species to drift into colder seas of 15°C or less, including NW Europe. Breeds in MS.

7. Kemp's Ridley Turtle *Lepidochelys kempii* **475**

Greyish heart-shaped/round carapace can be as broad as it is long. There are 5 costals, vertebral scutes are small and fairly uniform. Four inframarginals on the bridge (with pores on the hind border). Two claws on each flipper, four prefrontals, broad head and a beak that is not as robust as in the Loggerhead.World's rarest sea turtle, listed as endangered. Breeds in Gulf of Mexico, drifting N and W into colder nAO to 15°C, including NW Europe. 0.76m.

8. Olive Ridley Turtle *Lepidochelys olivacea* **475**

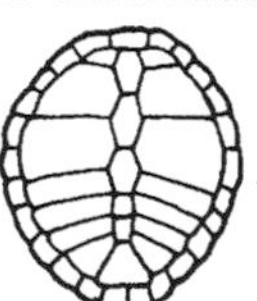

Like Kemp's Ridley but olive carapace with 6-9 costals that have a less uniform shape. IP. 76cm. Massed nesting sites on Central American and E Indian coasts.

SEA SNAKES

Sea snakes are found in the tropical Indo-Pacific (absent in AO) and have flattened rudder-like tails.

9. Yellow-lipped Sea Snake *Laticauda colubrina*
Yellow upper lip, body with black and blue-grey banding. Lays eggs in crevices on rocky shores and spends long periods out of the water. Enters the sea mainly to feed, diet consists mainly of eels. Bay of Bengal to wPO. 1.4m.

10. Stokes' Sea Snake *Astrotia stokesii*
A hydrophiine sea snake. The bulkiest of the sea snakes, adults being noted for their great girth (2m long specimens can be as thick as a human leg). IP.

11. Pelagic Sea Snake *Pelamis platurus* **475**
A hydrophiine sea snake. The only truly oceanic sea snake, feeding at the surface and carried around on surface currents. IP. 103cm.

1
6
6a
2
4
3
7
8
10
11
9

PLATE 25: PENGUINS AND SHEATHBILLS

Penguins are small to large thickset birds with an upright stance on land. All species are flightless, the wings modified as flippers for underwater propulsion. Most species are distinctive but the *Eudyptes* group requires great care and may be impossible to identify at sea. The *Spheniscus* group also appears similar but its species do not normally overlap in range. Size, colour of head and bill and range are important in identification. Sheathbills are rather chicken-like birds with affinities to both gulls and shorebirds. They are mostly terrestrial and feed by scavenging around penguin and seal colonies.

PENGUINS
Family: Spheniscidae

ANTARCTIC AND SUBANTARCTIC PENGUINS

1. King Penguin *Aptenodytes patagonicus*
Large, but smaller than Emperor, with different head pattern and orange neck patches. Immature browner with pale yellowish-white neck patches. Subantarctic only (not Antarctica). 94cm.

2. Emperor Penguin *Aptenodytes forsteri*
Largest penguin. Darker head and larger pale yellow neck patch than King Penguin. Immature paler than adult with whitish neck patches. Circumpolar Antarctica. 122cm.

3. Gentoo Penguin *Pygoscelis papua*
Distinctive; adults and immatures have white patch over eyes forming narrow band over crown. Bill, legs and feet orange-red. Circumpolar; subantarctic and Antarctic peninsula. 76cm.

4. Chinstrap Penguin *Pygoscelis antarctica*
Distinctive narrow black 'chinstrap' and white face. Eye red, bill black. Immature has greyish chin and sides of face. Almost circumpolar; Antarctica and subantarctic. 77cm.

5. Adelie Penguin *Pygoscelis adeliae*
Distinctive, black head and throat with conspicuous pale eye. Immature has white throat with black cap extending just below pale eye. Bill dark red. Circumpolar; Antarctica and subantarctic. 71cm.

6. Macaroni Penguin *Eudyptes chrysolophus* **476**
This species has an orange and yellow crest, joined across forehead but not reaching bill. Bill red with pinkish gape. Eye red. Juvenile has smaller crest and dusky throat. Subantarctic islands and Antarctic peninsula. 71cm.

7. Rockhopper Penguin *Eudyptes chrysocome* **476**
Smallest crested penguin. Similar to Macaroni but yellow crest does not meet on forehead. Hops (Macaroni waddles on land). Immature has whitish throat and shorter plumes. Widespread in subantarctic, not Antarctica. Northern form has longer crest and darker under-flipper pattern. 61cm.

PENGUINS OF THE NEW ZEALAND REGION

8. Yellow-eyed Penguin *Megadyptes antipodes*
Pale eye and distinctive yellow band through eye and across nape. Immature has broken yellow band and whitish chin. Breeds singly or in loose colonies in temperate forests and on grassy cliffs. Population vulnerable. South Is., New Zealand. 76cm.

9. Fiordland Penguin *Eudyptes pachyrhynchus* **476**
Broad yellow crest reaches base of bill and tends to lie flat when dry. Has diagnostic white stripes on cheeks and lacks fleshy gape. Immature has white chin and throat. Breeds in temperate rainforests. Breeds New Zealand, disperses to Tasmania. 67cm. (Other name: Fiordland Crested Penguin.) **Royal Penguin** *E. schlegeli.* Largest crested penguin. Very similar to Macaroni but more robust bill and whitish throat and face. Possibly a colour phase of Macaroni Penguin. Breeds on Macquarie Island. 73cm.

10. Snares Penguin *Eudyptes robustus* **476**
Like Fiordland but has prominent pink fleshy gape, black cheeks and partly raised crest. Only breeds on Snares Is., S New Zealand. 73cm. (Other name: Snares Island Penguin.)

11. Erect-crested Penguin *Eudyptes sclateri* **476**
Like Snares but with up-swept brush-like crest and less prominent fleshy gape. Under-flipper has broad dark margins. Immature similar to Rockhopper but has white chin. Breeds on subantarctic islands S of New Zealand. 68cm.

12. Little Penguin *Eudyptula minor*
Smallest penguin. Size and distinctive grey and white plumage diagnostic. S Australia and New Zealand. **'White-flippered Penguin'** *E. m. albosignata* has broad white flipper margins sometimes linked by band across centre and is possibly a separate species. Confined to Banks Peninsula, South Is., New Zealand. 40cm.

SPHENISCUS PENGUINS

13. Humboldt Penguin *Spheniscus humboldti* **476**
Single breast band. Large bill with pink base. Immature has greyer face and lacks breast band. Coasts of Peru and Chile. 70cm.

14. Magellanic Penguin *Spheniscus magellanicus* **476**
Two breast bands and smaller bill than Humboldt. Immature has whitish face and partial breast band. East and West coasts of S South America and Falkland Is. 71cm. **Jackass Penguin** *S. demersus* Similar to both Magellanic and Humboldt Penguins but confined to coasts of southern Africa where it is the only breeding penguin. Sometimes has partial second breast band. 68cm. **Galapagos Penguin** *S. mendiculus* Smaller than other species in genus and unmistakable. Two indistinct breast bands and narrow white line from eye to throat. Lower mandible mainly pale. Immature has all dark head. Confined to Galapagos Is. 49cm.

SHEATHBILLS
Family: Chionididae

15. Snowy Sheathbill *Chionis alba*
Chicken-like white shorebird. Yellowish bill, pinkish facial skin. Antarctic peninsula and subantarctic islands. Partially migratory, north to Falkland Is. and Patagonia. 41cm. (Other name: American Sheathbill.) **Black-faced Sheathbill** *C. minor.* Bill and facial skin dark. Subantarctic islands of IO. Sedentary. 39cm. (Other name: Lesser Sheathbill.)

1
2
3
8
4
7
5
6
9
11
12
13
14
15
10

PLATE 26: DIVERS/LOONS, GREBES AND PELICANS

Divers/loons are medium to large, sharp-billed diving birds. They have lobed (not webbed) feet and the legs are set well back. They swim low in the water and dive without up-ending. They have distinct breeding plumages but look superficially similar in immature and non-breeding plumages. Size, bill size and shape, and general plumage tones are important for identification. Confined to the northern hemisphere they are found in coastal waters during the boreal winter. Grebes are small to medium-sized diving birds with lobed feet. They have sharp bills and a tailless appearance. Most species have distinct breeeding plumages and several species occur in coastal waters in the non-breeding season. Size, bill shape and head pattern are important for identification. Pelicans are large, heavily built birds with long bills and a large distensible pouch which is used as a scoop. Though all species may occur in coastal waters only the two 'brown' species are truly marine. Size, pattern of upperparts, bill colour and range help with identification.

DIVERS/LOONS
Family: Gaviidae

1. Red-throated Diver/Loon *Gavia stellata*
Smallest diver. Bill slightly up-tilted, neck slim **(1a)**. In winter extensive white on face and around the eye creates paler appearance compared with Black-throated. Upperparts are paler and more mottled than other divers **(1b/c)**. Wingbeats fast. Circumpolar in N hemisphere, wintering south to about 30°N. 62/110cm.

2. Black-throated Diver/Arctic Loon *Gavia arctica*
Smaller and slimmer than Great Northern with narrower bill and more rounded head **(2a)**. In winter dark grey head and blackish upperparts contrast sharply with the white underparts. Note white patch on flanks **(2b/c)**. Breeds N hemisphere in Old World and W Alaska. 68/120cm. **Pacific Diver/Loon** *G. pacifica*. Formerly considered conspecific with Black-throated. Differs in breeding plumage in having green-glossed throat (purple-glossed in Black-throated), and paler grey head and hindneck. Indistinguishable in non-breeding plumage. Breeds NE Siberia and Nearctic, winters nPO.

3. Great Northern Diver/Common Loon *Gavia immer*
Large and bulky with thick neck **(3a)**. Dagger-like dark bill, paler in immatures but the culmen is dark. The angled crown and hind neck are usually darker than the back with an uneven light/dark divide and an extensive cap **(3b)**. Heavy goose-like flight **(3c)**. Breeds N Nearctic, Greenland and Iceland. Winters south to 30°N including coasts of W Europe. 76/138cm. **White-billed Diver/Yellow-billed Loon** *G. adamsii*. Largest diver, similar to Great Northern Diver but heavy, prominently up-tilted is bill ivory-yellow (including culmen). Sides of face, eye surround and neck paler. Breeds Arctic Siberia to N Canada, winters south to 50°N. 83/147cm.

GREBES
Family: Podicipedidae

4. Great Crested Grebe *Podiceps cristatus*
Longer-necked than divers. Distinct breeding plumage **(4a)**. In winter the distinct dark cap is separated from the eye by a white line **(4b)**. In flight shows a double wing-bar on leading and trailing edges of upperwing, extending to scapulars **(4c)**. Widespread in Old World. 50/88cm. **Western Grebe** *Aechmophorus occidentalis*. The largest grebe. Distinctive, long swan-like neck and long up-tilted bill. Long single wing-bar. Winters Pacific coast of N America. 65/90cm. **Clark's Grebe** *A. clarkii*. Formerly considered conspecific with Western Grebe. Differs in having lores and eye surround white (black in Western). Winters Pacific coast of N America. 65/90cm.

5. Red-necked Grebe *Podiceps grisegena*
Distinct breeding plumage **(5a)**. In winter greyer neck and ear-coverts than Great Crested with obvious white cheeks and a shorter, yellow-based bill. Black cap extends below eye **(5b)**. In flight double wing-bar does not extend to scapulars. Widespread in N hemisphere. 45/80cm.

6. Horned Grebe *Podiceps auritus*
Flat-crowned with straight pale-tipped bill **(6a)**. In winter distinct black cap with pale spot in front of red eye **(6b)**. In flight double wing-bar, broad on trailing edge, spot on forewing. Widespread in N hemisphere. 34/60cm. (Other name: Slavonian Grede.)

7. Black-necked Grebe *Podiceps nigricollis*
Compared with Horned smaller and shorter-billed with an up-tilted lower mandible and head more domed **(7a)**. In winter dusky on neck and ear-coverts with less distinct black cap **(7b)**. In flight single wing-bar along trailing edge. Mainly N hemisphere and scattered locations in the south. 30/58cm.

Other grebes that sometimes winter in coastal regions include the widespread **Little/Least Grebe** *Tachybaptus* spp. species complex, the small **Pied-billed Grebe** *Podilymbus podiceps* of the New World, plus several South American species including **Great Grebe** *Podiceps major,* **Silvery Grebe** *P. occipitalis,* **Hooded Grebe** *P. gallardoi* and **White-tufted Grebe** *Rollandia rolland.*

PELICANS
Family: Pelecanidae

8. Great White Pelican *Pelecanus onocrotalus*
Mainly white with black primaries and greyish secondaries. Also note bare part coloration. Widespread in Old World. Migratory, often in large flocks when it may be seen in coastal locations. 158/315cm. (Other name: Eastern White Pelican.)Two similar but less common species which overlap some of its range are **Dalmatian Pelican** *P. crispus* of E Europe and C Asia and **Spot-billed Pelican** *P. philippensis* of the Indian subcontinent and SE Asia. Both species have grey legs and feet and lack the distinct black and white underwing pattern of Great White. They are less likely at coastal locations. **Pink-backed Pelican** *P. rufescens*. Smaller and distinctly greyer than Great White Pelican with which it overlaps in Africa. Less black on secondaries and pinkish bill. Sub-Saharan Africa and Madagascar. Occasionally maritime and breeds in mangroves in Red Sea. 128/277cm.

9. Australian Pelican *Pelecanus conspicillatus*
Only pelican in Australia. Breeds at maritime as well as freshwater locations. 168/252cm. **American White Pelican** *P. erythrorhynchos*. Only New World 'white' pelican. Breeds N America, winters south to Gulf of Mexico and coasts of California and C America. 152/271cm.

10. Brown Pelican *Pelecanus occidentalis*
Marine pelican with greyish-brown plumage. Feeds by plunge-diving for fish. Breeds N and S America south to Brazil and Peru. 115/204cm. **Peruvian Pelican** *P. thagus*. Much larger than similar Brown Pelican and shows diagnostic pale rectangle on leading edge of upperwing. Humboldt Current coasts of S America. 152/228cm.

4c
3c
2c
1c
1b
1a
3a
3b
2b
2a
4a
4b
5a
5b
6a
6b
7a
7b
9
10
8
10

PLATE 27: GREAT, SOOTY AND PACIFIC ALBATROSSES

Albatrosses are large long-winged oceanic birds related to petrels and shearwaters. They have hooked bills, webbed feet and tubular nostrils either side of the bill. They generally glide on long, stiffly-held wings, but will flap in calm conditions. The 'great albatrosses' have a complex range of plumages depending upon age and it is necessary to be familiar with the various plumage stages to make an identification. The two dark *Phoebetria* species differ in structure from the others in having long, pointed tails. The southern and northern species do not normally overlap in range at sea. Besides the general pattern of the upper- and underwings it is useful to note bill colour.

GREAT ALBATROSSES

1. Wandering Albatross *Diomedea exulans*
Complex series of plumages related to age. Dark-bellied stage 1 distinctive **(1d)**. Stage 2 birds have pale mantles and less dark on belly **(1a/e)**. Plumage becomes progressively whiter with white on upperwing spreading outwards from a central white patch as stages progress, **(1b/c)** (see Royal Albatross). Note black on tail, only restricted in the most extreme form of the adult male, when easy to overlook. At least three races; older males of the 'Snowy' (southern) race reach stage 7 **(1c)** whilst the more northerly Gough and Antipodes races are smaller with less white on the wing, more like stage 5 **(1b)**. Females can breed in a stage 3 plumage, roughly intermediate in appearance between 1st and 5th stages. Circumpolar in southern oceans. 115/300cm.

2. Amsterdam Island Albatross *Diomedea amsterdamensis*
Rarest albatross, only 8 pairs known. Resembles dark stage 1 Wandering Albatross with a white belly, dark tip and cutting edges to mandibles and more extensive brown mark on leading edge of underwing. Breeds on Amsterdam Is. in IO; pelagic range unknown. Perhaps slightly smaller than Wandering Albatross. (Other name: Amsterdam Albatross.)

3. Royal Albatross *Diomedea epomophora*
Both subspecies lack brown underparts in stage 1 and quickly lose any black on the tail; crown mottled (in *epomophora* the head is white with some white on the upperwing) **(3a/c)**. Adults of the nominate southern race *D. e. epomophora* easy to confuse with Wandering Albatross but lack black in the tail after stage 1, and the white area on upperwing spreads backwards from the leading edge **(3b)**. The northern race *D. e. sanfordi* is unique with white mantle and wholly black upperwing; also shows diagnostic black mark on the underwing near carpal **(3d)**. Southern oceans, mainly sPO, also west coast of S America. 115/300cm.

SOOTY ALBATROSSES

4. Sooty Albatross *Phoebetria fusca*
Slender wings and long pointed tail. Unique with uniform dark plumage and white primary shafts. In immature and worn plumage has greyish collar and upper back. Bill black with orange-yellow sulcus. sAO and IO. 86/204cm.

5. Light-mantled Albatross *Phoebetria palpebrata*
Differs from Sooty in having a pale grey mantle extending to the lower back and rump and greyish-brown underparts. Bill black with purple or blue sulcus. Circumpolar in southern oceans. 84/216cm. (Other name: Light-mantled Sooty Albatross.)

PACIFIC ALBATROSSES

(Laysan Albatross is included on plate 28).

6. Short-tailed Albatross *Diomedea albatrus*
Largest and only white-bodied albatross in nPO; large pink bill diagnostic. Juvenile is similar to the Black-footed Albatross until the white belly develops, but bill pink and lacks white around base **(6a)**. Immature has dark cap and white in the wing **(6b)**. Rarely follows ships. Very rare. nPO, breeds on Torishima island, south of Japan. 90/222cm.

7. Black-footed Albatross *Diomedea nigripes*
Distinctive with mainly dark brown plumage and white at base of the dark bill **(7a)**. Immatures lack the white rump. Aberrant forms, possibly hybrids with Laysan Albatross, have paler body and bill colour and resemble Short-tailed Albatross **(7b)**. Widespread in nPO. 74/213cm.

8. Waved Albatross *Diomedea irrorata*
Whitish head, brown upperparts and white underwing-coverts unique in the restricted range. Does not follow ships. Breeds on Galapagos Is. and disperses to nearby coast of S America. 90/236cm.

1a
1b
1c
1d
1e
3a
4
3c
2
3b
3d
5
6a
6b
7a
7b
8

PLATE 28: MOLLYMAWKS AND GIANT PETRELS

'Mollymawk' is a term given to the smaller albatrosses of the genus *Diomedea*. As with the great albatrosses it is essential to note the underwing pattern and bill colour. Immature birds are distinct in plumage from adults though there are less intermediate stages than in the larger species. Giant Petrels are large, albatross-sized petrels with strongly hooked bills. The nostrils are fused into a single tube on top of the bill. The stiff-winged flight is less graceful than that of albatrosses and involves more flapping. While the white morph of Southern Giant Petrel is unmistakable, the dark morph is very difficult to separate from Northern Giant Petrel. The colour of the tip of the bill is diagnostic but can be difficult to observe at sea. They feed mainly by scavenging, often around seal colonies or trawlers at sea.

MOLLYMAWKS
Family: Diomedeidae

1. Black-browed Albatross *Diomedea melanophris*
White head with dark eyebrow and yellow bill. Underwing white with broad black margins (**1a/b**). Subadult has dusky band on breast and a dark tip to horn-coloured bill (**1c**). Juvenile has a darker bill. Widespread in southern oceans. 88/240cm.

2. Grey-headed Albatross *Diomedea chrysostoma*
Adult has grey head and black bill with bright yellow ridges above and below. Underwing similar to Yellow-nosed but shows broader black margins (**2a**). Juvenile has black bill, dusky grey head and very dark underwing (**2b**). Immature has more white on underwing and often shows grey breast band similar to juvenile Black-browed. Circumpolar in southern oceans. 82/220cm.

3. Yellow-nosed Albatross *Diomedea chlororhynchos*
Adult has black bill with orange-yellow culmen ridge and narrow black margins to mainly white underwing (**3a**). The head is white in the IO race *D. c. bassi*, but grey in adults of the Atlantic form *D. c. chlororhynchos*. Juvenile and immature of both races have white heads and black bills (**3b**). sAO and sIO. 76/192cm.

4. Shy Albatross *Diomedea cauta*
Largest mollymawk. Thin black margin to the underwing and black 'thumb mark' at the base of the leading edge of the underwing are diagnostic. Adult has white head with greyish cheeks and grey bill with yellow tip (**4a**). Juvenile has dusky breast band, like immature Black-browed Albatross, but with thin margins to the underwing and a black-tipped light grey bill. 99/256cm. (Other name: White-capped Albatross.) **Salvin's Albatross** *D. c. salvini*. Adult differs from Shy Albatross in brownish-grey head with a white forehead and more extensive black tip on underwing (**4b**). Juveniles only separable on underwing pattern. 94/251cm. **Chatham Island Albatross** *D. c. eremita*. Similar to Salvin's but has greyer head, lacks white forehead and bill is yellowish. Group fairly widespread in southern oceans. 90/220cm.

5. Buller's Albatross *Diomedea bulleri*
Adult similar to Grey-headed Albatross but has prominent white forehead, orange-yellow lower and upper ridges to black bill and more white on underwing. Immature similar but bill darker. sPO. 78/211cm.

6. Laysan Albatross *Diomedea immutabilis*
Both adults and immatures have white head with dusky cheeks, black-tipped yellow bill and distinctive dark mark on underwing-coverts. Feet project beyond tail in flight. Confusion possible with aberrant Black-footed Albatross or hybrids. Widespread in nPO. 80/200cm.

GIANT PETRELS
Family: Procellariidae

7. Northern Giant Petrel *Macronectes halli* **477**
Reddish tip to bill diagnostic. Adult dark brown above, paler below, with whitish face. Leading edge of underwing dark (**7a**). Juveniles of both species are completely dark brown and hard to separate, and the transition into the similar looking dark morphs makes identification very difficult unless the tip of the bill can be seen (**7b**). Circumpolar in southern oceans but not as wide ranging as Southern Giant Petrel. 88/190cm. (Other name: Hall's Giant Petrel.)

8. Southern Giant Petrel *Macronectes giganteus* **477**
White morph unique (**8a**). Dark morph like Northern but pale leading edge to underwing and greenish-tipped bill diagnostic in all stages; also whitish upper breast and head contrast with greyish-brown underwing and belly (**8b**). Circumpolar in southern oceans. 88/195cm. (Other name: Antarctic Giant Petrel.)

1a
1b
1c
2a
2b
3a
3b
4a
4b
5
6
7a
7b
8a
8b

PLATE 29: LARGER PETRELS, FULMARS AND SHEARWATERS

Petrels have a stiff-winged flight alternating with bouts of rapid wingbeats. They have rather heavy bills with the nostrils fused into a tube on the top of the upper mandible. Shearwaters are similar but have a more graceful flight and much more slender bills. The two black and white petrels are unlike any other seabirds in the southern oceans. The two fulmars occur in different hemispheres; Northern Fulmar has a dark morph which can be confused with other shearwaters and petrels but the heavy yellowish bill is a good feature. The dark petrels and shearwaters are a difficult group to identify. Flight and jizz are important in separating the genera. Attention should be paid tc the tail shape and both upper- and underwing patterns. Several species have restricted ranges while others have a virtually global distribution.

PETRELS AND SHEARWATERS
Family: Procellariidae

1. Antarctic Petrel *Thalassoica antarctica*
Brown upperparts contrast with broad, white rear edge of upperwing and dark-tipped white tail. Shows white hind-collar in worn plumage. Legs flesh. Circumpolar in Antarctic, usually close to pack ice. 43/102cm.

2. Cape Petrel *Daption capense*
Distinctive chequered brown and white plumage. *D. c. australe* is smaller and has less white in the wing. Legs dark. Circumpolar in southern oceans, gregarious, follows ships. 39/86cm. (Other name: Pintado Petrel.)

3. Northern Fulmar *Fulmarus glacialis* **477**
Superficially gull-like but distinctive flight, gliding and flapping on stiff wings. Heavy tubenose bill. Plumage variable; light morph with white head and pale grey upperparts in nAO, dark brownish forms in PO and high Arctic. Widespread in nPO, nAO and Arctic Ocean. 48/108cm. **Southern Fulmar** *F. glacialoides*. Paler than Northern Fulmar with darker wing tips. The only heavily built pale grey petrel with a white head and white underwing within its range. Circumpolar in southern oceans. 48/118cm. (Other name: Antarctic Fulmar.)

4. Bulwer's Petrel *Bulweria bulwerii*
Intermediate-sized brown petrel with pale diagonal bar on upperwing and long tapering tail. Weaving flight low over waves. Tropical and subtropical waters in AO, PO and IO. 26/67cm. **Jouanin's Petrel** *B. fallax* is larger, entirely blackish-brown and has a strong, sweeping flight high above waves. Arabian Sea and nwIO. 31/79cm.The localised **Fiji Petrel** *Pterodroma (Pseudobulweria) macgillivrayi*, recently rediscovered on the island of Gau, Fiji, is similar but lacks pale wing-bar. 30/?cm.

5. Great-winged Petrel *Pterodroma macroptera*
Large all dark petrel often with whitish feathers at base of stubby black bill. Legs black. Flight swift, sweeping in high arcs over waves. Southern oceans. 42/97cm.

6. Kerguelen Petrel *Pterodroma brevirostris*
Slate-grey plumage and reflective silvery leading edge and primary bases on underwing. Darker head gives hooded appearance. Black feet. Very fast swooping flight, often towering high above waves. Circumpolar in southern oceans. 36/81cm.

7. Herald Petrel *Pterodroma arminjoniana*
Polymorphic. All morphs have dark underwing with white leading edge and primary patch, and a dark 'M' mark on upperwing **(7a)**. Light morphs are greyish-white below, sometimes with darker breast band and flanks **(7b)**. Fragmented range in tropical PO, AO and IO. 37/96cm. **Kermadec Petrel** *P. neglecta*. Polymorphic. Resembles Herald Petrel but larger and heavier with more rounded tail and distinctive pale primary flash on upperwing. Legs pink. Tropical and subtropical PO. 38/92cm.

8. Providence Petrel *Pterodroma solandri*
Large, thickset greyish-brown petrel with conspicuous white flash on underwing. Compared with Kermadec, belly greyer, no white on leading edge of underwing and no white flash on upperwing. Legs black. wcPO. 40/94cm.

9. Murphy's Petrel *Pterodroma ultima*
Large, heavy greyish-brown petrel with indistinct 'M' mark on upperparts and pale base to primaries visible on underwing. Whitish mottling around base of bill visible at close range. The only round-tailed dark petrel of central tropical PO. 40/98cm. **Mascarene Petrel** *P. aterrima*. Medium-sized entirely dark petrel with stubby bill and high sweeping flight. Known only from a few sightings near Réunion, IO. 36/?cm.

10. Sooty Shearwater *Puffinus griseus*
Large and slim, with long narrow wings. Plumage sooty-brown with a variable silvery patch on underwing coverts. Global. 44/105cm.

11. Wedge-tailed Shearwater *Puffinus pacificus*
Large, rather slender-bodied with broad wings and long tail. Dark morph similar to Flesh-footed but bill dark **(11a)**. Light morph more scarce, similar to Pink-footed but head paler and underwing whiter **(11b)**. Legs pink. Flight buoyant, often with wings held forwards. Does not follow ships. IO and PO. 43/103cm. **Flesh-footed Shearwater** *P. carneipes*. Like dark Wedge-tailed with squarer tail and diagnostic pinkish bill and legs. Shows silvery primary bases on underwing. Slow, effortless flight. IO and PO. 43/103cm. **Short-tailed Shearwater** *P. tenuirostris*. Similar to Sooty Shearwater but has shorter bill and greyer underwing-coverts. Dark feet project beyond tail tip. In winter may show darker cap and whitish chin. Flight fast with rapid wingbeats followed by a glide. Transequatorial migrant in PO. 43/99cm. **Christmas Island Shearwater** *P. nativitatis*. Like small Sooty Shearwater but plumage entirely dark brown including underwing. Separated from Wedge-tailed by smaller size and dark legs. Tropical PO. 36/76cm. (Other name: Christmas Shearwater.)

12. White-chinned Petrel *Procellaria aequinoctialis*
Large size, blackish plumage, ivory bill and white chin diagnostic. White chin difficult to see unless head-on; some birds have white on sides of face. Legs dark. Circumpolar, southern oceans. 55/140cm. **Westland Petrel** *P. westlandica*. Separated from White-chinned by dark tip to pale bill and lack of white chin. Breeds Mar-Dec on South Is., New Zealand, where 1,000-5,000 breeding pairs. Disperses to cPO. 51/137cm. **Black Petrel** *P. parkinsoni*. Differs from Westland only in smaller size. Breeds Nov-Jun on islands off North Is., New Zealand, where less than 1,000 breeding pairs. Disperses through subtropical Pacific to coasts of Mexico and Galapagos. 46/115cm. (Other name: Parkinson's Petrel.)

1
2
3
4
5
6
7a
7b
8
9
10
11a
12
10
11b

PLATE 30: GADFLY PETRELS

The *Pterodroma* or 'gadfly' petrels are a highly pelagic group of small to medium-sized tubenoses found mainly in warmer waters, especially in the Pacific Ocean. They have a strong, purposeful flight and often tower high above the waves, particularly in high winds; most species do not follow ships. They present both taxonomic and identification problems. Important characters are upper- and underwing patterns and head markings. Some species are extremely rare and little-known.

GADFLY PETRELS
Family: Procellariidae

1. Soft-plumaged Petrel *Pterodroma mollis*
S Atlantic form *P. m. mollis* has a narrow breast band and an indistinct dark 'M' on upperwings **(1a)**. *P. m. dubia* of IO has a broad breast band. Rare dark morph has darker upperwing, brown underparts and lacks white leading edge to the wing **(1b)**. 34/90cm. Very similar **Cape Verde Petrel** *P. feae* has breast band reduced to dark smudges on sides of breast and heavier bill **(1c)**. Breeds Cape Verde Islands and Desertas Is. off Madeira, population about 1200 pairs. (Other names: Fea's Petrel, Gon-gon.) **Madeira Petrel** *P. madeira* is very similar but smaller, with narrower bill and paler forehead. Breeds Madeira, population only 20 pairs. (Other names: Zino's Petrel, Freira.) (All three forms may be conspecific.)

2. Atlantic Petrel *Pterodroma incerta*
Uniform dark brown including underwing and throat with white oval patch on breast and belly. May show paler head and dark eye-mask in worn plumage. Follows ships. sAO. 43/104cm. (Other name: Schlegel's Petrel.)

3. Phoenix Petrel *Pterodroma alba*
Like Atlantic with white undertail-coverts and pale band on underwing. White chin can be difficult to see, as can white leading edge to underwing. cPO. 35/83cm. **Tahiti Petrel** *P. rostrata*. Like Phoenix but with dark chin, paler upperparts and lacks white leading edge to underwing. wcPO. 39/84cm. **Beck's Petrel** *P. becki* is distinguished from Tahiti only by smaller size. Known from two specimens from New Guinea/Solomon Is. area. 29/??cm. **Magenta Petrel** *P. magentae*. Like a greyish-brown Phoenix with dark underwings. Virtually unknown, recently rediscovered on Chatham Is., New Zealand.

4. White-necked Petrel *Pterodroma cervicalis*
Black cap contrasts with white collar. Upperparts greyish-brown with darker 'M' mark and greyish rump. Underwings white with dark primaries and narrow black mark extending back from carpal. sPO, breeds off New Zealand. 43/95cm. **Juan Fernandez Petrel** *P. externa* lacks white collar and is paler on crown. Thinner black mark on underwing. sPO, breeds off Chile. 43/95cm.

5. White-headed Petrel *Pterodroma lessonii*
Distinctive white head and underparts and tail contrast with dark underwing and greyish upperwing with darker 'M' mark. Rarely follows ships. Circumpolar, southern oceans. 43/109cm.

6. Black-capped Petrel *Pterodroma hasitata*
Dark cap, mantle and upperwing contrast with white collar and rump. Thin black line on sides of breast. Underwing white with black margins and diagonal band from carpal. Atypical birds have reduced white on rump and collar. Rare, Caribbean. 40/95cm. **Bermuda Petrel** *P. cahow*. Similar to Black-capped but lacks white collar and rump patch; also see Soft-plumaged Petrel. Very rare, breeds Bermuda. 38/89cm.

7. Galapagos Petrel *Pterodroma phaeopygia*
Distinctive with black cap, entirely dark brown upperwing, rump and tail. Underparts white with thin black bar on underwing and a dark spot on axillaries. Sometimes shows white at sides of rump. Breeds Galapagos Is., disperses to S and C American coasts. 43/91cm. Very similar **Hawaiian Petrel** *P. sandwichensis* is indistinguishable at sea. Breeds Hawaii. 43/91cm.

8. Black-winged Petrel *Pterodroma nigripennis*
Distinctive with greyish head and patches on sides of breast. Underwing white with black borders and thick black diagonal mark from carpal. sw to cPO. 30/68cm. **Bonin Petrel** *P. hypoleuca*. Similar to Black-winged but darker on head and diagonal mark on underwing joins a broad black patch on primary coverts. nwPO. 30/67cm. **Chatham Islands Petrel** *P. axillaris*. Similar to Black-winged, but diagnostic underwing pattern with black line from carpal extending to axillaries. Very rare, probably less than 500 birds, restricted to Chatham Is., New Zealand. 30/67cm.

9. Mottled Petrel *Pterodroma inexpectata*
Distinctive grey belly and broad black diagonal bar on white underwing. Dark 'M' on greyish upperwing. Breeds on islands off New Zealand dispersing northwest across PO to Alaska. 34/74cm.

10. Cook's Petrel *Pterodroma cookii*
Pale grey upperparts with obvious dark 'M' on upperwing, narrow black diagonal bar on underwing and white sides to tail. Lacks a dark cap. Legs dark. wPO. 26/66cm. **Pycroft's Petrel** *P. pycrofti*. Very similar to Cook's but head darker giving capped appearance, dark eye-patch, less white in tail and browner on upperwing with less distinct 'M' marking. Less than 1,000 breeding pairs off North Is., New Zealand; probably disperses northwards. 26/66cm. In the ecPO the rare **Defilippe's Petrel** *P. defilippiana* is virtually indistinguishable from Pycroft's but has larger bill and less white in outer tail. Breeds off Chile, disperses in E Pacific. 26/66cm. **Stejneger's Petrel** *P. longirostris*. Similar to Cook's but darker crown and hindneck contrast with grey mantle. Tail uniform grey. Breeds off Chile and migrates across PO towards Japan and W United States. 26/66cm.

11. Gould's Petrel *Pterodroma leucoptera*
Blackish head and neck contrast with pale grey mantle. Black tip to tail. Pronounced dark 'M' on upperwing, extensive black margins to underwing and diagonal bar from carpal. Breeds off W Australia, disperses towards Galapagos. 30/71cm. **Collared Petrel** *P. brevipes* (possibly a race of Gould's) is very similar with variable dusky breast band and sometimes greyish underparts. Breeds swPO. 30/71cm. **Barau's Petrel** *P. baraui*. Only grey and white *Pterodroma* in IO. Black diagonal line on underwing-coverts. Breeds Réunion. 38/?cm.

1a
1b
1c
2
2
3
4
5
6
6
7
8
9
10
11

PLATE 31: SHEARWATERS

Shearwaters are long-winged, slender-bodied highly pelagic seabirds. They are nocturnal at the breeding colonies which are often vast. Most species are great wanderers and several are transequatorial migrants. They have slender, tubenose bills with a pronounced hook. The similarity of many species presents a number of taxonomic and identification problems. Flight action is important; the larger species tend to have an undulating flight, gliding on stiff wings interspersed with periods of rapid wingbeats. The smaller species tend to have a more fluttering flight on shorter, more rounded wings. Identification mostly requires careful observation of almost all plumage features and jizz is also important.

SHEARWATERS
Family: Procellariidae

1. Grey Petrel *Procellaria cinerea*
Large size, grey above with contrasting white underparts and yellow bill. Underwings uniform grey. Circumpolar in southern oceans and W coast of S America. 48/120cm.

2. Cory's Shearwater *Calonectris diomedea*
Large greyish-brown shearwater with whitish underparts and yellow bill. Lacks collar and has narrow white band on rump. MS, AO and sIO. 50/112cm. Cape Verde form *C. d. edwardsii* much smaller with dark bill (may be a full species). **Streaked Shearwater** *C. leucomelas.* Similar to Cory's but face whitish with darker streaks, scaly upperparts and broad dark margins to white underwings. nwPO dispersing to nIO. 48/122cm.

3. Great Shearwater *Puffinus gravis*
Large brown and white shearwater with distinctive black cap, white collar and rump patch. Dark belly patch diagnostic but hard to see. Dark diagonal bar on underwing. Dark bill. Often follows ships. Breeds sAO islands, transequatorial migrant to nAO. 47/109cm.

4. Pink-footed Shearwater *Puffinus creatopus*
Uniform greyish-brown upperparts, dusky white underparts and dark-tipped pinkish bill. 'Lazy' flight, slow wingbeats and low glides. Breeds sePO dispersing north to Alaska. 48/109cm.

5. Buller's Shearwater *Puffinus bulleri*
Large, slender, greyish shearwater with dark cap and striking dark 'M' on upperwing. Legs pink, graceful gliding flight. Breeds New Zealand dispersing to nwPO. 46/97cm.

6. Manx Shearwater *Puffinus puffinus*
Medium size, all dark upperparts sharply demarcated from white underparts. Black on face reaches below eye. Typical stiff-winged, shearing flight. Breeds nAO dispersing to S America and S Africa. 34/82cm. **Townsend's Shearwater** *P. auricularis.* Like Manx but with dark undertail-coverts and conspicuous white sides to dark-centred rump. ePO. 33/76cm. **Newell's Shearwater** *P. (a) newelli* is slightly larger with white or black-and-white undertail-coverts. Breeds Hawaii.

7. Mediterranean Shearwater *Puffinus yelkouan*
Similar to Manx but upperparts browner and less distinct dark underwing margins. Breeds in eMS. and Black Sea. 34/82cm. (Other name: Yelkouan Shearwater.) **Balearic Shearwater** *P. y. mauretanicus* is browner above than Manx with less distinct whitish underparts and dusky underwing and undertail-coverts; some individuals very dark below suggesting Sooty Shearwater. Breeds wMS. dispersing to nwAO and North Sea. 38/87cm.

8. Black-vented Shearwater *Puffinus opisthomelas*
Brown upperparts merge into whitish underparts, neck, upper breast and axillaries often dusky; lacks white sides to rump. c and ePO. 34/82cm.

9. Fluttering Shearwater *Puffinus gavia*
Upperparts brown to rusty-brown, whitish underparts. Flight low over waves with fluttering wingbeats. New Zealand-Australia. 33/76cm.

10. Hutton's Shearwater *Puffinus huttoni*
Like larger Fluttering but duskier on face, neck and breast with blackish-brown upperparts merging into whitish underparts. Dusky underwing-coverts and axillaries. Flight like Fluttering. New Zealand-Australia. 38/90cm.

11. Little Shearwater *Puffinus assimilis*
Small black-and-white shearwater with distinctive fluttering and gliding flight. Small bill and blue feet. Two forms, *P. a. assimilis* of the swPO and *P. a. boydi* of the AO, have white-faced appearance with dark cap not reaching eye, and *assimilis* has white undertail-coverts while *boydi* has brown undertail-coverts. *P. a. elegans* of the southern oceans is like a small Manx with blackish sides of face. 27/62cm. Includes **Bannerman's Shearwater** *P. bannermani* which is sometimes treated as a separate species. Breeds nwPO.

12. Audubon's Shearwater *Puffinus lherminieri*
Similar to Little, but browner upperparts and undertail-coverts, less white on underwing and pink legs. Tropical AO, IO and PO. 30/69cm. **Persian Shearwater** *P. persicus* is slightly larger with more brown on underwing and some streaking on flanks and axillaries. Occurs Arabian Sea. **Mascarene Shearwater** *P. atrodorsalis.* Newly described species (1995); similar in plumage to Manx but smaller, also sharing characters with Little and Audubon's. Bill relatively long and slender. wIO. **Heinroth's Shearwater** *P. heinrothi* is mainly sooty-brown with variable amount of white on underwing and sometimes whitish belly and chin. Tropical swPO off Solomon Is. 27/?cm.

1
3
2
6
6
2
4
5
7
10
11
8
9
8
11
12

PLATE 32: PRIONS, DIVING-PETRELS AND PHALAROPES

The unique Snow Petrel is placed in a monotypic genus and is possibly related to fulmars; it is typically associated with the Antarctic pack ice. The Blue Petrel is also monotypic and though it closely resembles prions it has some affinities with *Pterodroma* petrels. Prions are a confusing group of small, southern ocean petrels which have a fringe of lamellae on the upper mandible which allows them to filter plankton. They also have a gular pouch under the lower mandible which is used to hold food when transporting it back to feed chicks. The taxonomy of the group is still unresolved. Specific identification at sea is virtually impossible though an attempt can be made to distinguish those birds with a distinct head pattern from those which lack distinct head markings. Prions are often 'wrecked' on shore when specific identification of birds in the hand is possible. Diving-Petrels are equally difficult to identify at sea. They bear a superficial resemblance to the Little Auk of the northern hemisphere, but this is an example of convergent evolution and they are in fact true tubenoses. They differ from other tubenoses in having upward pointing nostrils. In the hand the bill shape is diagnostic. Phalaropes are the only shorebirds which regularly occur at sea. Two species are pelagic during the non-breeding season and may occur in large numbers, particularly in areas of ocean upwelling. Unlike other shorebirds they have lobed feet, an adaptation for swimming, and also possess well developed salt glands. They are also unusual in that the sexual roles are reversed with the duller-plumaged males taking responsibility for incubation and care of the chicks. They swim with jerky movements and will sometimes spin on the surface. Identification at sea of winter-plumaged birds requires care.

SMALL PETRELS
Family: Procellariidae

1. Snow Petrel *Pagodroma nivea*
Distinctive; only small all white petrel. Circumpolar in southern oceans, mostly near pack ice. 32/78cm.

2. Blue Petrel *Halobaena caerulea*
Resembles prions but has diagnostic white tip to tail and black cap extending to sides of breast. Circumpolar in southern oceans, dispersing north to Peru. 29/62cm.

PRIONS
Family: Procellariidae

All six species are extremely similar and very difficult to separate at sea.

3. Broad-billed Prion *Pachyptila vittata*
Has a diagnostic black bill and rather dark-headed appearance. sAO, IO and swPO. 28/61cm. **Antarctic Prion** *P. desolata* has a smaller, blue bill and grey mottling on sides of breast forming partial collar. Virtually circumpolar in southern oceans dispersing to about 10°S. 27/61cm. **Medium-billed Prion** *P. salvini* is virtually inseparable from Antarctic, but looks cleaner with less grey on sides of breast. sIO dispersing to S Africa and Australia. 28/57cm.

4. Fairy Prion *Pachyptila turtur*
Smallest prion. Pale head lacks streak through eye, broader black tip to uppertail. Virtually circumpolar in southern oceans. 25/58cm. **Fulmar Prion** *P. crassirostris* has stouter bill and more distinct 'M' mark on upperwing. sIO and swPO. 26/58cm.

5. Slender-billed Prion *Pachyptila belcheri*
Conspicuous dark eye-streak and long white supercilium. Upperparts paler than other prions with less distinct 'M' mark. Circumpolar in southern oceans. 26/56cm. (Other name: Thin-billed Prion.)

DIVING-PETRELS
Family: Pelecanoididae

All four species are virtually impossible to identify at sea but distribution is helpful.

6. South Georgia Diving-Petrel *Pelecanoides georgicus*
Has whitish tips to scapulars and white underwing-coverts. Blue legs have black rear edge and blackish webs. Seas around subantarctic islands. 20/32cm. (Other name: Georgian Diving-Petrel.) **Common Diving-Petrel** *P. urinatrix*. Like South Georgia Diving-Petrel but greyer underwing-coverts, plain brownish-black upperparts (lacking obvious tips to scapulars) and more mottled throat. Legs blue. Widespread in southern oceans. 23/35cm.

7. Magellanic Diving-Petrel *Pelecanoides magellani*
Distinctive plumage with white foreneck and crescent-shaped half-collar on sides of neck. Coasts of southern S America. 19/?cm. **Peruvian Diving-Petrel** *P. garnotii*. Range overlaps with Magellanic but darker foreneck and lacks white half-collar. Coasts of Peru and Chile. 22/?cm.

PHALAROPES
Family: Scolopacidae

8. Grey/Red Phalarope *Phalaropus fulicaria*
Distinctive breeding plumage (**8a**). In winter pale grey upperparts, darker eye-streak, pale base to blackish bill (**8b/c**). Circumpolar in arctic breeding range, winters south to Chile and S Africa. 20/37cm.

9. Red-necked Phalarope *Phalaropus lobatus*
Distinctive breeding plumage (**9a**). In winter black crown, darker slate-grey upperparts with white 'braces' on scapulars. Needle-like black bill (**9b/c**). Circumpolar in arctic breeding range, wintering south to S America, S Africa and New Guinea. 17/34cm.

10. Wilson's Phalarope *Steganopus tricolor*
Distinctive breeding plumage (not illustrated). In winter note diagnostic white rump, long thin bill and lack of wing-bar. Not pelagic in winter, but sometimes occurs in coastal waters off N and S America. 22/37cm.

1
2
3
4
3
5
5
6
7
10
9c
8c
8a
9a
8b
10
9b

PLATE 33: STORM-PETRELS

Storm-Petrels are amongst the smallest seabirds. With the exception of the Galapagos race of Wedge-rumped Petrel they are nocturnal at the breeding colonies. The species whose distribution is centred in the southern oceans tend to have rather shorter, more rounded wings than the northern ocean forms which are typically long-winged and often have forked tails. Southern ocean forms also tend to have longer legs which may project beyond the tip of the tail in flight, and patter on the surface more than northern species. Storm-Petrels are agile flyers and spend most of their time on the wing, rarely resting on the surface of the sea. Jizz, particularly flight action, is helpful when identifying similar species at sea. Though some species have rather restricted ranges, most are great wanderers and some are transequatorial migrants.

STORM-PETRELS
Family: Hydrobatidae

1. Wilson's Storm-Petrel *Oceanites oceanicus*
Small with U-shaped white rump extending broadly to undertail-coverts. Diagonal greyish bar on upperwing, indistinct greyish band on underwing. Feet project beyond square tail (yellow webs on feet are diagnostic but hard to see). Follows ships, 'walks' on water with legs trailing. Flight direct but skips over water when feeding. Worldwide, breeds in southern oceans. 17/40cm.

2. White-vented Storm-Petrel *Oceanites gracilis*
Small size; similar to Wilson's but pale underwing-coverts and variable white patch on belly. Feet project beyond square tail in flight. Follows ships. Little known, occurs ePO from Galapagos – Chile. 15/?cm. (Other name: Elliot's Storm-Petrel.)

3. Grey-backed Storm-Petrel *Garrodia nereis*
Like small White-bellied but rump, uppertail and diagonal upperwing-bar grey. Feet project beyond black-tipped tail. Southern oceans north to about 35°S. 17/39cm.

4. White-faced Storm-Petrel *Pelagodroma marina*
Medium size; grey and white storm-petrel with dark eye-patch and white supercilium. Feet project beyond forked tail. Rarely follows ships. Flight erratic and weaving, bounces off wave crests. Widespread (not nPO). 20/42cm.

5. Black-bellied Storm-Petrel *Fregetta tropica*
Medium size; black above with white rump joined to white flanks and underwing-coverts. Black breast and belly with narrow black line to undertail-coverts. Pale diagonal bar on upperwing, feet project slightly beyond tail. Flight distinctive, follows contours of sea with body twisting from side to side, bouncing off wave crests. Circumpolar in southern oceans dispersing north towards equator. 20/46cm. **White-bellied Storm-Petrel** *F. grallaria*. Very similar to Black-bellied including flight, but has paler greyish upperparts, more distinct upperwing bar and lacks black line on belly. Virtually circumpolar from 35°S north to tropics. 20/46cm.

6. Polynesian Storm-Petrel *Nesofregetta fuliginosa*
Largest storm-petrel, polymorphic. Dark morph uniform brownish with short pale bar on upperwing-coverts. Pale form has narrow white rump, white underwing-coverts and white throat and belly separated by dark breast band. Intermediates occur. Feet project beyond long forked tail. Distinctive flight with long glides on rounded wings then splashes down. Tropical PO. 25/?cm. (Other name: White-throated Storm-Petrel.)

7. European Storm-Petrel *Hydrobates pelagicus*
Small blackish storm-petrel with diagnostic white band on underwing. Square white rump does not extend to undertail-coverts, indistinct narrow diagonal bar on upperwing. Feet do not project beyond square tail. Follows ships. Fluttering bat-like flight. Breeds neAO and MS, winters in sAO. 15/37cm. (Other name: British Storm-Petrel.)

8. Least Storm-Petrel *Oceanodroma microsoma* (*Halocyptena microsoma*)
Smallest storm-petrel. Uniform sooty-brown apart from pale diagonal upperwing-bar. Tail wedge-shaped or rounded. Flight swift with deep wing beats. Pacific coast from California – Ecuador. 14/32cm.

9. Wedge-rumped Storm-Petrel *Oceanodroma tethys*
Distinctive large white rump patch extending to base of shallowly forked tail. Pale bar on upperwing-coverts. Rarely follows ships. ePO, California – Peru. 19/?cm.

10. Leach's Storm-Petrel *Oceanodroma leucorhoa*
Medium size; long-winged with forked tail. Distinct diagonal pale bar on upperwing, smudgy white rump usually divided by greyish centre, though some Pacific birds have dark rump. Does not follow ships. Erratic, bounding flight with deep wing beats and short glides. Widespread, breeds N hemisphere dispersing southwards. 20/46cm.

11. Band-rumped Storm-Petrel *Oceanodroma castro*
Medium size with slight fork to tail, pale diagonal bar on upperwing, broad curved white rump patch extends to sides of undertail-coverts. Does not follow ships. Buoyant zigzag flight with low glides. Tropical and subtropical PO and AO. 20/43cm. (Other name: Maderian Storm-Petrel.)

12. Ringed Storm-Petrel *Oceanodroma hornbyi*
Large and distinctive. Grey and white with diagnostic dark cap and breast band, broad pale diagonal bar on upperwing and forked tail. Pacific coast of S America. 22/?cm. (Other name: Hornby's Storm-Petrel.)

13. Ashy Storm-Petrel *Oceanodroma homochroa*
Medium size with long wings and forked tail. Uniform dark brown with diagnostic pale greyish underwing-coverts. Flight fluttering with shallow wing beats. Pacific coast from California – Ecuador. 20/?cm. **Black Storm-Petrel** *O. melania*. Like a large Ashy with a dark underwing. Flight deliberate with steady deep wing beats. Pacific coast from California – Peru. 23/48cm. **Markham's Storm-Petrel** *O. markhami*. Similar to Black but more extensive pale upperwing-bar and more deeply forked tail. PO from Mexico – Chile. 23/?cm.

14. Fork-tailed Storm-Petrel *Oceanodroma furcata*
Medium size; grey and white storm-petrel with blackish mask through eye and blackish underwing-coverts. Swimming bird could suggest phalarope at a distance. nPO. 22/46cm.

15. Swinhoe's Storm-Petrel *Oceanodroma monorhis*
Medium size, wholly dark brown with shallowly forked tail and diagonal pale upperwing-bar. nwPO, nIO, Red Sea and recently North Sea (vagrant). 20/45cm. **Matsudaira's Storm-Petrel** *O. matsudairae*. Like a large Swinhoe's with more deeply forked tail, paler upperwing-bar and diagnostic white bases to outer primaries forming patch on forewing. wPO to E African coast of IO. 24/56cm. **Tristram's Storm-Petrel** *O. tristrami*. Larger than Swinhoe's with more deeply forked tail, greyer plumage and more distinct wing-bar. nw to cPO. 24/56cm.

2
4
7
1
2
10
11
9
6
12
4
5
3
8
13
14
15

PLATE 34: CORMORANTS AND SHAGS

The terms Cormorant and Shag have been applied somewhat randomly to various species within the genus *Phalacrocorax* and are virtually interchangeable. They are medium to large aquatic birds; some species are virtually confined to freshwater habitats but the majority are marine, mainly occurring in coastal waters. Four almost exclusively freshwater species have been excluded below. Cormorants tend to up-end when diving, showing the longish tail as they dive.

CORMORANTS AND SHAGS
Family: Phalacrocoracidae

1. Great Cormorant *Phalacrocorax carbo*
Largest cormorant, black with green or blue gloss. White throat and yellow gular pouch. White thigh patch and filoplumes in breeding plumage **(1a)**. African form *P. c. lucidus* **(1b)** has white breast. Juveniles **(1c)** brown with whitish underparts. Widespread Old World species (coastal and inland), also east coast of N America. 90/140cm.

2. European Shag *Phalacrocorax aristotelis*
Adult black with a greenish sheen and in breeding plumage recurved crest. Duller in winter. Smaller than Great Cormorant with a thinner bill and small yellow gape. Juvenile brown with whitish underparts. Rocky coasts of neAO, MS and Black Sea. 72/97cm.

3. Socotra Cormorant *Phalacrocorax nigrogularis*
Large with glossy black plumage and white tuft on sides of head (absent in dull brownish winter plumage). Juveniles whitish below. Persian Gulf to S Red Sea, often in large flocks. 80/106cm. **Indian Cormorant** *P. fuscicollis.* Medium-sized with a long tail and long thin bill. Blackish-bronze with a white ear tuft when breeding. India to Indochina, also inland. 65/?cm.

4. Double-crested Cormorant *Phalacrocorax auritus*
Similar to Great Cormorant but smaller, has orange facial skin and lacks white patch on thighs. Duller in winter. Juvenile browner with whitish breast. Widespread (coastal and inland), N America to Caribbean. 84/134cm.

5. Pelagic Cormorant *Phalacrocorax pelagicus*
Smaller than very similar Red-faced but red facial skin does not join on forehead. Juvenile has less extensive facial skin. Pacific coasts of E Asia and N America. 68/96cm.

6. Brandt's Cormorant *Phalacrocorax penicillatus*
Similar to Double-crested but facial skin dark with yellowish feathers at base of gular pouch which is pale blue when breeding. Juveniles brown with paler 'V' on upper breast. Pacific coast of N America. 85/118cm.

7. Red-faced Cormorant *Phalacrocorax urile*
Black with greenish gloss and red facial skin joining across forehead. Breeding birds have double crest and white thigh patch. Juvenile has brownish-grey facial skin joining over bill. N Pacific coasts, Japan – Alaska. 84/116cm.

8. Japanese Cormorant *Phalacrocorax capillatus*
Very similar to Great Cormorant with greener iridescence and white on face extending under base of yellow gular pouch. Rocky coasts of NE Asia. 92/152cm.

9. Cape Cormorant *Phalacrocorax capensis*
All dark with yellow facial skin (duller in winter) **(9a)**. Browner in winter with whitish underparts (more extensive on juvenile). Occurs in huge numbers, flies in long skeins **(9b)**. Endemic to coasts of Namibia and S Africa. 63/109cm.

10. Bank Cormorant *Phalacrocorax neglectus*
Large, all dark cormorant with black facial skin and white rump when breeding. Some individuals have varying amount of white on face and neck. Endemic to coasts of Namibia and S Africa. 76/132cm.

11. Crowned Cormorant *Phalacrocorax coronatus*
Small, all dark glossy-green plumage with long tail, orange-yellow facial skin and red eye. Winter and juvenile plumage mainly brown, paler below on throat. Endemic to coasts of Namibia and S Africa. 50/85cm.

12. Neotropic Cormorant *Phalacrocorax brasilianus*
The only all dark cormorant in S. America. Diagnostic white border to yellow gular pouch. Breeding birds show white tuft on sides of head. Widespread S and C America (coasts and inland), also southernmost USA. 65/101cm. (Other name: Olivaceous Cormorant *P. olivaceus.*)

13. Guanay Cormorant *Phalacrocorax bougainvillii*
Dark above, white below. White throat separated from breast by dark band. Facial skin red, bill yellowish. West coast of S America. 76/?cm.

14. Red-legged Cormorant *Phalacrocorax gaimardi*
Distinctive pale grey plumage with white patch on neck and silvery wing-coverts forming wing patch in flight. Juvenile similar but browner. West coast of S America, also locally in S Argentina. 76/?cm.

15. Rock Shag *Phalacrocorax magellanicus*
Similar to Guanay but blackish bill and less white on throat. Breeding birds have white patch on sides of face. Non-breeding plumage entirely white below. Juvenile brown with white specks on belly. Coasts of southern S America and Falkland Is. 66/92cm.

16. Flightless Cormorant *P.* (*Nannopterum*) *harrisi*
Large, all dark flightless cormorant confined to Galapagos Is. 95cm. (Other name: Galapagos Cormorant).

17. Pied Cormorant *Phalacrocorax varius*
Larger than similar Black-faced with yellow-orange facial skin, yellowish bill and white on face extending above eye. Juvenile duller. Coastal and inland in Australia and New Zealand. 75/121cm.

18. Black-faced Cormorant *Phalacrocorax fuscescens*
Black above, white below with black facial skin and grey bill. Juvenile greyish on foreneck. Coasts of S Australia. 65/107cm. **Little Pied Cormorant** *P. melanoleucos.* Small polymorphic cormorant with short yellow bill and black legs. Dark morph all black with white face, intermediate morph has white on face extending to upper breast, pale morph has white face and underparts. Mostly inland but also coastal from Indonesia to Australia and New Zealand. 61/81cm.

19. Rough-faced Shag *Phalacrocorax carunculatus*
Black and white with white dorsal patch and double white wing-bar. Rare endemic in Cook Straits, New Zealand. 76/?cm. (Other name: New Zealand King Cormorant.) **Bronze Shag** *P. chalconotus.* Polymorphic. Pale morph greener than Rough-faced with one wing-bar and tiny dorsal patch, dark morph all oily-green, intermediate morph oily-green with variable white spotting on underparts. S. New Zealand. 68/?cm. (Other name: Stewart Island Shag.) **Chatham Islands Shag** *P. onslowi.* Like Rough-faced but velvety-blue/black above with larger white dorsal patch. Endemic to Chatham Is., off New Zealand. 63/?cm.

Continued on p.244

1a
1c
2
1b
3
5
4
9a
9b
8
6
7
10
11
12
13
16
17
18
14
15
19
22
20
21
23a
23b
23c

PLATE 35: TROPICBIRDS, GANNETS AND BOOBIES

The three species of tropicbirds are very distinctive medium-sized black and white birds with remarkably long tail streamers. They are amongst the most aerial of seabirds and are mainly pelagic, only visiting oceanic islands to breed. They hold their tail streamers raised on the rare occasions when they rest on the surface of the water. On land they are barely capable of standing due to the short legs which are placed well back on the body. Given good views the identification of adults is quite straightforward. Juvenile birds are much more similar, differing slightly in the upperwing pattern and the width of the black barring on the upperparts. Gannets and boobies are large narrow-winged seabirds with cigar-shaped bodies and long tapering bills which, with the exception of Abbott's Booby, end in a point. The external nostrils are closed as an adaptation for plunge-diving, and secondary nostrils above the gape are automatically closed by a flap of skin on entering the water. The two genera, *Morus* and *Sula*, are so closely related that many authors combine them all in *Sula*. They swim well and are excellent fliers.

TROPICBIRDS
Family: Phaethontidae

1. White-tailed Tropicbird *Phaethon lepturus*
Orange bill, upperwing pattern and white tail streamers distinctive in adult **(1a)**. *P. l. fulvus* of Christmas Is. has plumage washed gold/pink. Juvenile has a yellow bill, short eye-stripe, coarsely barred back and dark primaries **(1b)**. Throughout tropical and subtropical oceans. 78/92cm.

2. Red-tailed Tropicbird *Phaethon rubricauda*
Red bill, largely white plumage and red tail streamers diagnostic in adult **(2a)**. Juvenile similar to White-tailed but with much less black on upper primaries **(2b)**. Tropical and subtropical PO and IO. 78/107cm.

3. Red-billed Tropicbird *Phaethon aethereus*
Red bill, finely barred upperparts and long white tail distinctive in adult **(3a)**. Juvenile like White-tailed but denser barring and dark eye-stripe extends over neck to form collar **(3b)**. ePO, AO and nwIO. 98/105cm.

GANNETS AND BOOBIES
Family: Sulidae

4. Northern Gannet *Morus (Sula) bassanus* **477**
Adult white with golden head, black wing-tips and short gular stripe **(4a)**. Juvenile dark brown with white speckling **(4b)**. All gannets become progressively whiter with age until reaching adult plumage in 5th year. Third year plumage illustrated **(4c)**. nAO. 93/172cm.

5. Australasian Gannet *Morus (Sula) serrator* **477**
Adult differs from Northern Gannet by black secondaries and medium gular stripe; tail black with white sides. Australia and New Zealand; has occurred S Africa. 84/174cm.

6. Cape Gannet *Morus (Sula) capensis* **477**
Adult like Australian Gannet but tail all black and long gular stripe. Overlaps in range with Northern Gannet off W Africa. Breeds S Africa. 85/170cm.

7. Brown Booby *Sula leucogaster*
Adult has dark chocolate upperparts and breast clearly demarcated from white belly. Juvenile duller with short dark stripe on underwing-coverts. *S.l. brewsteri* of E Pacific has pale grey head and greyish bill. Widespread in tropical and subtropical oceans. 69/141cm.

8. Masked Booby *Sula dactylatra*
Adult resembles Cape Gannet but head white with black face mask, black tips to scapulars and yellow bill **(8a)**. Juvenile resembles Brown Booby but has white neck collar (becoming broader on immature) and longer dark line on underwing **(8b)**. Widespread in tropical and subtropical oceans. 86/152cm.

9. Blue-footed Booby *Sula nebouxii*
Adult has blue feet, speckled head and mantle barred brownish **(9a)**. At all ages has white spot at base of neck and clean white patch on axillaries. Juvenile differs from Brown Booby by white spot at base of neck and underwing pattern **(9b)**. Pacific coast of N and C America. 80/152cm.

10. Red-footed Booby *Sula sula*
Adults have diagnostic red feet. Complex range of morphs. Typical white morph has white tail and diagnostic dark carpal patch on underwing **(10a)**. Christmas Is. white morph similar but with yellow cast to plumage. Galapagos Is. white morph has a dark tail. Dark morph entirely greyish-brown as are juveniles of all morphs (which have yellowish feet). Other variants are white-tailed brown morph **(10b)** and white-headed, white-tailed brown morph. Various intermediate plumage types occur throughout range **(10c)**.Widespread in tropical and subtropical oceans. 71/152cm.

11. Abbott's Booby *Papasula (Sula) abbotti*
Distinctive; plumage all white with blackish upperwing and tail and dark-tipped pale bill. Juvenile duller. Endemic to Christmas Is., IO. 71/?cm.

12. Peruvian Booby *Sula variegata*
Resembles Blue-footed but head white, lacks spot at base of neck and lacks white patch at base of underwing. Juvenile similar to adult but plumage mottled. Coasts of Peru and Chile. 74cm/?cm.

CORMORANTS AND SHAGS
(continued from p.242)

20. Campbell Island Shag *Phalacrocorax campbelli*
Resembles Bronze Shag but foreneck black. Endemic to Campbell Is. off New Zealand where Little Pied is only other 'pied' cormorant. 63/105cm. **Auckland Islands Shag** *P. colensoi*. Resembles Campbell Is. Shag but foreneck white, sometimes with white collar. Resident Auckland Is. where only 'pied' cormorant. 63/105cm. **Bounty Islands Shag** *P. ranfurlyi*. Resembles Auckland Is. Shag but facial skin uniformly orange-red. Confined to Bounty Is. 71/?cm.

21. Spotted Shag *Phalacrocorax punctatus*
Distinctive grey-green above, pale grey below, double crest and white stripe from eye to sides of neck. Non-breeding plumage lacks crest and head greyish. Juvenile browner. Coasts of New Zealand. 69/?cm.

22. Pitt Island Shag *Phalacrocorax featherstoni*
Very similar to Spotted Shag but smaller, darker, lacks white stripe on head and neck in breeding plumage. Chatham Is., off New Zealand. 63/?cm.

23. Imperial Shag *Phalacrocorax atriceps*
Black and white dimorphic species with yellow caruncles, blue eye-ring, white wing (alar) bar and wispy crest. *P. a. atriceps* (Blue-eyed C.) has large white dorsal patch **(23a)** and *P. a. albiventer* (King C.) lacks dorsal patch **(23b)**. These two forms were formerly considered to be separate species. Southern oceans, S America and sIO. 72/124cm. Very similar **South Georgia Shag** *P. georgianus*, **Antarctic Shag** *P. bransfieldensis* **(23c)** and **Kerguelen Shag** *P. verrucosus* are now considered to be separate species.

1a
1b
2a
2b
3a
3b
4a
4a
4c
4b
5
6
7
8a
8b
9a
9b
10b
10a
10c
11
12

PLATE 36: SEA DUCKS

The term 'sea ducks' is loosely given to those species of ducks and geese which occur in coastal waters for at least part of the year. Some groups such as eiders are primarily marine while others such as scoters and mergansers are essentially freshwater species which utilise the marine environment during the non-breeding season. Most species are sexually dimorphic, the males being brightly coloured and the females dull or even cryptically coloured. The large, heavily built steamerducks occur in coastal waters of southern South America and the Falkland Islands. The four species are extremely similar and are the most difficult of all waterfowl to identify in the field. Only one species is able to fly, though not all individuals can do so. The three flightless species do not overlap in range so for identification purposes one has only to distinguish each one from the Flying Steamerduck. As many other species of waterfowl occur in coastal waters on occasions it is recommended that observers refer to a specialist work on wildfowl when faced with species which are not illustrated here.

DUCKS AND GEESE
Family: Anatidae

1. Kelp Goose *Chloephaga hybrida*
Male (not illustrated) entirely white with black bill and yellow feet. Female sooty-brown with white tail and vent, white barring on underparts, large white patches in wing and pink bill. Juvenile similar to female. Rocky coasts of Chile and Argentina, also Falkland Is. Four other species of *Chloephaga* are all non-maritime. 76cm. **Cape Barren Goose** *Cereopsis novae-hollandiae.* Large all grey goose with black primaries and tail and yellowish-green bill. Shoreline and coastal grasslands of S Australia. 90cm.

2. Crested Duck *Anas specularioides*
Both sexes have brownish plumage with dark eye-patch. Widespread in coastal areas of southern S America and Falkland Is. 65cm.

3. Flying Steamerduck *Tachyeres patachonicus*
Large, bulky, mainly grey duck of coastal Chile and Argentina; also Falkland Is. Wings longer than other species. Male has pale grey head and yellowish bill, female with browner head and grey bill. 69cm. Three other very similar flightless species are extremely hard to separate: **Flightless Steamerduck** *T. pteneres.* Both sexes have prominent orange-yellow bills. Chile and Argentina. 79cm. **Chubut Steamerduck** *T. leucocephalus* Chubut coast, Argentina. 68cm. **Falkland Steamerduck** *T. brachypterus.* Similar to Flying Steamerduck but larger. Falkland Is. 68cm.

4. Steller's Eider *Polysticta stelleri*
Male in breeding plumage distinctive. Female (not illustrated) dark brown with white borders to blue speculum and 'bump' on rear of head. Smallest eider. Arctic coasts, wintering south to Baltic Sea. 45cm.

5. King Eider *Somateria spectabilis*
Male in breeding plumage distinctive. Female (not illustrated) barred warm brown, very similar to female Common Eider. Arctic coasts dispersing south in winter. 58cm. **Spectacled Eider** *S. fischeri.* Breeding male black below, white above with green head and white 'goggle' patch around eyes. Female brown with pale 'goggles'. Bill cloaked with feathering. Arctic coast of NE Siberia and Alaska. 55cm.

6. Common Eider *Somateria mollissima*
Male in breeding plumage distinctive **(6a)**. Female barred brown **(6b)**. Widespread in Arctic and northern temperate waters S to 40°N. 61cm.

7. Surf Scoter *Melanitta perspicillata*
Male black with white head patches and distinctive bill. Female (not illustrated) sooty-brown with two pale patches on face and heavy bill. N America, vagrant in W Europe. 50cm.

8. Velvet/White-winged Scoter *Melanitta fusca*
Male black with small white patch below eye. Both sexes have a white wing patch (often obscured when swimming). Female (not illustrated) similar to Surf Scoter but bill smaller. Widespread in N hemisphere. 55cm.

9. Common/Black Scoter *Melanitta nigra*
Male all black with yellow base to bill. Female (not illustrated) dark brown with distinctive pale face and neck. Widespread in N hemisphere. 50cm.

10. Red-breasted Merganser *Mergus serrator*
Male distinctive **(10a)**. Both sexes have a crest. Female mainly grey with brownish head **(10b)**. Widespread N hemisphere. 55cm. **Goosander/Common Merganser** *M. merganser.* Male has pinkish-white underparts and greenish head lacking crest. Female differs from Red-breasted Merganser in clear demarcation between grey body and brownish head. Widespread N hemisphere, avoids open sea. 65cm. **Smew** *M. albellus.* Male distinctive, mainly white with black face mask; female grey with white throat and reddish head. Occurs locally in coastal waters. Widespread in Old World. 41cm.

11. Hooded Merganser *Lophodytes cucullatus*
Male distinctive. Female (not illustrated) dark brown with paler bushy crest. Coastal lagoons and estuaries. N. America. 46cm.

12. Common Goldeneye *Bucephala clangula*
Adult male has round white spot on bottle-green head **(12a)**. Brown-headed female appears all dark at a distance **(12b)**. Widespread N hemisphere. 50cm. **Barrow's Goldeneye** *B. islandica.* Differs from Common Goldeneye in steeper forehead and elongated white spot on purple-glossed head. Female very similar to Common Goldeneye, best distinguished by different head shape. N America, Greenland and Iceland. 53cm.

13. Bufflehead *Bucephala albeola*
Male has distinctive head pattern. Female (not illustrated) like small Goldeneye with a white patch behind eye. Widespread in N America. 35cm.

14. Harlequin Duck *Histrionicus histrionicus*
Male unmistakable. Female (not illustrated) brown with white face patches. N America, Iceland and E Asia. 41cm.

15. Long-tailed Duck *Clangula hyemalis*
Male distinctive in both summer **(15a)** and winter **(15b)** plumages. Female duller, lacks long tail; head all dark brown in summer, in winter **(15c)** sides of head white with dark patch. Widespread in Arctic and northern temperate waters. 40cm (male has 13cm tail projection).

10b
15c
6b
15b
12b
10a
12a
6a
8
5
4
1
9
2
3
4
5
7
6a
8
10b
9
10a
11
12b
12a
15a
13
14
15b

PLATE 37: FRIGATEBIRDS AND SKUAS

The frigatebirds are a group of five rather similar, highly aerial, piratical seabirds. They share a distinctive shape and structure having extremely long angular wings and a long, deeply forked tail which appears pointed when closed. They have long, thin, hooked bills and the males possess an inflatable gular pouch which can be blown up to form a huge scarlet ball during courtship display. Frigatebirds are barely able to swim and lack adequate waterproofing. The similarity between the species, together with sexual dimorphism and a range of immature plumages, makes identification very difficult. Careful observation of the pattern of the breast and belly, including any spur marks projecting onto the underwing, is necessary, but these details can be difficult to make out at times. The *Stercorarius* skuas (jaegers) all breed in the arctic or extreme northern temperate zone, sometimes far inland, and are largely pelagic in winter when they disperse as far as the southern oceans. Each species has a distinctive tail shape formed by elongated central tail feathers in breeding plumage, but these feathers are often lost or broken in winter. They are all piratical at times and are often found in association with other seabirds particularly gulls or terns. Although the identification of adults in breeding plumage is fairly straightforward, immatures and winter plumaged adults can be very difficult to separate; size and jizz are important aids. The larger skuas belong to the *Catharacta* genus which, with the exception of Great Skua, has a southerly or Antarctic distribution. The taxonomy of this group is rather confused which makes identification difficult. All are large, powerful gull-like birds with strongly hooked bills. Most species feed on carrion or by scavenging but they are also capable of snatching food from the surface of the water and, in the case of Great Skua, by harrying other seabirds, either to force them to give up their catch or (sometimes) to kill and eat them. Identification depends largely on plumage tones, but is often very difficult and may be impossible at times, particularly as some forms are known to hybridise. Distribution is a help but several species overlap during the non-breeding season.

FRIGATEBIRDS
Family: Fregatidae

1. Magnificent Frigatebird *Fregata magnificens*
Adult female black with white breast and collar and white-tipped axillaries forming 3-4 wavy bars on underwing. Adult male (not illustrated) entirely black, red pouch and blackish feet, lacking pale bar on upperwing. Tropical and subtropical coasts of America, rare off W Africa. 101/238cm.

2. Great Frigatebird *Fregata minor*
Adult male entirely black with red feet and pouch and pale bar on upperwing **(2a)**. Adult female differs from other frigatebirds in black axillaries **(2b)**. 2nd stage female has white breast and head **(2c)**. IO and tropical PO, also tropical eAO. 93/218cm.

3. Lesser Frigatebird *Fregata ariel*
Both sexes show white spur on axillaries at all ages. Adult female (not illustrated) also has white breast. IO and tropical wPO, also tropical eAO. 76/184cm. **Ascension Frigatebird** *F. aquila*. Adult male probably indistinguishable from Magnificent. Females dimorphic, all dark form like male but brownish collar and upper breast. Rare light morph has white patch on breast and belly. Endemic to Ascension Is., and the only frigatebird likely to occur in tropical wAO. 91/198cm. **Christmas Island Frigatebird** *F. andrewsi*. Adult male has diagnostic white lower belly, adult female has dark spurs at sides of breast and white spur on underwing. Breeds on Christmas Island disperses in IO around SE Asian coasts. 94/218cm.

SKUAS/JEAGERS
Family: Stercorariidae

4. Pomarine Skua/Jaeger *Stercorarius pomarinus*
Large, heavily built skua with blunt, spoon-shaped, twisted, central tail feathers in adults. Light phase has prominent breast band and some barring on flanks **(4a)**. Dark phase is uniform dark brown with conspicuous white wing flashes **(4b)**. Juvenile has heavily barred rump, underwing and undertail-coverts, diagnostic double white flash on underwing and pale bill with dark tip **(4c)**. Circumpolar arctic breeding range dispersing to southern oceans in winter. 56/124cm.

5. Long-tailed Skua/Jaeger *Stercorarius longicaudus*
Small, delicate, graceful skua with reduced white wing flash and very long tail streamers in adults. Adult is greyish-brown above with diagnostic dark trailing edge to upperwing, breast white, lacking dark band and merging into dusky lower belly and vent **(5a)**. Sooty-brown dark phase juvenile with heavily barred undertail and underwing-coverts, short tail projections and slim bluish bill **(5b)**. Light phase juvenile has greyish head and breast with white belly, distinctly barred like dark phase juvenile **(5c)**. Circumpolar arctic breeding range, dispersing to southern wPO and AO in winter. 54/111cm.

6. Arctic Skua/Parasitic Jaeger *Stercorarius parasiticus*
Medium-sized, very agile skua with short pointed tail projections in adults. Light phase usually shows darker breast band, flanks unbarred **(6a)**. Dark phase uniform dark brown sometimes with paler cheeks and medium-sized white wing flash **(6b)**. Juvenile variable, rusty-brown to sooty-brown with indistinct barring on rump, underwing and undertail-coverts **(6c)**. Circumpolar arctic breeding range dispersing to southern oceans in winter. 45/117cm.

7. Great Skua *Catharacta skua*
Large size, mainly brown plumage with golden shaft-streaks, indistinct darker cap and prominent white wing flash **(7a)**. Juvenile more uniform, lacking shaft-streaks **(7b)**. Breeds coasts of neAO, dispersing S to W Africa. 58/150cm. **Antarctic/Southern Skua** *C. antarctica*. Similar to Great Skua but more uniform greyer-brown plumage. Breeds Falkland Is., S Argentina, Gough and Tristan. 63/?cm. **Brown Skua** *C. lonnbergi*. Very similar to Antartic Skua (sometimes considered conspecific) but slightly larger with heavier bill and lacks dark cap. Juveniles have rufous-brown wash on underparts. Breeds Antarctic Peninsula and subantarctic islands dispersing north to about 30°S. 65cm. **Chilean Skua** *C. chilensis*. Like Antarctic Skua but usually darker with dark grey cap and cinnamon underparts and underwing-coverts. Juvenile similar to adult but underparts brick-red. Breeds Chile and Argentina, dispersing north to Peru. 58/?cm.

8. South Polar Skua *Catharacta maccormicki*
Polymorphic. Light morph has pale head, neck and underbody. Underwing-coverts are dark and contrast with rest of underwing **(8a)**. Juvenile similar but less distinct. Dark phase like Great Skua with finer pale streaking on neck **(8b)**. Intermediate morphs occur showing a pale collar on hindneck. Breeds Antarctica dispersing widely to nPO and nAO, vagrant in neAO and nIO. 53/127cm.

1
2a
2b
2c
3
4a
4b
4c
5a
5b
5c
6a
6b
6c
7a
7b
8a
8b

PLATE 38: GULLS I

Gulls are mainly medium to large birds with fairly long legs and powerful hook-tipped bills. Since many species occur inland and have a close association with man they are among the most familiar of seabirds. The large number of similar looking species is intimidating at first but the species fall into four broad groups: the smaller 'hooded' species, those with white heads and grey backs, those with white heads and black backs and a small group of dissimilar but highly distinctive species such as Swallow-tailed and Sooty Gulls. Distribution is often helpful in eliminating similar species, though the possibility of a vagrant should not be forgotten. Bill and leg colour together with the pattern of the wing-tips are the most important characters. The 'hooded' species lose their hoods in winter plumage. Immature gulls, particularly juveniles, are more difficult to identify though they are often found in association with adults of the same species. Most gulls are opportunistic feeders and mostly occur in inshore waters and inland.

GULLS
Family: Laridae

1. Pacific Gull *Larus pacificus*
Large black-backed gull with distinctive black tail band, massive red-tipped yellow bill and yellow legs. Upperwing lacks white mirrors. Immature distinguished from similar Kelp Gull by the heavy pinkish bill with a dark tip. SW Australia – Tasmania. 62/147cm.

2. Kelp Gull *Larus dominicanus*
Large black-backed gull with white tail, white mirrors at wing-tips and yellow bill with red spot on gonys (distinctive within normal range) **(2a)**. Distinguished from Lesser Black-backed Gull by dark irides and olive legs. Juvenile (not to scale) has black bill and brownish-pink legs **(2b)**. Southern ocean coasts from Australia to S America, S Africa and Antarctica. 58/135cm.

3. Band-tailed Gull *Larus belcheri*
Large black-backed gull with broad black tail band, dark-tipped yellow bill and yellow legs. Upperwing lacks white mirrors. Non-breeding birds have distinctive brown hood (browner second-winter has more extensive hood). Juvenile has dark legs and dusky brown on head extends to belly. Coasts of Peru and Chile. 52/124cm. **Olrog's Gull** *L. atlanticus* is very similar to Band-tailed Gull but ranges do not overlap. Rare breeder on coast of Argentina.

4. Slaty-backed Gull *Larus schistisagus*
Large black-backed gull with yellow bill and pink legs. Upperwing dark slaty with a broad white trailing edge, usually separated from black wing-tips by a narrow white band **(4a)**. First-winter mottled brown with whitish head and a dark tail band **(4b)**. Coast of Asiatic nPO. 64/147cm.

5. Western Gull *Larus occidentalis*
Very similar to Slaty-backed Gull (bare part colours same) but grey upperwing merges into black outer primaries. Northern birds have paler grey upperparts and dark eyes. First-winter is darker brown with greyish head and largely black tail. By the third winter grey mantle contrasts with a white rump. Pacific coast of N America. 64/137cm. **Yellow-footed Gull** *L. livens*. Very similar to Western Gull (sometimes considered conspecific) with a darker mantle and yellow legs. Gulf of California. 69/152cm. **Black-tailed Gull** *L. crassirostris*. Medium-sized dark-backed gull with black subterminal tail band and yellow bill with dark tip. Upperwing lacks mirrors. Only dark-backed gull with a tail band in the region. Juvenile has all black tail contrasting with white rump and mottled brown plumage. E Siberia and Japan winter south to Hong Kong. 47/120cm.

6. Dolphin Gull *Larus scoresbii*
Medium-sized dark-backed gull with diagnostic heavy blood-red bill and red legs. Adult assumes grey head in winter. Juvenile brownish above with black subterminal tail band and dusky bill. Coasts of Chile, Argentina and Falkland Is. 44/104cm.

7. Grey Gull *Larus modestus*
Medium-sized grey gull with white head and trailing edge to wing, black bill and legs. In winter head becomes brownish. Juvenile dark brownish-grey with paler edges to wing coverts forming pale patch on closed wing. W coast of S America. 46/?cm. **Lava Gull** *L. fuliginosus*. Large grey gull with a sooty hood, pale grey tail, white eye-crescents, black legs and black bill with reddish tip. Juvenile sooty brown with white crescent on rump. Endemic to Galapagos Is. 53/?cm.

8. Great Black-backed Gull *Larus marinus*
Largest black-backed gull with yellow bill and pink legs. Upperwing black with white trailing edge and large mirrors **(8a)**. Juvenile (not to scale) has dark bill and whiter head than similar species **(8b)**. Second-winter (not to scale) similar but has grey mantle **(8c)**. nAO (into polar seas), wintering south to MS and Florida coast. 75/160cm.

9. Lesser Black-backed Gull *Larus fuscus*
Large black- or grey-backed gull with yellow bill and legs. Black-backed form very similar to Great Black-backed Gull but differs in size and leg colour. Paler southern form has slaty-grey mantle and upperwing, contrasting with black wing-tips which show small white mirrors **(9a)**. Juvenile (not to scale) has pink legs and blackish bill. Head dusky, tail shows distinct dark band and upperwing has blackish trailing edge **(9b)**. Breeds NW Europe, dispersing south to E and W Africa, occasionally to E coast of USA. 56/140cm.

10. Sooty Gull *Larus hemprichii*
Distinctive medium-sized dark-backed gull with dark brown hood, dull yellow legs and dark-tipped yellow bill. Upperwing dark grey with narrow white trailing edge and black outer primaries. In winter upperparts become sooty-brown. Juvenile lacks dark hood, upperparts paler grey including upperwing-coverts, black tail contrasts with white rump and bill bluish with dark tip. Second-winter has black subterminal band on upperwing and indistinct tail band. Red Sea and nwIO south to Tanzania. 45/112cm. **White-eyed Gull** *L. leucophthalmus*. Smaller and paler than Sooty Gull with long black-tipped reddish bill and bright yellow legs. Hood black with striking white eye-crescents. Juvenile has a dark bill, greenish legs, dusky hood and a dark uniform brown upperwing. Endemic to Red Sea, some dispersal to NE Africa and Arabian Sea. 39/108cm.

11. Heermann's Gull *Larus heermanni*
Distinctive medium-sized dark-backed gull unmistakable within range. Adult has white head and underparts, black-tipped red bill and black legs **(11a)**. Juvenile entirely sooty-brown with black-tipped yellowish bill and dark legs. Recalls skua spp. but lacks white primary flashes **(11b)**. British Columbia to S Mexico, breeds Gulf of California. 49/130cm.

1
2a
2b
3
5
4a
6
4b
7
8a
8c
10
8b
9b
11a
9a
11b

PLATE 39: GULLS II

GULLS
Family: Laridae

1. Herring Gull *Larus argentatus*
Large grey-backed gull with black wing-tips showing small white mirrors. Bill yellow with red spot on gonys, legs pink, irides yellow **(1a)**. Underwing pattern helpful to separate adult Herring **(1d)** and Lesser Black-backed Gull **(1e)** in flight. In winter head streaked brown. Juvenile (not to scale) mottled brown with dusky head, blackish tail and dark bill. Separated from Lesser Black-backed Gull by pale 'window' on inner primaries and narrow dark trailing edge to upperwing **(1b)**. Second-winter (not to scale) has grey mantle and pale 'window' on inner primaries. **(1c)**. Third-winter like adult with more black on the wing-tips and trace of tail band. Complex racial differences across widespread range in N hemisphere. 61/147cm. **Yellow-legged Gull** *L. cachinnans* differs in yellow legs, darker mantle and slightly larger size. White 'mirror' reduced to single spot and underwing dusky grey. Head remains white in winter. Breeds MS – C Asia. **Armenian Gull** *L. armenicus* is smaller with more black on wing-tips, dark irides and a black band on shorter bill. Breeds inland SW Asia.

2. California Gull *Larus californicus* (not to scale)
Similar to Herring but smaller with yellowish legs, dark irides and (usually) black mark in front of red gonys. In winter head is spotted with brown. Juvenile has a black tip to pink bill, all dark tail and brownish rump. Breeds interior N America, winters on Pacific coast. 54/137cm.

3. Glaucous-winged Gull *Larus glaucescens*
Large grey-backed gull with long yellow bill with red spot on gonys, pinkish legs and slate grey tips to outer primaries. Juvenile has all dark bill and uniform light buffy-grey upperwing. Wing-tips barely project beyond tail at rest. Breeds nPO, wintering south to California. 65/147cm.

4. Glaucous Gull *Larus hyperboreus*
Large pale-backed gull with white outer primaries and sloping forehead. Legs pink, bill yellow with red spot on gonys **(4a)**. First-winter (not to scale) has distinctive pink bill with dark tip, and biscuit-coloured plumage **(4b)**. Wing-tips project slightly beyond tail at rest. Plumage becomes whiter in second year and by third year attains grey mantle. Circumpolar, mainly breeding north of Arctic Circle wintering south to California, Japan, Florida and France. 71/158cm. **Iceland Gull** *L. glaucoides*. Smaller and slimmer than Glaucous Gull with a rounded head and smaller bill. Plumage sequences identical. Juvenile has a dark bill which becomes pink at base by second winter. Wing-tips project well beyond tail at rest. Breeds Greenland, wintering south to NW Europe. 61/140cm. **Kumlien's Gull** *L. g. kumlieni*. Similar to Iceland Gull but shows grey markings near tips of outer primaries. Breeds Baffin Is., wintering south to Long Is., USA. 61/140cm. **Thayer's Gull** *L. g. thayeri*. Resembles Herring Gull but has dark irides and less extensive black on wing-tips. Underside of primaries whitish at all ages. Juvenile has dark bill and uniform greyish-brown primaries. Breeds Arctic Canada, winters south to Mexico. 59/140cm.

5. Laughing Gull *Larus atricilla*
Small dark-hooded, long-winged gull with drooping dark red bill and dull red legs. Upperwing dark grey with black wing-tips **(5a)**. In winter has smudgy partial hood. First-winter has dusky breast and flanks, brown flight feathers and black tail band **(5b)**. Eastern N America to Venezuela, winters south to Brazil and Chile. 40/103cm.

6. Franklin's Gull *Larus pipixcan*
Small dark-hooded gull with prominent white eye-crescents, red bill with black subterminal band and black legs. Upperwing grey with white band separating black outer primaries which have large white tips. In winter has dusky partial hood accentuating white eye-crescents. First-winter has similar head pattern, black tail with white outer tail feathers and white underparts. Breeds inland N America, wintering south to Peru and Chile. 35/90cm.

7. Grey-headed Gull *Larus cirrocephalus*
Medium-sized grey-hooded gull with red bill and legs and pale irides. Distinctive wing pattern **(7a)**. In winter shows trace of hood and faint ear-spot. First-summer has partial hood, dark 'M' mark on upperwing and dark tail spots **(7b)**. Juvenile has dusky head and complete black tail band. Breeds S America and Africa, mainly inland. 42/102cm.

8. Common/Mew Gull *Larus canus*
Medium-sized grey-backed gull. Adult has yellowish-green bill and legs, dark irides and prominent white 'mirrors' on upperwing **(8a)**. In winter head and breast streaked and variable dark band on bill. First-winter has dark grey mantle and clear-cut black tail band. Legs and bill pink, latter with dark tip **(8b)**. Juvenile browner. NW Europe to Siberia and western N America, wintering south to N Africa, Hong Kong and California. 49/124cm. **Ring-billed Gull** *L. delawarensis*. Similar but paler grey above with black band on heavier yellow bill, yellow irides and smaller white 'mirrors'. Spotted rather than streaked in winter. Immatures have black band on pink bill and less clear-cut tail band. Inland N America dispersing to coasts in winter. Vagrant in NW Europe. 45/124cm. **Audouin's Gull** *L. audouinii*. Large grey-backed gull, red bill with yellow tip and black subterminal band, greyish legs and dark irides. Upperwing grey with black primaries which have small white tips but no 'mirrors'. Juvenile has black tail band and dusky red bill. MS and NW Africa. 50/127cm.

9. Swallow-tailed Gull *Creagrus (Larus) furcatus*
Large, very distinctive dark-hooded gull. Striking wing pattern and forked tail unique. In winter head white with dark mark behind eye. Juvenile has scaly brown markings on mantle and upperwing-coverts and black terminal tail band. Mainly nocturnal. Breeds Galapagos Is. dispersing to coasts of western S America. 57/131cm.

10. Great Black-headed Gull *Larus ichthyaetus*
Large dark-hooded gull with long black-banded yellow bill and yellow legs. Distinctive wing pattern with large white tips to black-based primaries. In winter hood reduced to dark mark behind eye and streaks on hindneck. Juvenile has black tail band, grey mantle and pale wing panel. Breeds in C Asia, dispersing to NE Africa, SW Asia and India. 69/160cm.

11. Silver Gull *Larus novaehollandiae*
Medium-sized white-headed gull with red bill and legs and pale irides. Forewing white, primaries black with large white 'mirrors'. Juvenile has white head, dark bill and brown 'M' on upperwing. Australia, New Caledonia and New Zealand. 41/93cm. NZ form sometimes separated as **Red-billed Gull** *L.scopulinus*. **Black-billed Gull** *L. bulleri* is similar to Silver Gull but bill and legs black and white leading edge to upperwing. Juvenile has dark ear-spot and black-tipped pinkish bill. Endemic to New Zealand. 36/?cm. **Hartlaub's Gull** *L. hartlaubii*. Medium-sized gull with faint pale grey hood, red bill and legs and dark irides. Wing pattern similar to Grey-headed Gull. First-winter has blackish bill, little black on tail and no dark ear-spot. Endemic to coast of SW Africa. 38/91cm. (Other name: King Gull.)

1a
2
1b
1c
3
4a
5a
4b
5b
6
7a
9
7b
8a
10
11
8b
1d
1e

PLATE 40: GULLS III

This plate illustrates several somewhat similar 'hooded' gulls and some very distinctive species. Kittiwakes, Ross's, Sabine's and Ivory Gulls all breed in the arctic or temperate zone of the northern hemisphere and are largely pelagic outside the breeding season. The 'hooded' gulls mainly breed inland where they are associated with freshwater habitats. Most species disperse to coastal waters in the non-breeding season when they lose the dark hoods. Distribution is generally useful though a few species overlap in winter. The pattern of the upperwing is the most important feature to check when trying to identify any gulls in flight.

GULLS
Family: Laridae

1. Slender-billed Gull *Larus genei*
Medium-sized white-headed gull with a long dark red bill, red legs and pale irides. Forewing white with black primary tips. The long bill and sloping forehead give a distinctive jizz. Adult has underparts washed pink when breeding. In winter adult has very faint spot behind eye. Juvenile has a dark ear-spot, yellowish bill and legs and faint brownish 'M' on upperwing. MS, N Africa and SW Asia; breeds colonially in marshes, winters on coast and estuaries. 43/105cm.

2. Mediterranean Gull *Larus melanocephalus*
Medium-sized dark-hooded gull with conspicuous white eye-crescents, heavy deep red bill and red legs. Upperwing pale grey with white primary tips **(2a)**. In winter hood reduced to dark smudge behind eye. Juvenile has black legs, dark ear-spot, black tail band and brownish upperwing with pale central panel **(2b)**. Second-winter has white tail and black subterminal marks on outer primaries. English Channel and MS–Black Sea, dispersing south to NW Africa. 40/106cm. **Relict Gull** *Larus relictus*. Medium-sized dark-hooded gull with red bill and legs, conspicuous white eye-crescents and distinctive white wing-tips with black subterminal markings. First-winter has dark ear-spot, dark 'M' on upperwings and black tail band. Rare and little known. Breeds inland C Asia, occurs coasts of China and Vietnam in winter. 41/?cm.

3. Common Black-headed Gull *Larus ridibundus*
Small dark-hooded gull with slim dark red bill, red legs and white outer primaries with black tips **(3a)**. In winter hood reduced to dark ear-spot and faint crown markings. First-winter has dark-tipped pink bill, black tail band and brown 'M' on upperwing **(3b)**. N Eurasia dispersing south to Africa, India and SE Asia, and NE coast of USA. 38/104cm. **Brown-hooded Gull** *L. maculipennis*. Medium-sized dark-hooded gull closely resembling Common Black-headed Gull with less extensive black on outer primaries. S America from Brazil and Chile southwards. 37/?cm.

4. Brown-headed Gull *Larus brunnicephalus*
Medium-sized dark-hooded gull with red bill and legs, pale irides and black outer primaries with large white 'mirrors'. In winter hood reduced to dark ear-spot and patch on hindcrown. First-winter like winter adult but all black wing-tips and trailing edge to upperwing. Breeds inland C Asia. Disperses south to Asian coasts from Persian Gulf–Thailand. 42/?cm. **Andean Gull** *L. serranus*. Medium-sized dark-hooded gull with deep maroon bill and legs. Distinctive wing pattern with black band across white outer primaries. In winter hood reduced to dark ear-spot. First-winter has black tail band and dark 'M' on upperwing. Breeds High Andes, some dispersal to coasts of Ecuador–Chile in winter. 48/?cm.

5. Bonaparte's Gull *Larus philadelphia*
Small dark-hooded gull with thin black bill, red legs, white leading edge to wing and narrow black tips to primaries. In winter hood reduced to dark ear-spot and greyish patch on crown. First-winter like Common Black-headed Gull but darker carpal bar. Breeds inland N America, wintering south to Mexico. 31/82cm. **Saunders' Gull** *L. saundersi*. Small dark-hooded gull with short black bill, red legs and distinctive underwing pattern with dark wedge on outer primaries at all ages. In winter has small dark ear-spot and narrow band over crown. Rare and little known. Breeds China, in winter recorded from Vladivostok to Vietnam. 32/?cm. (Other name: Chinese Black-headed Gull.)

6. Little Gull *Larus minutus*
Smallest gull, with a black hood, black bill and red legs. Diagnostic dark underwing with white trailing edge **(6a)**. In winter hood reduced to dark ear-spot and hindcrown. First-winter has bold dark 'M' mark on upperwing, black tail band and pale underwing **(6b)**. Breeds inland, mostly N Eurasia and scarce in N America, wintering south to N Africa, MS and Japan. 27/64cm.

7. Black-legged Kittiwake *Rissa (Larus) tridactyla*
Medium-sized, white-headed, entirely marine gull with yellow bill, black legs and wholly black wing-tips. Underwing white with black tips **(7a)**. In winter has dusky ear-spot and greyish collar on hindneck. First-winter has black tip to forked tail, conspicuous dark 'M' mark on upperwings and black collar on hindneck **(7b)**. Almost circumpolar in Arctic, nPO and North Sea. Breeds colonially on rocky coasts, pelagic in winter. 41/91cm. **Red-legged Kittiwake** *R. (L.) brevirostris*. Differs in bright red legs, darker grey upperwing with broader white trailing edge and dusky underwing. First-winter lacks black tail band and collar. Upperwing grey with dark primaries, lacks dark carpal bar. Breeds Bering Sea, pelagic in winter in nPO. 38/85cm.

8. Ross's Gull *Rhodostethia rosea*
Distinctive small white-headed gull with black ring around head, black bill, red legs and diagnostic wedge-shaped tail. Upperwing pale grey with black outermost primary, underwing dusky with white trailing edge. Assumes pinkish tinge to underparts when breeding **(8a)**. In winter has dark crescent behind eye and lacks dark ring around neck. First-winter has dark 'M' mark on upperwing and black tip to tail **(8b)**. Breeds in high Arctic, E Siberia–Canada. Winters in arctic seas but vagrants occur south to Japan and NW Europe. 31/84cm.

9. Sabine's Gull *Xema (Larus) sabini*
Very distinctive small dark-hooded gull with black legs, yellow-tipped black bill, forked tail and distinctive tri-coloured wing pattern **(9a)**. In winter hood reduced to dusky patch or half-collar on hindneck. Juvenile has distinctive wing pattern and black tip to forked tail **(9b)**. Circumpolar in high Arctic, pelagic in winter south to western S America and SW African coast. 34/89cm.

10. Ivory Gull *Pagophila eburnea*
Very distinctive medium-sized entirely white gull with black legs and greyish bill with yellow tip **(10a)**. First-winter has dusky face and black spots on primaries, tail and upperwing-coverts **(10b)**. High Arctic, vagrant south to Newfoundland and NW Europe in winter. 43/110cm.

1
2a
3a
2b
5
3b
4
6a
6b
7a
8a
7b
8b
10a
9a
9b
10b

PLATE 41: LARGER TERNS I

Terns tend to be gregarious birds both in summer and winter. They often occur in mixed flocks and some breeding colonies contain several species, though usually in discrete areas. Most species feed by hovering above the surface before plunge-diving, though Gull-billed Terns also feed over land. They roost ashore often forming huge flocks on beaches. A few species are virtually confined to freshwater habitats. Terns generally have rather short legs, long pointed bills and deeply forked tails. Excluding the distinctive Inca Tern, all species have black caps, grey or blackish backs and white underparts. The main distinctions are size, the colour and shape of the bill and the pattern of the outer primaries, particularly on the underside. Those species which have shaggy caps or crests assume a non-breeding plumage where the black on the crown is greatly reduced in size.

TERNS
Family: Sternidae

1. Sooty Tern *Sterna fuscata*
Medium-sized tern, blackish above with white underparts and forehead. Juvenile entirely sooty-brown with pale tips to feathers of upperparts and white vent. Pantropical pelagic species, returning to land only for breeding. 43/90cm.

2. Bridled Tern *Sterna anaethetus*
Smaller and greyer than Sooty Tern. White forehead extends to form narrow line over eye and dark cap separated from mantle by pale collar. Juvenile has whitish underparts and less scaly appearance than Sooty Tern. Widespread in tropics and subtropics, less pelagic than Sooty Tern. 36/76cm.

3. Aleutian Tern *Sterna aleutica*
Distinctive black-capped tern with white forehead. Back, upperwings and underparts grey, rump and tail white. Juvenile has crown and mantle washed brown, rump and tail grey. Breeds Siberia and Alaska, winter quarters unknown. 36/78cm. **Grey-backed Tern** *S. lunata*. Similar to Bridled Tern but has grey mantle contrasting with brown upperwing and tail. Juvenile like Bridled Tern but upperparts more scaly. Tropical cPO. 36/74cm. (Other name: Spectacled Tern.)

4. Caspian Tern *Sterna caspia*
Very large tern with heavy black-tipped red bill and distinctive dark underside of outer primaries. Black cap becomes streaked in winter, sometimes reduced to an eye-streak. Immature has bill more orange and upperparts scaly brown. Widespread, except arctic and S America. Coasts and inland. 54/134cm.

5. Royal Tern *Sterna maxima*
Large tern with shaggy black cap and a large orange bill. Underwing shows dark tips to outer primaries not forming a wedge **(5a)**. Complete black cap only present for short period, otherwise has an extensive white forehead. Immature has dark carpal bar and pale panel in centre of upperwing **(5b)**. Breeds coasts of southern USA, Caribbean, S America and W Africa. 50/109cm.

6. Great Crested Tern *Sterna bergii*
Large tern with shaggy crest, a small white forehead (more extensive in winter) and a large drooping yellow bill **(6a)**. Juvenile has brownish upperwing with dark wedge on primaries and duller yellowish bill **(6b)**. Widespread on tropical and subtropical coasts from S Africa to cPO. 46/104cm. (Other name: Swift Tern.) **Chinese Crested Tern** *S. bernsteini*. Similar to larger Great Crested Tern but has black-tipped yellow bill. Virtually extinct, once occurred from China in summer to Philippines in winter. 38/?cm. **River Tern** *S. aurantia*. Similar to Great Crested Tern but has a complete black cap and red legs. Yellow bill is straight and upperwing dark grey. Breeds along rivers in India and SE Asia, occasionally strays to coast. 40/?cm. (Other name: Indian River Tern.) **Large-billed Tern** *Phaetusa simplex*. Very distinctive tern with huge drooping yellow bill, dark grey upperparts and black cap. Striking upperwing pattern with black primaries and grey forewing-coverts separated by diagonal white band. Freshwater, Columbia–Argentina. 37/92cm.

7. Lesser Crested Tern *Sterna bengalensis*
A medium-sized tern with shaggy crest and orange bill. Upperside of primaries silvery-white, rump grey. Juvenile similar to Royal Tern differing mainly in smaller size. sMS and Red Sea to S and W Africa, IO and swPO. 40/92cm.

8. Sandwich Tern *Sterna sandvicensis*
A medium-sized crested tern with shaggy black crest and a yellow-tipped black bill. Upperparts pale grey with white rump and tail. Dark outer primaries form wedge on upperwing **(8a)**. In winter crown mottled with white. Juvenile has all black bill, white forehead and dark barring on mantle and upperwing-coverts **(8b)**. Breeds N Atlantic coasts, also Black and Caspian Seas. Disperses to sAO, nwIO and Pacific coast of tropical America. 43/92cm. **Cayenne Tern** *S. s. eurygnatha*. As Sandwich Tern but bill usually all yellow, though sometimes black with yellow tip. Feet and legs sometimes yellow. Caribbean – Argentina. 42/96cm.

9. Gull-billed Tern *Sterna (Gelochelidon) nilotica*
Similar to Sandwich Tern but lacks crest and has heavy, gull-like, all black bill. Upperparts pale grey including rump and tail, primary tips darker grey **(9a)**. In winter head white with variable dark mark behind eye. Juvenile has prominent black mark behind eye and brownish mottling on mantle and upperwing-coverts **(9b)**. Feeds by hawking for insects over land, occasionally plunge-diving. Virtually cosmopolitan south of 50° N; occurs inland and on coasts. 39/94cm.

10. Elegant Tern *Sterna elegans*
Medium-sized tern with shaggy black crest and slim, decurved orange bill. Smaller and slimmer than Royal Tern with shaggier crest. In winter crest reduced to band from eye to hindcrown. Immature has yellowish bill and brownish mottling on upperparts. West coast of America from California – Chile. 41/86cm.

11. Inca Tern *Larosterna inca*
Very distinctive tern. Plumage entirely dark grey with blacker cap, white trailing edge to wing, red bill and legs and white facial plume. Juvenile and immature browner with dusky bill and facial plume reduced or lacking. Coasts of Ecuador – Chile. 41/?cm.

1
2
3
4
5a
6a
7
8a
11
9a
9b
8b
6b
5b
10

PLATE 42: LARGER TERNS II

Several species of medium-sized, black-capped terns are very similar in appearance. When at rest or seen at close range the shape and colour of the bill, leg length and length of tail streamers are important, but in flight the pattern of the outer primaries on both upper- and underside is more useful. Several species may occur together, particularly at roosts, and the situation is further complicated by distinct winter and juvenile plumages. Range is helpful in some cases but several species are long-distance migrants and could occur virtually anywhere. Some Arctic Terns reach the Antarctic during the austral summer but they tend to feed around the pack ice whilst the Antarctic Terns are breeding at this season and remain inshore. White-cheeked Tern poses a problem in winter when it overlaps with other similar species. The remaining species have a distinctive appearance or rather restricted ranges.

TERNS
Family: Sternidae

1. Roseate Tern *Sterna dougallii*
Medium-sized tern with black cap, dark red bill (appears blackish), bright orange legs and very long tail streamers. Thin dark wedge on upperwing formed by dark outer 2-3 primaries, underwing all white. Pinkish flush on underparts when breeding **(1a)**. In winter forehead white and bill darker. Juvenile has black bill and legs, dark barring on mantle, dark crown, obvious dark carpal bar and dark outer primaries on upperwing **(1b)**. Widespread nAO, IO and wPO. 39/78cm.

2. Common Tern *Sterna hirundo*
Medium-sized tern with black cap, long red bill with black tip, long red legs and tail streamers equal to wing-tips at rest. Dark wedge on upperwing formed by contrasting outer primaries. Narrow black tips to primaries merge into white underwing **(2a)**. In winter forehead white, bill darker (sometimes black) and dark carpal bar visible on upperwing **(2b)**. Juvenile has brownish mantle, white forehead, pale orange base to dark bill, orange-red legs, greyish secondaries and a dark carpal bar **(2c)**. Widespread in N hemisphere inland and coastal, also W Africa. In winter south to S America, S Africa and Australia. 36/80cm.

3. Arctic Tern *Sterna paradisaea*
Medium-sized tern with black cap, short dark red bill, short red legs and long tail streamers extending beyond wing-tips at rest. Underparts greyish contrasting with white cheeks. Upperwing entirely grey lacking dark wedge. Narrow clear-cut black tips to primaries contrast with white underwing **(3a)**. In winter forehead and underparts white **(3b)**. Juvenile has all dark bill, red legs, white forehead and underwing with clear-cut black tips to outer primaries. Faint carpal bar on rather white upperwing and brownish tinge on mantle **(3c)**. Widespread in N hemisphere south to about 50°N. In winter south through all oceans to Antarctic pack ice. 36/80cm.

4. Antarctic Tern *Sterna vittata*
Medium-sized tern with black cap, bright red bill and legs. In breeding plumage has dark grey underparts contrasting with white cheeks **(4a)**. Similar to Arctic Tern though attains breeding plumage in austral summer so confusion unlikely. In winter has white forehead and paler underparts **(4b)**. Juvenile has black bill, grey legs, mottled brown crown and mantle and conspicuous dark carpal bar on upperwing. Breeds islands in sAO, sIO and PO south of New Zealand. In winter disperses north to about 25°S off S America and S Africa. 41/79cm. **Kerguelen Tern** *S. virgata*. Very similar to Antarctic Tern though smaller with finer more dusky bill, darker grey overall with more contrasting white cheeks. In winter has mottled white forehead and black bill. Juvenile has dark brown cap and darker brown upperparts than Antarctic Tern. Sedentary, breeding on islands in sIO. 33/75cm. **South American Tern** *S. hirundinacea*. Medium-sized tern with black cap and bright red bill and legs. Upperparts pale grey, underparts white. Dark wedge on upperwing formed by outer primaries. Bill longer than that of Antarctic Tern and both upper- and underparts much paler. In winter has white forehead. Juvenile mostly dusky brown with mottled upperparts and whitish forehead. S American coasts from Peru – Brazil. 42/85cm.

5. Black-fronted Tern *Chlidonias* (*Sterna*) *albostriatus*
Small tern with black cap, bright orange bill and legs, greyish underparts and contrasting white cheeks. Flight and jizz closely resembles Whiskered Tern which does not overlap in range. Confined to New Zealand; breeds inland, dispersing to coasts in winter. 32/?cm.

6. White-cheeked Tern *Sterna repressa*
Medium-sized tern with black cap, black-tipped red bill and long red legs. Upperparts dull grey including rump and tail (all other similar species have white rump and tail). Underparts darker grey contrasting with white cheeks. In winter (illustrated) forehead white and underparts paler. Juvenile has black bill, yellowish-brown legs and extensive dark carpal bar on upperwing. Red and Arabian Seas, also coasts of E Africa and W India. 33/79cm. **Black-bellied Tern** *S. acuticauda*. Distinctive, rather small tern with black cap and belly, bright yellow bill and pale red legs. White cheeks contrast with black cap and greyish breast. Freshwater; Indian subcontinent–SE Asia. 31/?cm.

7. Forster's Tern *Sterna forsteri*
Medium-sized tern with black cap, black-tipped orange bill and long orange legs. Distinctive upperwing pattern with silvery inner primaries lacking any dark wedge. In winter (illustrated) bill mainly black and large black patch behind eye. Juvenile has brown crown, black patch behind eye and dark carpal bar on upperwing. Breeds N America dispersing south to Caribbean and C America. Vagrants reach NW Europe. 37/80cm.

8. White-fronted Tern *Sterna striata*
Medium-sized tern with heavy black bill, dark red legs and white forehead which merges into black cap. Upperparts very pale grey (appearing white at distance); underparts white including underwing which lacks dark trailing edge. In winter crown white. Juvenile has dark brown barring on mantle and upperwing-coverts and prominent dark carpal bar. Breeds in New Zealand dispersing to SE Australia. 41/76cm.

9. Snowy-crowned Tern *Sterna trudeaui*
Very distinctive medium-sized tern with almost white upperparts, dark patch behind eye, greyish belly and heavy yellow bill with black band near tip. Upper- and underwings virtually white lacking any dark markings. Juvenile has greyish sides of face and grey-brown mottling on mantle and upperwing. Inland and coastal, Uruguay, Argentina and Chile. 33/77cm. (Other name: Trudeau's Tern.)

10. Black-naped Tern *Sterna sumatrana*
Distinctive, white tern with black line from eye around hindneck. Bill and legs black. When breeding has pink flush on otherwise white underparts. Upper- and underwings almost white with black outerweb of outer primary. Juvenile has dark brown barring on mantle and dark carpal bar on upperwing. Tropical IO and wPO. 31/61cm.

1a
1b
2a
3a
2b
3b
3c
2c
4b
6
4a
5
7
8
9
10

PLATE 43: SMALL TERNS, NODDIES AND SKIMMERS

The 'marsh terns' *Chlidonias* spp. form a distinct group, differing from *Sterna* terns in having much shallower tail-forks, smaller bills and a distinctive manner of flight. During the breeding season they frequent inland freshwater habitats but on migration and in winter all three may occur at sea, though usually in sheltered inshore waters. In winter plumage they are more difficult to separate, particularly Whiskered Tern which closely resembles several species of *Sterna* terns. The Little Tern group is confusingly similar but most species do not overlap in range. Bill and leg colour and the amount of dark on the outer primaries are key features. The White Terns are extremely distinctive tropical species. Noddies differ from other terns in having wedge-shaped tails and, in contrast to *Sterna* terns, they have dark bodies and pale caps (Blue-grey Noddy excepted). The three dark species are extremely similar and difficult to separate. Size, overall plumage tone and the pattern of the cap and lores are the key features. Skimmers are very distinctive and have a highly specialised feeding technique which involves ploughing the surface of the water with the specially adapted lower mandible. The three species are allopatric and can be found on freshwater and sheltered coastal waters.

SMALL TERNS
Family: Sternidae

1. Black Tern *Chlidonias niger*
Distinctive in summer with blackish head and body contrasting with white underwing-coverts and undertail-coverts, grey upperwing, rump and tail. Bill blackish **(1a)**. In winter white below, uniform pale grey above with black hindcrown separated from mantle by white collar. Leading edge of upperwing is contrastingly dark grey **(1b)**. Juvenile similar, but has brownish mantle, dark carpal bar and conspicuous dark pectoral patches **(1c)**. Breeds N America, Europe and C Asia, dispersing to tropical coasts south to S Africa in winter. 23/66cm. **White-winged Tern** *C. leucopterus*. Very distinctive in summer with black body and underwing-coverts contrasting with whitish wings, rump, vent and tail. Short red bill. In winter like Black Tern but dark outer primaries and secondaries contrast with grey wings, often with some black on underwing-coverts. Juvenile differs from Black Tern by contrasting dark mantle and white rump; also lacks pectoral patches. Breeds from E Europe–China, dispersing south to Africa, SE Asia and Australia in winter, mainly on freshwater. 23/66cm. (Other name: White-winged Black Tern.)

2. Whiskered Tern *Chlidonias hybridus*
Most *Sterna*-like marsh tern. In summer has black cap, contrasting white cheeks, grey body and wings. Belly darker grey contrasting with pale underwing. Bill rather long, deep red **(2a)**. In winter uniform pale grey above with white underparts. Black crown flecked with white and no pale hind collar **(2b)**. Juvenile has brown mantle, smudgy dark cap which merges into mantle, and lacks dark carpal bar and pectoral patches **(2c)**. Breeds inland throughout most of Old World and Australia, dispersing south in winter; freshwater and coastal. 25/69cm.

3. Little Tern *Sterna albifrons*
Small *Sterna* tern with yellow legs, white rump and tail and dark outer primaries. In summer has black tip to yellow bill and white forehead which extends to form narrow stripe over eye **(3a)**. In winter bill blackish and crown mottled white **(3b)**. Juvenile has brownish mottling on upperparts. Rapid flight with fast wing beats and frequent hovering. Widespread on coasts and rivers from Eurasia to Australia and parts of Africa. 24/52cm. **Least Tern** *S. antillarum* is very similar but rump and tail are greyish and the call is higher and louder. N and C America, wintering south to Brazil. 23/51cm. **Peruvian Tern** *S. lorata*. Greyer overall than Little Tern including rump and tail. Coasts of Peru and Chile. 23/50cm. **Yellow-billed Tern** *S. superciliaris*. Like Little Tern but bill entirely yellow at all seasons. Inland eastern S America. 23/50cm. **Saunders's Tern** *S. saundersi* differs from Little Tern in darker outer primaries, brownish legs and lacks stripe over eye. nwIO 23/51cm. **Fairy Tern** *Sterna nereis*. Like Little Tern but orange bill and legs and white lores. Coastal W and S Australia. 25/50cm.

4. Damara Tern *Sterna balaenarum*
Small black-capped tern with black bill and dark legs. In winter (not illustrated) forehead and crown are white. Namibia to Gulf of Guinea coasts. 23/51cm.

5. Common White Tern *Gygis alba*
Very distinctive, all white tern. Bill slightly upturned, black with blue base. Juvenile has dark ear-spot and brownish cast to upperparts. Uniquely lays single egg on branch of tree. Pantropical. 31/80cm. (Other name: Fairy Tern.) **Little White Tern** *G. microrhyncha* breeds from Marquesas Is.–Easter Is. (recently separated as full species). 28/70cm.

NODDIES
Family: Sternidae

6. Blue-grey Noddy *Procelsterna cerulea*
Dark morph distinctive, uniform light blue-grey with darker wing-tips **(6a)**. Pale morph is mainly whitish-grey with greyer flight feathers. Juvenile has brownish wash and darker flight feathers **(6b)**. Tropical cPO. 27/60cm. (Other name: Grey Noddy.)

7. Brown Noddy *Anous stolidus*
Medium-sized brown tern with a whitish cap, sharply demarcated above the eye and separated from the bill by a band of dark feathers. Paler upper- and underwing-coverts contrast with darker flight feathers. Immature and juvenile have dark crowns, the latter with pale tips, and dark back and upperwing-coverts. Pantropical. 42/82cm.

8. Black Noddy *Anous minutus*
Smaller than Brown Noddy with finer bill, uniform blackish-brown plumage, whiter cap reaching base of bill and whiter hindneck. Upperwing and underwing uniform dark brown. Immature and juvenile show white forehead. Tropical and subtropical AO and PO. 34/76cm.

9. Lesser Noddy *Anous tenuirostris*
Smallest noddy with greyish-white cap reaching base of bill and extending below eye where it merges with brown plumage. Lacks contrast on upperwing and underwing. Juvenile similar but cap may be whiter. IO. Breeds Seychelles, Maldives and off W Australia. 32/60cm.

SKIMMERS
Family: Rynchopidae

10. Black Skimmer *Rynchops niger* **477**
Unmistakable. Juvenile has scaly upperparts. Coasts and rivers of America from N USA–Argentina, also Caribbean. 46/112cm. **African Skimmer** *R. flavirostris*. Like Black Skimmer but lacks dark tip to bill. Africa from S Egypt–Natal. 38/106cm. **Indian Skimmer** *R. albicollis*. Differs from other skimmers in white collar on hindneck. Iran–Indochina. 43/108cm.

1c
1b
2a
2b
2c
3a
3b
1a
4
5
6a
7
7
6b
8
8
9
10
10

PLATE 44: AUKS

A distinctive group of rather thickset dumpy birds with rapid wing beats, confined to the colder waters of the northern hemisphere. Identification is straightforward at breeding colonies but difficult in winter or at sea.

AUKS
Family: Alcidae

1. Razorbill *Alca torda*
Large heavily built auk with black head and upperparts, white below. Heavy blunt-tipped bill has diagnostic white stripes. Tail long and pointed **(1a)**. In winter plumage has white throat and sides of face **(1b)**. Breeds North Atlantic coasts, wintering south to MS. 43/64cm.

2. Black Guillemot *Cepphus grylle*
Medium-sized. Breeding plumage entirely black with white wing patches, red legs and gape **(2a)**. In winter barred grey above, white below **(2b)**. Circumpolar from Arctic to NE USA and NW Europe. Scarce in Bering Sea where range overlaps with Pigeon Guillemot. 33/58cm. **Pigeon Guillemot** *C. columba*. Very similar to Black Guillemot, but white wing patch broken by black wedge and underwing dusky. S Kamchatka form, *C. c. snowi,* has wing patch reduced to narrow white lines. Widespread nPO, wintering south to Japan and California. 32/58cm.

3. Spectacled Guillemot *Cepphus carbo*
Distinctive sooty-black breeding plumage with white 'spectacles' and post-ocular stripe. In winter underparts white including throat. Underparts barred in transitional plumage. nwPO from Kamchatka to Japan. 38/?cm.

4. Common Guillemot/Murre *Uria aalge*
Large auk with long pointed bill, chocolate brown head and upperparts, white below with dark streaks on flanks **(4a)**. A 'bridled' form with white eye-ring and post-ocular stripe occurs in nAO. In winter throat and sides of face white with diagnostic dark post-ocular stripe **(4b)**. nAO and nPO, wintering south to Japan, SW USA and MS. 42/71cm. **Brünnich's Guillemot/Thick-billed Murre** *Uria lomvia*. Upperparts darker than Common Guillemot. Bill thicker with white stripe on cutting edge, no streaks on flanks. In winter has dusky cheeks and lacks post-ocular stripe. Range similar to Common Guillemot but more northerly, including in winter when occurs south to Iceland, N Japan and S Alaska. 45/76cm.

5. Atlantic Puffin *Fratercula arctica*
Medium-sized chunky auk with distinctive multicoloured bill in summer **(5a)**. In winter face dusky and bill smaller and duller **(5c)**. Immature has slimmer, more pointed dark bill **(5b)**. Breeds N Atlantic coasts from N Greenland and Spitzbergen south to Maine and N France. Disperses in winter south to New York and Morocco. 32/55cm.

6. Tufted Puffin *Fratercula cirrhata*
Unmistakable in breeding plumage. In winter face mostly dark, ear tufts reduced and bill smaller and duller. Immature has greyish face and underparts and a yellow bill. Breeds N Pacific coasts; pelagic in winter south to 35°N. 38/?cm. **Horned Puffin** *F. corniculata*. Unmistakable within range. Like large Atlantic Puffin with red and yellow bill. Winter plumage and immatures have dusky face and bill is smaller and duller. Breeds N Pacific coasts becoming pelagic in winter south to California. 38/57cm.

7. Little Auk/Dovekie *Alle alle*
Tiny auk with stubby bill and whirring flight. Head and upperparts black with white streaks on scapulars; white below **(7a)**. In winter chin and throat white extending to sides of neck **(7b)**. Breeds Arctic and nAO, wintering south to E USA and France. Recently found off north coasts of Siberia and Alaska. 22/32cm.

8. Ancient Murrelet *Synthliboramphus antiquus*
Small auk with blue-grey upperparts. Face and throat black, indistinct white stripe behind eye and a pale bill. In winter throat whitish, stripe behind eye absent and white half-collar on sides of neck. Breeds N Pacific coasts, wintering south to Korea and California. 26/?cm. **Japanese Murrelet** *S. wumizusume*. Similar to Ancient Murrelet but has a short crest and larger white stripe behind eye extending to nape. Crest smaller in winter. Endemic to Japanese coast, dispersing in winter to Korea and Sakhalin. 26/?cm. (Other name: Crested Murrelet.)

9. Rhinoceros Auklet *Cerorhinca monocerata*
Large brownish auk with two white facial plumes and pale 'horn' at base of orange bill. In winter facial plumes are reduced, bill is yellowish and lacks 'horn'. Immature similar but lacks facial plumes. Widespread N Pacific coasts; nocturnal at breeding colonies. 37/?cm.

10. Marbled Murrelet *Brachyramphus marmoratus*
Small brownish auk with long pointed bill. Breeding plumage mottled brownish-grey **(10a)**. In winter blackish above with distinct black cap extending to below eye and white patch on scapulars **(10b)**. N Pacific coasts; breeding little known, probably nocturnal, non-colonial and nesting inland. 25/?cm. **Kittlitz's Murrelet** *B. brevirostris*. Similar to Marbled Murrelet but with tiny, short bill. Breeding plumage mottled sandy-brown. In winter sides of face white extending above eye. N Pacific coasts; range and breeding biology little known. 23/?cm.

11. Whiskered Auklet *Aethia pygmaea*
Very small greyish auklet with three white plumes and a long grey crest overhanging small red bill. In winter bill duller and plumes much reduced. Immature has three streaks on face and yellow bill. Breeds Aleutian Is., dispersing to Japan in winter. 20/?cm. **Crested Auklet** *A. cristatella*. Similar to Whiskered, but larger, with a single long white plume on face and a tall black recurved crest. Bill red with sheath extending towards eye in summer. In winter bill yellow and lacks sheath. Immature lacks facial plume and crest. nPO wintering to Japan. 27/?cm.

12. Least Auklet *Aethia pusilla*
Tiny auklet, dark above with white on scapulars, mottled brown and white below. Stubby red bill and white eye. In winter white scapulars more prominent and underparts entirely white. nPO, where abundant (some colonies over 1 million birds). 15/?cm.

13. Parakeet Auklet *Cyclorrhynchus psittacula*
Distinctive medium-sized auklet, black above, white below with white streak behind eye and stubby red bill. In winter throat white and bill duller. nPO, less gregarious than most auklets. 25/?cm.

14. Cassin's Auklet *Ptychoramphus aleuticus*
Small nondescript greyish auklet with white belly. At close range shows white crescent over pale eye and yellowish base to bill. Immature has whitish chin and throat. Breeds Aleutian Is. to Alaska, wintering south to California. 23/?cm.

15. Xantus's Murrelet *Synthliboramphus hypoleucus*
Medium-sized, long-billed auklet. Black above, white below with whitish underwing. Californian and N Mexican coasts. 25/?cm. **Craveri's Murrelet** *S. craveri*. Like longer-billed Xantus's Murrelet with dusky underwing and longer black breast spur. Californian and N Mexican coasts. 25/?cm.

1a
2b
2a
1b
4a
4b
5a
3
6
8
7a
9
10a
7b
11
13
12
14
10b
15
5c
5b

PLATE 45: GREAT WHALES I

Sperm Whale and Gray Whale lack a dorsal fin and have a knobbly hump in its place. Humpback is the only rorqual with a hump on which the dorsal fin lies. Body size, blow, surface profile, snout shape and the outline of the flukes (which are raised above the surface during diving in most species on this plate) are useful identification features. Visually inquisitive Gray, Humpback and Black Right Whales raise their heads above the water to look around.

RIGHT WHALES
Family: Balaenidae

Slow swimmers, no dorsal fin, deep, wide bodies and an arched upper jaw. Adult body weight estimated to attain 100 tonnes. Head rounded and broad. Solitary or in small groups, larger inshore congregations can occur. Rare.

1. Black Right Whale *Balaena glacialis*
Roughened patches of skin or callosities on head unique to this species, pattern is used in studies to sex and identify individuals. Body blackish with variable white belly and throat patches, calves can be entirely white. Females breed in coastal waters returning once every three years to give birth. V-shaped blow to 5m. Blackish baleen to 2.8m in length with fine fringes, feeds on zooplankton. Distribution cosmopolitan, seldom sighted south of Antarctic Convergence in S hemisphere. 18m. Listed as an endangered species. Total population in nPO numbered in hundreds only; at least three remnant populations (off Argentina, S Africa, W Australia) in S hemisphere. Total world numbers not known. S hemisphere population (?2,000) may be a separate species (*B. australis*); taxonomy needs clarification.

2. Bowhead Whale *Balaena mysticetus*
Large distinctive white patch on chin and sometimes on tail stock. Tail flukes can be entirely white. Head massive in bulk, more than half rest of body in length (about one third of total body length). Blackish body, head lacks callosities, blowholes lie high on head resulting a double-humped profile at the surface. V-shaped blow to 7m. Breaks breathing holes in ice up to 1m thick with its back; can utilise trapped pockets of air under ice. Longest baleen (blackish) of any whale, to 4.5m long with fine fringes. Full complement of plates averages 600 in number and 0.5 tonnes in weight. Feeds on zooplankton. Circumarctic distribution, living permanently amongst pack ice. Largest whale in this region, 20m. Population of 1000-3000 animals (though higher estimates have been made) in W Arctic (Alaska and Siberia), even rarer in E Arctic (polar nAO).

PYGMY RIGHT WHALE
Family: Neobalaenidae

3. Pygmy Right Whale *Caperea marginata* **478**
Similar rotund body shape to right whales (Balaenidae), curved upper jaw, flap-like lower lips and long, thin, pale-coloured baleen plates with fine fringes. Falcate dorsal fin set well back on body and two grooves on throat as in rorquals. Body colour grey, pale to white below with darker flippers. Smallest baleen whale, can be confused with Minke Whale (plate 46) at sea. Secretive, often only shows snout when it blows, which makes this species difficult to spot. Probable circumpolar distribution in southern oceans north of Antarctic Convergence; may make limited southerly migrations during austral summer. 6.4m.

GRAY WHALE
Family: Eschrichtiidae

4. Gray Whale *Eschrichtius robustus*
Irregular knobbly ridge on back, head V-shaped from above, upper jaw has a downward curve with projecting tip. Short grooves (generally two) present on throat. Mottled coloration of dark and light grey patches with barnacles and whale lice in clusters in the skin. Blow appears as single spout to 4.5m high. Mottled grey tail flukes shown when diving. Baleen pale-coloured with thick, coarse fringes, feeds on bottom-living invertebrates, mainly amphipods (*Amplisca, Atylus*). It scoops or sucks up mouthfuls of mud and sieves it through its baleen, leaving pits in the ocean floor. Visually inquisitive, frequently spyhopping. nPO; 15m. Two migratory populations: western (Okhotsk Sea to Korea) possibly extinct, though recent sightings off Sakhalin Is. of about 50 whales have been reported; eastern (E Siberia and Alaska to Mexico), c.20,000 whales, increasing at 3% annually. Extinct in nAO.

RORQUALS
Family: Balaenopteridae

5. Humpback Whale *Megaptera novaeangliae* **478**
Long flippers, one third of body length with scalloped leading edge. Scalloped, deeply notched flukes. Dorsal fin small, variable in both size and shape, with a hump beneath it. The broad head has numerous rounded wart-like projections, snout rounded at tip. Large lump ('cutwater') projecting at tip of lower jaw peculiar to Humpbacks. Colour blackish above, variably white on throat and belly. Flippers partly or entirely white above and below. Blow low and broad to 3m high, back arched high above the water when diving, flukes often raised high. The white pattern on fluke underside can be used to identify individuals. Most vocal of the baleen whales and agile, often breaching and spyhopping. Diet includes fish and zooplankton. Baleen plates black with coarse fringes. Evidence for geographically separate populations with specific migration routes between feeding and breeding areas. Cosmopolitan distribution; 16m. Most of world population (?10,000) in N hemisphere (nAO), but estimates of total numbers vary widely.

SPERM WHALES
Family: Physeteridae

6. Sperm Whale *Physeter catodon*
Unique rectangular head profile, cigar-shaped when seen from above. Low hump instead of dorsal fin with knobbly undulations (usually 5) on upper ridge of tail stock. Tail stock strongly keeled below. Body greyish, skin wrinkles along flanks, occasionally with whitish head, rarely whole body. Area around mouth and belly can be whitish, narrow lower jaw has curved conical teeth in two parallel rows, about 20 teeth in each. These fit snugly into recesses in the upper jaw. Upper jaw occasionally has small teeth. Single nostril close to tip of left side of head, S-shaped when closed, blow angled to height of 5m. Throat irregularly grooved with short pleats. Known to breach, but not easily observed since long periods (2hrs+) are spent underwater. Dives to 3,200m, foraging mainly for squid, but also takes fish. Most squid eaten are small (up to 1m mantle length) but may also capture Giant Squid *Architeuthis* sp. (7), known to reach 19.5m length (inc. tentacles). Circumglobal distribution, preferring open sea, seldom seen in shallow seas; only bulls venture into cold temperate and polar waters. Male 18m, female 12m. World population about 500,000, but no agreement on numbers.

1
2
3
4
5
7
6
1
2
3
4
5
6

PLATE 46: GREAT WHALES II

Rorquals have a grooved throat that expands enormously when feeding to engulf food. Sleek, fast whales with a small dorsal fin set well back and narrow pointed flippers. Females larger than males, maximum sizes recorded in S hemisphere. Double nostril forms one high blow. Baleen plates short compared to right whales. Most have V-shaped snouts from above, one snout ridge (except Bryde's) and generally do not show the flukes when diving (except Blue Whale). Fin whale has distinctive snout markings (asymmetrical). Rorquals are circumglobal over deep oceanic water, migrating between favoured feeding and breeding areas. Decimated by commercial whaling in the past, there are now large expanses of ocean where rorquals are rarely seen.

RORQUALS
Family: Balaenopteridae

Rorquals differ from right whales in the way their baleen plates are arranged at the front of their jaws; in rorquals the plates extend to the jaw tip, in right whales there is a space where the palate shows.

1. Sei Whale *Balaenoptera borealis* **478**
Typically a dark slate-grey with much scarring on tail stock but also reported to have a mottled bluish appearance at times. Head with central ridge from blowhole to snout tip, snout pointed in top view and upper jaw with down-turned tip. Throat with variable central white patch, sometimes anchor-shaped. Dorsal fin less than two thirds back on snout-tail distance, tall and falcate. Dorsal fin becomes visible almost simultaneously with blow (to 3m high). Rolls low at the surface, dives often shallow and short; can appear to sink below surface. Baleen blackish with fine greyish fringes. Diet mainly zooplankton, also fish and squid. Can skim feed at the surface like right whales. Observed to turn on its side when feeding like the Fin Whale (3). 'Shy' of boats. Cosmopolitan distribution. 18m. Common name pronounced 'say'.

2. Blue Whale *Balaenoptera musculus*
Body bluish-grey with pale mottling. Scattered white spots on throat region; underside of flipper and flipper leading edge white. Head broad and flat with a well rounded, U-shaped tip. Central ridge from blowhole to snout tip. Dorsal fin relatively tiny, variable in shape and set three quarters back on snout-tail distance. Tail stock deeply keeled. Blow to 9m high. Long and graceful roll at the surface, then dorsal fin appears well after the blow and shortly before the whale submerges. Flukes sometimes appear just above the surface, dives can last 30 min. Baleen blackish with coarse fringes, takes only larger zooplankton (krill) in polar waters, though feeding has also been seen in tropical waters. Yellowish hue on bellies of whales in colder waters is caused by a surface film of diatoms (*Cocconeis ceticola*). Cosmopolitan distribution. World's largest animal; longest correctly measured Blue Whale was 33.58m in length, heaviest weighed 190 tonnes. Listed as an endangered species; in nwAO (Gulf of St. Lawrence) 325 animals have been individually photo-identified from colour patterns and 785 animals identified in nePO (California/Mexico). Total S hemisphere population estimated at 450 animals in 1994. Total world population not known. A dwarf form is present in nIO and S hemisphere (named as the subspecies *brevicauda*, Pygmy Blue Whale); it does not differ in skin colour, and taxonomy needs clarification.

3. Fin Whale *Balaenoptera physalus*
Blackish-grey back, often with a dark line from the flipper to eye and a V-shaped (chevron) mark running from flipper to shoulder. Upper body coloration asymmetrical (extends further down flanks on left side of body) and sharply demarcated from whitish underside. Snout long and pointed, central ridge from blowhole to snout tip. Head with asymmetrical markings, lower jaw dark on left side, white on right. Baleen also white on right side (**3a**), otherwise brownish with coarse fringes. Dorsal fin two thirds back (not less) on snout-tail distance, falcate with a rounded tip. Leading edge of dorsal fin often less vertical than Sei and Bryde's Whales. Blow to 6m high, dorsal fin appears above surface a short while later. Feeds on krill, fish and squid. When feeding, may swim on its right side. Second largest whale and one of the fastest swimming Great Whales (24 knots burst speed). Can breach and leap. Cosmopolitan distribution. 27m.

4. Bryde's Whale *Balaenoptera edeni* **478**
Similar in colour to Sei Whale with which it is often confused, but head has three ridges (diagnostic) most clearly visible from above. Often inquisitive and will approach ships. Surfacing and dive profile similar to Fin Whale, but distinctive white marking on right side of Fin Whale snout absent in Bryde's Whale. Blow to 4m high. Baleen blackish with coarse fringes, broad diet including fish and squid. Pantropical and warm temperate distribution, not found in waters below 20°C, or more than 40° latitude. Local seasonal movements in tropics, not migrating to polar waters. 15.5m. Common name pronounced 'brooders'.

5. Minke Whale/Lesser Rorqual *B. acutorostrata* **478**
Smallest rorqual species. Bluish-grey above, sometimes with lighter stripes or blazes on sides and extending onto back, underside whitish. Flippers usually with diagnostic white patch (but absent in all S hemisphere Minke Whales except for dwarf form). Head small, snout sharply pointed, central ridge from blowhole to snout tip. Dorsal fin is nearer mid-body than tail notch, large and sickle-shaped. Visually inquisitive, often spyhopping around boats. Blow to 3m high, often diffuse and difficult to see, dorsal fin appears almost simultaneously with blow, roll high, humped and quick. Baleen whitish, posterior plates striped brown or black, with coarse fringes. Diet includes krill, squid and fish. Cosmopolitan, the rorqual species most likely to be seen in shallow polar and temperate coastal seas. 10.7m. There is a widely distributed dwarf form in S hemisphere which has a distinctive white shoulder patch extending onto flipper.

▼

1
2
3a
3
4
5
2
3
1
4
5

PLATE 47: BEAKED WHALES

These are the most secretive cetaceans (19 named species, four of these described since 1937, one species known from only two skulls). Adult male beaked whales are distinguished by a projecting pair of enlarged teeth in lower jaw (usually no teeth in upper jaw). Females and young males do not show this feature (except in *Berardius*) and often more difficult to separate to species level. Falcate dorsal fin is about two thirds back on body, flippers small, throat has a pair of V-shaped grooves and flukes generally lack notch. Most species are deep-bodied in side profile, some species reaching or exceeding the size of an adult Minke Whale. Females are usually larger than males. Characteristic scratch marks on skin due to rivalry between sexes (oldest males most scarred, young and females least); white oval scars probably from cookie-cutter sharks (*Isistius*). The genus *Mesoplodon* is the largest in the beaked whale family (13 currently recognised and named species). These are rarely observed species and their external features are poorly known; limited observations on skin colour have been made on stranded whales. Beaked whales are oceanic, perhaps diving to great depths in search of squid and deep-living fishes. Generally 'shy' of boats and spending long periods below the surface partly explains lack of sightings.

BEAKED WHALES
Family: Ziphiidae

1. Baird's Beaked Whale *Berardius bairdii* **478**
Body is slate-grey, heavily scarred particularly on back, belly and throat with white patches. Lower jaw projects markedly in front of upper and a pair of large teeth erupt near the lower jaw tip (in both sexes). A second pair of teeth located posterior to the first are hidden inside mouth. Tip of dorsal fin rounded. Flukes notched. Largest of the Ziphiidae, colder nPO. 12.8m. **Arnoux's Beaked Whale** *B. arnuxii* is very similar and occurs in S hemisphere (Antarctic Ocean and neighbouring waters). 9.75m.

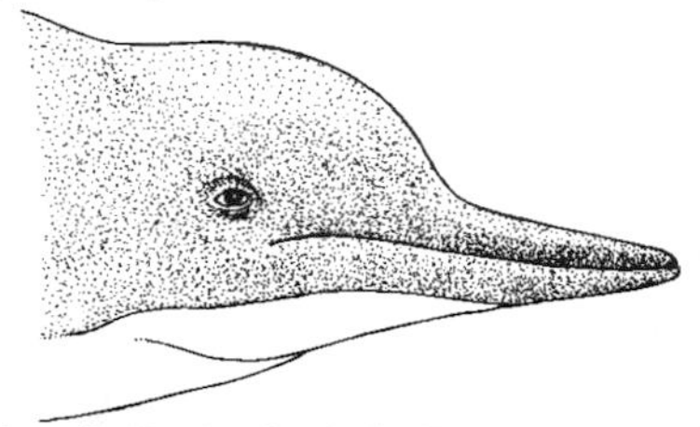

▲
2. Shepherd's Beaked Whale *Tasmacetus shepherdi*
Colour grey-brown, whitish below, with a light stripe along flank. Melon rises at a less steep angle than in Baird's. The only beaked whale with functional teeth inside both jaws, the larger pair of teeth at tip of lower jaw in males. Warm and cold temperate waters in S hemisphere (two possible sightings?). 7.0m.

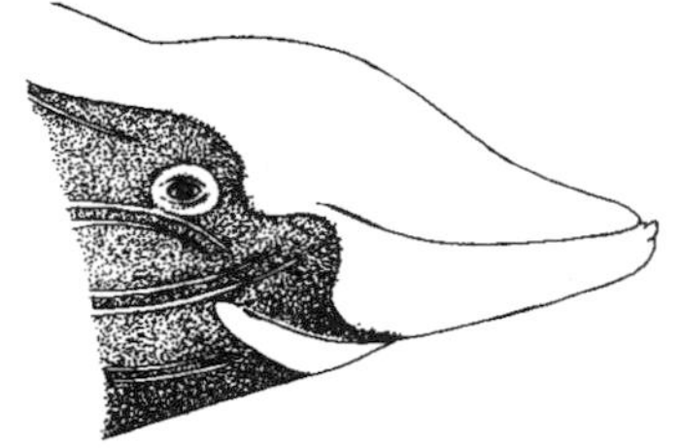

▲
3. Cuvier's Beaked Whale *Ziphius cavirostris*
Relatively small head and bird-like snout. Whole front of body whitish in older animals. Body bluish-grey colour, browner in juveniles, with scratches and white blotches. Dorsal fin distinctly hooked. Male has two projecting teeth at lower jaw tip. Wary of boats. Low blow slightly forward and to left. Circumglobal including polar waters of Bering Sea. 7.0m. Also known as Goose-beaked Whale.

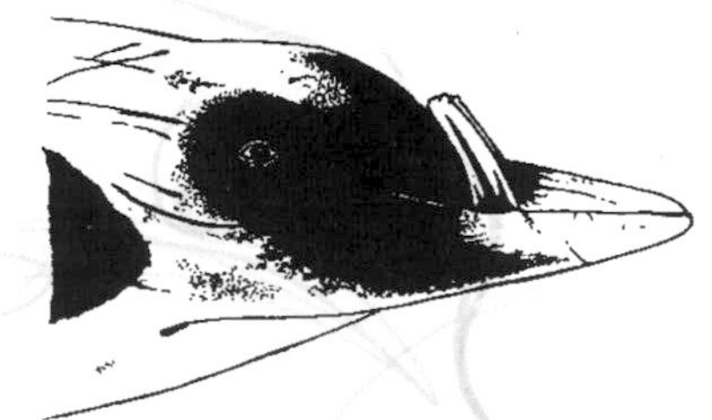

▲
4. Strap-toothed Whale *Mesoplodon layardii*
Dark grey with variable white patches on beak, head, back, genital slit and fluke tips. In males, long strap-shaped, tusk-like teeth erupt near midpoint of lower jaw. These can be up to 30cm in length, growing back over the upper jaw and restricting the gape width. Known mainly from more than 90 strandings on Australian and New Zealand coasts. Warm and cold temperate waters in S hemisphere. 6.15m.

5. True's Beaked Whale *Mesoplodon mirus*
Dark grey above, fading to light grey on sides and belly. Light patch on genital slit. In S hemisphere population, tail stock and dorsal fin are a striking white (as illustrated). In males, a pair of teeth erupt at the tip of the lower jaw **(5a)**. In N hemisphere, known only from nAO. In S hemisphere, warm temperate waters off S Africa and S Australia. 5.33m, 1.4 tonnes.

6. Hubbs' Beaked Whale *Mesoplodon carlhubbsi* **479**
Males are uniformly grey or black, females markedly lighter on back and sides with white belly. Snout tip whitish in both sexes. Males have a characteristically enlarged melon (marked by a striking white blotch) and tusk-like teeth **(6a)**. nPO. 5.32m, 1.4+ tonnes. Can be confused with Stejneger's Beaked Whale (head shown below) which also occurs in nPO; adult males can be identified using differences in shape of teeth, females and young only by using skull characteristics.

7. Blainville's Beaked Whale *M. densirostris* **479**
Bluish-grey above, belly, throat and lower jaw white. Distinctive flap-like upgrowths on lower jaw, especially marked in adult males **(7a)**, where tusk-like teeth are present. Most of tooth lies below gum, embedded in massive bony mandible. Skin of stranded animals rapidly darkens after death **(7b,c)**. Cosmopolitan in tropical and temperate waters. 5.8m.

▲
Stejneger's Beaked Whale ***Mesoplodon stejnegeri***

2
1
3
4
5a
5
6a
6
7a
7
7c
7b

PLATE 48: SMALL TOOTHED WHALES

This plate illustrates colour patterns and body shapes in small toothed whales. Note the differing profiles of the heads, dorsal fins, flippers and flukes (most are notched). False Killer Whales can be confused with both Pygmy Killer Whales and Melon-headed Whales; the False Killer differs from both in its distinctively curved flippers (see plate 49 for details).

DOLPHINS
Family: Delphinidae

1. False Killer Whale *Pseudorca crassidens*
Black with a greyish anchor-shaped patch on throat between flippers. Dorsal fin is high and falcate, tip pointed to rounded, placed behind mid-point of back. Flippers distinctive with arched outer margin and obvious outward pointing tips. Head rounded, similar to Pygmy Killer Whale and Melon-headed Whale, lips black. Inquisitive, one of the largest bow riding cetaceans. Common, oceanic, circumglobal except colder seas. 5.9m.

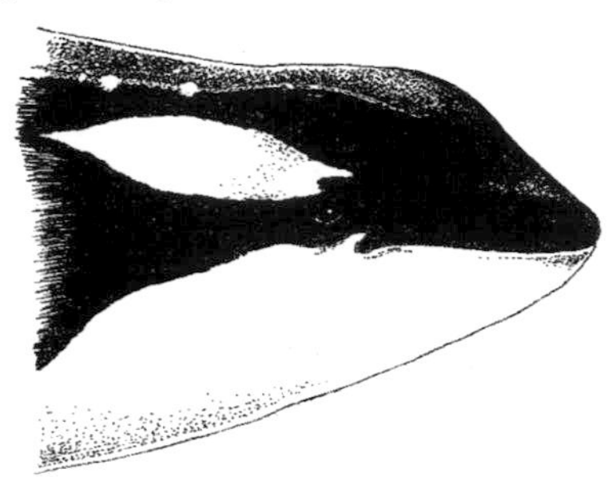

▲

2. Killer Whale/Orca *Orcinus orca*
Largest species of dolphin, striking appearance and coloration. Black above including large paddle-like flippers, dorsal fin and upperside of flukes. Grey 'saddle' behind dorsal fin. White belly, throat and underside of flukes extending onto rear flanks. Oval white patch behind eye. Slightly protruding beak. Large males have greatly enlarged dorsal and pectoral fins. Cosmopolitan distribution. 9.8m and about 10 tonnes, males larger. One of the fastest swimming whales, maximum recorded speed of 30 knots maintained for 20 minutes.

3. Long-finned Pilot Whale *Globicephala melas* **479**
Black with very distinct broad based dorsal fin placed forward of mid-body length. Grey saddle behind dorsal fin not always distinct. Obvious melon and long, pointed flippers (length up to 27% of body length) with distinct 'angle'. White band on underside from flippers to genital slit. Can be inquisitive but not a true bow rider, often indifferent to boats. Oceanic distribution, polar to warm temperate waters in nAO (+MS). Circumglobal in southern oceans. Extinct in nPO. Some overlap in range with Short-finned Pilot Whale (4). 6.7m, 2 tonnes, males larger than females.

4. Short-finned Pilot Whale *G. macrorhynchus* **479**
Similar in appearance and behaviour to Long-finned Pilot Whale. Melon often larger (even overhanging upper jaw) and squarish in adult males; white belly band less extensive. Main differences are the flippers, which are shorter (less than 23% of body length), curved rather than angled. Dorsal fin can be massively enlarged in adult males. Oceanic, circumglobal distribution in tropical and warm temperate seas (not MS). 6.1m, 3.6 tonnes, males larger than females.

NARWHAL AND BELUGA
Family: Monodontidae

Arctic whales, partly or entirely circumpolar in distribution. Dorsal fin lacking and well rounded heads with no obvious beak. Stocky, flippers paddle-like. Social, visually inquisitive. Slow swimmers, both species often encountered together.

5. Narwhal *Monodon monoceros*
The long spiral tusk of the male (origin of the unicorn myth) is used for sparring between males and display behaviour in competition for females. Note the ridge (4-5cm high) to the rear of the back and the profile of the flukes with highly curved rear margins and forward-angled tips. Body mottled greyish and white. Young grey all over, but white areas soon appear on underside becoming more extensive with age, often on flanks and back. Varied diet of fish, squid and shrimps. Males reach 5.0m excluding tusk (length up to 3m including root, generally present only in males, rarely paired), females smaller. Seldom seen in Alaskan and Siberian waters.

6. Beluga *Delphinapterus leucas*
Adults snowy white, young slate grey to reddish-brown. Adult colour, combined with the lack of a dorsal fin (replaced by low ridge), makes it almost unmistakable within its range, but confusion possible with aged Narwhals which can also be white. Visually inquisitive. Seen from above, has obvious neck constriction. Varied diet, including fish and bottom-dwelling prey. Dives to 440m in the wild. Frequents western Arctic of N America where Narwhal is rare. 5.5m, males larger than females.

BOTTLENOSE WHALES
Family: Ziphiidae

Large beaked whales, with an obvious, bulbous melon. Social, unlike many beaked whales, but older males may be solitary as in Sperm Whale.

7. Southern Bottlenose Whale *Hyperoodon planifrons*
Body grey, skin with scratch marks and white spots. Darker on back, lightening on belly in adults. Can appear distinctly brown (as illustrated) due to a covering of diatoms on the skin. Probable circumglobal distribution from 30°S to Antarctica, with possible widely scattered sightings in tropical PO. 7.8m. **Northern Bottlenose Whale** *Hyperoodon ampullatus*. Similar to Southern Bottlenose Whale in external appearance. The only large beaked whale with a bulbous melon in its range (flat-fronted melon when mature). Sometimes with entirely whitish body. Inquisitive, approaches boats. Colder waters of nAO up to Arctic pack ice edge, known as far south as MS. 9.8m (males larger than females).

1
2
5
7
6
3
4
7
1
6
3
4
2
5

PLATE 49: SMALL SPERM WHALES AND DOLPHINS

The toothed whales on this plate vary considerably in head and beak profile, as shown clearly in the top view of each species. The reaction to boats varies, small sperm whales (*Kogia*) do not bow ride. Schools of Right Whale Dolphins, Risso's Dolphins and Melon-headed Whales may be over one thousand animals strong.

SPERM WHALES
Family: Physeteridae

Few confirmed sightings, white bracket mark on neck very distinctive at sea. Widely distributed (circumglobal) in tropical and temperate waters.

1. Dwarf Sperm Whale *Kogia simus*
Dorsal fin tall (height more than 5% of total body length) and usually placed mid-back. Functional teeth in upper jaw (up to 3 pairs) and short throat creases characteristic. 2.7m.

2. Pygmy Sperm Whale *Kogia breviceps*
Similar to Dwarf Sperm Whale, but dorsal fin is lower and deeply hooked (less dolphin-like). Dorsal fin height less than 5% of total body length and usually placed well behind mid-back. Head proportionately longer than Dwarf Spem Whale. In some animals, tail ridge has low bumps and flank skin is wrinkled (compare to Sperm Whale, plate 45). 3.7m.

DOLPHINS
Family: Delphinidae

These dolphins can be inquisitive and bow ride. The Pygmy Killer Whale and Melon-headed Whale are externally similar and this can cause confusion. These beakless dolphins can be distinguished by a close view of the flippers (more pointed in Melon-headed Whales) and the head (greyish blowhole stripe, when present, only found in Melon-headed Whales).

3. Pygmy Killer Whale *Feresa attenuata*
Head rounded with white lips and small mouth. Tips of flippers rounded. Blackish grey, sides can be a lighter grey (dark cape below dorsal fin) with a limited white area on belly. Dorsal fin tall and falcate, tip broadly pointed, sometimes with irregular rear edge. Cicumglobal in tropical and warm temperate seas (+MS). 2.7m.

4. Melon-headed Whale *Peponocephala electra*
Blackish-grey, sometimes with darker cape under dorsal fin. Head rounded with white lips, lighter stripe runs from blowhole to tip of jaw. Dorsal fin falcate, flippers pointed. Social, range as Pygmy Killer Whale (3) (not MS). 2.7m. (Other name: Electra Dolphin.)

5. Risso's Dolphin *Grampus griseus* 480
A large, distinctive, beakless dolphin with blunt grooved head **(5a)** and tall falcate dorsal fin. Body often light greyish with a white head, throat and belly, and darker fins. Body covered in scratches and oval 'doughnut' scars, flippers sickle-shaped and pointed. Squid feeder. Spyhops at sea and in captivity. Cosmopolitan distribution except polar seas, often oceanic. 4m+.

▼

▲

6. Roughtooth Dolphin *Steno bredanensis*
Sleek snout lacking forehead-to-beak contour break. Dark bluish-grey saddle, lighter on flank, white belly (inc. lower jaw), with an irregular mottled dividing line. Circumglobal distribution in warmer seas, poorly known, appearing to favour deeper offshore waters (+MS). 2.8m.

▲

7. Bottle-nosed Dolphin *Tursiops truncatus* 480
Curving forehead well defined from short and stubby beak (longer in some tropical populations), lower jaw slightly protruding. Mouth curve forms a fixed 'smile'. Skin colour varies from light grey to dark grey or 'three tone' with a darker dorsal saddle, grey sides and light belly. Cosmopolitan distribution except polar seas. Usually coastal including turbid estuaries. Considerable range in length and bulk, up to 4m. This species is famous for its association with man (in the wild and in captivity), its intelligence and playful antics around swimmers.

8. Northern Right Whale Dolphin *Lissodelphis borealis*
No dorsal fin, profile very sleek with long tail stock. Body mainly black, including small sickle-shaped flippers and stumpy beak (lower jaw white-tipped). White underside. Flukes black or grey above (darker centrally), white below with dark trailing edge. Young are paler. In large schools, fast swimmer, burst speed of 24 knots. Cold temperate nPO, mainly oceanic. 3.1m.

9. Southern Right Whale Dolphin *L. peronii* 480
No dorsal fin, similar to Northern Right Whale Dolphin, but white pigmentation much more extensive reaching from underside high up onto flanks and melon. Flippers small, white except for dark trailing edge. Flukes variably black (dusky centrally) to white above, generally white below. Circumglobal distribution in southern oceans north of Antarctic Convergence, in large oceanic schools. Range extends northwards around S America as far as Peru on west coast and Brazil on east coast. 3.0m

1
3
2
4
5
6
7
8
9
1
2
3
5
4
6
9
8
7

PLATE 50: DOLPHINS I

Some of the dolphin species shown on this plate are inquisitive and frequently bow ride (3,4,5,6); others tend to be shyer. Widespread, in coastal seas and offshore in both hemispheres. Close inspection of complex body colour patterns important for identification purposes. To assist identification, the species are divided according to geographical locality where possible; all species stocky with short, robust snouts.

NORTH ATLANTIC

1. White-beaked Dolphin *Lagenorhynchus albirostris*
Large and robust with a short, broad white beak. Dorsal fin high and strongly falcate. Between the dorsal fin and the dark flipper there is a white or greyish variably shaped foreflank patch, often seen when surfacing. There is a more diffuse rear flank patch, sometimes forming one long streak, separating the dark flukes and tail stock from the lighter trunk. Sometimes bow rides. Cold temperate waters. 3.2m.

2. Atlantic White-sided Dolphin *Lagenorhynchus acutus*
Dark dorsal area continuous from upper beak to flukes, sharply demarcated from the grey flanks, and these in turn demarcated from the white belly and lower jaw. Black eye-bridle with grey flipper line. Long patch on rear half of the flank is whitish in front and brown behind. Tail stock distinctly keeled. Rarely bow rides, usually indifferent to boats. Similar range to the White-beaked Dolphin (1), but not as far north (absent from N Norway and rarely seen in Baltic Sea). 2.8m.

NORTH PACIFIC

3. Pacific White-sided Dolphin *L. obliquidens* **480**
Both the dorsal fin (hooked with a round tip) and the flippers can be a 'two tone' black and greyish-white, indistinctly divided. Belly white with a thin black line dividing it from flanks, which are dark with extensive pale patches. Inquisitive, often surfs and bow rides. Temperate waters. 2.5m. A dwarf form is found in northern part of range (nwPO).

SOUTHERN HEMISPHERE

4. Dusky Dolphin *Lagenorhynchus obscurus* **480**
Coloration similar to Pacific White-sided Dolphin (3), both species share 'two-tone' dorsal fin and thin white tail streak trailing over back towards blowhole. Inquisitive, often bow rides. Distribution in temperate southern seas, mainly coastal in habit (New Zealand, southern Africa, S America). Confusion possible with Peale's Dolphin (6) in S American waters. 2.1m.

5. Hourglass Dolphin *Lagenorhynchus cruciger*
Distinctively marked species with pattern of two white flank patches contrasting with black dorsal area, flanks and beak. Belly white. Note the highly falcate dorsal fin. Bow rides. Possible circumglobal distribution in colder southern oceans, and Antarctic waters to edge of pack ice. Most southerly known small cetacean. 1.83m.

6. Peale's Dolphin *Lagenorhynchus australis*
Colour pattern similar to Pacific White-sided Dolphin (3), black chin and black 'face' distinguish this species from all other southern *Lagenorhynchus*. Coastal southern S. America and Falkland Is. Bow rides. Possible sighting well out in tropical PO (Palmerston Atoll). 2.16m.

OCEANS WORLDWIDE

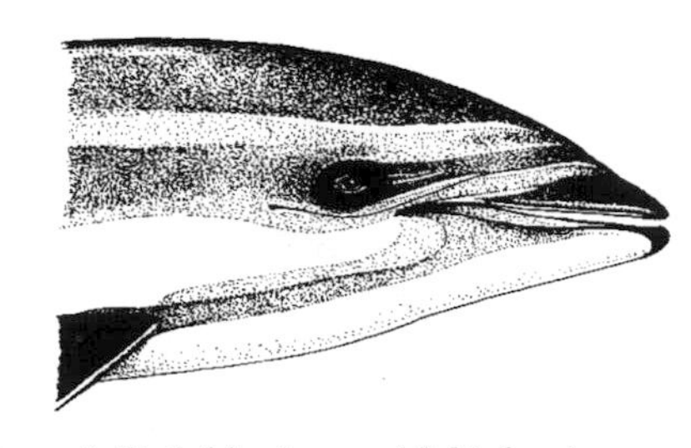

▲

7. Fraser's Dolphin *Lagenodelphis hosei* **481**
Body colour grey with distinctive (in many but not all individuals) horizontal stripe from melon to base of tail. Upper belly yellowish, lower belly white. Flippers, dorsal fin and flukes relatively small. Can be inquisitive. Observations on bow riding have been made in some areas of its range. Can be confused with Striped Dolphins (plate 51) especially as some populations of Fraser's Dolphins have similar spinal blaze (not shown on illustration here). The short-snouted Fraser's Dolphin can, however, be easily distinguished when the head is seen at close range from the long-snouted Striped Dolphin. Circumglobal distribution in tropical seas, also warm temperate waters in nwPO north to Honshu Is., Japan. 2.7m.

▲

Striped Dolphin ***Stenella coeruleoalba***

▲

Hourglass Dolphin ***Lagenorhynchus cruciger***

1
2
3
4
5
6
7
1
2
3
4
6
5
7

PLATE 51: DOLPHINS II

These dolphins are generally oceanic and social, living in schools that can be hundreds strong; there is an unexplained association between schools of Spinner and Spotted Dolphins and tunas (Yellowfin and Skipjack, see plate 15). The long slender jaws and numerous pointed teeth of the dolphins shown here are adapted for grasping the small, soft-skinned fish and squid on which they feed. All species illustrated here approach boats and bow ride, appearing to 'jostle' each other for the best positions on the pressure wave and turning on their sides to look up at people on the ship. Some species are energetic in air, displaying spectacular acrobatics (leaps, spins and backward somersaults). The Spinner Dolphin derives its name from the habit of spinning during leaps above the water; each spin can have up to 4 rotations on the body axis, up to 3m above the surface. Off Hawaii, spinning is seen most frequently just before dusk when Spinner Dolphins move offshore to their feeding grounds.

DOLPHINS
Family : Delphinidae

1. Striped Dolphin *Stenella coeruleoalba* **481**
Cape dark bluish-grey with a notch or 'spinal blaze' running up to dorsal fin. This distinctive feature enables Striped Dolphins to be separated from the closely similar Common Dolphin (6). Also a distinctive stripe through eye. Belly white. Circumtropical and warm temperate (+MS, where it is considered common), but also entering cooler waters. 2.6m.

2. Atlantic Spotted Dolphin *Stenella frontalis*
Cape black with spinal blaze, obscured to varying degrees by an overlay of spots, especially in larger animals. Tail stock uniformly grey. Belly white or white-spotted. Calves unspotted, subadults spotted **(2a)**, adults have dark spots on belly and light spots on dorsum **(2b)**. Confusion likely with Pantropical Spotted Dolphin (3), detail of cape differs. Tropical and warm temperate wAO (east to W African coast?). 2.3m.

3. Pantropical Spotted Dolphin *Stenella attenuata*
Cape black curving deeply below dorsal fin, obscured to varying degrees by an overlay of spots, especially in larger animals. Tail stock dark above, keeled below in adult males. Belly white or white with grey spots, becoming entirely grey as spots merge during growth. Snout tip white. Calves unspotted, spots appear first on belly, then on back in subadults **(3a)**; adults are sometimes so heavily spotted that they appear whitish above (silverbacks, **3b**). Circumtropical. 2.6m. May form mixed schools with Spinner Dolphins.

4. Spinner Dolphin *Stenella longirostris*
Cape dark grey. No spots on body. 'Three tone' colour pattern formed by dark upper cape, light grey flank stripe and white belly is most characteristic for this species **(4c)**. In some regions (ePO), cape is not apparent and skin colour is dark grey overall except for lightening on belly. Males and females also with strongly marked physical differences in ePO; in males dorsal fin triangular and forward canted, tail with lower keel **(4a)**, less obvious in females **(4b)**. Circumtropical, living close to oceanic islands in PO, but pelagic in ePO. 2.4m. A dwarf form has been described from the Gulf of Thailand.

5. Clymene Dolphin *Stenella clymene*
Cape dark grey, curving down onto flanks below dorsal fin and on forehead above eye. Belly white. Short snout and cape distinguishes this species from Spinner Dolphin (4). At closer range, the dark line running from each eye to upper jaw and onto snout is visible ('moustache'). Little-known species, recently reported to spin during leaps above water's surface. Tropical AO. 2.0m.

6. Common Dolphin *Delphinus delphis*
Cape black forming a 'saddle pattern' with point below dorsal fin. Flank patch is a striking golden-yellow colour (sometimes brownish or greyish-green) above the pectoral fins which distinguishes it from Clymene Dolphin (5). Circumtropical and warm temperate (+MS), but also entering cooler waters. 2.6m.

Common, Spotted and Spinner Dolphins all have a distinctive dark-coloured back or 'cape' which is starkly contrasted against the light-coloured flanks. The exact size, shape and position of this cape (shown below) is different in each species and is a useful aid to identification.

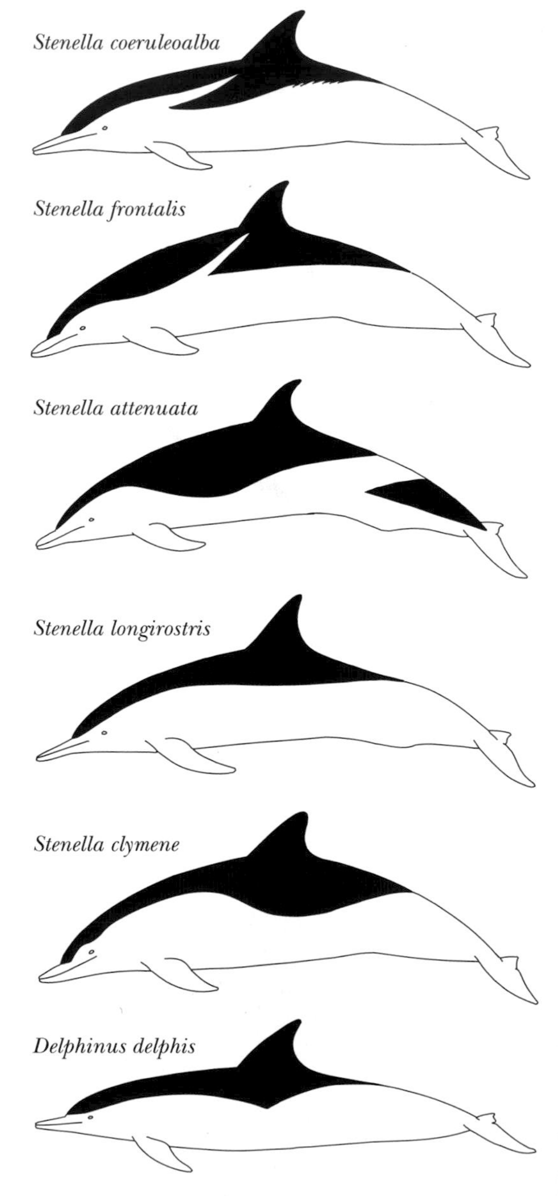

4c
1
2a
2b
3a
3b
4b
4a
4b
5
6
1
2b
3a
3b
4a
4c
5
6

PLATE 52: DOLPHINS III

***Cephalorhynchus*, with four species, is a genus of small, tubby dolphins with black, grey and white skin patterns. All are found in localised regions in S hemisphere. *Sousa*, a genus with three species, is widely distributed in tropical inshore waters from W Africa to Australia. Colour of body lead grey, spotted or pinkish-white, dorsal fin lies on a variably pronounced raised ridge or 'hump'. *Orcaella*, a genus with a single species, is at home in freshwater, frequenting the Mekong and Irrawaddy river systems (and other tropical rivers), well above tidal regions. It also inhabits marine coastal waters.**

DOLPHINS
Family: Delphinidae

CEPHALORHYNCHUS

Short snout, flippers grey or black and belly white. Three streaks of white on base of tail, shorter central band in midline (absent in Commerson's Dolphin). Dorsal fin distinctly rounded (except in Heaviside's Dolphin). World's smallest marine delphinid.

1. Heaviside's Dolphin *Cephalorhynchus heavisidii*
Low triangular dorsal fin. Black cape extending forward as narrow band to blowhole. Throat black with white bib. Coastal SW Africa, in waters of Benguela current. 1.7m, 74kg.

2. Chilean Dolphin *Cephalorhynchus eutropia* **481**
Body with dark grey back, lighter grey sides and white belly. Melon with light grey marking extending behind blowhole. White ovoid patch on axilla and throat region. Flukes, flippers and dorsal fin uniformly dark grey except for lighter leading edge on flippers. Prominent flipper to eye stripe. SW coast of S. America between Valparaiso and Navarino Island, Tierra del Fuego. 1.67m, 63kg.

3. Hector's Dolphin *Cephalorhynchus hectori*
Body mainly light grey. Wide black mask from flipper to the eye, narrowing along the mouth to the jaw tip. Coastal waters of New Zealand. Most of world population of Hector's Dolphin (estimated at 3000-4000 animals in 1988) is found in coastal waters of South Island. 1.53m, 57kg.

4. Commerson's Dolphin *C. commersonii* **481**
Striking black and white body. Black patch over genital slit and black head mask narrowing to flippers. White tear drop marking on throat. SE coast of S America and around Tierra del Fuego. Also Falkland and S. Shetland Is. 1.47m, body weight less than 45kg. This dolphin also occurs around Kerguelen Is. Adults there reach longer body lengths than in mainland waters (1.74m, 86kg) and have greyish rather than white back and flanks.

SOUSA

Stout build, head and body proportions similar to the Bottle-nosed Dolphin (plate 49). Coastal in habit, often in brackish waters (river mouths, mangroves). The limits of the ranges of these dolphins are poorly known and it has been suggested that all Indo-Pacific *Sousa* belong to the single variable species *S. chinensis*.

5. Atlantic Humpback Dolphin *Sousa teuszii*
Lead-grey skin colour. Dorsal fin variable in shape, raised on marked hump. Wide, obviously keeled tail stock. Coast of W Africa from S Morocco (W Sahara) to Cameroon and possibly N Angola. Mouths of Senegal and Gambia rivers, also reported from Niger River. 2.5m.

6. Indian Ocean Humpback Dolphin *Sousa plumbea*
Lead-grey skin colour, prominent back ridge and tail keels. Animals in some regions show heavy spotting (**6a**). Coastal waters of southern Africa, The Gulf and east to India. 2m.

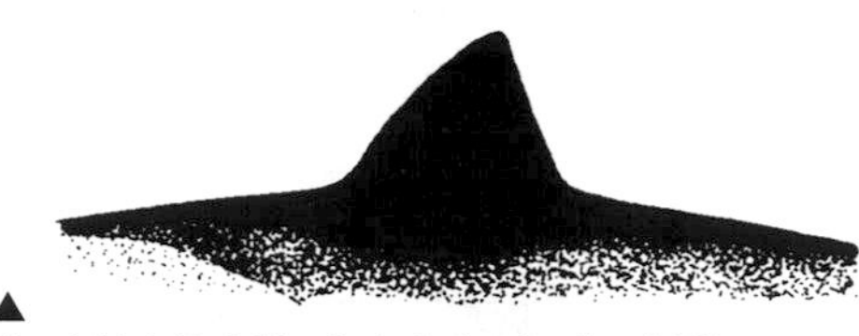

▲
Heaviside's Dolphin ***Cephalorhynchus heavisidii***

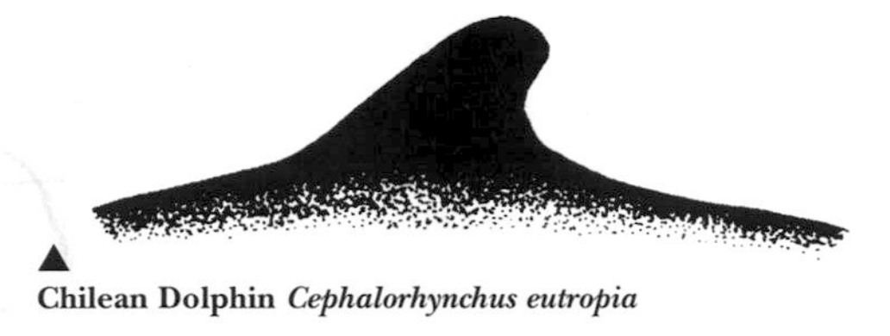

▲
Chilean Dolphin ***Cephalorhynchus eutropia***

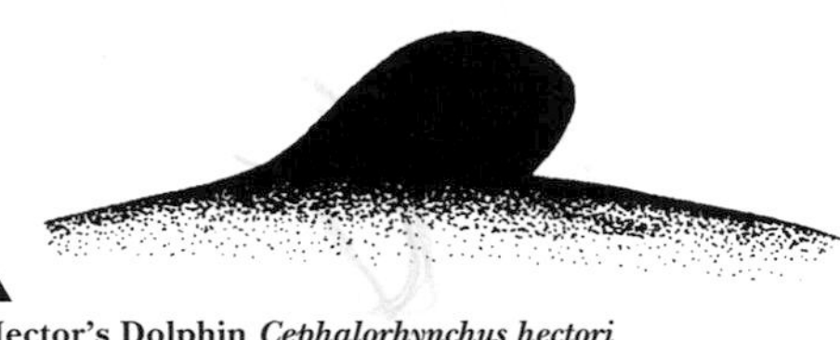

▲
Hector's Dolphin ***Cephalorhynchus hectori***

▲
Commerson's Dolphin ***Cephalorhynchus commersonii***

7. Pacific Humpback Dolphin *Sousa chinensis*
Whitish skin colour, back ridge and tail keels not prominent. Distribution from N Australia, Indonesia and north through Philippines to S China, in coastal waters. 2.8m.

ORCAELLA

Beakless, blunt-headed dolphin with small dorsal fin behind mid-point of body, rounded flippers and reverse blowhole slit. Distinct neck region, obvious in side view.

8. Irrawaddy Dolphin *Orcaella brevirostris* **483**
Greyish skin colour without any prominent markings. Diet composed mainly of fish. Capable of ejecting a stream of water from mouth, said to be used as feeding technique to concentrate fish. Coastal waters and larger freshwater rivers from E coast of India to N Australia. Could be confused with Finless Porpoise (plate 53) and Dugong (plate 56), though both these species lack a dorsal fin.

1
2
3
4
5
6
7
8
1
2
3
4
6a
5
7
8

PLATE 53: 'RIVER DOLPHINS' AND PORPOISES

This plate illustrates two species of 'river dolphins' which are the smallest of the world's cetaceans; the little Tucuxi from the central Amazon barely exceeds 50kg in adult body weight and 1.5m in length. Only the Vaquita may be smaller still, but too few specimens have been examined to be certain. Information on the S hemisphere porpoises (5,7) from sightings or strandings is currently very limited.

'RIVER DOLPHINS'

Seafarers in S American waters may encounter two dolphin species (La Plata Dolphin and Tucuxi) that occur in both freshwater rivers and in tropical, marine waters of coastal wAO. The La Plata Dolphin is classified with a small group of dolphins that are found mainly in freshwater (River Dolphins, Platanistidae); the Tucuxi is classified with the (largely marine) true dolphins (Family Delphinidae).

1. La Plata Dolphin/Franciscana *Pontoporia blainvillei*
Beak long and slender in adult, stubby in young **(1a)**. Broad flippers, eyes small and dorsal fin triangular. Body colour variable shade of grey-brown, lighter below. Brazil to Argentina from Doce River in the north to Peninsula Valdez in the south. 1.77m, 52kg.

2. Tucuxi *Sotalia fluviatilis*
Similar to the Bottle-nosed Dolphin (plate 49) but smaller. Grey above, lighter and often pink below, with dark flipper band. Amazon river system, coastal Central and S America from Panama to Florianopolis, Santa Catarina State, Brazil, 2.1m. Smaller in Amazon, 1.52m.

PORPOISES
Family: Phocoenidae

Small, compact cetaceans lacking an obvious beak (except Dall's Porpoise), dorsal fin (where present) is generally triangular, small flippers. The teeth are spade-shaped. Porpoises (except Dall's Porpoise, which bow rides) are indifferent to boats, rolling low and quietly in the water. Easily missed, giving the impression to observers that they are 'busily preoccupied'.

3. Vaquita *Phocoena sinus*
Grey skin colour with dark eye-patch, chin patch and lips. Dorsal fin high, pointed and quite unlike that of other porpoises. Probably restricted to the uppermost Gulf of California and seldom seen. 1.5m, 47kg. Listed as an endangered species, population estimated in low hundreds.

▲

4. Common/Harbour Porpoise *P. phocoena* **482**
Dark above, white below, with dark pigment line to mouth. Light flank patch above flipper of variable size and intensity. Low dorsal fin and blunt-tipped, paddle-like flippers; both these appendages have small lumps ('tubercules') on their leading edges. Widespread in coastal waters in temperate nAO and nPO; distribution extends into tropical and polar waters in some areas of range, isolated population in Black Sea. 2m, females larger than males. Historical records from MS (no strandings on MS coast of France this century), present size of population there uncertain. Confusion possible with Dall's Porpoise (8) in nPO, but difference in shape and colour of dorsal fins makes accurate identification possible.

5. Spectacled Porpoise *Australophocaena dioptrica*
Black above, including dorsal fin and flukes, sharply demarcated from white ventral area, which curves up at rear of tail stock. Lips dark, with grey line (not always present) to white flippers. Females **(5a)** have a low triangular dorsal fin with a rounded tip, males **(5b)** have higher rounded dorsal fins with more extensive whitening on the upper tail stock. S America, Tierra del Fuego to Uruguay. Also reported from subantarctic islands, including Falkland and Auckland Is. Seldom seen. 2.3m.

6. Finless Porpoise *Neophocaena phocaenoides* **482**
Dorsal fin absent. Colour uniformly bluish-grey, lighter below. Paired dark grey stripes on melon. A dark stripe from eye to mouth and also from flipper to tip of jaw. In animals from eastern part of range (China and Japan), a low ridge running most of length of back is present. Ridge surface covered with tubercules. In rivers, estuaries and shallow seas from Persian Gulf to Japan (Honshu Island), south to Irian Jaya. 1.9m.

7. Burmeister's Porpoise *Phocoena spinipinnis* **481**
Distinctive flat-fronted dorsal fin is located behind body mid-length, leading edge has tubercules along its length ('spines', from which the species name is derived). Colour slate-grey with dark eye-patch, lips and flipper line, lighter below. S American coast from N Peru (5°S) south to Patagonia and around to S Brazil (28°S). Little-known species. 2.0m.

8. Dall's Porpoise *Phocoenoides dalli* **482**
Small head, stocky build, strongly keeled tail, the only beaked porpoise. Dorsal fin distinctly pointed and tip often white. Much variation in colour, all black **(8c)** and all white specimens have been reported. Size of white flank patch distinguishes the western Pacific form **(8b)** from the widely distributed, commoner central and eastern Pacific colour variant **(8a)**. Mixed schools of Dall's Porpoise containing both these types (8a, 8b) have been seen off the east coast of Japan. Readily bow rides, burst speeds up to 27 knots. Forms a characteristic plume of spray at the surface and zig-zags energetically at the bow. Distribution in nPO including Sea of Japan, Sea of Okhotsk and Bering Sea. 2.4m.

1
1a
3
2
4
5a
6
5b
7
8a
8b
8c
1
2
3
4
5
6
7
8

PLATE 54: EARED SEALS: SEA LIONS AND FUR SEALS

The two types of eared seals are the sea lions and the fur seals; they are classified in separate subfamilies. Marked differences exist between the sexes in eared seals and each species pair on this plate shows the larger bull with characteristic mane beside the smaller female; pups are brownish or blackish. Claws are well developed on the three central toes of the hindflippers and are used for scratching and grooming; the soles are well supplied with blood vessels and sweat glands. Seals resting at the surface in a distinctive flipper-out pose (3a) lose heat from their soles to cool themselves using fanning movements in very warm weather. The seals illustrated are shown with dry coats, when species are easier to separate; identification at sea is much more difficult. No eared seals occur in the North Atlantic.

EARED SEALS
Family: Otariidae

I SEA LIONS

Fur short and stiff. Hindflipper digits are unequal in length, middle ones shorter. Ear flaps relatively short and close alongside head. Five species in five genera, all of which are illustrated here.

▲

1. Steller's (Northern) Sea Lion *Eumetopias jubatus*
World's largest sea lion. Buff-yellow to brown coat, lightening to grey when wet. Flippers blackish. Bulls muscular and bulky at the neck with rounded crown. Social, haul-outs can number thousands of animals. Polygynous, males defending territories against rivals. Feeds mainly on fish and squid. Coastal, N. Japan to California. Range similar to Northern Fur Seal (7) from which it can be distinguished by differing head shape. 3.3m, 1 tonne.

2. Hooker's (New Zealand) Sea Lion *Phocarctos hookeri*
Bulls dark brown to blackish with blunt well rounded face and very bulky neck. Adult females and subadults of both sexes light coloured with dark foreflippers. Polygynous, males defending territories against rivals. Feeds on wide variety of prey including fish, squid, crustaceans and occasionally penguins. Subantarctic Is. of New Zealand, mainly Auckland Is, Campbell Is. and Snares Is. 3.3m.

3. Californian Sea Lion *Zalophus californianus*
Males dark chocolate-brown, females uniform tan. Steep forehead and raised crown on bull, muzzle and area around eyes becomes lighter with age. Flippers black, subadults and females with more dog-like snouts. Feeds on fish and cephalopods. Polygynous, males defend territories on land and in water. Coastal, breeds in California and Mexico where population estimated at 160,000. Also found in Galapagos Is. 2.4m.

4. South American (Southern) Sea Lion *Otaria byronia*
Distinctive darker bulls with contrasting lighter underparts, thick mane and bulky neck. Muzzle curiously upturned at tip. Females with yellowish coats. Polygynous, males defending territories against rivals. Opportunistic feeder, taking both vertebrate and invertebrate prey items. Usually on sandy or shingle beaches, also rocky shores. Coastal South America: N Peru around to S Brazil, including Juan Fernandez Is. and Falkland Is. 2.5m.

5. Australian Sea Lion *Neophoca cinerea*
Bulls dark brown with creamy crown and nape, paler chest and throat. Young males have grey sheen. Females silvery grey. Foreflippers darker above. Polygynous, males defend territories and actively herd females (harems) within them. Feeds on shallow water benthic prey, including fish and squid. Non-migratory, confined to beaches from N of Perth to Adelaide. Population c.10,000. 2.5m.

II FUR SEALS

Dense coat with thick underfur is a primary distinction from sea lions. Hindflipper digits roughly equal in length. Ear flaps relatively larger and more obvious. Muzzle narrower. Nine species in two genera, four species are illustrated here. The genus *Arctocephalus* contains eight externally similar species; all but one are found in the S hemisphere. The genus *Callorhinus*, with a single species, is confined to nPO. Fur seals are polygynous, males aggressively defend territories containing shifting groups of females, but do not actively collect females with the possible exception of the Northern Fur Seal.

▲

6. Northern Fur Seal *Callorhinus ursinus*
Bulls are grey, brown or black with grey or yellowish frosting on mane. Adult females and subadults greyish with light 'bib'. Short, blunt face compared to *Arctocephalus*, with downturned muzzle; fur on foreflipper also differs in extending only to wrist level, ending in an abrupt line. Feeds mainly at night on fish and squid. nPO, Japan to California. Main breeding colony on Pribilof Is. (E Bering Sea), c.1.3 million animals. 2.1m.

7. Antarctic (Kerguelen) Fur Seal *Arctocephalus gazella*
Bulls dark grizzled grey with dark gingery ventral surface. Females and subadults grey and paler below. Females have uniform pale belly which can extend to muzzle rather than an obvious 'bib'. Unusual white form occurs infrequently. Feeds on krill, takes fish in Antarctic summer. Generally occurs S of Antarctic Convergence and as far south as pack ice in Antarctic winter. Main breeding colony on NW South Georgia. 2.0m.

8. Subantarctic Fur Seal *Arctocephalus tropicalis* **484**
This is the only southern fur seal with a colour pattern. Chest and face is a bright yellow or creamy hue, the belly is light brown and the back is grey, brown or black. Adult females have lighter coloured backs than adult males. A topknot of long guard hairs in adult males. Diet of fish, squid, krill and penguins. Widely distributed in S hemisphere. Breeds on subantarctic Is. (Gough, Prince Edward, Marion, Amsterdam, St. Paul). Main breeding colony on Gough Is. (c.200,000) animals. 1.8m.

9. South American Fur Seal *A. australis* **484**
Bulls are blackish or brownish-grey. Adult females and subadults are brownish to grey above and a lighter shade (brown, tan or grey) below with a pale neck. Light areas on muzzle and ear region. Diet of fish, molluscs and crustaceans. Can be confused with South American Sea Lion which occurs in same range; can be separated by differing head shape and habitat (fur seal is found only on rocky shores). Coast of central Peru around Patagonian coast to S Brazil, including Falkland Is. 1.9m.

3a
7
1
3
2
4
6
5
8
9

PLATE 55: NORTHERN TRUE SEALS

The eight species of true seals illustrated here are placed in the subfamily Phocinae or northern true seals; the Monachinae (southern true seals) are descibed on plate 56. In the Arctic, the commonest seal is the Ringed Seal (5) which is found on fast ice as far as the North Pole where there is open water amongst the ice. There are distinct differences in coat markings between pups and adults in the species illustrated; pups of Harp (2), Ribbon (4), Ringed (5), Grey (6) and Larga Seals (8) are white-coated. In the Hooded Seal (1) and Common Seal (7), the pup's white coat is generally shed in the mother's uterus before birth.

TRUE SEALS
Famliy: Phocidae

1. Hooded Seal *Cystophora cristata* **483**
Large seal, silvery-grey coat marked with irregular black patches. Male has black sac or 'hood' which hangs in front of mouth; when inflated strikingly increases apparent size of head. An inflatable red membrane can also be blown out of the left nostril (**1a**). Both are used during visual and vocal display in aggressive encounters between males. Breeds on pack ice, well away from land. Pups called 'bluebacks' (**1b**), coat with blue-grey back sharply demarcated from white belly. Feeds on fish and squid. Newfoundland to Spitsbergen, wandering individuals have been found as far south as San Diego, California. 2.6m, males larger.

2. Harp Seal *Phoca groenlandica*
Distinctive adult coat pattern, silvery-white with dark 'harp' marking on back and black face patch. Pups are born with a pure white coat **(2a)** but this soon changes after first moult to a grey colour with dark patches. The adult pattern does not appear until the seals become sexually mature (about 5 years of age), though some adults may retain juvenile spotting and an incomplete harp pattern (called 'spotted harps'). A few individuals retain all their spots with dark streaks over a grey background ('sooty harps'). Found mainly on pack ice often at edge, breeding at margins, may move north in summer to high Arctic. Wide diet includes fish and planktonic crustacea. Gulf of St. Lawrence to E Siberia. 1.9m, males larger.

3. Bearded Seal *Erignathus barbatus* **483**
Large seal with brownish to light or dark grey coat, darker above, often with scattered spots. Head smallish with large and distinctive display of whiskers (**3a**). Snout and areas around eyes lighter often with dark patch on crown extending forwards onto muzzle. Newborn pups are dark-coloured (usually brown) with white patches on back, crown, face and flippers. Tends to be solitary, normally curious but wary out of water, with heads near water's edge. Can be found asleep vertically in water. Wide diet of benthic prey species, feeds on molluscs like the Walrus (plate 56) and both these seals have big patches of prominent whiskers. Circumpolar. 2.5m, females larger.

4. Ribbon Seal *Phoca fasciata*
Striking markings on coat. Whitish to yellowish bands on reddish brown to black coat, females lighter with less distinct bands. Pups have whitish woolly coat. Solitary for much of their lives, wintering on southern edge of the pack ice, pups born on ice floes. Moves into open water of Bering Sea during summer. Diet of fish, cephalopods and benthic invertebrates. N Japan to Alaska. 1.8m.

5. Ringed Seal *Phoca hispida* **483**
Coat marked with pale-ringed dark spots and blotches, background hue varies, often grey above and lighter below. Little or no spotting on underside distinguishes it from other spotted seals, especially Common Seal (7) and Larga Seal (8). Breeds on fast ice where females excavate lairs; pups have whitish woolly coats. All other polar seals give birth on the exposed ice surface. Wide diet includes fish and planktonic crustaceans. Population estimated at 6-7 million animals. Preyed on by Polar Bears and taken by native peoples; pups have also been found in stomachs of the Greenland Shark (plate 10). Circumpolar. 1.65m. Two isolated populations Caspian Seal, *Phoca caspica* and Baikal Seal *Phoca sibirica*, are related to the Ringed Seal but are considered different species.

6. Grey Seal *Halichoerus grypus*
Long snout with flat top to muzzle. Nostrils set wide apart and almost parallel unlike Common Seal with which it can be confused. Adult males more thickset, up to 1.5 times longer than females, older bulls with high, arched muzzles (**6a**) giving a 'Roman-nosed' appearance. Coat individually variable, dark grey or brown to creamy-brown, often darker above with irregular blotches on sides. Underside and neck can be orange or reddish, males tend to darken with age. Pups have creamy-white coat (**6b**). Generally gregarious when hauled out on rocky shores for breeding and moulting. Grey Seals are polygynous, but without territories (as in Walrus) or harems (as in elephant seals). Visually inquisitive, often 'bottling' with only head exposed at the surface. Diet includes fish and invertebrates, known to take seabirds. Coastal distribution in cold temperate and subarctic waters. Population in British and Irish waters estimated at 115,000 (c.50% of world population). 2.3m, males larger.

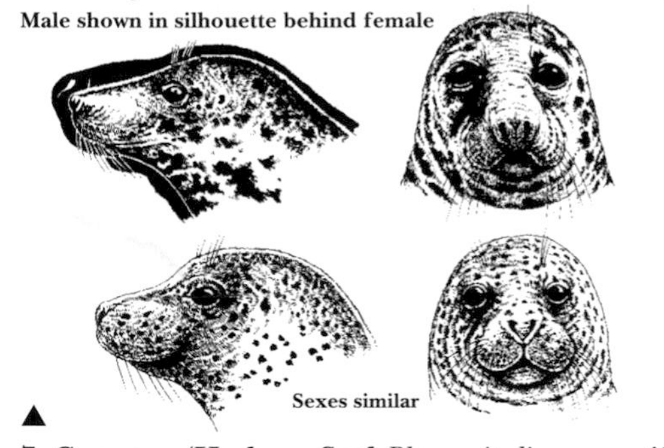

7. Common/Harbour Seal *Phoca vitulina* **483**
Small seal with large head and short body. Nostrils in wide 'V', face rounder, flatter and more endearing than adult Grey Seal. Coat variable, often brownish to grey with dark spots and blotches, paler with sparse spots below. Pups usually have adult markings on coat from birth. Birth may occur on sandbanks, offshore rocks or even in the water. Often seen in 'banana posture' on haul-out sites with head and tail raised. Diet includes fish and invertebrates. Widespread species in the N hemisphere in coastal temperate and polar waters, using sandbanks, rocky terraces and ice for hauling out. British and Irish population estimated at 30,000+ (c.5% of world population). 1.9m, males larger.

8. Larga Seal *Phoca largha* **483**
Once considered as a subspecies of the Common Seal, now recognised as distinct. Breeds on pack ice, not on beaches and sandbars. Coat dark above and lighter below. Adult coat covered in even scattering of dark spots all over body, unlike Common Seal. Face and snout are also darker than in Common Seal. Pups have white woolly coat. Feeds on benthic fish, crustaceans and cephalopods. Yellow Sea to Alaska. 1.7m, males larger.

1a
2
2a
1
1b
3a
3
4
6
5
6b
6a
7
8

PLATE 56: SOUTHERN TRUE SEALS, WALRUS AND SEA COWS

Southern phocids, placed in the subfamily Monachinae, include the elephant seals, monk seals and the Antarctic seals. This subfamily contains the largest of all the world's seals, the elephant seals, and the most abundant, the Crabeater Seal. The latter is probably the world's most numerous, non-domestic, large mammal after Man.

TRUE SEALS
Family: Phocidae

I ANTARCTIC SEALS

1. Weddell Seal *Leptonychotes weddellii*
Short muzzle, head looks small relative to body, especially outside spring breeding season when body bulk is at a maximum. Coat greyish-white below, dark bluish-grey above, with variable blotches. Can fade to browner colour prior to moulting. Pups with silvery-grey woolly coat. Not social but will congregate around breathing holes in ice. In winter, ice holes are kept open by scraping away newly formed ice with teeth. Dives deeply, up to 700m and for up to 82 minutes. Varied diet, mainly fish, which are hunted under the ice using keen eyesight and swift swimming. Most southerly breeding seal, in fast ice areas as well as pack ice and islands on Antarctic Peninsula. 3.2m, females larger.

2. Ross Seal *Ommatophoca rossii*
Obvious brown or reddish streaks on thickset neck. Hind flippers very long, about 20% of body length (measured belly up from snout tip to tail). Pups dark above, pale below, with throat streaks. Feeds on cephalopods, fish and krill.Widespread but solitary. Will rear up when approached by humans on the ice and 'sing'. Inhabits dense pack ice, poorly known species. 2.4m, females larger but few seals have been measured.

3. Leopard Seal *Hydrurga leptonyx*
Sleek body, large head with reptilian appearance and distinctive neck. Long foreflippers. Characteristically large gape (**3a**), with impressive teeth. Adult coat dark above, light below and variably spotted. Woolly coat of pups similar colour to adults but with dark stripe down back, pale below. Mostly inactive when out of water (**3b**). Wide diet includes krill, penguins, seals and occasionally carrion. Can appear to behave aggressively near boats and their occupants, reported to bite and sink moored inflatable craft. Widespread but uncommon. Antarctic and subantarctic, vagrants occur well to north. 3.4m, females larger.

▲

4. Crabeater Seal *Lobodon carcinophagus*
Sleek with long muzzle. Coat uniform silvery-grey to brown, rich sheen when recently moulted, paler on belly. This seal is also known as White Seal, as the coat fades all over to a pale colour in older animals in summer (**4a**). Often individuals have long dark scars, attributed to Leopard Seal attacks when young. Darker blotches around foreflippers. Pups with greyish-brown woolly coat, flippers darker. Feeds on krill mainly at night; ornately cusped interlocking teeth which are particularly marked in this seal, sieve prey from water. Dives recorded to 430m. Found around pack ice often in large groups, unlike other southern phocids. Vagrants occur well to north. Most abundant seal, perhaps numbering 11-12 million animals. 2.6m.

II ELEPHANT SEALS

5. Southern Elephant Seal *Mirounga leonina* **484**
On land, colour is light to dark grey when newly moulted, though normally rusty-brown due to discoloration from sand, mud and excrement. Bulls are huge, heavily scarred with inflatable sac or proboscis on snout. Females much smaller with broad muzzle and large-eyed appearance. Pups with black woolly coat. Elephant seals are polygynous, males display aggressively during the breeding season for the right to mate with groups of females (harems). A bull may control up to about 50 females in a harem. Breeding displays are usually decided without bloodshed but occasionally a pair of bulls will fight inflicting lacerations on each other with their sharp canine teeth. Remarkable divers, females recorded at depths of over 1700m and underwater for up to 2 hours. Young have a varied diet; adult diet consists of fish and cephalopods. Widely distributed in S hemisphere, breeds on islands on both sides of Antarctic Convergence. Males up to 5m and 4 tonnes, females much smaller (3m) and lighter (800kg). **Northern Elephant Seal** *M. angustirostris.* Similar to Southern Elephant Seal, but male smaller with larger proboscis. Population once down to 100 animals on Guadalupe Is. off Mexican coast after near extermination by sealers; now breeds in good numbers. Enormous migratory range from Gulf of Alaska to Baja California, straying as far west as Japan.

III MONK SEALS

Robust, long-bodied seals with well rounded faces and upwardly angled nostrils on top of muzzle. Monk seals are characteristically lethargic when hauled out. Both illustrated species are listed as endangered; the **West Indian Monk Seal** *Monachus tropicalis* is probably extinct, last reported in 1952. Its coloration was said to be brown grading to white below. A captive breeding programme for the Mediterranean Monk Seal using animals from the Cap Blanc peninsula (W Sahara), where up to 130 seals may be present, is planned.

6. Mediterranean Monk Seal *Monachus monachus*
Colour varies from reddish-brown to black with pale belly patch. Pups have woolly black coat with light belly patch. Isolated non-migrating groups exist in MS (5 relict populations) and eAO (2 relict populations). There are less than 300 animals alive today. Secretive in its habits, known to haul-out and breed in caves with hidden or underwater entrances in MS. 2.4m.

7. Hawaiian Monk Seal *Monachus schauinslandi*
Grey to brownish above, paler below. Pups with woolly black coat. Occurs in leeward Hawaiian Islands, occasionally on the main islands. Population c.1000. 2.4m, males smaller.

WALRUS
Family: Odobenidae

8. Walrus *Odobenus rosmarus*
The unique tusks of this species are elongate canine teeth up to 1m long, each with a weight of up to 5kg, which occur in both sexes. They are straighter, longer and more heavily built in males. Tusks are used to haul-out on to beaches and ice floes, for social interactions and to maintain breathing holes in the ice. The Walrus is

2
3b
3
3a
8
4
4a
5
7
11
10
9

polygynous, with males forming small aquatic territories where social interactions occur (vocalising, displaying) and males fight amongst themselves. Pups are grey-coated with black flippers; a white coat is shed before birth in the mother's uterus. The Walrus is vocal in air and produces eerie bell-like calls underwater. Found in shallow coastal waters associated with pack ice in AO, PO and Laptev Sea where its food (bivalves) is abundant. 3.2m, 1.2 tonnes, females smaller.

SEA COWS
Order: Sirenia

Fully aquatic marine and freshwater herbivores. Four living species, three partly or fully marine species (illustrated here) and one entirely freshwater species.

9. Dugong *Dugong dugon* **482**
Skin colour grey or brown, lighter on belly. Whale-like tail and paddle-like forelimbs without nails. IP distribution, in areas with sea grass beds. 3.3m, estimated weight 420kg at 3m body length.

10. West Indian Manatee *Trichechus manatus*
Greyish or brownish skin, often with greenish colour from attached algae. Tail rounded and nails present on paddles. This illustration shows female with calf nursing at axillary nipple. Coastal Caribbean, Gulf of Mexico, Florida, tropical wAO coast of S. America, middle and lower reaches of Orinoco River. 3.9m, 1.5 tonnes.

11. West African Manatee *Trichechus senegalensis*
Similar in colour, length and build to West Indian Manatee, but blunter snout and more 'pug-faced'. Coasts and freshwater river systems of tropical W Africa.

Evolution of the Vertebrates

by

Geoffrey Waller

INTRODUCTION

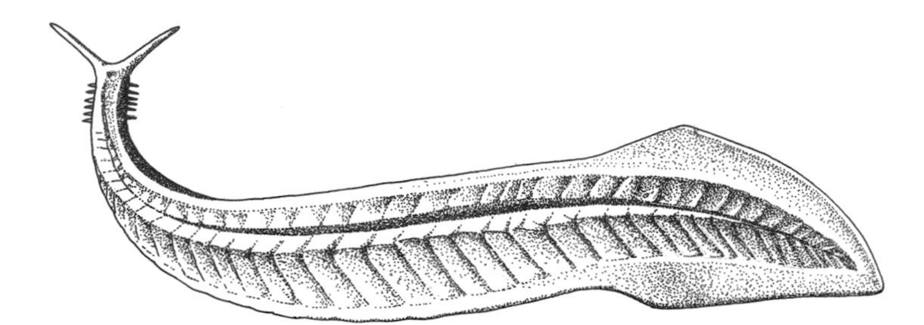

Figure 179. Reconstruction of the fossil chordate *Pikaia* (length about 5cm) found in rocks of Middle Cambrian age.

Most of the living vertebrate species are aquatic; these are primarily fishes (over 20,000 species) and many of the 4000 amphibian species. Among the terrestrial vertebrates, the birds (about 9,000 species) and the squamate reptiles (lizards and snakes, about 5,700 species) are the most successful groups in terms of species numbers. The marine tetrapods (reptiles, birds and mammals) number about 500 species with about 80 species of reptiles and 300 species of birds. Mammals are poorly represented in the marine environment. There are some 4600 mammalian species of which about 2.5% are found in the marine ecosystem. Marine mammals, which number about 120 species, are distributed in three of the 26 orders within the class Mammalia, namely Carnivora (carnivores), Cetacea (whales and dolphins) and Sirenia (sea cows). Some 'marine' mammals occur only in freshwater, such as certain river dolphins (Platanistidae), the Baikal Seal *Phoca sibirica* and the Amazonian Manatee *Trichechus inunguis.*

The subphylum Vertebrata (vertebrates) traditionally comprises five classes of backboned animals (fishes, amphibians, reptiles, birds and mammals). The Vertebrata is one of three subphyla making up the phylum Chordata (table 8). Two of these subphyla, the tunicates and the lancelets, are termed the 'protochordate' group of organisms since they show certain structural features found in vertebrates and are probably closely related to vertebrates in their evolutionary history.

THE CHORDATES AND VERTEBRATE ANCESTRY

Characters that broadly define vertebrates include anterior paired sense organs, complex muscle systems for active movement and hard phosphatic skeletal tissues for support. All vertebrates share the following common features:

1) Gill-bearing pharyngeal slits
2) Skeletal supporting rod or notochord
3) Hollow dorsal nerve cord
4) Tail

Two protochordate groups (table 8) share some of these features with the vertebrates. Some members of

PHYLUM : HEMICHORDATA

Class : Enteropneusta (acorn worms)
Class : Pterobranchia (*Cephalodiscus, Atubaria, Rhabdopleura*)

PHYLUM : CHORDATA

Subphylum : Urochordata (tunicates) } Protochordata
Subphylum : Cephalochordata (lancelets, amphioxus) } Protochordata
Subphylum : Vertebrata
Class : Agnatha (hagfishes and lampreys)
Class : Chondrichthyes (sharks, rays and chimaeras)
Class : Osteichthyes (bony fishes and tetrapods)
Subclass : Actinopterygii (ray-finned fishes)
Infraclass : Cladistia (bichirs and Reedfish)
Infraclass : Actinopteri
Chondrostei (sturgeons, paddlefishes)
Neopterygii (gars, Bowfin) and modern ray-finned fishes (Teleostei)
Subclass : Sarcopterygii (lobe-finned fishes)
Infraclass : Actinistia (coelacanths)
Infraclass : Rhipidistia
Dipnoi (lungfishes)
Porolepiformes*
Osteolepiformes*
Tetrapoda (land vertebrates)
Lissamphibia (modern amphibians)
Amniota (all other tetrapods, i.e. reptiles, birds and mammals)

Table 8. A higher classification of extant (and extinct*) major chordate and near-chordate groups. The two extinct groups of the infraclass Rhipidistia were fishes with either lungfish-like features (Porolepiformes) or primitive amphibian features (Osteolepiformes).

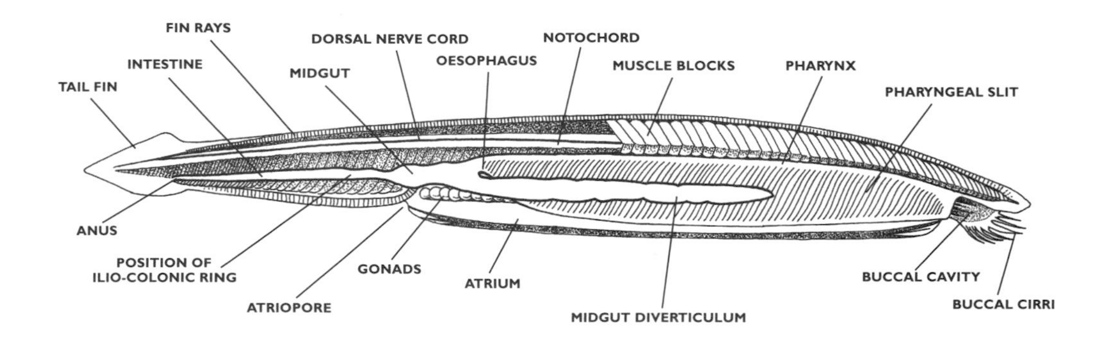

Figure 180. Organisation of amphioxus (atrial wall and most muscle blocks removed on right side). The body is dominated by the enormous pharyngeal chamber which is used to filter food from sea water. There are no paired fins and only limited head-tail differentiation.

the first group, the tunicates (subphylum Urochordata), possess a tadpole-like larva in which some vertebrate features are apparent (notochord, cranial sense organs, neural tube and muscular tail). The evolution of vertebrate ancestors by the process of neoteny (premature development of adult reproductive organs) occurring in a tunicate-like larva has been suggested as a possible evolutionary route linking invertebrates and vertebrates, though without implying the direct evolution of vertebrates from urochordates. The chordate-like features of the sea squirt larva are shown in fig.182. The second group, the lancelets (subphylum Cephalochordata) look much like fish larvae (fig.180), and amphioxus *Branchiostoma* has several anatomical features (mode of formation and relations of muscle blocks or myotomes, notochord, neural tube, digestive tract and vascular system) that are typically vertebrate. However, other features such as formation of mesoderm during development, sperm structure and innervation of muscles (fig.181) are typically echinoderm in nature.

During vertebrate evolution, numerous types of body form have developed and gills and tails may have been lost. Structural modifications of the notochord (e.g. the calcified vertebral column of bony fishes and most tetrapods) have also developed. Not all the structural features listed above may be evident in advanced vertebrates but they are always evident in larval or embryonic stages.

It is traditionally held that the early vertebrates had a free-swimming form similar to amphioxus or an ammocoete larva (see page 304). This larva may have had a 'tongue' or endostyle, ciliated pharyngeal slits and a segmentally arranged muscle system innervated by segmental nerves. In the case of an ammocoete-like ancestor there would also have been paired cranial sense organs (eyes, ears) and an olfactory apparatus. Some of these features are seen in the tadpole-like larva of the sea squirt *Ciona* which has a free-living planktonic life of 6 - 36 hours before attaching itself the substrate. Here it metamorphoses into a sessile form and develops both sets of reproductive organs (*Ciona,* like most tunicates, is hermaphroditic). Unlike the adult sea squirt, this tadpole-like larva has a distinct head-tail axis. This important feature indicates that polarity of the body axis is an ancient trait in evolution and not a new 'invention' of the vertebrates.

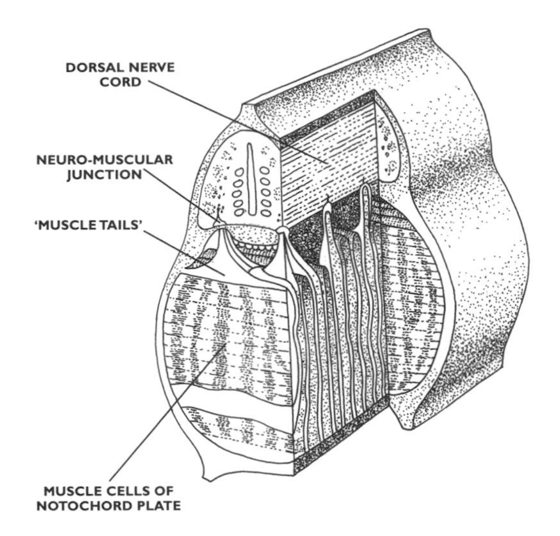

Figure 181. Transverse section of the nerve cord and notochord of amphioxus. The notochord muscles are not innervated by peripheral nerves but instead make direct electrical connection to the nerve cord by processes or 'muscle tails'. Muscle tails occur in the segmented muscle blocks of the body wall. This type of nerve-muscle connection is also found in echinoderms and nematodes.

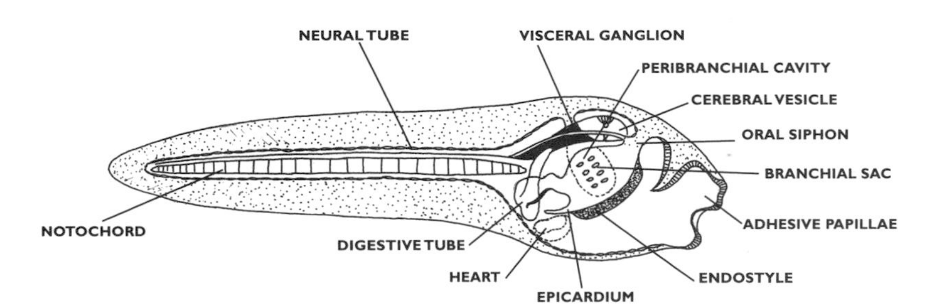

Figure 182. Late larval stage of the sea squirt *Clavelina* showing advanced structural organisation characteristic of tunicate larvae.

The earliest fossils that are unquestionably vertebrate are the ostracoderms; the first known examples of these fishes come from the Middle Ordovician period. Ostracoderms are thought to have been suspension feeders as they lacked both teeth and jaws. They may therefore have been relatively inactive and ecologically comparable to present-day amphioxus and larval lampreys. The heavy armour of bony scales and plates present in ostracoderm fishes may have been linked to defence and this observation suggests that the vertebrate skeleton evolved primarily as a protective device. In this scheme, teeth evolved from bony scales that migrated to the mouth in a process that lasted 100 million years. The change from suspension feeding to active predation was therefore a gradual process in which tooth formation was secondary to the evolution of the skeleton. However, the presence of fossil teeth from an enigmatic group of animals called conodonts in Upper Cambrian deposits suggests that teeth may have been present much earlier in the evolution of the vertebrates. Some authorities consider the conodonts to be vertebrates which would suggest that active predation evolved before suspension feeding.

The neoteny theory of evolution proposes that the ancestral vertebrate condition is represented by a tunicate larva that developed reproductive organs and lost the sedentary stage in its life cycle. Neoteny is well known in a species of salamander (Axolotl *Ambystoma mexicanum*) where it results in the suppression of the normal adult stage in the life cycle. In this respect, the derivation of an early vertebrate ancestor by neotenic development of the tunicate larva is an attractive explanation where the adult stage is a 'degenerate' sessile form. Indeed, in some tunicates (class Larvacea) the tadpole-like larval form persists and develops reproductive and other organs enabling it to survive as an 'adult larva'.

The acorn worms or enteropneusts have pharyngeal slits and in some species a hollow nerve cord is present. These burrowing, filter-feeding organisms have been classified as chordates in the past but they lack a true notochord and differ from chordates in other important ways. These 'prechordates' are placed here in a separate non-chordate group (phylum Hemichordata, table 8). The hemichordates comprise two classes: the class Enteropneusta (acorn worms) includes *Saccoglossus* and *Balanoglossus*; the class Pterobranchia (pterobranchs) comprises *Cephalodiscus, Atubaria* and *Rhabdopleura*. Some species of acorn worms have a free-living larva (tornaria) that closely resembles the larva of starfish (bipinnaria). Other features of the coelom of enteropneusts may also indicate a close relationship with echinoderms and, indirectly, an evolutionary connection between echinoderms and chordates.

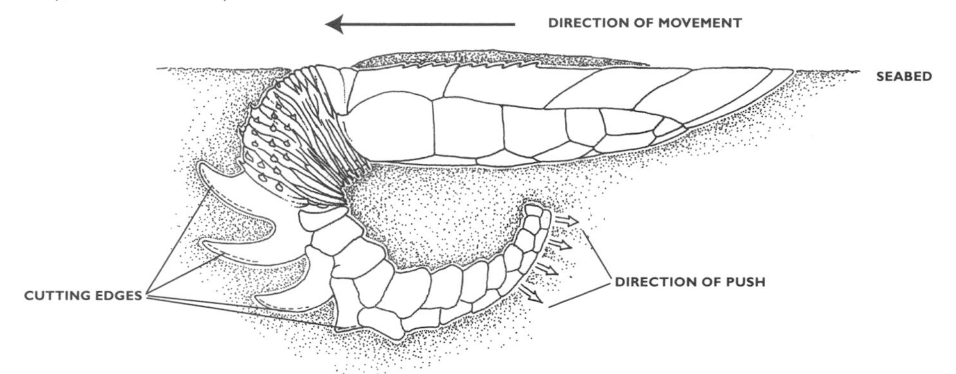

Figure 183. A calcichordate *Chinianocarpos thorali* in side view showing how the tail was used in locomotion on the seabed to move the animal backwards. This species is from Lower Ordovician sediments (head length about 10mm). (After Jefferies 1986).

The calcichordates are fossil animals known from the Cambrian to the Devonian period that have often been grouped with the echinoderms. These curious marine animals have an asymmetrical body shape and an echinoderm-like external skeleton. However, the calcichordates also show evidence of typical chordate features (notochord and segmentally-arranged muscular tail). They seem to have benthic animals that probably used their heavily armoured and muscular tails for moving along the seabed (fig.183). The calcichordates are considered by some authorities to hold a pivotal position in the evolution of early vertebrates from hemichordate ancestors.

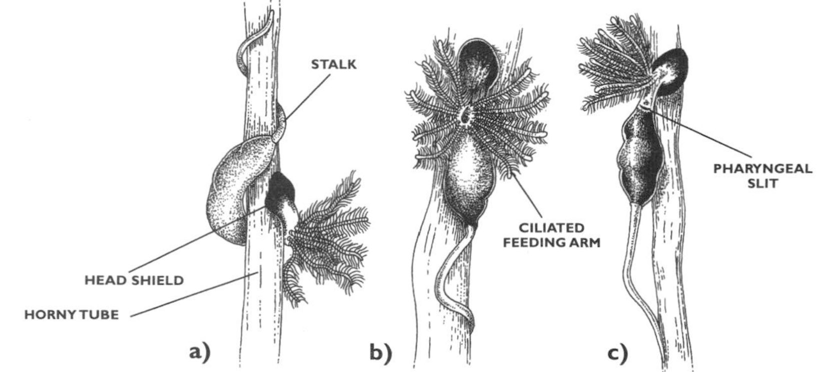

Figure 184. The pterobranch *Cephalodiscus* secretes a horny tube attached to the substrate inside which it shelters. When it moves out of the tube, it creeps using the head shield as a grasping organ and the stalk as a prehensile 'tail'. *Cephalodiscus* is 5-15mm long excluding the stalk.

These ancestors are thought to have been close to the extant pterobranch hemichordate *Cephalodiscus*. This pterobranch has a distinct head-tail axis and is capable of independent movement amongst the horny tubes in which it lives (fig.184). It has ciliated feeding arms and a single pair of pharyngeal slits that probably aids the entry of food material into the digestive tract. Pharyngeal slits are also found in the related pterobranch *Atubaria* (fig.185). There are numerous similarities in the body plan of pterobranchs and calcichordates which suggest a close evolutionary relationship between these two groups, and in turn, with stem chordates.

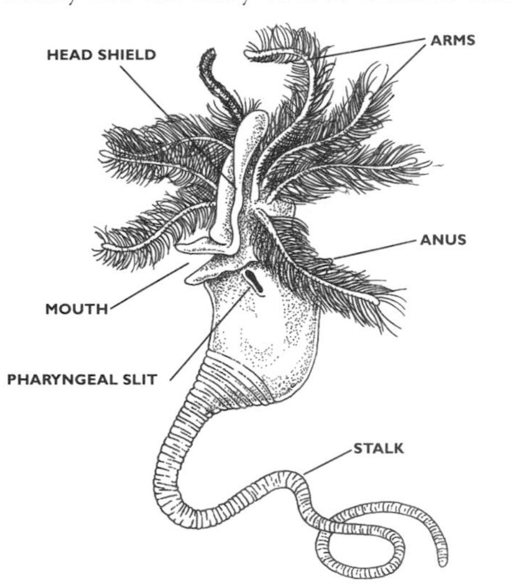

Figure 185. The pterobranch *Atubaria* has a pair of pharyngeal slits that lie below the mouth. It lacks a horny tube and lives as solitary individuals on hydroid colonies.

The earliest known chordate, *Pikaia*, comes from the Cambrian shales of Canada of about 550 million years of age (fig.179). *Pikaia* had some features that are similar to amphioxus; both have V-shaped muscle blocks along their sides supported by a notochord and in *Pikaia* there is a pair of sensory tentacles at the front end. Behind this is a row of projections which may be similar to the buccal cirri of amphioxus. The presence of primitive chordates in sediments of this great age suggests that the chordate lineage may have originated at about the same time as many of the invertebrate phyla (early Cambrian period).

Whilst there is much discussion about the merits of different invertebrate groups as potential chordate ancestors, the weight of evidence presently available shows unequivocally that the extant invertebrate group most closely related to the chordates is the phylum Echinodermata.

EARLY PALAEOZOIC VERTEBRATES

Fishes have an ancient ancestry extending back about 500 million years. The oldest known fossil records of vertebrates are small fish-like animals found in rocks from Bolivia and Australia of Middle Ordovician age. These early jawless fishes (ostracoderms) evolved into a variety of different forms, though most became extinct by the end of the Devonian. They possessed either external armour or small bony plates and an internal skeleton of cartilage. One group, the anaspids, possessed structural features similar to those of modern lampreys. The anaspid *Jamoytius,* which was a marine fish, had a terminal sucking mouth and a row of circular gill openings behind the eye (fig.186).

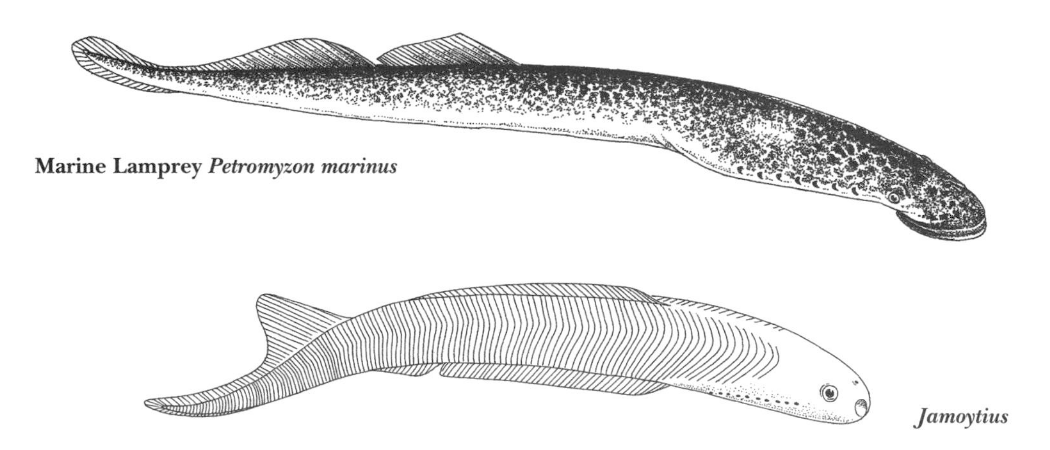

Figure 186. Marine Lamprey *Petromyzon marinus*, and the extinct anaspid *Jamoytius*

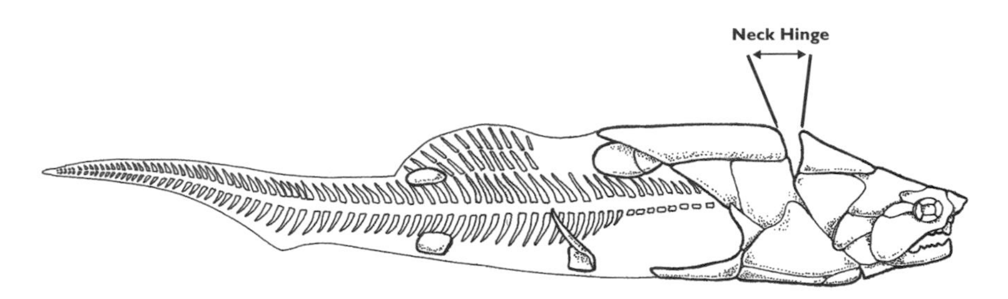

Figure 187. Body outline and skeleton of a placoderm of the order Arthrodira.

The first recognisable fossil lampreys (*Hardistiella* and *Mayomyzon*), similar to today's modern forms, were discovered in marine sediments from the Carboniferous period and they probably have an ancestry among the anaspid fishes. By contrast, there is little fossil evidence of hagfishes (a more primitive animal than the lamprey) which probably evolved as a much earlier offshoot from the early vertebrate line. The hagfishes and lampreys are traditionally grouped together in the Agnatha (jawless fishes).

Fossil records show that the earliest known vertebrates with jaws (gnathostomes) appeared during the Silurian period. In addition to primitive jaws, early gnathostomes differed from most ostracoderms as they possessed paired, mobile fins. There were also differences in the arrangement of the semicircular canals in the ear. Primitive gnathostomes are represented by two distinct groups, the Acanthodii and Placodermi. Both groups lived at about the same time in geological history and probably occupied different habitats which may have reduced competition for food. During their evolution both groups diversified, with the evolution of mid-water and bottom-dwelling forms. The placoderms comprised heavily armoured (fig.187) or scale-plated species that varied in length from 30cm to 6m. The acanthodians were small, marine or freshwater animals living in rivers, lakes and swamps. Their bodies were covered with small scales which became enlarged as small plates on the head. Early placoderms were mostly marine and most later forms were freshwater species.

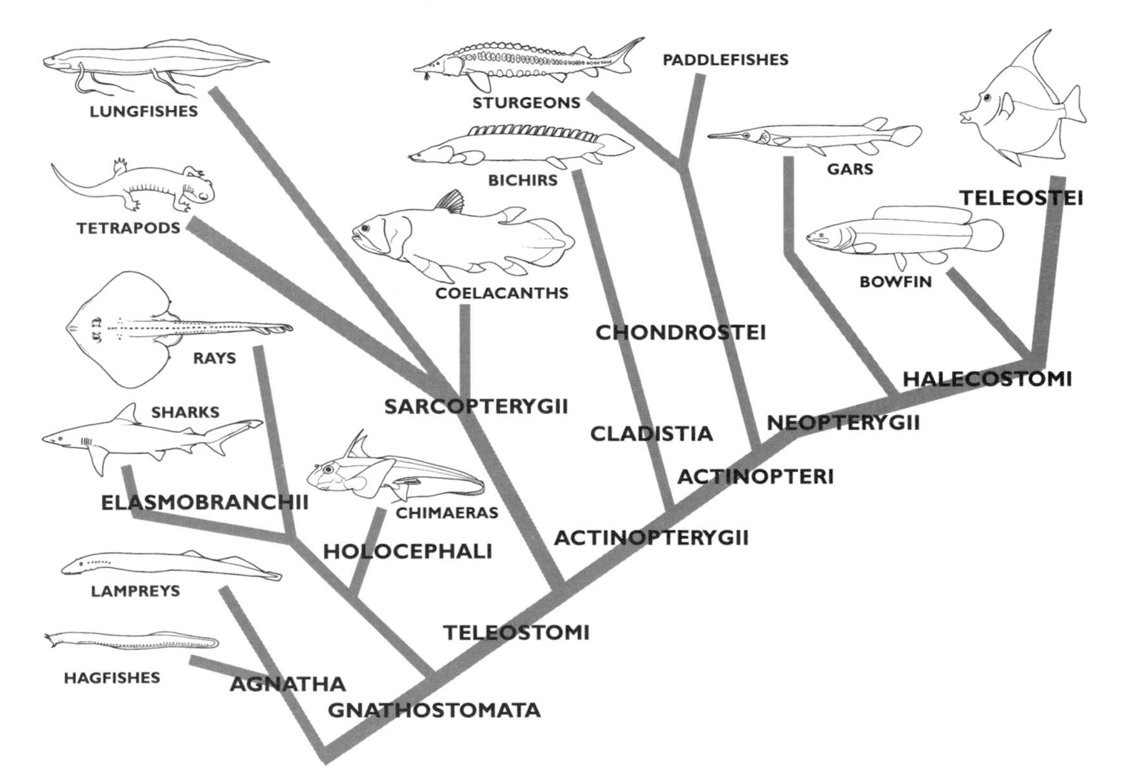

Figure 188. Relationships of living fish groups. The extinct acanthodian fishes (not shown) are assumed to be the earliest known bony fishes (teleostomes).

The early acanthodians are typified by the genus *Climatius.* Although this species was only a few centimetres in length it had the typically recognisable fish shape with a body that tapered from snout to tail. In addition to the heterocercal tail there were a number of median and paired fins (fig.192). The body was completely covered by diamond-shaped rhombic scales and the eyes were well developed and positioned forwards. This suggests that eyesight was of sensory importance to the fish. *Climatius* had five gill arches covered by a large external operculum similar to the more advanced fishes of today. It also had an arrangement of ear stones (otoliths) similar to modern ray-finned fishes. As the acanthodians evolved and diversified, some species became long and slender, while others developed deeper bodies. The fin structures and arrangements also changed. By early Devonian times the acanthodians reached their peak of evolutionary development and by the end of the Palaeozoic they had disappeared.

The legacies of the acanthodians and placoderms were the evolution of fish-like body forms, paired fins and most importantly the development of jaws. Without jaws and paired fins the vertebrates could not have completely abandoned the predominately bottom-dwelling lifestyles of the primitive ostracoderms or taken full advantage of new food sources. The development of paired fins was also fundamental in the later evolution of the vertebrates when fishes moved from water to land. The closest ancestors to the tetrapods are the lobe-finned fishes which evolved a strong fin skeleton with a central supporting column of bones. This type of skeleton is found in two groups of living fishes, the lungfishes and the Coelacanth *Latimeria chalumnae.* Modern-day lungfishes use their limbs when moving along the bottom in the same way as amphibians walk on land. There is much discussion about which of these two groups of fishes is most closely related to tetrapods and three alternative evolutionary relationships are shown in figure 189.

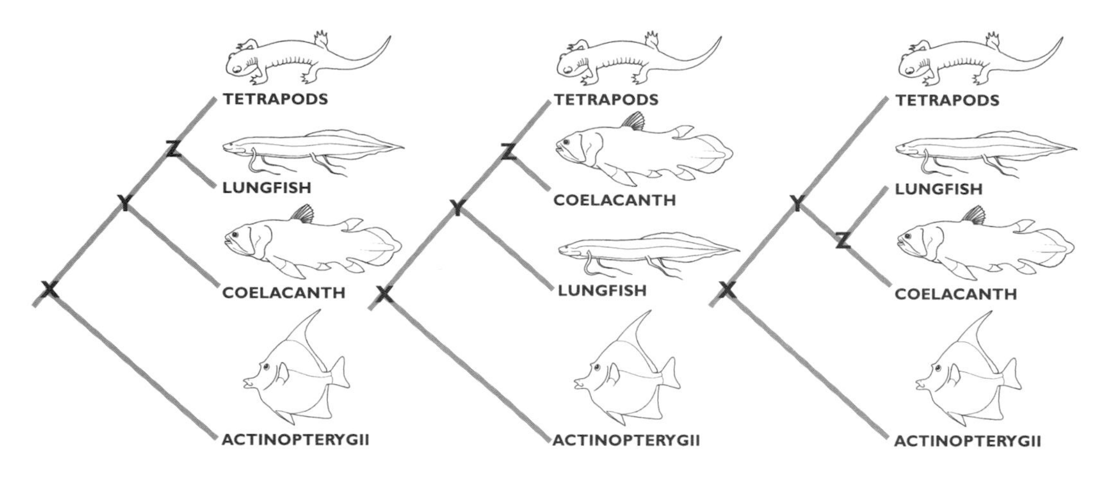

Figure 189. Hypothesised relationships among the major groups of living bony fishes (Teleostomi) showing three different interpretations. X, Y and Z denote branch points in each tree.

JAW EVOLUTION

Although the ostracoderms lacked jaws, they did develop hard tissues that supported the gill pouches. These were internal pockets that opened into the pharynx and were lined with fine blood vessels for gaseous exchange. Later these gill pouches developed external openings or gill slits. This enabled the flow of water to enter the pharynx via the mouth opening and pass out through the gill slits to the surrounding water. In the extinct cephalaspids (ostracoderms), an anterior mouth opened into the pharynx that contained a series of ten pairs of gill slits. These slits opened along each side of the underside of the head and each gill opening was supported by a rod of hard tissue. Such supporting rods are thought to have developed over time into the gill skeleton (branchial gill arches) which evolved in later fishes.

The upper and lower jaw bones of the early gnathostomes probably developed from the cartilaginous supports of the first gill arch. The jaws were initially suspended by ligaments and buttressed against the floor and sides of the skull (autostyly), though this a hypothetical condition since the earliest known gnathostome fossils show a progression beyond this stage. In these fossil fishes the jaw suspension incorporated the hyomandibula cartilage (one of the elements of the upper portion of the second or hyoid gill arch). This provided the jaws with posterior support. This primitive type of jaw suspension where the jaws were attached both to the skull and hyomandibula is termed amphistyly (fig.190a). Later a more advanced type of jaw suspension developed involving only the hyoid arch (direct articulation with the skull was lost) and this is termed hyostyly (fig.190b). In most modern fishes the jaw suspension is of the hyostylic type.

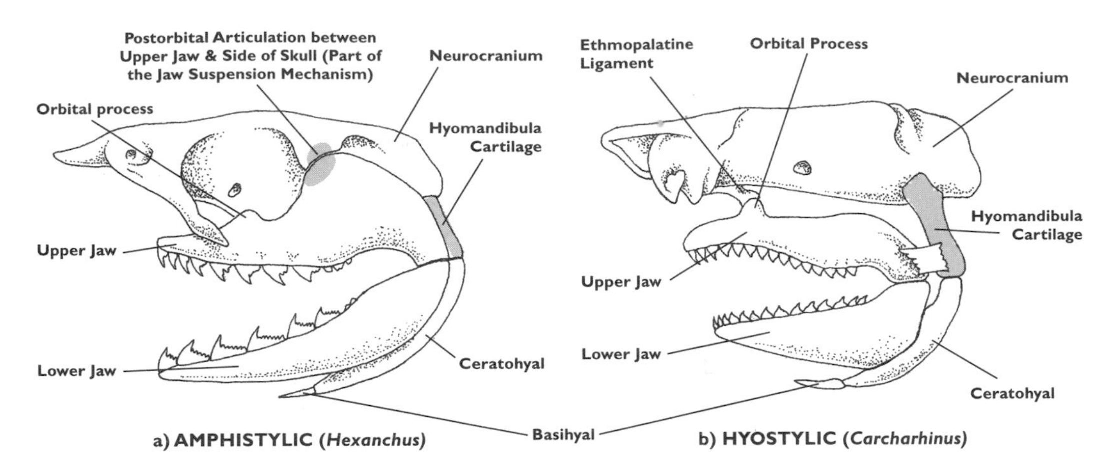

Figure 190. Comparison of a) amphistylic and b) hyostylic jaw suspension in sharks.

The spiracle is the only remnant of the gill slit originally associated with the hyoid arch. Therefore gill arches first developed to support gill slits and were later modified as jaw bones and the jaw suspension mechanism. Finally, connection with other branchial arches provided the multi-functional hyoid apparatus of modern fishes that is used for both feeding and breathing. The transformation of an anterior pair of gill arches into paired upper and lower jaws was a progressive evolutionary development, perhaps the simplest possible solution to the basic evolutionary problem of developing jaws.

EVOLUTION OF CARTILAGINOUS FISHES

The ancestry of both the cartilaginous fishes (Chondrichthyes) and the bony fishes (Osteichthyes) can be traced back to the Silurian period. During this time sharks developed rapidly and continued their expansion throughout the Carboniferous and Permian periods. Most of the early sharks were marine forms.

Although ancient and modern sharks mostly lack bone, it is presumed that sharks had a common ancestry with the early bony fishes. From fossil records one of the better known of ancient sharks is *Cladoselache* (fig.191). This shark grew to a length of about one metre and had a typical shark-like appearance. It had a large heterocercal tail, two dorsal fins, pectoral and pelvic fins and a pair of small, horizontal fins (one on either side of the tail base).

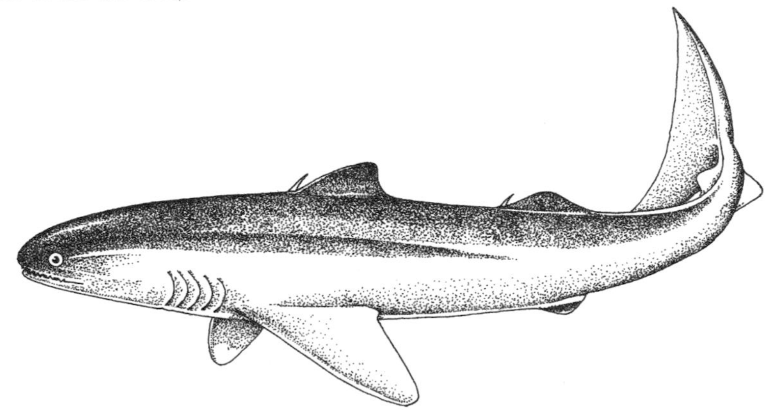

Figure 191. The Devonian shark *Cladoselache*.

Cladoselache had small eyes and well developed nostrils. The jaw suspension was primitive (amphistylic) and the teeth had long central cusps with low lateral cusps on each side. It was probably an active and efficient predator on other fishes. However *Cladoselache* is considered too specialised to have been related to modern sharks. The ancestral group of all the modern forms of sharks may have been the ctenacanthid sharks of which *Bandringa* was an example (fig.192). From the Ctenacanthidae evolved the hybodont sharks (*Hybodus*, fig.192) which are abundant in marine Triassic and Jurassic sediments. The present-day horn sharks (*Heterodontus* spp.) may be near relatives of the hybodonts.

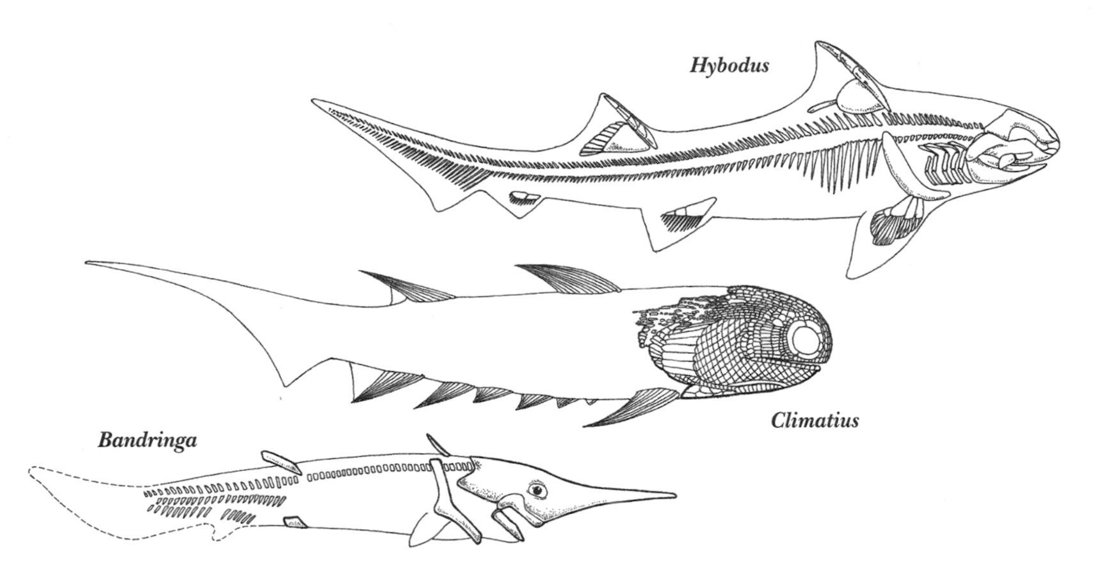

Figure 192. Body outlines and skeletons of *Hybodus*, *Climatius* and *Bandringa*.

When early sharks appeared in the Devonian period, they rapidly radiated and displaced the mid-water acanthodian and placoderm fishes that occupied the same habitats. At about the same time the chimaeras (Holocephali) became common in the habitats of the bottom-dwelling placoderms and acanthodians to become the major, bottom scavengers and mollusc feeders. They probably evolved from shark-like ancestors and are known from the Upper Devonian; indeed some Carboniferous holocephalan-like fishes had shark-like teeth and dermal denticles in the skin (Iniopterygiformes). The chimaeras reached their peak of diversity during the Permian period but by the Mesozoic era they had been largely displaced by rays. Today there are only three surviving families of chimaeras which are only distantly related to modern sharks.

Sharks reached their peak of diversity and expansion during the upper Palaeozoic era, but by the end of the Permian period many forms were extinct and the succeeding members less numerous. However, whilst today's living sharks are successful predators at the top of the marine food chain, they are represented by relatively few species. The reasons for their demise in the Permian are uncertain, though it is possible that reduction of suitable prey may have played a role. By the Jurassic period, fossil records indicate that some groups of sharks evolved into bottom-living forms and the skates and rays of today probably evolved from a shark-like ancestor. The modern rays represent a highly specialised type of adaptation to a bottom-dwelling existence, though some species have re-adapted to a mid-water lifestyle.

Although sharks have been evolving for the past 400 million years in one form or another, all modern lineages of sharks and rays were present by the beginning of the Cretaceous period, about 140 million years ago. Some modern species can be traced back in the fossil records to similar fossil forms from the Jurassic period such as guitarfishes (Rhinobatidae), cow sharks (Hexanchidae), wobbegongs (Orectolobidae), catsharks (Scyliorhinidae), and angel sharks or monkfishes (Squatinidae).

EVOLUTION OF BONY FISHES

One of the most important events that occurred in the early evolutionary history of the vertebrates was the radiation of the bony fishes during the Devonian period. This influx of comparatively advanced fishes was crucial in leading to the decline and subsequent disappearance of the dominant primitive vertebrates of that time such as the ostracoderms and placoderms. From the Middle Devonian, the bony fishes evolved with great rapidity and variety, first as freshwater types and later as marine fishes.

During the early stages of their evolution, the bony fishes differentiated to comprise two separate lineages, the lobe-finned fishes (Sarcopterygii) and the ray-finned fishes (Actinopterygii). During the Palaeozoic era, both lineages of bony fishes were found in small but similar numbers which suggests they inhabited different environments which reduced competition. It is thought the lobe-finned fishes initially diversified in the sea and returned to freshwater by the late Devonian period. By contrast, the ray-finned fishes may have diversified first in freshwater and remained there until the Mesozoic era when the break up of the supercontinent (Pangaea) created new habitats. The strong skeletons and fleshy fins of lobe-finned fishes may have been used for walking on the bottom of shallow seas. The lighter and more manoeuvrable ray-finned fish limbs (in which the fins are composed of a web of skin supported by many slender, horny rays) were possibly used to control body movements in fast-flowing rivers and lakes.

Most lobe-fins became extinct by the close of the Palaeozoic era though a few sarcopterygian fishes still survive to the present day. These include six species of freshwater lungfishes (Dipnoi) that are represented by three genera from Africa, Australia and South America. All living species are specialised bottom-dwellers, eating detritus and crushing shellfish with their large tooth-plates. The more advanced African (*Protopterus*) and South American (*Lepidosiren*) lungfishes have two lungs and are able to survive periods of drought, cocooned in moist mud burrows of dried out river beds. However, they must maintain the moisture levels in the burrows to survive and they do this by secreting mucus. The Australian lungfish (*Neoceratodus*) cannot do this and only uses lung breathing when oxygen levels fall below the minimum for gill respiration. Throughout fish evolution, use of the gas-bladder as a functioning lung has evolved more than once, as it is found among widely separated groups of fishes (evolutionary convergence).

The only other lobe-finned fish that survives today is the Coelacanth *Latimeria chalumnae* which is a rarely caught species found only in the Indian Ocean off east Africa (fig.193). It shows remarkable similarities to lobe-finned fishes (Actinistia) from the Devonian period with its large scales and trilobed tail fin, and is the only surviving member of this infraclass.

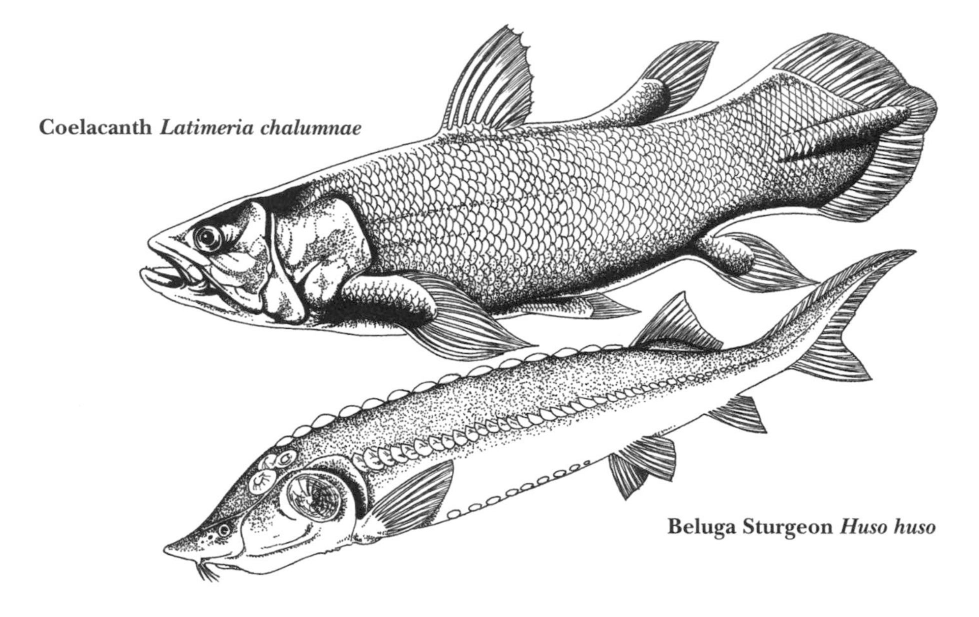

Figure 193. Coelacanth *Latimeria chalumnae*, and Beluga Sturgeon *Huso huso*.

The bony fishes, as their name implies, differ from the cartilaginous fishes in several important ways. They possess true bony (ossified) internal skeletons that support the body tissues and there is a well formed neurocranium covered by flat dermal bones that are fused together to form a rigid skull unit. Most bony fishes also have external scales covering the head and body. In primitive fishes scales were heavy and generally rhombic in shape.

They were of two basic types, the cosmoid scales of the lobe-finned fishes and the ganoid scales of the ray-finned fishes (fig.194). Bony fishes also have a single flap or operculum covering a single gill exit in contrast to separate gill openings in sharks and rays. The spiracle is also reduced or lost in the bony fishes. A swim-bladder is present in the majority of bony fishes and this has evolved by the modification of the primitive fish lung. Eyes are generally large and are of primary sensory importance, whereas the olfactory sense is secondary. The opposite is true for sharks and rays; modern chondrichthyes rely more on olfaction and electrical sensitivity than eyesight.

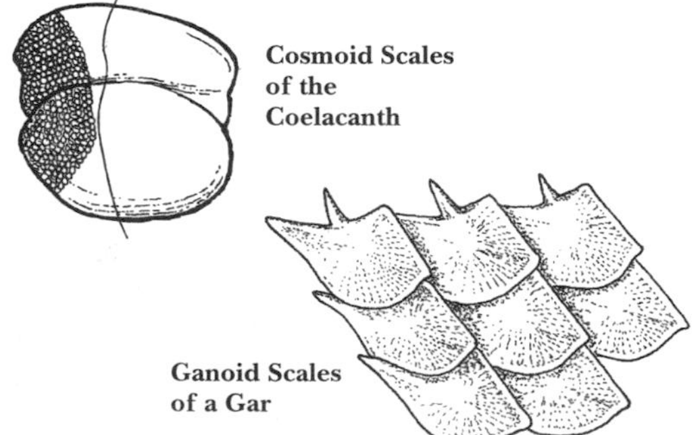

Figure 194. Cosmoid and ganoid scales.

Early ray-finned fishes known from the fossil record belong to the order Cheirolepiformes. The extinct chondrostean fish *Cheirolepis* which belongs to this order, exemplifies the primitive ray-finned fish form known from the Devonian period. This fish had large eyes and a mouth that extended the full length of the skull. The fusiform body tapered back to a heterocercal tail and the vertebrae were not completely ossified. There was an anal fin, paired pectoral and pelvic fins and a single dorsal fin that was positioned well back on top of the body. It is thought the main evolutionary line of bony fishes, the ray-finned Actinopterygii, evolved from a *Cheirolepis*-like fish.

The evolution of the ray-finned fishes can be traced through three major stages of development from primitive through intermediate and then to advanced forms. This progressive sequence from Chondrostei to Neopterygii and finally to the advanced neopterygians or Teleostei can be linked using not only fossil records, but the study of the few living examples of these ancient groups of fishes that still survive today.

The order Polypteriformes is currently represented by ten species of freshwater bichirs (*Polypterus* spp.) and the Reedfish *Calamoichthys*. These polypterid fishes are found in African rivers and have numerous primitive features (e.g. functional spiracles and a lung-like swim-bladder) but their taxonomic position is not certain. They are placed here in the infraclass Cladistia and are considered an evolutionary offshoot from primitive ray-finned fishes.

The order Acipenseriformes, represented by the living sturgeons (Acipenseridae) such as the Beluga Sturgeon *Huso huso* (fig.193) and the curious freshwater paddlefishes (Polyodontidae) from China and USA, also has a long evolutionary history. It is the only surviving group of the Chondrostei which was once widespread during the Palaeozoic era. The chondrostean fishes reached a peak of expansion during the Triassic period with evolutionary advances characterised by shortening of the jaw, reduction of heavy scales and a less pronounced heterocercal tail. The chondrostean fishes declined considerably by the end of the Jurassic period and were largely replaced by early neopterygian fishes. Living relatives of these early neopterygians include the freshwater gars (Lepisosteidae) of North and Central America and the unusual freshwater Bowfin *Amia clava* of north-eastern N America.

The neopterygian fishes evolved further specialisations of features inherited from their chondrostean ancestors. These include a more abbreviated heterocercal tail, refinements in jaws and skull, increased ossification of vertebrae, structural reduction in scales, less fin rays and loss of the spiracle. With the advance of the Cretaceous period, the near ancestors of modern bony fishes (Teleostei) evolved. The teleost fishes have shown an explosive increase in species diversity and structural complexity. The most important anatomical advancements in the development of the teleosts were the replacement of heavy rhombic scales by thinner, rounded ones which allowed the fish greater flexibility of movement. Their skeletons have become completely ossified and the primitive heterocercal tail structure has been replaced by the homocercal condition (with more or less symmetrical lobes and the vertebral column ending near the middle of its base). Primitive fish lungs have been replaced by a modern swim-bladder that is usually completely hydrostatic in function. Pelvic fins have gradually moved forwards during the evolution of modern teleosts. Recent teleosts have radiated widely and adapted to occupy virtually all types of aquatic habitats. Their considerable evolutionary success has established them as masters of the aquatic world.

The most striking feature apparent from the evolution of fishes is succession and replacement of one large group by another in the fossil record. This characteristic pattern is evident from the replacement of the Acanthodii and Placodermi by the Chondrostei. The Chondrostei were then replaced by the early Neopterygii and finally, these too were replaced by advanced Neopterygii (Teleostei). Throughout their evolution, fishes have progressively lost their heavy armour, increased internal ossification and become more deep-bodied. The complexity of the jaw and suspensory mechanism has also advanced with time. However, it was the modification of the primitive fish lung into a hydrostatic swim-bladder that allowed fishes to take full advantage of new habitats. At present the fishes have reached a zenith in their evolutionary history as they now occupy nearly all aquatic habitats and greatly outnumber all other aquatic and terrestrial vertebrate groups.

EVOLUTION OF OTHER VERTEBRATES

In the Devonian period, fishes were abundant in marine and freshwater habitats. It has been suggested that at some time during this period, some sarcopterygian fishes came out of freshwater to live on land. Geological evidence suggests that cyclical drought conditions caused streams to run dry in some regions in the Devonian. Falling water levels may have had an important influence on the evolution of fishes inhabiting these streams as they would have been forced to move over dry land to new water bodies. Devonian rhipidistians and actinistians had unusual fins supported by movable stalks or lobes and may have been able to move on land. It is paradoxical that evolution of limb-like fins may have enabled these fishes to continue the same way of life in changing climatic conditions, but at the same time this condition represents an important pre-adaptation to the type of semi-terrestrial way of life seen in certain amphibians. On the other hand, the earliest tetrapods may have been entirely aquatic evolving initially to exploit a shallow water environment, their limbs perhaps facilitating movement in weed-filled channels.

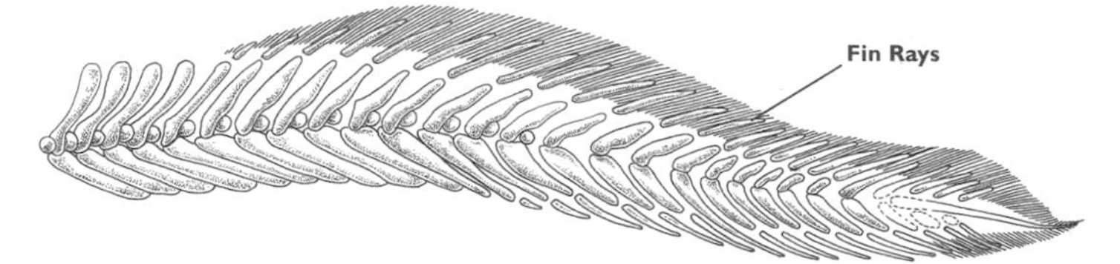

Figure 195. Tail of the primitive amphibian *Ichthyostega*.

The earliest known tetrapods are fossil amphibians (*Acanthostega* and *Ichthyostega*) of the Upper Devonian which had long, fish-like tails (fig.195) and a lateral line system. This type of tail structure is also found in an extinct family of lobe-finned fishes (panderichthyids) which had a bizarre mosaic of fish-like and amphibian features. This family is thought to be the group most closely related to the ancestor of the first amphibians. Modern amphibians (Lissamphibia) differ in many respects from stem amphibians and their ancestry is still the subject of debate. Primitive reptiles (cotylosaurs) probably evolved from a group of amphibians known as labyrinthodonts which may themselves have been direct descendants of the ancestral ichthyostegids. With the evolution of the shelled egg, primitive reptiles were released from their last links with freshwater to become, as their evolutionary history suggests, the dominant terrestrial vertebrates of the Mesozoic era (Age of Dinosaurs). Birds evolved from dinosaurs, possibly from among the coelurosaur group which comprised slender, bipedal, carnivorous reptiles with long necks, tails and fingers. Many modern bird orders appeared in the Tertiary period. Mammals are thought to have evolved from mammal-like reptiles (therapsids) towards the end of the Triassic period but the great radiation of the placental mammals probably did not begin until more than 100 million years later with the demise of the dinosaurs.

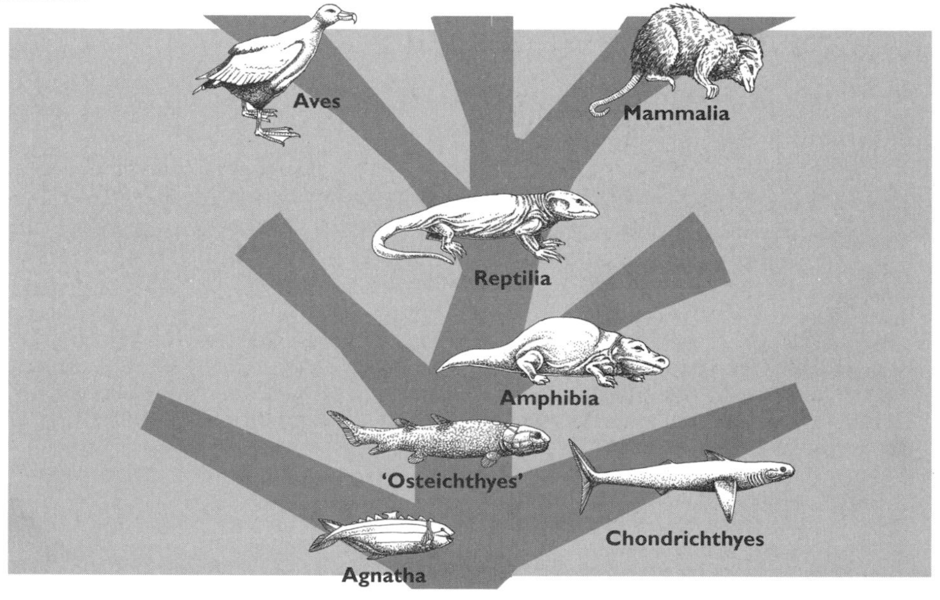

Figure 196. Evolutionary relationships of the classes of vertebrates.

BIBLIOGRAPHY

Colbert, E. H. & Morales, M. 1991. (4th ed.). *Evolution of the Vertebrates.* Wiley-Liss, New York.

Jamieson, B. G. M. 1991. *Fish Evolution and Systematics: Evidence from Spermatozoa.* Cambridge University Press, Cambridge.

Jefferies, R. P. S. 1986. *The Ancestry of the Vertebrates.* British Museum (Natural History) and University of Cambridge.

Romer, A. S. & Parsons, T. S. 1986 (6th ed.). *The Vertebrate Body.* HRW International Editions.

Wake, M. H. (ed.). 1979 (3rd ed.). *Hyman's Comparative Vertebrate Anatomy.* University of Chicago Press, Chicago.

Young, J. Z. 1962 (2nd ed.). *The Life of Vertebrates.* Oxford University Press, Oxford.

Marine Fishes

by

Michael Burchett

INTRODUCTION

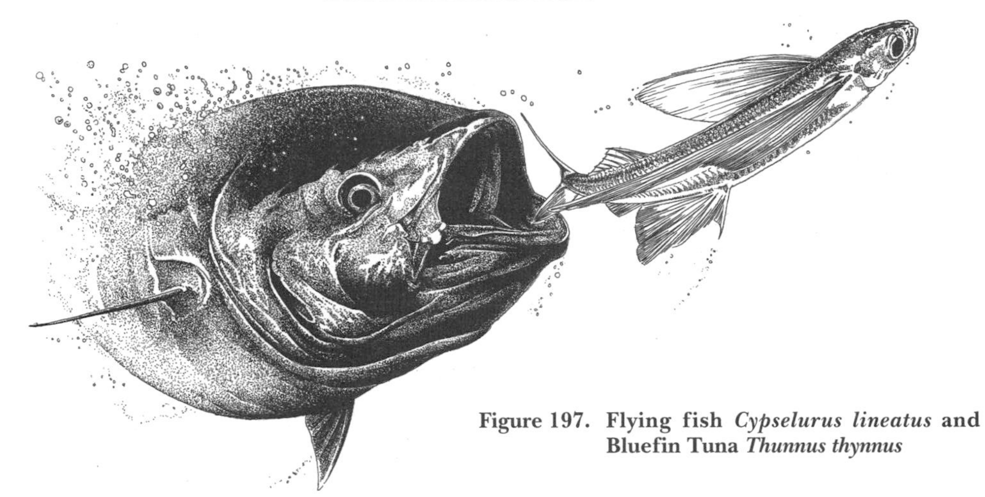

Figure 197. Flying fish *Cypselurus lineatus* and Bluefin Tuna *Thunnus thynnus*

Fishes are the most successful vertebrates that have ever lived on the Earth. There are over 24,000 species known at present and they total almost 50% of all living vertebrates (fig.180). Marine fishes (about 14,000 species) account for about 60% of all known fishes and they vary in size from less than 10mm at maturity (e.g. Indian Ocean Pygmy Goby *Eviota* sp.) to over 12m in length (Whale Shark *Rhincodon typus*). The life span of fishes is variable. Some species live less than one year and others, such as the long-lived sturgeons (Acipenseridae), may survive for over 100 years in the wild. Fishes have evolved and diversified to occupy nearly all known aquatic habitats including the deepest oceans at about 10,000m in depth and lakes at an altitude of 5,000m above sea level.

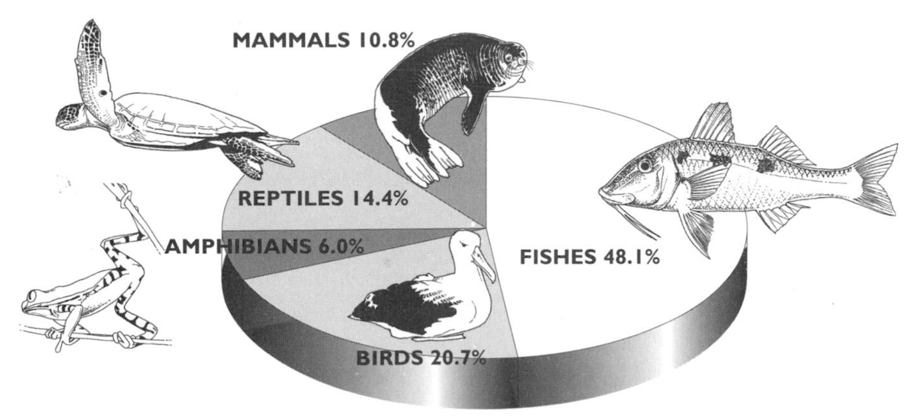

Figure 198. Species composition of the various groups of living vertebrates.

Fishes of the genus *Trematomus* (Nototheniidae) inhabit polar waters around Antarctica where sea temperatures regularly fall below freezing to -1.9°C. At the other extreme some species of *Tilapia* (Cichlidae) have adapted to life in hot springs and soda lakes where temperatures may reach 44°C. Some fishes can live in habitats where there is little or no natural light. About 50 different species of fishes have lost their eyes and inhabit subterranean caves, underground waterways and the deepest oceans. Many deep-sea fishes live in near total darkness where the only form of light is from bioluminescence produced by many living animals including some fishes. Salmon and Sea Trout (Salmonidae) and some eels (Anguillidae) live different parts of their life cycle in both sea water and freshwater. Fishes may have free-living lifestyles, or may form symbiotic or parasitic associations. Some marine fishes (e.g. mudskippers, Gobiidae) are adapted to life out of water for short periods and flight has evolved in both marine and freshwater fishes (e.g. flying fishes, Exocoetidae and flying gurnards, Dactylopteridae).

JAWLESS OR CYCLOSTOME FISHES (Agnatha)

GENERAL FEATURES

These fishes lack jaws, vertebrae, true fin rays, paired fins and scales. In lampreys (Petromyzontidae) the body is eel-like and there is a funnel-shaped mouth with horny teeth. Hagfishes (Myxinidae) are worm-like in form. The skin has many mucus pores which in hagfishes are arranged in a series of slime glands along the sides of the body. The gills are enclosed in gill pouches that open externally through paired lateral pores (lampreys and some hagfishes) or into common gill tubes which have two openings placed well back along the trunk (fig.199). There is a caudal fin which may extend partially along the dorsal and ventral surfaces of the body; lampreys have one or two dorsal fins (fig.200).

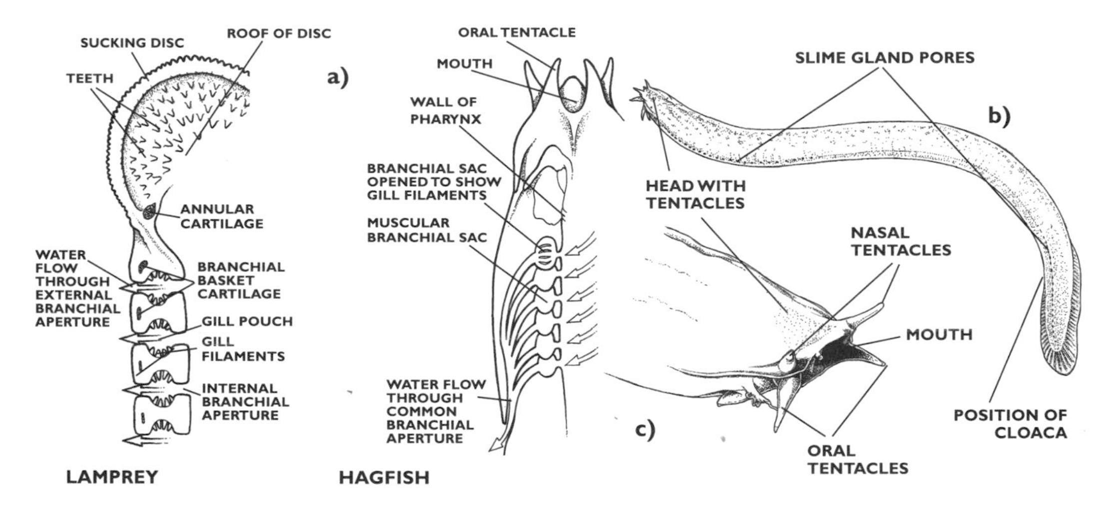

Figure 199. a) Gill structures of a lamprey and hagfish shown in horizontal section (based on Lagler *et al.* 1977). b) External features of a hagfish. c) Detail of the head of a hagfish.

Only adult lampreys possess a well formed pair of external eyes and a pineal eye (light-sensitive organ) situated between them. Its function is to control hormone release by detecting changes in levels of light intensity. The skin colour of lampreys changes from light during the night to dark during the day and pineal activity controls this daily change of coloration. Hagfishes have reduced eyes hidden below the surface of the skin and the skin itself is sensitive to light. Hagfishes have paired tentacles around the mouth and a single nostril, but in lampreys there is a muscular, toothed, sucker-like disc surrounding the mouth. The internal structure of these fishes is relatively simple (fig.200). There is no swim-bladder and a tubular notochord provides support. There is a single gonad which lacks a duct to the exterior.

Hagfishes such as *Myxine glutinosa* from the temperate north Atlantic use the single nostril as a channel to draw water into the six pairs of gill pouches for respiration though they are also able to absorb oxygen directly through the skin. The blood volume is double that of lampreys and there are large vascular sinuses below the skin. In adult lampreys, the muscular, valved gill sacs draw in water for respiration; the single nostril has no connection with the gills. There are seven pairs of gill openings which are supported by an elaborate cartilaginous branchial basket.

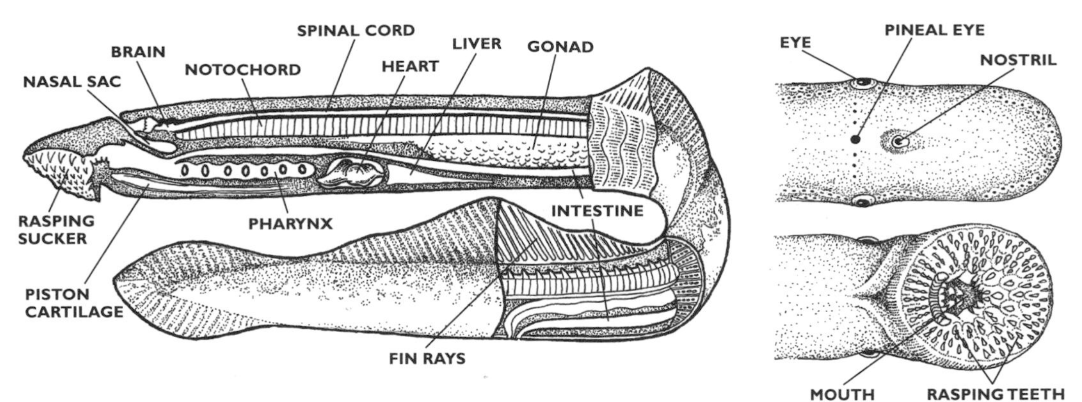

Figure 200. Anatomy of a lamprey.

Lampreys and hagfishes lay eggs which are externally fertilised. Young lampreys go through a process of metamorphosis from larvae (ammocoetes) to adults. Hagfishes have direct development without a larval stage and hatch as miniature adults. Adult lampreys are either marine or freshwater in habit but all species breed in freshwater. Hagfishes are strictly marine. There are about 43 species of hagfishes and 41 species of lampreys; about half of all lamprey species are parasitic.

The evolutionary relationship between hagfishes and lampreys is uncertain. According to one view, they separated only recently; a second view suggests they may have separated early in cyclostome history and have undergone a long period of independent evolution. The fact that hagfishes are the only vertebrates whose body fluids have an osmotic pressure similar to that of sea water suggests they never had a freshwater phase in their history, and this supports the latter view.

BIOLOGY OF AGNATHA

All agnathan species are structurally primitive and yet have specialised, unusual lifestyles. Lampreys may be parasitic on other fishes or have a free-living lifestyle (e.g. brook lampreys). Hagfishes have benthic, often scavenging lifestyles.

Hagfishes and marine lampreys are found throughout the world in cooler waters, normally over continental shelves. There are nine sea lamprey species that feed parasitically on other fishes. In northern Atlantic waters, Sea Lampreys *Petromyzon marinus* can be found attached to the sides of Basking Sharks *Cetorhinus maximus* or Sperm Whales *Physeter catodon,* on which they hitch rides. They also attach themselves to Atlantic Cod *Gadus morhua,* Haddock *Melanogrammus aeglefinus,* Atlantic Salmon *Salmo salar,* and other fishes on which they feed.

The larval life history of lampreys takes place in freshwater (fig.201). The ammocoete burrows in sediment at the bottom of rivers and feeds by extending its head out of the burrow into the current, straining the water with an expanded upper lip or oral hood. Mucus exudes from a 'tongue' or endostyle at the base of the gill chamber (pharynx) and traps food particles from the water current entering the mouth. Food-laden mucus then passes back to the gut along a chain of mobile ciliary hairs. Ammocoetes lack eyes but have light-sensitive receptors in the tail. The eyes and disc-like mouth form during metamorphosis (at five or more years of age). Many features of this extended larval development in lampreys are unique among fishes.

Feeding behaviour in both groups is unusual. In lampreys, the adults of parasitic forms attach themselves to the underside of fish using their sucking mouths. Once attached, a ring of horny teeth rasps away the flesh until blood starts to flow. Anticoagulants secreted from special glands help keep the blood flowing and the host's blood provides the lamprey with nourishment. However there is much variation; the non-parasitic freshwater Brook Lamprey *Lampetra planeri* of northern Europe does not feed at all in the adult stage and parasitic species show delayed metamorphosis leading to smaller non-parasitic adults when host fishes of suitable size are scarce.

Hagfishes locate their prey with six sensitive tentacles which are arranged in pairs on the head (fig.199). *Myxine glutinosa* feeds on benthic invertebrates, primarily crustaceans. It also scavenges dead or dying fishes and although vision is limited or absent, the senses of smell and touch are well developed. When a fish is located, the toothed tongue bores into it and with strong wriggling motions the hagfish enters inside the body of the prey. It may also enter through the mouth. Once inside, the hagfish speedily and voraciously consumes the internal tissues of the prey, often leaving just the skin and bones. Both hagfishes and, to a lesser extent, lampreys exude copious quantities of slimy mucus from secretory glands in the skin. This mucus helps to protect the animals from attack, making them unpalatable to many predators. It may also help with movement by reducing friction and may possibly reduce water loss through the skin.

Lampreys lay their eggs in freshwater in shallow excavated pits (redds) between small stones and gravel. Stones are moved using the sucker-like mouth. About 24,000 to 240,000 eggs (in *Petromyzon marinus)* are laid over a period of several days and these are fertilised by the males. The eggs are covered and hatch about two weeks later into larvae. The sexually mature female develops a large anal fin which may help to cover the eggs during the spawning movements. The male develops a cloacal papilla ('penis') when sexually mature. Lampreys die after spawning.

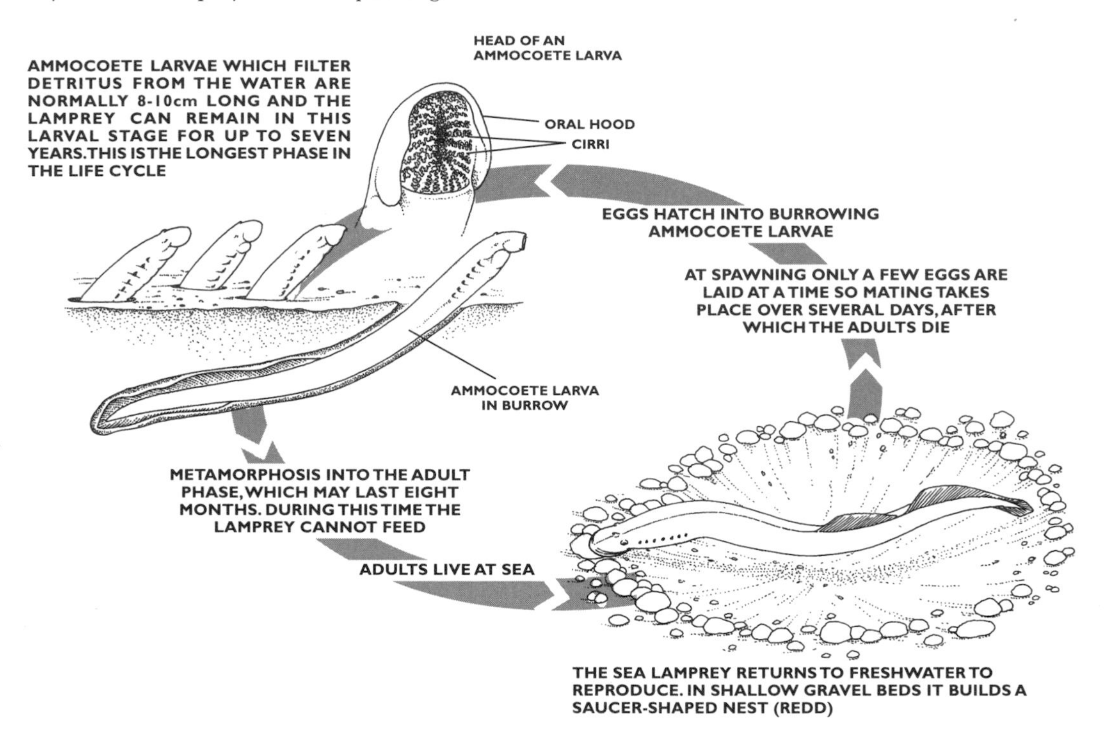

Figure 201. Life cycle of the Sea Lamprey *Petromyzon marinus.*

Less is known about hagfish reproduction but fewer eggs are produced and these are laid on the seabed. Only about 30 eggs are laid at any one time and these are about 20mm in length, each attached to its neighbour to form a rosette-shaped cluster. The eggs hatch after two months or more. Hagfishes are thought to reproduce throughout the year in deep waters of the oceans (some hagfish species are hermaphrodite), but little else is known. Finds of European Hagfish *Myxine glutinosa* eggs are rare and nothing is known about its development.

CARTILAGINOUS FISHES (Chondrichthyes)

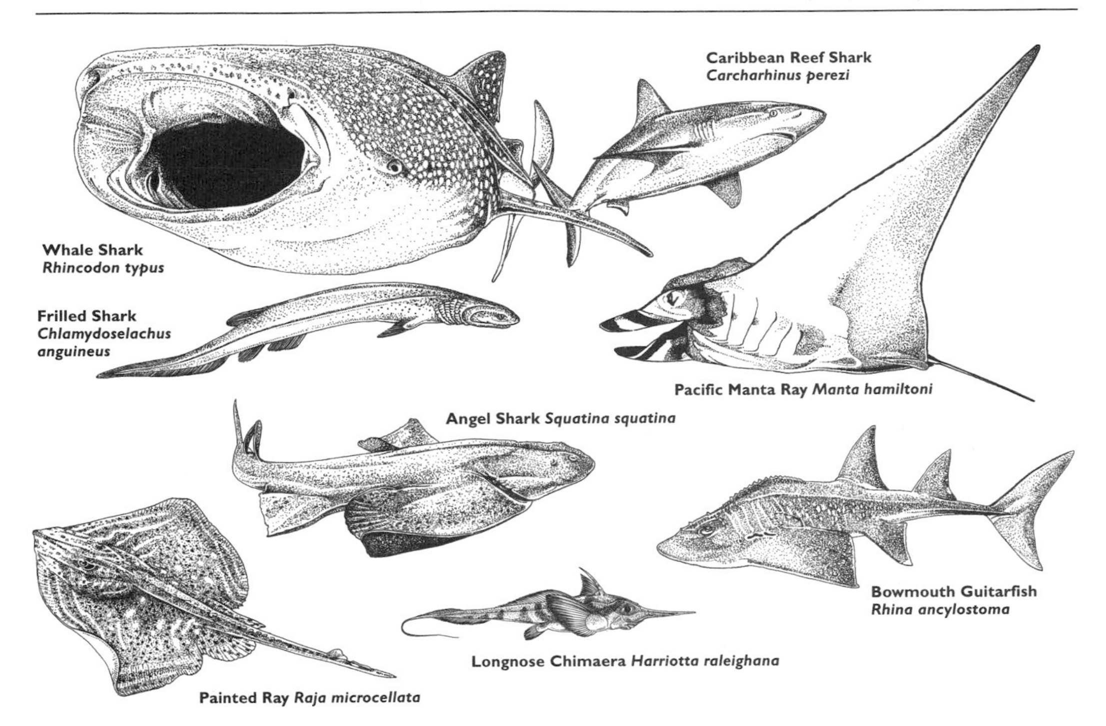

Figure 202. Adaptive radiation of cartilaginous fishes.

Sharks and rays (elasmobranchs) and chimaeras are caught throughout the world's oceans and about 30 species have evolved to spend their entire lives in rivers. They have adapted to occupy nearly every type of marine habitat from shallow littoral waters of the surf zone, to open oceans and deep-water margins of the continental shelves below 3,000m. Many sharks and rays are bottom-dwellers, while others are pelagic, mesopelagic or live just above the bottom (benthopelagic). A few species are even semi-parasitic. Chimaeras are generally fishes of deep waters.

There are about 900 species of cartilaginous fishes (Class Chondrichthyes) which are divided into two subclasses. The subclass Holocephali has one order, the Chimaeriformes that has three families of living chimaeras with about 30 species (page 320). The subclass Elasmobranchii has eight orders of living sharks (about 30 families, 375 species) and one order of living rays (Rajiformes) with 12 families. The rays are the most successful Chondrichthyes in terms of numbers with about 450 species.

SHARKS AND RAYS

GENERAL FEATURES

Most modern sharks and rays have characteristic long snouts and a mouth on the underside of the head that is set well back from the snout tip. The jaws in most species are of modern design (hyostylic suspension, fig.190). Teeth are continually replaced as if on a conveyor belt, and are arranged in many rows, one behind the other. Pelagic shark teeth vary enormously in shape and may be long and dagger-like or broad-based and triangular in shape. The benthic rays generally have flat button-like teeth that may be arranged as crushing plates. Fins are generally stiff and fleshy and, unlike the fins of bony fishes, cannot be folded alongside the body or change shape. In sharks, the tail fin is asymmetrical with the upper lobe usually better developed than the lower lobe (heterocercal form). The skin of cartilaginous fishes is tough and covered by 'dermal denticles' and these have a hard enamel covering similar to teeth. Denticles are not true scales but are in fact modified teeth. Along each side of the body, a lateral line system is present and external gills are arranged in five to seven pairs on each side of the head. The gills lack gill covers. Some sharks and all rays possess a pair of openings known as 'spiracles' that are located behind the eyes. These help with breathing when the fish lies on the bottom. Inside the body, cartilage is present instead of bone. However, there may be some salt deposition for strengthening. Lungs and swim-bladders are absent, the liver is large and the gut is short with a distinct spiral valve (fig.214). These fishes tend to produce few large yolky eggs or live young. Fertilisation is internal and sperm is transferred to the female using paired penes or claspers. Young cartilaginous fishes are well developed on hatching and have a body form similar to the adults.

The behaviour of cartilaginous fishes is not well understood and bad publicity surrounding a few potentially dangerous species has generally given sharks a poor reputation. However, as more is learnt about their behaviour and other aspects of their biology, many of the popular misconceptions can be addressed more objectively.

BODY FORM

Body form varies considerably and not all sharks are streamlined for leisurely, energy efficient cruising and fast hunting. In the fastest cruising sharks, the body shape conforms closely to the hydrodynamically ideal fusiform (torpedo) shape. Here, the body is ovoid in cross section and reaches a maximum girth about one third from the anterior end (e.g. Porbeagle *Lamna nasus*, plate 12). The body tapers evenly towards the tail to reduce friction and drag. The head end of many cruising sharks is slightly flattened and a curved upper surface gives some lift during swimming. This line of evolution is taken to its extreme in the hammerhead sharks (Sphyrnidae). As lifestyles become increasingly sluggish or bottom-living, body forms become flattened in the vertical plane (e.g. wobbegongs, angel sharks and rays). The Frilled Shark *Chlamydoselachus anguineus* is unusual as it is the only shark to have become attenuate or eel-like during its evolution (fig.202); it lives in mid-water or just above the bottom in deep water.

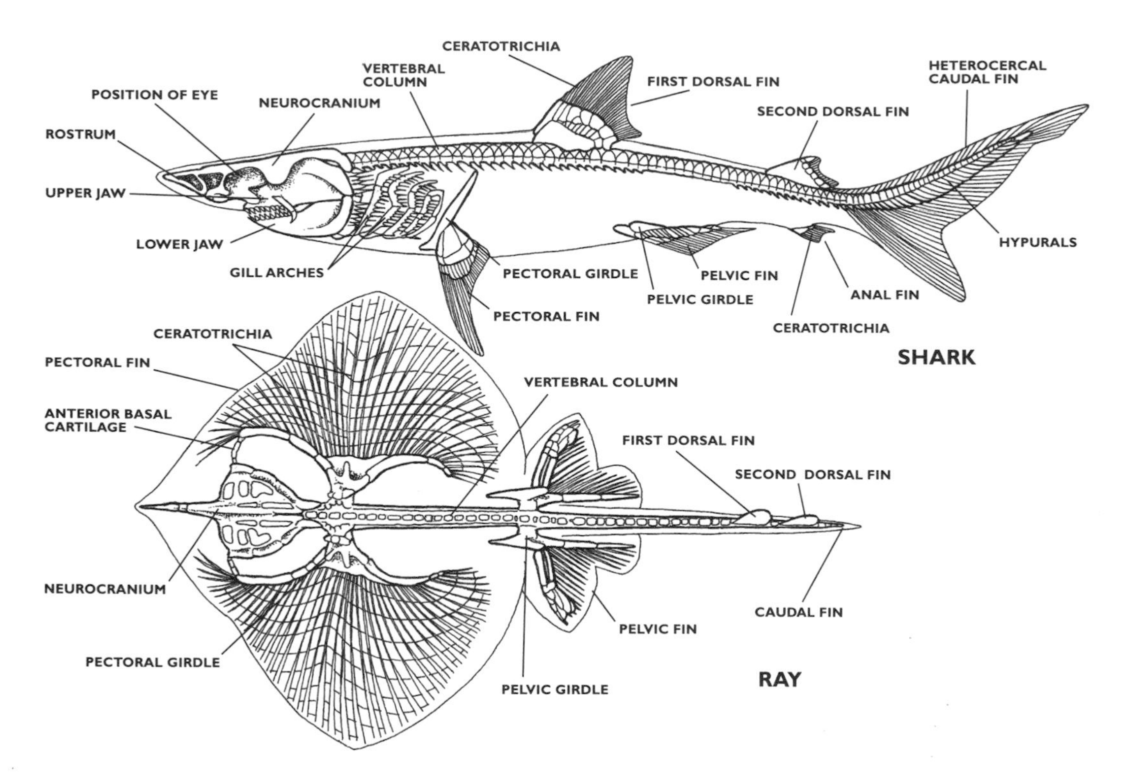

Figure 203. Cartilaginous skeleton of a shark (side view), and ray (from above, branchial basket not shown).

SKELETON

Cartilaginous fishes, as their name suggests, do not possess skeletons of bone, but instead their skeletons are formed from cartilage which is a tough, flexible, gristle-like material. True bone is not entirely absent but is confined to the dentine of the teeth, skin denticles and spines. The shark 'skull' is a cartilaginous box (neurocranium) that surrounds the brain. The neurocranium comprises olfactory, visual and auditory sections. The jaws of sharks are loosely attached by ligaments to the underside of the skull and also to a short, separate rod or strut of cartilage (hyomandibula) at the back of the neurocranium (fig.190). Flexibility of the mobile upper jaw allows a shark to thrust this part of its jaws forward when biting prey (fig.217). In rays the jaws have lost all connection with the skull and are suspended from the hyomandibula. The gill arches of sharks are hooped cartilaginous rods that contour the sides of the body. Most species have five pairs of lateral gill slits, but a few have six or seven pairs (e.g. cow sharks, Hexanchidae). The vertebral column is highly flexible and each segment consists of an articulated base (centrum) with a dorsal arch that protects the delicate spinal nerve cord. Although sharks do not have a rib cage, small unsupported processes from each centrum are structurally equivalent to reduced ribs. The vertebral column extends almost to the end of the upper lobe of the tail fin, which forms the characteristic heterocercal tail shape. The large, thick fins of cartilaginous fishes are different from those of bony fishes and are stiffened internally by struts of cartilage at the base and hundreds of horny filamentous rods (ceratotrichia) peripherally. The pectoral fin skeleton attaches to a strong pectoral girdle. The pelvic girdle supports the pelvic fins and is embedded in the ventral layers of body muscle.

The anterior skeleton of rays is more complex and strut-like rays of cartilage fan out to form a flexible framework that supports the broad wing-like pectoral fins. The pectoral girdle is attached to the vertebral column and the anterior bases of the pectoral fins link with the skull (fig.203). Between the skull and pectoral girdle, the vertebral column is fused to form a stiff tube and this arrangement supports the broad pectoral fins and head (fig.203).

SKIN

The skin of cartilaginous fishes is extremely tough and flexible. Together with the cartilaginous skeleton, it has an important role in maintaining body shape, particularly in the head region. The skin consists of an outer layer of epidermis and below this, a thick dermis with connective tissue that has tough, flexible fibres arranged at angles of up to 70° to each other. This arrangement, which is similar to a carpet weave, gives the skin of sharks and rays its strength. There are no true scales covering the bodies of cartilaginous fishes, but instead there are hardened placoid scales (denticles). These are small tooth-like structures that are closely packed with their tips facing backwards to give elasmobranch skin a rough, abrasive feel when rubbed against the direction of water flow (fig.204). This characteristic was used commercially for many hundreds of years to produce non-slip grips on handles of knives and swords.

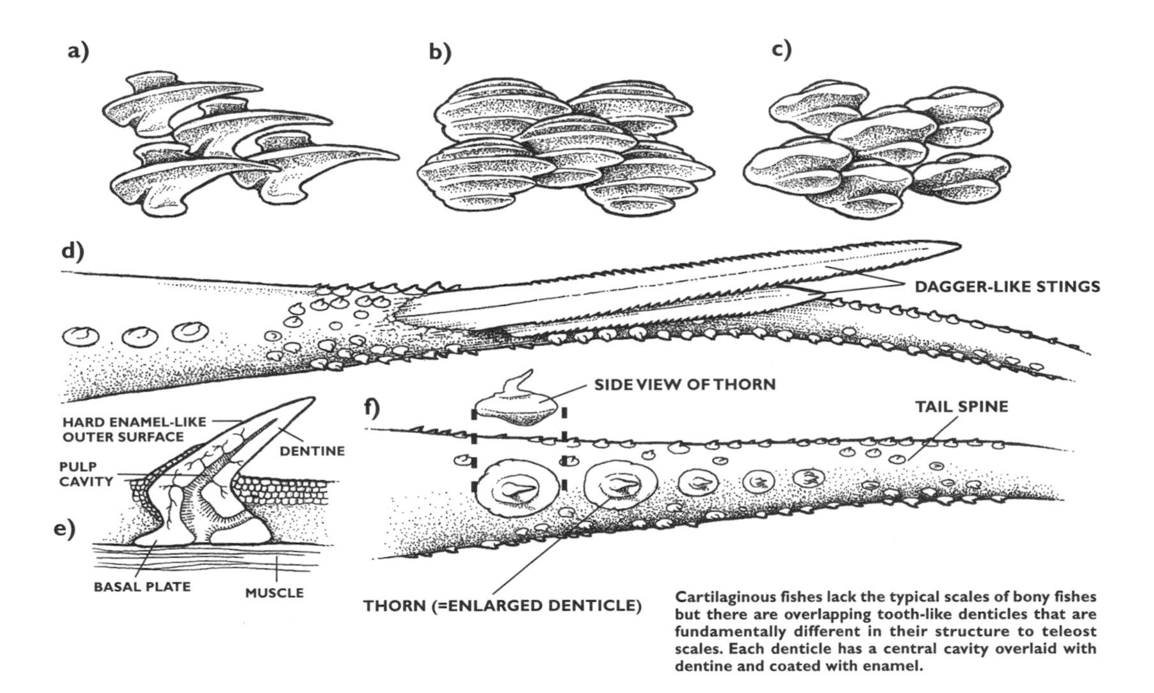

Figure 204. Denticles covering the skin of various elasmobranch fishes and tail spines of a stingray. a) dermal denticles of the Whale Shark *Rhincodon typus*, b) requiem shark (*Carcharhinus* sp.) and c) Tiger Shark *Galeocerdo cuvier*; d) stings of a stingray *Dasyatis* sp.; e) cross section of a denticle showing the tooth-like structure; f) thorn of the Thornback Ray *Raja clavata*.

There are many variations to the basic denticle shape and some of these may have evolved for specific purposes. It is apparent that dermal denticles of cartilaginous fishes closely resemble the structure and form of their teeth, but the skin denticles are usually smaller. These placoid scales may become enlarged to form vicious thorn-like structures along the ridges, tails and dorsal surfaces of many rays (e.g. Thornback Ray *Raja clavata*, fig.204). The main function of denticles and thorns is for protection against predators, but in some fast-cruising sharks the denticle covering may have a hydrodynamic function that reduces drag. However some fast swimming eagle rays have lost their denticle covering altogether. If denticles become damaged the skin quickly replaces them and new ones are formed during skin growth. Some elasmobranchs have enlarged clasper denticles and claws that help the male to grip the female during copulation.

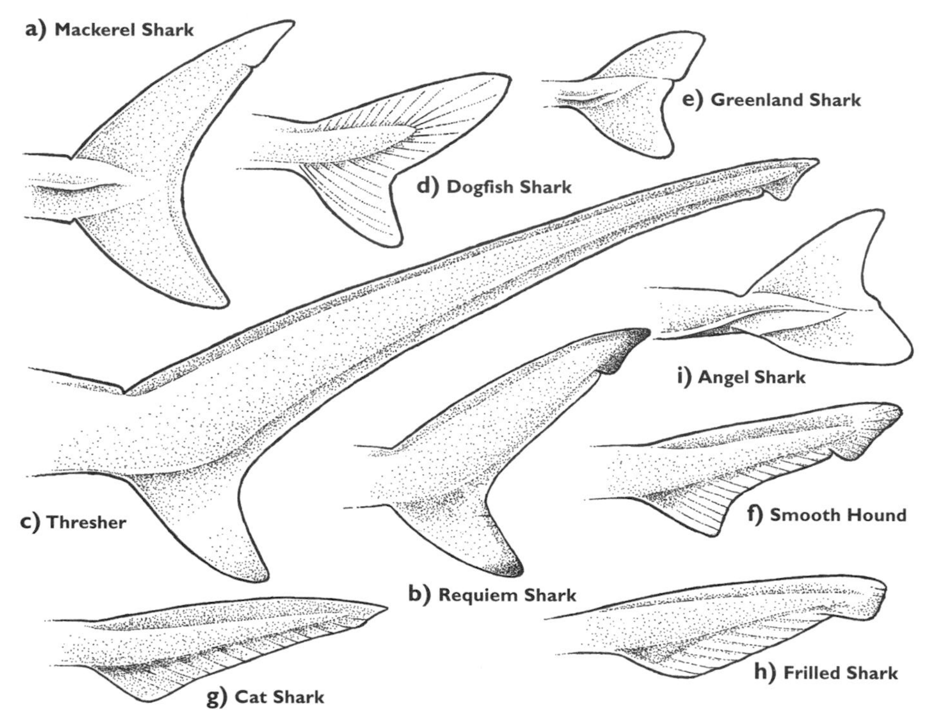

Figure 205. Tail shapes of sharks. The lifestyle of sharks is often reflected by their caudal fin shapes and in turn this is indicative of swimming ability. Fast swimmers such as a) mackerel sharks (Lamnidae) have sickle-shaped tail fins with upper and lower lobes almost symmetrical. b) requiem sharks (*Carcharhinus* spp.) and c) thresher sharks (*Alopias* spp.) have asymmetrical tails and are also powerful swimmers. Less powerful tails of d) dogfish sharks (Squalidae) and e) the Greenland Shark (*Somniosus*) still show lower lobe development and active-swimming habits. Among sluggish sharks such as f) smooth hounds (Triakidae), the lower lobe is reduced and it is virtually absent in many bottom-dwelling sharks including g) catsharks (Scyliorhinidae) and h) the mesopelagic Frilled Shark *Chlamydoselachus anguineus*. Bottom-dwelling i) angel sharks (Squatinidae) have the lower lobe of the tail fin larger than the upper lobe.

Skin coloration is primarily due to the presence of pigments produced by special chromatophore cells in the dermis. Most sharks and rays are quite drab and subdued in colour compared to teleost fishes. Although some sharks and rays have bold patterns, the usual shades are based upon tones of brown, grey and black. Pelagic cartilaginous fishes have dark grey or blue dorsal surfaces with electric tints and these lighten laterally towards the creamy-white ventral surface. The colorations provide camouflage from both predators and prey, as the dark back of the fish blends with the deeper waters when viewed from above and the pale underside becomes less distinct against the lighter surface. This is called countershading (see page 332). The pigmented dorsal surface may also give some protection against the harmful effects of the sun when the fish is swimming near the surface. Bottom-dwelling elasmobranchs match their skin colour to resemble the seabed upon which they are resting (e.g. collared carpet sharks, Parascyllidae and wobbegongs, Orectolobidae, plate 9). Wobbegongs also have projecting flaps on the sides of the head to increase the cryptic effect by breaking up the outline of the body. Camouflage with the seabed offers both protection from predators and easier capture of prey.

Deep-water sharks (*Etmopterus, Centroscyllium,* Dalatiidae) and some other sharks (*Isistius, Squaliolus*) have light-producing photophores in their skin. In *Etmopterus* they are numerous on the underside, sparse on the flanks and almost absent on the back. Light from photophores is thought to eliminate the shadow caused by surface light striking the shark's body. This 'luminescent countershading' acts as an anti-predator device and is also used by many species of deep-water bony fishes.

Poison glands are present in the skin of some sharks in the form of modified mucus glands and the function of the venom is probably defensive.

FINS, BODY SHAPE AND MOVEMENT

Fins are present on all fishes in one form or another. They may be multi-functional in their use, or of more specific design for a particular purpose. Most fins function in a number of ways: they propel the fish through the water, aid stabilisation of the body, control direction of movement, provide lift and act as brakes during swimming. Other fins are used for defensive and courtship purposes. Among the teleost fishes, the complex diversity of body shapes varies a great deal and this will be discussed later. In sharks and rays, where body form is less variable, the use of fins in association with body shape and movement is more easily understood. The basic principles of fin use and function can be applied to most other fishes with similar body forms.

All cartilaginous fishes have paired and unpaired fins that are internally supported by ceratotrichia. Cartilaginous fishes have a pair of symmetrical fins low down on the sides of the body that are equivalent to the hands, wings or forelimbs of terrestrial animals. A pair of pelvic fins on the lower sides correspond to the hind limbs of land animals. Sharks have a caudal fin on the end of the tail which may be reduced or missing in rays. A single anal fin between the pelvic fin and tail may be present but this is absent in certain sharks and all rays. One or two dorsal fins are present in sharks but both have been lost in some rays. Some sharks have fin spines on the leading edges of their dorsal fins and spines are also found on the tails of some rays.

The caudal fin and posterior portion of the body of sharks provide the thrust to move the animal forwards through the water. The tail fin also acts as a rudder to steer the body while the fish is moving forwards. Pectoral fins of rays have become enlarged, muscular swimming structures that replace the caudal fins of sharks and their tails may be reduced and slender (e.g. devil rays *Mobula* spp.). Tail shape varies widely among sharks and its form appears to be related to swimming abilities and lifestyles. The fastest swimming sharks (e.g. makos *Isurus* spp.) have a near symmetrically-shaped tail similar to many fast-cruising bony fishes (fig.205). Sharks that lead a sluggish, benthic lifestyle have reduced lower tail lobes. The exceptions are the angel sharks (Squatinidae) which have evolved an enlarged lower lobe of the tail fin. The heterocercal tail is taken to extreme lengths in the thresher sharks (*Alopias* spp.) where the upper lobe may equal the length of the body and can be used for stunning prey (fig.205) .

Figure 206. Swimming sequence of a dogfish (Scyliorhinidae) seen from above. The S-shaped waveform starts at the head, passes smoothly along the body and becomes more pronounced as it reaches the tail. The tail-beat frequency dictates the swimming speed of the fish.

Sharks use their tails and body muscles for locomotion, and swim with snake-like undulations formed by a series of S-shaped curves (sinusoidal wave motion) that travel along the body (fig.206). These actions are accomplished by alternating serial contractions of the segmental muscle blocks (myotomes) on each side of the body, that flex the fish (or successive parts of the fish) first one way and then the other into a waveform. Each wave starts at the head and passes smoothly along the body towards the tail, where it becomes more exaggerated as it reaches the tail tip. The tail-beat (the frequency at which waves travel along the body) will dictate the swimming speed. The whole action of rhythmical body movements gives the shark a graceful, sinuous swimming motion.

Fins of sharks are quite stiff and cannot be folded along the body in the manner of teleost fishes. The stiff fins of the shark help to stabilize the body and reduce 'roll' (fig.207b). Adjustments to the angle of the pectoral and pelvic fins control 'pitching' (up and down motion) and these work in a similar way to the hydroplanes of a submarine or the elevators of an aircraft. Yawing (side to side motion) is more difficult to control and a combination of fine fin and tail adjustments accomplish this. Even then, many bottom-dwelling sharks tend to have a slight lateral nod of the head as they swim through the water.

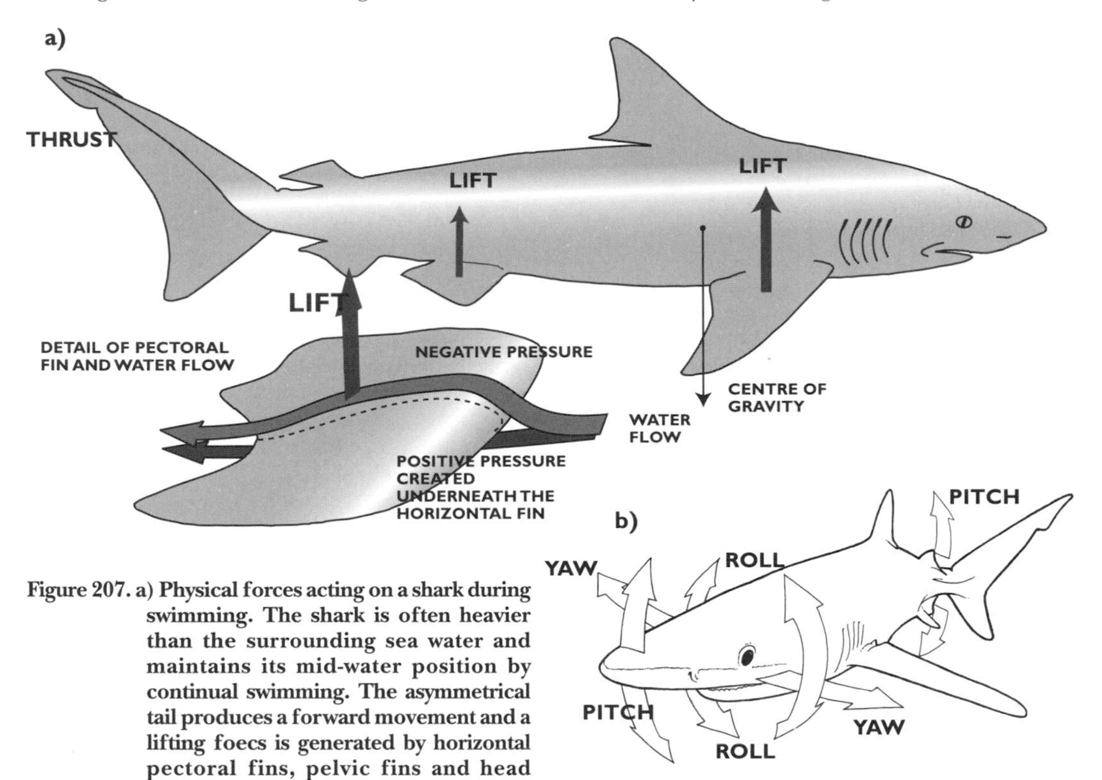

Figure 207. a) Physical forces acting on a shark during swimming. The shark is often heavier than the surrounding sea water and maintains its mid-water position by continual swimming. The asymmetrical tail produces a forward movement and a lifting foecs is generated by horizontal pectoral fins, pelvic fins and head region. b) Movement of a shark showing roll, pitch and yaw.

Cartilaginous fishes do not have swim-bladders and they slowly sink when forward swimming motion stops. With the help of lift from the head and paired fins, and a large, buoyant liver, sharks can control and maintain their level in the water column even at low cruising speeds. Therefore another function of lateral fins is to provide lift while the fish is swimming forwards. The cross section of the paired fins is similar to a section of an aircraft wing. The fin shape forces water to travel further over the top surface. This has the effect of reducing pressure on top of the fins and increasing pressure underneath, thus generating a lifting force (fig.207a). Tilting the angle of the pectoral fins further increases this lifting force, but too great an angle will cause stalling and a decrease in speed. This effect is useful as a braking system. The leading edges of all the fins are smooth and round and gently taper back towards the trailing edge. This design is hydrodynamically efficient in reducing drag and the profile is similar to the designs used for boat keels and rudders.

Unlike many teleosts, cartilaginous fishes cannot swim backwards as they are unable to reverse the muscular wave motion. Compared to sharks, rays use a slightly modified method of swimming. The S-shaped, undulating wave motion does not travel along the body but along the enlarged pectoral fins (fig.208a). The eagle rays, manta rays and devil rays (Myliobatidae and Mobulidae) have the largest pectoral fins of all rays. Their fins have a higher aspect ratio than other rays and their pectorals do not meet at the front of the head. The whole fin is moved up and down with a beat sequence similar to that of a bird's wing and it provides both upward lift and forward propulsion (fig.208b). By contrast, teleost fishes achieve mobility and manoeuvrability using small, thin, flexible fins.

Figure 208. a) Swimming sequence of a Thornback Ray *Raja clavata*. Most rays swim using the wing-like pectoral fins. Undulating waves pass along the pectoral fins from front to back. Some rays such as sawfishes (Pristidae), electric rays (*Torpedo* spp.) and a few guitarfishes (Rhinobatidae) are unable to do this, but instead scull with their tails.
b) Swimming sequence of the Devil Ray *Mobula diabolus* adapted from a tracing of an underwater film (from Alexander 1975). I-VIII downstrokes; IX-XV upstrokes.

SENSES

Cartilaginous fishes have a sophisticated array of senses including sight, hearing, vibration, smell, taste, temperature, touch (barbels), and even electric receptors. Therefore fishes use many different senses to monitor the environment and several senses may be in use at the same time. Some senses such as olfaction are useful at long range whilst others are only effective at closer ranges (e.g. touch, vision). Together they form an integrated sensory 'picture' of the three-dimensional marine environment.

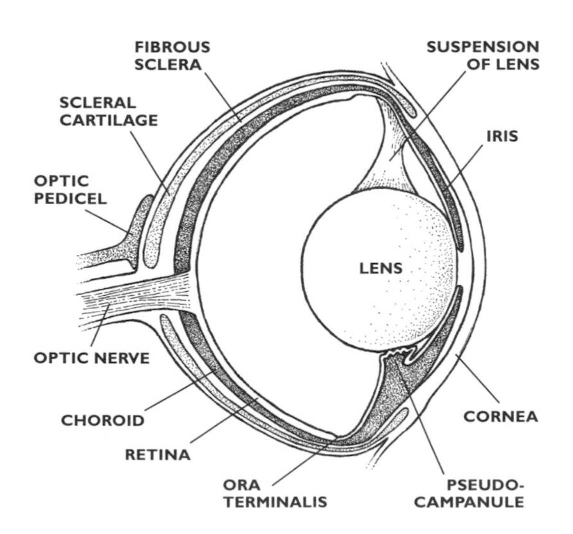

Figure 209. Vertical section through the eye of a shark.

Sight Sharks and rays have well developed eyes (fig.209). In those species with eyes positioned near the front of the head, binocular vision may be possible. The presence of cone cells in the retinas of some sharks suggests that colour vision may be present. Eyes of bottom-dwelling sharks and rays may be less developed or even obsolete in some electric rays (narcinids) but other senses are more acute. By contrast, many nocturnal sharks and deep-water species have eyes sensitive to low light levels. These species may have a layer of cells at the back of the eye called the tapetum lucidum that helps with low light vision. This layer double-reflects the light back into the retina and maximises the use of any available light. During the day when the fish is near the surface, a layer of pigment covers the sensitive tapetum like a protective shield. An interesting feature in the eyes of some large oceanic sharks (e.g. Carcharhiniformes and Lamniformes) is the extra eyelid or nictitating fold. This structure is moved across the eye to protect the cornea just before biting prey. Therefore, in the last seconds before biting, the shark cannot see its prey and may rely on other senses such as electroreceptors at close range.

Hearing Internal or inner ears of sharks can detect low frequency sounds below 1,000Hz. Sharks will often seek out and investigate distant sources of sounds, such as those coming from dying fishes. However, the main function of the inner ear with its three semicircular canals is spatial orientation. This structure senses which way up the body is in relation to the three-dimensional aquatic environment (fig.210).

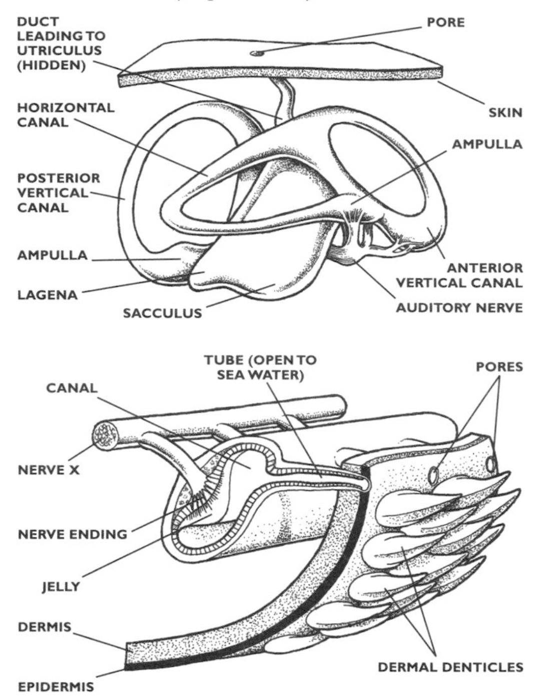

Figure 210. Inner ear of a shark (one of a pair). The inner ear comprises three saclike structures with three associated canals arranged at right angles to each other. The sacs include an upper utriculus and lower sacculus and lagena, each with a sensitive spot (or macula). The macula contains the ear stones (otoliths) and associated sensory cells which register orientation and speed changes in the shark. The ampulla with the internal gelatinous mass (cupula) and sensory epithelial cells detect movement in any one of three spatial planes.

Figure 211. Enlarged view showing the lateral line of a shark. Presure waves in the sea water around the shark vibrate the liquid in the canal. The vibration is transmitted through the jelly to the sensory nerve endings that relay the message via nerves to the brain. Therefore the shark is alerted to the presence of movement in the water which it can avoid or investigate (based on Stoops & Stoops 1994).

Vibration Along each flank of a cartilaginous fish is the lateral line canal (acoustico-lateralis system) which can detect water movements. In low visibility water or at night, it may help the fish to detect predators and prey, and to avoid other moving objects. The system consists of a network of fluid-filled canals below the surface of the skin (fig.211). It has several branches in the head region and a single line running back along the flank to the tip of the upper lobe of the tail. Short tubes link the canals to the outside of the body where they open externally as a line of pores. The walls of the internal canals have sensory organs called neuromasts. Each consists of a gelatinous mass (the cupula), with clusters of sensory hair cells connected to nerve fibres that run to the brain. Single pore detectors (pit organs) may also be present over the entire body in some sharks.

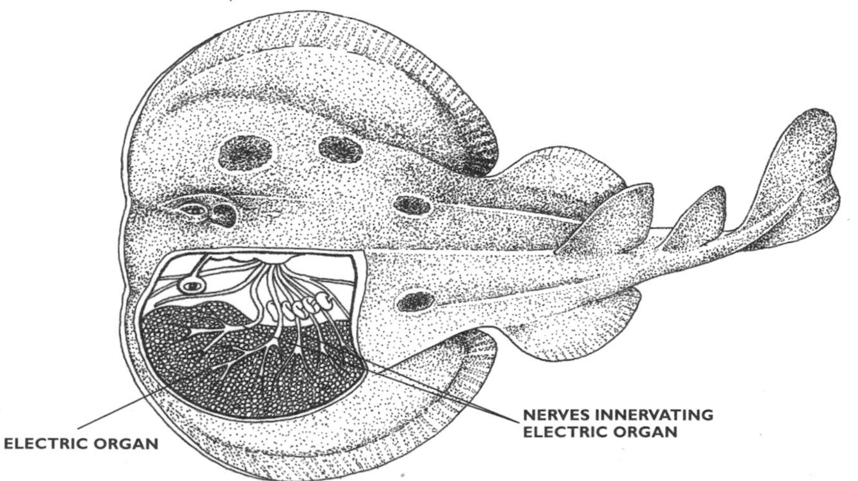

Figure 212. Eyed Electric Ray *Torpedo torpedo*, dissected on the left side to show one of the paired electric organs with associated nerve supply (after Young 1973).

Olfaction and Taste The sense of smell (olfaction) is well developed in cartilaginous fishes and is of major sensory importance for sharks and rays. Water flows in through the nostrils and passes to the nasal sac where odours excite sensory cells in the sac lining. Some sharks can detect minute traces of substances (one part per million, or equivalent to one drop of blood in a large bucket of water) therefore olfaction is an important sense for sharks. The mouths of cartilaginous fishes also have taste buds and these are present on the touch feelers (barbels) of many bottom-dwelling species (e.g. nurse sharks, Ginglymostomatidae). They are used to detect prey such as shrimps and molluscs that are buried below the surface of the sand.

Electric Sensitivity Cartilaginous fishes have an electro-receptor system. It consists of large numbers of elongate, blind-ending tubes containing a jelly-like fluid and a sensory nerve cell at the inner end. Each tube opens on to the skin as a pore and clusters of pores are most noticeable around the underside of the snout. These 'ampullae of Lorenzini' are sensitive to weak electric fields. Sharks can detect living animals that produce very small electrical signals, even when they are hidden under loose bottom substrates.

The ability to discharge electric currents has evolved in some rays and this has developed from specialised muscular tissue derived from the gill muscles. The electric organs of electric rays (Torpedinidae and Narcinidae) are kidney-shaped structures located in the pectoral wings (fig.212). Electric organs of these fishes are primarily defensive but they can also be used to stun prey. A shock of up to 220 volts at 8 amps can be produced by larger electric rays (e.g. *Torpedo nobiliana*). Work on skates has shown that they also possess weak electric organs. These organs are found in the tail but do not become functional until sexual maturity which suggests a possible role in breeding behaviour.

GILLS AND RESPIRATION

The gills of sharks and rays work in a similar way to those of the bony fishes and there are many structural similarities. Sharks swim with their mouths open to allow water to flow in to the gill chambers. All sharks and rays can actively pump water into the mouth through the gills and out of the body via the gill slits (fig.213a). This pump functions by lowering the floor of the mouth through muscular action and expanding the walls of the pharynx using the branchial arches and associated muscles. This causes a rush of water into the mouth which is then closed, the floor of the mouth is raised and the water is expelled through the gill (branchial) apertures. Gill arches bear the gill filaments (fig.213c) and the leaflike folds (lamellae) increase the surface area for gaseous exchange. Blood flows through the lamellae in the opposite direction to the flow of water. As it leaves the gills, the blood contacts the high oxygen concentrations and low carbonic acid levels of the water. Gaseous exchanges take place and the heart pumps re-oxygenated blood around the body of the fish through the arterial system. Gill arches may also support many comb-like structures (gill rakers) and their function is to strain the passing water of small food particles. Such gill rakers are present in the plankton-feeding sharks and rays (plate 9).

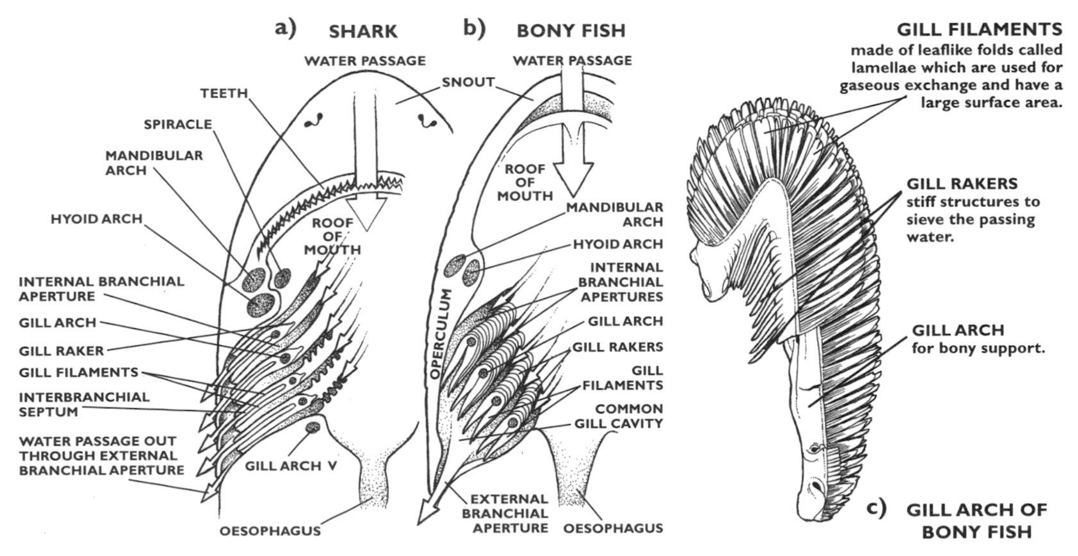

Figure 213. Gill structures of a) an elasmobranch and b) a bony fish shown in horizontal section (based on Lagler *et al.* 1977). c) A single gill arch from a bony fish showing the arrangment of gill filaments and gill rakers.

Bottom-dwelling rays and angel sharks cannot breathe effectively by drawing water in through the flattened ventral mouth while lying buried in sediment on the seabed. They have evolved a pair of openings just behind the eyes known as spiracles. The spiracles allow oxygenated water to be drawn in from above the fish even when the animals partially bury themselves.

Nostrils are present in both cartilaginous and bony fishes and they are especially well developed in many bottom-living species (e.g. horn sharks Heterodontidae). Complex connections (nasoral grooves) between the nostril region and mouth enable water to circulate into the nasal sacs during respiratory movements. However, unlike land vertebrates, the nostrils of cartilaginous fishes are used only for olfaction and they do not have a respiratory function.

WATER BALANCE

Water balance in aquatic organisms depends on the physical process of osmosis. Osmosis describes the effect of water transfer along an invisible gradient from a fluid with low salt concentration to a fluid of high salt concentration. This gradient is established as soon as an animal enters an aquatic medium and it occurs over all parts of the body surface across their semipermeable membranes. For physiological reasons, fishes must maintain the salt concentrations in their body fluids at about one-third that of sea water, but in so doing they become osmotically out-of-balance.

Almost all cartilaginous fishes are marine and they have solved their osmotic problems in an interesting way. By maintaining elevated concentrations of organic compounds (mostly urea) in the body fluids, the total osmotic concentration of their blood equals or slightly exceeds the osmotic concentration of sea water. The urea is a waste product coming from the breakdown of protein and because of its toxicity to the body, it is normally excreted by the kidneys in other vertebrates. A shark's body also needs a powerful excretory system to expel the excess salts. This is partly accomplished through the kidneys and also by the rectal gland which opens via a duct into the rectum and acts like a third kidney. With these effective excretory structures, cartilaginous fishes can excrete a fluid that has a higher salt content than the surrounding sea water and therefore body fluids stay in near osmotic equilibrium with their surrounding environment. Unlike bony fishes, elasmobranchs do not have to actively drink large amounts of sea water to maintain their osmotic balance.

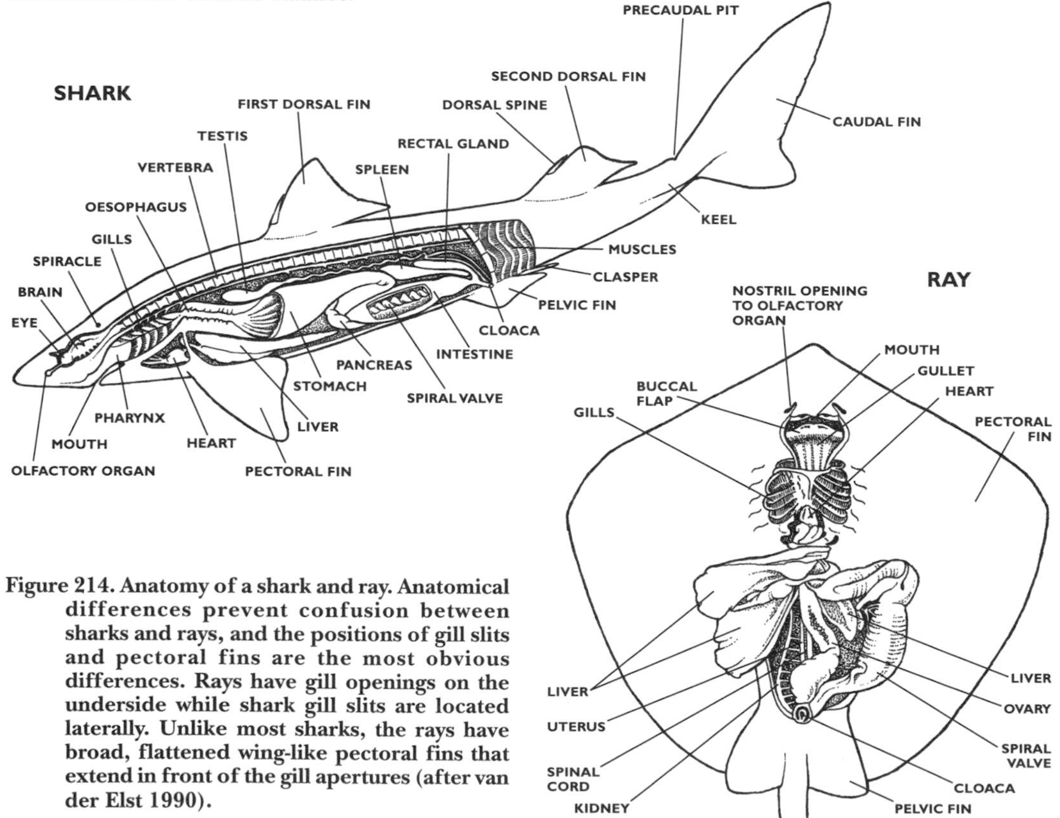

Figure 214. Anatomy of a shark and ray. Anatomical differences prevent confusion between sharks and rays, and the positions of gill slits and pectoral fins are the most obvious differences. Rays have gill openings on the underside while shark gill slits are located laterally. Unlike most sharks, the rays have broad, flattened wing-like pectoral fins that extend in front of the gill apertures (after van der Elst 1990).

DIET AND FEEDING BEHAVIOUR

Stomachs of sharks are large, sac-like structures and their ability to stretch enables large quantities of food to be eaten and stored in a single meal. This may be important to wandering, oceanic species where prey may be few and far between. The intestines of sharks and rays are shorter than those of bony fishes. There is also an unusual corkscrew-shaped structure known as the 'spiral valve' (fig.214) associated with the intestine. This structure helps digestion by slowing the passage of food through the gut and increasing the surface area for nutrient absorption.

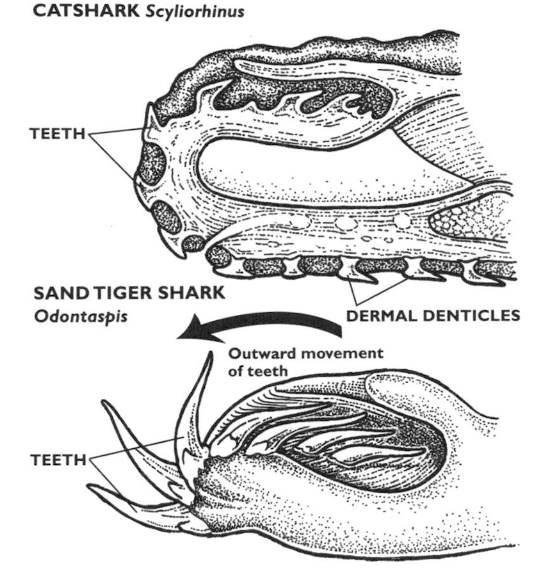

Figure 215. Cross section through the upper jaw of a shark showing succession of tooth replacement. New teeth form on the inside of the jaw and move to the top when old teeth are lost or worn (after Young 1973).

Many sharks and most rays live on or near the seabed, feeding on small fishes, molluscs and crustaceans. Some species such as the smooth hounds (Triakidae) and catsharks (Scyliorhinidae) are good swimmers and regularly move off the bottom. Benthic species including angel sharks (Squatinidae), sawsharks (Pristiophoridae), skates (Rajidae) and electric rays (Torpedinidae) rarely leave the bottom, but may swim or glide over it. Most bottom-living sharks blend in with the surrounding background, (e.g. collared carpet sharks, Parascyllidae and wobbegongs, Orectolobidae), whilst many benthic species have a habit of partially burying themselves in loose sand or gravel. By lying still and being well hidden, prey may wander by unaware of any danger; thus the predator need expend little energy in searching for its prey. Some species of sharks rest in caves (e.g. some nurse sharks, Ginglymostomatidae) or under rocks during the daytime and hunt at night. These nocturnal species often return to the same places to rest during the daytime. Oceanic Blue Sharks *Prionace glauca* perform continuous vertical movements between the surface and the deeper water in a regular and systematic manner to sample the water column for olfactory signals.

Many coastal sharks and rays show strong behavioural responses to tidal cycles. They will move into

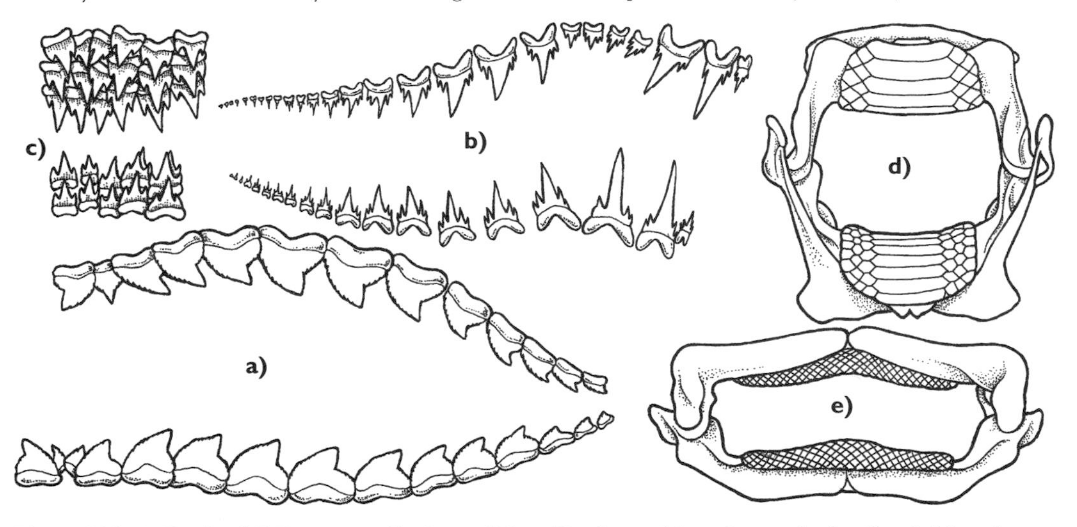

Figure 216. a) Teeth of different cartilaginous fishes. Teeth provide a clue to the feeding habits of many fishes and most teeth can be divided into five major types: a) serrated triangular teeth for 'cutting' (e.g. Tiger Shark *Galeocerdo cuvier*); b) dagger-like tri-cusped teeth for 'piercing and tearing' (e.g. Smalltooth Sand Tiger Shark *Odontaspis ferox*); c) teeth for 'clutching' (e.g. Chain Dogfish *Scyliorhinus rectifer*); d) teeth for 'grinding' (e.g. eagle rays, Myliobatidae) and e) teeth for 'crushing' (e.g. Thornback Ray *Raja clavata*). Grinding and crushing teeth are found in bottom-dwelling species while the more pointed teeth are found in the free-swimming hunters.

shallow bays or inlets with the flood tide and then move into deeper waters on the ebb tide. Within a coastal area there may be several different species of sharks and rays present. Each species is integrated into the local ecosystem and occupies a different niche with its own food resource. An example of this can be found in the temperate shallow waters off the California coast, where the cartilaginous fish fauna differs both spatially and seasonally. Different species of smooth hounds (Triakidae) occupy the muddy shallows with the eagle rays (Myliobatidae) and skates (Rajidae). The smooth hounds prey on clams, worms, mud shrimps and small fishes, while in the deeper channels, eagle rays crack open oysters with their crushing teeth. Cruising in the waters above the muddy bottom are the larger sharks in search of bigger prey such as bony fishes and cephalopods (e.g. squid and octopus). By occupying these different habitats with differing food resources the competition is reduced between the different cartilaginous fish species which coexist in the same areas.

Tooth shapes of cartilaginous fishes are not only a useful indicator of the species they came from, but can also give information about diet and lifestyles. Teeth are arranged in rows and they are conveyed towards the front of the mouth on special folds of skin that form the gums (fig.215). The teeth move forwards and continually replace worn or damaged teeth on the leading edge of the jaws. During their lives, sharks may form many thousands of teeth. Sharks and rays are active predators that prey upon other animals, sometimes larger than themselves. They will also scavenge dead animal carcasses. Other food sources may include small planktonic animals that are sieved through the gill rakers of plankton-feeding elasmobranchs. Seabirds, turtles, seals and large fishes form part of the diet of the Tiger Shark *Galeocerdo cuvier* and Great White Shark *Carcharodon carcharias*. The teeth of the Great White Shark and many requiem sharks (*Carcharhinus* spp). are triangular in shape and heavily serrated for seizing, cutting and tearing. The angled comb-like teeth of the Tiger Shark (fig.216) have backward facing points to hold the prey more securely. The Sand Tiger Sharks (Odontaspididae) use dagger-like, tri-cusped teeth for grasping and piercing prey. Smooth hounds (Triakidae) *Varium* and the Zebra Shark *Stegostoma* have flat-topped teeth for crushing hard-shelled invertebrates. Horn sharks (Heterodontidae) have a double dentition in their jaws in the form of pointed front teeth and flat-topped rear teeth which are effective in dealing with sea urchins.

Loose jaw connections to the skull enable many shark species to push their mouths well forward to get a better bite while attacking prey (fig.217). Violent head shaking or twisting of the body also helps to break off or cut out large lumps of flesh when tackling large prey. Among the bottom-dwelling sharks and rays, 'table manners' are a little better. Diets mostly consist of shellfish (e.g. clams, mussels, lobsters, crayfish, crabs and urchins), octopuses and small fishes. There may still be preferential feeding behaviour among these bottom feeders, but many species are essentially opportunistic and consume a wide range of prey species in their diets. To cope with hard-shelled prey, small crushing teeth form hard 'tooth-plates' and powerful jaw muscles have developed; the Spotted Eagle Ray *Aetobatus narinari* is a mollusc specialist which has teeth in the form of fused, flat-topped crushing plates. Some skates (Rajidae) have been observed using their jaws to 'pluck' invertebrate animals from the bottom. They may also use their flexible, muscular snouts to root out buried crustaceans and molluscs from the seabed.

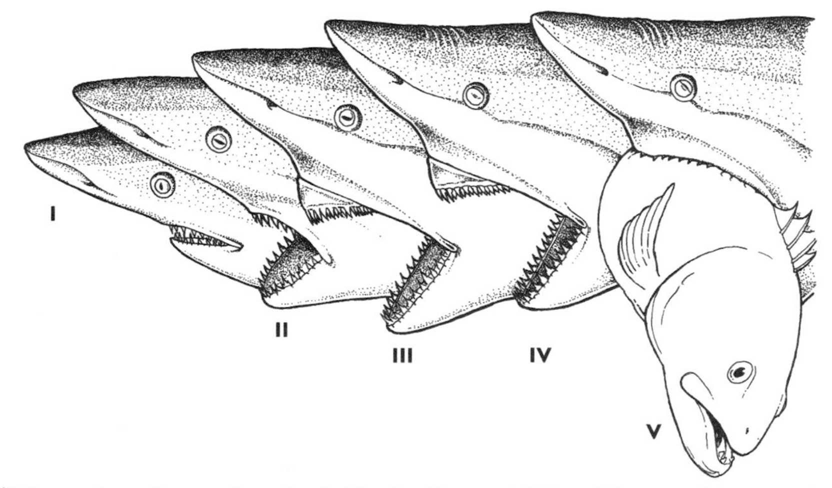

Figure 217. Jaw action of a requiem shark (*Carcharhinus* sp.) biting. The mouth is opened wide and the moveable upper jaw is thrust forwards to expose the teeth (I-III). Just prior to biting many sharks cover the eyes with a skin fold or nictitating membrane to protect the cornea (IV) and are unable to see during the final stages of feeding (based on van der Elst 1990).

Two 'antisocial' species of sharks include the semiparasitic cookie-cutter sharks (*Isistius* spp.) and the sawfishes (Pristidae). The Cookie-cutter Shark *Isistius brasiliensis*, which is less than 0.5m long, preys on large fishes, whales, dolphins and seals. It attaches itself onto the skin with its large sucker-like lips and, with its single row of fused, overlapping lower teeth, the shark bites into the skin. It then swivels round cutting free and swallowing an oval-shaped plug of flesh. The benthic sawfishes (*Pristis* spp.) have elongate snouts with rows of sharp teeth embedded in the sides. This snout is used in a sideways slashing motion to disable or kill fish. It may also be used like a rake to dislodge crustaceans and molluscs from the seabed. The food is then sucked in through the small, slot-like mouth.

Figure 218. Threat display of Grey Reef Shark *Carcharhinus amblyrhynchos* (Johnson and Nelson 1972).

Social feeding behaviour in sharks can be complex and is not well understood. Some sharks may hunt in packs (e.g. lantern sharks *Etmopterus* spp. and Spurdog *Squalus acanthias*) and herd their prey before feeding together. The Sandbar Shark *Carcharhinus plumbeus* of Australia may also hunt in a pack and drive schools of anchovies close to the shore before feeding. On occasions, when large numbers of sharks may congregate to feed on a single food source (e.g. on schooling fishes or squid), 'feeding frenzies' may take place with sharks darting in all directions and biting at whatever they can including, on occasion, their own species. This contrasts with the behaviour of solitary oceanic sharks which are often cautious when approaching potential prey. They may swim about in the vicinity and make several close passes before deciding to move in to begin feeding. However, dead whales floating at sea are often surrounded by large groups of feeding Blue Sharks. By contrast, the Great White Shark *Carcharodon carcharias* either stalks its prey on the surface or close to the seabed. It has been observed swiftly swimming from the depths to ram and incapacitate unsuspecting surface prey such as seals and sea lions. Following a kill, several sharks may feed on the same carcase as a group. However, a social order of dominance is first established through body language. Once established, the largest sharks tend to feed first, but the dead prey is still approached with caution.

Sharks may exhibit an aggressive type of threat display which has been observed when swimmers approach certain requiem sharks (*C. amblyrhynchos* and *C. galapagensis*). The behaviour includes exaggerated swimming movements, lowering of the pectoral fins, back arching, lifting and swinging of the head (fig.218). The shark may sometimes swim around in a horizontal spiral pattern or in a series of loops as it slowly approaches its target. If the swimmer continues approaching the animal, the threat display may culminate in a rapid rush by the shark with a fierce slashing attack. Although the attack is potentially fatal to humans, it is not the result of a feeding response.

AGE, SIZE AND GROWTH

About half of all adult sharks and rays reach a maximum length of less than one metre. Only 5% of all species exceed four metres in length. The largest of all cartilaginous fishes are the plankton-eating Whale Shark *Rhiniodon typus* (which can reach a length of 12m), Basking Shark *Cetorhinus maximus* (up to 9m) and Manta Ray *Manta birostris* (body spans up to 7m). Certain species of dogfish sharks (Squalidae) and finback catsharks (Proscylliidae) are amongst the smallest of shark species and become sexually mature at lengths of only 15-25 cm.

The ages of sharks and rays can be determined by counting the bands or rings in the vertebral centra or layering in the fin spines. This method is similar to counting the annual growth rings of trees. Different species may take between 6-20 years to reach sexual maturity. Cartilaginous fishes generally seem to be long-lived, and grow relatively slowly, though at present age is known in only a small number of species. Longevity differs widely between species; the Tiger Shark can live between 7 and 10 years while the Great White Shark may live up to 23 years. The maximum age for the Spurdog is at least 25-30 years and maximum recorded age for any shark or ray species in captivity was 75 years.

REPRODUCTION

Outside the mating period adults of many shark species spend much of their life in sexually segregated groups. It is believed there are certain areas or periods during the year when large numbers of sharks gather to mate (e.g. some hammerheads *Sphyrna* spp.). However, observations on mating behaviour in elasmobranchs are rare, even in captivity.

Mating behaviour is poorly understood and varies between species. In large sharks, close parallel swimming often precedes mating. Males (often the smaller of the pair) may hold on to the female by gripping the thickened skin just behind the head using their teeth. Alternatively, males may hold on to the trailing edge of the pectoral fins (fig.219b). Female Blue Sharks *Prionace glauca* have skin four times thicker than males, presumably as protection from the bites they receive from males during mating. In small sharks (e.g. catsharks, Scyliorhinidae) the copulatory behaviour differs; here the male coils his body around the female in a 'python-like' squeeze (fig.219a). In the Round Ray *Urolophus halleri* mating occurs in a belly-to-belly position.

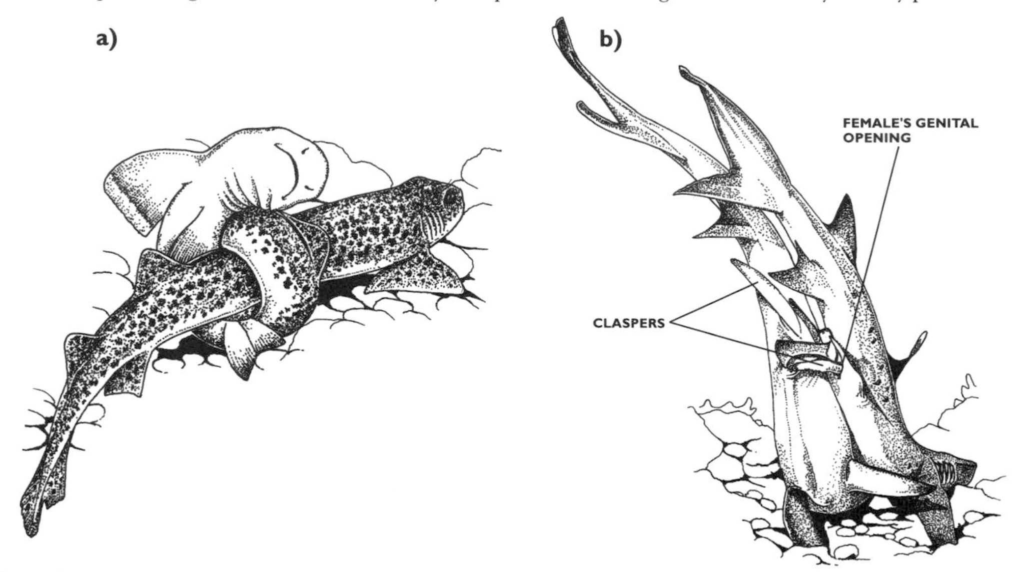

Figure 219. Mating postures in sharks: a) Lesser Spotted Dogfish *Scyliorhinus canicula* with the male coiled around the female, and b) White Tip Reef Shark *Triaenodon obesus*.

All male sharks and rays have a pair of elongate clasper organs formed from the inner part of the pelvic fins and these can be erected by muscular action during mating. When mating in a side-to-side posture, claspers fill with seminal fluid and one or both claspers may be inserted into the female's vent (fig.219b). In the Basking Shark *Cetorhinus maximus*, sperm is transferred to the female in golf ball-sized packages called spermatophores which protect the sperm inside until it is required. Female Basking Sharks may store spermatophores in their oviducts so that eggs can be fertilised sometime later, when they pass down from the ovary. It seems likely that other sharks may also be able to store sperm, perhaps for several years at a time.

Cartilaginous fishes have three different types of reproductive strategy (fig.220). They may be oviparous and lay external eggs with leathery cases (fig.220b). The young develop inside the egg cases, feeding off their yolk sacs. In the ovoviviparous sharks, the egg shells are thin and are retained inside the female's body (fig.220c). The yolk sac nourishes the young but once the yolk supply is finished they 'hatch' inside the mother and are soon born.

A modified type of ovoviviparity occurs in lamniform sharks. The young of the Bigeye Thresher Shark *Alopias superciliosus* feed on bundles of unfertilised eggs produced inside the mother's reproductive system. The female Sand Tiger Shark *Carcharias taurus* has a more bizarre strategy for its developing young. The first foetus to develop in the oviduct eats any other foetuses that are coming down the oviduct behind it (uterine cannibalism) before reverting to eating unfertilised eggs.

The third reproductive strategy is viviparity, which is also common among many higher vertebrates including mammals (fig.220a). In viviparity there is little development of the egg and the embryonic yolk is quickly used up. At this stage the yolk sac becomes connected to the maternal uterine wall to form a yolk-sac placenta. A placental cord allows the transfer of nutrients to take place from the maternal circulatory system to the developing foetus.

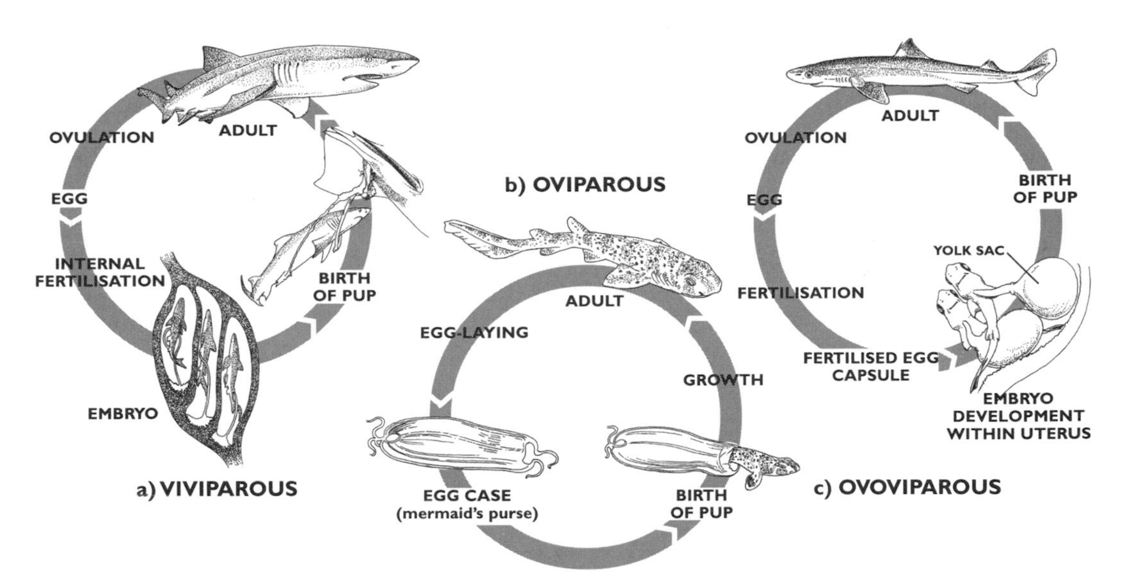

Figure 220. Typical life cycles showing the reproductive strategies of cartilaginous fishes.

In viviparous rays, the oviduct walls secrete a nutrient substance (uterine 'milk') which is ingested by the developing foetus. Viviparous sharks and rays may have long gestation periods ranging from a few months to nearly two years (some of the longest for any vertebrate). Litter sizes vary from a few young to over 130 offspring in the Blue Shark, but the average litter size is about 20 'pups'. A large size at hatching gives the young sharks and rays a better chance of survival.

About 40% of cartilaginous fishes lay eggs. Some ground sharks (Carcharhiniformes) including many of the catsharks (Scyliorhinidae), all horn sharks (Heterodontiformes) and, in the rays, all known skates (Rajidae) are egg layers. These species lay small numbers of large, yolky eggs which are fertilised inside the female. The leathery egg case (mermaid's purse) varies in shape between species (fig.221). The egg case may be cone-shaped, spindle- or sac-like with protruding horns, tentacles, lateral fins or spiral flanges that help attach it to the seabed or seaweed. Some sharks have specific pupping areas which they repeatedly use. A few species have even been observed repositioning their egg cases for better protection, but none are known to guard their eggs. Many empty egg cases may be found washed up on beaches along the strandlines and they make interesting finds. After the eggs have been laid incubation periods vary from one to fifteen months . Once birth or egg laying is complete there is no further parental care and in many cases sharks may even feed on the eggs of their own species.

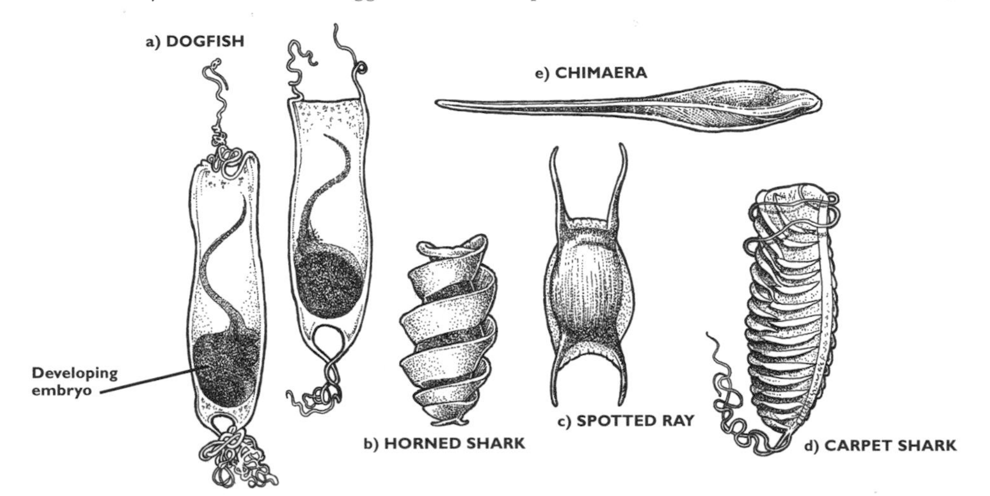

Figure 221. Egg cases of cartilaginous fishes: a) Lesser Spotted Dogfish *Scyliorhinus canicula*; b) Port Jackson Shark *Heterodontus portusjacksoni*; c) Spotted Ray *Raja maculata*; d) Tasmanian Carpet Shark *Parascyllium multimaculatum* and e) a chimaera *Chimaera phantasma*.

WARM-BLOODEDNESS

Warm-bloodedness or endothermy is a characteristic of birds and mammals. In fishes, endothermy is the capacity to raise body temperature above the surrounding sea water and maintain it at a constant level even when sea water temperature fluctuates. Most fishes are cold-blooded. However, some mackerel sharks including the Shortfin Mako *Isurus oxyrinchus*, porbeagles *Lamna* spp., Great White Shark *Carcharodon carcharias* and some thresher sharks have complex vascular arrangements to recycle heat generated by the active swimming muscles. In other sharks and rays this heat is lost through the gills, but normally most excess heat is lost from the body surface. The evolution of warm-bloodedness in these fishes seems to be related to large body size (low surface area for heat loss) and efficient hydrodynamic swimming. Similar endothermic mechanisms have evolved independently in some bony fishes (see plate 15).

CHIMAERAS

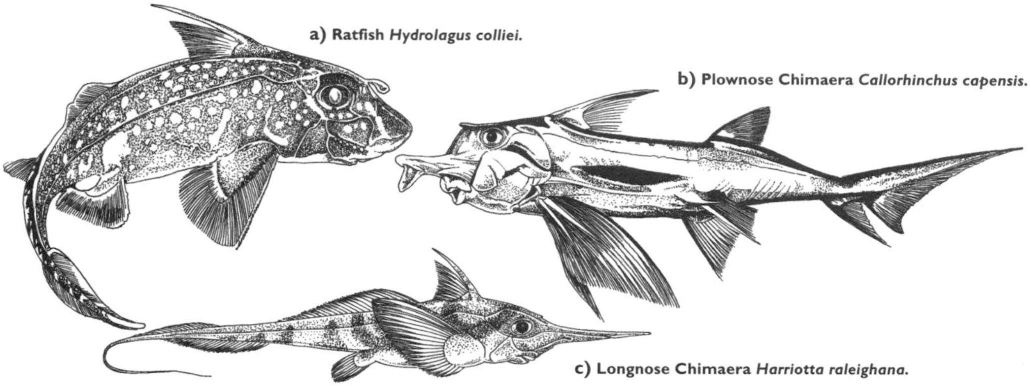

a) Ratfish *Hydrolagus colliei.*

b) Plownose Chimaera *Callorhinchus capensis.*

c) Longnose Chimaera *Harriotta raleighana.*

Figure 222. Representatives of the three main families of chimaeras: a) shortnose (Chimaeridae), b) plownose (Callorhynchidae) and c) longnose (Rhinochimaeridae).

The Chimaeras (order Chimaeriformes) are unusual, uncommon, generally deep-water fishes that grow to lengths of about one metre and weigh about 2.5kg (fig.222). They are only distantly related to the sharks and rays and the biology of this poorly known group is discussed separately here.

There are three different families of chimaeras, all of which are marine. Chimaeras are found in the world's temperate and polar regions. At lower latitudes they seek out the deeper, cooler waters and have been recorded at depths of 2,400m (8,000ft). The fishes are slow, clumsy swimmers and usually stay close to the bottom where they have been observed motionless and resting on the tips of their fins. Fertilisation is internal and copulation is performed using special 'clasper' organs. However, male chimaeras also have a unique clasper in the centre of the forehead to help hold the female during the mating process. Longnose (Rhinochimaeridae) and plownose (Callorhynchidae) species have fleshy snout projections (rostra) which are rich in chemical and electrical receptors. As longnose chimaeras live in deep, dark waters their receptors are most likely to be used for detection of prey and possibly a mate. By contrast the plownose chimaeras live in shallower waters, including bays and estuaries, where they probe the sand for invertebrate prey. Dentition is unusual as teeth are fused together into three pairs of continuously growing tooth-plates. This forms a solid beak-like structure that is used for crushing prey such as hard-shelled crustaceans and molluscs. The gut of chimaeras is a simple structure and the stomach is less well-defined than in elasmobranchs with only a limited spiral valve in the intestine. The skin of young fish has a covering of dermal denticles, but most adults lose them at maturity.

The first dorsal fin has a free mobile spine in front of it with a poison gland that produces venom. The shortnose and longnose species have long, tapering, whip-like tails whilst the plownose species have a shark-like tail fin. Unlike other cartilaginous fishes, chimaeras take in water through large nostrils and channel the water through special ducts directly to the gills. The four pairs of gills are, unusually, protected by soft gill covers, similar to the opercula of bony fishes. All female chimaeras lay large yolky eggs between 150 and 250mm in length that have leathery shell cases which are attached to the seabed (fig.221e). The female produces only two eggs at a time (one from each uterus) and the young hatch about 9-12 months later (six to eight months in plownose chimaeras), looking like small editions of their parents.

BONY FISHES (Osteichthyes)

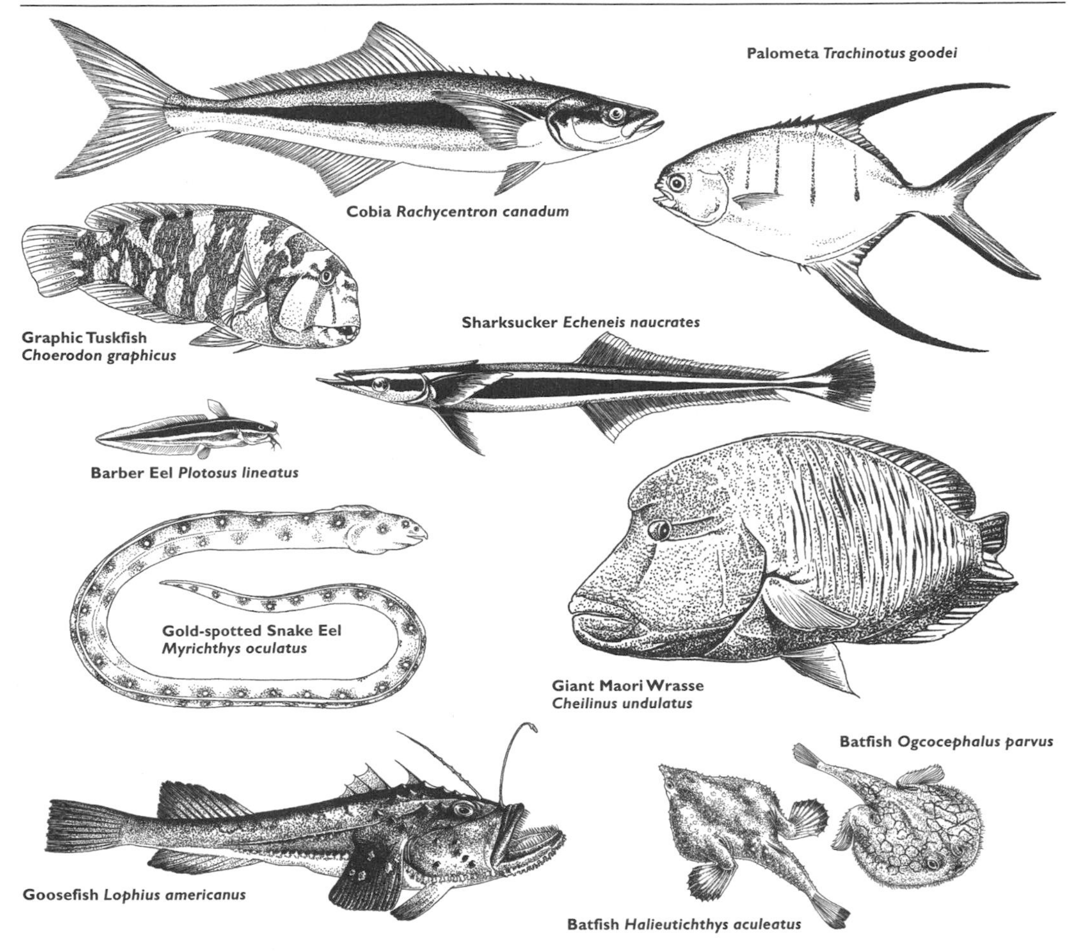

Figure 223. Ten bony fishes illustrating diverse adaptive radiation of the Osteichthyes.

Bony fishes or Osteichthyes (as defined here) comprise two separate subclasses, the ray-finned fishes and the lobe-finned fishes. The lobefins (Sarcopterygii) include the freshwater lungfishes (Lepidosireniformes), Coelacanth *Latimeria chalumnae,* and many extinct forms some of which are thought to be closely related to the early land vertebrates. Ray-finned fishes (Actinopterygii) contain, amongst other groups, the Chondrostei (sturgeons and paddlefishes) and the Teleostei (teleosts), both important groups in the sea.

Two teleost groups are ecologically dominant. These are the Ostariophysi, which account for 64% of all freshwater fish species, and the Percomorpha which is the most numerically important of all the vertebrate groups present in the marine environment. There are nine orders of fishes that make up the Percomorpha of which the perch-like fishes (Perciformes) is the largest (about 9,000 species). Pufferfishes, flounders, scorpionfishes, seahorses, dories, squirrelfishes and whalefishes are a few examples from the other marine percomorph orders. During their evolution, the perch-like fishes have undergone an explosive adaptive radiation, often with the development of unique structural modifications seen nowhere else amongst the fishes.

In modern teleostean orders there are both less advanced and more advanced forms, with many fishes exhibiting characteristics between the two types. Teleost diversity ranges from the primitive freshwater bonytongues, through eels, herrings and catfishes at one end of the scale, to the advanced cod-like fishes, mullets and pufferfishes at the other. The more primitive teleosts include the tarpons (Megalopidae) and herrings (Clupeidae). These fishes have characteristics that include: streamlined, elongated bodies; a single dorsal fin; pelvic fins located near the posterior part of the body; fin supports of flexible, branching, soft fin rays instead of spines; cycloid scales covering the tail and trunk but not extending onto the head, and a pneumatic duct between the swim-bladder and intestine (e.g. open or physostomous type, fig.242b).

The most advanced teleosts are the spiny-rayed fishes (Acanthopterygii) – see Factsheets for comparison with lower teleosts. They include mullets (Mugilidae), wrasses (Labridae), dories (Zeidae), blennies (Blenniidae), anglerfishes (Lophiidae), gurnards (Triglidae) and flatfishes (Pleuronectiformes). Major characteristics of spiny-rayed fishes include: two dorsal fins with the first dorsal fin supported by a stiff spine and the second dorsal fin having flexible, unbranched rays; the position of the pelvic fins are shifted forwards along the body to a position beneath the pectoral fins; spines are present in the anterior border of the pectoral fins; ctenoid scales extend onto the head and opercula (gill covers) and there is a closed swim-bladder (physoclistous type) with no pneumatic duct connection between the bladder and intestine.

Among the unusual modern teleosts are the eels (Anguilliformes) which have snake-like bodies and no pelvic fins. Eels are believed to be closely related to the less advanced tarpons (Megalopidae), as both groups have similar ribbon-shaped, transparent larval stages known as leptocephalous larvae (fig.253). Marine fishes exhibiting intermediate characteristics between the more primitive soft-rayed and advanced spiny-rayed forms include the sticklebacks (Gasterosteidae). Pipefishes and seahorses (Syngnathidae) are probably related to the sticklebacks. Flying fishes (Exocoetidae) also belong to this intermediate group of fishes.

GENERAL FEATURES

There are several characteristics that set apart the bony fishes from the cartilaginous fishes. These include the presence of true scales on the skin, a flap or operculum covering the gills and the presence of movable rays in the fins and tail.

These fishes have a bony (ossified) skeleton that supports the body tissues. There is a well-formed neurocranium around the brain that is sheathed externally by a fused outer covering of bone (dermal bones); the neurocranium and dermal bones together form the ossified skull. Within the skull are three pairs of ear stones (otoliths) that form part of the auditory and balance organs of the fish. Jaws are of the modern hyoid suspension type and are not rigid in ray-finned fishes; in lobe-finned fishes the jaw system is firmly attached to the skull. The paired gill chambers are protected by a hinged, external flap (operculum). The anatomy of a bony fish is shown in figure 224 and 226.

Teeth are present in most fishes and may line the jaws, bones inside the mouth (e.g. vomerine teeth) and also the pharynx (pharyngeal teeth). The oesophaguses of many fishes can distend considerably and therefore large prey can swallowed whole. The stomachs and intestines of bony fishes exhibit adaptations that differ according to their diets and some herbivorous fishes do not have stomachs. However, loss of the stomach among the carnivorous fishes also occurs but is generally rare. Numerous pyloric caeca behind the stomach increase the surface area of the gut for food absorption. The liver is relatively small by comparison with the elasmobranchs.

Paired fins are present on all known teleost fishes. They are thin structures, strengthened by fin-rays that can be either articulated and branching (e.g. soft rays) or 'unjointed' and stiff (spiny or hard rays). Fin rays of the caudal fin are directly supported by projections of the last caudal vertebrae and they lack any intermediate supporting bones. Unlike the fleshier fins of the lobe-finned fishes, ray-finned fishes can fold the thin, membranous fins and lay them flat against the surface of the body. In fast-swimming ray-finned fishes the fins are folded into grooves in the skin. Most teleosts are partially or completely covered in true scales although a few fishes do not have any at all. Many bony fishes possess a swim-bladder. However, it is often absent in bottom-dwelling species and in some of the mid-water species.

Reproduction normally occurs by the production of large numbers of small eggs that are externally fertilised and hatch into larvae. Larvae are transported by currents and a larval stage is important for dispersal of many fish species. A few species are viviparous with internal fertilisation and have developed complex arrangements for nutrient transfer between the female and her developing young.

BODY FORM

Evolution of the gas bladder and its subsequent development into a hydrostatic organ (swim-bladder) enabled fishes to maintain their positions at any depth in the water column with little effort. Therefore fishes no longer needed to continually swim to maintain depth. Slow swimming, highly manoeuvrable and agile free-swimming fishes were able to evolve and this was accompanied by changes in body forms. At the same time, fin shapes became modified and together these changes enabled fishes to expand into new habitats.

The theoretically ideal body form for mid-water living and swimming is a streamlined, torpedo (fusiform) shape that is ovoid in cross section. The widest part lies about one-third back from the snout and the trunk gently tapers towards the tail.

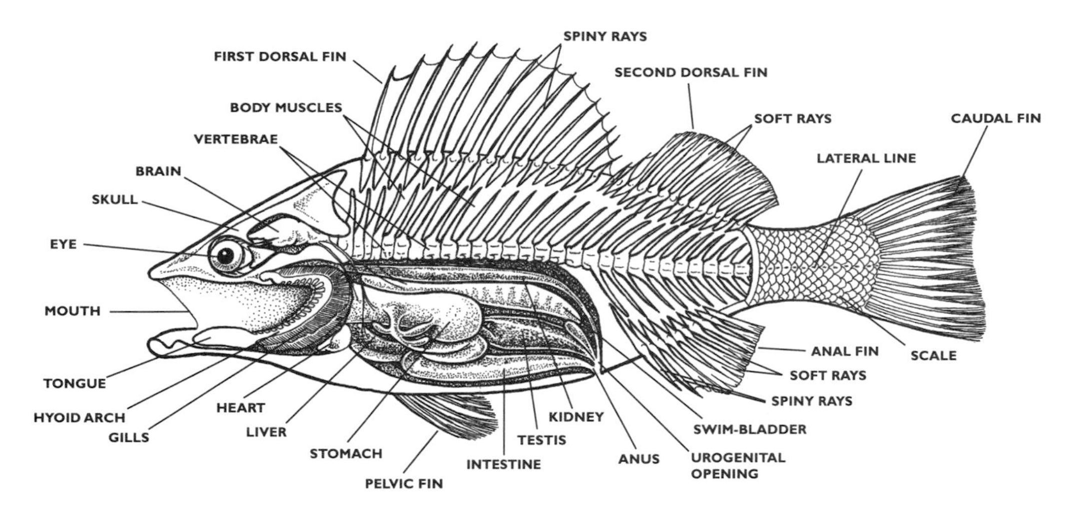

Figure 224. Anatomy of a bony fish showing the principal organs.

Fishes that have a fusiform shape include the tunas (Scombridae) and billfishes (Xiphiidae) and many pelagic species. However, the shapes of other fishes are very varied: some are globe-shaped (globiform, e.g. pufferfishes, Tetraodontidae); serpentine (anguilliform, e.g. eels, Anguillidae); thread-like (filiform, e.g. snipe-eels, Nemichthyidae) or elongate (attenuate, e.g. needlefishes, Belonidae). Some fishes are flattened from top to bottom (depressed, e.g. anglerfishes, Lophiformes) or flattened from side to side (compressed, e.g. dories, Zeidae; butterflyfishes, Chaetodontidae and flatfishes, Pleuronectiformes). Other species appear shortened from front to rear (truncate, e.g. triggerfishes, Balistidae, ocean sunfishes, Molidae and boxfishes, Ostraciontidae).

Within these types of body form, the degree to which the change from a basic fusiform shape has evolved will vary between species and even different parts of the life cycle of the same species. For example, the pelagic larvae of many fishes are fusiform in shape. However, if they develop and settle to the bottom to lead a demersal lifestyle, they may then become flattened (depressed). Despite such variations in body form, most fishes (except adult flatfishes) exhibit bilateral symmetry (left and right halves of the body look the same). Departure from the typical fusiform body shape normally has an adverse effect on swimming speed but does have the advantage of increasing the manoeuvrability of the fish.

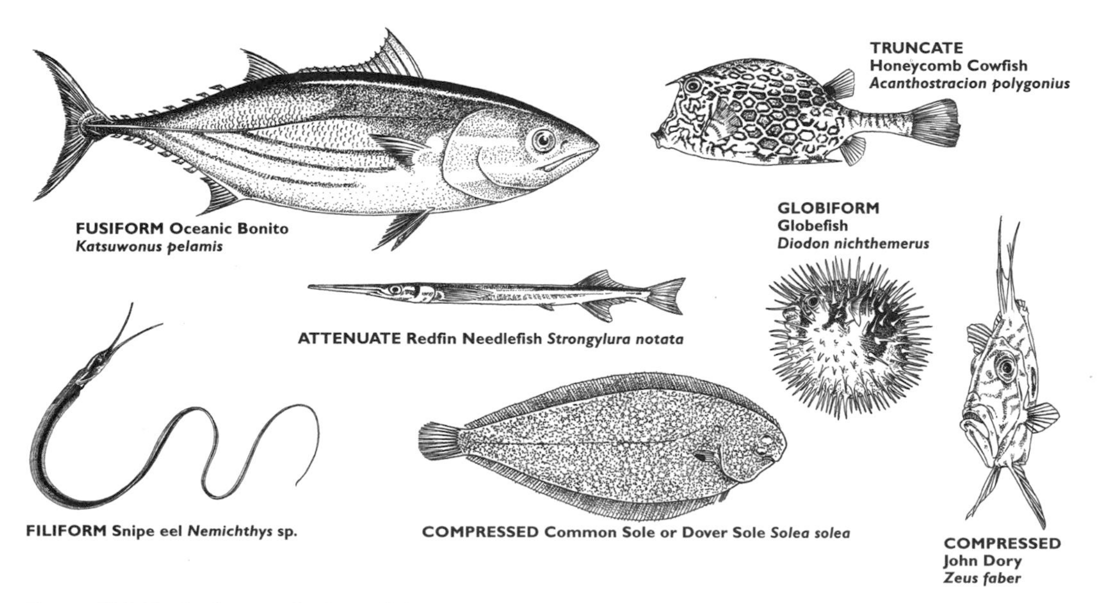

Figure 225. Body shapes of teleost fishes.

SKELETON

A tough bony skeleton gives the fish body protection and provides a supportive framework for muscle block attachments. The fish skeleton can be divided into three main parts: 1. the skull (cranium) which encapsulates the brain, supports the eyes and suspends the jaws and gill arches; 2. the backbone of vertebrae with associated projections and ribs; and 3. a fin skeleton of bones that anchors and supports the fins and tail (fig.226). The bones are either joined by ligaments or fuse to form a solid structure (e.g. the skull). The major lateral muscles attach firmly to the skeleton of the fish and are used primarily for body locomotion (fig.230). The backbone design is ideal to withstand the alternating tension and compression forces of the contracting locomotory muscles yet still retains the necessary degree of flexibility. Other groups of muscles may be used for movements of individual parts of the skeleton such as jaws and fins. A variety of protective functions of bones include ribs that surround vital organs, a skull that encloses the brain and the vertebral arches of the backbone which form a column to surround and protect the delicate spinal nerve cord. Fish bones either develop from cartilage that later becomes calcified into bone during growth, or develop from the dermal layers in the skin (e.g. scales). Study of the fish skeleton provides an understanding of how fishes swim and may also be used in classification and age determination.

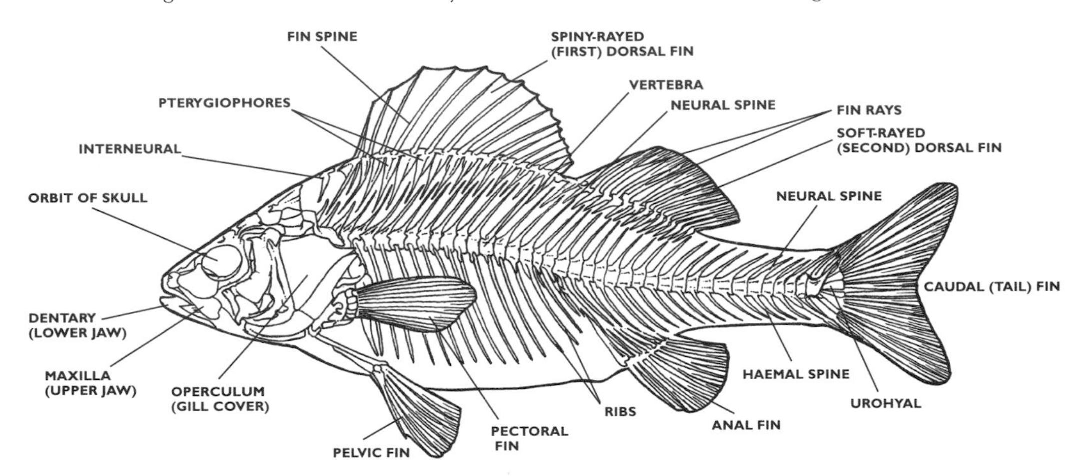

Figure 226. Bony skeleton of a higher teleost fish.

SKIN

A common form of protection for most marine teleosts is a thin, flexible 'suit of armour' consisting of overlapping scales. This provides defence against infection and protection from predators. Scales are easily replaced if they become worn or lost. They may also be modified to form spines, bony scutes or thorns (fig.228).

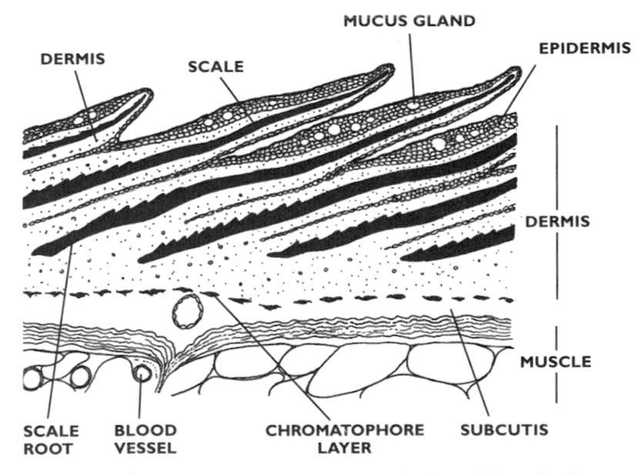

Figure 227. Cross section through the skin of a bony fish.

The skin envelope around the body of a fish forms a complex organ that has several different functions. It provides a physical protective barrier against fungal and bacterial attack, and also has respiratory, excretory, secretory and osmoregulatory roles. The skin also helps with streamlining by providing a smooth and sometimes slimy surface that is covered by scales and mucus. Skin colouration provides defence, warning, display, camouflage and sex recognition through a variety of characteristic patterns and markings. The skin may have several different glands including mucus glands, poison glands and a variety of receptors to detect taste, touch and vibrations. Outgrowths of skin called barbels may be used to feel for prey hidden below the surface of soft sediments. They may also carry taste and smell receptors to sample the surrounding water.

Skin consists of two main layers, the outer epidermis and underneath, the much thicker dermis (fig.227). The dermal layer of the skin contains blood vessels, nerves, connective tissue and cutaneous sense organs to monitor the environment.

The structure and arrangement of scales on the body differs considerably among fishes (fig.228). They usually form a continuous layer of overlapping plates similar to the arrangement of roof tiles on a house. Scales are often thin and flexible and being lightly attached at one end, they are able to move independently. This allows the body to flex easily. Scales also have colour pigments and if scales become worn, damaged or lost they are easily replaced. When fishes hatch they have naked skin, but scales start to appear as their bodies develop. Fishes that have large scales often develop them early on, but in other species scales may never develop (e.g. moray eels, Muraenidae; catfishes, Siluriformes and jawless fishes, Agnatha) and these fishes remain 'naked'. Partially scaled marine fishes include sturgeons (Acipenseridae) which have lateral and dorsal rows of scutes (thickened scales) whose function is probably protective. Many modern bony fishes have a complete covering of scales and this is a feature of most marine teleosts.

The shapes of scales and their structural modifications vary considerably. There are severel scale types: 1. diamond-shaped or ganoid type (e.g. primitive bichirs, Polypteridae; gars, Lepisosteidae; sturgeons Acipenseridae and paddlefishes, Polyodontidae); 2. comb-like or ctenoid type with the posterior margin bearing small teeth (e.g. most spiny-rayed acanthopterygian fishes); 3. round or cycloid type with smooth edges (e.g. most soft-rayed fishes including Clupeiformes, Gadiformes and lungfishes) and 4. cosmoid type (e.g. the lobe-finned Coelacanth fig.193). A particular scale type may not be confined to one order alone and more than one type may appear on a single species of fish (e.g. freshwater bass, *Micropterus* spp.). However, in most fishes the scales will be predominantly of one type. Normally scales of modern fishes consist of two plate-like layers (leptoid type) comprising a bony layer and a thin, fibrous layer. The front end (anterior) of the scale is embedded in a pocket of the dermis while the other end (posterior) is exposed and quite free. Most scales exhibit many fine growth rings which increase in number as the fish ages (fig.210a). Biologists can often determine the age of a fish by counting the light and dark bands of the scale, of which each pair (a light and dark band) represents one year's growth. This technique is similar to counting the growth rings in trees.

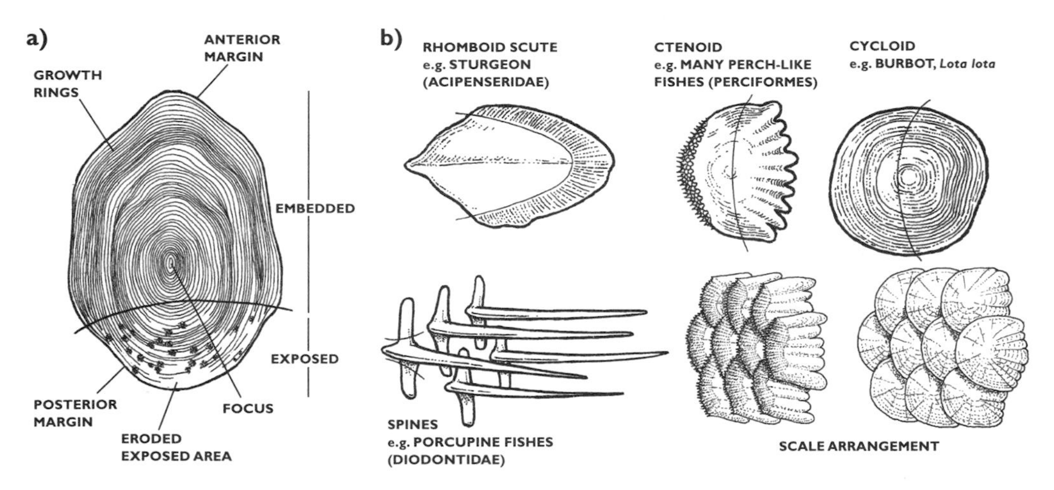

Figure 228. a) Dorsal view of a scale, and b) scale types of different teleost fishes. Scales form a tough outer layer to the skin which helps protect the animal from infection, predation and loss of body fluids.

POISON GLANDS

Poison glands are present in many fishes and are formed from modified mucus glands in the skin. Venom produced from the poison glands varies in toxicity amongst different species and in some extreme cases, it may even be fatal to other animals. In bony fishes, many reef fishes have toxic flesh, for example: porcupine fishes (Diodontidae), pufferfishes (Tetraodontidae), trunkfishes (Ostraciidae), some mackerels and tunas (Scombridae) and the moray eels (Muraenidae). Humans and other higher predators are susceptible to 'ciguatera-type' poisoning. This is a nervous reaction where the victim feels alternately hot and cold and in 7% of cases, the reaction is fatal. Many fish families are periodically involved in this type of poisoning. They include trunkfishes (Ostraciidae); surgeonfishes (Acanthuridae); barracudas (Sphyraenidae); parrotfishes (Scaridae); wrasses (Labridae); breams and porgies (Sparidae); jacks (Carangidae); snappers (Lutjanidae); basses (Serranidae); goatfishes (Mullidae); anchovies (Engraulidae); herrings (Clupeidae) and bonefishes (Albulidae).

Venom produced from a stinging apparatus (e.g. fin spines) has evolved independently in many fish families and these stings can be painful and even fatal to humans in certain circumstances. Some of the more dangerous families are scorpionfishes (Scorpaenidae), stargazers (Uranoscopidae) and weever fishes (Trachinidae). Non-fatal but painful stings come from dragonets (Callionymidae), surgeonfishes (Acanthuridae), and toadfishes (Batrachoididae). The following is a more detailed description of the poison apparatus and the fish species involved:

1. Stargazers (Uranoscopidae) have shoulder stingers with venom glands at their bases.
2. Rabbitfishes (Siganidae) have anal, dorsal and pelvic fin spines with poison glands (e.g. *Siganus* spp.).
3. Dragonets (Callionymidae) have fin spines with venom (e.g. *Callionymus* spp.).
4. Surgeonfishes (Acanthuridae) have stingers on each side of the caudal peduncle. Venom glands are in the sheath of the spine (e.g. *Acanthurus* and *Naso* spp.).
5. Catfishes have spines on dorsal and pectoral fins with glands beneath the skin that open through pores at the bases of spines (e.g. sea catfishes, *Galeichthys felis* and *Bagre turus,* Indo-Pacific catfishes *Heteropneustes, Clarias* and *Plotosus* spp., and bullheads *Ictalurus* spp.).
6. Toadfishes (Batrachoididae) have opercular stingers and dorsal fin spines with glands at their bases (e.g. *Batrachoides, Thalassophyryne* and *Opsanus* spp.).
7. Scorpionfishes, rockfishes and lionfishes (Scorpaenidae) have dorsal, anal and pelvic fin spines with venom glands in their grooves (e.g. scorpionfishes *Scorpaena* spp. bullheads *Notesthes* spp. lionfishes *Pterois* spp. and stonefishes *Synanceia* spp.).
8. Weever fishes (Trachinidae) have opercular stingers and spines on dorsal fins with venom glands in their grooves (e.g. *Trachinus* spp.).

FINS, BODY SHAPE AND MOVEMENT

Most bony fishes use a series of 'sinusoidal' waveform contractions, starting at the head and passing back along the body, to move themselves forward through the water. The myotome blocks are the main swimming muscles which are arranged in a complicated pattern along the flanks of the fish (fig.229 and 230). Each contraction of a muscle block will affect a considerable section of the body. In the fast swimming species, each myotome may overlap as many as 19 vertebrae and the more overlap there is, the more powerful the contraction can be. Prominent blocks of lateral trunk muscles are clearly seen when a fish is skinned or cut vertically into steaks for cooking. After cooking, it becomes apparent there are two types of swimming muscles in different positions of the body. These are distinguishable by their colour. Both 'red' and 'white' muscles are clearly visible in a cooked Atlantic Herring *Clupea harengus.* The white muscle constitutes the bulk of the muscle blocks, but there is a lateral strip of red muscle underneath the skin, positioned on the flanks of the fish. It has been discovered that the red muscle is used for slow, sustained swimming while the white muscle is used for sudden bursts of speed. Red muscle is particularly well developed in the oceanic cruising fishes such as the tunas and mackerels (Scombridae) and billfishes (Xiphiidae).

The fins of many teleosts no longer provide lift during swimming, but instead the swim-bladder is used to control the buoyancy and vertical position of the fish in the water column, even when the animal is stationary. The presence of a swim-bladder has freed teleosts from a bottom-living existence and has enabled fins to become modified and adapted for different uses (e.g. fine movement control). In the more advanced teleost fishes, fins have been modified for various uses (fig.231) including gliding (e.g. pectoral fins of flying fishes, Exocoetidae and flying gurnards, Cephalacanthidae); defence (e.g. spines of weeverfishes, Trachinidae; stonefishes, Syngnathidae and triggerfishes, Balistidae); suction (e.g. dorsal fin of remoras, Echeneidae and pelvic fins of the Lumpsucker *Cyclopterus lumpus* and Clingfish *Lepidogaster gouami);* display and courtship (e.g. male Common Dragonet, *Callionymus lyra*); signalling a warning (e.g. lionfishes *Pterois* spp.); walking (e.g. gurnards, Triglidae); touch (e.g. threadfin fishes *Polynemus* spp.), hopping (e.g. mudskippers, Periophthalmidae) and as a lure (e.g. the fleshy, mobile dorsal fin ray of anglerfishes, Lophiidae).

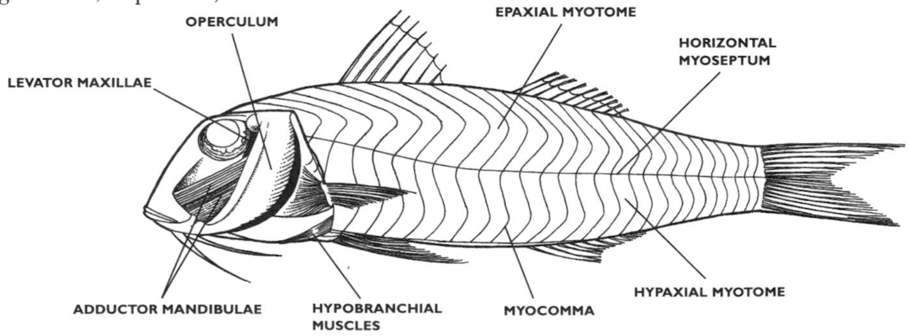

Figure 229. Muscle structure in the Red Mullet *Mullus surmuletus* (based on Young 1973).

The streamlined body form in many diverse groups of cruising fishes has evolved independently to produce a similar shape (evolutionary convergence). Cruising speeds associated with these fishes are dependent on a mathematical relationship between speed and overall body length. Cruising speeds which are maintained by free-swimming fishes are in the order of three to six times the body length per second. Fins of many fast-cruising teleosts have become reduced and thin, and they can often be flattened against the body or folded away into grooves to reduce drag during swimming. Fast fishes (plate 15) have narrow, forked tails with high aspect ratios. However, there is always some compromise in shape as the tail fin must still function reasonably well at lower speeds (fig.232). At low speeds, dorsal and anal fins may be raised to reduce rolling (e.g. sailfishes *Istiophorus* spp.).

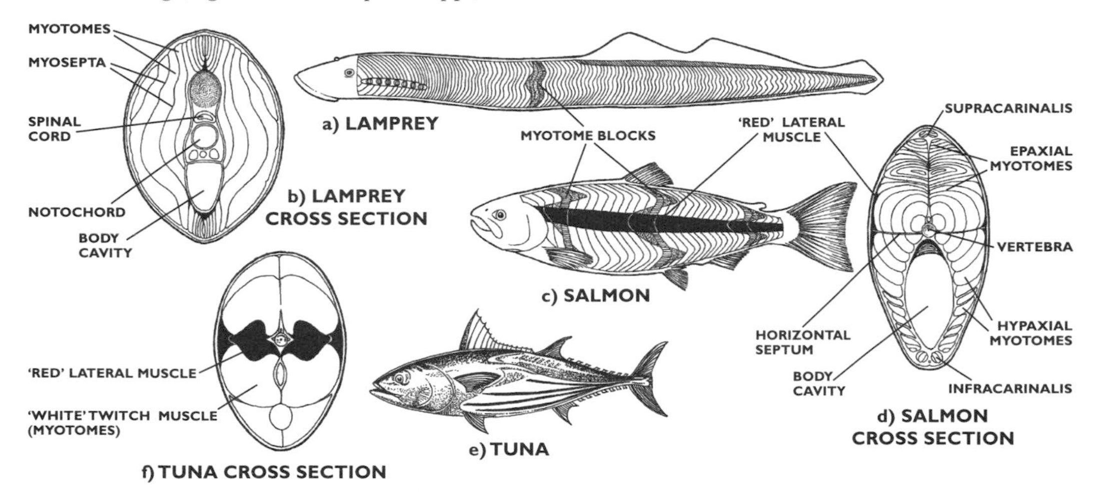

Figure 230. Muscle arrangement in three fishes. a) The lamprey *Lampetra tridentata* myotome pattern and b) the lamprey's simple body musculature in cross section. c) The salmon's reduced myotome number and overlying red lateral (tonic) muscle and d) the Chinook Salmon's *Oncorhynchus tshawytscha* more complex white (twitch) musculature cross section. e) The Skipjack Tuna *Katsuwonus pelamis*, and f) its substantial proportion of red lateral (tonic) muscle which is used during steady cruising.

In the advanced, slow-swimming teleosts, pectoral fins are positioned high on the sides of the body and pelvic fins are usually well forward. The broad pectoral fins found in slow-swimming fishes allow them to be used as effective brakes for deceleration (fig.233b). This happens when they are thrust out laterally from the body. The action often has a side effect as it causes a slight rise of the body (pitching) during high speed deceleration. However, this is counteracted by tilting the leading edges of the pelvic fins downwards. Alteration of pelvic fin positions can control the rise and dive of the body (similar to aircraft elevators) but by turning one of them outwards, they can also produce a rolling motion when required. With the increase in delicate control of movements, many other improvements in slow-swimming techniques have taken place.

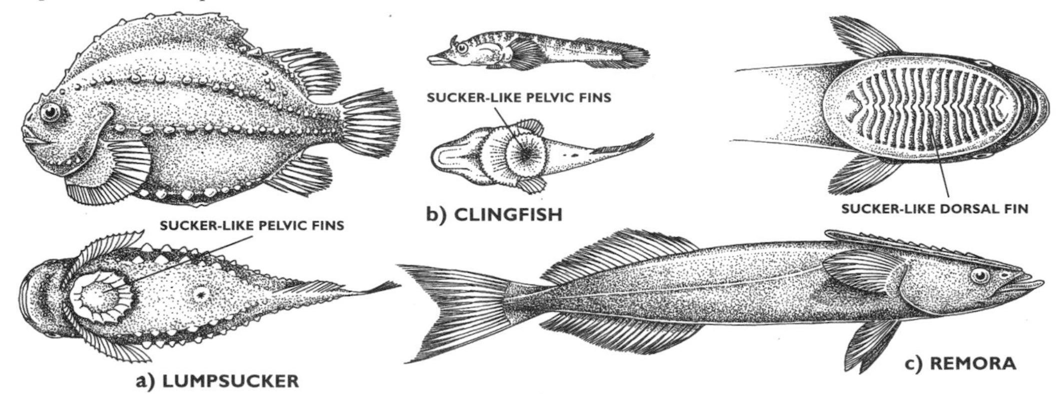

Figure 231. Fin modifications among marine teleosts. a) Sucker-like pelvic fins of the Lumpsucker *Cyclopterus lumpus*, and b) the Clingfish *Lepidogaster gouami*. c) Sucker-like dorsal fin of a remora (Echeneidae).

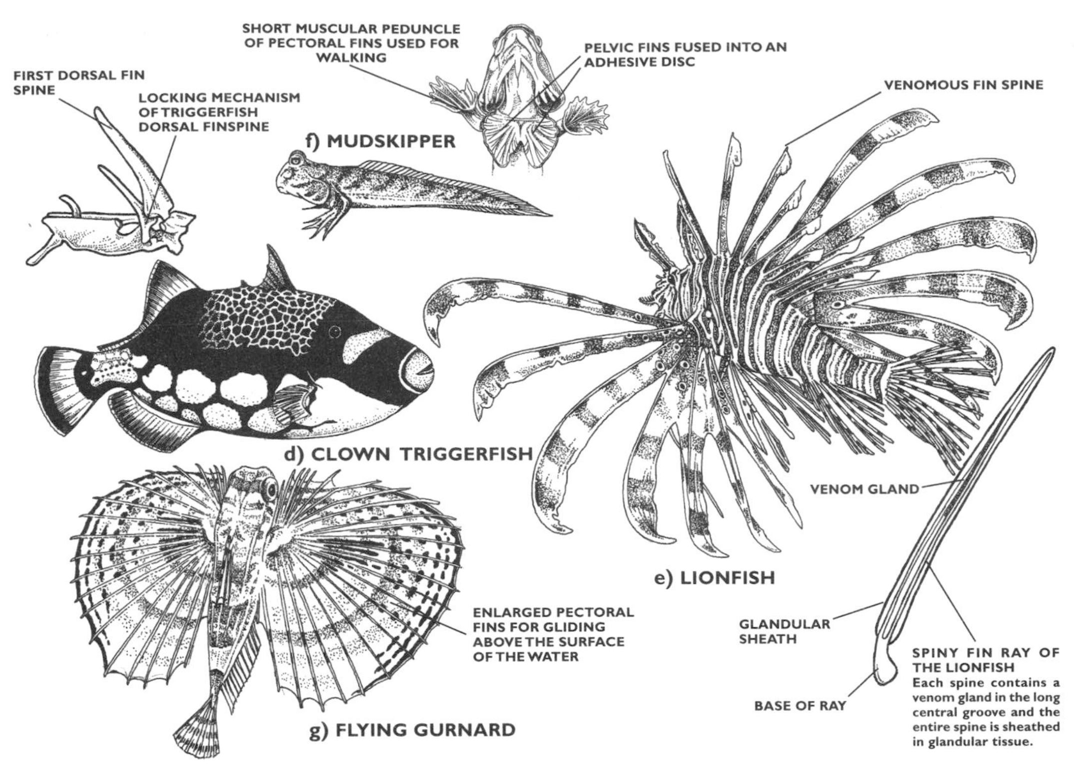

Figure 231 (cont.) Fin modifications among marine teleost fishes. d) Fin spines of triggerfishes (Balistidae) e.g. Clown Triggerfish *Balistoides conspicillum*, and e) lionfish *Pterois* sp. f) Leg-like pectoral fins of mudskippers (*Periophthalmodon* spp.). g) wing-like pectoral fins for gliding (Cephalacanthidae) e.g. Flying Gurnard *Dactylopterus volitans*.

More unusual modifications of fins are evident in the trunkfishes, ocean sunfishes, porcupinefishes, pufferfishes and seahorses. Their bizarre body shapes do not allow efficient body undulations for normal locomotion. Instead, the reduced dorsal, pectoral and pelvic fins perform a 'sculling' action to move the fish in any direction, including backwards. Caudal (tail) fins have changed considerably during teleost evolution towards the symmetrical homocercal shape. This is an efficient design for directional turning of the body without creating tail lift. Broad, flattened tails are common in slow-swimming and bottom-dwelling teleosts where manoeuvrability is more important than cruising speed (fig.232). The tails of reef-dwelling butterfly fishes (Chaetodontidae) and angelfishes (Pomacanthidae) only just extend beyond the posterior margins of the body (plate 22). The whole rear end of these fishes functions as a mechanism for increased lateral mobility. By contrast, the tail fins of eels have become pointed, and the dorsal and anal fins extend along the upperside and underside of the body. The snake-like movements of the Common Eel *Anguilla anguilla* and its pencil-shape makes this fish an efficient long-distance swimmer, capable of migrating many thousands of miles across oceans (see page 63). The S-shaped waves of undulations which pass along the body can be reversed to enable the eel to swim backwards. Their sinuous and flexible bodies are ideal for movements between rocks, cracks, crevices and through holes. Seahorses have completely lost the swimming action of the rear portion of their body. Instead the posterior region has evolved into a prehensile-like tail similar to that of a monkey. The seahorse secures itself to weed or coral by coiling its tail around the object.

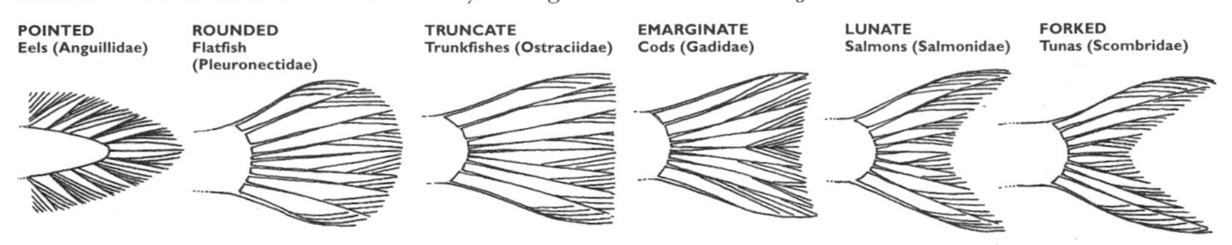

Figure 232. Tail shapes of teleosts (after van der Elst 1990). Fast-swimming pelagic fishes have forked tails that are hydrodynamically efficient. Fishes that are slower swimmers or bottom-dwelling have larger and more rounded tails for greater manoeuvrability.

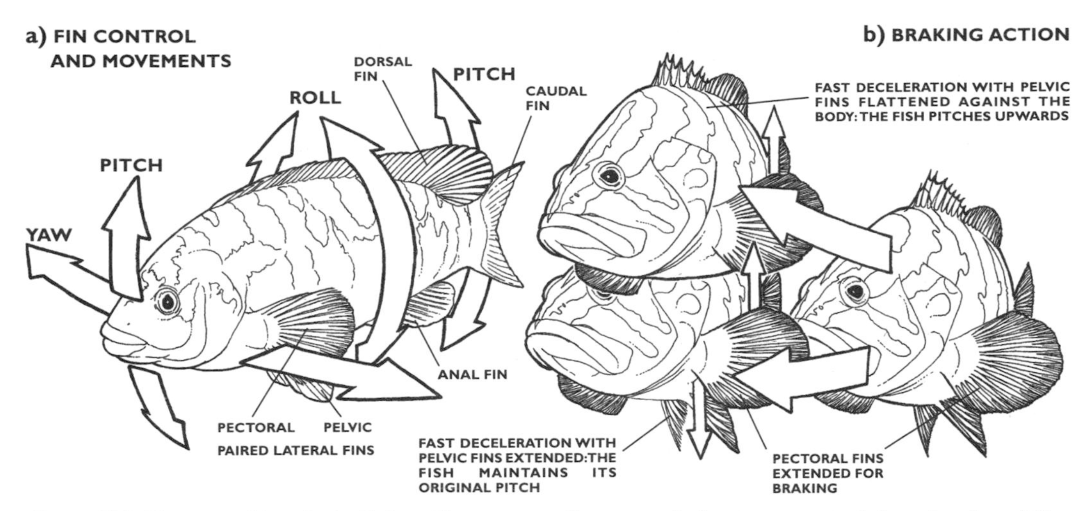

Figure 233. Fin control in teleost fishes. Fins serve as fine controls for movement: a) dorsal and anal fins prevent 'rolling'. The paired lateral fins control rising and diving and can be used to produce 'rolling' movements. The caudal fin is used as an efficient rudder to steer the animal and a combination of subtle fin actions control 'yaw'. b) Braking in teleosts is normally produced by laterally extending the pectoral fins. This often causes the fish to pitch upwards during fast deceleration. To prevent this, the pelvic fins are positioned so as to produce a counter-balancing downward thrust, thus maintaining the fish's vertical position when braking.

Departures from the streamlined body forms of fast-swimming fishes are numerous in many bottom-dwelling species. Many of these animals do not have swim-bladders. Dorso-ventral flattening is far less common among teleosts than it is in cartilaginous fishes. Some of the most flattened teleosts include the benthic anglerfishes (Lophiidae) which are common in cool temperate coastal waters. These fishes are heavy and their swimming is ungainly and slow, reflecting their sluggish lifestyles. Their skin coloration and body shape make them difficult to detect on the seabed and they usually wait motionless for prey to pass by within striking distance. Some species have a modified dorsal fin forming a 'tassel' or a 'luminous' lure to attract prey. Once the prey is near enough to strike at, the huge mouth shoots forwards and engulfs it. Stargazers (e.g. *Uranoscopus* and *Astroscopus)* also have dorso-ventrally flattened bodies and these fishes wait for prey by burying themselves in sand with only their eyes showing and their open mouths facing upwards. A red patch in the open mouth attracts small inquisitive fishes close enough to bring them within striking range.

Flatfishes have unusual life cycles in which they pass through several body changes as they develop or metamorphose from larvae to adults (fig.234). The pelagic larval stage has bilateral symmetry, with an eye on each side of the head. However, as the larval body becomes thin (compressed), the eye on one side moves round over the top of the head to a final position close to the eye on the opposite side. This leaves one side of the fish temporarily 'blind', but when the young fish settles to the bottom, the blind side becomes the underside of the body. The turbots, brills, topknots and megrims (Bothidae) are all 'left-eyed' (left side uppermost, plate 18); most flounders, plaice, halibuts (Pleuronectidae) and soles (Soleidae) are 'right-eyed' (right side uppermost, plates 18 & 23).

Schooling Many fish species swim together in formations or schools. Schooling is a behavioural mechanism that appears to be partially genetic (i.e. inherited), with the finer points being learnt during the early larval and 'fingerling' stages of life. Schooling is not just a randomly organised group of individual fishes but a social organisation to which each fish is 'bound' by behavioural and physical constraints. True schools of fishes are able to move, turn and flee together in common directions yet still maintain remarkable group cohesion. This peculiar organisation does not have 'leaders'. Fishes swimming at the leading edge of the school frequently change places with those behind when changing direction; the lead becomes the flank and the flank becomes the rear as the school moves around. When the fishes move together, they maintain spatial distances between each other and only break ranks when frightened or disturbed. They then flee in all directions but quickly re-group. Schools of fishes may consist of mixed sex groups or single sex groups, but this varies according to the species. Fishes within schools are often of similar size and shape and are likely to come from one age group. It is believed that sight or acoustic vibrations monitored by the lateral-line system are the main senses that keep the school organised. Schools do not break up at night or in poor visibility, therefore sight cannot be the only sense used for maintaining group cohesion.

The advantages of schooling are not always easy to understand. Some species school for life while others may only come together for specific purposes such as feeding, breeding or migration to other areas. There may also be safety in large numbers as few predators can catch and eat large numbers of fishes at once. Many pairs of watching eyes are also better than one, and predators may be confused by fishes darting in many directions during a chase. Within mixed school groups, finding a breeding partner at the right time is less of a problem. Maintaining formation during continuous swimming may be the most energy efficient method for schooling fishes to cover long distances. Whatever the reasons for schooling, this behavioural strategy is successful and allows fishes to survive in large numbers to maturity. This may be especially important in open waters where there is little natural cover.

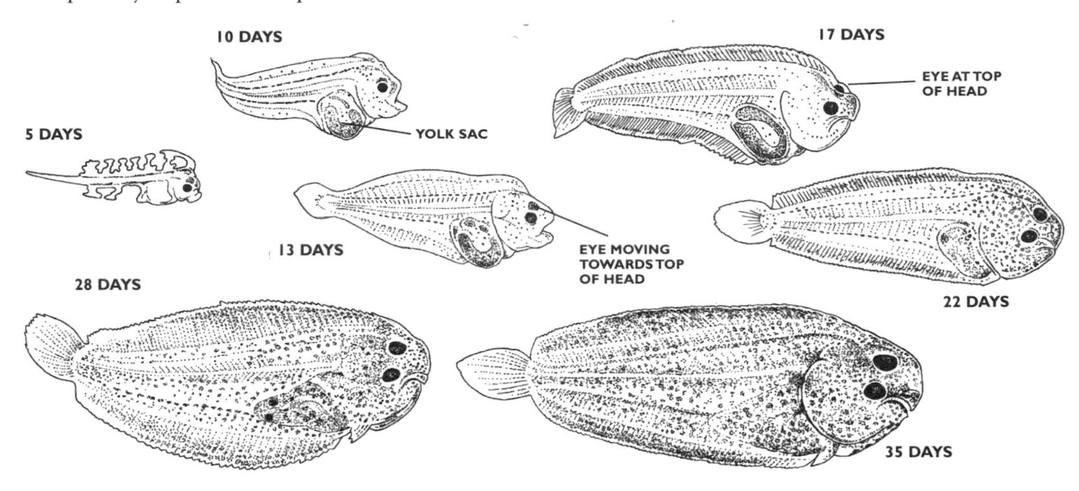

Figure 234. Stages in the early development of the Common Sole *Solea solea*.

5 days (3.5mm). The larva develops like a normal fish larva and lives pelagically near the surface where it hatched from the floating egg. At this stage it lives off the yolk sac nutrients.

10 days (4.0mm). Most of the fins have developed and the mouth can be opened and closed.

13 days (5.0mm). Most of the yolk sac food has been used up and the larva starts to feed on microscopic plankton. The fish starts to become asymmetrical in shape as the left eye moves towards the top of the head and the skull and jaws twist slightly. Fins become more distinct and fin rays start to develop.

17 days (6.5mm). The left eye has migrated to the top margin of the head. Skin pigment increases and the larval fish continues to actively feed on larger plankton.

22 days (8.0mm). Both eyes are now on the right side of the body and the post-larva is blind on the left side. The adult body shape is well defined and pigment increases in the head region and along the margins. The remnants of the yolk sac have virtually disappeared.

28 days (9.0mm). Fin rays, backbone and ribs are clearly defined and pigment increases over the body.

35 days (10.0mm). Metamorphosis is nearly complete and the skull increases in size and strength. At this stage the young sole has migrated from the spawning areas to the shallow coastal waters. From now on the skin pigment increases and by about 50 days most young fishes have settled to the seabed to become bottom-dwelling. Within the next few months the scales will develop and the skin coloration will alter to match the bottom substrate which will help to conceal the fish. When they become bottom-dwelling their diet changes to worms and small crustaceans.

COLORATION

One of the most striking features of fishes is the brilliant variety of colours, markings and patterns that are apparent to human eyes. However, there are some common colour elements recognisable amongst the bewildering array of designs. Many free-swimming, pelagic fishes have a similar pattern with white bellies, silvery flanks and dark iridescent, blue or green backs (plates 15, 16).

In the clear coastal waters of the tropics many fishes have bright colours and distinct patterns. Some of the most colourful fishes include the groupers (Serranidae), damselfishes (Pomacentridae), wrasses (Labridae), angelfishes (Pomacanthidae), butterfly fishes (Chaetodontidae), parrotfishes (Scaridae), surgeonfishes (Acanthuridae), squirrelfishes (Holocentridae) and triggerfishes (Balistidae) (plates 19, 20, 21, 22).

In the colder and often murky waters of temperate and polar regions, the colours of fishes are generally more subdued with shades of green, brown and yellow dominating. These fishes often merge with the background habitats they live in. There may also be dimorphic (sex) differences of coloration, with the male fish often being the more vivid and colourful partner while the female appears quite dull by comparison (plate 18). The bright colours of males are often used for courtship displays to attract a suitable mate. In many tiny fishes and the larval stages of larger species there may be no pigment present. The body appears glass-like and transparent except for the fine bones and the intestine. This uniform look may make the fish difficult to see during early stages of the life cycle and so increases its chances of survival.

There are two types of cells that impart colour to fishes. Chromatophores are found in the dermis of the skin either outside or beneath the scales and these cells provide true colour pigment. The second type of cell, the iridocyte or mirror cell, contains reflective material. These produce an iridescent appearance if the cell material is on the outside of the scales. However, if the material is inside the scales it imparts a white or silvery hue. Chromatophores are branched cells that contain pigments of either black (melanophores), red or orange (erythrophores), yellow (xanthophores) or white (leucophores). Spacing of chromatophores and iridocytes produces interference effects and therefore a wide range of subtle spectral hues are possible. A dispersed mixture of yellow and black chromatophores imparts brown or green hues and blue is usually an interference colour.

The use of colour by fishes can be complex and the reasons are often not clearly understood. All humans perceive colour in a similar way, but qualities of image and brightness are not necessarily the same as those seen through the eyes of fishes. Many fishes see colours but some are colour-blind; others have excellent low level light vision. Sea water preferentially reflects, scatters and absorbs light of different wavelengths. Therefore the typical wide colour spectrum seen at the surface (e.g. on a shallow tropical reef) will not be evident deeper down. What is important to the human eye may be less significant to fishes. Therefore perceiving colour is likely to be associated with tones and patterns, rather than true colours.

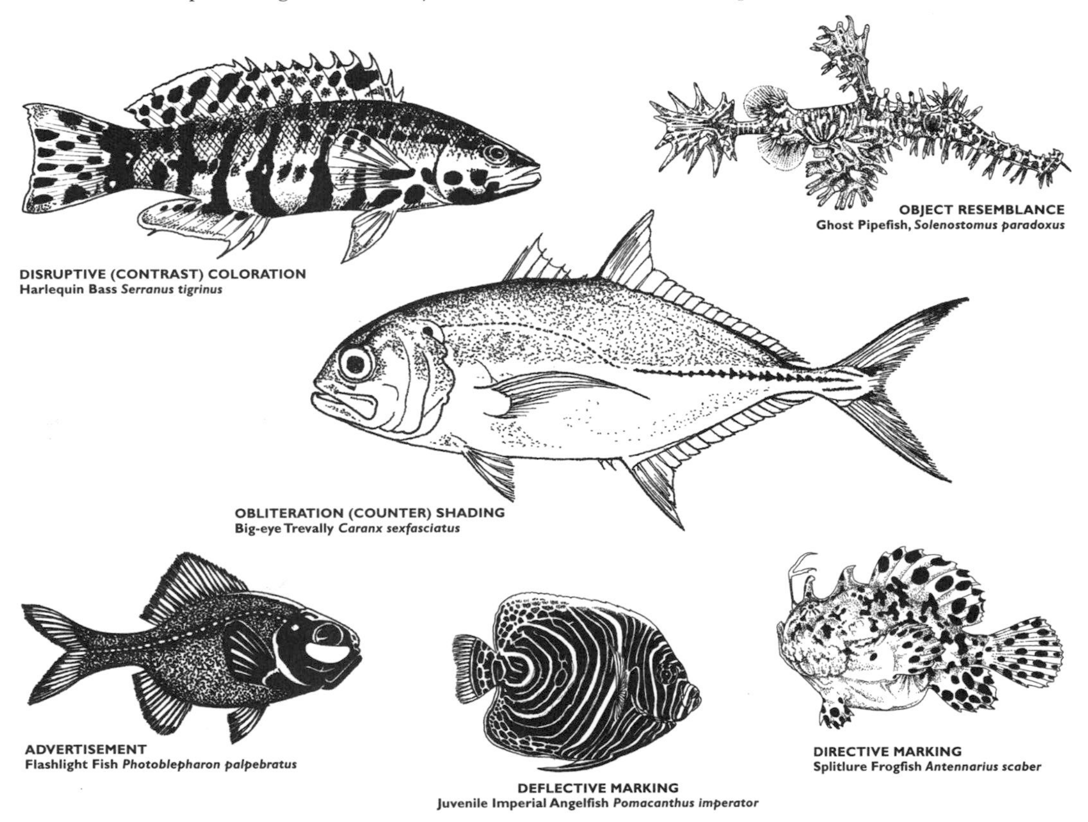

Figure 235. Concealment and camouflage of marine teleosts.

Coloration of fishes may be useful for a variety of purposes including sex recognition, warning, threat, distress signals, concealment, disguise and display. Concealment colours are imparted in different ways through: 1) resemblance; 2) obliteration shading, and 3) disruptive coloration.

1. **Resemblance.** Many coastal fishes try to match the hues of their habitat backgrounds. Therefore colourful coral reefs often have colourful fishes associated with them, while subdued backgrounds mostly have subdued-looking fishes. Fishes living among weeds or on the bottom will try to blend in with the substrate by creating similar body hues, tones and patterns. If fishes move to a different background they may change to match it. Therefore, the colorations of most fishes are not constant throughout their life. There are both short-term and long-term changes to help the fish adapt to its habitat. Mechanisms for body colour changes are both physiological and morphological. Morphological changes are normally slow and involve additions of pigment and chromatophores. Physiological changes may be in response to visual or non-visual stimuli. They are controlled through hormonal or nervous activities and cause a contraction of black melanophores or redistribution of pigment granules within the chromatophores.

   Many fishes show behaviour that mimics an inert object or water movements. Some fishes resemble floating weed, twigs or logs in their posture, form and coloration. Seahorses such as the Leafy Sea-dragon *Phycodurus eques* (fig.236) and Sargassum Fish *Histrio histrio* (plate 14) have many loose, leaflike skin projections (trailers) to disguise the true body form and these animals are difficult to distinguish from the weed they are inhabiting. Pipefishes (*Syngnathus* spp.) also adopt a body attitude similar to the weed and they will even sway with the algae as it moves with the current. The Atlantic Spadefish *Chaetodipterus faber,* (plate 14) resembles the seed pod of the red mangrove *Rhizophora* spp. and the Lumpsucker (plate 18) appears like a floating capsule (pneumatocyst) of brown alga.

Figure 236. Leafy Sea-dragon *Phycodurus eques.*

   Some coral reef blennies (Blenniidae) mimic other fishes; *Plagiotremus laudandus* looks similar to the inoffensive blenny *Meiacanthus atrodorsalis* so that it can approach its prey and nip off pieces of fins, skin and scales.

2. **Obliteration shading**. This concealment mechanism is based upon countershading which acts to reduce the clarity and outline (contour) of the body. Most fishes exhibit some form of countershading. The simplest form is a pale belly and toned flanks that become increasingly darker towards the dorsal surface of the animal. Light from above brightens the back and shadows the underneath of the body. The fish appears to lose its contrast and contours to merge with the surrounding water. This mechanism is common among the free-swimming pelagic fishes.

3. **Disruptive (contrast) coloration.** Dazzle-shading is a concealment mechanism that uses patterns, banding and patches of irregular contrasting tones or hues. They distort the true contours of the fish to draw attention away from the shape or an area of the body. Therefore a predator looks at the pattern instead of the general body form of the fish. Disguise is another related form of concealment that may be partially morphological and partially behavioural.

Deflective or **directive** markings are found on many fishes and they function to draw the attention of a predator. Deflective marks draw the attention of an attacker away from the vulnerable body parts (e.g. an ocellus or eyespot on the dorsal fins or flanks of many fishes, plates 14, 18, 22). The vulnerable head region and eyes may be hidden by a head stripe and this is a common feature among the coral reef fishes (spadefishes, plate 14; wrasses, plate 20; butterfly fishes, plate 22). Directive markings draw the attention of prey towards a part of the body through a combination of colour and body form (e.g. the lures of anglerfishes (Lophiidae) and bright mouth patches of the stargazers (Uranoscophidae).

Advertisement colours of some fishes function to reveal the presence of the fish instead of concealing it. The purpose may be sexual recognition and display, but for many fishes its main purpose is warning (e.g. black dorsal fin rays indicate danger of venomous fin spines in weeverfishes). The distinctive coloration of turkeyfishes advertises the presence of their deadly spines (plate 14). Furthermore, non-toxic fishes mimic toxic species to gain protection from predators (e.g. Filefish *Paraluteres prionurus,* plate 23).

SENSES

Like the cartilaginous fishes, bony fishes have a range of senses to scan their three-dimensional aquatic environment. Teleost senses include sight, hearing, vibration, smell and taste.

Sight

Eyes vary widely in structure and function and their adaptations will reflect the lifestyles and types of habitats that fishes occupy. Deep-sea fishes may be highly sensitive to light and have large eyes or they may be completely blind; turbid water inhabitants may rely little upon eyesight whereas clear water species may rely on eyesight as their main sensory input. Many fishes rely upon eyesight for capturing prey, for mating and display purposes, to receive warnings and distress signals, for fleeing and to find shelter.

Colours of the visual spectrum are reflected, scattered or quickly absorbed with increasing water depth. In turbid waters there may be little light penetration below 5-10m depth (see p.37). Research has shown that many shallow-water fishes have good colour vision and they can detect many spectral hues. Many fishes have evolved eyes that are sensitive to the wavelength (colour) of light. However, it is intensity (brightness) and tonal contrast that reveal movement and shape of underwater objects and these parameters are of most importance to fishes.

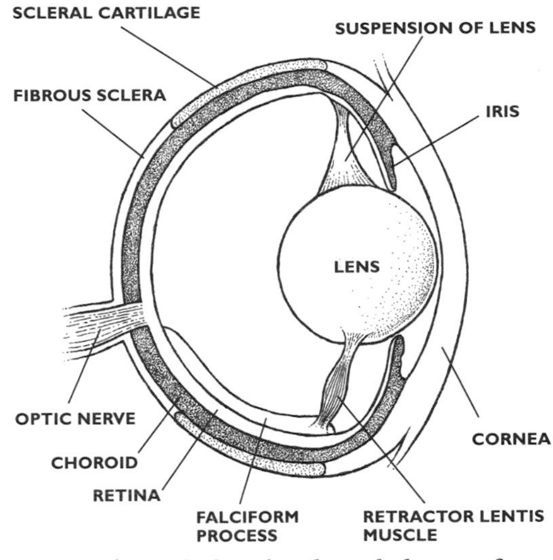

Figure 237. Vertical section through the eye of a teleost fish.

The rapidity with which colour fades with depth will depend on the clarity of the water and the entry angle of the light. Red and orange colours are quickly absorbed, blue light penetrates deepest but little visible light remains at 1,000m, even in the clearest ocean waters. Fish eyes are especially sensitive to yellow, green and blue light as these colours have short wavelengths that are absorbed least by sea water. These are also the colours most often produced through bioluminescence. Most fishes are capable of focusing an image on the retina by muscular movement of the eye lens (fig.237). Some fishes also have good binocular vision with varying angles of overlap in different species. This is important for judgement of distances and depth of field. Many fishes can also move their eyes independently and adjust the portions of the visual field for each eye. Stream fishes can fix their body position in the flowing current by visually marking a stone or some other fixed object. Eyes are often used in association with the balance organs of the inner ear to help maintain the fish in the proper position during swimming or at rest.

Hearing

The inner ears of fishes provide them with a receptor organ that allows both reception of waterborne sounds, and signals the position of the body in relation to gravity. This enables the animal to set its fins and eyes in appropriate positions for movement and orientation. During movement, the semicircular canals (that are enclosed in the 'otic' bones of the skull) also signal body accelerations and the 'otoliths' (ear stones) may respond to sonic (sound) vibrations. The ear sac is subdivided into the three semicircular canals and three chambers: the sacculus, utriculus and lagena (fig.210). In each chamber there is an otolith named the sagitta, lapillus and asteriscus respectively.

In fishes that hear well there is often a perilymphatic space that forms a connection between the swim-bladder and ear. This connection may occur either directly, where the sac of the bladder extends forwards (e.g. Clupeidae) or indirectly, through a chain of paired modified vertebrae called the 'Weberian ossicles' (fig.238).

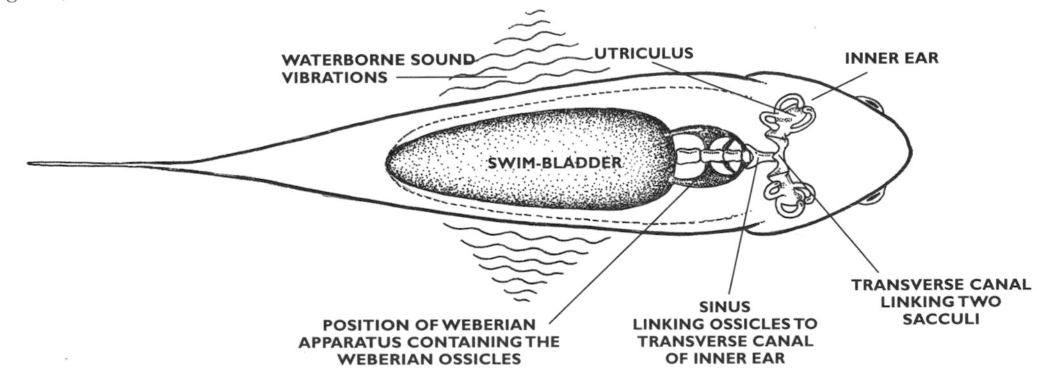

Figure 238. Auditory organ of freshwater Cypriniformes. Horizontal section through a fish to show the position of the Weberian ossicles between the swim-bladder and the ear. These bones transmit the sounds from the swim-bladder to the inner ear.

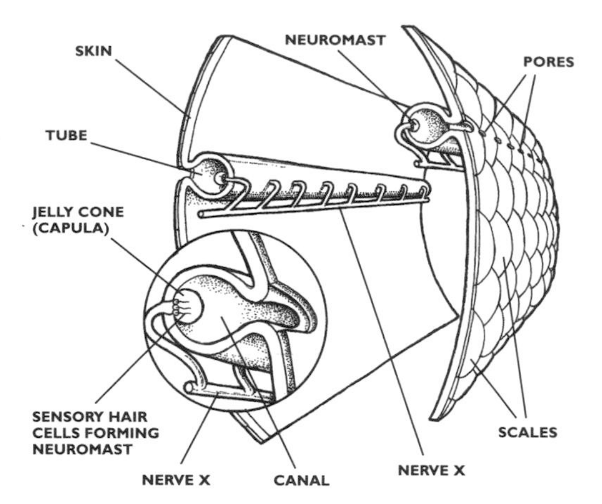

Figure 239. Enlarged view showing the internal lateral line structure and its position in a teleost fish. This sense organ transmits information about movements of water. It consists of a fluid-filled canal with pores opening to the surrounding water. The sensory receptor (neuromast) behind each pore is structurally similar to that of the cartilaginous fishes (fig.211) and operates in a similar way.

In freshwater Cypriniformes (e.g. dace, roach and goldfishes) the inferior part of the ears (sacculus and lagena) and the sagitta, in association with the special wing-like projections, are suspended in an arrangement that amplifies small sounds. Those fishes with acute hearing have sound discrimination abilities approaching that of the human ear and yet they lack any organ comparable to the human cochlea.

Many fishes can produce sounds that are often audible to the human ear. Sounds travel well in water and can be heard considerable distances away from their sources. Fishes may produce sound for the purposes of schooling (spacing between individuals), to warn of danger, for sex discrimination and possibly for courtship displays. Noise-making fishes include the drumfishes (Sciaenidae), many gurnards (Triglidae) and many catfishes (Siluridae). In deep-water rattail fishes (Macrouridae), only males have sound producing structures.

The mechanism for the production of sound varies considerably between species and may include: 'stridulation' (plucking) of the vertebrae (e.g. some Siluridae), operculum (e.g. bullheads, Cottidae), the pectoral girdle (e.g. triggerfishes, Balistidae), the teeth (e.g. some mackerels, Scombridae, and deep-water morid cods, Moridae) or 'phonation' of the gas-bladder. The advantages of underwater sound production have led to the evolution of similar mechanisms in several different groups of fishes (convergence).

Vibration

The 'lateral line' organs occur partly as rows of distinct pits and partly in canals that connect with the body surface through pores in the scales. The lateral lines of bony fishes are similar in structure to those of the cartilaginous fishes (fig.239) detecting water movements and vibrations from moving objects (and also stationary ones). This system is important for nocturnal fishes, or those living in deep-sea habitats, caves and low visibility waters.

Olfaction and Taste

These senses detect chemical substances that are dissolved in water. Smell (or olfaction) is usually the 'distance' sense that guides the fish to its prey, whereas taste is used to sample the chemical source just before ingestion and during swallowing. Taste buds may occur on the tongue, barbels, and sometimes all over the surface of the body. The sensitivity to taste and olfaction varies widely amongst fishes that live in low visibility habitats where chemical reception is likely to be important. Based upon human perceptions of tastes such as bitter, sweet, sour, salty and acid, many fishes have varying levels of awareness to these flavours. However, the importance of taste in fishes may reflect particular chemical sensitivity instead of specific types of taste.

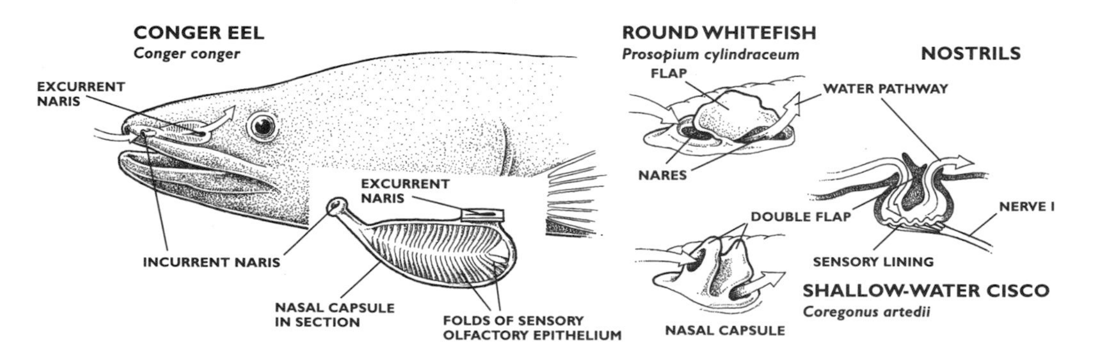

Figure 240. The olfactory organs of some bony fishes (based on Lagler *et al.* 1977).

Olfactory organs of many bony fishes are their main sense organs although they are quite small by comparison to the equivalent organs of cartilaginous fishes. Teleost olfactory organs have several adaptations for the intake and exchange of water (fig.240). The organ has either a single 'nasal' opening (naris) or a pair of apertures (nares). Nasal apertures lead directly or indirectly through a short tube to the nasal capsule that contains folds of sensory olfactory epithelium. Many fishes possess both incurrent and excurrent nares that allow water to be drawn in, pass through and exit by different openings (e.g. Common Eel *A. anguilla*). Freshwater sculpins (*Cottus* spp.) and sticklebacks (*Gasterosteus* spp.) have only one nasal aperture. This single nasal pouch alternatively fills and empties using the breathing cycles of the animal.

Particularly well developed olfactory sensitivity is known in eels which hunt by smell, and salmon which migrate long distances during their life cycles. Adult salmon and Sea Trout can 'home-in' on the characteristic 'smells' or chemical signatures of the original stream they left as young fishes. Their acutely sensitive olfactory system plays a part in guiding them back from the sea to the coastlines, then into the correct estuaries and finally up the original tributary or stream where they were born. It is believed there is a chemical 'imprint' which is learnt during the early stages of the fish's river life and this imprint is not forgotten during its lifetime.

LIGHT ORGANS

Bioluminescence (cold light produced by animals) is well known in marine invertebrate groups and in many marine fishes. There are no known freshwater animals that exhibit light production. Possible uses of bioluminescence may include lures to attract prey, to confuse or deter predators, to advertise or disguise body outlines and communication between individuals or potential mates.

Most luminescence is blue or green in colour and fishes may produce the light in two ways, either from modified mucus glands called photophores or from luminescent bacteria contained in special organs (fig.241). Production of luminescence may come from the tissues of the body (intracellular luminescence) or by the discharge of luminescent secretions (extracellular luminescence). Some photophores have lenses to reflect luminescent light to the exterior whilst in some other fishes photophores may be intermittently covered up by a structural mechanism to produce flashes (e.g. the Flashlight Fish *Photoblepharon palpebratus*). This is similar to the Aldis lamp principle where shutters are opened or closed while the lamp stays on.

Fish families that exhibit bacterial bioluminescence include the cods (Gadidae); grenadiers (Melanonidae), anamolopsids (Anamolopsidae), pinecone fishes (Monocentridae), cardinal fishes (Apogonidae), slipmouths (Leicognathidae), basses (Serranidae), swallowers (Saccopharyngidae) and anglerfishes (Lophi- formes). Many of these species have light organs associated with the ventral body wall, anus or digestive system. They have internal reflectors and translucent gelatinous tissue to diffuse the light.

The largest numbers of self-illuminating teleosts belong to the deep-sea scaly dragonfishes (Stomiatidae), lanternfishes (Myctophiformes), toadfishes and midshipmen (Batrachoididae). They have light organs on the ventral surface of the body that beam light downwards and these are under direct nervous control.

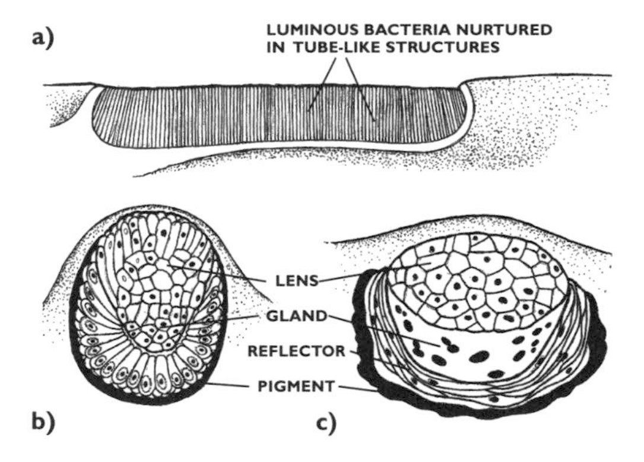

Figure 241. Light organs: a) with luminous bacteria, b) self-luminous with pigment layer and c) self-luminous with pigment and reflector layers.

GILLS, RESPIRATION AND SWIM-BLADDERS

Most teleosts use their gill mechanisms to extract oxygen from the surrounding water where it is 30 times more dilute than it is in air. By using the muscular pharynx and enlarging the buccal (mouth) cavity, water is drawn in through the mouth and then squeezed out through the gill apertures. During its passage through the mouth water passes through a fine mesh of gill filaments (lamellae), that are arranged in double rows on the gill arches to extract the oxygen (fig.213b). Gills of bony fishes differ from cartilaginous fishes as the gill arches have reduced septa (dividing walls) and open into a common gill chamber. The external opercular flaps are kept closed as water is drawn into the mouth but are forced open by the pressure of the expelled water. In fast-swimming teleosts, the flow of water passing into the mouth reduces the need to actively pump in water, but stationary fish must do so.

Flatfishes partially bury themselves in loose bottom substrates leaving only their mouths, eyes and upper operculum uncovered. This presents the fish with a breathing problem since the lower operculum exit is on the underside of the body and pumping water through the sand or gravel is not an energetically sensible option. In practice only the upper opening is used when the flatfish is partially buried or resting on the bottom. Both opercula keep moving to flush water over the gills. However, water pumped over the lower gills travels through a passage connecting the two opercular cavities, to emerge through the upper operculum opening.

The swim-bladder is an important organ in bony fishes and its main purpose is hydrostatic (buoyancy control). In a few ancient species the bladder still partially acts as a lung (e.g. lungfishes) but it may have also evolved in some other fishes for the secondary purpose of receiving and transmitting sound (a hydro-acoustic organ). The gas-bladder modified as a buoyancy compensator has enabled fishes to become complete masters of the aquatic environment.

The swim-bladder is a thin-walled sac in the body cavity of the animal that develops from the foregut (small intestine). It fills with oxygen from an adjacent gas secretory gland, but in many deep-sea fishes nitrogen gas may replace oxygen. With the aid of the swim-bladder, fishes can maintain their position in the water column at any given depth with little movement or energy expenditure. By contrast, fishes that do not possess a swim-bladder must keep swimming or they will slowly sink. Sea water is more dense than freshwater, therefore marine fishes can float more easily. Their freshwater counterparts have developed larger swim-bladders (about 7% by body volume compared to 5% in marine fishes). Since gas has the physical property of compression and expansion with pressure, a given volume of air will decrease with increasing depth (the volume of gas halves for every 10m increase in depth). These problems are overcome by the ability of the fish to secrete or absorb gas from the gas gland which is supplied with a dense network of blood vessels. However, fishes must still avoid ascending to the surface too quickly, as the expanding gas cannot be absorbed quickly enough and over-inflation may cause damage to the swim-bladder. Some fishes do make considerable vertical migrations in the water column and the mesopelagic hatchetfishes (Sternoptychidae) may travel 500m up and down in 24 hours.

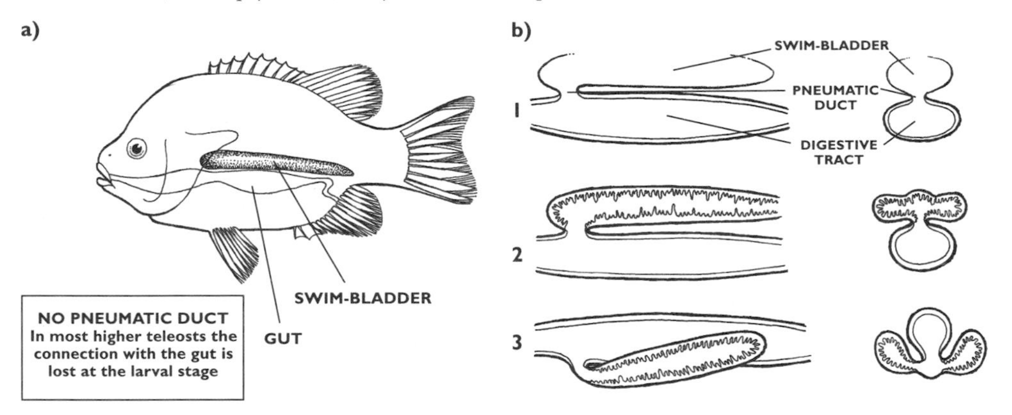

Figure 242. a) Position of the swim-bladder in many higher teleost fishes. b) Variations of bladder form and their connections with the gut in lower groups of fishes: 1. Sturgeons (Acipenseridae) and most lower teleosts. 2. Gars (Lepisosteidae) and the bowfin (Amiidae). 3. African and South American lungfishes (Lepidosireniformes).

It is important that the position of the swim-bladder is correct in the body. The ideal position is around the balance point along the length of the fish. If its position is too far forward or backwards the ends of the body would lift. Therefore many fishes have elongated swim-bladders to compensate for this tendency. The swim-bladder is often situated in the top half of the body above the gut, otherwise the fish would tend to turn upside down (fig.242a). The detailed structure of the swim-bladder differs in fishes. In the less advanced ray-finned teleosts the bladder may consist of a single, cigar-shaped chamber (e.g. salmon and trout). In most of these soft-rayed fishes a functional pneumatic duct connects the bladder with the pharyngeal region of the gut. This condition is open or physostomatous (fig.242b). However, herrings (Clupeidae) have an odd arrangement with a pneumatic duct that communicates with the exterior through a pore near the anus. In the spiny-rayed fishes (acanthoptygians) the pneumatic duct has been lost and this is known as the closed or physoclistous condition.

Many species of bony fishes can live out of water for a short time and in freshwater fishes the swim-bladder may function as an accessory breathing organ. It has been observed that freshwater Common Eels *Anguilla*

anguilla make considerable journeys across damp fields between ponds and streams. These eels have no special structures apart from normal gills, but oxygen is taken in directly through the moist, slimy skin. By contrast, mudskippers which spend much of their time out of water on exposed mud at low tides, can actively gulp air to increase their oxygen intake.

WATER BALANCE, BLOOD AND BODY FLUIDS

The living cells of the body need to have the right amounts of water and salt ions if they are to function properly. All animals have to regulate their body fluids but it is more difficult for those creatures living in aquatic environments, or those that have 'semi-permeable' tissue surfaces. Fishes often have skins that are semi-permeable and large gill areas where exchange takes place between the water and blood. The sea has a higher salt concentration (hypertonic) than the body fluids and tissues of fish. By contrast, freshwater is lower in salt ions (hypotonic) compared to the fish's body fluids. Therefore semi-permeable areas of the fish's body are continually 'leaking' and are either gaining or losing water. Unless this water movement is controlled by the body, the fish would quickly die. A good way of regulating this passage of water is by adding or eliminating salt from the body.

Marine fishes are osmotically more dilute than the water in which they live and they lose body water to the surrounding sea water through their gills and in the form of urine. To compensate for this water loss, marine fishes usually drink substantial amounts of sea water. The ingested salts (sodium and chloride) are absorbed in the intestine and then actively eliminated through the gills by active transport. Magnesium and sulphate ions are excreted by the kidneys. Body fluids of freshwater teleosts are osmotically more concentrated than the surrounding water. They suffer a constant osmotic influx of water, mostly through the gill membranes and into the blood system. To compensate for this, excess water is excreted as urine and lost salts are actively taken up through the gills. Teleosts that spend parts of their life cycles in each of the different environments (e.g. salmon, Sea Trout, and eels) must spend sometime in brackish waters as an acclimatisation period.

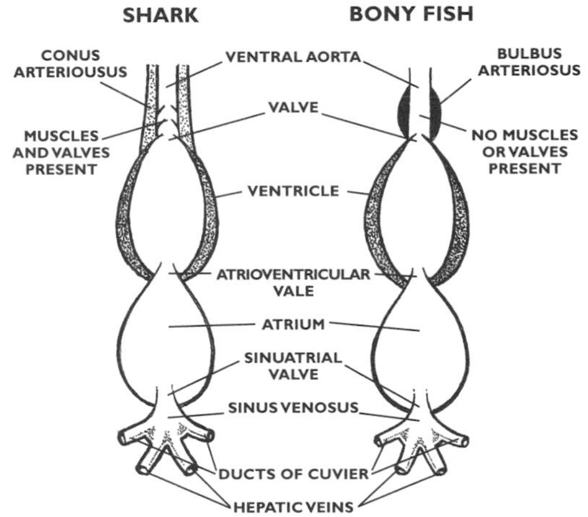

Figure 243. Generalised schemes of the heart of a shark and bony fish (after Lagler *et al.* 1977).

In proportion to body weight, the volume of blood in fishes is considerably less than that of land vertebrates, and freshwater teleosts tend to have more blood than marine teleosts. Active cruising fishes also have greater volumes of blood. In less active species, blood accounts for approximately 2% of the body weight (in humans it is near 10%). An important component of blood are the red cells that contain the pigment haemoglobin, which has an affinity for oxygen. The blood circulatory system carries the oxygen

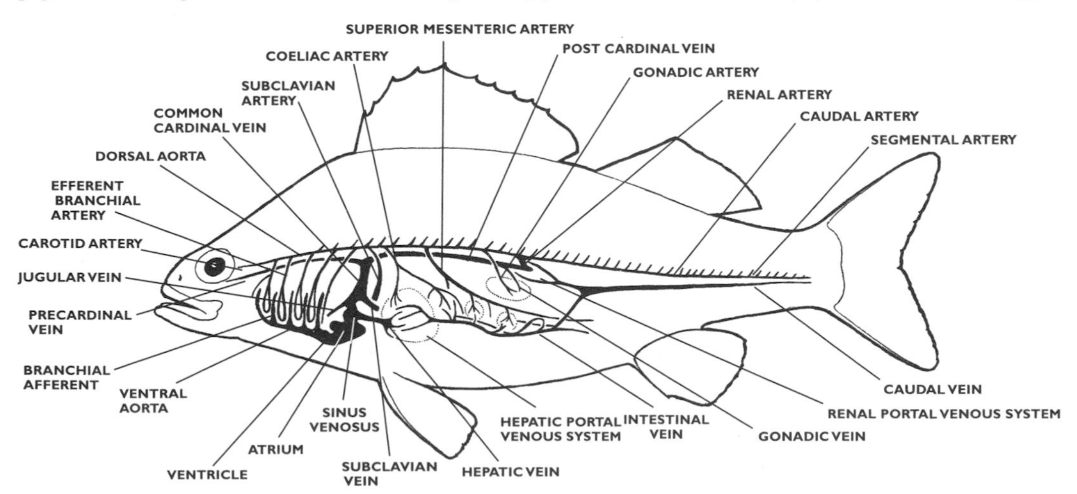

Figure 244. Generalised blood circulation system of a teleost fish (after Lagler *et al.*1977).

rich blood from the gills to the tissues of the body (fig.244). The waste products (e.g. carbon dioxide) are carried away in the blood and excreted via the gills. Blood is pumped around the body by contractions of the heart. The fish heart consists of a single auricle and ventricle, but the structure does vary between fishes (fig.243).

Marine teleosts inhabiting polar regions have to overcome the environmental hazard of freezing. Sea water contains salt (sodium chloride) that lowers the freezing point of water. Therefore sea water will not freeze until it reaches a temperature of -1.9°C. Deep polar waters never reach this critical temperature, so teleosts living at depth allow their body fluids and saturated tissues to 'supercool' to the temperature of the surrounding water. However, fishes living near the surface of the water may come into contact with ice, and ice crystal formation in the blood would normally cause rapid death. Surface-living fishes have evolved 'antifreeze additives' (glycopeptides) in the body fluids to stop their tissues freezing. One advantage of cold water is the high concentration levels of dissolved gases such as oxygen. The pallid Antarctic icefishes (Chaenichthyidae) have completely lost their red, oxygen-carrying cells from the blood. Instead, the fluid plasma of the circulatory system carries some oxygen and the rest is supplied by direct absorption through the surface of the skin from the supersaturated water. The sluggish, bottom-dwelling lifestyles of many of these fishes also helps to reduce oxygen demand.

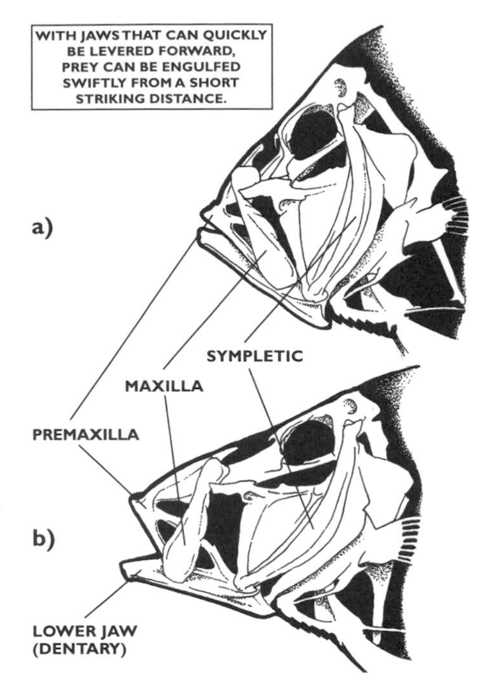

Figure 245. Lateral views of the head and jaw skeleton of the John Dory *Zeus faber*. a) the mouth closed with the premaxillae retracted and b) the mouth open with the premaxillae extended.

PREDATION, DIET AND FEEDING BEHAVIOUR

Fishes use a number of behavioural mechanisms to avoid being eaten. When danger threatens, a quick retreat to a nearby crack, crevice, hole or burrow will often suffice. In the case of the clownfishes (Pomacentridae) the tentacles of a stinging anemone are home, but to avoid being eaten by the anemone these fishes secrete a skin mucus that 'anaesthetises' the stinging cells of the host. Burrowing into loose substrates of sand and gravel is another form of concealment. Many benthic fishes and reef fishes have developed concealment to a high level, using body colour patterns and modified body shapes. Flatfishes exhibit this form of behaviour to lessen the chances of detection, by blending their body colour patterns with the background substrates. This helps to break up the outline of the body. Compression of the body either dorsally (e.g. flatfishes) or laterally (e.g. John Dory *Zeus faber*, plate 19 and fig.225) also makes fishes difficult to see from some positions. The John Dory with its thin, slow moving body is difficult to see from certain angles and therefore it can approach its prey with less chance of being detected. When it is within striking distance, the extendible jaws shoot forward to suck in the victim. These movements are made possible by special articulations of the premaxillae and other jaw bones (fig.245). Avoidance of detection by being well hidden has an extra advantage, as it allows the fish to wait for passing prey to come within striking distance. This saves time and energy as active hunting methods are not required. However, patience and stealth with sudden short bursts of speed are needed for successful capture of prey. The juvenile Twin-spotted Wrasse *Coris aygula* has an unusual hiding strategy; it can wriggle its way below loose gravel with a quick flick of the body and tail. The flicking motion of the body and fins piles the loose stones on top of its body and the fish 'instantly disappears'.

Some species (e.g. surgeonfishes) have barb-like blades positioned laterally near the tail that are used in a defensive slashing action. Venom is another powerful type of defence. Many percomorph fishes have spiny dorsal and pectoral fins that make them difficult to catch and swallow. Some fishes go even further and have poison glands at the base of such spines (fig.231e). This can be injected into predators through a hollow fin ray that is similar to a hypodermic needle. Other defence mechanisms may include teeth for defensive displays and biting predators. Some formidable toothed fishes include the barracudas (Sphyraenidae, fig.246g) and wolf-fishes (Anarhichadidae). Among the deep-sea fishes, the production of bioluminescent light flashes from special organs (photophores) or symbiotic partners may serve to confuse an enemy during a chase. Pocupine fishes (Diodontidae) and pufferfishes (Tetraodontidae) have a remarkable defence mechanism associated with the stomach. Both groups can inflate themselves with large quantities of water or air when threatened. They assume large, inflated ball shapes and the distention of the skin erects the surface spines on the scales of porcupine fishes (e.g. *Diodon histrix,* plate 14). Water is pumped in and out of the stomach using the muscles of the stomach walls.

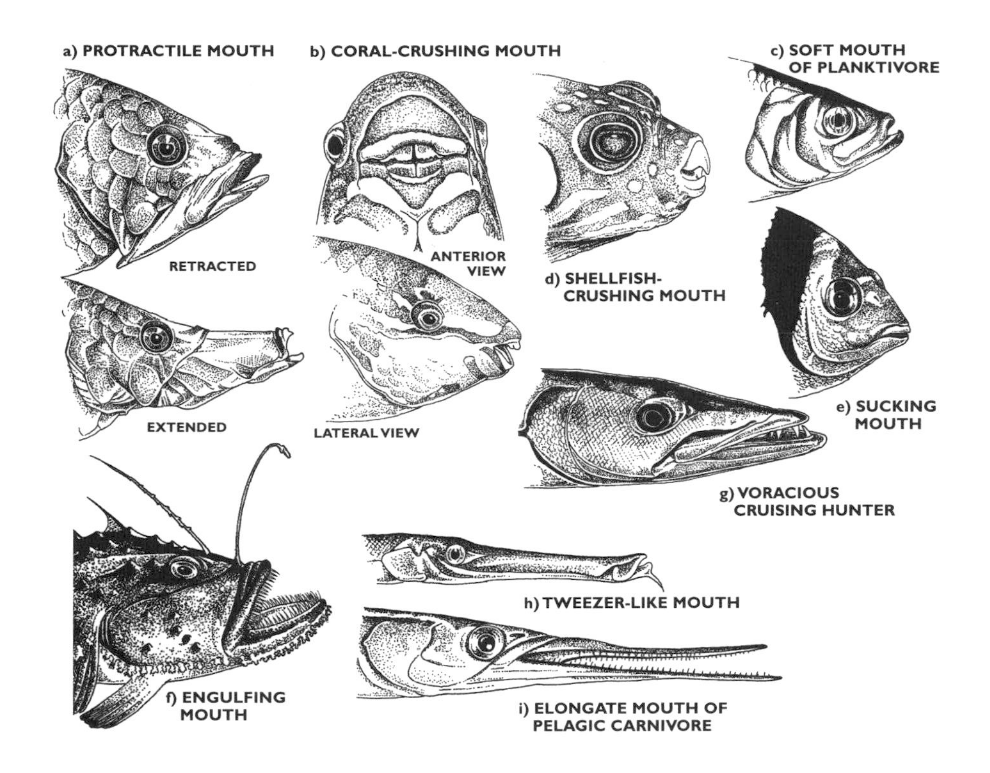

Figure 246. Head profiles of teleost fishes showing different mouth types. a) Mouth extension in the jaw mechanism of the Slingjaw Wrasse *Epibulus insidiator;* b) coral-crusher (e.g. Yellow-sided Parrotfish *Scarus psittacus*); c) planktivore (Atlantic Herring *Clupea harengus*); d) shellfish-crusher (e.g. White-spotted Puffer *Arothron hispidus*); e) sucking mouth (Two-banded Bream *Diplodus vulgaris*); f) engulfer (e.g. Goosefish *Lophius americanus*); g) cruising hunter (e.g. Great Barracuda *Sphyraena barracuda*); h) tweezer-like mouth (e.g. Trumpetfish *Aulostomus maculatus*) and i) pelagic carnivore (e.g. Garfish *Belone belone*).

Many open water fishes will rely on their fast-swimming abilities, good sensory receptors and camouflage skills to out-manoeuvre a predator during a chase. Some prey fishes will confuse their chasers by leaping out of the water and changing direction as soon as they re-enter the water. Flying-fishes (Exocoetidae) can leave the water for considerable distances but they do not actively fly (fig.197). Using powerful flicks of the tail they glide between waves using their enlarged, wing-like pectoral fins. However, some predators can visually track the glide path of these fishes through the air-water interface.

Elongated snouts are characteristic of cruising carnivorous hunters that roam the open waters. Barracudas (Sphyraenidae) are typical examples which have dagger-like teeth to seize and hold prey. The position of the mouth often differs and a superior mouth is typical of slow-swimming fishes. An inferior mouth is characteristic of many bottom-dwelling fishes that ambush their prey. A terminal mouth at the tip of the snout is often seen on fishes that probe and nibble food from surfaces (e.g. parrotfishes, wrasses and butterfly fishes).

Diet and feeding habits vary widely in bony fishes. Heads, jaw shapes, teeth and digestive systems have evolved to take advantage of many food sources (fig.246). Most fishes are predatory and eat animals that are smaller or weaker than themselves. More than one type of tooth may be present in the dentition which is often modified to cope with either soft prey (e.g. worms) or hard prey (e.g. shellfish). The intestine may also show modifications to cope with carnivorous, herbivorous or omnivorous diets (fig.247). Among the sturgeons and mullets, stomachs are modified into grinding organs, similar to the gizzards of birds, that help break up the coarse food. Some fishes do not have stomachs at all. Herbivorous-grazing parrotfishes (Scaridae) have long intestines that replace the stomach for digestion and absorption of plant materials. Some carnivorous zooplankton-consumers including seahorses and pipefishes have also lost their stomachs.

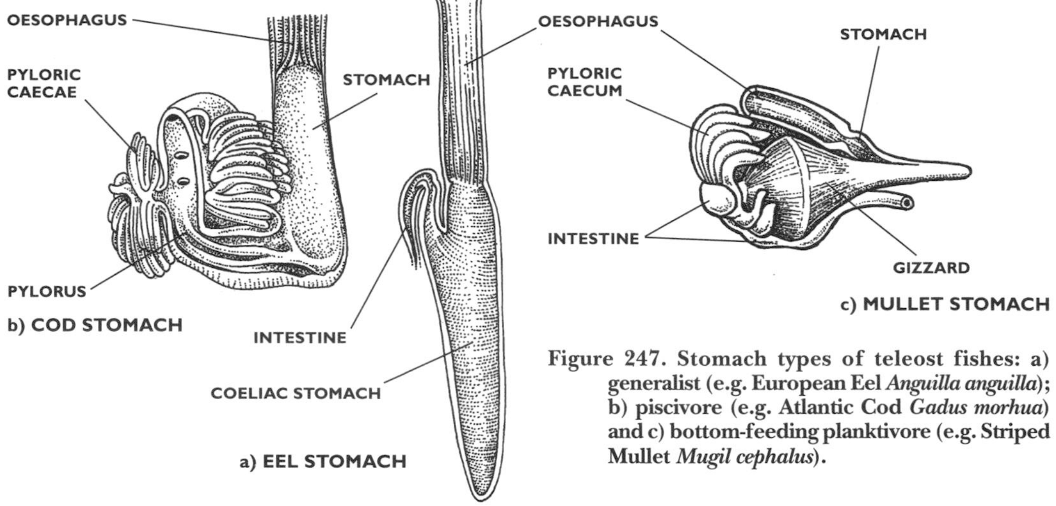

Figure 247. Stomach types of teleost fishes: a) generalist (e.g. European Eel *Anguilla anguilla*); b) piscivore (e.g. Atlantic Cod *Gadus morhua*) and c) bottom-feeding planktivore (e.g. Striped Mullet *Mugil cephalus*).

Teeth and jaws have developed to cope with many types of prey and tooth shapes can provide some clues to the dietary habits of many fishes (fig.248). Bony fishes may have teeth in the mouth and pharynx and on the jaw bones. Within the mouth, teeth may be present on the top surface of the tongue and in the roof of the mouth. Pharyngeal teeth may develop as tooth-like plates or pads on various gill arch elements. These have several uses including grasping, tearing and grinding food.

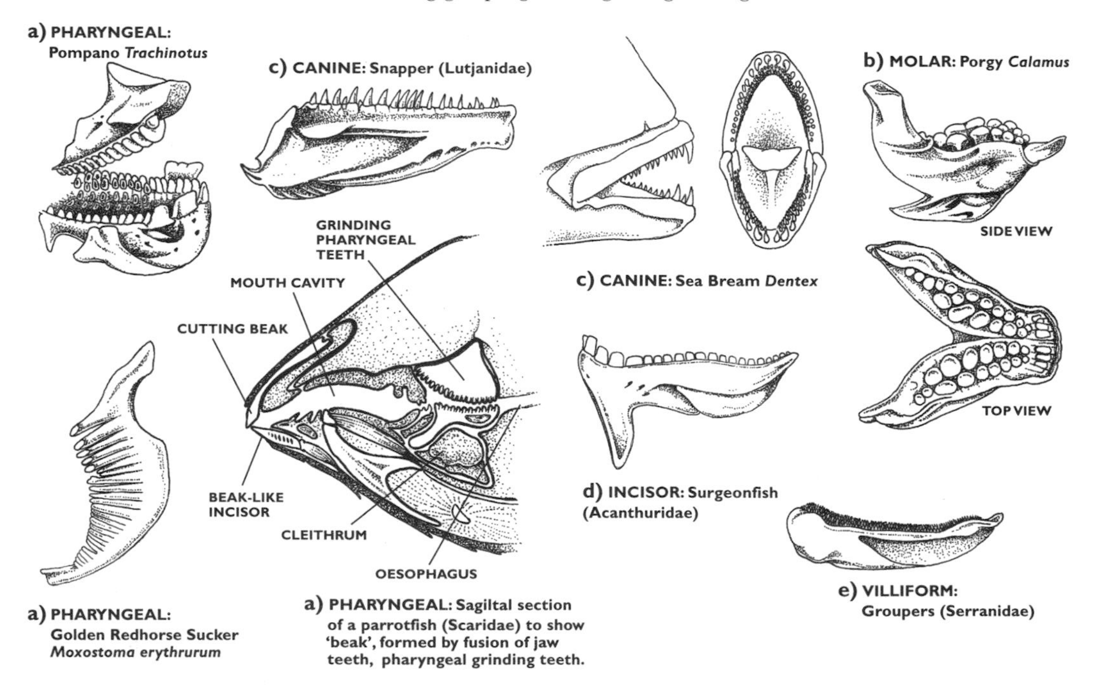

Figure 248. Teeth of teleost fishes and their locations on the jaws. Tooth shape may give an indication of the eating habits of some fishes. Some species may possess more than one type of tooth. Teeth types include: a) pharyngeal teeth for sieving, grinding, holding or tearing food within the gullet; b) molars for crushing hard prey such as shellfish; c) canines to seize, hold and pierce prey; d) incisors for slicing and cutting; e) villiform teeth to hold prey more securely .

Jaw teeth vary in form and function (fig.248). Cardiform teeth are numerous, short, fine pointed, and often form in multiple rows. This arrangement is common among the perches (Percidae) and many basses (Serranidae). Villiform teeth are fine teeth typical of the needlefishes (*Belone* spp.) and lionfishes (*Pterois* spp.). In groupers (Serranidae), the sandpaper-like villiform teeth are used in association with the powerful jaws to grip and hold the prey. Canine-type teeth are fang-like teeth used for piercing and holding prey, which may point backwards for better grip. They are common among many predators including snappers (Lutjanidae), snooks (Centropomidae), barracudas (Sphyraenidae) and walleyes (Percidae). Moray eels (Muraenidae) and the lancetfishes (Alepisauridae) have hinged canines so that the jaws can be locked closed to give a vice-like hold from which the prey has little chance of escape. Incisor-type teeth are sharp-edged cutting teeth which work by a shearing action. Here, two surfaces overlap and pass close to each other to slice the prey. Examples of these teeth are seen in surgeonfishes (Acanthuridae) and sea bream (Sparidae). Parrotfishes (Scaridae) have horny beak-like incisors at the front of the head which are formed from many fused teeth. These fishes scrape off thin layers of algae from the rocks and coral. The food is then crushed and ground to a powder using the strong pharyngeal plates and the soft tissues of the prey are then extracted. Molar-type teeth are used for crushing and grinding. These teeth often have broad, flattened and occluded surfaces. They are common among bottom-dwelling fishes (e.g. some drumfishes, Sciaenidae) which have a diet of hard-shelled crustaceans and gastropods. Here, the shells are crushed and broken before the soft tissue can be eaten.

Many voracious predators that eat soft-bodied prey do little more than seize it and swallow it whole. A highly muscular pharynx also helps to hold the prey and to squeeze it into an appropriate shape for swallowing. The bottom-living soles (*Solea* spp.) have an overhanging, hook-like snout which divides the mouth into left and right halves. The connection between the jaws is loose enough to allow the oddly shaped lower jaw to rotate and open only the lower side of the mouth for feeding. The upper part of the mouth works independently for respiration, and thus it is possible for respiration and feeding to take place together. Plankton-feeding fishes (e.g. herrings, Clupeidae and anchovies, Engraulidae) may have small teeth, or no teeth at all, and often have weak mouths. Instead they may have a filtering system of branched gill rakers that sieve the passing water for small zooplankton (fig.213c). Where gill rakers are present in a reduced form, their main function is to protect the fine-meshed gill structures from coarse sand and ingested food.

REPRODUCTION

Bony fishes have evolved a variety of breeding methods and behaviour. As a group their reproductive strategies are part of the evolutionary success of bony fishes. Most fish species have about equal numbers of males and females in a population. Through the process of reproduction, new genetic material passes from one generation to the next (fig.231). This allows the species to survive and adapt so that at least two individuals succeed the parents. Most modern teleosts accomplish this by laying large numbers of small eggs. Sperm is released externally to fertilise the eggs. A few species of fishes do have individuals that are hermaphrodite where each fish carries both male and female sex organs (e.g. some trout *Salmo* spp. and black basses *Micropterus* spp.). Some basses (Serranidae) and wrasses (Labridae) are unusual as individuals can be either male or female at first, but later change to the opposite sex. Very few bony fishes are viviparous.

Eggs vary widely in sizes, shapes and numbers produced. The eggs of most marine fishes have a certain amount of yolk that consists of protein and fat in the form of oil droplets. This yolk is the food source for the embryo that is developing inside the egg. Most bony fishes lay at least 30-100 eggs (e.g. marine sticklebacks, Gasterosteidae). However, the large ocean sunfishes (Molidae) and ling (*Molva* spp.) may lay as many as 28 million eggs during a single season. Atlantic Cod *Gadus morhua* and Turbot *Psetta maxima* lay about 10 million eggs in a season.

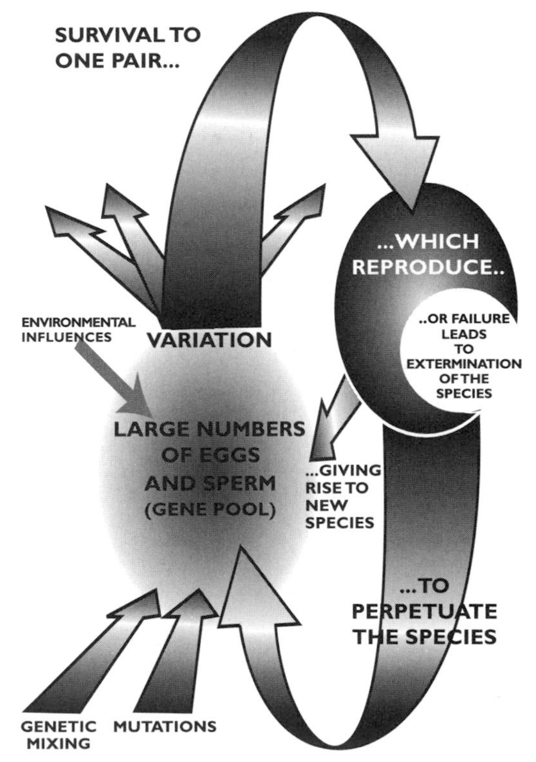

Figure 249. Role of reproduction in the survival and evolution of fishes.

Once fish eggs are released, they either float (pelagic) or sink to the bottom (demersal). Most marine teleosts lay buoyant eggs (e.g. cods, Gadidae and hakes, Merlucciidae) and the pelagic eggs may be widely dispersed by surface water movements. By contrast, the pelagic Atlantic Herring *Clupea harengus*, salmon and Sea Trout *Salmo* spp. (which spawn in freshwater) lay heavy demersal eggs that sink to the bottom. Most demersal eggs are adhesive or temporarily adhesive to help them stay in place on the bottom. The flying fishes (Exocoetidae) have egg cases with external filamentous hairs that become entangled on the seabed.

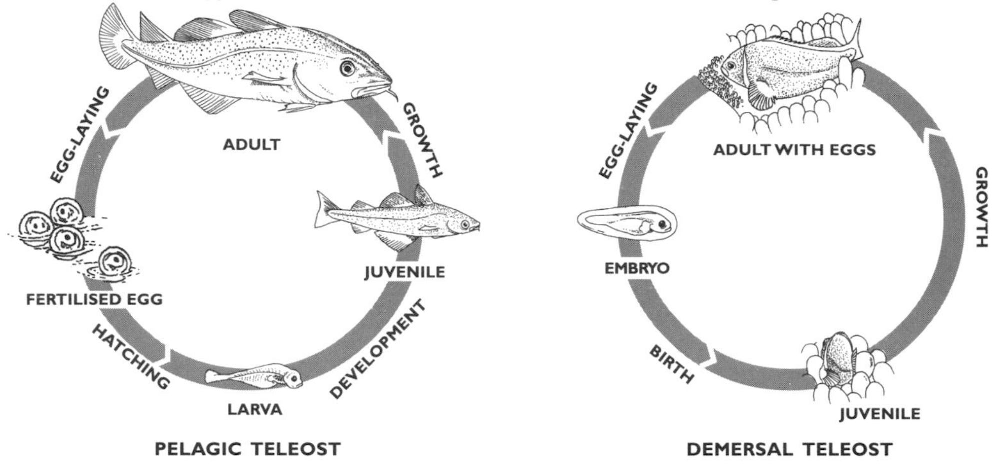

Figure 250. Major types of reproductive life cycles in marine teleost fishes.

Differences between the external appearances of male and female adults may not be obvious in most species. However many non-pelagic species do have sex-related (dimorphic) differences that may become apparent during breeding periods (fig.251). A 'papilla' (or fish penis) may be present and this is very noticeable in one group of sculpins (Oligocottinae). Many female fishes may also appear more rotund or pot-bellied during the spawning period when they are heavily laden with eggs. Often, coloration differences between sexes are evident and generally males have brighter and more intense hues than females. Bright coloration of males is often associated with behavioural courtship displays to attract a suitable female mate before spawning (e.g. marine sticklebacks, Gasterostedae and dragonets, Callionymidae). The males of some breeding salmon (e.g. Pink Salmon *Oncorhynchus gorbuscha)* may develop a characteristic hump back and hooked kype on the upper and lower jaws (fig.251a).

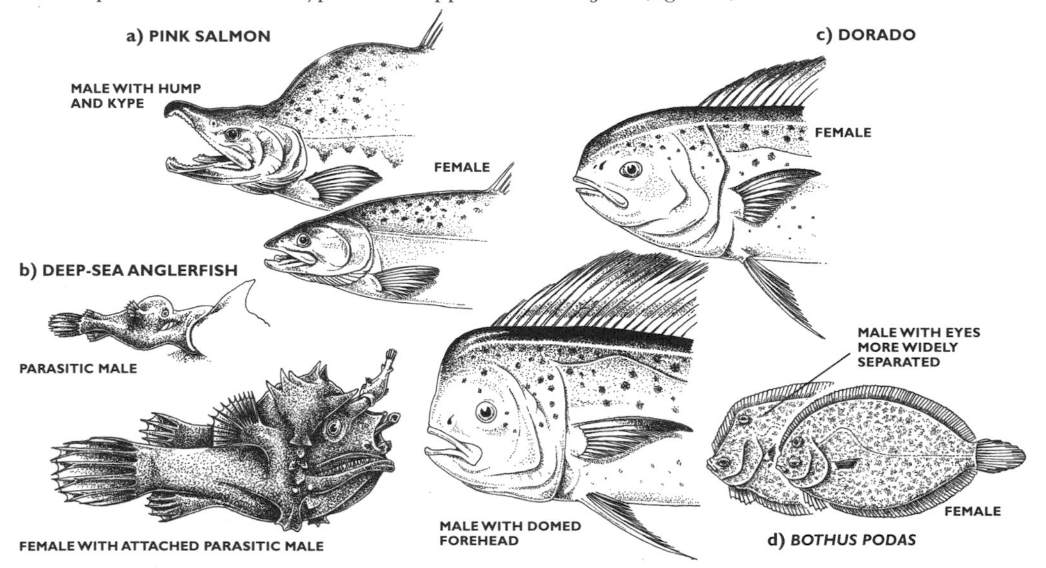

Figure 251. Sexual dimophism of selected teleost fishes. a) Humped back and hooked jaw (kype) of breeding male Pink Salmon *Oncorhynchus gorbuscha*; b) parasitic male of the deep-sea anglerfish *Photocorynus spiniceps*; c) domed forehead and anterior position of the dorsal fin in the Dorado *Coryphaena hippurus* and d) eye positions of the male and female flatfish, *Bothus podas*.

One bizarre form of sexual dimorphism occurs in the deep-sea anglerfish *Photocorynus spiniceps* (Linophrynidae). The diminutive male is a fraction the size of the female. He lives parasitically attached to the head of the female. The male latches on by his mouth to a protuberance on the female that provides him with nourishment. In return the male provides sperm to fertilise the female's eggs (fig.251b).

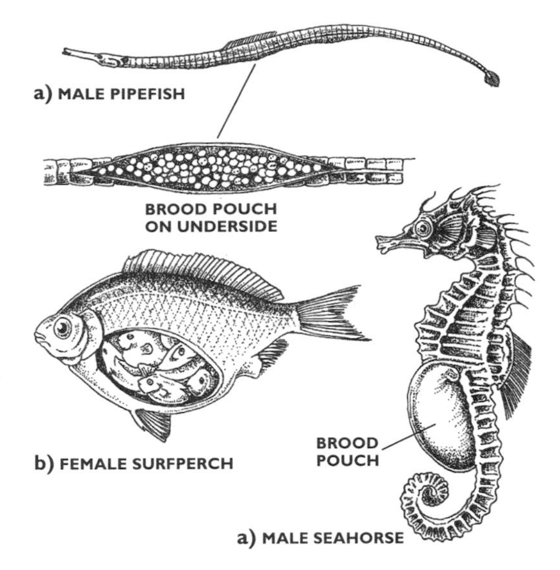

Figure 252. Brood care among marine teleost fishes. a) male pipefishes and seahorses (Syngnathidae) have brood pouches on the ventral surface; b) surfperches (Embiotocidae) are viviparous and give birth to well developed, sexually mature young.

The age at which fishes reach sexual maturity varies and there may also be differences between males and females of the same species. Some species such as the viviparous Surf Perch *Micrometrus aurora* are sexually mature at birth. However, the females will not give birth to live young for a few years. Most species of teleosts tend to mature between two and fifteen years of age. Sunfishes (*Lepomis* spp.) and Salmonidae mature early, whereas the European Eel *Anguilla anguilla* and sturgeons (Acipenseridae) may take up to 14 years to reach full maturity. In many kinds of fishes it is the male that matures first.

Fishes have developed a host of different ways to ensure breeding success. Many pelagic schooling fishes and cold water species have short breeding seasons, therefore males and females will reach maturity together. These species often migrate to specific spawning areas on, or near, the edges of the continental shelves. Environmental and biological signals may trigger mass spawning with eggs and sperm being released in quick succession. Larval mortality is generally high, but by producing large numbers of eggs a few young fishes will survive and eventually reach sexual maturity. High fecundity is a successful reproductive strategy among marine fishes. This is evident from the huge populations of some pelagic and demersal fish species such as cod, ling, hake and herring.

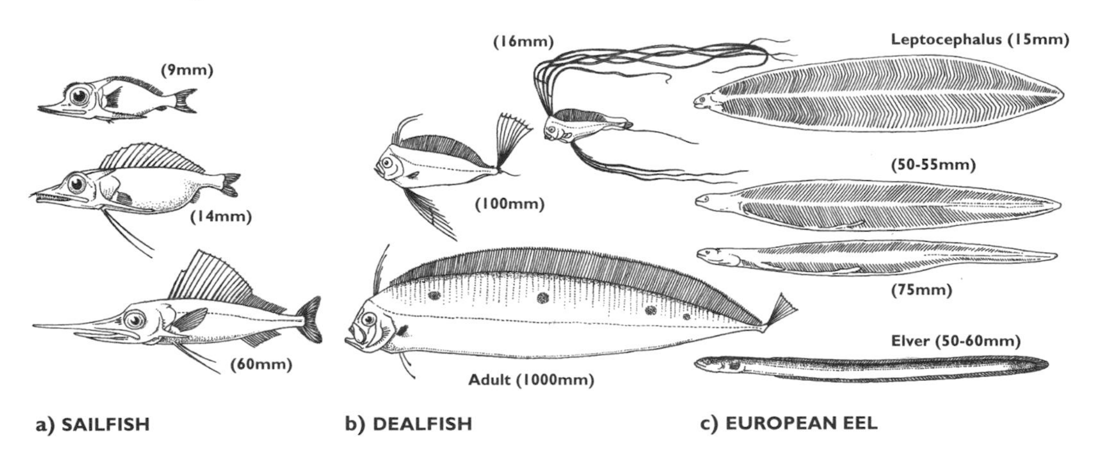

Figure 253. Developmental stages in three marine teleost fishes: a) Sailfish *Istiophorus* sp.; b) Dealfish *Trachypterus arcticus* and c) European Eel *Anguilla anguilla*, from the leptocephalus to the elver (glass eel) stages.

The breeding habits of freshwater fishes are more diverse than the marine teleosts. A greater variety of freshwater species take care of eggs and young after they hatch, whereas marine fishes show little parental care. However, there are a few demersal marine teleosts that do have an inclination to protect the brood. Unusual examples are the seahorses and pipefishes (Syngnathidae) that have evolved various methods of incubation. The female of the species places the eggs into a special brood pouch located on the male, who then carries them around until they hatch (fig.252). Once hatched, the young leave the brood pouch to be dispersed by the water currents.

Many fishes may have successive broods in the same season, because the total numbers of eggs stored by the female cannot all develop together at the same time. Different species will spawn small numbers of large eggs, or large numbers of small eggs. Large eggs normally produce larger, more fully developed young which have a better chance of survival in the race to grow and mature. Smaller eggs provide the developing embryo with little stored food and the supply is soon depleted. Larvae that hatch at a smaller size are limited in the types of foods they can consume. The larvae and young of most fishes are heavily preyed upon by many larger predatory creatures, so mortality is high in the early stages of their life cycles. By the post-larval stage of their development the yolk sac disappears and skin pigment starts to increase. By the juvenile stage, the adult body form is well advanced and many species have juveniles that resemble the adults, in all except size and reproductive maturity.

Some demersal fish species have pelagic larvae or juvenile stages that only later change their behaviour and body form to resemble the adults (e.g. many flatfishes, fig.234). A few species have reduced or eliminated the precarious larval stage of their life cycles altogether. Surfperches (Embiotocidae) are viviparous fishes that give birth to live young which are sexually mature when they are born (fig.252b). Senescence among fishes also varies. Generally the smaller, faster growing fishes and those living in the warm waters of the tropics do not live as long as those inhabiting the waters of cooler regions. In cold temperate and polar regions, fish have slower growth rates and may live more than 30 years (e.g. *Dissostichus mawsoni*, Nototheniidae). The coldwater, deep-sea Orange Roughy (Holocentridae) has an estimated maximum life span of 150 years.

BIBLIOGRAPHY

Alexander, R. McNeill. 1975. *The Chordates.* Cambridge University Press, Cambridge.

Bannister, K. 1993. *The Book of the Shark.* Quintet Publishing, London.

Bent, J., & Dahlstrom, P. 1981. *Collins Guide to the Sea Fishes of Britain and North-Western Europe.* Collins, London.

Carcasson, R. H. 1977. *A Field Guide to the Coral Reef Fishes of the Indian and West Pacific Oceans.* Collins, London.

Colbert, E. H. 1969 (2nd ed.). *Evolution of the Vertebrates.* John Wiley & Sons, New York.

Compagno, L. J. V. 1984. *FAO Fisheries Synopsis* No. 125, Vol.4, Parts 1 and 2. Sharks of the World. FAO, Rome.

Elst, R. van der. 1988 (2nd. ed.). *A Guide to the Common Sea Fishes of Southern Africa.* Struik, Cape Town.

Hutchins, B., & Swainston, R. 1986. *Sea Fishes of Southern Australia.* Swainston Publishing, Australia.

Johnson, R. & Nelson, D. R. 1972. Agnostic display in the Grey Reef Shark, *Carcharhinus menisorrah* and its relationship to attacks on man. *Copeia* 1: 76-84

Joseph, J., Klawe,W., & Murphy, P. 1988. *Tuna and Billfish - Fish without a Country.* Inter-American Tuna Commission, La Jolla, California.

Lagler, K. F. *et al.* 1977 (2nd. ed.). *Ichthyology.* John Wiley, New York.

Luther, W., & Fiedler, K. 1976 *A Field Guide to the Mediterranean Seashore.* Collins, London.

Marshall, N. B. 1954. *Aspects of Deep Sea Biology.* Hutchinson, London.

Marshall, N. B., & Marshall, O. 1971. *Ocean Life.* Blandford, London.

Masuda, H., Amaoka, K., Araga, C., Uyeno,T., & Yoshino, T. (eds). 1984. *The Fishes of the Japanese Archipelago.* Tokai University Press.

Nelson, J. S. 1994. *Fishes of the World* (3rd. ed.). John Wiley, New York.

Norman, J. R. 1958. *A History of Fishes.* Ernest Benn, London.

Rhodes, F. H. T. 1976. *The Evolution of Life.* Pelican, London.

Smith, M. M., & Heemstra, P.C. (eds). 1986. *Smith's Sea Fishes.* Springer Verlag, Berlin.

Stoops, E. D., & Stoops, S. 1994. *Sharks.* Sterling Publishing, London.

Wheeler, A. 1978. *Key to the Fishes of Northern Europe.* Frederick Warne, London.

Young, J. Z. 1973. *The Life of Vertebrates.* Oxford University Press, Oxford.

Cartilaginous Fishes Factsheet

MAJOR EXTERNAL FEATURES OF A TYPICAL SHARK

Lateral view

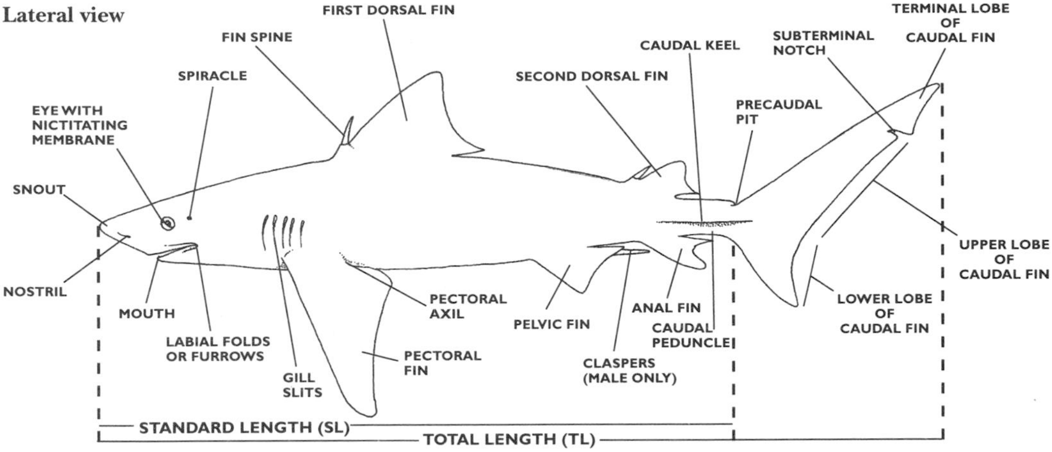

FIRST DORSAL FIN
FIN SPINE
SPIRACLE
EYE WITH NICTITATING MEMBRANE
SNOUT
NOSTRIL
MOUTH
LABIAL FOLDS OR FURROWS
GILL SLITS
PECTORAL AXIL
PECTORAL FIN
PELVIC FIN
SECOND DORSAL FIN
CAUDAL KEEL
PRECAUDAL PIT
SUBTERMINAL NOTCH
TERMINAL LOBE OF CAUDAL FIN
UPPER LOBE OF CAUDAL FIN
LOWER LOBE OF CAUDAL FIN
ANAL FIN
CAUDAL PEDUNCLE
CLASPERS (MALE ONLY)
STANDARD LENGTH (SL)
TOTAL LENGTH (TL)

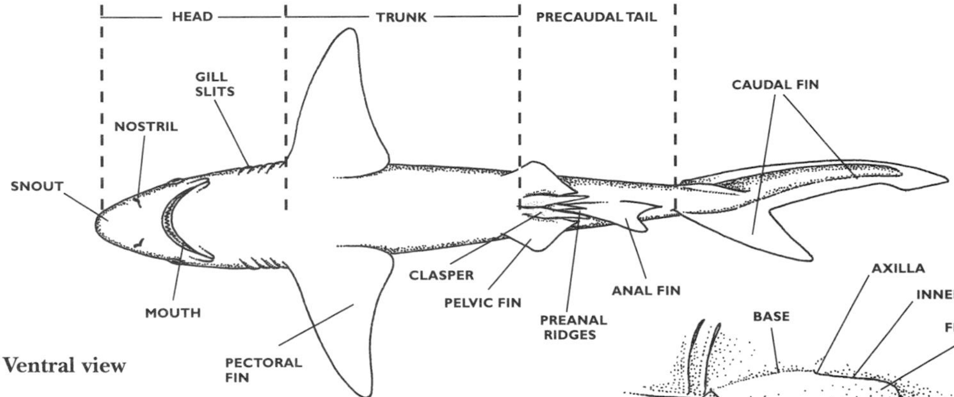

HEAD
TRUNK
PRECAUDAL TAIL
GILL SLITS
NOSTRIL
SNOUT
MOUTH
CAUDAL FIN
CLASPER
PELVIC FIN
ANAL FIN
PREANAL RIDGES
PECTORAL FIN

Ventral view

AXILLA
INNER MARGIN
FREE REAR TIP
BASE
ORIGIN
POSTERIOR MARGIN
ANTERIOR MARGIN
APEX

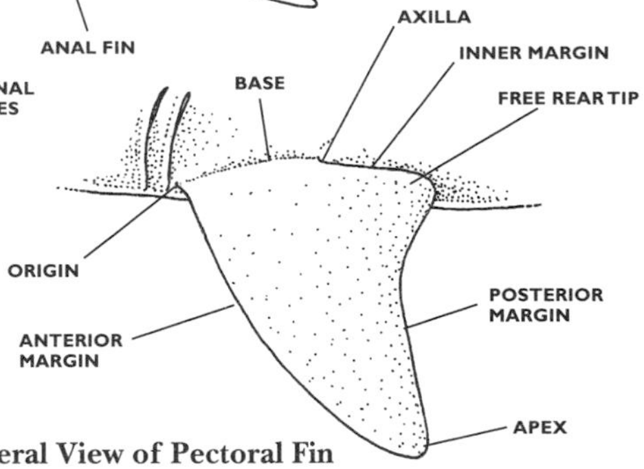

Lateral View of Pectoral Fin

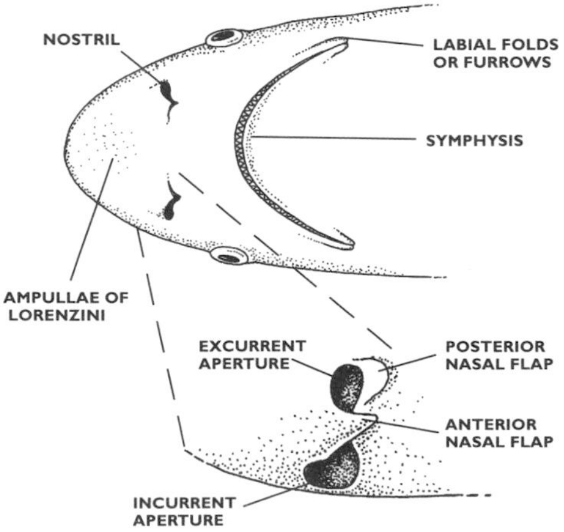

NOSTRIL
LABIAL FOLDS OR FURROWS
SYMPHYSIS
AMPULLAE OF LORENZINI
EXCURRENT APERTURE
POSTERIOR NASAL FLAP
ANTERIOR NASAL FLAP
INCURRENT APERTURE

Ventral View of Head of Pelagic Shark

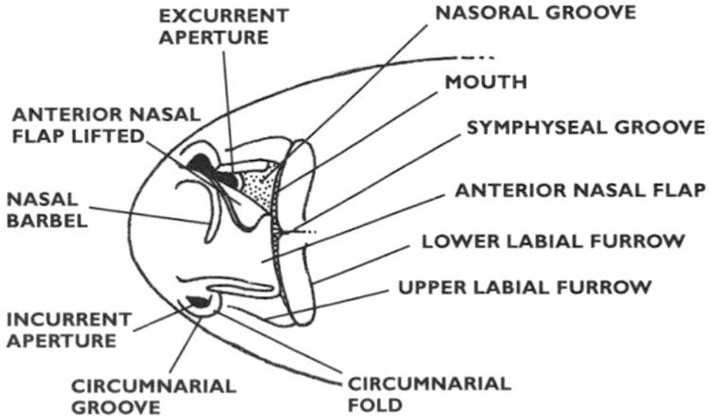

EXCURRENT APERTURE
NASORAL GROOVE
MOUTH
ANTERIOR NASAL FLAP LIFTED
SYMPHYSEAL GROOVE
ANTERIOR NASAL FLAP
NASAL BARBEL
LOWER LABIAL FURROW
UPPER LABIAL FURROW
INCURRENT APERTURE
CIRCUMNARIAL GROOVE
CIRCUMNARIAL FOLD

Ventral View of Head of Bottom-dwelling Shark

MAJOR EXTERNAL FEATURES OF TYPICAL RAYS (RAJIFORMES)

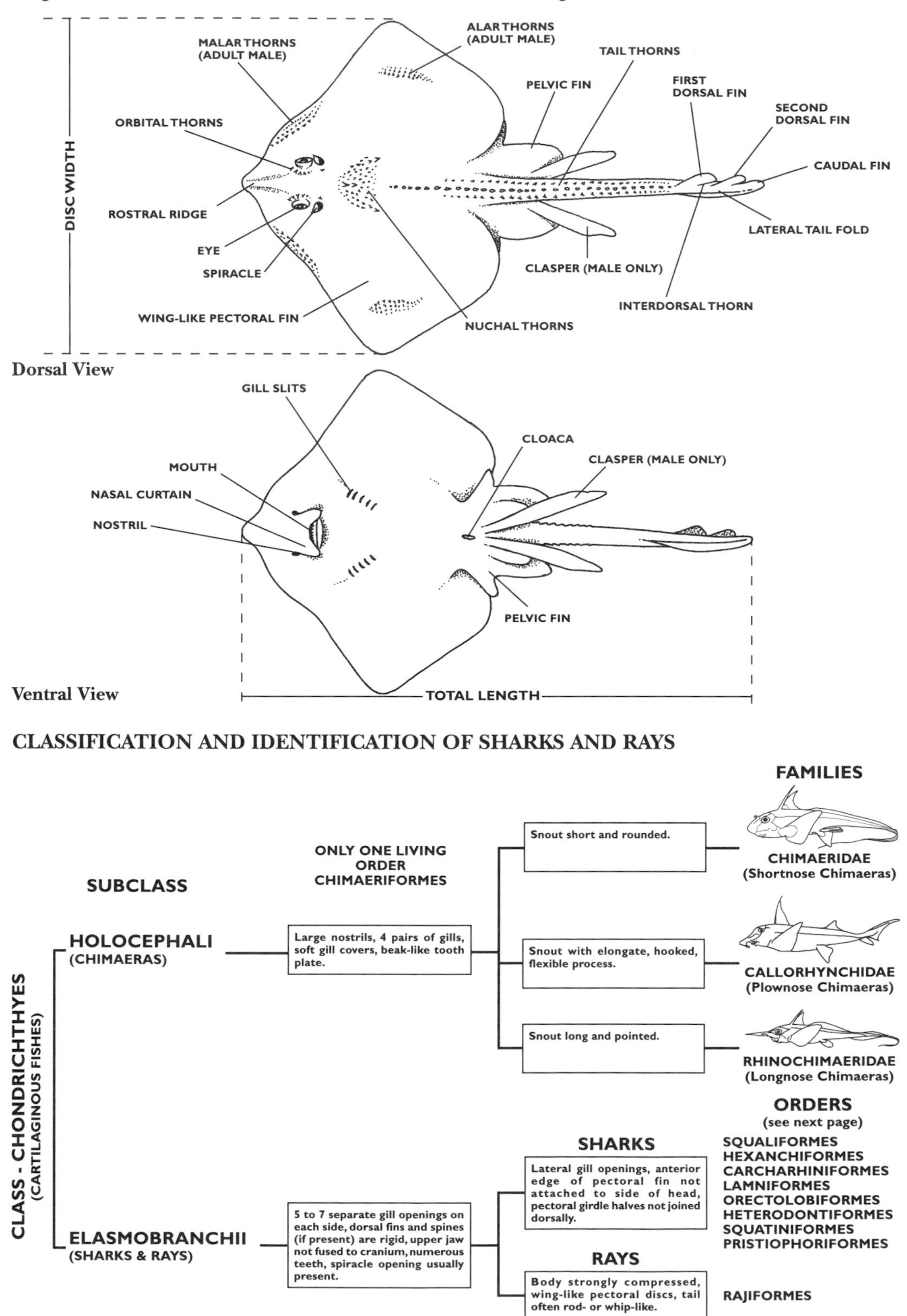

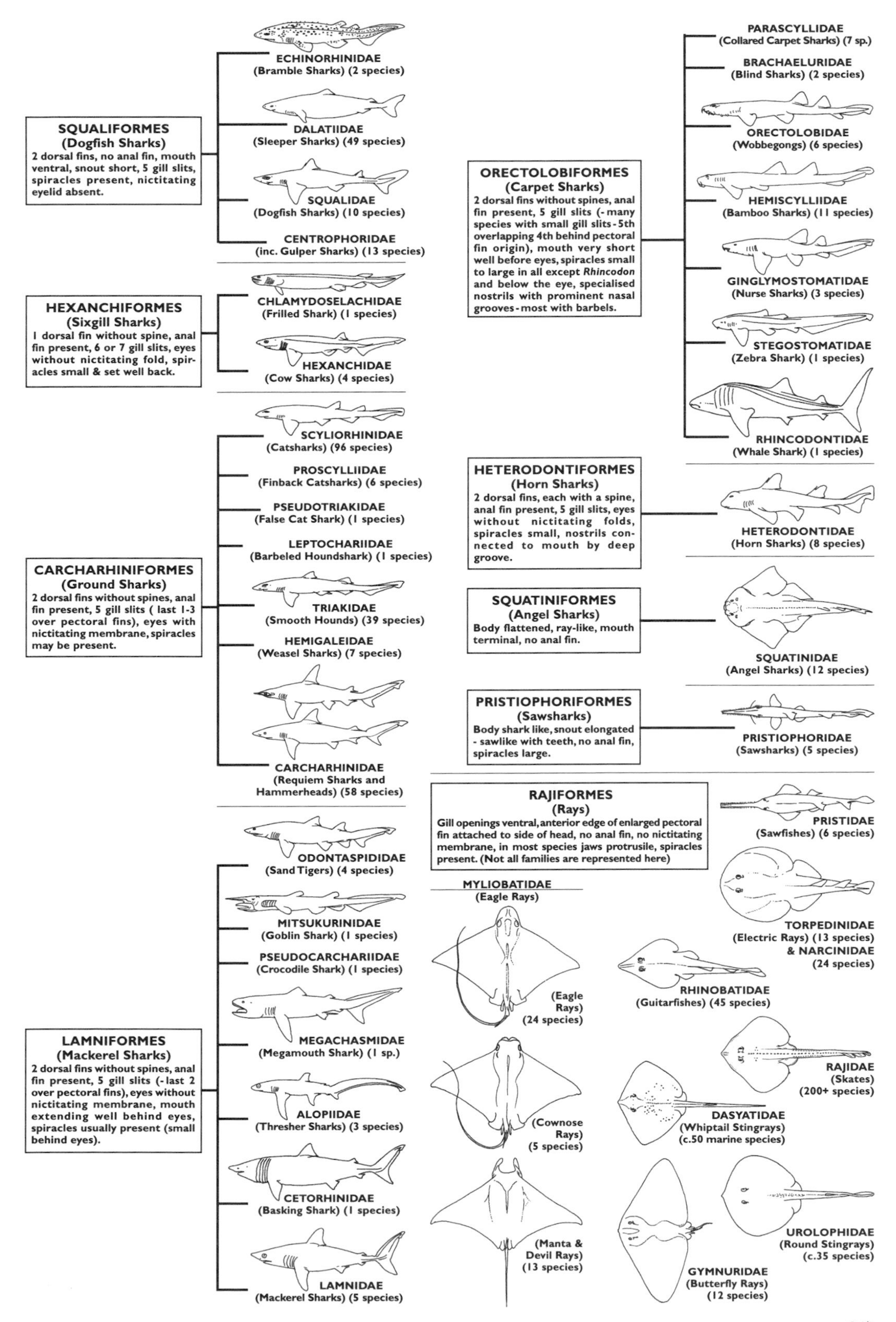
SQUALIFORMES
(Dogfish Sharks)
2 dorsal fins, no anal fin, mouth ventral, snout short, 5 gill slits, spiracles present, nictitating eyelid absent.
ECHINORHINIDAE
(Bramble Sharks) (2 species)
DALATIIDAE
(Sleeper Sharks) (49 species)
SQUALIDAE
(Dogfish Sharks) (10 species)
CENTROPHORIDAE
(inc. Gulper Sharks) (13 species)
HEXANCHIFORMES
(Sixgill Sharks)
1 dorsal fin without spine, anal fin present, 6 or 7 gill slits, eyes without nictitating fold, spiracles small & set well back.
CHLAMYDOSELACHIDAE
(Frilled Shark) (1 species)
HEXANCHIDAE
(Cow Sharks) (4 species)
CARCHARHINIFORMES
(Ground Sharks)
2 dorsal fins without spines, anal fin present, 5 gill slits (last 1-3 over pectoral fins), eyes with nictitating membrane, spiracles may be present.
SCYLIORHINIDAE
(Catsharks) (96 species)
PROSCYLLIIDAE
(Finback Catsharks) (6 species)
PSEUDOTRIAKIDAE
(False Cat Shark) (1 species)
LEPTOCHARIIDAE
(Barbeled Houndshark) (1 species)
TRIAKIDAE
(Smooth Hounds) (39 species)
HEMIGALEIDAE
(Weasel Sharks) (7 species)
CARCHARHINIDAE
(Requiem Sharks and Hammerheads) (58 species)
LAMNIFORMES
(Mackerel Sharks)
2 dorsal fins without spines, anal fin present, 5 gill slits (-last 2 over pectoral fins), eyes without nictitating membrane, mouth extending well behind eyes, spiracles usually present (small behind eyes).
ODONTASPIDIDAE
(Sand Tigers) (4 species)
MITSUKURINIDAE
(Goblin Shark) (1 species)
PSEUDOCARCHARIIDAE
(Crocodile Shark) (1 species)
MEGACHASMIDAE
(Megamouth Shark) (1 sp.)
ALOPIIDAE
(Thresher Sharks) (3 species)
CETORHINIDAE
(Basking Shark) (1 species)
LAMNIDAE
(Mackerel Sharks) (5 species)
ORECTOLOBIFORMES
(Carpet Sharks)
2 dorsal fins without spines, anal fin present, 5 gill slits (-many species with small gill slits-5th overlapping 4th behind pectoral fin origin), mouth very short well before eyes, spiracles small to large in all except Rhincodon and below the eye, specialised nostrils with prominent nasal grooves-most with barbels.
PARASCYLLIDAE
(Collared Carpet Sharks) (7 sp.)
BRACHAELURIDAE
(Blind Sharks) (2 species)
ORECTOLOBIDAE
(Wobbegongs) (6 species)
HEMISCYLLIIDAE
(Bamboo Sharks) (11 species)
GINGLYMOSTOMATIDAE
(Nurse Sharks) (3 species)
STEGOSTOMATIDAE
(Zebra Shark) (1 species)
RHINCODONTIDAE
(Whale Shark) (1 species)
HETERODONTIFORMES
(Horn Sharks)
2 dorsal fins, each with a spine, anal fin present, 5 gill slits, eyes without nictitating folds, spiracles small, nostrils connected to mouth by deep groove.
HETERODONTIDAE
(Horn Sharks) (8 species)
SQUATINIFORMES
(Angel Sharks)
Body flattened, ray-like, mouth terminal, no anal fin.
SQUATINIDAE
(Angel Sharks) (12 species)
PRISTIOPHORIFORMES
(Sawsharks)
Body shark like, snout elongated - sawlike with teeth, no anal fin, spiracles large.
PRISTIOPHORIDAE
(Sawsharks) (5 species)
RAJIFORMES
(Rays)
Gill openings ventral, anterior edge of enlarged pectoral fin attached to side of head, no anal fin, no nictitating membrane, in most species jaws protrusile, spiracles present. (Not all families are represented here)
PRISTIDAE
(Sawfishes) (6 species)
MYLIOBATIDAE
(Eagle Rays)
(Eagle Rays) (24 species)
(Cownose Rays) (5 species)
(Manta & Devil Rays) (13 species)
TORPEDINIDAE
(Electric Rays) (13 species)
& NARCINIDAE
(24 species)
RHINOBATIDAE
(Guitarfishes) (45 species)
RAJIDAE
(Skates)
(200+ species)
DASYATIDAE
(Whiptail Stingrays)
(c.50 marine species)
UROLOPHIDAE
(Round Stingrays)
(c.35 species)
GYMNURIDAE
(Butterfly Rays)
(12 species)

BONY FISHES FACTSHEET

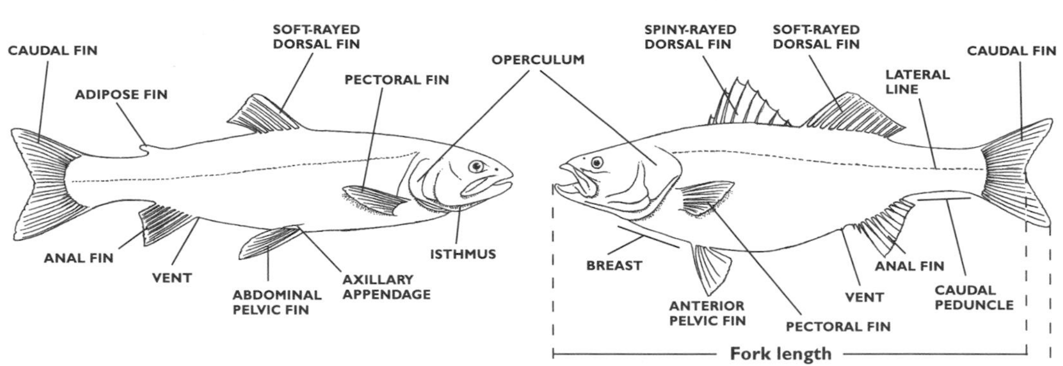

e.g. Salmons (Salmonidae)

No fin spines

No second dorsal fin, but fleshy ('adipose') fin may be present

Pelvic fins abdominal

Upper jaw bones have limited mobility, small gape

Scales usually cycloid

Swim-bladder duct present (physostomous)

e.g. Basses (Serranidae)

Fin spines in dorsal and pelvic fins

Second dorsal fin present

Pelvic fins anterior

Mobile upper jaw bones enlarged, large gape

Scales usually ctenoid

Swim-bladder duct absent (physoclistous)

LOWER TELEOST ORDERS

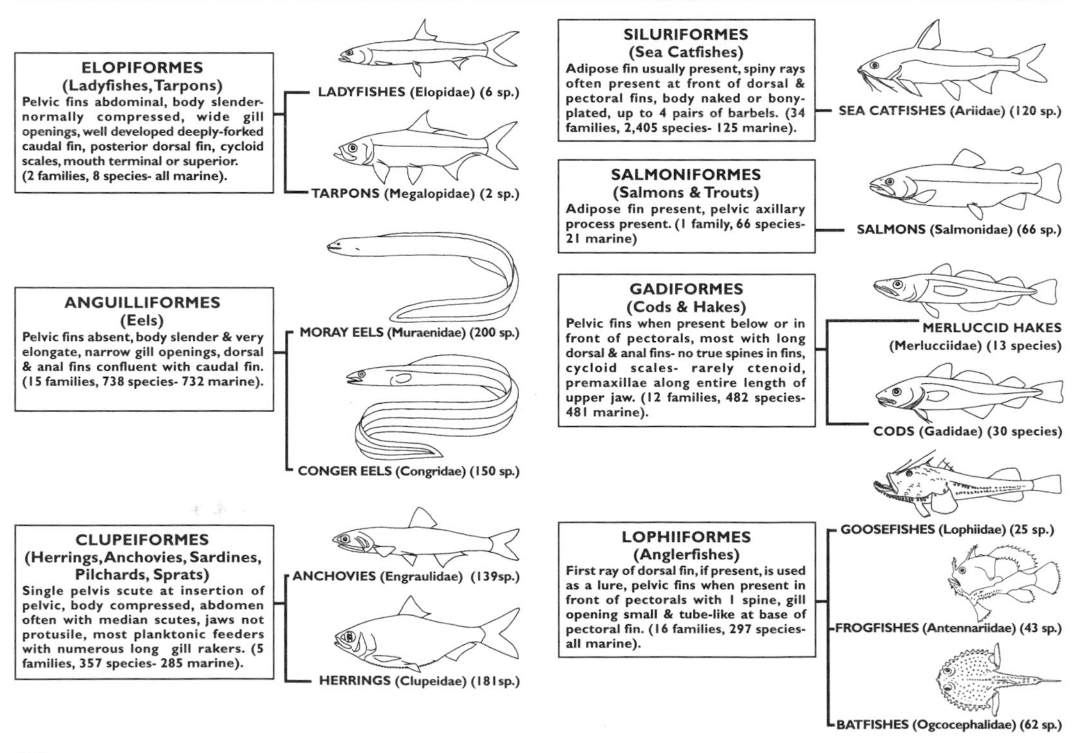

MUGILIFORMES
(Mullets)

Pelvic fins sub-abdominal, gill rakers long, lateral line absent or very faint, widely separated dorsal fins (spiny-rayed with 4 spines), ctenoid scales except *Myxus*, mouth of moderate size.
(1 family, 66 species- 65 marine).

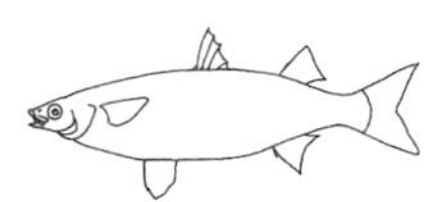

MULLETS (Mugilidae) (66 sp.)

BELONIFORMES

Fixed or non-protusile upper jaw, lower caudal fin lobe with more rays than upper lobe. (5 families, 191 species- 140 marine).
BELONIDAE- Both upper & lower jaws elongate, dorsal, anal & pelvic fins placed well back.
SCOMBERESOCIDAE- Mouth opening small, jaw length varies fom long & slender to only lower jaw slightly extended, 5 to 7 finlets behind posterior dorsal & anal fins.
HEMIRAMPHIDAE- Upper jaw much shorter than lower, pectoral & pelvic fins short.

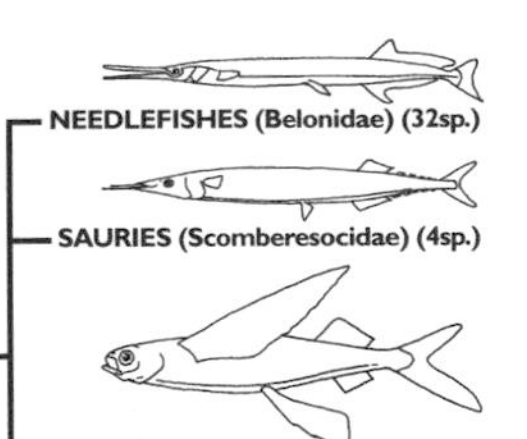

NEEDLEFISHES (Belonidae) (32sp.)

SAURIES (Scomberesocidae) (4sp.)

FLYINGFISHES (Exocoetidae) (52sp.)

HALFBEAKS (Hemiramphidae) (85 sp.)

BERYCIFORMES

(7 families, 123 species- all marine).
ANOMALOPIDAE- Luminous organ beneath eye with rotational or shutter mechanism for controlling light emission.
HOLOCENTRIDAE- Pelvic fin with one spine & c.7 soft rays, long dorsal fin with spiny portion, caudal fin forked, large eyes, operculum spine, usually reddish.

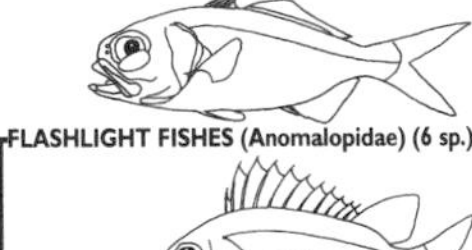

FLASHLIGHT FISHES (Anomalopidae) (6 sp.)

SQUIRRELFISHES (Holocentridae) (65 sp.)

GASTEROSTEIFORMES
(Includes Sand Eels, Sticklebacks, Tubesnouts, Snipefishes & Shrimpfishes)

Body often with dermal plates, mouth usually small.
(11 families, 257 species- 238 marine).

SEA MOTHS (Pegasidae) (5 species)

PIPEFISHES & SEAHORSES (Syngnathidae) (215 species)

TRUMPETFISHES (Aulostomidae) (3 sp.)

CORNETFISHES (Fistulariidae) (4 sp.)

SCORPAENIFORMES
(Includes Flying Gurnards, Velvetfishes, Pigfishes, Flatheads, Sablefishes, Oilfishes, Poachers & Snailfishes)

'Mail-cheeked' fishes, head & body tend to be spiny or with bony plates, pectoral fin rounded, caudal fin rounded or truncate, membrane between lower rays incised.
(25 families, 1,271 species- 1,219 marine).

Scorpionfishes (Subfam. Scorpaeninae 150sp.)

Stonefishes (Subfam. Synanceinae 10sp.)

SCORPIONFISHES (Scorpaenidae) (388 species)

GREENLINGS (Hexagrammidae) (11 sp.)

GURNARDS (Triglidae) (100 species)

LUMPFISHES (Cyclopteridae) (28 sp.)

SCULPINS(Cottidae) (300species)

PERCIFORMES

Most diversified of all fish orders, largest of all vertebrate orders.
(148 families, 9,293 species- 7,371 marine).

Suborder PERCOIDEI

Spines present in dorsal, anal & pelvic fins, 2 dorsal fins, no adipose fin, thoracic pelvic fin which has 1 spine & 5 soft rays, pectoral fin bases are laterovertical, 17 or less caudal fin rays, ctenoid scales.
(71 families, 2,860 species- 2,522 marine).
Includes- Snooks, Temperate Basses, Basslets, Dottybacks, Jawfishes, Bigeyes, Cardinal fishes, Tilefishes, Bluefishes, Cobia, Moonfish, Slipmouths, Tripletails, Emperors, Threadfins, Sweepers, Armourheads, Sea Chubs & Hawkfishes.

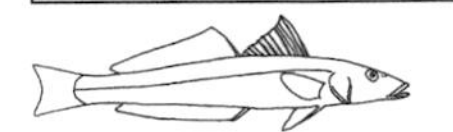

SILLAGOS (Southern Whiting) (Sillaginidae) (31 species)
Body elongate, mouth small, 2 dorsal fins with little or no gap, soft-rayed with 1 slender spine, anal fin long with 2 spines.

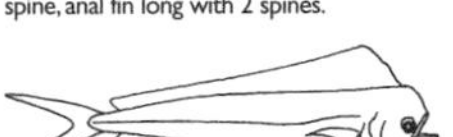

DOLPHIN FISHES (Coryphaenidae) (2 species)
Forehead prominent- particularly steep & high in males, dorsal fin continuous, originating on head, dorsal & anal fin with no spines, caudal fin deeply forked.

SNAPPERS & FUSILIERS (Lutjanidae) (125 species)
Mouth terminal, moderate to large, usually with enlarged canine teeth, jaws same length or lower slightly projecting, dorsal fin continuous with shallow notch, anal fin with 3 spines, pelvic fins just behind pectoral base, caudal fin truncated to deeply forked.

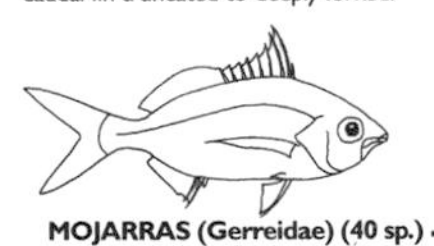

MOJARRAS (Gerreidae) (40 sp.)
Mouth highly protrusile, mouth shape illustrated very typical & distinct, head scaly, upper surface smooth, scaly sheath at base of dorsal & anal fins, caudal fin deeply forked.

DRUMS (Sciaenidae) (270 sp.)
Bony flap above gill opening, dorsal fin long with deep notch, soft dorsal fin with 1 spine, anal fin up to 2 spines, caudal fin slightly emarginated or rounded, sometimes with small barbels, pores on snout & lower jaw.

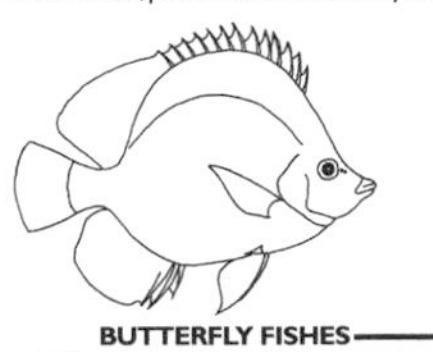

BUTTERFLY FISHES (Chaetodontidae) (114 species)
Body strongly compressed, no spine on operculum, well developed pelvic axillary process, dorsal fin continuous- sometimes slight notch, anal fin with 3-5 spines, dorsal & anal fins with elongated extensions, caudal fin rounded or emarginated, scales extending onto dorsal & anal fins.

BASSES (Serranidae) (449 sp.)
Operculum with 1 central main spine & 2 smaller spines, dorsal fin usually continuous & may be notched, caudal fins normally rounded, truncate or lunate, no scaly axillary pelvic process, lateral line complete & continuous not extending onto caudal fin. 3 subfamilies- Serraninae, Anthiinae & Epinephelinae.

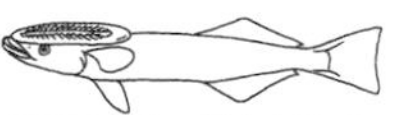

REMORAS (Echeneididae) (8 sp.)
Body elongated, head flattened, lower jaw projecting past upper jaw, small cycloid scales, dorsal fin & anal fin with no spines, transformed spiny-rayed dorsal fin forming sucking disc on head.

JACKS & POMPANOS (Carangidae) (96 species)
Body normally compressed- ranging from deep to fusiform, generally small cycloid scales, many species with lateral line scales modified into spiny scutes, dorsal fin & anal fins variable- sometimes with up to 9 detached finlets behind dorsal & anal fins, caudal peduncle slender, & fin widely forked.

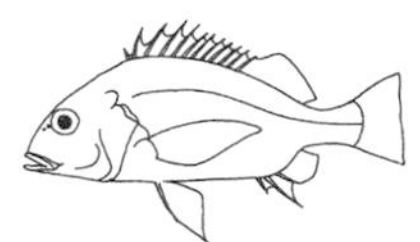

GRUNTS (Haemulidae) (150 sp.)
Dorsal fin continuous, anal fin with 3 spines, mouth small, upper jaw usually projects beyond the lower, usually cardiform teeth, generally have enlarged chin pores.

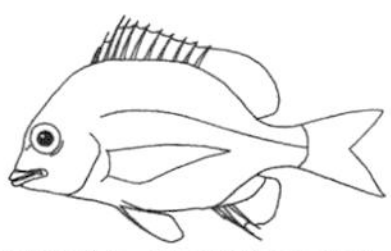

BREAMS (Sparidae) (100 species)
Dorsal fin continuous, anal fin with 3 spines, mouth small, & horizontal, teeth are stout, maxillae covered with sheath when mouth closed, no spines on operculum, posterior nostrils are large & elongated.

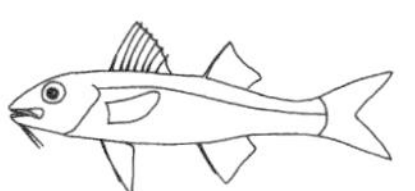

GOATFISHES (Mullidae) (55 sp.)
Body elongated, 2 long movable hyoid barbels, 2 widely separated dorsal fins- first with 6-8 spines second with soft rays, soft dorsal fin shorter than anal which has up to 2 spines, caudal fin forked.

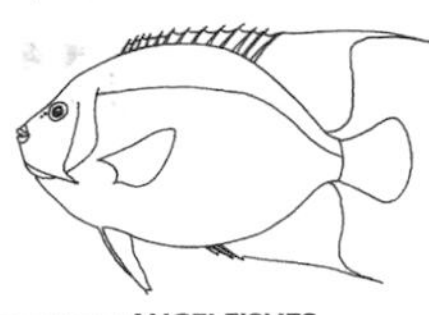

ANGELFISHES (Pomacanthidae) (74 species)
Body strongly compressed, strong spine on operculum, no well developed axillary process, dorsal fin continuous, anal fin with 3 spines, dorsal & anal fins with elongated extensions, caudal fin rounded or lunated.

PERCIFORMES (cont.)

Suborder LABROIDEI

Most species of labrids & scarids can change sex; the majority of this suborder are freshwater fish, the cichlids (c.1275).
(6 families, c.2,235 species, c.960 marine). Includes- Surfperches.

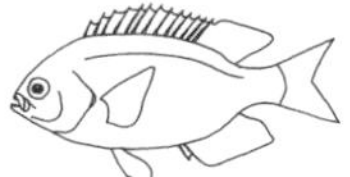

DAMSELFISHES
(Pomacentridae) (315 species)
Body usually high & compressed, single nostril on each side of snout, mouth small, lateral line incomplete or interrupted, dorsal fin continuous- spine-rayed part longer than soft-rayed, anal fin with 2 spines.

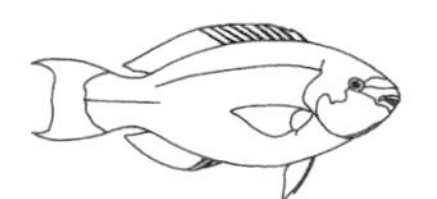

PARROTFISHES
(Scaridae) (83 species)
Body moderately compressed, mouth nonprotractile, jaw teeth usually fused to form a pair of beak-like dental plates on each jaw, continuous dorsal fin with 9 spines & 10 soft rays, anal fin with 3 spines, pelvic fins with 1, scales large & cycloid.

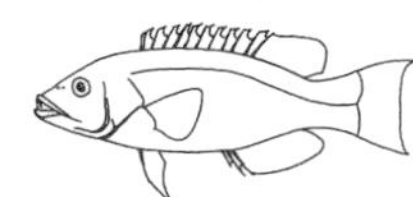

WRASSES
(Labridae) (c.500 species)
One of the most diverse groups of fish, terminal mouth that is protractile, maxillary not exposed on cheeks, lips are thick, jaw teeth mostly separate usually projecting out, cycloid scales, dorsal fin continuous with slender spines.

Suborder ZOARCOIDEI

All have a single nostril, but no other characteristic external features.
(9 families, 318 species, all marine). Includes- Pricklebacks, Gunnels, Wolfishes.

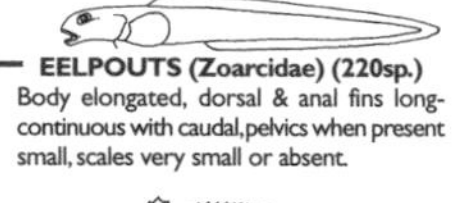

EELPOUTS (Zoarcidae) (220sp.)
Body elongated, dorsal & anal fins long- continuous with caudal, pelvics when present small, scales very small or absent.

Suborder NOTOTHENIOIDEI

Pelvic fins each with 1 spine, one nostril each side, usually 2 or 3 lateral lines, 3 plate-like pectoral fin radials.
(5 families, 122 species, all marine). Includes- Icefishes, Antarctic Dragonfishes, Sandfishes, Sand perches, Sand Lances & Stargazers.

WEEVERS (Trachinidae) (4 species)
Body elongated, first dorsal fin with 5-7 spines, pelvic fins in front of pectorals, poison glands associated with first dorsal spine & gill-cover spine.

Suborder BLENNIOIDEI

Pelvic fin usually with 1 embedded spine & 2-4 soft rays which are placed ventrally in front of pectorals- normally long rather than deep, anal fin with less than 3 spines & all soft rays not branched, 2 nostrils on each side of head, cirri often on head.
(6 families, 732 species, all marine). Includes- Triplefin Blennies, Kelpfishes & Pikeblennies.

SAND STARGAZERS
(Dactyloscopidae) (41 species)
Body elongated, mouth oblique- lips usually fringed, upper margin of gill cover with many thin extensions, eyes dorsal & may be on stalks, pelvic fins 1 spine, dorsal fin long.

COMBTOOTH BLENNIES
(Blenniidae) (345 species)
Body elongated & naked, head usually blunt, jaws with comb-like teeth, first dorsal fin with 5-7 spines, anal fin with 2 spines, many species mimic other fish species.

Suborder ICOSTEOIDEI

Highly compressed & limp, no spines in fins, adults naked. (1 species, marine).

RAGFISHES (Icosteidae)

Suborder GOBIESOCOIDEI

Ventral sucker present.
(1 family, 120 species, mostly marine).

CLINGFISHES
(Gobiesocidae) (120 species)
Body elongate & naked, pelvic fins modified into thoracic sucking disc, 1 dorsal fin with no spines.

Suborder CALLIONYMOIDEI

Head broad & depressed, body naked, mouth small, 2 dorsal fins, pelvic fin with 1 spine. (2 families, 91 species, all marine).

DRAGONETS
(Callionymidae) (84 species)
Body elongated, gill opening small on upper side of head, gill cover with spine, normally high dorsal fins- especially in male as generally sexually dimorphic.

Suborder GOBIOIDEI

Body elongated, lateral line system reduced, barbels on some fish species, gill membranes joined to isthmus, pelvic fins below pectorals, pelvics with 1 spine.
(8 families, 2,121 species, 1921 marine). Includes- Gobies, Sleepers, Sandfishes & Wormfishes.

GOBIES (Gobiidae) (1,875 species)
More marine species than any other fish family & one species is smallest vertebrate. Pelvic fins united- usually forming sucking disc, spiny dorsal (when present) separate from soft-rayed dorsal fin- first 2-8 spines flexible, 2nd dorsal fin & anal fin with weak initial spine, caudal fin usually rounded.

PERCIFORMES (cont.)

Suborder ACANTHUROIDEI

Body deeply compressed, mouth small, gill membranes broadly joined to isthmus, usually lunate caudal fin.
(6 families, 125 species, all marine). Includes- Spadefishes, Scats, Rabbitfishes, Louvar & Moorish Idol.

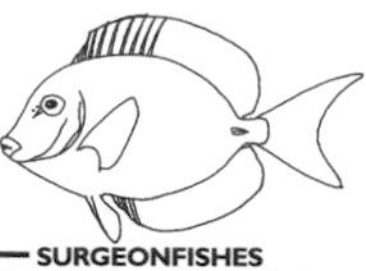

SURGEONFISHES
(Acanthuridae) (72 species)
Pelvic fin with 1 spine, dorsal fin 4-9 spines, anal fin 2-3 spines; Unicornfishes with 1 or 2 plates on caudal peduncle; Sugeonfishes with 1+ spine(s) on caudal peduncle which can extend & produce deep lacerations.

Suborder SCOMBROIDEI

Species in this suborder are probably fastest swimming fish, with the evolution of endothermy. Body shape elongated or fusiform, upper jaw not protrusile.
(5 families, 136 species, all marine). Includes- Snake Mackerels & Cutlassfishes.

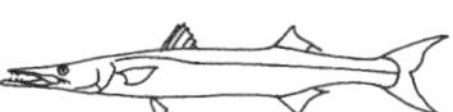

BARRACUDAS (Sphyraenidae) (20 sp.)
Body elongate, mouth large with jutting lower jaw & fang-like teeth, lateral line well developed, 2 widely separated dorsal fins, pectoral fins relatively low.

TUNAS & MACKERELS
(Scombridae) (49 species)
Body fusiform, 2 well spaced dorsal fins- depressible into grooves, with 5-12 finlets behind 2nd dorsal & anal fins, pelvic fins placed below high pectoral fins, slender caudal peduncle with 2 keels.

BILLFISHES (Xiphiidae) (12 species)
Fusiform body with elongated premaxillary bill, mouth inferior, dorsal fin origin over back of head, 2 anal fins, pectoral fins low on body, pelvics reduced or absent, finlets absent, 1-2 keels on narrow caudal peduncle.

Suborder STROMATEOIDEI

Characteristic internal feature with toothed saccular outgrowths in gullet behind last gill arch. (6 families, c.65 species, all marine). Includes- Driftfishes, Squaretails & Butterfishes.

MEDUSAFISHES
(Centrolophidae) (27 species)
Pelvic fins in adult, dorsal fin continuous, anal fin with 3 spines & usually long.

PLEURONECTIFORMES (Flatfish)

Adults not bilaterally symmetrical, body highly compressed, one eye moving to other side of head- both eyes protude above body surface, dorsal & anal fins long, very distinctive group.
(11 families, 570 species- 566 marine). Includes- Psettodids, Citharids, Southern Flounders & American Soles.

LEFT-EYE FLOUNDERS
(Bothidae) (115 species)
Eyes sinistral, pelvic fin base on blind side is short-based, first ray distinctly posterior to first ray on ocular side.

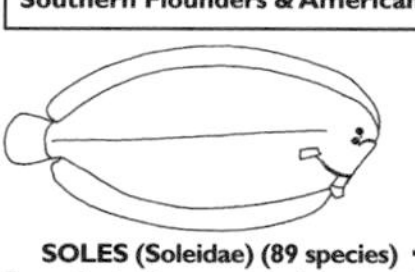

SOLES (Soleidae) (89 species)
Eyes dextral, preoperculum margin completely concealed, head small.

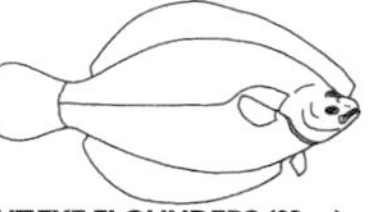

RIGHT-EYE FLOUNDERS (93sp.)
(Pleuronectidae)
Eyes almost always dextral, both pelvic fins generally short-based & roughly in line.

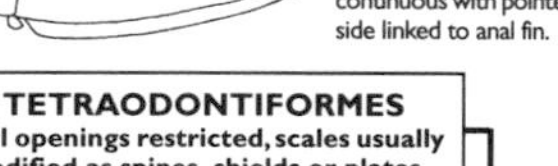

TONGUEFISHES (Cynoglossidae) (110sp.)
Eyes sinistral, preoperculum concealed, dorsal & anal fins continuous with pointed caudal fin, sometimes pelvic fin of blind side linked to anal fin.

TETRAODONTIFORMES

Gill openings restricted, scales usually modified as spines, shields or plates
(25 families, 1,271 species- 1,219 marine). Includes- Spikefishes, Triplespines, Filefishes & Three-toothed Puffer.

TRIGGERFISHES (Balistidae) (c.40 sp.)
Body usually compressed, no pelvic fin, first dorsal spine with locking mechanism- 3 dorsal spines, upper jaw protuding, 4 incisor teeth, plate-like scales.

PUFFERFISHES
(Tetraodontidae) (121 sp.)
Body inflatable, skin tough, body naked with only short prickles, single dorsal fin posterior, anal fin in similar lateral position.

BOXFISHES (Ostraciontidae) (33sp.)
Body encased in bony carapace with polygonal plates, single dorsal fin posterior.

SUNFISHES
(Molidae) (3sp.)
Appear to be all head, 2 fused teeth in jaws, no caudal peduncle, caudal fin absent or formed by few rays derived from dorsal & anal fin rays, high dorsal & anal fins.

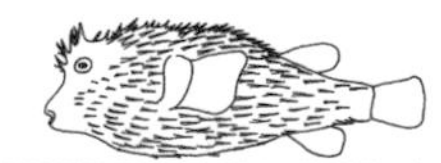

PORCUPINE FISHES (Diodontidae) (19 sp.)
Body inflatable covered with well developed sharp spines, 2 fused teeth in jaws.

MARINE REPTILES

by

Colin McCarthy

INTRODUCTION

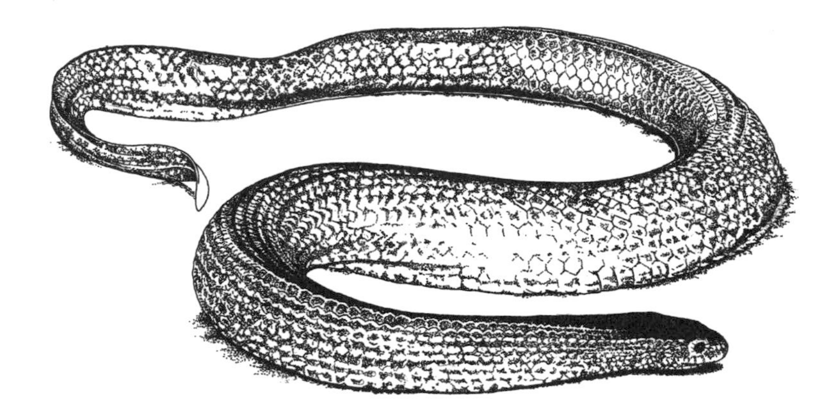

Figure 254. Spine-bellied or Hardwick's Sea Snake *Lapemis curtus hardwickii.*

The sea is by no means a typical habitat for reptiles. Of about 6,000 known species of living reptiles only about 80 (1.3%) occupy marine or brackish environments; sea snakes comprise the largest single group of marine reptiles in terms of number of species. Amphibians are more poorly represented in the marine ecosystem than reptiles. The Crab-eating Frog *Rana cancrivora* from coastal mangrove swamps of South-East Asia is one of the few amphibians that can tolerate such conditions. All marine reptiles have salt-excreting glands which are located in various parts of the head. These remove excess salt (taken in with their food) which the reptilian kidney is unable to handle.

The class Reptilia has traditionally been divided into six subclasses, three of which are now extinct. Thus, among living reptiles, turtles, terrapins and tortoises (Chelonii) belong to the **Anapsida**, lizards, snakes, worm-lizards (Squamata) and the Tuatara (*Sphenodon*) are in the **Lepidosauria**, and crocodilians (Crocodylia) are placed in the **Archosauria**. However, the artificiality of this arrangement has long been apparent. For instance, modern classifications now consider birds (Aves) to be archosaurians, and archosaurs appear more closely related to lepidosaurs than they are to chelonians. The simplified classification below expresses these relationships and highlights the living marine reptiles and their approximate number.

CLASS: REPTILIA					No. of marine species
SUBCLASS: ANAPSIDA		Order	**Chelonii**	Turtles	7
SUBCLASS: SAURIA					
	ARCHOSAURIA		(**Aves**	Birds)	
		Order	**Crocodylia**	Crocodiles	1
	LEPIDOSAURIA	Orders	**Rhynchocephalia**	(*Sphenodon punctatus*)	-
			Squamata	Suborder: **Amphisbaenia** (worm-lizards)	-
				Suborder: **Lacertilia** Lizards	1
				Suborder: **Serpentes** Snakes	71

CROCODILES

Crocodilians are essentially freshwater creatures. However, through their 200 million year history there have been many true marine specialists. None of the truly marine forms have survived to the present but at least two crocodile species favour estuarine habitats and individuals are sometimes found far out to sea. The American Crocodile *Crocodylus acutus* is found in coastal areas, rivers and lakes from southern Florida, several of the Caribbean islands, parts of Central America and northern South America. Perhaps the most striking example of seagoing is provided by the Estuarine or Indo-Pacific Crocodile *Crocodylus porosus* which is widely distributed in brackish waters through tropical Asia and the Pacific.

Crocodilians have aquatic adaptations that enable them to colonise food-rich shallow waters at the land edge. The tail is flattened from side-to-side for swimming and there is a well developed web between the toes of which there are five on the front limbs and four on the hind limbs. The mouth can be opened underwater without flooding the lungs because the airway between the nasal passages and lungs is closed by a valve at the back of the throat. The nostrils can be closed by muscular flaps and the eardrum is protected by movable scales when the animal dives. There is a diaphragm-like structure associated with well developed lungs which assists in effective gas exchange - crocodilians are capable divers.

Salt glands in crocodiles are modified salivary glands. They occur on the tongue in the Estuarine Crocodile and they are also found in a number of crocodiles which do not normally occur in salt water (but not in the American Alligator or caimans). There is evidence to suggest that some of the freshwater crocodile species in the Indo-Pacific region might extend into more brackish habitats were it not for competition from such a dominant species as the Estuarine Crocodile.

Figure 255. Salty secretions of the salt gland in the tongue of an Indo-Pacific Crocodile *Crocodylus porosus.*

Estuarine or Indo-Pacific Crocodile

Figure 256. Estuarine or Indo-Pacific Crocodile *Crocodylus porosus.*

This is the largest living reptile, dubiously reported to have reached 9m in length but certainly known to achieve 7m and captive weights of over 1,000 kg. Males are sexually mature at about 3.2m (16 years), females at 2.2m (10 years).

It is typically found in brackish habitats (mangroves) but also extends into freshwater rivers and swamps. Wandering Estuarine Crocodiles have sometimes been found at sea well outside the normal breeding range for the species. The ability of the species to survive long sea journeys has enabled it to reach remote islands such as Cocos (Keeling) and Fiji.

Estuarine crocodiles feed opportunistically, the type and size of prey relating to the age and habitat. Young animals often eat crustaceans (crabs and shrimps) and insects. As they get bigger the proportion of vertebrate prey increases, larger juveniles taking a wide range of fish and sometimes snakes and birds; progressively larger prey is tackled including turtles, sharks, cattle, horses and humans. Mammals are frequently taken at the water's edge, occasionally being knocked into the water by a blow from the crocodile's tail. The formidable conical teeth interlock and are so arranged that the lower ones bite slightly inside the upper ones. They are used for gripping (as in crocodiles generally), not cutting or chewing.

Nesting time varies across its wide range but generally coincides with the start of the annual wet season. The nest is constructed from vegetable debris and mud. A freshwater environment is usually chosen but in coastal swamp areas the nest may be built on a floating vegetation mat.

On average there are 50 eggs in a clutch and incubation takes 80-90 days. The female often remains on guard near the nest during the incubation period. She assists the hatchlings by opening the nest and protects them by carrying them from the nest to the water in her throat pouch.

TURTLES AND TERRAPINS

Figure 257. Ornate and Northern forms of the Diamondback Terrapin *Malaclemys terrapin*.

Sea turtles are the best known of the marine-adapted turtles and terrapins (Chelonii). However some other predominantly freshwater terrapins and turtles also live in estuarine habitats and regularly use coastal beaches and neighbouring banks and dunes as nest sites.

Diamondback Terrapins *Malaclemys terrapin* live in coastal habitats along the Atlantic coast of North America (from Massachusetts southwards) and are confined to salt or brackish water. Freshwater Slider Turtles, *Trachemys scripta venusta*, from Costa Rica, frequently enter the Caribbean Sea to be carried south on the longshore current to their beach nest sites. This seasonal activity occurs from January to March and only involves the females, the males remaining in freshwater habitats. The hatchlings emerge in May and June. They appear not to enter the sea but move overland to reach swamps and waterways.

In Asia, River Terrapins *Batagur baska* and Painted Terrapins *Callagur borneoensis* nest on islands and coastal beaches. Some mainly freshwater softshell turtles (such as *Pelochelys bibronii* from Asia and *Trionyx triunguis* from the eastern Mediterranean) may spend extended periods at sea.

Sea Turtles

Turtles and tortoises (Chelonii) have an extensive fossil record, the oldest being known from the Triassic of Germany. Sea turtles themselves have a long history, and a range of fossil genera (and two living genera *Caretta* and *Chelonia*) have been found in the Upper Cretaceous of Europe and North America. The hard-shelled sea turtles (family Cheloniidae) are regarded as being most closely related to the Leatherback Turtle (family Dermochelyidae) and several extinct sea turtle families (Plesiochelyidae, Protostegidae and Toxochelyidae).

Seven living species of sea turtle are recognised; see plate 24, and the turtle factsheet for their diagnostic characters.

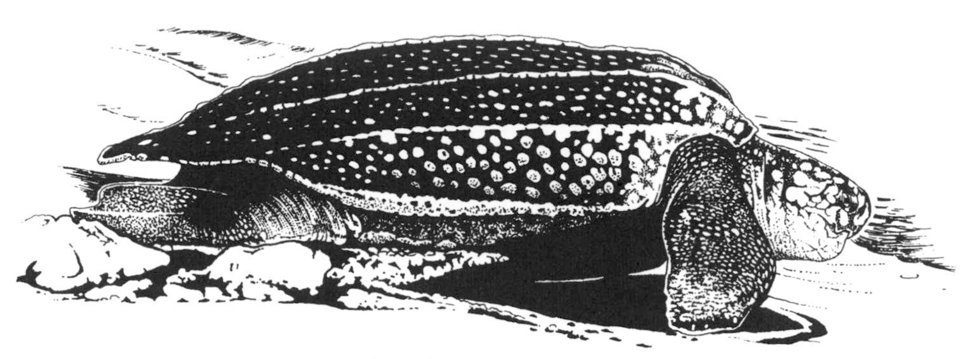

Figure 258. Leatherback Turtle *Dermochelys coriacea.*

Marine turtles have streamlined shells and long flippers which make them well adapted for swimming. Buoyancy is enhanced by a relatively reduced bony skeleton. However, out of water they are cumbersome and essentially defenceless. On land, if turned on their backs they generally find it impossible to right themselves. They are unable completely to retract the limbs and head within the shell, thus also making them prone to shark attack. The chest is rigid as the ribs support the bony carapace and breathing occurs by contraction of the abdominal muscles. Excess salt is excreted by eye (orbital) glands. This is the reason for the copious gluey tears shed by these reptiles; the secretion also protects and lubricates the front of the eyeball.

Turtles are most commonly associated with warmer seas; in cold conditions they usually become lethargic. Loggerheads have been known to hibernate when body temperatures reach 15°C, burying themselves in mud. However, Leatherback Turtles, in cold water, have the unusual ability to maintain deep body temperatures up to 18°C above their surroundings. This attribute, due to a combination of their large size, thick blubber, high oil content of the flesh, and special arrangement of the blood vessels, enables the Leatherback Turtle to range further north than other sea turtles, into waters as cool as 8°C.

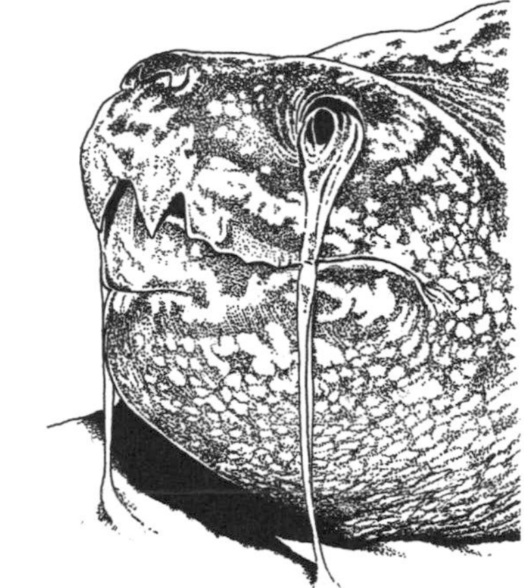

Figure 259. 'Tear-like' salt secretion in the Leatherback Turtle *Dermochelys coriacea.*

Although they spend most of the time in the sea, turtles start life on the shore, hatching from eggs buried in sand. The males spend their whole lives at sea. The females mate offshore and at night they climb out onto the beach to dig a hole for the eggs with their back flippers. The soft leathery-shelled eggs are round. Commonly around 100 eggs are laid per clutch and the females may return to lay more eggs several times during the season.

If the eggs are laid in sand of a suitable temperature (about 29°C) and survive predation from enemies, including humans, they hatch after about two months. Emergence occurs one to seven days (mean two and a half days) after hatching. In most cases, emergence occurs at night when the sand temperatures are falling, although a small percentage may emerge in the early morning or later. The incubation temperature in the nest affects the sex ratio of the brood; in Loggerhead Turtles it has been shown that the percentage of female hatchlings increases with incubation temperature. After the hatchlings have dug themselves out of the nest chamber, they often face a hazardous run to the sea and may be attacked on the way by birds, crabs and ultimately fishes. Hatching at night eliminates exposure to diurnally active predators and hot surface sand which could be lethal to young turtles. Those that survive this frenzied period swim out to sea to spend a period drifting at the surface, often associated with floating mats of seaweed and feeding on small invertebrates.

Once they grow to about 35cm shell length many move inshore to defined feeding grounds (Leatherback Turtles, however, have a more diffuse range in that they follow drifts of jellyfish). Maturation is long, frequently around thirty years, after which migration to breeding areas (rookeries) occurs. The female completes the life cycle by laying several clutches of eggs. The adult turtles then return to their feeding areas. Migration to rookeries, rather than being an annual event, frequently occurs at two to three year intervals, probably to allow for fat reserves to build up in order to sustain the journey and egg production.

Sea turtles may travel up to 1600km or more between feeding and nesting sites. Green Turtles tagged on Ascension Island have been recovered 2240km away off the coast of Brazil. They return to the same small nesting areas of the island. The methods that adult turtles use to navigate over these long distances are not known, although it is most likely that the homing process is a composite one employing different senses (vision, smell, hearing, temperature, wind/wave motion and current sensitivity) based on a multiplicity of cues.

Recent studies on hatchlings have shown that the Loggerhead Turtle has a light-compass sense and a magnetic sense which are used in orientation. Newly hatched Loggerhead Turtles find their way to the sea using vision with which they detect light from the moon or stars reflected off the sea surface. Once in the surf zone, the turtles head for the open sea by sensing wave motion. Of particular interest was the finding that when the sea surface was calm, hatchlings used a magnetic sense to guide their movement away from the shore. Strong evidence for this was found in laboratory tests where the effects of artificially reversing the magnetic fields produced startling changes in swimming direction. Since starlight and moonlight reflect more brightly from the sea than the land, it has been suggested that the initial exposure to light at hatching sets the turtles' internal magnetic compass. This takes over when the turtles reach the sea and is used in combination with their wave motion sense. Confusing signals from beachside buildings and motorway lights disrupt the turtles' light-compass and have the effect of causing movement inland away from the sea. Adult Loggerhead Turtles may also use a magnetic compass sense to navigate during their five to seven year migration across the North Atlantic.

Marine Lizards

Figure 260. Marine Iguana *Amblyrhynchus cristatus*.

A few lizards are known to forage on the sea shore. For instance small skinks, *Cryptoblepharus*, on islands in the Indian Ocean seek out insects and crustaceans in the intertidal region. However, only one lizard can be described as truly marine, the Marine Iguana *Amblyrhynchus cristatus*, of the Galapagos Islands; these are the only living lizards to gain the major part of their food requirements from the sea.

The Marine Iguana is thought to be most closely related to the Galapagos Land Iguana *Conolophus subcristatus*, but the relationship of these two iguanas to the other iguanine lizards remains uncertain. There is just one species of Marine Iguana in the Galapagos, but seven subspecies (races), which differ from each other in size and coloration, are distributed on the various islands. Adult males can reach 140cm in total length and up to 12.5kg in weight. Their short snouts and jaws lined with three-cusped teeth are skull adaptations for feeding on marine algae attached to rocks. Adult males engage in head-butting contests, interlocking prominent pyramid-shaped scales which cover the head. The tail is rather longer than the body and quite flattened from side-to-side. There is a well developed crest of pointed scales on the neck, back and tail.

Marine Iguanas live in large groups on the lava fields and rocks adjacent to the sea. They are herbivorous and as the tide retreats they move into the intertidal zone to feed on the exposed algae. Their feet have long claws which enable the iguanas to cling firmly to the rocks while being pounded by surf. The sea temperature is appreciably cooler than their preferred body temperature, hence their frequent reluctance to get wet. However, they are capable swimmers and divers. They swim holding their limbs tightly back against their slowly undulating bodies. They dive up to 15m to feed on algae growing on the seabed. Although they can survive submergence for half-an-hour or more, diving periods usually last a few minutes.

Surplus salt is excreted by salt glands which open into the nose; the salty secretion is sneezed out through the nostrils.

Breeding of Marine Iguanas usually starts during November and December. At this time the males, with display behaviour and fighting, attempt to defend territories consisting of a few square metres of lava. They assume breeding coloration, often developing red mottling on their sides. Fighting is highly ritualised and may last for up to five hours. Mating occurs when females enter the territory. Egg laying occurs about five weeks later. The females move to traditional nest sites where they dig burrows in sand up to 80cm deep and lay two or three eggs. They then seal the burrow entrance and level the ground. The nests are guarded for several days before the females leave the area to feed. The eggs hatch in 89 to 120 days. When the hatchlings emerge they run rapidly to the nearest available cover; eventually they move to the shore.

MARINE SNAKES

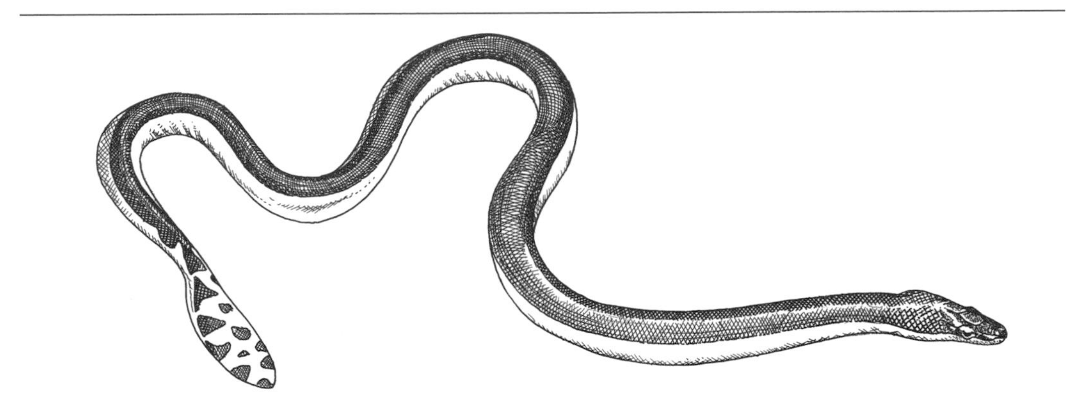

Figure 261. Yellow-bellied Sea Snake *Pelamis platurus*. This is the most widely distributed of all sea snakes and is the world's most abundant reptile. The venom of this species is extremely toxic.

The dominant groups of marine snakes are the 'true' sea snakes (hydrophiines, 56 species) and sea kraits (laticaudines, five species) and these are the main subjects of this section. They are venomous, front-fanged, aquatic snakes most closely related to the mainly terrestrial elapid group (cobras and their relatives). However, a number of other snake species enter the sea, though most of these occupy relatively shallow water.

Among colubrids (one of the largest groups of mainly terrestrial species), the homalopsines, rear-fanged aquatic snakes from the Indo-Australian area, comprise 34 species in ten genera. Most live in freshwater but some (seven species) also occur in estuaries and coastal areas. Several North American natricine snakes (*Nerodia*) have races that live in saltmarshes and estuaries but these enter the water mainly to feed.

Wart snakes (family Acrochordidae) comprise three highly aquatic species which occur in the Indo-Pacific region. Two of these are freshwater species but the Asiatic File Snake *Acrochordus granulatus*, from the coasts of India through to northern Australia and the Solomon Islands, occupies a wide range of habitats including freshwater ponds and rivers, estuaries, mangroves and the sea.

IDENTIFICATION

Identification of sea snakes is frequently difficult. External features (shape, coloration and scalation) observed on living specimens can be unreliable; for certain identification it is often necessary to perform scale counts, microscopic study and dissection to study internal features. Illustrated regional guides are helpful for provisional identifications, especially in areas where a fairly restricted number of species occurs. It should be noted, however, that range extensions frequently occur and sea snake species often turn up outside their previously known distribution.

A summary of characters that distinguish sea kraits and hydrophiines is given below:

Sea Kraits - nostrils on side of snout and enlarged scales on belly (like land snakes). Egg layers. Partly terrestrial.

Hydrophiines - nostrils on top of snout and reduced scales on belly. Young are born alive. Entirely marine, do not venture on land except for a few species that forage on tidal flats in Australia.

ADAPTATIONS

Sea snakes have a number of features which help them to thrive in the marine environment. Nasal valves and close-fitting scales around the mouth keep out water during diving. Flattened, paddle-shaped tails provide extra propulsion when swimming. Many sea snakes have small, ventral belly scales. In terrestrial snakes these scales are enlarged and assist in locomotion by catching on surface irregularities. Those sea snakes that spend part of the time on land (sea kraits, *Laticauda*) also have large belly scales but most other sea snakes have very little power of terrestrial locomotion and are effectively helpless if accidentally washed-up on the shore.

Buoyancy seems at least partly controlled by the amount of air in the lung which is an exceptionally long structure in sea snakes, extending from just behind the head to the vent. The terminal part of the lung consists of a muscular sac which acts as an air reservoir. The lung, of course, also plays an important role in respiration (as in land snakes), but sea snakes are additionally able to exchange gases across their skin; indeed it has been estimated that up to about a fifth of the total oxygen needs can be absorbed and almost all of the carbon dioxide produced can be eliminated through that route. Sea snakes frequently remain submerged for about half an hour but occasionally have been found to remain underwater for almost two hours.

The salt gland in sea snakes is located in the mouth; the salty fluid is expelled as the snake protrudes its tongue.

DISTRIBUTION AND ECOLOGY

Sea snakes are mainly found in the tropical seas of South-East Asia, northern Australia and adjacent parts of the Pacific. One species, the Yellow-bellied Sea Snake *Pelamis platurus* (plate 24 & fig.243) is, however, extraordinarily widespread being found from the Cape (South Africa) eastwards through the Indian Ocean, Asiatic waters, and the Pacific to Central America.

The greatest concentration of sea snakes is in Sout-East Asia; in the area of the Strait of Malacca alone, no fewer than 28 species have been recorded. It seems that the comparatively recent geological history of the area might well account for this, with past fluctuations in sea level splitting and subsequently reuniting sea snake populations. Quite how so many different species manage to coexist in virtually the same habitat without over-exploitation of resources is an intriguing question.

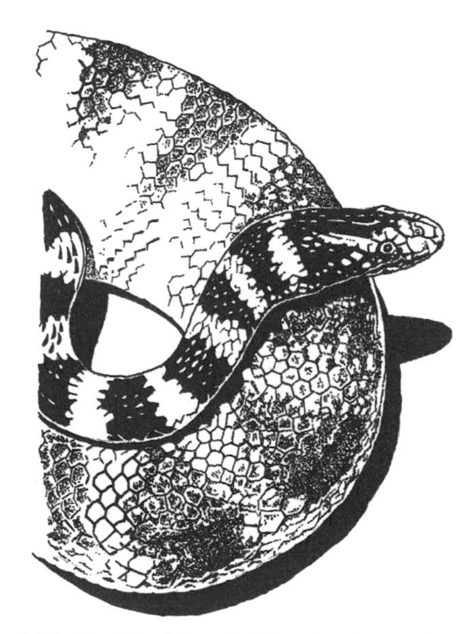

Figure 262. The Black-headed Sea Snake *Hydrophis coggeri* has a tiny head for reaching into crevices whilst foraging for prey.

Studies have categorised the various species of sea snakes according to depth and/or feeding preferences. The only truly pelagic (surface feeding) sea snake is *Pelamis*. Among the remaining species, in any one area, there are often distinct shallow-water, intermediate and deep-water specialists. Feeding categories include eel-eaters, burrowing goby-eaters, goby-eaters, egg-eaters, catfish-eaters and generalists.

The prey diversity of most sea snake species is low with a single prey species comprising as much as 50% of the diet. However, generalist feeders seem capable of exploiting a wide spectrum of prey. One example is *Lapemis*, in which 21 fish families, as well as invertebrates such as cuttlefish and squid, have been recorded in its diet.

Sometimes sea snakes occur in almost incredible numbers. In 1932 millions of intertwined sea snakes (*Astrotia stokesii*, plate 24) in a line three-metre wide and 100km long, were recorded in the Malacca Strait.

While the majority of sea snakes live in salt water, at least some are tolerant to brackish or even freshwater. A few species that are commonly found around estuaries sometimes make their way up rivers and one species occurs in Lake Taal in the Philippines. In the brackish Lake Te-Nngano in the Solomon Islands there are two species of sea krait: *Laticauda crockerii*, an endemic found only in this lake,) and *Laticauda colubrina*, a widespread species found in marine habitats through much of South-East Asia and the western Pacific (plate 24).

VENOM

Sea snakes are regarded as being among the most potentially dangerous snakes on account of their highly toxic venom (for example the Beaked Sea Snake *Enhydrina schistosai* can produce a lethal dose sufficient to kill 53 people). It seems likely that this is related to their need to rapidly subdue relatively resistant prey such as eels. However, human fatalities are not as common as these facts may imply. Although a few species have a reputation for being ill-tempered, others rarely bite even when severely provoked and fish egg-eating specialists are not regarded as dangerous because they produce tiny quantities of weak venom. Most fatal bites occur among South-East Asian fishermen who wade in muddy estuaries where sea snakes are common, treading on them accidentally or being bitten repeatedly when extracting the snakes from nets.

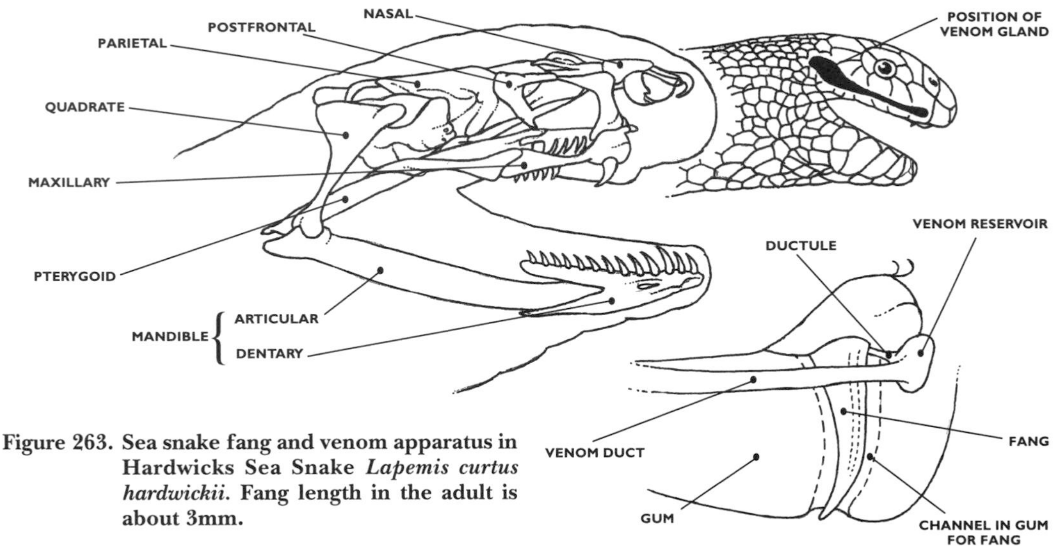

Figure 263. Sea snake fang and venom apparatus in Hardwicks Sea Snake *Lapemis curtus hardwickii*. Fang length in the adult is about 3mm.

BIBLIOGRAPHY

Dunson, W. A. (ed.) 1975. *The Biology of Sea Snakes.* University Park Press, Baltimore and London.

Ernst, C. H., Lovich, J. E. & Barbour, R. W. 1994. *Turtles of the United States and Canada.* Smithsonian Institution Press, Washington.

Groombridge, B. 1982. *The Amphibia-Reptilia Red Data Book, Part I: Testudines, Crocodylia, Rhynchocephalia.* IUCN, Switzerland.

Halstead, B. W. (ed.) 1970. *Poisonous and Venomous Marine Animals of the World.* Government Printing Office.

Heatwole, H. 1987. *Sea Snakes.* New South Wales University Press.

Mrosovsky, N. 1983. *Conserving Sea Turtles.* British Herpetological Society, London.

Perry, R. (ed.) 1984. *Key Environments: Galapagos.* Pergamon Press, London.

Pritchard, P. C. H. 1979. *Encyclopedia of Turtles.* TFH Publications, New Jersey.

Ross, C. A. (ed.) 1992. *Crocodiles and Alligators.* Blitz Editions, Leicester.

Turtle Factsheet

This factsheet shows the structure of the shell and the variation in head shape.

PARTS OF A SHELLED TURTLE

The turtle shell consists of two parts, an upper carapace and a lower plastron. These are illustrated in the Loggerhead Turtle below.

The plastron of Ridley Turtles differs from all others in the presence of inframarginal pores. This is an important diagnostic feature - it is sometimes difficult to distinguish the Loggerhead (above) from Kemp's Ridley (below).

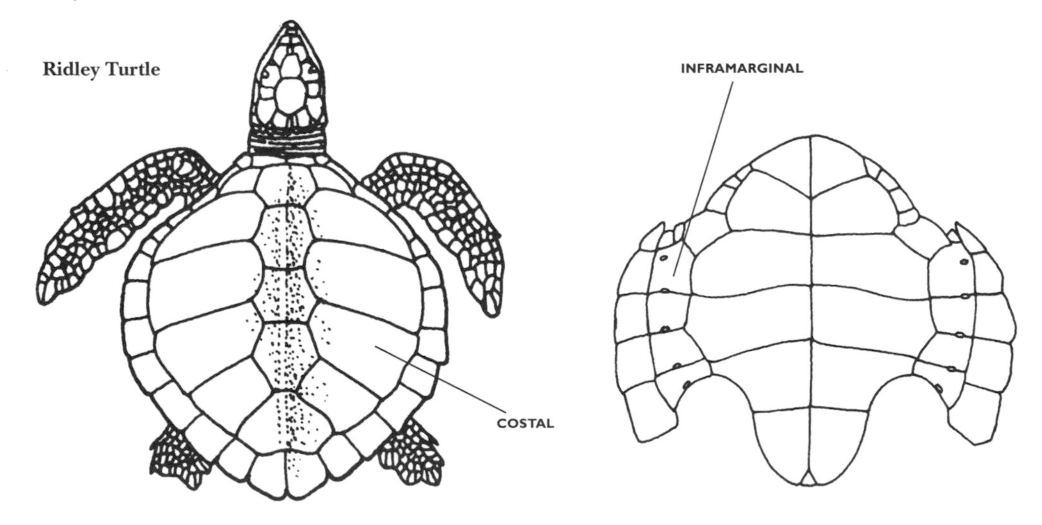

The Leatherback Turtle differs from all other sea turtle species in that the carapace and plastron lack horny shields but are covered with leathery skin.

HEADS OF SEA TURTLES

In cases where shells are damaged or covered with encrusting organisms, useful identification features can be found by examining the heads.

Between the eyes there are variations in the size and number of scales; these are called prefrontal scales.

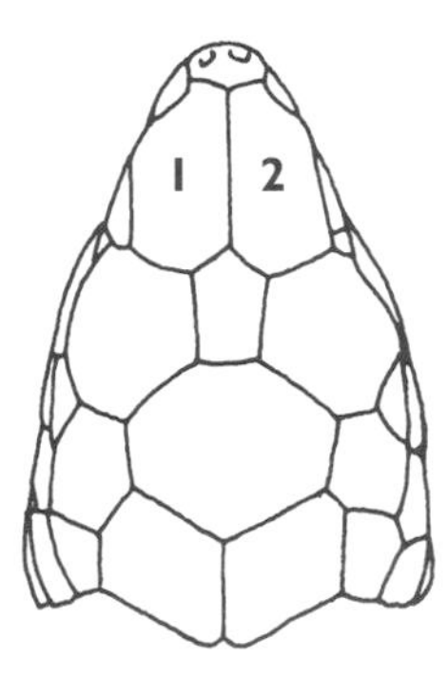

Green Turtle Head
Chelonia mydas
2 Prefrontals
Nostrils on beak

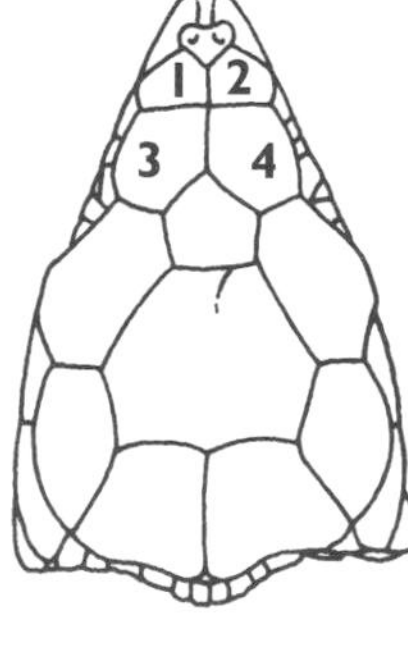

Hawksbill Turtle Head
Eretmochelys imbricata
4 Prefrontals
Nostrils on separate scale

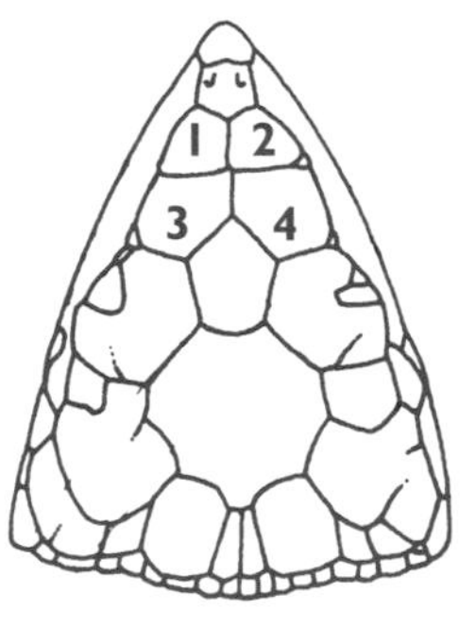

Kemp's Ridley Turtle Head
Lepidochelys kempi
4 Prefrontals
Nostrils on separate scale

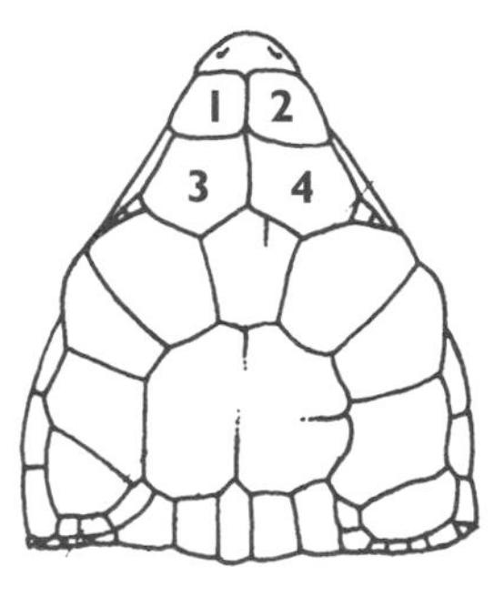

Loggerhead Turtle Head
Caretta caretta
4 or more Prefrontals
Nostrils on beak

Additional useful features are shape of the jawline, position of the nostril and the overall dimensions of the head. These features are shown in the side views below.

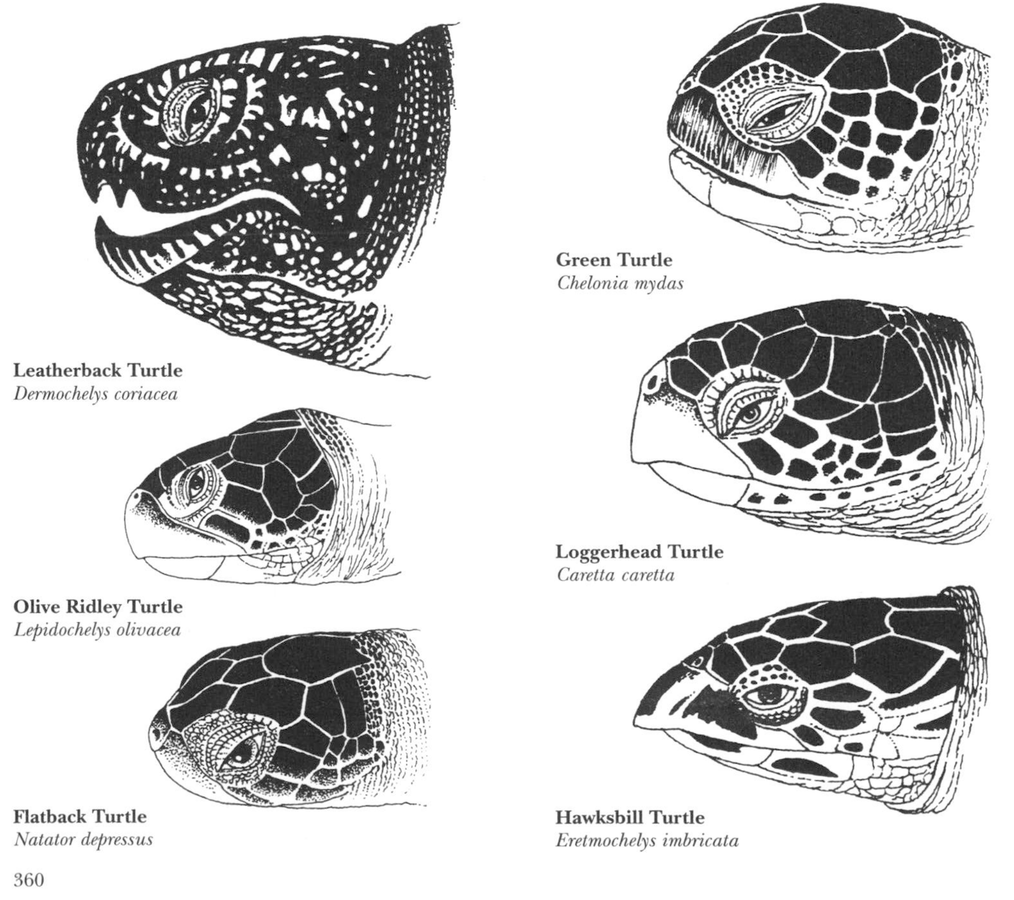

Leatherback Turtle
Dermochelys coriacea

Green Turtle
Chelonia mydas

Olive Ridley Turtle
Lepidochelys olivacea

Loggerhead Turtle
Caretta caretta

Flatback Turtle
Natator depressus

Hawksbill Turtle
Eretmochelys imbricata

SEABIRDS

by

Iain Robertson & Michael Burchett

INTRODUCTION

Figure 264. Adelie Penguins *Pygoscelis adeliae* diving off the ice edge.

Birds constitute the class Aves and with about 9000 living species are the second most numerous group of vertebrates after fishes. All birds possess a unique feature, the presence of feathers, which is a diagnostic characteristic of the class. The class is traditionally divided into 29 orders, seven of which contain all the world's seabirds. In common with most reptiles, all birds lay eggs and most birds have the power of flight, although this has been lost in the order Sphenisciformes (penguins) and in a few species of Podicipediformes (grebes) and Pelecaniformes of the family Phalacrocoracidae (cormorants).

The definition of a 'seabird' is somewhat imprecise; in this work it is taken to include those species of birds which are wholly or largely dependent upon the marine environment, together with some closely related species which are generally associated with freshwater habitats but occasionally occur in coastal waters.

Those seabirds which can be considered to be exclusively marine belong to the families **Diomedeidae** (albatrosses), **Procellariidae** (shearwaters and petrels), **Hydrobatidae** (storm-petrels), **Pelecanoididae** (diving-petrel), **Phaethontidae** (tropicbirds), **Sulidae** (gannets and boobies), **Fregatidae** (frigatebirds) and **Alcidae** (auks). The remaining seabirds are: the **Gaviidae** (divers or loons) which are largely marine during the non-breeding season and the **Podicipedidae** (grebes) which are mainly confined to freshwater but a few species regularly utilise the marine environment during the non-breeding season. Most of the **Phalacrocoracidae** (cormorants) are largely marine, indeed some are exclusively so, but a few species are primarily associated with freshwater as are the **Pelecanidae** (pelicans). The **Anatidae** (ducks and geese) include a number of species such as eiders and scoters which have become secondarily marine, as well as some species which occur in salt water habitats during the non-breeding season. The vast majority of this large family, however, are freshwater species. Among the **Scolopacidae** (shorebirds) only two species of phalarope are pelagic outside the breeding season but most of this large family will utilise the littoral environment at times. The **Chionididae** (sheathbills) mainly inhabit the littoral zone though some make long migratory flights across the oceans. The **Stercorariidae** (skuas/jaegers) are primarily marine though some species are more terrestrial during the breeding season. The majority of **Laridae** (gulls) and **Sternidae** (terns) are marine though a few species are largely associated with freshwater. The **Rynchopidae** (skimmers) are largely freshwater birds but they regularly feed in brackish and inshore waters.

Amongst the other types of bird which are sometimes associated with coasts are some species of flamingo, many herons and egrets, certain raptors notably Osprey *Pandion haliaetus*, Black Kite *Milvus migrans* and sea-eagles, and several species of crows. These are not considered in this work.

The sea covers about 70% of the earth's surface but seabirds comprise only 3.8% of the world's bird species. However, many of these species are extremely numerous and consequently play a major role in the ecology of the marine environment.

TAXONOMY

Seabird species are arranged by Order and Family; the more traditional classification is as follows:

			No. of species
ORDER SPHENISCIFORMES			
	Family		
	Spheniscidae	Penguins	17
ORDER GAVIIFORMES			
	Family		
	Gaviidae	Divers or Loons	5
ORDER PODICIPEDIFORMES			
	Family		
	Podicipedidae	Grebes	22
ORDER PROCELLARIIFORMES			
	Family		
	Diomedeidae	Albatrosses	14
	Procellariidae	Petrels and Shearwaters	75
	Hydrobatidae	Storm-Petrels	21
	Pelecanoididae	Diving-Petrels	4
ORDER PELECANIFORMES			
	Family		
	Phaethontidae	Tropicbirds	3
	Pelecanidae	Pelicans	8
	Sulidae	Gannets and Boobies	9
	Phalacrocoracidae	Cormorants and Shags	36
	Fregatidae	Frigatebirds	5
ORDER ANSERIFORMES			
	Family		
	Anatidae (part)	Ducks and Geese	22
ORDER CHARADRIIFORMES			
	Family		
	Scolopacidae (part)	Shorebirds	2
	Chionididae	Sheathbills	2
	Stercorariidae	Skuas or Jaegers	8
	Laridae	Gulls	50
	Sternidae	Terns	44
	Rynchopidae	Skimmers	3
	Alcidae	Auks	22
	Total number of seabird species		**372**

An alternative treatment based on DNA x DNA hybridisation techniques has been proposed by Sibley & Ahlquist (1990) as follows:

ORDER ANSERIFORMES

Family
Anatidae

ORDER CICONIIFORMES

Family
Scolopacidae
Chionididae
Laridae
Podicipedidae
Phaethontidae
Sulidae
Phalacrocoracidae
Pelecanidae
Fregatidae
Spheniscidae
Gaviidae
Procellariidae

The general characteristics of each family are given in the Factsheet and specific information is included in the plate captions.

While there are two views of taxonomy at order and family level, there are also differing views on taxonomy at the species level. The situation is not helped by the fact that the definition of a 'species' is still not fully resolved. Two methods generally used by taxonomists are the biological species concept, where species are defined as "groups of actually or potentially interbreeding natural populations which are reproductively isolated from other such groups", and the phylogenetic species concept which considers that "a species is the smallest diagnosable cluster of individual organisms within which there is a parental ancestry and descent". Those who follow the biological species concept agree on about 9000-9500 different species of birds whilst followers of the phylogenetic concept would recognise about 25,000 species.

Modern taxonomic studies include the analysis of tape recordings of vocalisations and DNA comparison techniques which were not available to earlier workers in the field. Some of the most problematical and controversial groups are amongst the *Pterodroma* petrels, *Puffinus* shearwaters, *Catharacta* skuas and the larger *Larus* gulls, particularly the Herring Gull/Iceland Gull/Lesser Black-backed Gull complex.

In recent years there have been great advances in identification techniques which has led to more and better trained observers. An increasing interest in seabirds (and ornithology in general) has encouraged studies by amateur and professional ornithologists with the result that there is now a much greater understanding of many species.

DISTRIBUTION AND MOVEMENTS

Seabirds are not evenly distributed throughout the oceans. The greatest numbers of both individuals and different species are found in the most productive areas. Cold waters are far more productive than warmer waters and this is reflected by relative abundance of seabirds in the temperate and subpolar regions compared with the tropics and subtropical zones.

Regional distribution of species

South Pacific	128	North Pacific	107
South Atlantic	73	North Atlantic	74
Indian Ocean	73	Mediterranean	24
Antarctica	44	Arctic	31
	318		**236**

Some of the greatest concentrations of seabirds are found in areas where the rich cold water currents from the Antarctic flow northwards along the west coasts of South America and South Africa. The natural upwelling areas of the oceans also make these areas particularly rich in food supply and therefore also rich in birds. The North Pacific is another highly productive area with a great diversity of species, particularly auks.

Migration is generally a feature of the higher latitudes. For many Antarctic species the rigours of the winter are avoided by moving north. A few species such as Adelie Penguin, and Snow and Antarctic Petrels migrate to the edge of the pack ice, whilst most move to latitudes between the subantarctic and subtropical convergences. Some species are transequatorial migrants; these include Sooty Shearwater *Puffinus griseus*, Great Shearwater *Puffinus gravis* and Wilson's Storm-Petrel *Oceanites oceanicus* which reach the waters off southern Greenland during the non-breeding season. The Short-tailed Shearwater *Puffinus tenuirostris* makes a figure-of-eight migration from its breeding grounds in southeast Australia, northwards to Japan and Sakhalin, across the Bering Sea to Alaska, then south off the west coast of North America and back across the south Pacific in the region of Oceania.

In the opposite direction the Arctic Tern *Sterna paradisea* leaves its northern breeding grounds and can travel as far south as the edge of the Antarctic pack ice during the austral summer. Long-tailed Skuas *Stercorarius longicaudus* and Sabine's Gull *Xema sabini* leave their high Arctic breeding grounds to winter off the coasts of western South Africa and South America. Many species have much less spectacular migrations; Common Guillemots *Uria aalge* and Razorbills *Alca torda* from British breeding colonies disperse to the coasts of France and Spain. Amongst those species of seabird which breed inland, particularly gulls, movements tend to involve a general dispersal to the coast outside the breeding season.

The speed at which migration takes place can be very impressive; a Short-tailed Shearwater flew 16,000 kilometres from Australia to the Bering Sea in six weeks, while a Manx Shearwater *Puffinus puffinus* was recovered in Brazil only 16 days after being ringed at a colony in Wales, giving it a minimum speed of 740 kilometres per day.

Not all seabird movements are migratory, and some species make considerable journeys to feed. Manx Shearwaters from colonies in Wales regularly visit the Bay of Biscay on feeding trips while gannets, auks and kittiwakes frequently travel at least 150 kilometres from the breeding colonies in search of food.

The movements of many seabirds are still largely unknown, particularly those species which spend the non-breeding season far out at sea. Recent studies using satellite tracking of Wandering Albatrosses fitted with small radio transmitters have produced extraordinary results. Immature albatrosses travel thousands of kilometres over the open oceans, only returning to land when they reach maturity. Furthermore, the sexes appear to circumnavigate the southern oceans in different directions.

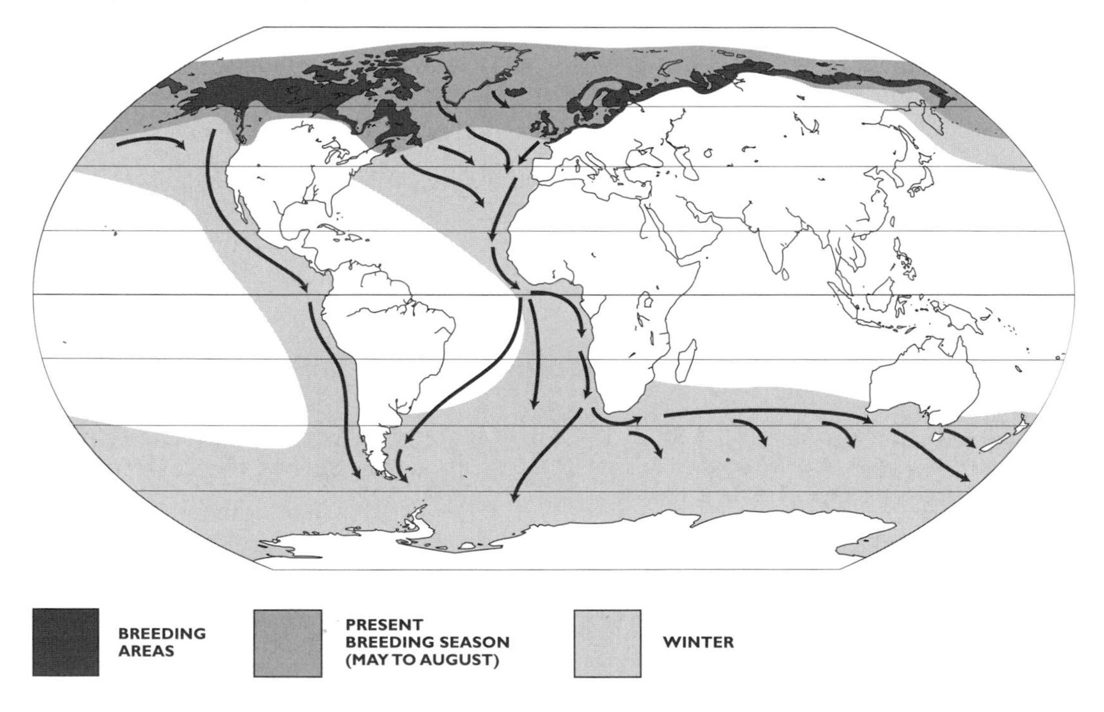

Figure 265. Distribution and migration routes of the Arctic Tern *Sterna paradisaea*.

PHYSIOLOGY AND ADAPTATIONS

The development of true flight enables birds to range far and wide in search of suitable food and habitats. Flight opens the whole planet to birds and the envelope of air with its associated wind systems provides direct flight paths around the Earth. Birds are the most successful group of vertebrates in terms of species numbers to inhabit the land, sea surface and air and they are found from the tropics to the polar regions.

Birds probably evolved from bipedal diapsid ancestors that used their hind legs for walking (unlike quadrupeds that use all four legs). This enabled them to develop two independent and entirely separate movement systems, namely; posterior legs which are adapted for walking, swimming or both; and wings which are adapted for flight. Some birds such as penguins (Spheniscidae) have lost the ability to fly and their wings are modified as flippers for underwater swimming. Many birds have legs that are further modified for grasping (e.g. perching passerine birds) or food capture and manipulation (e.g. Steller's Sea Eagle *Haliaeetus pelagicus*).

At first sight, birds appear to vary both in size and shape and exhibit many types of bills, feet and plumages. However, a closer look will reveal a far more uniform morphology, anatomy and physiology compared to mammals. These similarities are a reflection of the restrictive requirements needed for flight and relatively few flying species deviate from stable aerodynamic designs. Birds are able to fly as a result of several evolutionary developments including a powerful heart and high metabolic rate, a unique breathing system, light, hollow bones, feathers, warm bloodedness (endothermy), a streamlined body form and elongated wings with powerful pectoral muscles (table 9). These flight modifications can be considered under two headings: 1) those that help reduce weight and 2) those that enhance power. All birds must have a high power to weight ratio if they are to fly. In birds that have lost the ability to fly, keeping weight down is not as important.

Weight-reducing adaptations	**Power-promoting adaptations**
Thin, hollow bones	Warm bloodedness and high metabolic rate
Extremely light feathers	Plumage for insulation
Elimination of teeth and heavy jaws	An energy-rich diet and high blood sugar levels
Elimination of some bones and extensive fusion of others	Rapid and efficient digestion
A system of branching air sacs	Highly efficient respiratory system
Absence of sweat glands	Air sacs for efficient cooling during muscular flight activities
Oviparity rather than viviparity	Breathing movements synchronised with wing beats
Atrophy of gonads in non-breeding periods	Large heart and high pressure circulation
Eating energy-concentrated foods	
Efficient and rapid digestion	
Excretion of uric acid instead of urea	

Table 9. Avian adaptations for flight (After Welty 1975).

SKELETON

The skeleton of a bird is essentially an animal airframe which combines strength with lightness (fig.266). In some sections of the skeleton elasticity is required while in other parts rigidity is more important. Since most birds walk, swim or dive with their hind limbs and fly with their front ones, the rearrangement of bones to cope with these mechanical stresses is seen principally in the pectoral (breast and shoulder) and pelvic (hip) regions. These regions are known as the pectoral and pelvic girdles respectively and they function to support the limbs and to provide anchorage points for the muscles that power them. Although girdle bones are large, they are not heavy and the thin bones are strengthened by ridges to prevent distortion when they are stressed by muscles working under load. The pectoral girdle is firmly attached to the breastbone, therefore the body is efficiently suspended from the wings during flight. This is made possible by the extra development of the coracoid bones which run from the upper side of the keel (sternum) nearly to the shoulders. The coracoids are essentially compression struts that not only bridge the pectoral girdle and wings, but stop the whole structure collapsing from the compressive forces generated by the powerful wing muscles (pectoral muscles). The partially fused, elastic wishbones (furcula) that are situated in front of and between the coracoids, help to prevent the chest region collapsing, especially during flight. The pelvic girdle is similarly strengthened and designed so that the legs can effectively carry the weight of the bird while on the ground, perched, climbing or on the water. The pelvic girdle must also be able to withstand the stresses of take-offs and landings and therefore function as an efficient shock-absorption system.

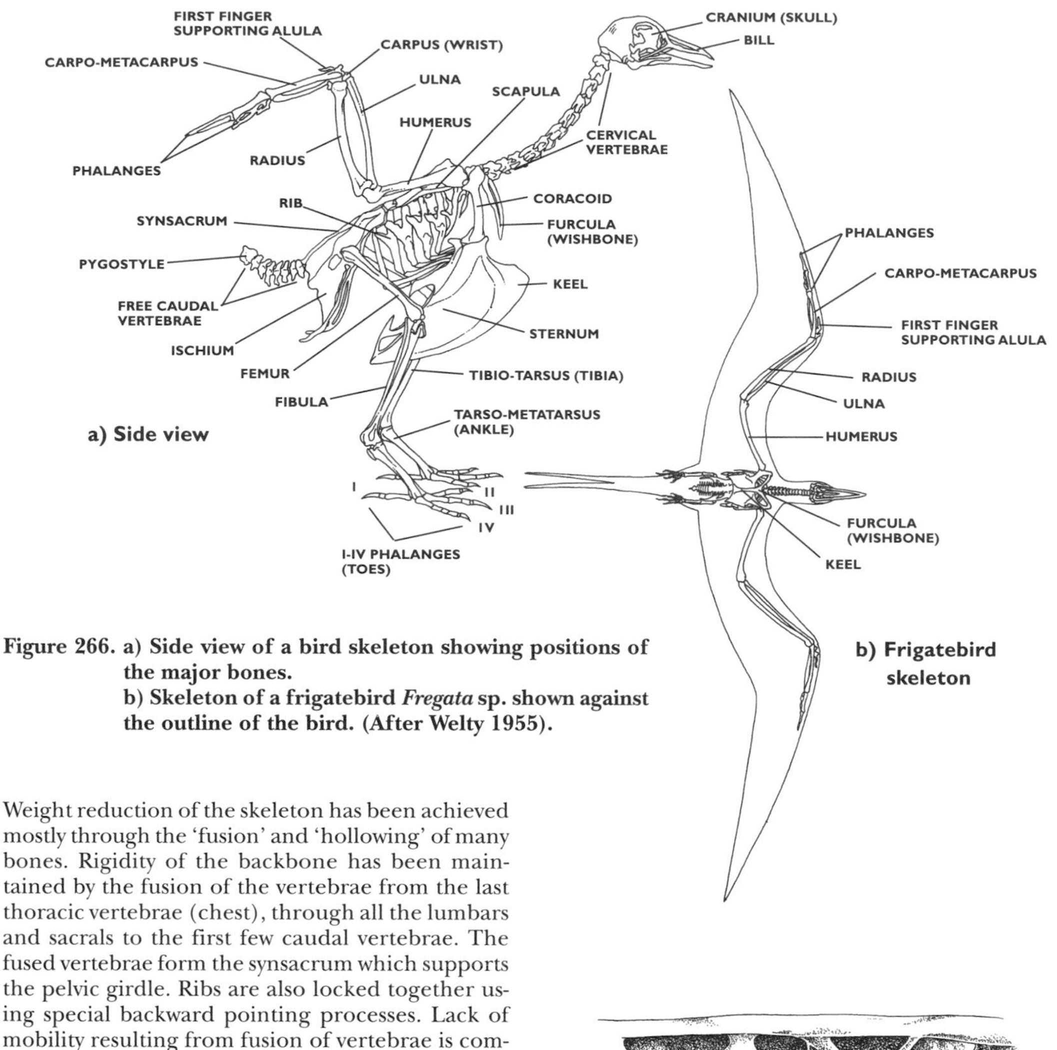

**Figure 266. a) Side view of a bird skeleton showing positions of the major bones.
b) Skeleton of a frigatebird *Fregata* sp. shown against the outline of the bird. (After Welty 1955).**

Weight reduction of the skeleton has been achieved mostly through the 'fusion' and 'hollowing' of many bones. Rigidity of the backbone has been maintained by the fusion of the vertebrae from the last thoracic vertebrae (chest), through all the lumbars and sacrals to the first few caudal vertebrae. The fused vertebrae form the synsacrum which supports the pelvic girdle. Ribs are also locked together using special backward pointing processes. Lack of mobility resulting from fusion of vertebrae is compensated by a highly articulated neck and many birds can rotate the head and neck through 180°.

Many of the long-bones of the legs and wings are hollowed or honeycombed, or moulded into curved, thin plates in such a way that they are strong enough for the tasks they have to perform. A tubular structure is not only lighter but stronger than a solid rod of equivalent size. Further internal strengthening of hollow bones is possible through the formation of struts arranged in triangular formation (fig.267). 'Triangulation' is a well known engineering principle to gain maximum strength and rigidity from a minimum amount of bracing. Many large, soaring birds (e.g. albatrosses, Diomedeidae and gannets, Sulidae) have hollow long-bones and their internal spaces are extensions of the air sacs. The skull of a bird is another area that is highly modified to form a light, streamlined structure. This is necessary as a heavy weight at the end of a long, thin neck would require stronger, larger and therefore heavier muscles and tendons to hold

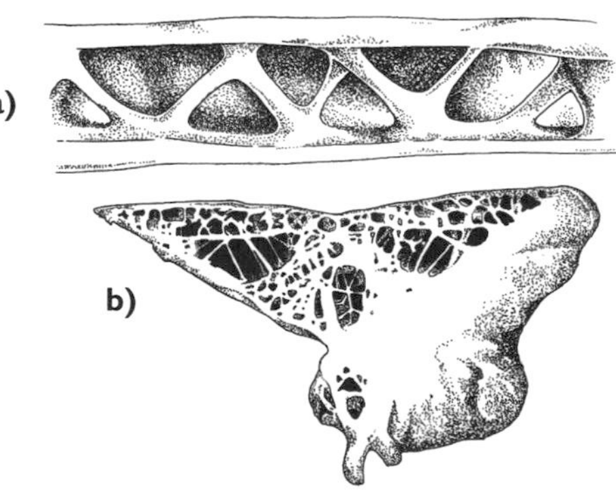

Figure 267. a) Longitudinal section showing the internal structure of the metacarpal wing bone of a large soaring bird. The triangular arrangement of struts allows great strength and rigidity with a minimum amount of bracing. b) Longitudinal section through the frontal bone of a bird skull to show the internal hollowing and bracing. (After Welty 1955).

it in position. To decrease weight in the head region, the bill replaces teeth and many bones are once again fused and are thin or hollow (fig.267).

The leg skeleton of birds is much modified from the basic vertebrate plan. The bones of the lower leg and tarsal bones of the foot are elongated and fused to provide an additional joint, the tarso-metatarsus. Usually, the toe flexor muscles originate above the knee and their tendons pass in front of the knee, behind the ankle and beneath the toes. When a bird (e.g. a sea eagle) lands on a perch and the legs are flexed, the animal's own body weight automatically closes the toes, enabling it to perch passively without having to use the muscular energy of contraction. In addition, small projections on the tendons function like ratchets and grip the tendon sheaths to prevent slipping. These arrangements ensure that the toes grip tightly to a perch even when the bird is asleep. Many seabirds are unable to perch because of their webbed feet.

MUSCLES

Weight for weight, bird muscles are able to generate more power compared to the muscles of most other vertebrates. This is only made possible through their higher body temperatures (muscles work more efficiently when warm) and elevated metabolic levels. If wing surfaces and their associated covering of specialised feathers give birds their lift and motive force, then muscles provide the power, equivalent to an avian engine. Muscles that move the wings up and down are large and account for 20-30% of a bird's total body weight. These muscles are especially well developed in smaller birds that fly mostly by flapping. Aerodynamic stability is achieved if the centre of gravity of a flying bird is at or below the level of the wings. For this reason the bulk of the muscle used to elevate the wings is positioned below them.

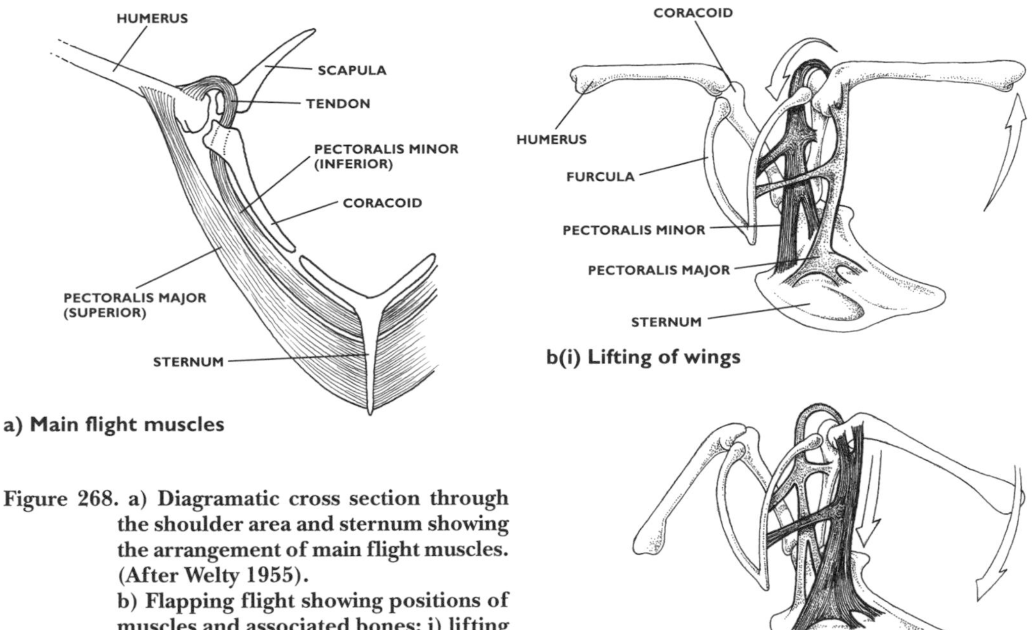

Figure 268. a) Diagramatic cross section through the shoulder area and sternum showing the arrangement of main flight muscles. (After Welty 1955). b) Flapping flight showing positions of muscles and associated bones: i) lifting of wings and ii) downbeat of wings. (After Beazley 1976).

The powerful downward depression of the wings is produced by the massive 'pectoralis major' muscles which are attached to the humerus bone. At the other end they form the outer layer of muscles on the keel bone which in turn projects down from the sternum (fig.268). The pectoralis muscles are often dark red in colour as they are packed with myoglobin, a pigment which has a high oxygen affinity. Oxygen is therefore immediately available for the muscles to use when active. Elevator muscles (pectoralis minor or supracoracoideus) which raise the wings are situated next to the pectoralis majors on the keel. This second set of muscles end as tendons that pass dorsally over the shoulder joints and attach onto the upper side of the humerus bones. Through a pulley action, elevator muscles that are positioned well below the wings are able the raise them during flight. Other arm muscles operate to fold or extend the wings and alter their positions and attitudes during flight. The positioning of individual flight feathers is controlled by an elaborate system of muscles and tendons running along the back of the hand. Albatrosses have breast muscles that are greatly reduced in size but instead, well developed tendons and ligaments hold the wing in position for long periods of gliding with little muscular effort.

AERODYNAMICS

A wing that flaps is a complex structure as it provides lift, motive force and stability all at the same time. Therefore an understanding of fundamental aerodynamics is required to gain an appreciation of bird flight. In soaring birds that maintain their wings fully extended with little flapping, the aerodynamics can be likened to the fixed wing of an aircraft. An object with a large surface area that moves through the air at an inclined angle is called an aerofoil. Although a bird's wing moves forward through the air, it is easier to think of it as stationary with the air flowing past it (as it would be in a wind tunnel). If the upper surface of a contoured wing is convex and the underside is concave, a 'cambered' aerofoil is created. A cambered aerofoil causes the air pressure flowing over the two surfaces to become unequal because the air has to travel quicker and farther over the top surface relative to the ventral surface (fig.269). The result is a reduction of air pressure over the top of the wing which, in turn, produces a lifting force. Drag is another important force that tends to slow down or stop motion in any direction. Both lift and drag forces are proportional to the square of the speed, therefore for sustained flight in still air, a bird or aircraft needs to generate sufficient speed to produce a lifting force equal to its own weight before it becomes airborne.

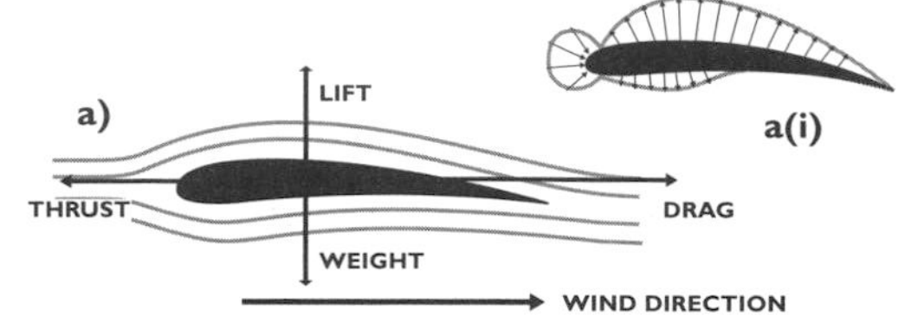

a) Wing level to show forces acting on the aerofoil; a(i) arrows indicate pressure levels.

b) Wing angle of attack is too high for a given air speed. Turbulence causes air to break away from the wing's top surface and it stalls.

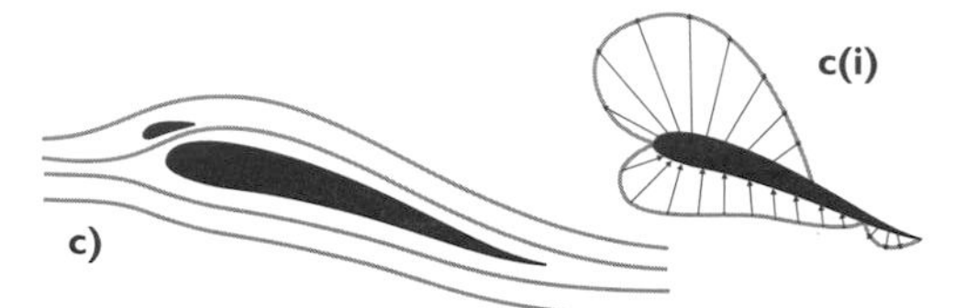

c) Laminar flow is restored by placing a slot on the leading edge of the wing; c(i) arrows indicate increased pressure differences.

Figure 269. Effect of airflow passing an aerofoil wing under various conditions.

If a wing is inclined slightly upwards (increasing the angle of attack) the pressure differences can be further increased to provide extra lift which is useful if a body is flying at lower air speeds. However, if the angle of attack is increased beyond the critical point (in birds, between 15° and 20°) for a given air speed, the air flow over the top of the wing ceases to be laminar (smooth) and turbulence is created. When this happens, air breaks away from the surface of the top wing and the lifting force is destroyed, causing the whole structure to lose altitude. A bird or aircraft will also lose altitude when its speed falls below a critical point and the aerofoils no longer produce sufficient lift to overcome the weight of the flying body. The smooth flow of air over the top surface of the wing can be restored at high angles of attack by placing a slot on the leading edge of the wings. Large aircraft often have forewing slots that can be opened or closed to assist take-offs and landings at reduced air speeds. In birds (especially larger ones) each forewing has a leading 'alula' (bastard wing) to prevent the wings from stalling at high angles of attack, a necessary adaptation to keep them operating at low air speeds. Primary flight feathers also function as a series of independent, overlapping aerofoils, each smoothing the flow of air over the one behind.

Another characteristic of an aerofoil is the outward flow of air from under the wings and an inward flow over them (fig.270). At the tips of the wings, 'vortex' eddies form which is known as induced-drag. As the air speed increases, little vortices along the trailing edge of the wings move outwards towards the tips and join the larger wing-tip vortices. Long, thin wings reduce this effect as the vortex disturbances at the tips are widely separated and there is proportionally more wing area over which the air can flow smoothly. The slot principle also works at the wing tips of soaring birds and at low air speeds they prevent the wing tips stalling. Resistance due to 'form' (profile) drag occurs when the body of the bird (equivalent to the fuselage of an aircraft) forces its way through an oncoming air stream. This is reduced as far as possible through the streamlined body having a pointed anterior end (e.g. head region), a long, cylindrical torso and a thin, tapering posterior. Feet are often trailed aft, folded in or tucked away to enhance the streamlined profile of the animal.

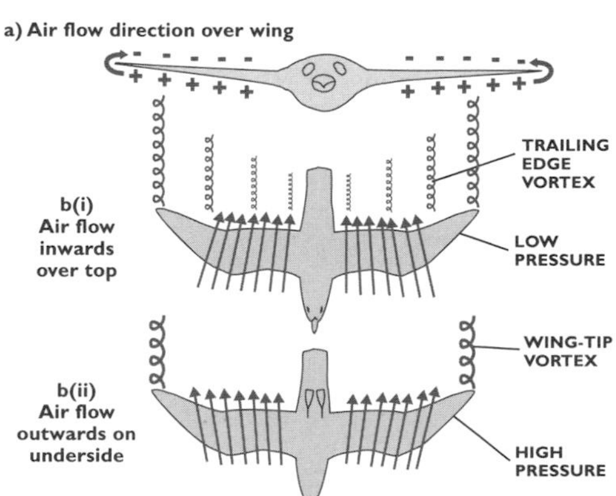

Figure 270. a) Two airflow directions- outwards underneath and inwards over the top of the wing. b) Cross flowing air streams cause trailing edge vortices which move outwards towards the wing tips to form longer vortices.

WINGS

Wing shape is important in determining flight capabilities. Different shapes of wings are suitable for different types of flight and the major factors to consider with wing shapes are:

a) wing area and loading
b) aspect ratio (wing length/breadth)
c) wing taper and outline
d) wing camber
e) slots (gaps between individual flight feathers)

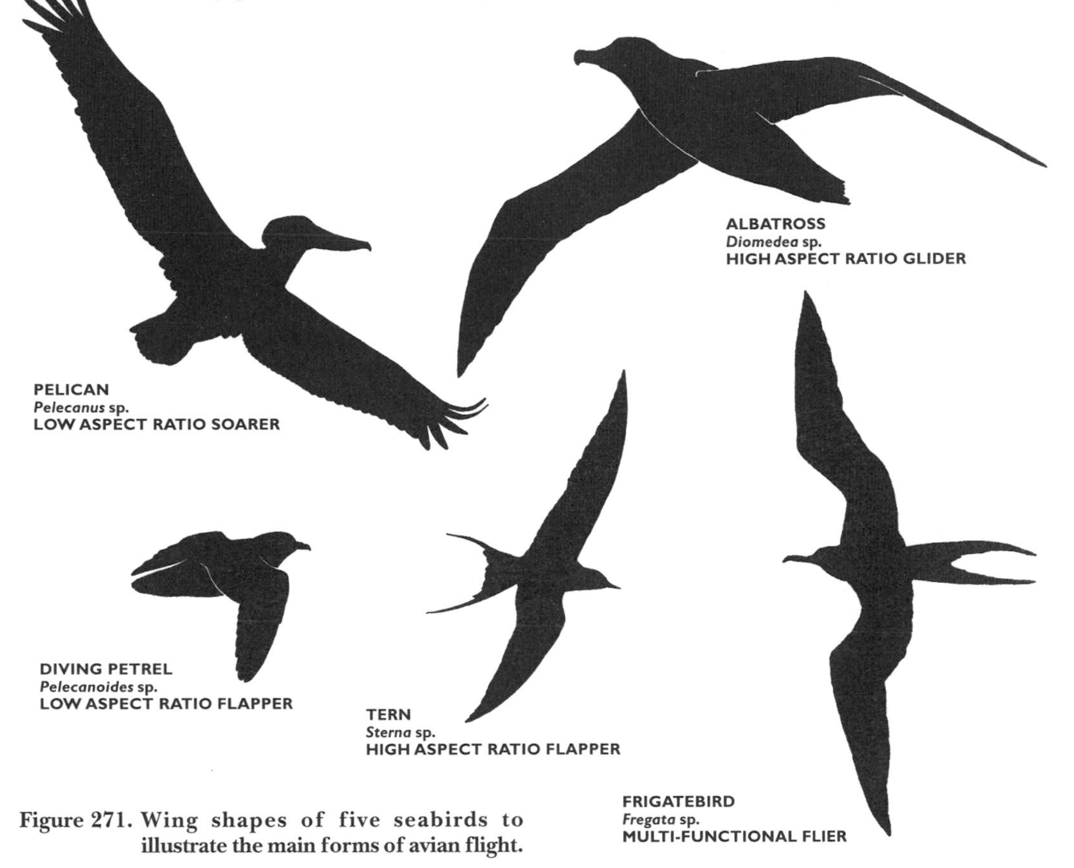

Figure 271. Wing shapes of five seabirds to illustrate the main forms of avian flight.

A relatively small wing area is required for high speed flight. For birds that produce their motive force by 'flapping', a small body size is advantageous for fast flight. By contrast, a large wing area enables birds to fly more slowly and provides proportionally more lift for the larger, heavier birds such as pelicans (Pelicanidae). Therefore larger birds tend to have larger wings compared to smaller birds. 'Wing-loading' (span loading) is the weight of the bird divided by the surface area of the wings and this in turn has an effect on induced drag. Essentially the lighter the wing loading, the less the power that is required to sustain flight.

There are four basic forms of avian flight:

1) forward flapping flight (e.g. auks, Alcidae and diving petrels, Pelecanoididae)
2) hovering by flapping (e.g. terns, Sternidae)
3) static soaring (e.g. pelicans, Pelecanidae)
4) dynamic soaring (e.g. albatrosses, Diomedeidae and gannets, Sulidae)

Frigatebirds, Fregatidae are unusual in that they exhibit several different forms of flight.

Birds that spend little time flapping usually gain altitude by soaring or gliding and maintain height using three types of air movements - ascending air currents (e.g. updraughts or rising thermals); variations in wind velocities at one particular level (e.g. wind gusts) and differences in wind velocities at various heights. A 'soaring' wing has a broad, highly cambered profile that generates large amounts of lift and performs well at low flying speeds. However, low aspect wings produce more drag which reduces the air speed. With broad, soaring wings the birds weight is spread over a large surface area, therefore the wing loading is comparatively low. This gives a slower sinking rate (best gliding angle) in still air. Providing these birds find air rising faster than their sinking speed, they will gain height with minimum effort. Slots on both the pinions and the forewings (thumbs) help to keep the air flow laminar and this enables the wings to fly at low air speeds. This is a most important factor at the wing tips, as soaring birds need to be able to perform tight turns to keep inside the thermal envelope. When a bird executes a tight turn, the inside wing will be moving more slowly than the outside one, and the slotted pinions help to prevent the inner aerofoil stalling. Surfaces such as the tail are often broad and manoeuvrable in order to control stability and promote lift at low air speeds. Gusty wind conditions are favoured by manoeuvrable fliers such as gulls (*Larus* spp.) that use the ever changing pressures and eddies to provide lift and forward momentum.

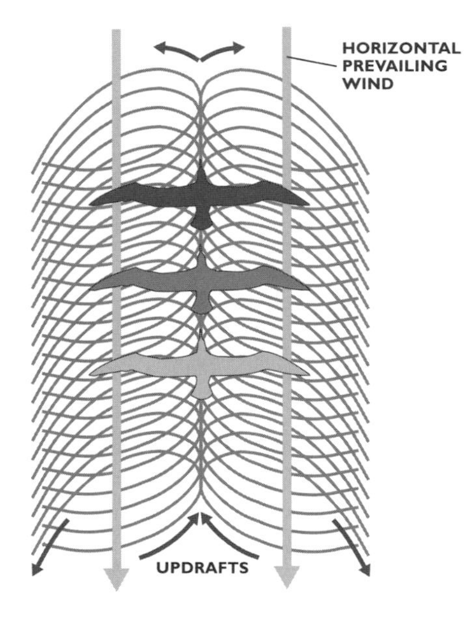

Figure 272. Seabirds can sometimes soar on updrafts (small arrows) of prevailing winds (large arrows). These cylinders of warmer air rise from the surface of the sea. If winds increase in strength, the cylinders lie parallel to the sea surface. (After Storer 1952).

Many birds that fly fast often make prolonged migrations, attack prey during flight or have heavy wing loadings as part of an activity (e.g. diving into water from height). These birds often have high aspect ratio wings (e.g. albatrosses 25:1, gulls 11:1, puffins 10:1). High aspect ratio wings have evolved similar (convergent) shapes. There is reduced camber (flatter wings) and the wings are long, narrow and swept back to a tapering point. Air resistance (drag) is reduced due to a smaller wing surface area. This wing design is similar for both fast flapping birds (e.g. puffins) and those that glide fast enough to obtain enough kinetic energy to convert into altitude (e.g. albatrosses, gannets and loons). Slotted pinions are less evident in the high aspect ratio wings of large oceanic gliding birds as they have little need to move in tight, spiralling circles.

Although high aspect ratio/low area wings are aerodynamically efficient with low sink rates, they stall at relatively low speeds. Consequently, in calm air, large soaring and gliding birds find it difficult to generate sufficient lift to take off. Because of their size and inertia, large wings are also tiring and energy consuming to flap, therefore large birds are quickly exhausted. Large birds tend to stay on the ground or water and only become airborne when conditions are right. It is impossible for large albatrosses to take off from the water in still air conditions. Not surprisingly, many of these birds and some large petrels (e.g. Southern Giant Petrel *Macronectes giganteus*) inhabit the oceanic Trade Wind belts or are associated with the windswept, stormy regions of the southern oceans.

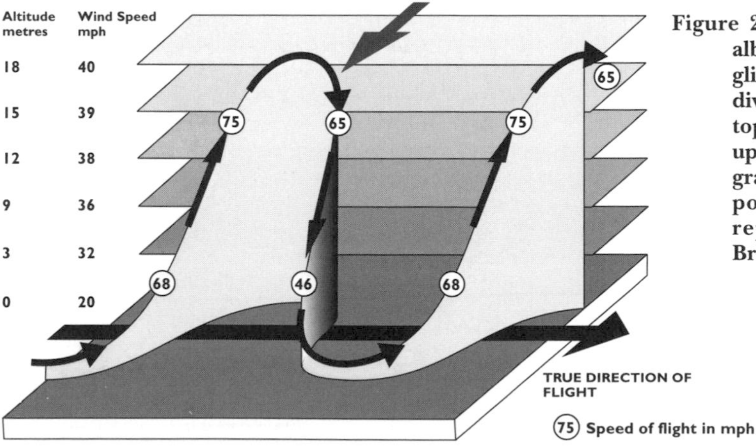

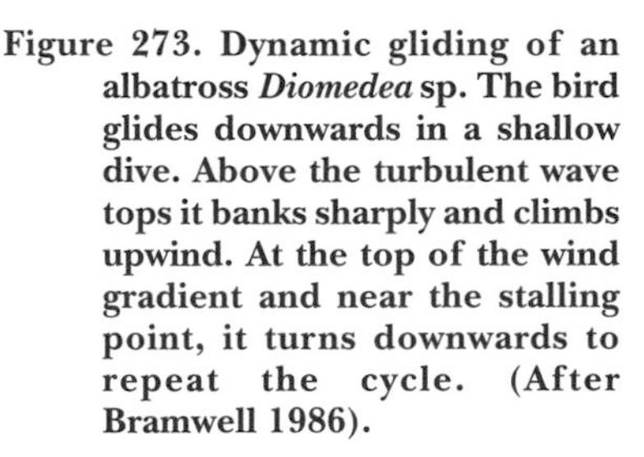
Figure 273. Dynamic gliding of an albatross *Diomedea* sp. The bird glides downwards in a shallow dive. Above the turbulent wave tops it banks sharply and climbs upwind. At the top of the wind gradient and near the stalling point, it turns downwards to repeat the cycle. (After Bramwell 1986).

The Wandering Albatross *Diomedea exulans* is the classic ocean wanderer that seeks out the gale torn waters that surround Antarctica, where the circumpolar winds of the West Wind Drift circle the continent in a clockwise direction. A combination of wind and waves produces a vertical gradient of moving air where the bottom 15 to 20m layer is slowed by friction as it passes over the irregular ocean surface. To harness the energy of these natural forces, the albatross speeds downwind on outstretched wings in a shallow dive until it meets the turbulent air above the wave tops. As it nears the surface it turns sharply into the wind and gains altitude. Because the bird is flying into wind of increasing velocity as it rises, the loss of air speed is not as great as its loss of ground speed. Consequently the bird does not stall until it has reached the top of the wind gradient. At this point the albatross once more turns downwind to repeat the cycle to drift with the wind (fig.273). In this way, these birds are known to range up to 8,000 miles from their nesting sites and it is thought that they are capable of circumpolar flights. Nesting sites for many species of albatrosses are on open ground on slopes or ridges near sea cliffs. The prevailing winds and the air deflected up and over the cliffs assist the birds with take-off, and cliff soaring is a common sight during the breeding season.

The frigatebirds (Fregatidae) have unusually long, narrow wings, adapted for soaring and gliding as well as powered flight, and the three segments of the forelimbs are about equal in length. The gliding albatrosses have developed wing lengthening to an extreme and the humerus, has become thinner and lighter (fig.274).

Elliptical-shaped wings are highly cambered with a low aspect ratio which gives good lift generation. They are associated with birds that need to manoeuvre around obstructions or fly in confined spaces, where speed is of secondary importance. Birds with elliptical wings are mostly associated with forests and woodland habitats, therefore few seabirds have this type of wing. However, many birds have wings that are intermediate between elliptical and high speed designs in order to suit a particular habitat.

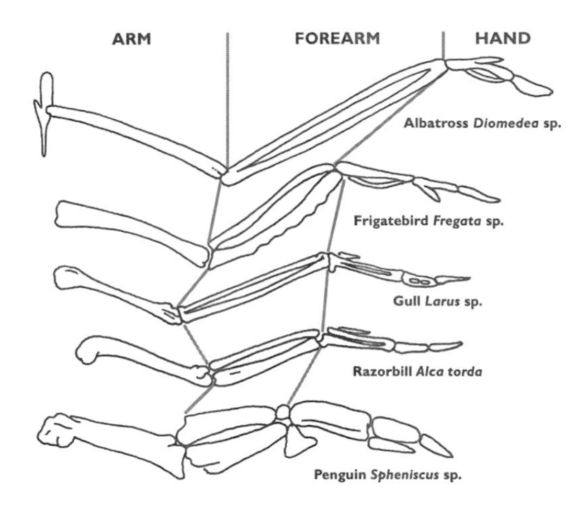

Figure 274. Comparison of five seabird wing bones drawn at about the same size to show relative proportions of proximal (arm), middle (forearm) and distal (hand) portions.

SKIN AND FEATHERS

The skin of birds tends to be thin, loose, dry and without sweat glands. The only epidermal gland present is the 'uropygial' gland (preen gland) at the base of the tail. The papilla of the gland often has a tuft of modified, brush-like, down feathers which help the transfer of an oily secretion from the gland to the bill during preening. The bill is the instrument used to apply the oil to the plumage. The uropygial gland is well developed in aquatic birds and the oily dressing is used to clean and waterproof the feathers. Without oil, feathers soon become waterlogged or matted and this leads to a reduction of thermal insulation and flight efficiency.

The three main functions of feathers are to conserve heat, provide lift and produce motive force. It is thought that feathers are homologous with reptilian scales and that they evolved from them. However, there is little evidence to support this hypothesis. Feathers are produced from the papillae in the epidermal layer of the skin. Feathers are made from a keratin-type material (similar to scales, hair and nails of other animals) which is light, durable and resistant to degradation. Birds keep their plumage in top condition not only by preening regularly but by moulting the feathers at frequent intervals in their life cycles (e.g. after the breeding season). New feathers are produced from the old papillae.

Feathers provide a high level of insulation, and weight for weight they have higher thermal retention than mammalian hair or almost any other natural or synthetic material. Penguins spend about half their lives in the cold waters of the southern hemisphere, and therefore need to be especially well insulated against loss of heat. Apart from the bill, feet and brood patch, penguins have a complete covering of small feathers (about 12 per square centimetre of body surface). Small tufts of down (aftershafts) grow out from the base of each feather to form an underlayer which traps air close to the skin thus preventing further heat loss. In addition, the skin is relatively thick and beneath this, a fatty tissue layer provides both a food store and added insulation. The penguin's body shape, large size (small animals lose heat faster) and insulation is so effective that the Emperor Penguin *Aptenodytes forsteri* happily lives at air temperatures of -50°C. However, if temperatures rise or the body is active, overheating may become a problem. Under these circumstances the birds pant and the webbed feet are splayed to act as cooling radiators. If conditions become too hot the birds return to the water for a cooling dip. Most birds are able to pant and lose heat in this way, but the massive build up of heat during muscular flight is controlled by convection cooling as air passes through the respiratory system. To conserve heat in cold conditions, most birds fluff out their feathers to trap more air and they are also capable of shivering. Through their high metabolic activities and heat produced from vigorous flight, birds normally have enough body heat to maintain themselves in cold conditions.

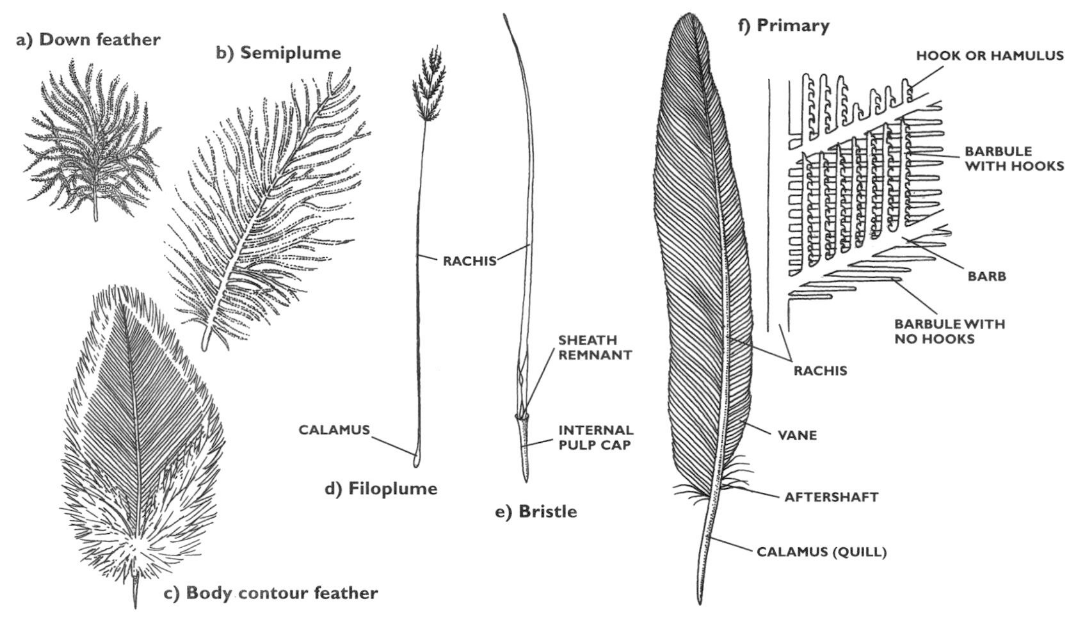

Figure 275. Different types of feathers showing main features, including detailed structure of a primary.

There are several different types of feathers specialised for different functions. They include: contour feathers (e.g. coverts, secondaries and primaries); semiplumes; filoplumes; down and bristles (fig.275). 'Contour' feathers provide an aerofoil covering to the wings, protect the downy undercoat, shed water, reflect or absorb solar radiation and may also have visual and auditory functions. 'Primary' flight feathers are highly modified contour feathers of the outer wing that produce the motive forces through flapping flight. Outer primaries may be further modified by being tapered in shape or notched to form conspicuous gaps or slots when they spread. This specialisation helps to reduce drag. 'Secondary' feathers of the inner wing provide the large surface areas required for lift. The loss of a number of secondary feathers still enables a bird to fly but the loss of even a few primaries greatly alters, or entirely prevents, flight. Covert feathers give much of the wing its overall shape and outline, and help fill many of the spaces. Tail feathers (rectrices) are large, stiff contour feathers that are modified to provide the bird with an additional surface. Thus the tail region helps with lift, stability and manoeuvrability. 'Semiplumes' are feathers with a structure intermediate between contours and down feathers. They are situated along the margins of contour feathers or are in their own tracts where they are mostly hidden below the contour feathers. Again they help to fill out the contours of the bird's body and also provide thermal insulation. 'Filoplumes' are fine, hair-like feathers with a thin shaft and a few short barbules at the tip. Filoplume feathers are thought to have a sensory role and help the operation of other feathers. In comorants (Phalacrocoracidae), the filoplumes grow out over the contour feathers and enhance the appearance of the plumage, but most filoplumes are generally concealed below other feathers. 'Down' feathers are fine, fluffy structures with a rachis that is shorter than the longest barb, or entirely absent. These feathers provide insulation and completely cover young birds at or soon after hatching. Most natal down feathers (neossoptiles) are moulted as other feathers develop, especially prior to fledging. 'Powder' feathers have a similar structure to down but they produce a fine, white keratin powder that is shed into the body plumage. This powder is believed to be another form of waterproof dressing. 'Bristles' are stiff, specialised feathers that may or may not have barbs on their outer ends. They are often found adjacent to contour feathers and commonly occur around the base of the bill, around the eyes and, in some cases, on the head and even the toes. They may help certain bird species to capture insect prey while on the wing and provide protection around the eyes. The detailed structure of a typical feather is most easily seen on a contour feather (fig.275). The short tubular base of the feather (quill or calamus) remains firmly embedded into a skin papilla until a moult occurs. A long, thin tapered midrib (rachis) bears closely spaced side branches (barbs) in a single plane.

Each barb has a central rib (ramus) with barbules that project out on either side. The barbules on the side of the barbs away from the body have hooks (hamuli) that catch on the unhooked barbules of the next barb. This arrangement has a marvellous ability to withstand damage and unhooked barbules can be quickly reconnected to reform an unbroken surface (a vane). An intact vane is not quite airtight and the controlled seepage of air through the vane provides the high lift properties of flight feathers. The two vanes (one either side of the rachis) of a contour feather may be asymmetrical or symmetrical in outline; Flight feathers are asymmetrical and tail feathers are symmetrical. At the base of the contour feather vanes, tufts of soft down may be present. The contour feather is therefore a multi-functional structure, as it forms part of an aerofoil surface while the protected downy base provides insulation.

FLIGHT

For birds to take off and fly, they must have high power to weight ratios, therefore they must be light and strong. For example, a frigatebird wing that has a wing span of 1.5m has a skeleton weight of only 114 grams (fig.266b). Lengthening the wings to provide an aerofoil lifting surface may seem simple, but it does create an awkward engineering problem. The upward lifting forces on a wing (which must balance the bird's weight) take place away from the body and this causes a large bending movement at the wing root. Flight forces produced on the outer wings (for example from flapping) are of little use if the wings bend or fold up under stress. Wings need to transfer the forces of flight to the shoulder joints and finally to the body. Therefore, although wings have to be aerodynamically thin, they need to be stiff enough to resist bending.

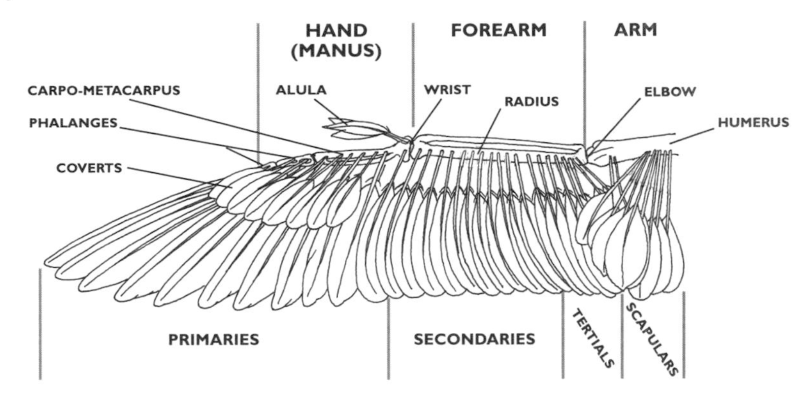

Figure 276 . Dorsal surface of a gull *Larus* sp. wing with most coverts and skin removed to show flight feather positions and insertions into the bones. The inner wing (shoulder to wrist) produces lift, the outer wing (wrist to wing tip) produces motive force and provides control. (Based on Storer 1952).

Feathers that make up the surface of the wing are stiff enough to need supporting at one end only. Feather shafts radiate out from the radius and manus (hand) and the forces developed on the feathers are transmitted to these bones and ultimately concentrated at the head of the humerus (fig.276). The primary feathers of the hand are not fixed immovably into the bones but are held by a wide flexible membrane that allows each feather considerable movement. Freedom of movement is critical if the primaries are to develop motive force.

The wings of a bird are more complex than aircraft wings as they often generate both lift and power. A bird also has the ability to alter the area, position and shape of the wings in relation to its body attitude. Changing the aerodynamic properties of wings during flight helps the bird to manoeuvre, change direction, land and take-off.

Each wing can be thought of as having an 'outer' wing which moves separately at the wrist halfway along the structure, and an 'inner' wing that operates from the shoulder joint. The secondary flight feathers of the inner wing are arched and together they form a large cambered aerofoil that provides most of the lift. During flight, the inner wing is held quite rigidly at a slight upward angle. A group of feathers (the alula or bastard wing) provides a small aerofoil surface at the front edge of the wrist where the inner and outer wings join. The alula forms a slot to smooth the air flow over the wings when they are held at an increased angle of attack (fig.269c). Re-establishing laminar air flow prevents turbulence and drag, therefore wings can be held at higher inclined angles to increase lift without stalling. The prevention of stalling and increasing lift are especially important at slow flight speeds. Without an alula many birds (especially large ones) would find it difficult or impossible to take off and land properly.

During flapping flight the outer wing and its associated primary feathers provide the power to move the bird forwards through the air. At take-off, when the the air speed over the inner wing is too slow to generate much lift, the outer wing is brought swiftly forwards to create enough lift to raise the bird off the ground (fig.277). The motive force is also generated by the outer wing as it moves backwards on the recovery stroke. The effort of taking off is often very tiring and it can only be maintained for a short period. Large birds may have to run a considerable distance before they generate enough motive power to become airborne.

Figure 277. Seabird launching from and landing on a sea cliff. Sequence a) shows wing beat movements following take-off. Sequence b) shows wing beat movements for landing. (After Orr in Sparks 1972).

Primary feathers are superbly designed to produce the propulsive forces that ultimately move the bird forwards through the air. This is made possible as the front vane of the primary feather (on the forward side of the rachis) is narrower than the rear vane, therefore it has a smaller surface area. During the downbeat, air pressure on the larger vane of the asymmetrical primary feather twists each one of them upwards and backwards along their length until they take the shape of a propeller blade. The degree of twist (pitch) and shape of the feather is controlled mostly by the design of the quill (calamus) and shaft (rachis) which is quite rigid at its base but flattened and more flexible towards its tip. A powerful downbeat often elevates the feather tips almost at right angles to the rest of the wing and in the direction of flight. The twisted shape is held only momentarily and during this phase the feathers are pressed firmly together by the air pressure. However, unlike an aircraft propeller which rotates around a central pivot in one direction, the feather propeller moves rapidly up and down to produce the thrust. On the upstroke the primaries are deflected only a little or not at all. In some birds, the barbs of the feathers are so arranged that they open like the vanes of a blind when under pressure from above, but shut when the pressure is from below. The basic requirement needed to make fast flapping flight possible is that the magnitude of the net aerodynamic forces should be greater on the downstroke than on the upstroke. Throughout the entire cycle of the wing beat the primary feathers are constantly changing their position and shape, adjusting automatically to the air pressure and movements of the wing. While the bird is flapping steadily, only the tips of the primary feathers may twist. If the bird requires more power and flaps vigorously, the whole outer wing may be twisted by the increased air pressure so that it performs as one large motive propeller. Thus, gross movements of the wings are downwards and forwards on the downstroke with a quicker upwards and backwards motion on the upstroke (fig.278). This produces a complex figure of eight flapping cycle. The amount of backwards and forwards movement will vary with the speed of the

Figure 278. Powered flapping flight cycle of a sea duck. Primary feathers twist into individual propellers on the down stroke. Gross wing movements are downwards and forwards on the downstroke with a quicker upwards and backwards motion on the upstroke.

wing beat. If the wings beat fast (e.g. on take-off) the increased pressure forces the wing tips forwards on a more horizontal path. By contrast, during normal flight the wing movement is nearly vertical.

Other parts of the body surface of a bird play a more passive role in flight. The tail is often depressed and spread during take-offs and landings. This helps the bird to increase its overall lift and improve control at low flying speeds. In some species with long, forked tails (e.g. terns *Sterna* spp. and frigatebirds, Fregatidae), a spread tail enhances their slow flight and hovering capabilities. A tail is also important to steer and balance a bird. It steers by turning the tail up, down or sideways. However the balance of a bird is mostly accomplished by the wings; if the body tips to one side, the bird can restore itself to an even keel by increasing the lift of that wing, either by beating more strongly with it or by changing the wing angle.

In many seabirds that have webbed feet, the downwardly splayed position allows the feet to be used as effective airbreaks during descents. This is especially important when a steep angled glide path is required during a controlled descent. In the auks (Alcidae) the tail is short and the large webbed feet can also be used as supplementary lifting surfaces that are added to the sides of the tail in slow flight. Just before touchdown the feet are rotated downwards and forwards from the lateral position to absorb the shock of landing (fig.279).

Figure 279. Puffin *Fratercula* sp. landing. Webbed feet are opened to provide air breaks and they twist forward prior to landing. The tail feathers are spread for extra lift and control as flight speed is reduced. The alula feathers are raised and primary feather slots open to smooth airflow over the wings and prevent stalling. (Drawn from a photograph sequence in Sparks 1972).

RESPIRATION

Birds have an unusual respiratory system with a large air capacity consisting of a normal pair of small, non-distensible lungs connected to an extra, extensive system of thin-walled, non-respiratory air sacs (fig.280). These air sacs ramify throughout the viscera and branches may extend into the hollow, long bones of the wings and legs, sometimes even into the small toe bones. Like mammals, the respiratory system of birds is generally tidal, but unlike mammals which have dead-end structures (alveoli) bird lungs have air channels (parabronchi) which are continuous. The route taken by air through the respiratory system is uncertain, but it is partly under the control of the bird. This may be important in diving birds where the air may be passed backwards and forwards through the lungs for more effective extraction of oxygen. In the diving penguins the lungs are protected from collapsing through increased pressure during a dive. This is accomplished by the compression of the ramifying air sacs forcing air into the lungs to maintain their volume.

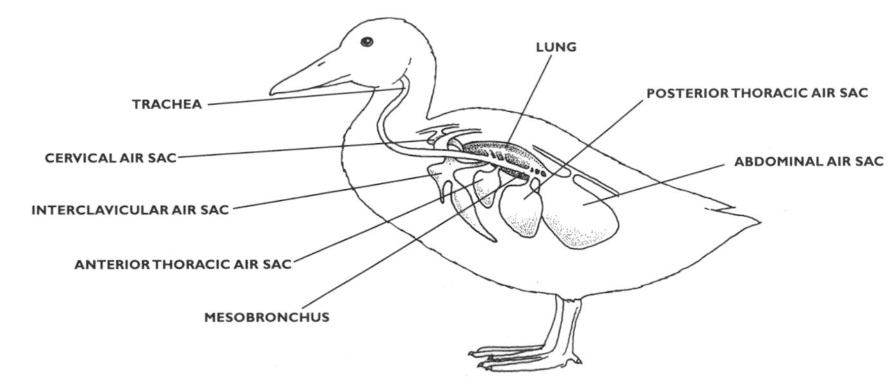

Figure 280. Respiratory system of a sea duck showing the lungs and associated air sacs. (After Schmidt-Nielsen 1971).

Air flows through the respiratory system into all but the smallest of tubes and gaseous exchange takes place regardless of whether air is being inhaled or exhaled. While flying, wing beats compress the rib cage to automatically expel the air. Therefore instead of running out of breath, incoming fresh air continually fills the lungs and air sacs during flight. Air sacs are normally kept at their maximum volume and the lungs are stabilised by the elasticity of the lung tissues. The fact that birds are generally able to extract about 6.5% of the available oxygen from respired air (about the same level as many mammals) suggests that the complexity of the respiratory system in birds is related to other functional features (see table 9 on p.365).

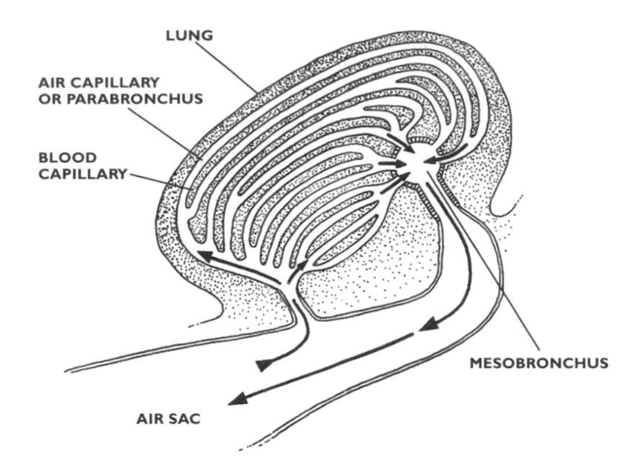

Figure 281. Longitudinal section through the lung and air sac of a bird. Air passes through the lungs, into and out of large air sacs. Air flowing in capillaries travels in the opposite direction (counter-current) to blood flow in the blood capillaries, therefore efficient gaseous exchanges take place. (After Clements 1962).

All birds are covered by feathers that provide a high degree of insulation, but they do not possess sweat glands with which to cool themselves. With their high metabolic rate and heat generated from vigorous muscular activities, birds would quickly overheat without an efficient means of thermoregulation. This is achieved mostly through 'convection cooling' where a passing fluid (air or water) draws heat away from the body. Birds that live most of their lives in the water have little problem as water is a good conductor of heat; on the contrary, the situation may arise where the birds actually need to conserve heat, especially those living in the cold temperate and polar latitudes. During flight the lungs and ramifying air sacs are exposed to a rapid and continuous stream of air and this through-flow of air draws large amounts of heat out and away from the body. Even in the hottest climates the passing air is likely to be below the bird's blood temperature, especially if the bird is flying at higher altitudes.

BLOOD SYSTESM AND METABOLISM

Birds require an efficient blood circulation system and metabolism as large amounts of blood sugar and oxygen need to be supplied at high rates to the flight muscles. The blood system has no special anatomical features and compared to non-flying vertebrates, the blood system is not significantly richer in haemoglobin. Birds, like mammals, have a four-chambered heart and a double circulation system where blood passes through the lungs for gaseous exchange before it circulates through the body again. However, relative to their body size, birds' hearts are about twice as large as mammalian hearts because blood needs to be pumped around the blood-circulatory system at a higher rate, especially during periods of prolonged flight or high levels of activity. As may be expected, their blood pressure is generally higher, blood sugar levels are twice that of mammals, and body temperatures may exceed 40°C. Therefore birds (especially the smaller species) burn their metabolic 'candles' at both ends and as a result they live relatively short lives.

It is often thought that prolonged exertion or activity must inevitably lead to exhaustion. However, this is now known not to be true provided the oxygen and blood sugar requirements, and thermoregulatory needs are within the metabolic constraints of the body. There is evidence that small passerine species can fly continuously for 50 hours or more, while swifts (Apodidae) can fly day and night, only landing for breeding purposes.

DIGESTION

The bird intestine is comparatively short, wide and thin-walled in fruit- and meat-eating species, but longer and more developed in the seed-eaters. Swallowed food may be stored in a large oesophageal bag (crop) which is well developed in grain-eating birds. The true stomach is divided into two separate compartments, a glandular proventriculus and a muscular gizzard. Macerated food from the crop passes to the proventriculus where peptic enzymes are added to chemically break down the food. The food then enters the gizzard where it is mechanically broken down, sometimes with the help of swallowed grit and small stones. Gizzards are reduced or absent among fruit-eaters, insectivores and many carnivores but are large in many birds that consume fish, shellfish, vegetation and seeds. One difference between the gut of birds and mammals is the presence of blind-ending tubes (caeca) which are outgrowths of the posterior end of the small intestine. They are longest in grain-eaters and their function may include the production of certain vitamins or fermentation of bacteria used in the breakdown of cellulose. The cloaca is the posterior part of the gut that forms a common opening for the ureter and the reproductive system.

The high quality, high energy foods that birds consume and the elevated metabolic rates of birds makes the passage of food through the gut quite rapid. Food consumption is generally high and may reach 30% of the body weight per day in small bird species. Therefore birds defecate often and food passage times vary from 20 minutes to several hours, depending on species. Bird faeces are a combination of waste materials from the kidneys and the digestive tract. The bird excretory system is designed to reduce body weight by eliminating nitrogenous waste with minimum amounts of water. The kidneys reabsorb most of the fluids and excreta are mostly in the form of an insoluble, pasty substance called uric acid. In addition, birds have no bladder and this makes them more compact and lighter - an obvious advantage for flight. Waste products that cannot be easily digested (e.g. bones, feathers and hair) are often regurgitated as a ball of waste material (a pellet). This practice is common among raptors and carnivorous birds.

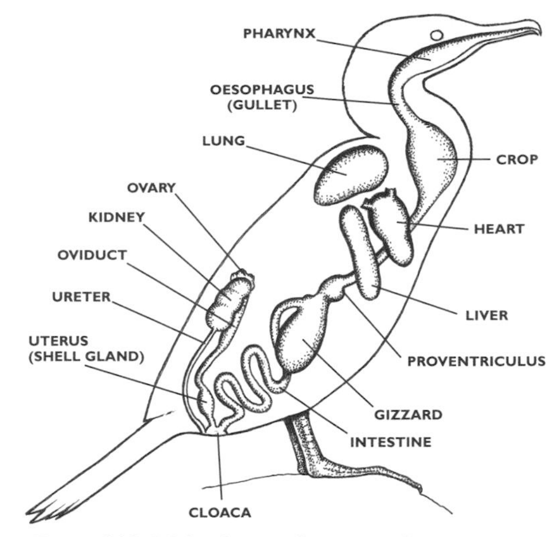

Figure 282. Major internal organs of a cormorant *Phalacrocorax* sp., including its digestive system. (After Lofgren 1984).

SALT EXCRETION

Most air-breathing vertebrates are unable to drink large quantities of sea water without dehydration and kidney damage. This is because sea water contains sodium chloride and other salts at three times the concentration of the blood and body fluids of animals. Under normal circumstances, the vertebrate kidney functions in a regulatory capacity to eliminate salts and to produce complex waste products for excretion. However, seabirds are capable of drinking large quantities of sea water without ill effect, yet they have kidneys that are no more efficient than other vertebrates. Seabirds and some other marine vertebrates (e.g. the marine iguana and turtles) possess special salt secreting glands in the head, normally close to, or above the eyes. These glands eliminate salts far more rapidly than the kidneys.

Unlike kidneys, salt glands only function to excrete salt from the body fluids and these glands work only when required. The fluid produced by the salt glands is nearly five times as salty as the bird's body fluids or blood and the fluid excreted is an almost pure 5% solution of sodium chloride, which is nearly twice as salty as sea water. The glands can perform a large amount of osmotic work in a short time and in one minute can produce up to half their own weight of concentrated salt solution (kidneys can produce about 1/20th their weight of urine per minute).

The structure of the salt gland is relatively simple(fig.283). It comprises many parallel, cylindrical lobes each containing thousands of branching tubules that radiate from a central duct. It is cells lining these ducts that actively secrete the salty solution. Blood in a network of capillaries flows adjacent to the tubules in a counter-current flow. This arrangement allows more efficient and quicker transfer of salts to the cells of the salt gland

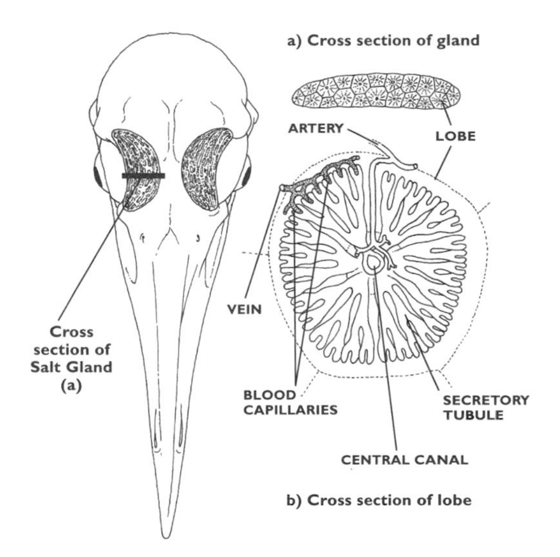

Figure 283. Salt gland of a gull *Larus* sp. showing its position and detailed structure. a) Cross section of a gland. b) Cross section of a lobe showing blood capillaries and secretory tubules. Blood flows in through the arteries and exits via the veins in a counter-current flow to the fluid in the secretory tubules. A central connecting canal collects the salty fluid for eventual secretion. (After Schmidt-Nielsen 1959).

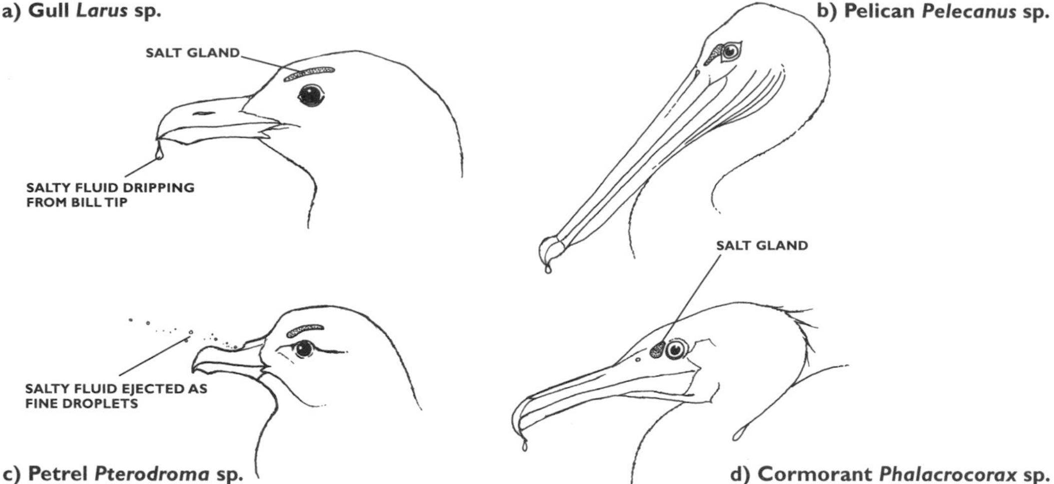

Figure 284. Locations of salt glands in different seabirds. (After Schmidt-Nielsen 1959).

Although the structure and function of salt glands is virtually identical in all seabirds, its position does vary (fig.284). The duct from the gland opens into the nasal cavity and, in most bird species, the salty solution flows out through the nostrils and drips from the tip of the bill. However, this is not always the case; for example, pelicans have a pair of grooves in the longer, upper bill which channels the fluid down to the tip. This avoids the salty fluid entering the bill pouch and being swallowed. In gannets and cormorants, the nostrils are covered with skin and therefore are not functional. Instead, the salty solution exits through the internal nostrils in the roof of the mouth and the fluid flows to the bill tip. Petrels which spend long periods flying at sea have a curious adaptation to remove the salty fluid. These birds physically expel the fluid outwards through a tube on the upper mandible as a tiny stream of droplets.

SOME ADAPTATIONS FOR SWIMMING AND DIVING

Although birds are not completely aquatic (even penguins come ashore to breed), nearly 400 species are adapted for swimming. Almost half of these species also dive underwater in search of food or to escape predators. Most adaptations for swimming involve modifications to the hind limbs, although other changes include widening of the body to increase underwater stability, enhanced waterproofing of feathers, dense plumage and a fat layer for increased buoyancy and insulation. Legs are most often placed near the posterior end of the body which is the best position for streamlining, propulsion and steering (fig.285). Aquatic birds generally bathe and preen themselves far more regularly than terrestrial birds. By bathing, preening and oiling their plumage, the feathers do not dry out or become waterlogged.

To function efficiently the feet of aquatic birds have developed webs or lobes between the toes (fig.285). Webbing between the three forward toes has independently evolved (convergence) in the waterfowl, petrels, penguins, gulls, auks and divers (loons). Webbing between all four toes (totipalmate condition) has only evolved in the pelicans, comorants and gannets. While swimming on the surface, many birds move their feet back and forth, but during the backward (power) stroke, the webbed feet are spread apart. On the recovery stroke the webs are closed and the toes bend to decrease water resistance.

Diving requires further modifications to limbs and body. Hind limbs either become better adapted for underwater swimming or wings are used as flippers that enable the birds to literally fly underwater. Specialised foot-propelled diving species include the comorants and divers (loons). Wing-propelled divers include the diving-petrels, gannets, penguins and auks. Ducks may be adapted for either wing- or foot-propulsion.

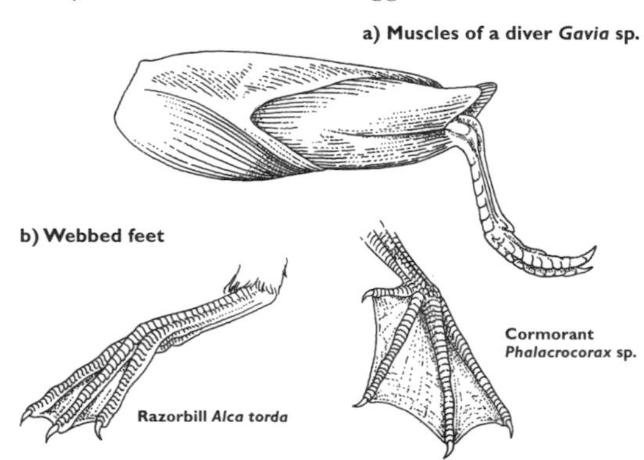

Figure 285. Adaptations for swimming and diving. a) Muscles and leg position of a diver (loon) *Gavia* sp. b) Feet sho0wing webbing between three toes of the Razorbill *Alca torda* and between four toes (totipalmate condition) of a cormorant *Phalacrocorax* sp.

Swimming and diving birds have a high level of buoyancy due to air trapped in feathers, air sacs and even some bones. Diving birds have partially overcome these problems by expelling air from the plumage prior to diving, reducing the air cavities in their bones and breathing out before a dive. Metabolic adaptations include reduction of blood flow and heart rate to reduce oxygen demand while underwater. The oxygen carrying capacity of the blood and tissues is also greater and CO_2 tolerance is higher.

The most specialised underwater swimmers are the wing-propelled penguins. Being flightless they are able to have large body sizes and reduced pectoral muscles. Their feathers, metabolism and body shape have become further modified. Penguins often swim near the surface and porpoise out of the water to take in quick breaths of fresh air. During fast swimming, wing flippers are mostly used to maintain directional control. On the surface, penguins paddle without the use of flippers and instead use their webbed feet.

Many aquatic birds are less mobile on land than in the water. Large webbed feet make perching difficult or impossible, therefore these birds choose a reasonably wide surface on which to stand. Well developed, hooked claws help to maintain a grip on slippery surfaces such as cliff ledges. Penguins walk upright and waddle with a rolling gait. Stiff tail feathers help them to maintain balance while walking uphill and provide a third point of balance when standing. Penguins will also toboggan down snow slopes as this is often an easier and quicker method of returning downhill. The tail and claws are especially important for balance and grip when they leap out of the water onto ice or rock. With potential danger from pounding surf and predators, slipping back into the water may be fatal, especially when Leopard Seals patrol the areas where they come ashore. Adaptations for feeding and breeding among seabirds are discussed in the following sections.

EYES AND VISION

Birds depend more on their eyes than other senses. Eyelids and nictitating membranes protect the eyes of a bird. They have keen vision and possess both rods (for low light vision) and cones (for colour vision). The cones are concentrated in retinal areas of high visual acuity called the foveae. In eagles (e.g. Sea Eagles), two foveae are present, one for forward binocular vision and the other for monocular side vision. The bony rings (scleral plate, fig.268) that encircle the eyes of a bird help them keep their shape.

Many night flying birds and diurnal predators have a double method of active accommodation (focusing) for near vision. This is made possible by the use of the anterior (Crampton's) and posterior (Brücke's) ciliary muscles. The Brücke's muscle draws the lens forwards into the anterior chamber, but because the eye shape is fixed by the sclerotic plates, the lens becomes increasingly curved and thus accommodates for near vision. Contraction of the iris sphincter helps this action. During this process the Crampton's muscle pulls on the cornea and shortens the corneal radius to further help this accommodation process. Among diving birds, the cornea is of little use for image formation underwater, therefore the Crampton's muscle is reduced or absent. However, the Brücke's muscle is often large and in the cormorant eye the powerful iris muscles assist the ciliary muscles to enable impressive shape changes of the soft lens which are necessary for underwater vision (fig.286).

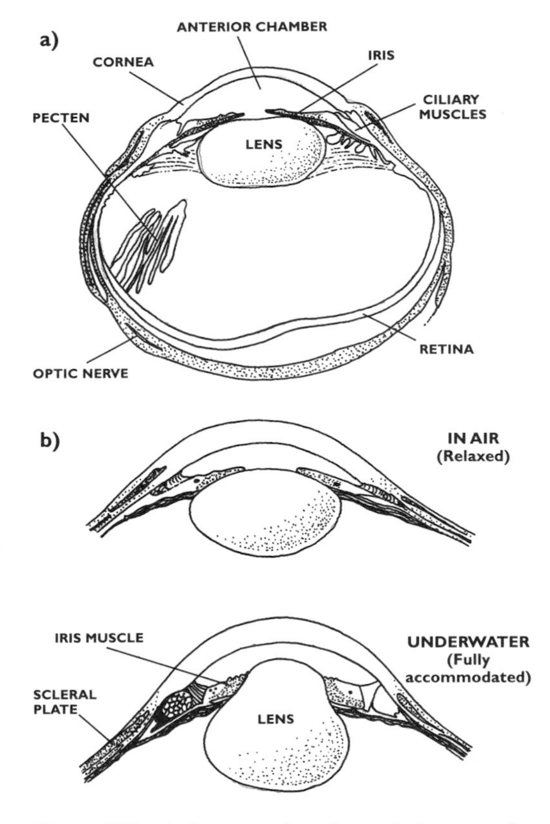

Figure 286. a) Cross section through the eye of a Great Cormorant *Phalacrocorax carbo*. b) Muscular accommodation in a cormorant's eye. (After Walls 1942).

MORPHOLOGY AND PLUMAGE

The size, shape and structure of seabirds varies enormously. Albatrosses possess the longest wings of any bird, whilst penguins are flightless and have the wings reduced to flippers. Storm-petrels may weigh as little as 25 grams while a Royal Albatross *Diomedea epomophora* can reach 12 kilograms. This great diversity has resulted in a range of modifications to the basic avian skeleton and musculature. Size and structure are correlated with various aspects of ecology and behaviour. Within families, size tends to reflect the size and type of prey that is taken. Bill structure is extremely varied amongst seabirds and shows adaptation according to the prey type and the method of feeding.

Seabirds lack the wide range of colourful plumages that are found amongst landbirds and there is also much less sexual dimorphism. With a few exceptions, most seabirds are black, white, brown or grey. There are three main plumage types: (a) mainly dark plumage with any white plumage restricted in area (b) mainly dark above and white below and (c) mainly white or pale-coloured. It has been demonstrated that fish react less to birds with white underparts than to those which have dark underparts, and the various plumage types may be adaptations resulting from the method of feeding. Many of those species which feed by active underwater pursuit of fish have darker plumage than those which plunge-dive or pick food from the surface. Species which swim on the surface often have white confined to the parts of the bird which are below the waterline whereas species which search for food on the wing and plunge-dive or swoop down to the surface tend to have a white frontal aspect comprising the head, breast and underwing. The brilliant white upperparts of gannets and some terns, which is very conspicuous against the sea, may be useful in attracting the attention of other individuals to productive fishing areas. Abrasion is the effect of wear upon the vanes of the feathers and the preponderance of black or brown in the plumage of many seabirds may also be important for anti-abrasive reasons, as melanin, the dark pigment in the feathers, helps to strengthen the feathers. In penguins and some species of auks, the dark plumage of the upperparts is also well designed for absorbing heat from the sun thus increasing body temperature.

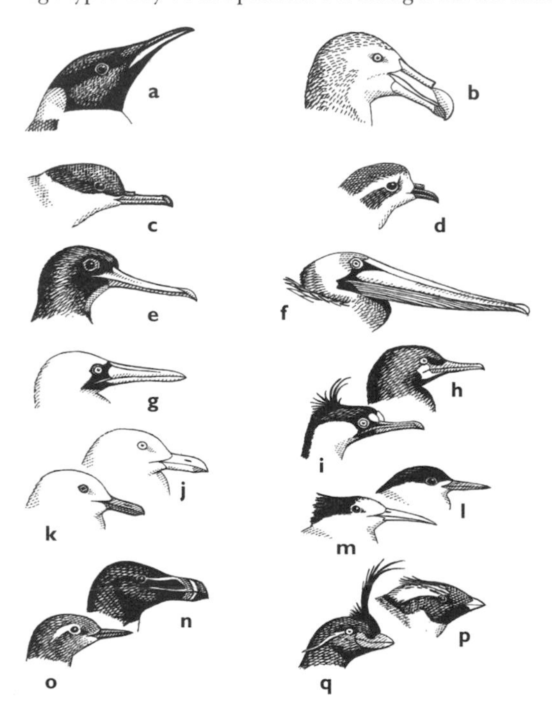

Figure 287. Variation in heads and bill shapes. a) Emperor Penguin *Aptenodytes forsteri*, b) Northern Giant Petrel *Macronectes halli*, c) Great Shearwater *Puffinus gavis*, d) White-faced Storm-Petrel *Pelagodroma marina*, e) Christmas Island Frigatebird (female) *Fregata andrewsi*, f) Brown Pelican *Pelecanus occidentalis*, g) Masked Booby *Sula dactylatra*, h) Cape Cormorant *Phalacrocorax capensis*, i) Imperial Shag *Phalacrocorax atriceps*, j) Glaucous Gull *Larus hyperboreus*, k) Iceland Gull *Larus glaucoides*, l) Common Tern *Sterna hirundo*, m) Crested Tern (adult winter) *Sterna bergii*, n) Razorbill *Alca torda*, o) Spectacled Guillemot *Cepphus carbo*, p) Japanese Murrelet *Synthliboramphus wumizusume* and q) Crested Auklet *Aethia cristatella*.

A few species of seabirds do show some brighter colours but these are decorative features and are mostly confined to the head, including the facial skin and bill, and to the feet. The brightly coloured bills of puffins are used in display and are reduced in both size and colour outside the breeding season. In a few species, notably some gulls and terns, the white underparts develop a rosy or pinkish tinge during the breeding season. Most seabirds which have dark hoods or caps lose these features in winter, as do those species which have ornamental crests or head plumes. The *Eudyptes* penguins are an exception as they keep their brightly coloured plumes throughout the year.

The plumage of juvenile birds tends to be drabber than that of adults and often shows some cryptic coloration such as the brown tips to the feathers on the upperparts of many gulls and terns. This makes these birds less conspicuous to predators. The downy young of gulls and terns are generally mottled or blotched in shades of brown, black and grey making them highly camouflaged at an age when they are most vulnerable to predators.

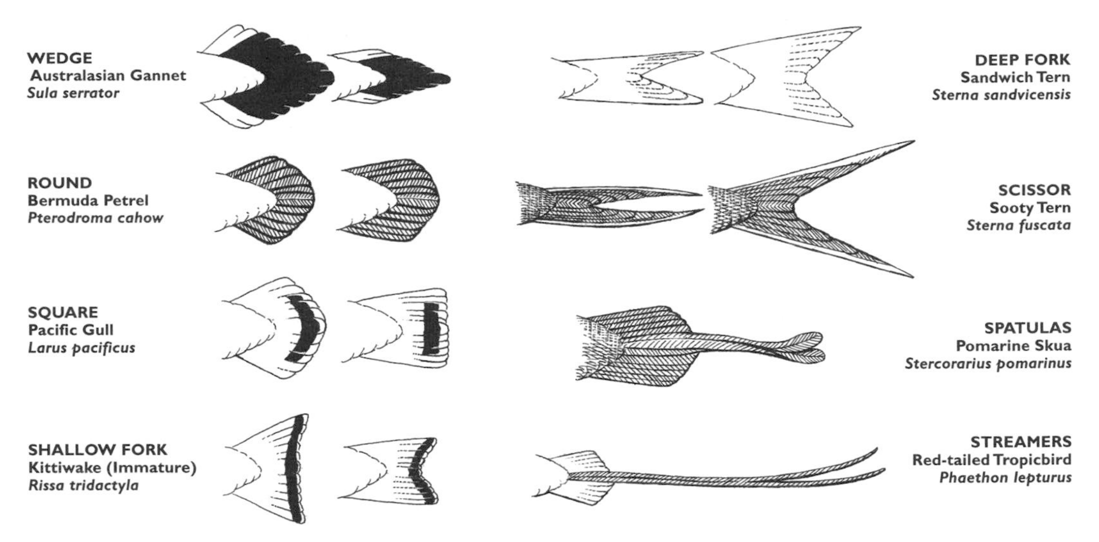

Figure 288. Variation in tail shapes of eight seabirds.

FOOD AND FEEDING

About half of all species of seabirds are fish eaters; these include some penguins and shearwaters, divers and grebes, gannets and boobies, cormorants, pelicans, some ducks, many gulls and terns, skimmers and many species of auks. The second largest group of seabirds feeds mainly on crustacea and other marine invertebrates; these include some penguins, many storm-petrels, diving-petrels, many shearwaters, prions and petrels, and a few auks. The third main group are those species which prey on cephalopods; these include most albatrosses, several *Pterodroma* petrels, some storm-petrels, many shearwaters, and some gulls, terns, frigatebirds and tropicbirds.

Seabirds have developed many different feeding strategies in order to exploit the varied riches of the marine environment. These differences result from the type and size of prey and also the depth at which the prey is located.

The main methods of feeding are: (a) picking food from or just below the surface while flying, (b) swimming on the surface and obtaining food by pursuit-diving in the upper layers of the sea, (c) plunge-diving to obtain food at deeper levels, (d) diving from the surface and pursuit-swimming at depth, (e) piracy or klepto-parasitism, and (f) scavenging and opportunistic feeding, for example taking discarded fish from trawlers.

Some species of seabirds are highly specialised feeders whilst others employ a variety of feeding techniques to exploit a greater variety of prey. The availability of prey changes according to the time of day. Plankton and cephalopods rise to the surface at dusk and through the night, which accounts for the mainly nocturnal feeding habits of many shearwaters and petrels and such specialised feeders as the Swallow-tailed Gull *Creagrus furcatus* which, due to its dependence on cephalopods, is the only nocturnal gull. Amongst the most specialised feeders are the skimmers *Rynchops* spp. which plough the surface of the sea with their specially adapted lower mandible, snapping the bill shut when a food item is detected. Many gulls and skuas are omnivorous taking almost any available food item including fish, small rodents, young birds and eggs, stranded carcasses, road kills, discards from fishing boats, food scraps from waste tips, earthworms and other invertebrates disturbed by ploughing or silage cutting etc. The diets of adults and chicks often differ as do the diets of sympatric species.

Kleptoparasitism (fig.271) is an unusual feeding method in that there are no species which are solely dependent upon this method of obtaining prey, though there are probably some individual birds which are. Piracy is chiefly practised by skuas, some gulls and the frigatebirds.

Penguins living in the Antarctic and subantarctic zones tend to feed mainly on krill (*Euphausia*), while those that live further north are largely dependent on fish. The depth to which some birds will dive in search of prey is remarkable; Emperor Penguins *Aptenodytes forsteri* have been recorded to dive to a depth of 265 metres.

The abundance of food at sea is dependent on a number of factors; the phenomenon known as El Niño, which is a periodic anomaly of certain marine currents caused by an influx of warm water from the north causing the thermocline to sink taking the nutrients down with it, has produced extreme effects. These have been most marked on the coasts of Chile and Peru where an estimated 28 million seabirds in 1955 dropped to a mere 6 million after the El Niño of 1957/58.

The impact that seabirds have upon the resources of the marine environment is significant. It has been estimated that a population of around 16 million seabirds (mostly Guanay Cormorants *Phalacrocorax bougainvillii*) breeding on the Peruvian coast annually consumed 2.5 million tons of fish, almost entirely anchovies *Engraulis ringens*, and the 5 million Adelie Penguins on Lawrie Island were estimated to consume 9000 tonnes of krill and larval fish each day.

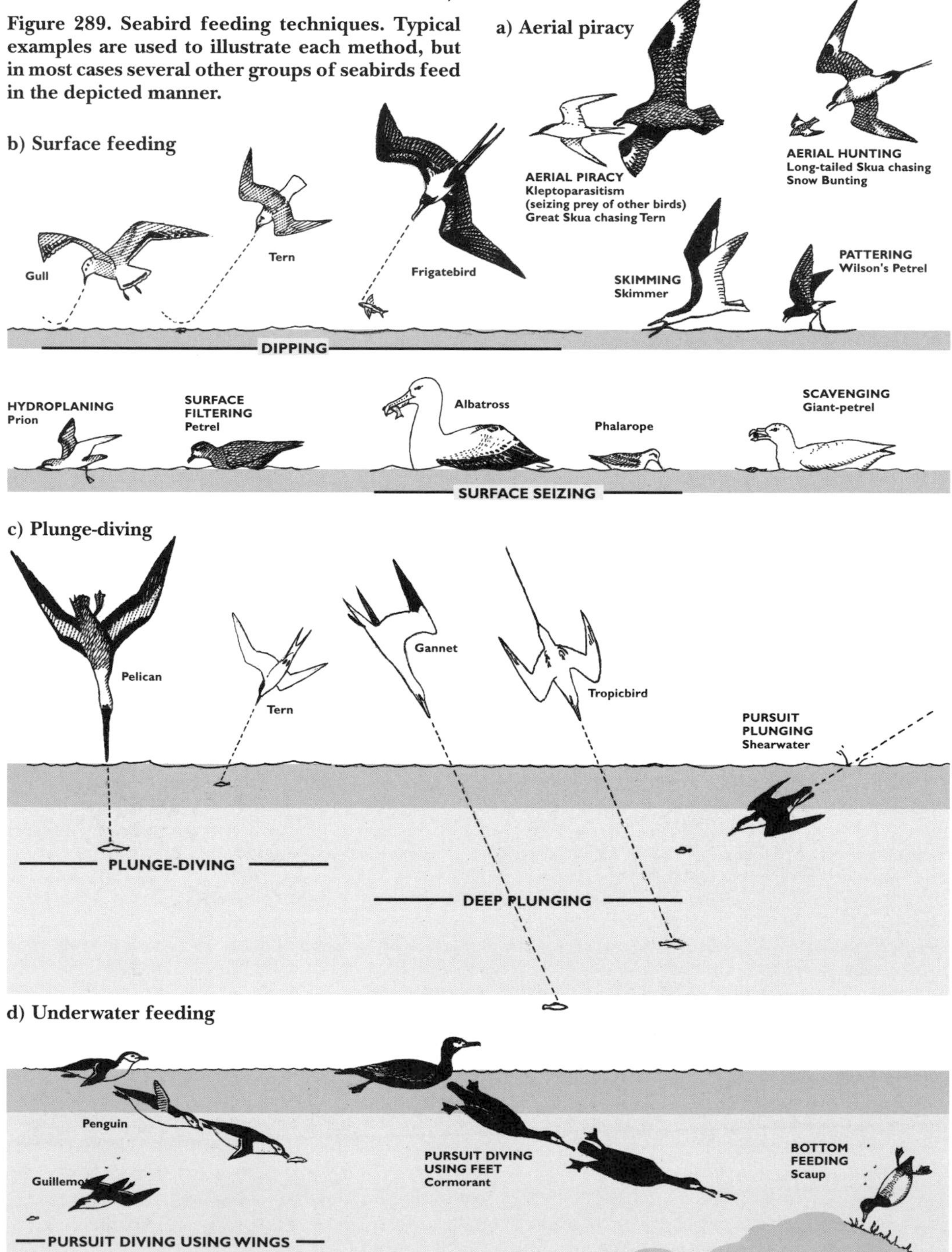

Figure 289. Seabird feeding techniques. Typical examples are used to illustrate each method, but in most cases several other groups of seabirds feed in the depicted manner.

BREEDING

All seabirds lay their eggs on land and therefore must come ashore to breed for part of the year. The breeding cycles and strategies of seabirds are remarkable for their diversity and are a fascinating subject to study.

Seabirds utilise a wide range of habitats during the breeding season; these range from ledges on sea cliffs to sand or shingle beaches, holes or crevices amongst boulders or on screes (often far inland), in burrows, on open ground and even on ice (fig.290). Some species have adapted to tree-nesting, ranging from the White Terns *Gygis* spp. which lay a single egg on a branch without a nest to support it, to the more elaborate nests of frigatebirds *Fregata* spp., and even tree-hole nesters like the Christmas Island race of White-tailed Tropicbird *Phaethon lepturus*. Some seabirds such as albatrosses *Diomedea* spp.,cormorants *Phalacrocorax* spp. and kittiwakes *Rissa* spp. build well constructed nests using mud and vegetation, but most species merely make a shallow scrape to hold the eggs. Guillemots (murres) *Uria* spp. lay pyriform (pear-shaped) eggs to reduce the chance of the eggs rolling off the rock ledges.

About 95% of all seabird species are colonial and require the presence of other members of the same species to provide the social stimulation necessary to breed. These colonies may range in size from a handful of pairs to vast numbers such as the 10 million Chinstrap Penguins *Pygoscelis antarctica* which breed on Zavodevski Island in the South Sandwich group.

Seabirds must reach physical maturity before they are able to breed. The age of first breeding varies from 10 years in the Wandering Albatross *Diomedea exulans* and Royal Penguin *Eudyptes schlegeli*, to only 2 years in the Antarctic Tern *Sterna vittata*.

The period of incubation is correlated with the relative size of the egg. The egg of Royal Albatross *Diomedea epomophora* requires 79 days of incubation, European Storm-Petrel *Hydrobates pelagicus* 41 days and Little Tern *Sterna albifrons* about 20 days. The length of the fledging period is similarly variable from between 263-303 days for a Wandering Albatross *Diomedea exulans* to about 25 days for an Arctic Tern *Sterna paradisaea*. The young of some species of auks leave the nest site before they can fly. In Craveri's Murrelet *Synthliboramphus craveri* this takes place only 2-3 days after hatching while in Common Guillemot *Uria aalge* the chicks leave the breeding ledges (mainly at night to avoid the attention of predators) after 18-25 days. The young are then cared for at sea by the parents until they become independent.

Birds breeding at high latitudes have the shortest breeding seasons and those species involved have highly synchronised breeding. Almost every species of seabird has a well defined breeding season though there are exceptions such as the Red-footed Booby *Sula sula* in the Galapagos Islands which shows no particular preference in the population as a whole.

Sooty Terns *Sterna fuscata* on Ascension Island have a closely synchronised season, breeding about every 9.6 months. Bridled Terns *Sterna anaethetus* in the Seychelles have an 8-month cycle and Audubon's

Figure 290. Five different nesting strategies of seabirds. a) Frigatebird, b) Puffin, c) Kittiwake, d) Little Tern and e) Wandering Albatross.

Shearwater *Puffinus lherminieri* in the Galapagos Islands breeds every 9-10 months. Some species like Brown Booby *Sula leucogaster* and frigatebirds *Fregata* spp. have cycles of more than one year. It is likely that the period between breeding seasons is the least amount of time that is required for the birds to breed, moult and return to breed again, independent of the annual cycles of the food supply. These non-annual cycles are quite common amongst seabirds though they are rare in landbirds. Close synchrony may reduce predation and may also be beneficial in that individuals may learn about good feeding areas from other individuals within a colony.

A few seabirds such as Wandering Albatross *Diomedea exulans* have longer breeding cycles which can take two years to complete. In these cases parental care usually takes much longer than one year and so individuals breed about every second year.

The King Penguin *Aptenodytes patagonicus* has a unique breeding cycle amongst birds. It lays a single egg in November which develops into a chick by the following winter. It is dependent upon the parents for food throughout the winter but does not develop quickly and may even lose weight during this time. The following spring it begins to develop more quickly and becomes independent by December. The parents then undergo a moult and are not ready to breed again until January, about 14 months since the previous egg was laid. If the chick from this later season's laying survives it remains small and takes the whole of the following summer to reach independence in March. Successful King Penguins can raise two young in a three year cycle. If a chick is lost they will revert to laying in November and will raise only one chick in two years.

By contrast, the closely related Emperor Penguin *Aptenodytes forsteri* breeds annually, laying eggs in midwinter on the sea-ice. The single egg is balanced on the feet and incubated by the male for about nine weeks, during which time he will lose about half his body weight. The chicks hatch at the start of the Antarctic spring and take the whole summer to develop, reaching independence in late summer when they leave the parents. In general, older birds are in better breeding condition and return to breeding colonies earlier in the season, occupying the best sites and producing larger eggs and bigger clutches which give rise to more and healthier young compared to younger breeding birds which have recently been recruited to the colony.

STATUS AND CONSERVATION

Seabirds include some of the commonest as well as some of the rarest birds in the world. Wilson's Storm-Petrel *Oceanites oceanicus* has often been cited as the most numerous bird in the world. In the Antarctic there are single colonies of Adelie Penguin *Pygoscelis adeliae* of over 1 million birds. In the northern hemisphere auks are amongst the most numerous seabirds and millions of pairs of Little Auks *Alle alle* and guillemots *Uria* spp. form vast colonies. Forty-four species of seabirds are considered to be globally threatened. Amongst these are the Madeira Petrel *Pterodroma madeira* which has a known world population of 20-30 pairs while the total population of the closely related Bermuda Petrel *Pterodroma cahow* is only 35 pairs. In 1986 on the Japanese island of Torishima the total population of 146 adult Short-tailed Albatrosses *Diomedea albatrus* raised 77 young. Some species are still virtually unknown; Beck's Petrel *Pterodroma becki* is only known from two specimens taken off Papua New Guinea and the Solomon Islands, while the Mascarene Petrel *Pterodroma aterrima* is known from four specimens and a few recent, sight records, mostly off Réunion.

Seabird populations can be affected by both natural and man-made conditions (fig.291). Climatic factors can influence breeding success, either directly though chick loss due to chilling in cold or prolonged periods of wet weather, or indirectly by poor weather hampering the feeding efforts of parents or delaying the onset of breeding so that the young hatch and fledge at less than opportune times.

Figure 291. A gull entangled in a plastic beer can holder.

Seabirds are vulnerable to a wide range of pollutants through external contact and by ingestion. Oil pollution is a major threat throughout the world. Blowouts from offshore rigs or large spills resulting from accidents can cause massive local mortality, while some seabird populations have suffered a gradual decline resulting from chronic pollution caused by recurring discharges from oil tankers washing out their tanks at sea and from oil escaping into the sea from rivers and coastal industrial sites. A summary of 17 years of beached bird surveys in the Netherlands revealed that of 100,000 birds found dead on the shore, 68% were oiled. Major oil tanker disasters such as the wrecks of the Torry Canyon, Amoco Cadiz and Christos Bitas hit the headlines and became infamous. In one recent accident after the Exxon Valdez went aground in Prince William Sound, Alaska more than 30,000 dead birds of 90 species were recovered and it was estimated that between 100,000-300,000 birds were killed. Besides direct mortality it is known that oiled birds can transfer oil onto their eggs during incubation with adverse effects on hatching success.

Industrial chemicals such as organochlorine compounds, pesticides and heavy metal residues have been implicated in reductions in breeding success. Concentrations of heavy metals in seabird tissue are typically much higher than those found in landbirds.

Another cause for concern is the overfishing of food stocks such as capelin, sand-eels and anchovies. In Peru the anchovy became the base of the world's largest fishing industry in the 1960s. After the 1971 El Niño the fish stocks collapsed, with a huge reduction in the population of seabirds notably the Guanay Cormorant *Phalacrocorax bougainvillii.* Overfishing of pilchards and sardines off South Africa reduced the populations of Cape Cormorant *Phalacrocorax capensis* and Cape Gannet *Morus capensis.* In the North Sea there was a dramatic breeding failure of many seabirds after a period of intensive industrial fishing of sand-eels in the 1980s. Many seabirds also perish as a result of becoming entangled in netting and other waste from fishing boats, while thousands of seabirds are drowned in fishing nets, particularly monofil drift nets.

Other threats to seabirds include predators such as introduced rats, feral cats and mink which have had disastrous local effects on island seabird colonies with the result that many species are now virtually confined to predator-free islands. Habitat destruction by introduced domestic animals such as goats have destroyed the breeding ground of petrels on islands off New Zealand. Pigs from a shipwreck on Clipperton Island in the eastern Pacific reduced a huge population of frigatebirds, Sooty Terns *Sterna fuscata* and boobies (possibly the largest colony of Masked Boobies *Sula dactylatra* in the world) to a few non-breeding pairs. The pigs were eventually destroyed and ten years later the population of boobies had increased to over 4000 pairs.

Direct persecution by man can also have an affect on populations. The hunt of Brünnich's Guillemots/Thick-billed Murres *Uria lomvia* off the Newfoundland/Labrador coast takes between a 0.5-1 million birds annually. The eggs of some species of tropical terns are collected in vast quantities for food, and in West Africa wintering terns from Europe are snared for food. Seabirds have also been killed for their feathers. Five million seabirds were killed on islands off Japan between 1887-1903. Practically the entire population of 3000 Short-tailed Albatrosses *Diomedea albatrus* was killed in 1922-23 and even now, following protection since 1957, the population has only recovered to 300-400 individuals. The exploitation of guano has damaged breeding colonies at a number of locations around the world.

Fortunately many important seabird colonies are now protected as reserves, and organisations such as BirdLife International are taking a lead in the global conservation of seabirds.

BIBLIOGRAPHY

Beazley, M. (ed). 1976. *The Natural World.* Colour Library Books, Godalming.

Brooke, M. 1990. *The Manx Shearwater.* T. & A. D. Poyser, London.

Campbell, B. & Lack, E. (eds). 1985. *A Dictionary of Birds.* T. & A. D. Poyser, Calton.

Collar, N. J. & Andrew P. 1988. *Birds to Watch.* ICBP, Cambridge.

Clements, J. A. 1962. Surface Tension in the Lungs. *Readings from Scientific American; Vertebrate Adaptations.* W. H. Freeman, San Francisco.

Cramp, S. (ed). 1977-85. *The Birds of the Western Palearctic.* Vols. 1-4. Oxford University Press, Oxford.

del Hoyo, J., Elliott, A. & Sagartal, J. 1992. *Handbook of the Birds of the World.* Vol 1. Lynx Edicions, Barcelona.

Furness, R. W. 1987. *The Skuas.* T. & A. D. Poyser, Calton.

Harris, M. P. 1984. *The Puffin.* T. & A. D. Poyser, Berkhamsted.

Harrison, P. 1985. *Seabirds: an identification guide.* Christopher Helm, London.

Harrison, P. 1987. *Seabirds of the World: a photographic guide.* Christopher Helm, London.

Lockley, R. M. 1974. *Ocean Wanderers.* David & Charles, Newton Abbot.

Lofgren, L. 1984. *Ocean Birds.* Croom Helm, London.

McFarland, W. W., Pough, F. H., Cade, T. J. & Heiser, J. B. 1979. *Vertebrate Life.* Macmillan. New York.

Monroe, B. L & Sibley, C. G. 1993. *A World Checklist of Birds.* Yale University Press, New Haven & London.

Nelson, B. 1992. *Seabirds, their biology and ecology.* Hamlyn, London.

Nettleship, D. N. (ed). 1994. *Seabirds on Islands: Threats, Case Studies and Action Plans.* BirdLife Conservation Series No. 1, BirdLife International, Cambridge.

Pennycuick, C. 1972. *Animal Flight.* Studies in Biology No.33. Edward Arnold, London.

Schmidt-Nielsen, K. 1959. Salt Glands. *Scientific American.* 200: 109-116.

Schmidt-Nielsen, K. 1971. How Birds Breathe. *Readings from Scientific American: Birds* (1980). W. H. Freeman, San Francisco.

Sibley, C. G. & Ahlquist, J. E. 1990. *Phylogeny and Classification of Birds: A Study in Molecular Evolution.* Yale University Press, New Haven & London.

Sparks, J. (ed). 1972. *Animal Design.* BBC Publications, London.

Storer, J. H. 1952. Bird Aerodynamics. *Readings from Scientific American: Birds* (1980). W. H. Freeman, San Francisco.

Walls, J. 1942. *The Vertebrate Eye.* Cranbrook Institute of Science, Bulletin 19. Michigan.

Warham, J. 1990. *The Petrels: their ecology and breeding systems.* Academic Press, London.

Welty, J. C. 1955. Birds as Flying Machines. *Readings from Scientific American: Birds* (1980). W. H. Freeman, San Francisco.

Welty, J. C. 1975 (2nd ed.). *The Life of Birds.* Saunders, Philadelphia.

Young, J. Z. 1973 (2nd ed.). *The Life of Vertebrates.* Oxford University Press. London.

SEABIRD FACTSHEET

Watching Seabirds

As might be expected, the greatest variety of seabirds is to be found at sea! For those who want to observe seabirds in their element there is no substitute for an ocean voyage, a pelagic birdwatching trip or a few days aboard a working fishing boat. Organised pelagic trips to watch seabirds now take place regularly in many parts of the world and there is usually an expert guide aboard to help with identification. Birds are attracted by the use of a technique called 'chumming' when a mixture of fish offal, oil and popcorn is poured over the stern of the boat.

Other ways to watch seabirds include 'sea-watching' which usually takes place from a promontory or headland, mostly during the spring or autumn migration periods. The land-based observer has the advantage of a stable position where a telescope can be used, unlike the rolling deck of a ship. Under certain weather conditions, and depending upon the locality, many species of seabirds can be seen from land, often at quite close range. Breeding colonies afford a wonderful opportunity to observe seabirds at close range and to study their behaviour. Although some colonies are on remote, inaccessible islands there are plenty of sites on mainland coasts where access is straightforward and the birds can be observed with little effort. Some seabird colonies are protected reserves which may offer facilities for visitors. The sight, sound and smell of a seabird colony at the height of the breeding season is an unforgettable experience. Many species of divers (loons), sea ducks, gulls, terns and skuas breed inland at marshes, lakes, rivers or on the arctic tundra. Most species of gulls enjoy a close relationship with humans, some following the plough, and others congregating at fishing ports or at rubbish dumps.

Identification

Seabirds include some of the most distinctive birds in the world, easily identified by even the most inexperienced observer. Seabirds also include some species which are amongst the most difficult birds to identify, proving a real headache even for experienced ornithologists. It is often impossible to come to a conclusion about the identity of some individuals and the seabird watcher must be prepared to let some birds 'get away'. The concept of 'jizz' which can be described as a combination of characters including size, shape and flight action is something which can only be learned through experience. Once familiar with the 'jizz' of a particular species the experienced observer can often confidently identify a bird at long range when the finer details of plumage or bare part coloration cannot be seen.

When faced with identifying a particular species it is essential to be able to recognise the family to which the bird belongs. This greatly simplifies the rest of the process. A list of the most important characteristics of the families and key features for the subgroups is given below.

Distribution can be a great help in some circumstances. For example a black auk with white wing patches in the North Atlantic will undoubtedly be a Black Guillemot as no other similar species occurs in the region; in the northwest Pacific a similar looking bird will be a Pigeon Guillemot but off the coast of Alaska both species occur and greater care will be needed to achieve a correct identification. Some species can be found in almost any ocean; for example, Sooty Shearwaters can be seen off the coast of New Zealand and off the coast of Greenland, though normally at different times of year. Many seabirds are great wanderers and are prone to vagrancy outside their normal range. Several species of North American gulls regularly occur on the other side of the Atlantic from Northwest Europe to the coasts of West Africa. However, these vagrants are usually lone individuals which often stand out amongst the more common resident species of the area. Although one should always be aware of the possibility of a vagrant, remember that rare birds are seldom encountered.

Weather conditions have an important role in the identification of birds at sea. Shearwaters and petrels fly quite differently in calm conditions compared with flying head-on into a gale. Strong sunlight can make birds appear much paler than they really are, while foggy conditions often make birds appear both larger and darker than normal. After periods of strong winds or tropical storms some seabirds often appear inshore or even inland. In the polar regions the ever changing size of the icecaps has an effect on the distribution of some species.

Size can be helpful in identification and both total length and wingspan (where known) is given for all species on the plate captions. However, it is notoriously difficult to assess the size of lone or distant birds. Size is most useful when there are several species of birds together so that direct comparison can be made, preferably with a common species with which the observer is familiar.

Age is also an important consideration when identifying seabirds. Most species have a distinctive juvenile plumage, the first set of feathers (rather than down) that they grow. Between juvenile and adult plumage there may be as many as seven stages, all quite different from each other. Some species attain adult plumage in their first year while others may take as long as eight years. It is beyond the scope of this work to describe every plumage stage for every species of seabird, so the descriptions usually refer to adult, juvenile and perhaps one or more intermediate stage. It is necessary to refer to a more specialised work to understand the complex range of plumages that occur in some species. Many seabirds also have distinct breeding and non-breeding plumages (sometimes called 'summer' or 'winter' plumage) and these are described in the text and in some cases illustrated. Many species attain 'breeding' plumage while they are still in their winter quarters or while on migration, and birds in 'non-breeding' plumage may be observed at breeding colonies, usually early or late in the season.

Sex The majority of seabirds are not sexually dimorphic - males and females generally differ only slightly in size, but some groups such as frigatebirds have distinct male and female plumages which, when combined with age differences, make some individuals very difficult to identify. Most sea ducks are sexually dimorphic, as are phalaropes in breeding plumage. Most species of ducks undergo a complex sequence of moults and the full range of plumages is not described here. Reference to a specialist work on wildfowl is recommended.

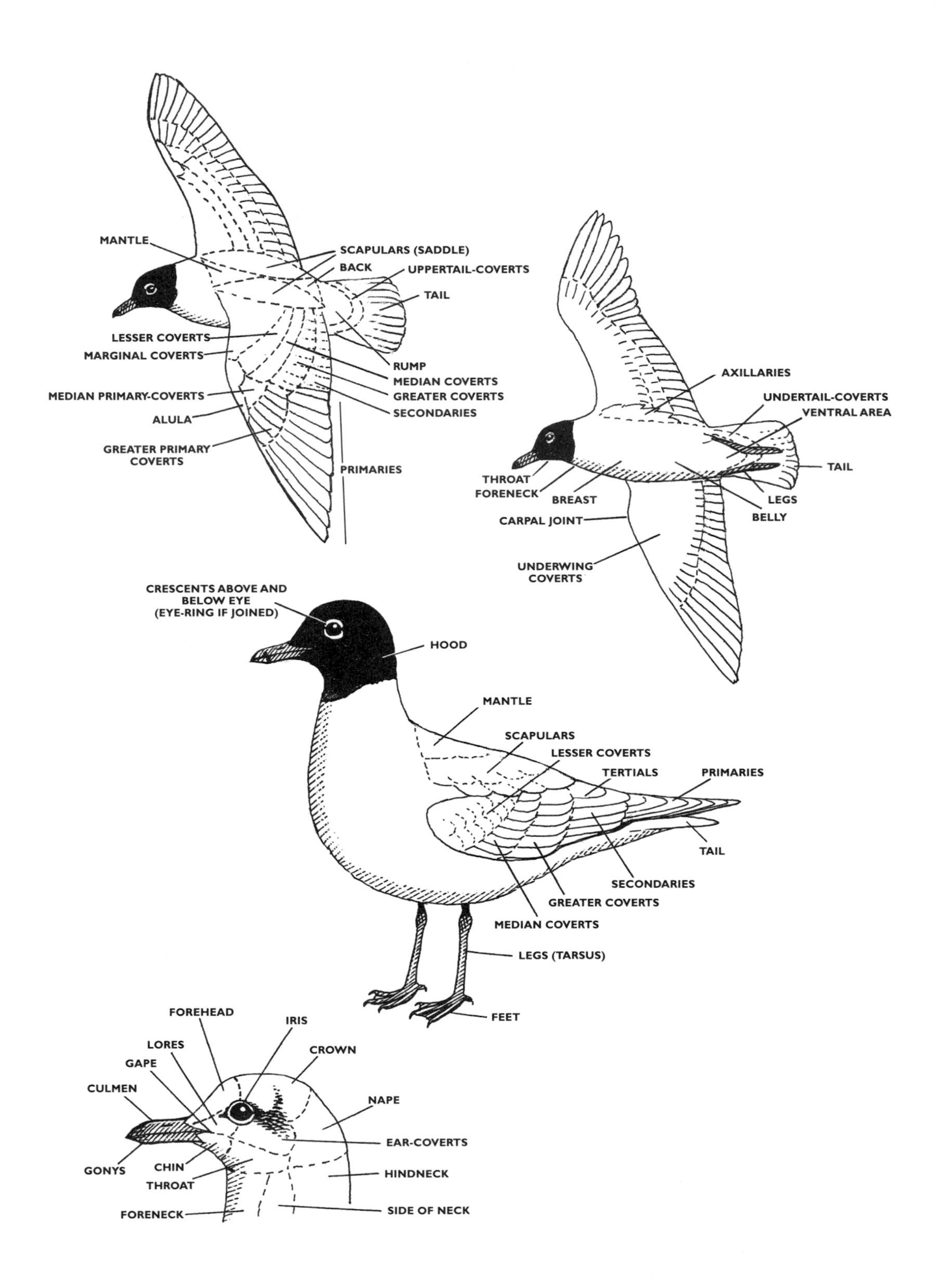

Figure 292. Topography of a Mediterranean Gull *Larus melanocephalus*.

Family Identification

PENGUINS Sphenisicidae

These are amongst the most familiar of birds, even to those who have never seen one in life. Most species are very distinctive and are not likely to prove a problem to identify, at least on land. At sea penguins are much more difficult to identify as they have a low profile in the water and generally dive at the approach of a boat. All penguins are flightless and are rather clumsy on land. In the water they become supreme swimmers and divers using their rigid flippers for propulsion. Some species can dive to great depths. The important features to check are the colour of the bill and any facial patches or facial adornments such as plumes or crests. This is particularly important for the *Eudyptes* group as similar looking species overlap in range in some areas. The *Spheniscus* group look superficially alike but can normally be separated on range.

Chinstrap Penguin *Pygoscelis antarctica* chasing fish

White-billed Diver (adult winter) *Gavia adamsii*

DIVERS or LOONS Gaviidae

These are generally distinctive in breeding plumage but in non-breeding plumage they are superficially alike. Size, bill shape and colour, and contrast between pale and darker parts of the plumage are important. Juveniles resemble non-breeding adults but have pale tips to the feathers of the upperparts which give a scaly appearance at close range. They can be distinguished from cormorants by their pointed bills which lack a hooked tip and tailless appearance which is noticeable in flight or when diving.

GREBES Podicipedidae

Most species are distinctive in breeding plumage but in non-breeding plumage it is important to check size, bill and head shape, and distribution of colour on the head and neck. Juvenile grebes have distinct striped plumages but this is normally lost before the birds move to the coast, by which time they resemble non-breeding adults. The larger species superficially resemble loons or cormorants but are distinguished by their pointed dagger-like bills, thin necks and tailless appearance.

Little Grebe *Tachybaptus ruficollis*

Black-browed Albatross *Diomedea melanophris*

ALBATROSSES Diomedeidae

These huge seabirds fly effortlessly on long stiff wings, soaring over the ocean and sometimes dipping so low into the troughs between waves that their wing tips appear to cut the wave crests. In calm conditions they have great difficulty in taking off from the surface of the ocean and appear clumsy with a heavy flapping flight. They are great wanderers, only coming ashore to breed. At sea they will often follow ships, sometimes for days on end. Identification requires a careful study of both upper- and especially underwing patterns, head pattern and bill colour. Apart from Wandering Albatross they are not sexually dimorphic in plumage. All species of *Diomedea* albatrosses have distinct juvenile and immature plumages. Wandering Albatross has seven different plumage stages, Royal Albatross has four and most of the smaller species of 'mollymawks' have three or four stages. Reference to a specialist work on seabirds is recommended when faced with the intermediate stages of the 'great' albatrosses. The two species of *Phoebetria* albatrosses differ from all others in having pointed tails and largely dark plumage at all ages. In general the southern ocean species do not overlap in range with those of the northwest and central Pacific.

PETRELS and SHEARWATERS Procellariidae
This is a diverse group containing some of the largest and smallest species of seabirds. All have the nostrils fused into a single tube along the top of the bill and all are pelagic, only returning to land to breed.

Fulmars include the two large giant-petrels of the southern oceans which superficially resemble albatrosses. They are heavier, less graceful birds with a rather stiff-winged flight and feed primarily by scavenging. The only diagnostic feature is the colour of the bill tip. The two 'true' fulmars are considerably smaller, superficially gull-like but differ in their stiff-winged flight. Prions are a confusing group; several closely related forms are variously considered to be full species or merely races. All are small blue-grey petrels with a prominent 'M' mark on the upperparts. Identification at sea is almost impossible, though with experience the more distinctive forms can be separated by a combination of head pattern, bill size and the extent of the black tail-tip.

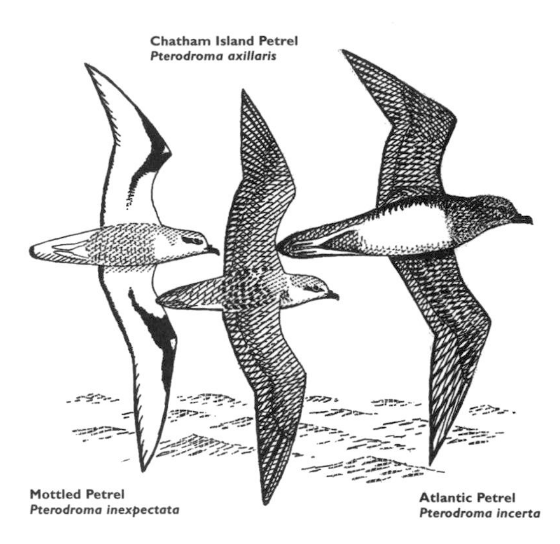

Pterodroma or Gadfly Petrels are an interesting group. Highly pelagic, they rarely follow ships and are often seen only at long range. They have rather short, stubby bills, a strong, purposeful flight on angled wings and some species have a tendency to tower high above the waves, particularly in strong winds. Key features are the head markings, and upper- and especially underwing patterns. Some species are polymorphic, occurring in pale and dark morphs, sometimes with confusing intermediate forms. Several species are extremely rare and little-known and new information on identification criteria is continually adding to our knowledge of their distribution.

The all dark *Bulweria* petrels are distinctively long-tailed and are more likely to be confused with storm-petrels than 'true' petrels.

The large *Procellaria* petrels are a difficult group, differing from Gadfly petrels in having long, thin bills and a rather stiff-winged flight. Given reasonable views, a combination of bill colour, overall size and proportions are aids to identification, but the largely dark-plumaged species require great care and are often impossible to identify with certainty.

The true shearwaters *Calonectris* and *Puffinus* include a number of very similar and confusing forms. As knowledge of the identification criteria improves, the taxonomy of this group changes. Furthermore, new species are still being discovered. The larger species have a typically 'shearing' stiff-winged flight, alternatively showing underside then upperside as the bird follows an undulating course above the waves. The smaller species have a more fluttering flight interspersed with periods of gliding. The 'jizz', particularly the manner of flight, is particularly important in identifying shearwaters. Other key features are the colour and pattern of the head and face, in particular the presence of a cap or collar, the colour of the rump and undertail-coverts and in some cases the pattern of the underwing. Bill and leg colour can also be useful though these are more difficult to record accurately at sea. At breeding colonies most species are largely nocturnal.

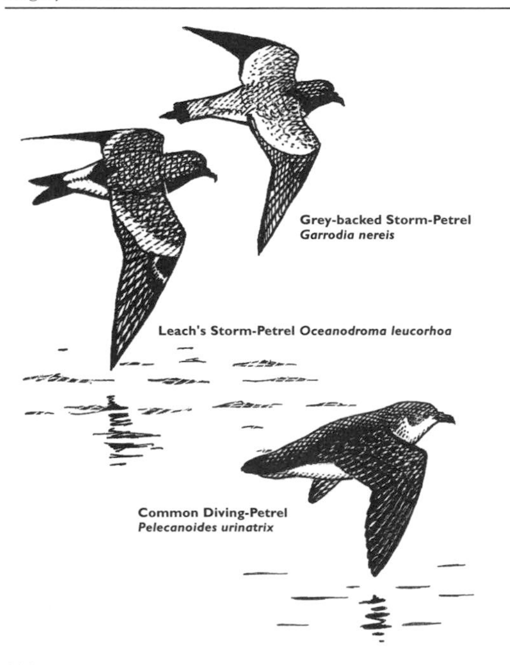

◀ **STORM-PETRELS Hydrobatidae**
Amongst the smallest of seabirds this group presents some great challenges in identification. Critical features are the colour and shape of the rump patch (formed by the rump, uppertail-coverts and lateral undertail-coverts), presence or absence of a diagonal bar on the upperwing, the pattern of the underwing-coverts, tail shape and leg length. Most southern species have rather rounded wings while the northern hemisphere *Oceanodroma* have longer, more pointed wings. Jizz is also important, particularly feeding behaviour and flight. Some species habitually follow in the wake of boats and most can be attracted to 'chum'. At breeding colonies they are largely nocturnal.

◀ **DIVING-PETRELS Pelecanoididae**
This small group is virtually impossible to identify at sea with the exception of the reasonably distinctive Magellanic Diving-Petrel. Even in the hand it is sometimes impossible to identify some individuals. They are rather small, dumpy and thickset with a superficial resemblance to the northern hemisphere auks. They have a distinctive flight, fast and low with rapid wing beats, and they enter the water in full flight, simply flying into the wave crests. They are nocturnal at breeding colonies and are sometimes attracted to the lights of a ship, particularly in misty conditions. Reference to a specialised work on seabirds is essential but most sightings will be impossible to resolve.

TROPICBIRDS Phaethontidae

These are amongst the most beautiful and graceful of seabirds and generally present no identification problem if seen well. Key features are the colour of the bill and tail streamers and the pattern of the upperwing. Juveniles are more difficult to identify and observers should look for a nuchal collar, the extent of barring on the upperparts and the amount of black on the outer primaries. As their name suggests they are confined to tropical and sub-tropical waters.

Red-billed Tropicbird ***Phaethon aethereus***

Pink-backed Pelican
Pelecanus rufescens

PELICANS Pelecanidae

Even the most inexperienced observer will have no difficulty in recognising the familiar shape of a pelican. Distribution is helpful as few similar looking species occur together. The largely white pelicans are best distinguished by a combination of bill and leg colour and by the wing pattern in flight. The two brownish pelicans only overlap on the northwest coast of S America. They are best distinguished by a combination of size and head pattern.

GANNETS and BOOBIES Sulidae

These have distinctive cigar-shaped bodies, long narrow wings and pointed tails. They have a direct flight with periods of flapping followed by a glide, and often fly in line formation. Most species feed by plunge-diving in a spectacular manner. Some species are polymorphic and all have distinctive juvenile and immature plumages that are quite complex. The inexperienced observer may confuse the darker immature stages with albatrosses or larger shearwaters but attention to structure and flight should avoid this pitfall. The breeding distribution of gannets does not overlap. However, off West Africa two species occur together and vagrants may occur elsewhere. Gannets are best distinguished by a combination of wing and tail pattern or, at close range, by the length of the gular stripe. Boobies are distinguished by a combination of plumage features, particularly head and underwing patterns, and foot colour is diagnostic in two species.

Red-footed Booby (Galapagos white morph) ***Sula sula***

European Shag ***Phalacrocorax aristotelis***

CORMORANTS and SHAGS Phalacrocoracidae

This large group appears to present many identification problems. However, distribution is often helpful as most similar-looking species do not occur together. All have rather long necks, hooked bills, elongated bodies and long tails. They feed by diving and pursuing fish underwater. Many species are gregarious throughout the year. Their feathers are not completely waterproof and consequently birds come ashore to dry their plumage, often with wings held in a spread-eagle posture. The most important features are the colour of bare facial skin, bill and legs, the distribution of white on face, underparts, and rump, and the presence and shape of facial plumes or crests. Some species are largely confined to freshwater and are unlikely to be seen away from coastal lagoons or estuaries.

FRIGATEBIRDS Fregatidae
These are spectacular, highly aerial long-winged seabirds with deeply forked tails. They do not land on water but soar above the sea swooping down to pick food from the surface or to chase other seabirds in order to rob them of their catch. All are sexually dimorphic and have a range of confusing immature plumages which may last as long as six years. Great care is needed to distinguish the different species and particular attention should be given to the exact pattern of white on the underparts and underwing. In some plumages head colour and breast band are also important features. Reference to a specialised work on seabirds is almost essential to separate the species and to understand the complex plumage sequences.

Ascension Frigatebird *Fregata aquila*

Three species of ducks which can be found on inshore waters in the non-breeding season: Smew *Mergus albellus* a) male and b) female. Scaup *Aythya marila* c) male and d) female. Tufted Duck *Aythya fuligula* e) male and f) female.

SEA DUCKS Anatidae
Ducks and geese comprise a huge family, a few of which regularly occur in the maritime environment where they are generally confined to inshore waters, particularly during the non-breeding season. Besides the species illustrated on plate 36, many other species of ducks and geese occur on inshore waters at times. Most species are sexually dimorphic and have distinct first-winter plumages which mostly resemble that of the female. Most species undergo a moult after breeding; in males the resulting plumage stage is called 'eclipse' plumage and generally resembles that of the female. Most species do not assume adult plumage until the second year. As plumage, size and shape differ greatly between species it is difficult to generalise about specific identification features. Head pattern and bill shape are often important as is the wing pattern of flying birds. The steamerducks are notoriously difficult to identify but distribution can be helpful. With all species of sea ducks and geese, reference to a specialised work on wildfowl is recommended.

PHALAROPES Scolopacidae
Two species of these diminutive shorebirds spend the non-breeding season at sea where they are gregarious, often forming large flocks along an area of ocean upwelling. They swim jerkily, bobbing about on the surface picking at minute items of food and occasionally spinning as they do so. The colour and pattern of the upperparts including the face and the proportions of the bill are key features for identification. A third species, Wilson's Phalarope *Steganopus tricolor* which winters in S America, does not normally occur at sea.

Red-necked Phalarope (female summer) *Phalaropus lobatus*

Snowy Sheathbill *Chionis alba*

SHEATHBILLS Chionididae
This distinctive family comprises just two species. Both occur only in the southern hemisphere, breeding in the Antarctic and subantarctic, with one species moving north to Patagonia outside the breeding season. The plumage of both species is largely white and they can be distinguished by bare part coloration. On land they are inquisitive and rather chicken-like, with sturdy legs (for running) and unwebbed feet. In flight they are more pigeon-like. They are notorious scavengers and are often found around penguin and seal colonies.

Arctic Skua (adult light morph) *Stercorarius parasiticus*

◀ **SKUAS or JAEGERS Stercorariidae**

The larger *Catharacta* skuas are large gull-like birds with strongly hooked bills, largely brown overall with prominent white wing flashes. The species are very similar and difficult to separate, but a combination of overall plumage tones, a cap or collar and, in some cases, underwing pattern may help. The taxonomy of the group is confusing and some hybridisation may occur. They feed mainly by scavenging, chasing other seabirds to rob them of their catch and also, on occasion, to drown them. The *Stercorarius* group are smaller, more agile birds. They are polymorphic (though dark Long-tailed Skuas may be melanistic examples) and have distinctive barred juvenile plumages. In breeding plumage adults have diagnostic tail shapes, though these may be lost in winter when plumages resemble those of immature birds. Important features are size and structure, tail shape, the size of the white wing flash and the distribution of barring (particularly in winter or immature plumages). *Stercorarius* skuas feed mainly by harrying terns and gulls in order to rob them of food though in the breeding season rodents, young birds and carrion are also eaten. They are swift and powerful fliers which show great agility when chasing other seabirds.

GULLS Laridae

Gulls, many of which breed inland, are amongst the most familiar of seabirds. With a few notable exceptions they can be divided into the larger species which have white heads and a contrasting grey or black mantle and the smaller species which have dark brown or blackish hoods during the breeding season. The dark hood is generally reduced to a dusky patch behind the eye (ear-spot) during the non-breeding season. All have distinct juvenile plumages and the larger species have a series of intermediate stages until they attain adult plumage in the fourth year. Key features are the colour of the bill and legs, the tone of the mantle and the wing pattern, in particular the exact distribution of black, white or grey on the primaries. In hooded species, the colour and extent of the hood are important as are the size and shape of white eye crescents. In winter plumage, the distribution of darker patches on the sides of the head or crown are useful. Juvenile and immature gulls are more of a problem to identify. The pattern of the upperwing is important as are size and structure, and in particular bill size and shape, and the presence or absence of a tail band. A knowledge of how to age a bird and familiarity with the various plumage stages associated with age are important and reference to a specialised work on gulls is recommended. The taxonomy of some of the larger species, particularly the Iceland/Herring/Lesser Black-backed Gull complex is confusing. Hybrids occur and identification may be impossible in some cases. The large number of similar-looking species is at first confusing but distribution is often helpful though many gulls are prone to vagrancy and great care should be taken when faced with an unusual individual.

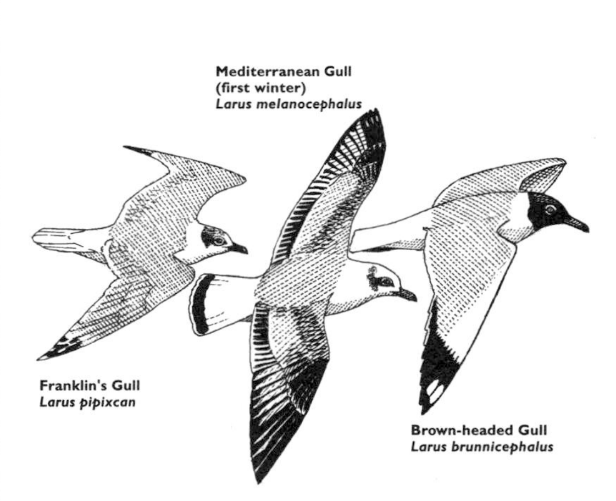

Mediterranean Gull (first winter) *Larus melanocephalus*

Franklin's Gull *Larus pipixcan*

Brown-headed Gull *Larus brunnicephalus*

TERNS Sternidae

Terns are amongst the most graceful of seabirds and are colloquially called 'sea-swallows'. They can be divided into several groups. The small *Chlidonias* or 'marsh terns' are mainly associated with freshwater habitats and have a distinctive flight action which, to an experienced observer, allows them to be identified on jizz alone. The breeding plumage is generally contrasting with white, dark grey or black; identification in this plumage is quite straightforward. Juveniles and birds in non-breeding plumage are much more difficult and particular attention should be paid to the extent of the dark cap, presence of breast markings or contrasting 'saddle', upperwing

Sooty Tern *Sterna fuscata*

pattern and the colour of the rump and tail. At close range, bill structure and colour are also useful. The typical *Sterna* terns can be subdivided into two groups; those that are mostly grey and white with a black cap in breeding plumage and those which have dark brown or grey upperparts and contrasting white underparts. The 'capped' terns vary considerably in size from the large Caspian Tern *Sterna caspia* to the diminutive Least Tern *S. antillarum.* Key features are size and structure (particularly at rest when bill shape, leg length and the length of the projection of the tail beyond the wing tips should be checked), bill and leg colour, the pattern of the outer primaries on both upper- and underwing, and the colour of the outer tail feathers. In winter plumage the black cap is lost or reduced to a dark mark behind the eye, sometimes continuing to form a band on the hind crown. Juveniles have distinct plumages usually with some brownish scaling on the upperparts and a pattern on the upperwing which often shows a carpal bar. Adult plumage is not attained until the second or third year and the confusing intermediate plumage stages present a great challenge even to experienced observers. The darker-backed group of terns can be identified by noting the colour of the upperparts, including contrasting collar if present, and the shape of the white forehead.

SKIMMERS Rynchopidae

Skimmers are large, distinctive birds with specialised feeding habits. Although superficially similar to terns, they are not very closely related. They have long, narrow bills with the lower mandible longer than the upper. The wings are long and pointed, but the tail and legs are rather short. Skimmers feed low over water, ploughing the surface with the longer lower mandible; when the latter touches a prey item, the bill snaps shut. Feeding takes place mainly around dawn or at dusk. The three species do not overlap in range and identification is straightforward. Two species are frequent in coastal locations; Indian Skimmer is largely a freshwater species (see map on p.477).

Black Skimmer
Rynchops niger

AUKS Alcidae

This distinctive group is confined to the colder waters of the northern hemisphere where they are the ecological counterparts of penguins. They are mostly rather thickset, dumpy birds with a whirring flight and rapid wing beats. They dive and swim underwater using their wings for propulsion. Most species are black and white or greyish, often with distinctive head ornaments and bill shapes. They have distinct non-breeding plumages when facial ornaments are lost. Distribution is helpful as some species are confined to certain regions such as the North Atlantic or Northwest Pacific. Particular attention should be paid to overall size, the shape and colour of the bill, facial adornments such as crests or plumes, and the general distribution of black, white or grey, particularly around the head which is especially important in non-breeding plumage. Identification is usually straightforward at breeding colonies but in winter or at sea it can be very difficult to separate similar species.

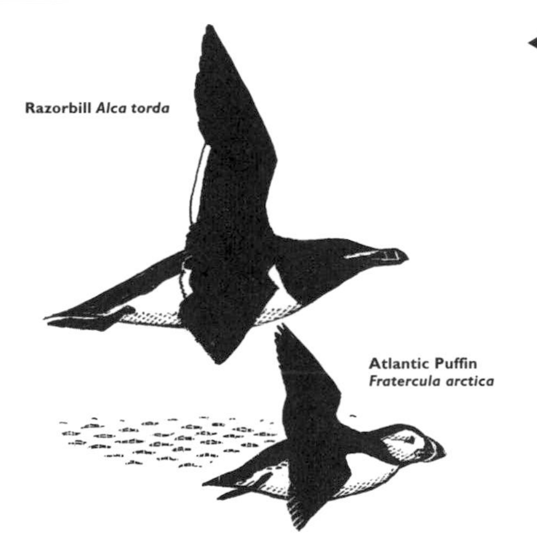
Razorbill *Alca torda*

Atlantic Puffin
Fratercula arctica

CETACEANS

by

Sara Heimlich-Boran

INTRODUCTION

Figure 293. Humpback Whale *Megaptera novaeangliae* breaching.

Whales, dolphins and porpoises are known as cetaceans and are placed in the order Cetacea (from the Greek 'ketos' and the Latin 'cetus', both meaning a large marine creature or sea monster). Cetaceans are mammals, yet they live like fish, conducting their entire lives, from birth to mating and death, underwater. With their wholly aquatic lifestyles the typical features of mammalian existence – breathing air, maintaining a constant body temperature, giving birth to live young and nursing offspring – present many difficulties. Of the few groups of mammals that have successfully returned to the sea, cetaceans exhibit the most profound and marvellous anatomical and physiological adaptations for surviving and thriving in a medium which is essentially hostile to mammalian life. The wide variety of habitats in which cetaceans live and their diverse systems of structural and social organisation are testimony to their remarkably complete secondary adaptation to the aquatic environment.

In the past, the only way of studying whales was by the analysis of dead specimens collected from whaling activities and strandings. Our knowledge of cetacean biology remains surprisingly scant; even in the 1990s, new species have been described and the taxonomy and evolutionary history of Cetacea is still being pieced together. Whales and dolphins are becoming familiar from oceanaria, films and television exposure. Yet despite their popularity and wide distribution throughout the world's oceans, few cetacean species are known in any detail. The oceanic lifestyle of many cetaceans makes them difficult to study and this inaccessibility imposes many restrictions on the type of scientific work that can be attempted.

EVOLUTION

Cetaceans evolved from the same ancestral mammals that gave rise to modern even-toed ungulates (deer, camels, pigs, cows and hippos). The study of cetacean evolution relies on fragmentary fossil remains and inevitably there are large gaps in the fossil record. For example, although certain unique features of the anatomy of the Gray Whale *Eschrichtius robustus* suggest one of the most ancient of origins, it is paradoxical that Gray Whale fossils date only to 100,000 years ago. These fossil bones are virtually indistinguishable from the bones of Gray Whales living today.

Primitive whale-like mammals (archaeocetes) evolved during the Eocene period in the brackish estuaries of the Tethys Sea, in the region of what is now the Mediterranean-Arabian Gulf area. The oldest known fossil archaeocete is *Pakicetus* which lived about 50 million years ago. The archaeocetes became extinct at the end of the Eocene period (37 million years ago). During the following Oligocene period two specialist groups of whale-like mammals evolved. Squalodontids, kentriodontids and other odontocete groups developed the means to catch other animals, their long toothy jaws giving them the ability to hunt fast-moving prey. The cetotheriids (early mysticetes) developed an enlarged mouth and exploited swarms of zooplankton, becoming slow-swimming filter feeders as sieve-like baleen gradually replaced teeth. However, all these groups became extinct by the end of the Tertiary period.

During evolution, a streamlined body shape developed with the migration of the nostrils to the top of the head, accompanied by a pronounced elongation ('telescoping') of the skull in which the jaws and face bones extended far forward of the bony entrance of the nares (nostrils) in the skull. Hair was lost, and teats and genitals became concealed in slits in the body wall. The external ear flaps disappeared, leaving only a pinhole opening. The two forelimbs became flattened paddle-like flippers (still containing the finger, hand, and lower and upper arm bones) which rotated at the shoulder joint for hydrodynamic stability and steering. The hind limbs and pelvic girdle disappeared almost completely, leaving only small bony remnants embedded in the side muscles (fig.294). Fibrous, horizontally-flattened tail flukes developed for propulsion.

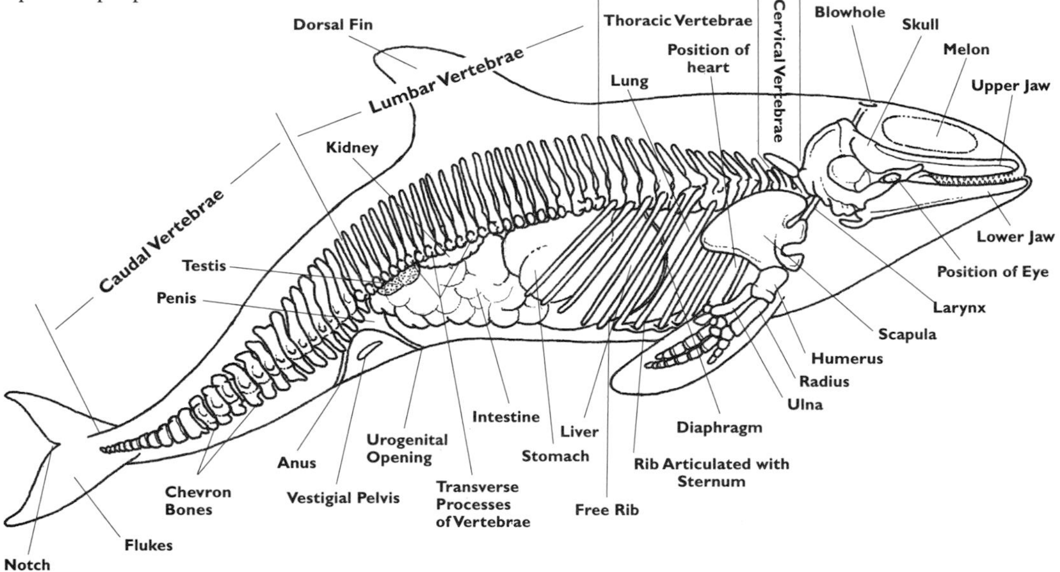

Figure 294. Dolphin anatomy.

Through the millennia of the Tertiary period, continental drift and the opening and closing of seaways permitted the spread of whales around the globe. Split by oceanographic or physical barriers, populations evolved different characteristics. As habitats changed, further structural modifications and adaptations evolved.

Whale fossils with recognisably modern features of present-day cetaceans can be traced back 23-25 million years in the early Miocene period, including sperm whales (Physeteridae) river dolphins (Platanistidae), beaked whales (Ziphiidae) and right whales (Balaenidae). Modern forms of the Humpback Whale *Megaptera novaeangliae*, Blue Whale *Balaenoptera musculus*, Fin Whale *B. physalus* and porpoises (Phocoenidae), can be traced back 10-11 million years in the late Miocene period. Dolphins (including the Killer Whale *Orcinus orca*, and pilot whales *Globicephala* spp.), the Beluga *Delphinapterus leucas*, and Narwhal *Monodon monoceros* are some of the most recently evolved cetaceans.

All present-day cetaceans existed about 5 million years ago in the Pliocene period, but their distributions at that time were still in the process of change. Porpoises probably originated in the cooler regions of the North and South Pacific, radiating from there to other parts of the world as recently as 4 million years ago. The fossil remains of the earliest known Narwhals and Belugas have been found in California and Mexico, yet today they have mainly Arctic distributions.

TAXONOMY

The majority of cetacean species evolved and became extinct long before the appearance of modern humans. There are about 78 species of living cetaceans known at present which are divided into two taxonomic groups or suborders (table 10), based primarily on the type of feeding apparatus in the mouth. There is a third suborder comprising the extinct fossil relatives of modern whales (Archaeoceti).

ORDER: CETACEA

Suborder: Mysticeti

Family: Balaenidae
- *Balaena* - Bowhead Whale and Black Right Whale

Family: Balaenopteridae
- *Balaenoptera* - Minke (Lesser Rorqual), Sei, Bryde's, Blue and Fin Whales
- *Megaptera* - Humpback Whale

Family: Eschrichtiidae
- *Eschrichtius* - Gray Whale

Family: Neobalaenidae
- *Caperea* - Pygmy Right Whale

Suborder: Odontoceti

Family: Delphinidae
- *Cephalorhynchus* - Commerson's, Chilean (Black), Heaviside's and Hector's Dolphins
- *Delphinus* - Common Dolphin
- *Feresa* - Pygmy Killer Whale
- *Globicephala* - Short-finned and Long-finned Pilot Whales
- *Grampus* - Risso's Dolphin
- *Lagenodelphis* - Fraser's Dolphin
- *Lagenorhynchus* - Atlantic White-sided, White-beaked, Peale's, Hourglass, Pacific White-sided and Dusky Dolphins
- *Lissodelphis* - Northern and Southern Right Whale Dolphins
- *Orcaella* - Irrawaddy Dolphin
- *Orcinus* - Killer Whale (Orca)
- *Peponocephala* - Melon-headed Whale (Electra Dolphin)
- *Pseudorca* - False Killer Whale
- *Sotalia* - Tucuxi
- *Sousa* - Pacific, Indian Ocean and Atlantic Humpback Dolphins
- *Stenella* - Pantropical Spotted, Clymene, Striped, Atlantic Spotted and Spinner Dolphins
- *Steno* - Rough-toothed Dolphin
- *Tursiops* - Bottle-nosed Dolphin

Family: Monodontidae
- *Delphinapterus* - Beluga
- *Monodon* - Narwhal

Family: Phocoenidae
- *Australophocaena* - Spectacled Porpoise
- *Neophocaena* - Finless Porpoise
- *Phocoena* - Common (Harbour) Porpoise, Vaquita, Burmeister's Porpoise
- *Phocoenoides* - Dall's Porpoise

Family: Physeteridae
- *Kogia* - Pygmy and Dwarf Sperm Whales
- *Physeter* - Sperm Whale

Family: Platanistidae
- *Inia* - Amazon River Dolphin (Boto)
- *Lipotes* - Yangtze River Dolphin (Baiji)
- *Platanista* - Ganges and Indus River Dolphins
- *Pontoporia* - La Plata Dolphin (Franciscana)

Family: Ziphiidae
- *Berardius* - Arnoux's and Baird's Beaked Whales
- *Hyperoodon* - Northern and Southern Bottlenose Whales
- *Mesoplodon* - Sowerby's, Andrews', Hubbs', Blainville's, Gervais', Ginkgo-toothed, Gray's, Hector's, Strap-toothed, True's, Pygmy, Stejneger's and Longman's Beaked Whales
- *Tasmacetus* - Shepherd's Beaked Whale
- *Ziphius* - Cuvier's Beaked (Goose-beaked) Whale

Table 10. A classification of living Cetacea adapted from the arrangement of Mead and Brownell (1993). Species are listed for each genus using common names (alternative names in brackets).

The Mysticeti ('moustached' or baleen whales) contain 10 or 11 species of baleen-bearing, filter-feeding whales. The Odontoceti (toothed whales) comprise about 68 living species of whales, dolphins and porpoises which have teeth in their jaws. The further taxonomic subdivisions in both suborders are still debated. There are 10 to 15 or more proposed groups (families), some of which contain as few as a single species. The largest family (Delphinidae) contains over 30 species of dolphins.

Whereas there are clear-cut anatomical differences between porpoises and dolphins which necessitate grouping them in two separate families, anatomical differences between dolphins and the larger Killer Whale *Orcinus orca*, False Killer Whale *Pseudorca crassidens*, Pygmy Killer Whale *Feresa attenuata*, and pilot whales *Globicephala* spp. are relatively minor; it appears difficult to justify separating these larger dolphins into a family of their own (Globicephalidae of some authors). Intermediate-type species that have external features of both groups, such as the Melon-headed Whale *Peponocephala electra*, also complicate such divisions.

The terms whale, dolphin and porpoise can be useful as a general guide to size as well as giving taxonomic information. Cetacean species with body lengths of 3m or more are usually called whales, and those species smaller than this generally fall into one of two categories, dolphins or porpoises, the latter being the smaller of the two. However, the Spectacled Porpoise *Australophocaena dioptrica* is larger than some dolphins (*Cephalorhynchus, Sotalia* and some *Lagenorhynchus* species) and the adult Bottle-nosed Dolphin *Tursiops truncatus* can be larger than the adult Dwarf Sperm Whale *Kogia simus*. Therefore it is evident that as far as length is concerned, there is an overlap where fixed terms are not exactly applied.

The baleen whales can be placed in four families:

1. Balaenidae (containing the right whales, *B. mysticetus* and *B. glacialis*)
2. Balaenopteridae (containing rorquals such as the Blue Whale *B. musculus* and Humpback Whale *M. novaeangliae*)
3. Eschrichtiidae (containing a single species, Gray Whale *E. robustus*)
4. Neobalaenidae which contains a single species (Pygmy Right Whale *Caperea marginata*)

The toothed whales can be placed in six families:

1. Delphinidae (containing the dolphins and including the Killer Whale and pilot whales)
2. Monodontidae (containing the Beluga and Narwhal)
3. Phocoenidae (containing the porpoises)
4. Physeteridae (containing the sperm whales)
5. Platanistidae (containing the river dolphins and the La Plata Dolphin *Pontoporia blainvillei*)
6. Ziphiidae (containing the beaked and bottlenose whales)

The external characteristics of marine odontocete and mysticete species are described in plates 45-53.

HABITATS AND DISTRIBUTION

Cetaceans can be found in many types of riverine, estuarine and marine habitats, from muddy coastal shallows to clear tropical oceanic waters and icy polar seas. The type of habitat occupied has a profound influence on lifestyle, impacting every aspect of cetacean biology from feeding to breeding and geographical distribution.

Riverine and coastal species have fairly predictable sources of food and fixed calving or nursing sites. They seldom undertake long seasonal movements, and have relatively small home ranges. River dolphins living in large continental river systems and Heaviside's Dolphin *Cephalorhynchus heavisidii* of the cold coastal Benguela current are examples of species that have become highly adapted to particular habitat types with restricted geographical distributions.

Offshore species forage opportunistically, move great distances on a daily basis and have broad 'home' ranges. Some, like the Sperm Whale *Physeter catodon* and pilot whales, favour deep areas off the continental shelf, over oceanic canyons and near volcanic islands. Spinner and Spotted Dolphins, *Stenella* spp., and Rough-toothed Dolphins *Steno bredanensis*, are typical oceanic dolphins with a worldwide (cosmopolitan) distribution, keeping seaward of the continental shelf and often limiting coastal visits to deep-water oceanic islands.

Many species have a 'pantropical' distribution and live in warm equatorial waters around the globe. Others are 'circumpolar', having a continuous distribution around the world at high latitudes. Almost all rorquals have a continuous distribution from polar waters to the equator. However, the equator acts as a barrier between populations because the seasons in the two hemispheres are six months out of phase. Of the species which can be found in all oceans, the most ecologically adaptable and truly cosmopolitan are the Killer Whale and the Bottle-nosed Dolphin. Both species occur from high latitudes to the tropics in stable coastal groups, or in open ocean roaming groups.

MIGRATION

Cetaceans respond to seasonal changes in water temperature and the availability of prey. The extent of seasonal movements is quite variable among different species. Bowhead Whales *Balaena mysticetus*, Narwhals and Belugas synchronise their movements with the annual formation, break up and drift of Arctic pack ice. The seasonal relocation of Humpback and Gray Whales away from the polar winter occurs through the open ocean and along coastlines over thousands of kilometres. These two species are the only cetaceans that have truly predictable migrations, apparently following the same routes each year.

Generally, populations of baleen whales divide their year between high latitude, cold water feeding grounds and low latitude, warm water breeding grounds (fig.295). In both hemispheres, feeding is concentrated in the short polar summer months when plankton blooms fuel the food chain. Here the pregnant females gorge themselves and foetal growth rapidly increases. The inability of newborn calves to survive in the cold water of the feeding grounds necessitates the migration of females to warmer waters where birth occurs. This is the case for all migratory species except the Bowhead Whale which is the only mysticete that calves in polar seas. Adult males also migrate to gain access to adult females for breeding once they have given birth. Different reproductive and age classes leave and arrive at different times and a 'population' may spread over vast distances. Little or no food is consumed during the six to nine months between round trips to the feeding areas; if adequate food is found on the way, some migrants will stay and not complete the journey to high latitudes.

Odontocetes usually do not make these long annual migrations. Their seasonal responses tend to be onshore-offshore movements. However, in both hemispheres Sperm Whales are the exception. Most mature males move to colder seas during the spring and summer, in bachelor herds or as individuals (lone bulls). Females and calves remain in warm waters all year round.

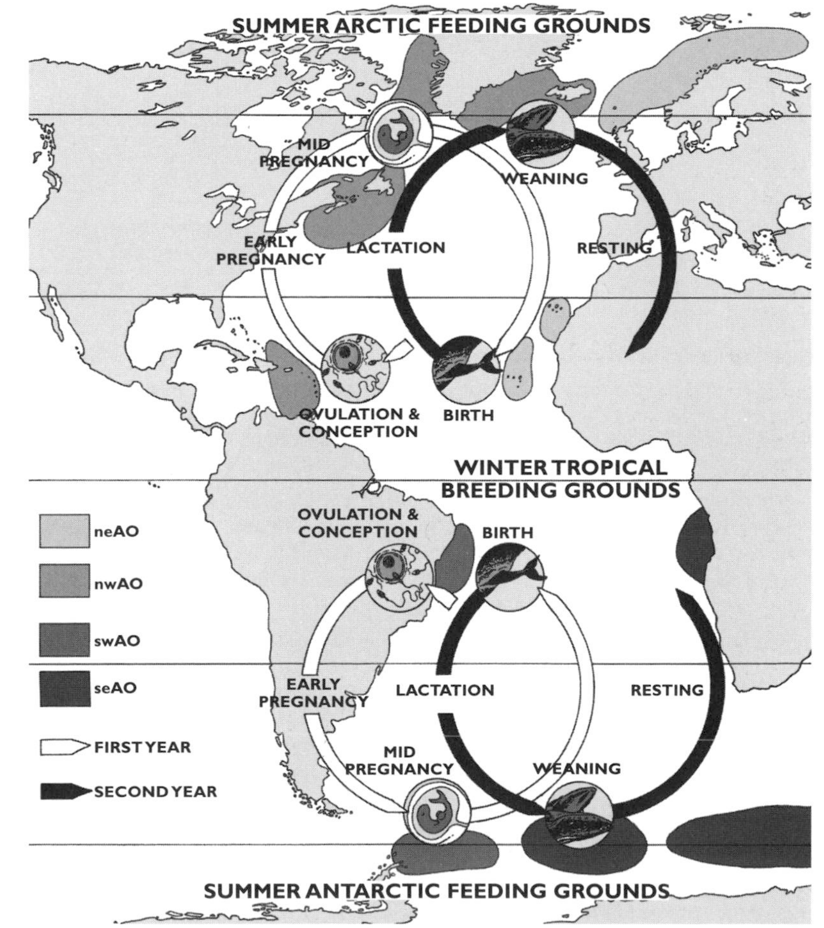

Figure 295. Humpback Whale *Megaptera novaeangliae* migration and breeding cycle in four Atlantic Ocean populations.

FOOD AND FEEDING

Cetaceans have a body core temperature of 37°C. The rigours of seasonal water temperature changes, rapid movement into colder water during deep diving, metabolic changes occurring during reproduction and patchy food resources all affect the daily requirements for energy. Prey consumption averages from 1.5 to 14% of body weight per day. Smaller cetaceans need to consume more in order to compensate for a higher rate of heat loss (a consequence of their higher body surface area to body volume ratio).

Unlike odontocetes, migrating baleen whales experience extreme but regular annual oscillations in food intake. Food representing a year's energy requirements is consumed and stored as blubber within a 4-5 month period on the polar feeding grounds. They feed little, if at all, in the tropics and the 40% weight gain in polar waters is lost gradually over the rest of the year, as the blubber energy store is used up.

Odontocete teeth are generally cone-shaped and adequate for grasping prey, which is then swallowed whole. Porpoise teeth are spade-shaped, and exclusive squid-eaters (such as beaked whales) have few or no visible teeth. The first set of teeth are retained throughout their lives (unlike other mammals which shed their milk teeth). The teeth of some species are so highly modified that they are no longer used in feeding (fig.296, plates 47, 48).

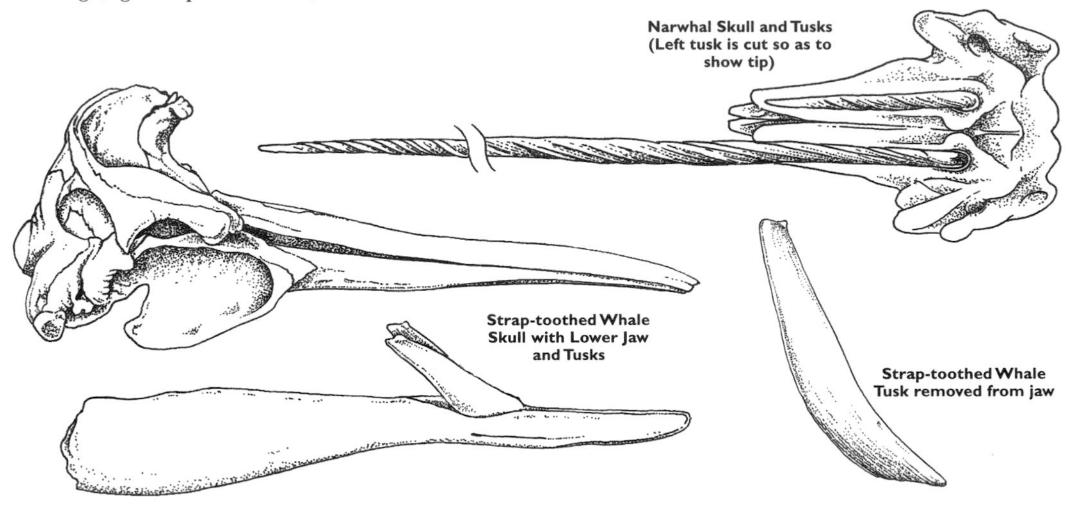

Figure 296. Skulls and tusks of Narwhal *Monodon monoceros* and Strap-toothed Whale *Mesoplodon layardii*.

The diet of most odontocetes reflects the relative abundance of available prey and they are essentially opportunistic feeders. Most eat a mixture of fish, squid and, to a lesser extent, crustaceans. Killer Whales, False Killer Whales and Pygmy Killer Whales are the only species known to hunt warm-blooded prey; the Killer Whale diet includes a wide variety of warm-blooded prey such as other cetaceans, seals and penguins.

Odontocetes use vision and sonar for locating prey, combined, in some species, with co-operative hunting behaviour. Oceanic dolphins surround and trap fish schools against the sea surface, and then take turns gorging themselves on the dense ball of prey. Killer Whales have been known to tip up ice flows to seize penguins or seals, surf up onto steep beaches to snatch seals, and stun herring by lashing them with their tails. Belugas have some lip mobility and may simply suck their prey off the bottom.

Mysticetes (except Gray Whales) eat krill and other zooplankton, schooling fish or small squid depending on the size, flexibility and density of their baleen plates. Formed from springy tissue similar to human fingernails, baleen plates are triangular in shape and arranged with their bases embedded in the roof of the mouth like dishes stacked in a drying rack (fig.297). There are up to 480 plates on each side of the upper jaw.

The hair-like fringe on the inner edge of each plate overlaps with those of neighbouring plates to form a dense fibrous mat which traps food as water passing through is pushed out of the mouth. Food is rasped off and swept into the throat with the tongue. As the tip of the baleen plate wears down, new tissue is added at its base in the gum where the plate is continuously forming.

Right whales have long, elastic baleen plates (the favoured material in Victorian corsets) hanging from highly arched jaws that provide the largest surface area for filtering. The fine inner fringes of the plates sieve tiny zooplankton (copepods) from the water. Right whales are 'skimmers', swimming along with the mouth slightly open, allowing a continuous flow of water in at the front, through the baleen and out at the sides. Skimming is not limited to the surface, but is also used at varying depths in the water column.

Rorquals (Balaenopteridae) have shorter, coarse-fringed baleen and generally feed on larger prey. The highly elastic throat pleats and tissues that cover the entire underside from the jaw tip to the umbilicus in some species permit the intake of large quantities of water. Rorquals are 'lunge and gulp' feeders, filling the throat pouch by lunging through the water with their mouths wide open (the gape may be as much as 90°) and forcing the water out through the baleen between the jaws by contracting muscles in the throat pleats. However, Sei Whales *Balaenoptera borealis*, which have finer baleen fringes, are also skim feeders.

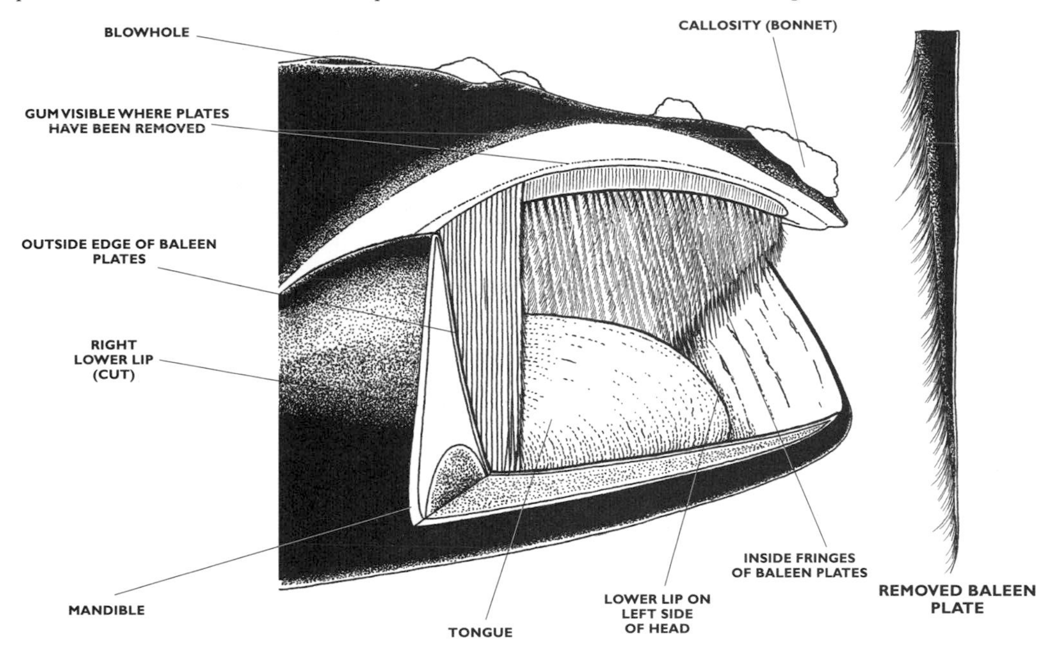

Figure 297. Section through the mouth of a Black Right Whale *Balaena glacialis.*

Gray Whales have short, coarse baleen and feed on gammarid amphipods, molluscs, crabs and polychaete worms which live in mud on the sea floor. They are the only benthic-feeding mysticetes. They 'plough' the substrate, sucking mud into the mouth with the large muscular tongue and squeezing out the sediment with the tongue pressed against the palate. The prey is trapped by the baleen sieve.

Other mysticetes employ a wide range of complex techniques to assist feeding. Fin Whales use the white marking on the right side of their heads to scare prey into tight schools as they circle to the surface. Humpbacks blow bubble clouds, curtains or circular 'nets' to concentrate prey. The rising air is a physical barrier through which the prey are unwilling or unable to pass. Humpbacks then lunge feed individually, in pairs or in groups on the concentrated prey.

Most cetaceans feed at dusk, when zooplankton rises nearer to the surface, causing an upward shift in the entire marine food chain. The daily onshore/offshore movements of localised populations of dolphins are also often related to the vertical migrations of their prey in the water column.

SWIMMING

Propulsion is provided by the broad, flat tail flukes. The flukes are a boneless, extremely tough, fibrous, rubber-like horizontal paddle moved by muscles in the tail stock (that part of the tail which lies just in front of the fluke). The muscles anchor into the spine which runs along the middle of the tail into the flukes. The flukes have an efficient hydrofoil shape (in cross section the fluke looks like an aircraft wing), reducing water drag and providing lift during the upstroke. The tail stock is flattened from side-to-side which also reduces drag. It is swept up and down (not sideways, as in fishes) and involves the entire posterior part of the body. The dorsal fin (where present) is also fibrous and provides stability, as do the bony pectoral flippers.

Laminar flow is improved by the skin structure and the tissues lying underneath it. Cetacean skin can develop temporary areas of small ridges that dampen potential turbulence before it develops. The skin also contains lubricants which may reduce friction. Superficial skin cells exude tiny oil droplets and the outer layer of skin is shed up to 12 times a day.

Surface behaviour and activities vary between species and these provide useful species identification clues. Some cetaceans keep a low profile at the surface and seldom display any exuberant swimming behaviour. Species of the genus *Stenella* (Spotted, Spinner, Clymene and Striped Dolphins), Bottle-nosed Dolphins (*Tursiops truncatus*), Common Dolphins (*Delphinus delphis*), and Atlantic and Pacific White-sided Dolphins (*Lagenorhynchus acutus* and *L. obliquidens*) are some of the most surface-active species. Twists, somersaults, backward flips and high leaps are common acrobatics, but it is only the Spinner Dolphin and Clymene Dolphin which spin on the long axis of their bodies. 'Breaching' is a particularly forceful full-body thrust out of the water, followed by a resounding splash with plenty of noise as the cetacean's body falls back into the water. It is especially spectacular in larger species, although it is only the Humpback (fig.275) which breaches with any regularity. Splashing is also achieved with 'tail-lobbing', where the tail flukes are raised out of the water and slapped with huge force onto the sea surface. Most species also 'spyhop', raising the head out of the water to have a look around.

Many smaller species of dolphins 'bowride' in front of boats and larger whales. They ride in the pressure wave formed at the front as the larger object moves at speed through the water. They position themselves in the small region of forward force where they can stop active swimming. Dolphins will continuously flicker about in the wave, trying to find the exact spot for an effortless ride. The pressure wave tends to break down at slow speed or when the bow thrusts out of the water.

Specific activities such as travelling, feeding and defence affect swimming speed, surface behaviour, group size and spread. Travelling dolphins often cluster in tight groups or in a line abreast (echelon formation) and swim at a steady speed. Feeding groups disperse as individuals or form into small, loose bunches of animals which move randomly and with variable speed. When not on feeding or breeding grounds, baleen whales disperse as mother-calf pairs or form trios. When they group together for specific activities such as defence, Black Right, Bowhead, Gray and Sperm Whales cluster in protective circles with flailing tails pointed outwards. Cetaceans may be dispersed over large areas and seem to remain acoustically 'linked'. The amount of group spread can be observed from a ship and is generally a useful clue to species type and behaviour.

RESPIRATION AND DIVING

Cetaceans inhale and dive with their lungs full of air. Nasal plugs inside the blowhole of odontocetes provide a watertight seal, opening only by muscular contractions controlled by voluntary or conscious action. The lungs are very efficient at transferring oxygen from the air into the liquid medium of blood and cetacean lungs collapse completely during diving below 100m depth. Oxygen is stored in blood and muscle, where it is most needed. Cetacean blood is high in haemoglobin, which carries dissolved oxygen around the body and to the brain. The muscles are rich in myoglobin, a protein that binds with oxygen from the blood and stores it for metabolism during the dive. Myoglobin gives the characteristic dark coloration to cetacean flesh; the deep-diving Sperm Whale has muscle that looks almost black.

Cetaceans also possess unusual networks of blood vessels called the retia mirabilia ('wonderful nets'). These complexes of contorted, spiralled blood vessels form blocks of vascular tissue on the inside wall of the chest cavity near the backbone and in other areas. A number of functions have been suggested for the retia: a trap for nitrogen gas which bubbles out of the blood on surfacing from deep dives; blood flow regulators which even out blood pressure differences as water pressure changes during diving or surfacing; temporary reservoirs of oxygenated blood for the brain used during diving; blood temperature regulators that hold large stores of blood at body core temperature and blood stores used in the redistribution of blood during deep diving.

Cetaceans inhale and exhale about 4 times per minute when at rest: humans normally breathe about 15 times per minute when resting and become acutely distressed after three minutes without air. As the lungs compress during a dive, air is pushed into the relatively impermeable windpipe, reducing the amount of nitrogen absorbed into the blood; high nitrogen levels in the blood can cause gas embolism or decompression sickness when surfacing from a dive. On exhalation, air can have as little as 1.5% oxygen after prolonged breath holds and the proportion of lung volume exchanged is 80-90%, compared to 15% in humans.

During prolonged dives, the heart rate becomes markedly reduced (bradycardia) and in combination with vascular adjustments, blood flow is limited to all areas except the heart and brain. Heat loss is also reduced. Blood supplies to the blubber layer are minimal, reducing heat loss from the body's surface. A countercurrent heat exchange system, in which veins serving the body's periphery (skin and fins) form intricate networks with arteries, ensures that body core heat given up by outflowing blood is recovered by inflowing blood.

Male Sperm Whales dive to at least 1,000m and have been tracked to 2,800m during dives lasting over two hours; females and calves forage for about 40 minutes at less than 500m. Northern Bottlenose Whales

Hyperoodon ampullatus can dive for as long as two hours. Both Sperm and Bottlenose Whales have greatly enlarged 'foreheads' which contain spermaceti oil, but it is only in the Sperm Whale that a spermaceti organ is present (fig.298). Smaller dolphins can dive to depths of 280m and may remain submerged for over eight minutes. Baleen whales dive to about 100m, and a normal surfacing pattern has a typical sequence of 10-15 blows at 15-second intervals, followed by eight minutes below the surface; dives in excess of 30 minutes are rare.

Baleen whales have two blow holes that produce twin columns of vapour and in most species these merge to form a single spout. The Blue Whale has the most conspicuous blow, rising to 9m as a tall, thin cone of smoke-like vapour. The Fin Whale also has a tall, thin blow. The Humpback's blow is low and slightly bushy. Right whales make two separate bushy blows that diverge on either side of the head. The Sperm Whale, with its left-sided blow hole, produces a low bushy plume thrust forward to the left. All odontocetes have only one blow hole and the blow in most of the smaller species is usually low and brief and, especially in warm air, not visible.

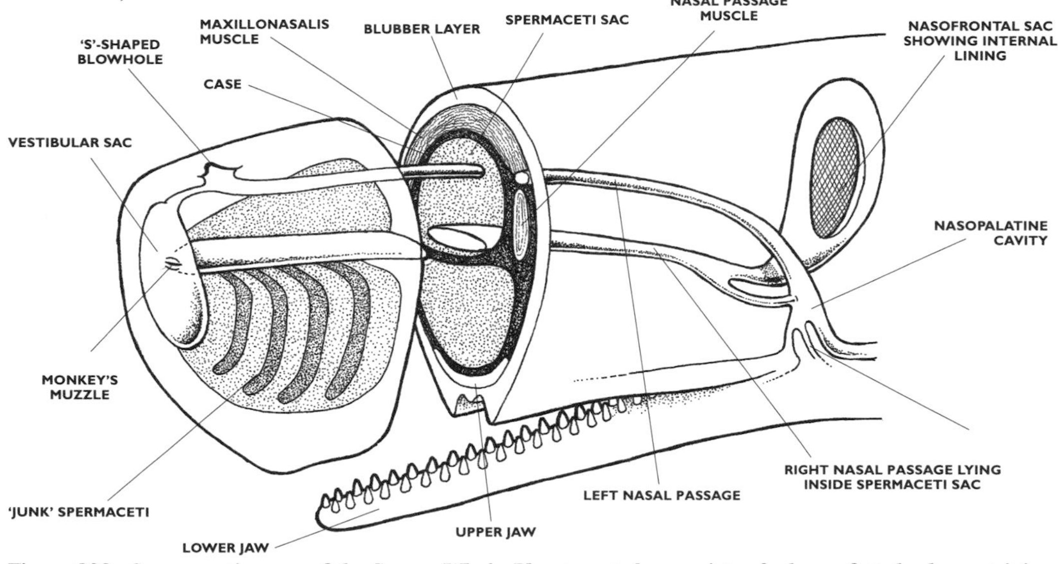

Figure 298. Spermaceti organ of the Sperm Whale *Physeter catodon* consists of a lower fatty body containing alternating bands of fibrous and oily tissue ('junk') and an upper sac-like structure containing spermaceti oil. This organ may help control buoyancy during deep dives, but its primary function, in conjunction with special adaptations in the nasal passage, is probably sound production.

SENSES

Whales have all the senses generally found in terrestrial mammals with the exception of olfaction (smell) which is very reduced or absent. Taste provides a means of monitoring the marine environment through sampling water-borne chemicals in the immediate surroundings. Taste buds are present in the tongue of some cetacean species and may be used in feeding. They may also be used to 'taste' the reproductive or emotional state of other individuals. Taste perhaps provides information about more distant surroundings from chemicals borne along in currents. Cetacean skin is sensitive, especially the area around the blowhole; captive dolphins object to being touched there. Hairs or whisker pits along the top of the head and on the chin may be tactile organs, helping to sense the air-water interface. Touch seems to maintain order within social groups. Stroking and touching are part of courtship rituals in most species. Mothers and calves frequently reaffirm bonds with touch.

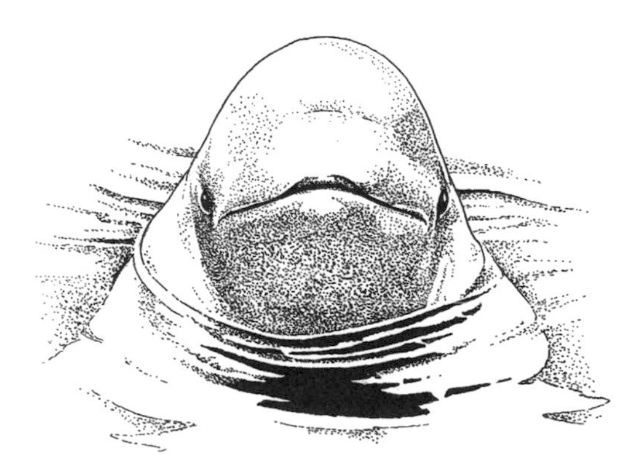

Figure 299. Binocular vision in the Beluga *Delphinapterus leucas*, which can see forward in front of its snout with both eyes.

Figure 300. Binocular vision in the Black Right Whale *Balaena glacialis*, which can see forward below its head with both eyes.

Dolphins see well in both water and air. When feeding, Bottle-nosed Dolphins are able to track airborne flying fish, apparently seeing fairly well through the air-water interface. Small odontocetes with short snouts or pointed beaks have good binocular vision both forward and also downward (fig.299). Less is known about vision in large whales. In those species with blunt heads such as the Sperm Whale, the eyes are positioned near the widest part of the head and far back from the tip of the jaws, affecting binocular vision. They are not able to see immediately in front of their jaws and have predominantly monocular vision, each eye seeing a separate part of the total visual field. However, the Black Right Whale has an area in front of and below the snout where binocular vision occurs (fig.300). Cetaceans have an asymmetrical eyeball and use different areas of the retina for seeing in air and underwater. These areas have different lens-to-retina distances. This visual mechanism relies on the principle that light is more strongly refracted by the eye in air and less strongly in water. Vision in water is more useful to species inhabiting clear tropical seas than those living in turbid environments. Indus and Ganges River Dolphins (*Platanista* spp.) have reduced eyes that probably only sense light direction and intensity.

Sound travels 4.5 times faster underwater than it does in air and the ear structure of cetaceans differs from terrestrial mammals. The middle and inner ear (tympanic bulla) is acoustically isolated from the skull by foam-filled air spaces, an adaptation that helps whales to discern the direction of underwater sound. How cetaceans hear sounds is not fully known. In odontocetes, the lower jaw is filled with oily tissue and it is thought that sounds, especially higher frequencies, pass along this sound-conducting medium in the lower jaw to the tympanic bulla where sounds are received. Lower frequency sounds may reach the bulla by a different route through the outer ear canal (fig.301).

Most cetacean species are vocal underwater and some have a sophisticated range of sounds. Baleen whales produce sounds with subsonic frequencies, rarely above 5 kHz. Song-like, melodic patterns of moans and groans have been recorded from Bowhead and Humpback Whales. Different Humpback populations sing different songs, and these songs also change with time. Mysticetes lack the anatomical structures in the blowhole and lower jaw present in odontocetes for producing and receiving the ultrasound clicking noises associated with sonar. However, Minke *Balaenoptera acutorostrata* and Gray Whales make click-like sounds but how these are produced is not yet known.

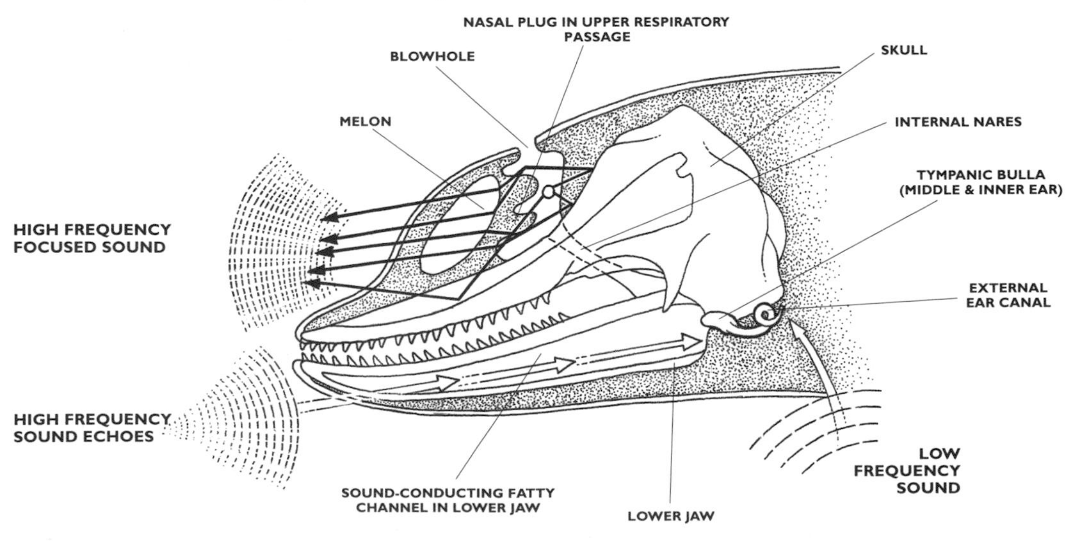

Figure 301. Sound pathways in a dolphin head.

Odontocetes produce a wide variety of calls for communication, including whistles, creaks and squeaks. The Sperm Whale uses pulsed sounds for communication underwater between the members of a school (fig.284). The Beluga is known as the sea canary for its melodious calls. Killer Whales use distinct calls or 'dialects'. Some are apparently unique to each group or pod, but others are shared with other pods living in the vicinity. In captivity, Killer Whales may learn 'foreign' dialects.

Odontocetes make ultrasound clicks (up to 300kHz in frequency) which are used for echolocation. Echolocation clicks are thought to be made by the paired nasal plugs in the upper nasal passage between the skull and the blowhole (fig.301). These clicks are very short (less than one millisecond) in duration and are repeated many times a second, sometimes sounding like a creaking door; the frequency of most of the sound is above the upper limit of human hearing (20kHz). Broad frequency clicks cover a longer range and give a low detail acoustic 'picture'. Clicks with a narrower frequency are used to give acoustic detail at closer range. Control of the sound beam 'width' is thought to occur in the fatty melon. The shape of the melon can be altered by muscles in the head and this is thought to change the characteristics of the sound passing through it. The frequency of the returning echoes and their loudness provide odontocetes with information about their environment including size, shape, density, distance and movement of objects in it. Man-made sonar systems cannot yet match the elaborate and finely tuned bioacoustic sense of the smaller toothed whales.

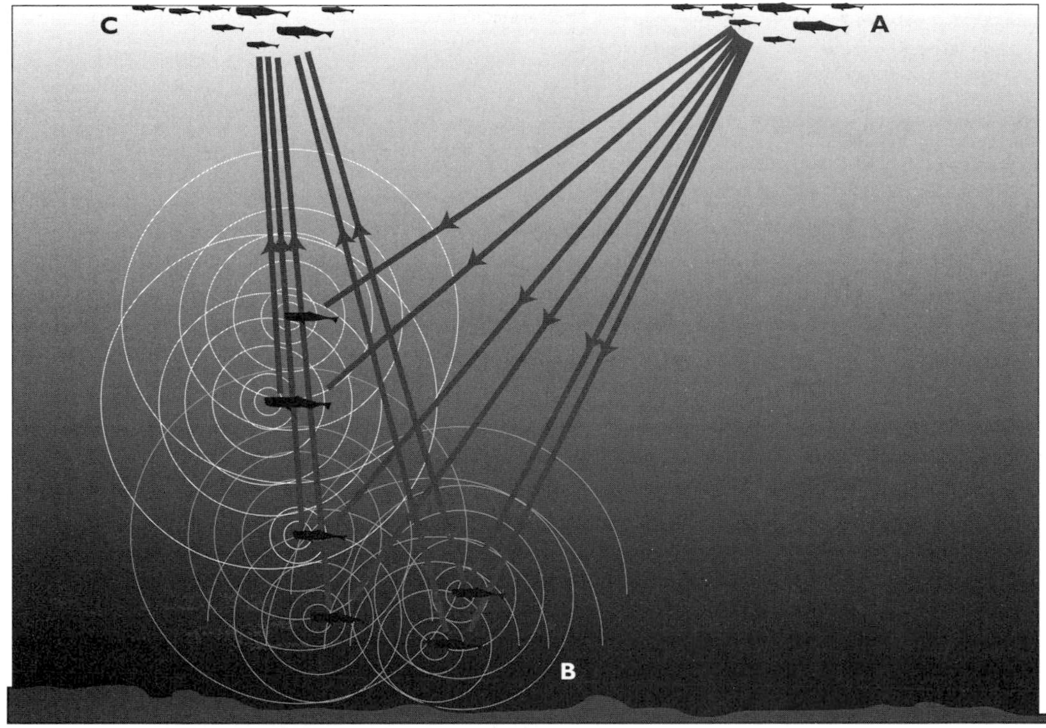

Figure 302. Acoustic communication in the Sperm Whale *Physter catodon.* a) School of Sperm Whales swimming at the surface. b) Sperm Whales disperse at depth, each individual emitting a 'signature' click-train that permits the whales to recognise the identity of all the other members in the school and determine their relative positions. c) Sperm Whale school re-forms at the surface.

LIFE HISTORY

Odontocetes are generally gregarious and some species seem to have a well defined social organisation. Among these, Killer Whales and Sperm Whales exhibit strong social bonds. In the Killer Whale, group members remain together for life and pods are stable from one generation to the next. Pod size varies from 4 to 40 and larger aggregations occur when several pods gather temporarily. The Sperm Whale is polygynous and forms different school types which vary with season and location: nursery schools of mothers and calves; harems overseen by one sexually mature male; calves and juveniles of both sexes; bachelor schools of young adult males and bull schools of sexually active males. When one member of a school is wounded, the others form a protective cluster around it with their flailing tails pointing outwards (fig.303).

Baleen whales do not form complex groups and social units are not fixed. In most species, the young return to the feeding grounds of their mothers. Individuals may gather in a group one day, then be alone or with a different group on the following day.

Figure 303. Marguerite formation of Sperm Whale *Physter catodon*, from below.

Most cetaceans have an average life span of 20 - 40 years. Some species are considered aged at 10-15 years old, whereas others can reach an estimated 80 years of age. Cetaceans have comparatively low reproductive rates and long calving intervals. Ovulation, conception, pregnancy, birth and lactation ranges from a year in porpoises, about two years in the larger whales and three or four years in the Delphinidae. Females in most species remain sexually active until death but the frequency of calving tends to diminish as the female ages. Killer Whales and possibly pilot whales may be the exceptions, as older females in these species enter a post-reproductive phase when ovulation stops.

Pregnancy varies between 8 and 16 months. The Blue Whale and Bottle-nosed Dolphin are examples of species that gestate for just under 12 months. Calves are born tail first (fig.304), the blowhole emerging last to minimise the risk of inhalation of water before the first breath. Body size at birth is generally larger than in terrestrial mammals which may be important in reducing heat loss, as larger animals have a relatively lower surface area per unit volume. Growth rates of calves are remarkably high. Newborn Blue Whales are about 7m long and by eight months, they average 15m and will have gained 90kg per day. Such rapid growth is possible because the mother's milk has a fat content of 16-46% (compared to 3-5% in humans and cows); daily quantities of milk can be up to 100 litres in the largest whales.

Mothers nurse their young for at least four months. Calves in social species may take milk from several females who are probably all relatives, such as aunts or grandmothers. In less social species, calves stay close to their mothers throughout the nursing period, and the mother's dives are probably restricted during early stages of calf dependency. Sperm and Gray Whale mothers aggressively defend their young if threatened and these species were called 'devilfish' by open-boat whalers who found calves dangerous to harpoon.

Cetaceans are generally promiscuous in their reproductive behaviour. Mating in cetaceans is unusual among mammals for it occurs belly-to-belly. Sexual activity in the more gregarious species occurs frequently and in all seasons, suggesting it has a wider social meaning. Courting Humpback males produce songs as part of their breeding ritual.

Sexual maturity occurs anywhere from 2 to 20 years, depending on the species and sex. In some species, male sexual maturity is delayed long after that of females: female Sperm Whales mature at nine years of age, but males mature at 20; first ovulation in female pilot whales is at seven years, but males mature between 15 and 20 years. In Killer Whales, the minimum age of sexual maturity in females is seven years but in males it is 10-12 years; males begin to show growth of their dorsal fin at about seven years but it does not attain full size (about 2m high) until 14 years of age.

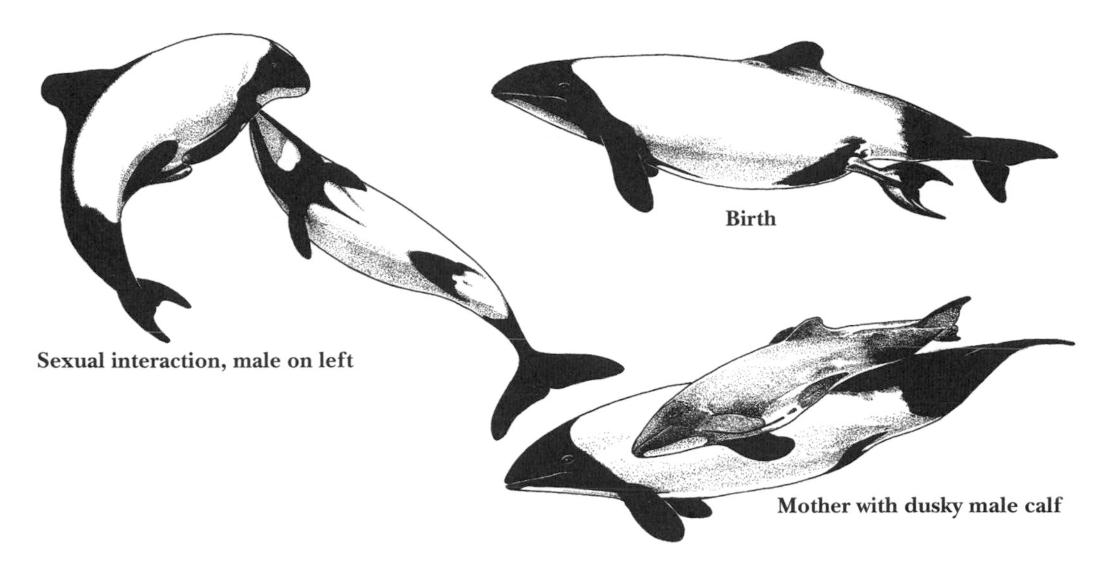

Figure 304. Reproduction in Commerson's Dolphin *Cephalorhynchus commersonii.*

STRANDING

Although the causes of whale strandings are still not known, injury, parasitic infestation, disease, marine toxins in the food chain, social factors and errors in navigation are some of the better known explanations. Storms, increased water noise and suspended sediments near coastlines may also be contributory factors especially where they cause interference with the sonar of odontocetes. Their sonar may also be disrupted by internal dysfunction of the ear due to parasitic infestation of the tympanic cavity. This has been suggested as a direct cause of strandings but parasites have been found in the ears of dolphins and porpoises caught at sea. Cetaceans may follow magnetic contours and simply blunder into unknown waters, where they panic in the shallows and strand as the tide drops. However, these animals are likely to be visually aware of entering shallow water, if only from the increased turbidity encountered in nearshore waters.

Cetaceans may be washed ashore singly or in groups (mass strandings). Coastline shape and substrate seem to be important features of mass stranding sites. The most common stranding sites are sandy bays and estuaries. Some 20 cetacean species mass strand, of which only 10 are more frequently involved; Sperm Whales, Killer Whales, False Killer Whales, pilot whales, Bottle-nosed Dolphins and certain beaked whales are the most frequently stranded animals. Odontocete species that normally live in large, tightly-knit groups seem to be more susceptible. There have been relatively few mass strandings of any of the baleen whale species.

Coping with live strandings requires common sense and a practical approach. Every stranding is different and requires its own solution, ranging from the disposal of already dead animals, to rescue and refloating of living animals or removal into captivity if sick animals are present and, in rare cases, humane killing (under veterinary supervision).

Stranded cetaceans may perish quickly from physiological stress. They need to be kept moist and cool (covering with wet blankets is effective) and the blowhole kept clear, while curious onlookers should be kept away. The nearest maritime authority should be informed and expert help found. The local and national museum may keep stranding records and should be notified. If refloating is an option in a mass stranding, attempts should be made to release as many animals as possible simultaneously, otherwise social bonds may cause whales already refloated to return to those remaining on the beach only to strand a second time.

Smaller species can be carried back to the sea in slings. Larger species need to be assisted down a channel dug back to the sea at high tide. A more risky option (because of causing damage to fins or flukes) is a gentle roll towards the water. Rescues are more successful if the animals are supported in shallow water and 'walked around' until they show signs of recovery.

Detailed information should to be gathered from each stranding: species; number of stranded animals; body length and sex of each animal; date and location of the stranding, with a map reference if possible;

number refloated and other cetaceans (if any) seen at the time of the stranding. Further details are given in the factsheet.

WHALES AND MAN

Commercial whaling activities have caused a catastrophic reduction or, in some cases, near extinction of the majority of the world's populations of large whales. However, with the decline in whaling, the fishing industry now presents the greatest threat to cetaceans, especially the smaller species. Inadvertent entrapment of dolphins and porpoises in fishing gear causes more cetacean deaths per year than whaling did at its peak. The impact on cetaceans of declining fish stocks due to overfishing is also of growing concern.

Environmental degradation also affects cetaceans. The ever-increasing flotsam and jetsam of man-made objects such as balls, balloons, discarded nets, plastic bags and non-degradable refuse become caught around, or are swallowed by, cetaceans, often resulting in death. Oil spills not only poison cetaceans unlucky enough to be caught in them, but also poison the marine environment on which they depend by affecting the entire food chain. Damming, mining, logging and other developments on land may have deleterious effects by causing erosion and destroying watersheds important to freshwater-spawning fish, a major food source for riverine and coastal cetaceans. Cetacean populations with localised distributions, such as coastal species, are especially vulnerable to habitat disturbance.

Toxic chemical compounds find their way into the marine environment, deposited directly from ships, near shore drainage, and rivers. Heavy metals, pesticides and radioactivity accumulate in the marine food chain. Low-level contamination of smaller prey animals low in the food chain becomes concentrated in the tissues of larger predators which feed on them. This bio-accumulation builds up to a peak in marine mammals. A cetacean foetus absorbs toxins into its bloodstream through its mother's placenta. After birth, the calf receives a further toxin load from its mother's milk during nursing. Thus a concentration of toxins is present in the young before they start feeding on their own.

The contamination of odontocete tissues can be high but the wider effects on metabolism are difficult to detect. Belugas living in the St. Lawrence River, Canada, have high levels of toxins including polychlorinated biphenyls (PCBs), the pesticide Mirex, DDT and heavy metals. The main sources of toxins found in the Belugas are thought to be from feeding on contaminated American Eels *Anguilla rostrata* and from toxins transferred during pregnancy and suckling. Calving rates for the St. Lawrence Belugas are lower than in Alaskan Belugas and it has been suggested that toxins, especially organochlorines, are the cause. Numerous unexpected diseases have been found in Beluga carcasses washed up on the banks of the St. Lawrence and it is likely that the continued existence of the remaining population of about 500 Belugas there is threatened by the effects of chemical pollution.

Sound pollution affects cetaceans. Disturbance responses include longer dives, movements away from sound sources, and temporary displacement from feeding or breeding areas. Before regulations were imposed, Gray Whale mothers and their calves shifted between lagoons in Baja California in response to disturbance from low-flying aircraft and speedboats. In Hawaii, the number of Humpbacks present varies inversely with increasing amounts of daily boat traffic and activity on the military bombing range. Marine seismic surveys are also sources of potentially harmful sound. Tests are often conducted at frequency ranges and decibel levels which may interfere with the hearing of both baleen and the larger toothed whales.

BIBLIOGRAPHY

Carwardine, M. 1994. *On the Trail of the Whale.* Thunder Bay Publishing, Guildford.

Evans, P. G. H. 1987. *The Natural History of Whales and Dolphins.* Facts on File, New York.

Evans, P. G. H. 1994. *Dolphins.* Whittet Books, London.

Harrison, R. J. & Bryden, M. M. (eds). 1988. *Whales, Dolphins and Porpoises.* Merehurst Press, London.

Hoyt, E. 1990. *Seasons of the Whale.* Mainstream Publishing, Edinburgh.

Hoyt, E. 1984. *The Whale Watcher's Handbook.* Doubleday, New York.

Leatherwood, S. & Reeves, R. 1983. *The Sierra Club Handbook of Whales and Dolphins.* Sierra Club Books, San Francisco.

Martin, A. R. 1990. *Whales and Dolphins.* Salamander Books, London.

May, J. (ed). 1990. *The Greenpeace Book of Dolphins.* Century Editions, London.

Mead, J. G. & Brownell, R. L. 1993 (2nd ed.). Order Cetacea. In: *Mammal Species of the World* (eds. D. E. Wilson & D. M. Reeder). Smithsonian Institution Press, Washington.

WHALE FACTSHEET

1. IDENTIFYING WHALES AT SEA

When a cetacean surfaces to breathe, most of the body remains below the waterline (fig.287). This restricts the possibilities for identification to features visible above the sea surface as the cetacean rolls. If it appears close enough to the boat, it will also be possible to see at least part of the body beneath the water. It can be difficult to be sure of the size of a cetacean from its surface appearance without a reference point (e.g. other whales of different ages).

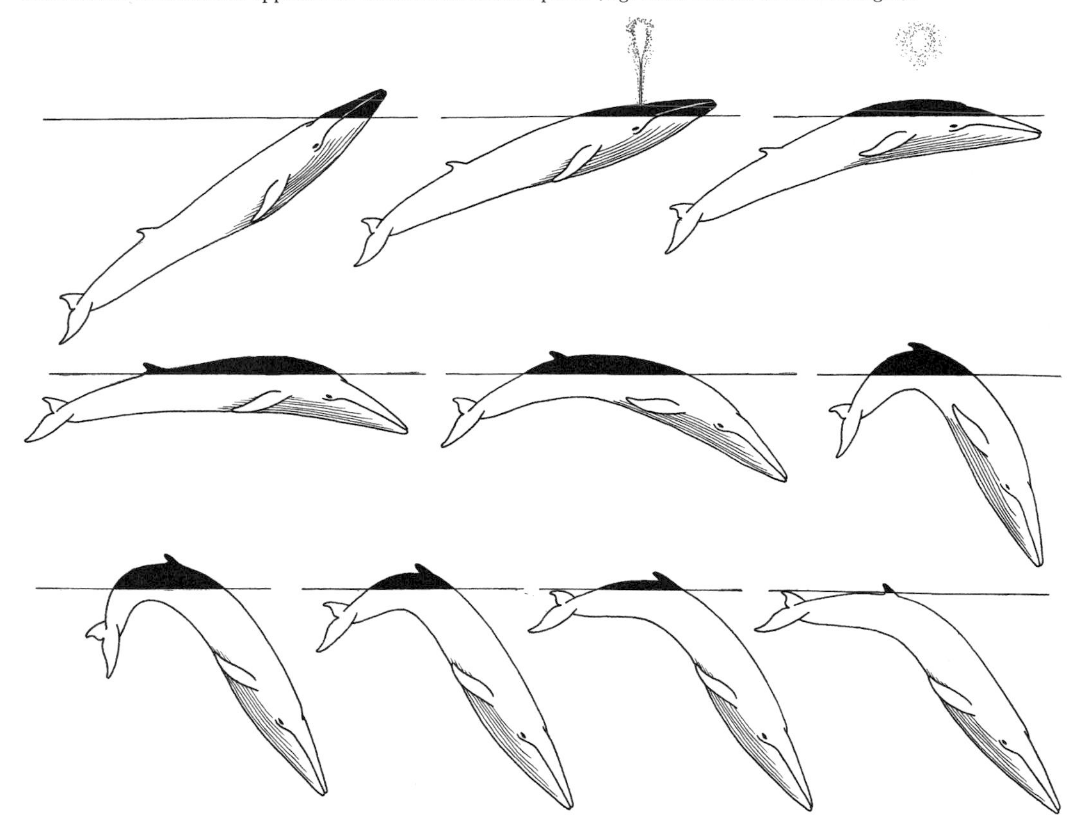

Figure. 305 Fast swimming Fin Whale *Balaenoptera physalus* surfacing to breathe. The size of the back (relative to the total body length) observable above the waterline during the roll is shown.

The following useful features should be looked for when attempting identification of large whales at sea. It should be noted that wind strength, sea condition, sea surface glare, cloud cover and distance from sighting are factors that affect the way whales appear to the human eye at sea. It is important to make allowances for these factors.

The detailed identification of cetaceans at sea is beyond the scope of this book; the features given below are generally applicable to baleen whales and larger odontocetes.

1) Whale's blow
Shape, size and direction of blow
Shape of the blowhole region, the head itself and the head surface (e.g. presence of ridges, markings and callosities)

2) Whale's back
Shape of body exposed above surface during 'roll'
Distance between blowhole and dorsal fin
Characteristic features of tail stock (area between dorsal fin or hump and flukes)

3) Whale's flukes
Appearance (if any) of flukes above surface
Width, shape and colour of flukes

4) Whale's flippers
Length, shape and presence or absence of markings

5) Whale's dive
Degree of back arching seen above surface and appearance (if any) of flukes
Angle and style of descent below surface

These characteristics are highlighted in the illustrations below, showing the surface appearances of four whale species.

BLACK RIGHT WHALE

Blow: Wide, V-shaped, appearing as two spouts
Back: Rotund, lacking dorsal fin
Flukes: Distinctly concave with deep notch
Dive: Low roll, flukes thrown into air, then slip vertically beneath surface

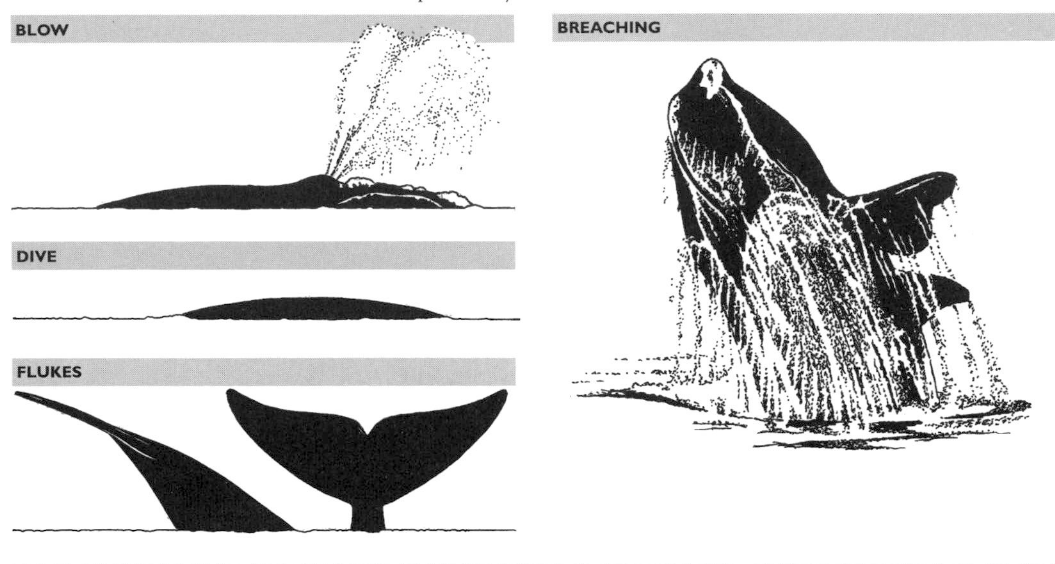

SPERM WHALE

Blow: Single spout projected obliquely forward from left side of snout tip
Back: Skin obviously wrinkled, prominent hump instead of a dorsal fin, bumpy ridge on tail stock
Flukes: Broad relative to width, nearly triangular, deeply notched
Dive: Arched, wrinkled back; flukes show above surface

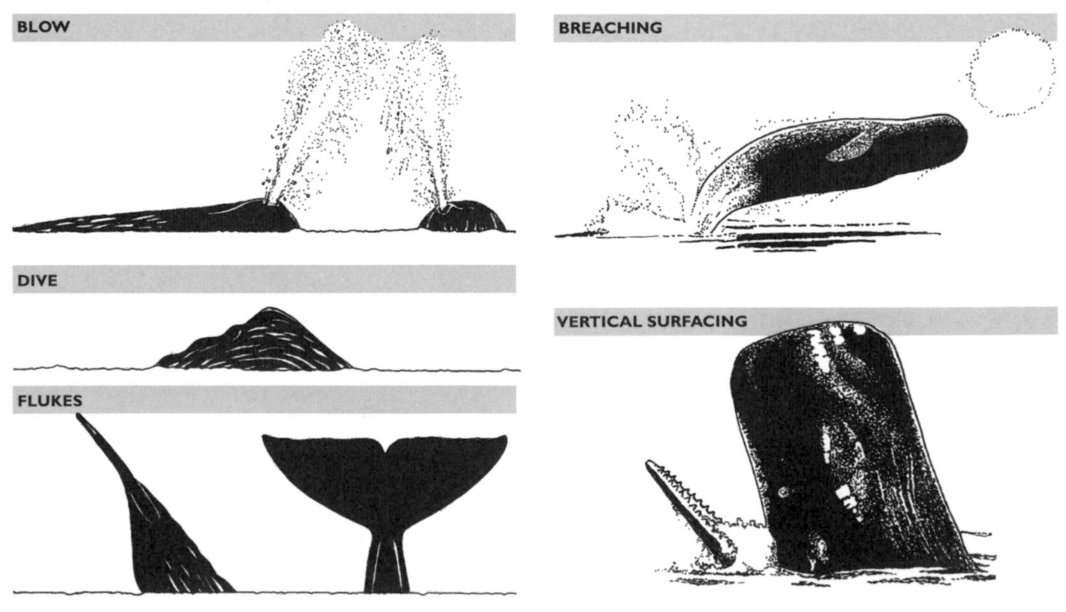

HUMPBACK WHALE

Blow: Single spout, shaped like a helium balloon, wider in relation to its length than other rorquals
Back: Small dorsal fin, variable in shape, placed on a step or hump on last third of snout-to-tail length
Flukes: Often white underneath (each individual has unique pattern), concave, scalloped rear margin
Flippers: Very long with scalloped front margin; may be held above surface
Dive: High roll, distinctly arched back, flukes raised high above surface

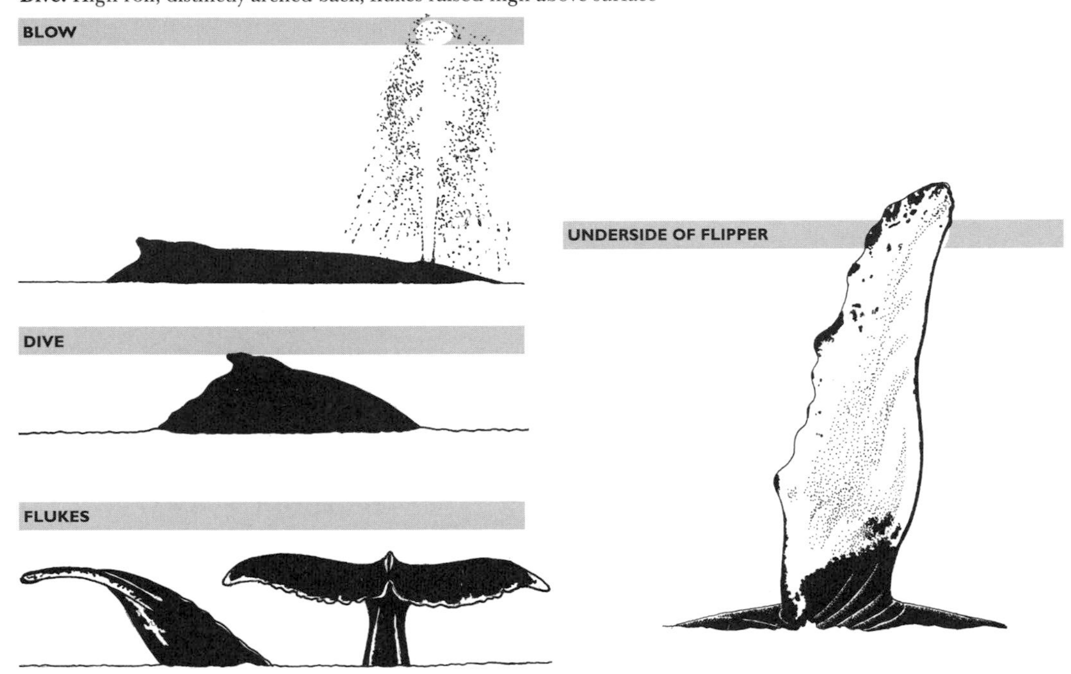

GRAY WHALE

Blow: Bushy, heart-shaped
Back: Low hump instead of dorsal fin on last third of snout-to-tail length; bumpy ridge on tail stock
Flukes: Broad with convex rear margin and pointed tips, often barnacle encrusted
Dive: Low roll, flukes raised high into air

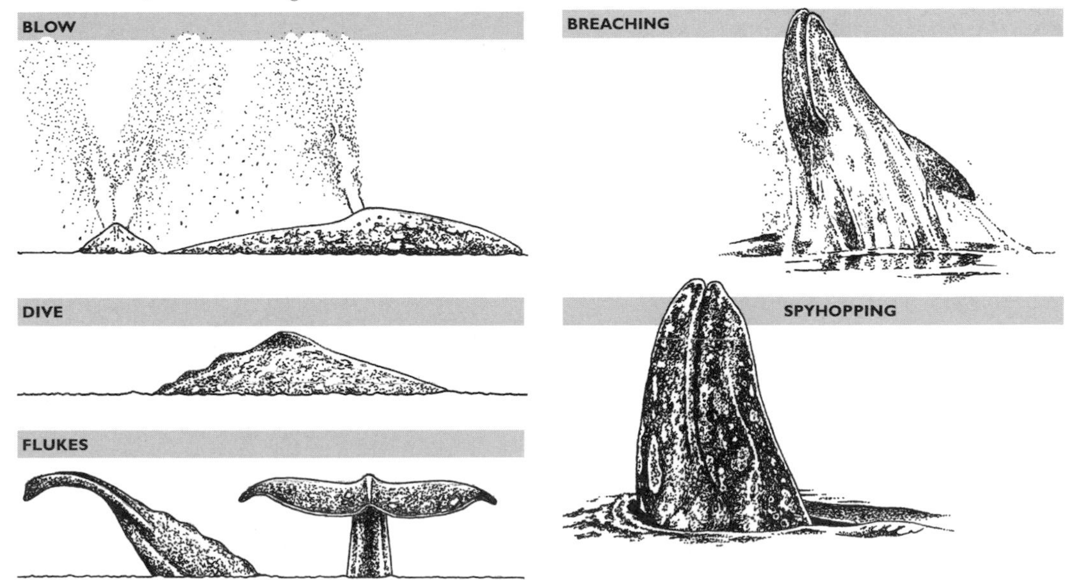

2. USEFUL GUIDELINES FOR STRANDED CETACEANS

A. Determine the species if possible, condition of the animal(s), number of animals, group composition (one large group, clusters, or scattered individuals), and the distance between individuals or groups. A photograph or sketch is useful.

B. Take basic body measurements (fig.288). Measurement numbers 1, 4, 6, 8 and 12 are the most useful for large whales. These should always be made point-to-point in a straight line except for the girth measurement.

1. Length from tip of upper jaw to the deepest point of the notch between the flukes. If there are no flukes, measure to the farthest point noting where your measurement ended (you can make a sketch).
2. Length from tip of upper jaw to the front edge of the dorsal fin. If there is no dorsal fin, make a note.
3. Length from tip of upper jaw to the blowhole.
4. Length from tip of upper jaw to the centre of the eye.
5. Length from tip of upper jaw to the corner of the mouth.
6. Length from tip of upper jaw to front of the pectoral flipper, where it joins the body wall.
7. Length from tip of upper jaw to the anus, measured along the belly or side.
8. Length of the pectoral flipper from the tip to the body wall, measured along the front edge.
9. Maximum width of the flipper (note if damaged or missing).
10. Maximum girth of the body (in front of dorsal fin).
11. Height of dorsal fin from the base to the tip (note if damaged or missing).
12. Width of flukes (note if damaged or missing).

C. Determine the sex of the animal(s) (fig.289) and the number of calves and adults, if possible.

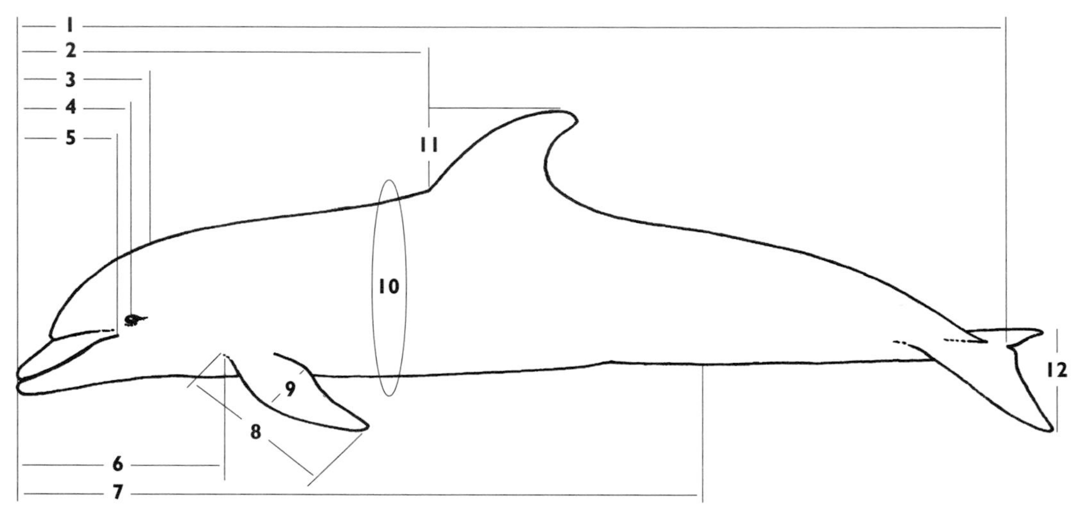

Figure 306. Measurements used for recording total body length and body proportions for a cetacean.

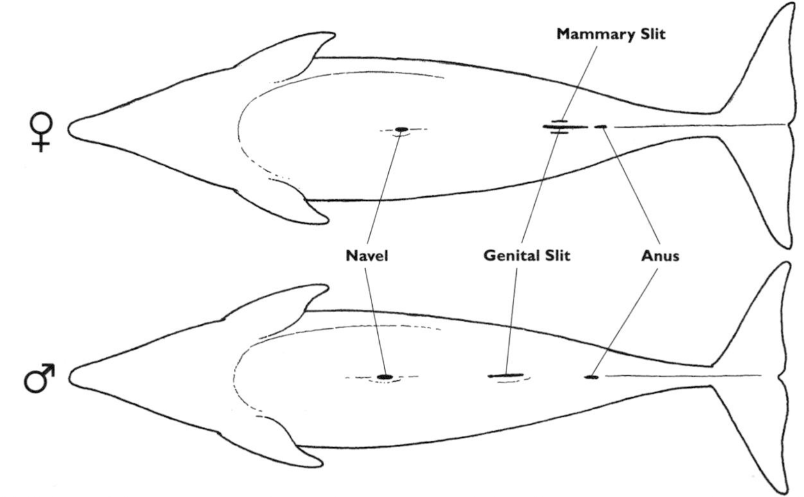

Figure 307. External features used for sexing cetaceans. The distance between the genital slit and anus is larger in the male.

SIRENIANS

by

Geoffrey Waller

INTRODUCTION

Figure 308. The West Indian Manatee is a docile herbivore living in coastal waters and rivers of the western Atlantic from Virginia, USA to Espirito Santo, Brazil.

Manatees and the Dugong (also known as sea cows or sirenians) are the only herbivorous mammals living in the sea. Although manatees move between the sea and freshwater river systems or are wholly restricted to freshwater, the Dugong is entirely and strictly marine. The Dugong has the widest distribution of any living sirenian, its range extending over much of the Indo-Pacific between latitudes 30°N and 30°S. Manatees are confined to Atlantic coastal waters and associated river systems (fig.308).

Sea cows are docile, slow moving animals showing none of the athletic exuberance so characteristic of dolphins and seals. A dorsal fin is lacking and hind limbs are absent (fig.312). The front limbs (or more correctly 'paddles') move at the elbow, the upper arm being enclosed within the flank. The tail is flattened for propulsion and the tail lobes are supported only by a central row of caudal vertebrae as in whales. The nostrils open at the top of the muzzle as a pair of circular nasal openings and are closed by anteriorly hinged valves.

The teats are located at the 'armpits', just behind the paddles of the female. In elephants the mammae have a similar location behind the front limbs.

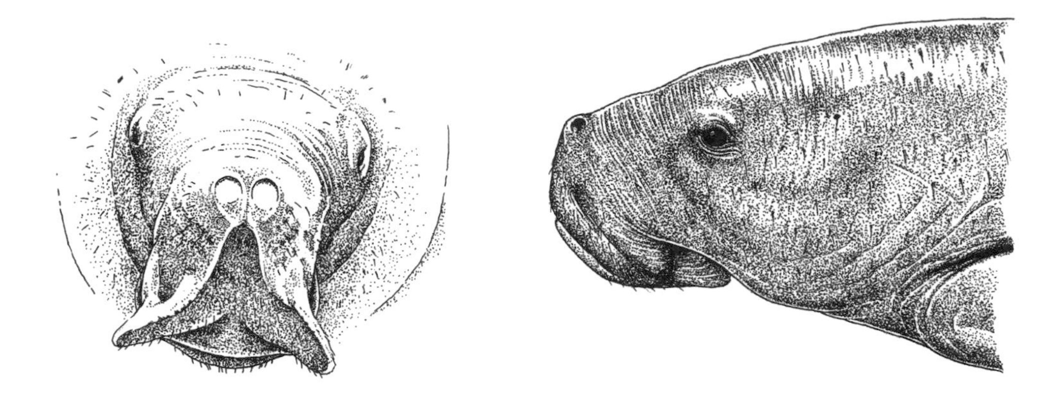

Figure 309. Head of Dugong (front view and side view). The ear opening lies just behind the eye.

The thick skin is tough but not entirely naked; sparsely distributed hairs are evident on close inspection. The head is distinctly flat-faced with a heavy-jowled, densely bristled appearance. The two enlarged lobes of the upper lip hang down, one each side of the mouth, in a manner similar to the pendulous lips of a bloodhound. There is a prominent and bristly chin. There are no external ear flaps and the ear canal opens at the skin surface behind the eye (fig.309). The eyes are relatively small without well defined eyelids, and are protected by a heavy tear secretion.

Manatees and Dugongs differ in a number of important external features (table 11, fig.310), and because of these and other important differences in their internal anatomy, they are classified in separate families of the order Sirenia.

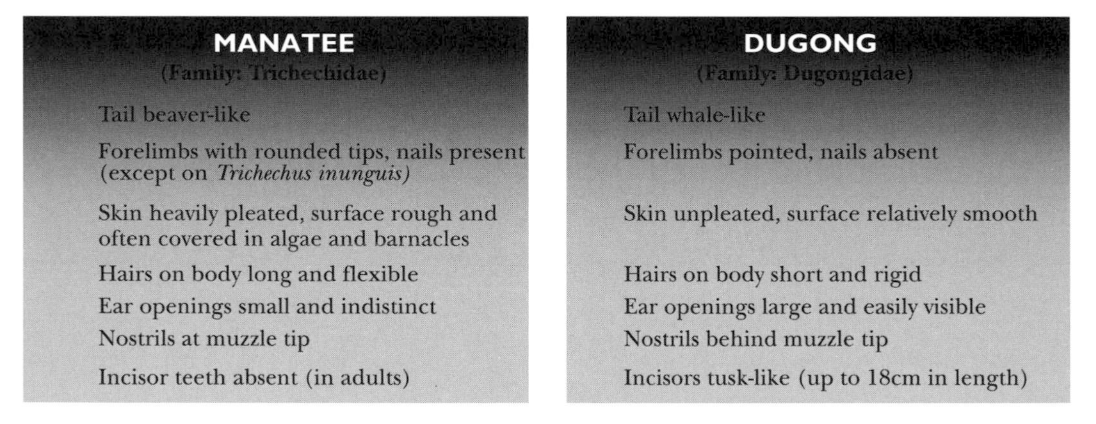

MANATEE (Family: Trichechidae)	DUGONG (Family: Dugongidae)
Tail beaver-like	Tail whale-like
Forelimbs with rounded tips, nails present (except on *Trichechus inunguis*)	Forelimbs pointed, nails absent
Skin heavily pleated, surface rough and often covered in algae and barnacles	Skin unpleated, surface relatively smooth
Hairs on body long and flexible	Hairs on body short and rigid
Ear openings small and indistinct	Ear openings large and easily visible
Nostrils at muzzle tip	Nostrils behind muzzle tip
Incisor teeth absent (in adults)	Incisors tusk-like (up to 18cm in length)

Table 11. External differences between manatees (*Trichechus* spp.) and the Dugong *Dugong dugon.*

TAXONOMY AND DISTRIBUTION

Manatees and the Dugong belong to the order Sirenia, a name derived from one of three mythical sea nymphs of the Greek tales whose song lured unsuspecting seafarers to their deaths. Sirenians have ungulate features in their teeth, skull, skeleton and soft tissues, and are distantly related in evolution to elephants (Proboscidea). The Eocene fossil sirenian *Protosiren* from Egypt had more numerous teeth than modern sea cows and small hind limbs were present. Incisors, canines and two types of molars are discernible in the upper jaw of *Protosiren.* In present-day manatees only molar teeth are present in adults.

There are three species of manatees: the West Indian Manatee *Trichechus manatus* is found throughout the tropical coastal Caribbean Sea and in coastal or riverine areas of the western Atlantic seaboard; the West African Manatee *T. senegalensis* is found in coastal waters, rivers and lakes of tropical West Africa from northern Angola to southern Mauritania; the Amazonian Manatee *T. inunguis* is the only strictly freshwater sea cow species and is confined to the Amazon river system of Brazil, Columbia, Ecuador, Guyana and Peru.

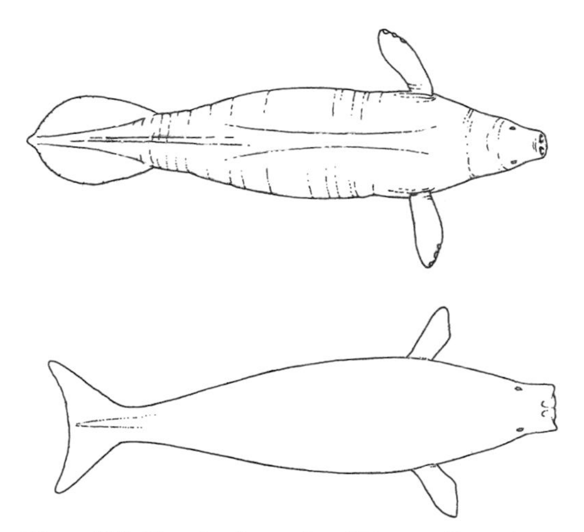

Figure 310. Manatee (upper) and Dugong (lower) seen from above. There are differences in the snout, paddles and tail (see table 11 for details).

There is a single species of Dugong *Dugong dugon*. This is essentially an animal of tropical seas with a wide though presently discontinuous Indo-Pacific distribution from the head of the Gulf of Suez (32°E) to the New Hebrides (170°E). Steller's Sea Cow *Hydrodamalis gigas*, a giant sea cow and recently extinct relative of the Dugong, was confined to the north Pacific basin.

The distribution of modern sea cows is probably limited by their sensitivity to low water temperature. There is evidence to suggest that the ability of sea cows to raise their metabolic rate in cold water is limited. The Dugong has relatively thin blubber and lacks heat exchangers in the skin. These structures are found in the skin of seals and whales and function to reduce heat loss from the body surface in cool water. However, Steller's Sea Cow lived year-round in the Bering Sea where it was well adapted to survive the severe winter conditions.

SWIMMING AND DIVING

The main thrust in swimming is produced by the tail which is moved up and down with powerful strokes. When changing direction of movement, the tail acts as a horizontal rudder by twisting on its axis whilst stroking is stopped. The paddles are used for guiding turns and decelerating; in fast swimming the paddles are raised and held against the sides. When stationary, a manatee can make abrupt movements to the left or right using its paddles. To turn to the left, it pushes backward with its left forelimb whilst pulling forward with the right. The paddles are always feathered to present maximum surface area on the power stroke and minimum resistance on the return stroke (fig.311). Left and right paddles can be used alternately to 'walk' along the bottom; when alarmed, manatees can rapidly pull themselves along the bottom, working their paddles in unison to increase forward momentum. Manatees can also swim backwards using their paddles. In the Dugong, paddles are used for sculling at low speed, or for pivoting the body above the bottom during resting or feeding.

Sea cows are slow swimming when undisturbed. During fast swimming, burst speeds of 10-12 knots have been recorded for the Dugong and 13 knots for the West Indian Manatee. These speeds can only be maintained over short distances (< 100m in the latter species).

Sea cows dive with their lungs full of air and are similar to whales in this respect but differ from seals. A large manatee can stay submerged for about 20 minutes when resting and the Dugong can dive for about half this time. The amount of air renewed when rising to the surface to breathe is about 90% of lung volume in manatees; this is a similar figure to whales but much higher than in humans (15%

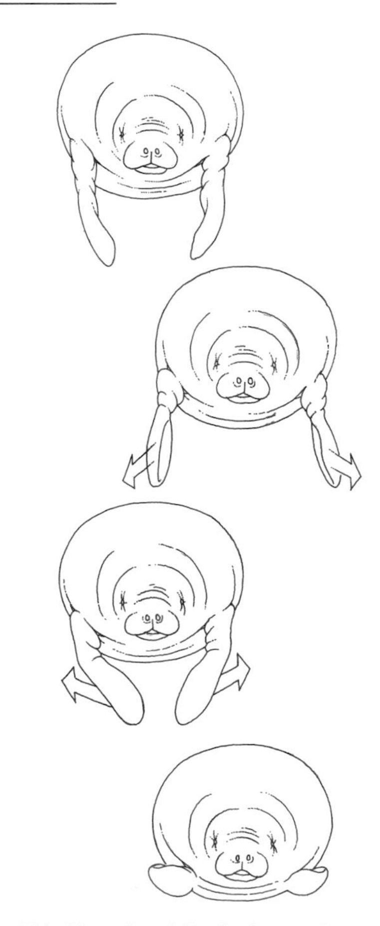

Figure 311. Use of paddles in forward movement in the West Indian Manatee.

Figure 312. A group of Dugongs swimming at the surface.

of lung volume is renewed in a single breath). Oxygen consumption in manatees is low in comparison to other marine mammals and is only one-ninth that of humans under true resting conditions.

Resting manatees are able to ascend and descend effortlessly through the water column, for example when surfacing to breathe. This movement is achieved partly through a remarkable modification of the lung cavity of these mammals (fig.313). The lungs usually have a forward position in front of the gut in mammals, but in sea cows the lung cavity extends along most of the whole length of the animal's back. The lungs themselves, which reach as far back as the kidneys, have large air spaces within them. It is thought that the lung volume can be reduced to cause the manatee to sink or increased to cause it to rise in the water column, using voluntary muscular compression and relaxation. Compression of gas decreases the lung volume and therefore buoyancy is reduced. Certainly the muscular diaphragm covering the lungs seems well suited to assist such a function in the control of buoyancy.

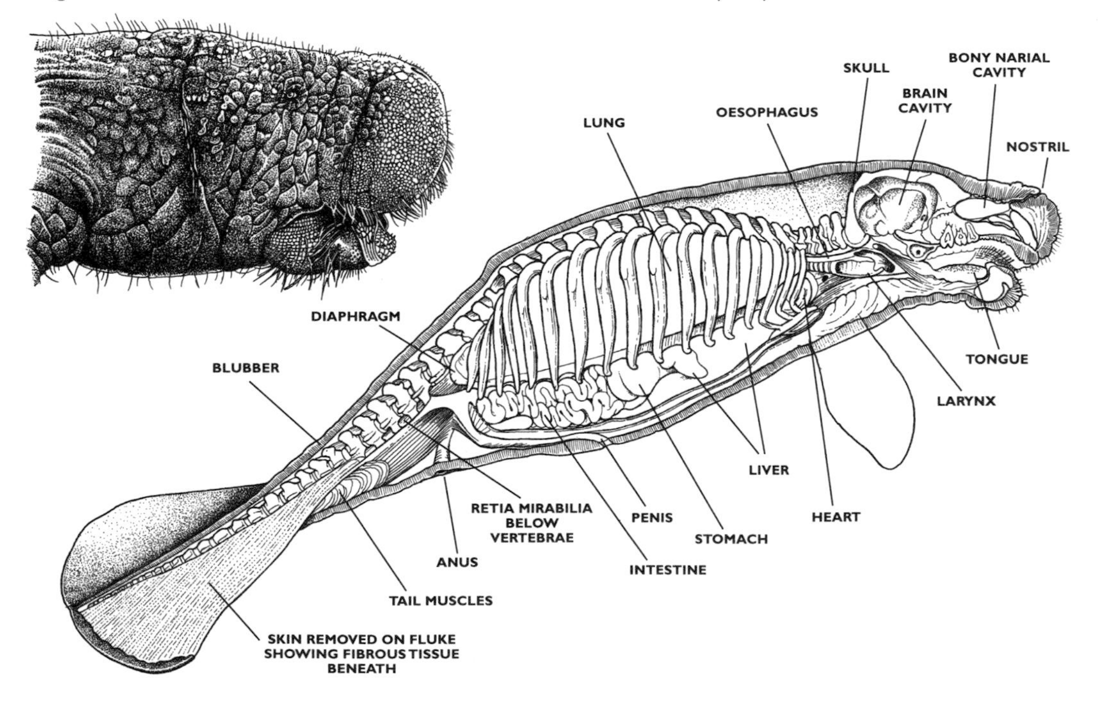

Figure 313. Side view of head and internal body plan of the West Indian Manatee. The long body cavity is almost filled by the large lungs which are important in buoyancy regulation. The horizontal diaphragm forms a partition that divides the body cavity into two unequal halves. The stomach and liver lie in the lower half.

The bones themselves are dense and surprisingly heavy. This is due to the complete loss of the internal marrow cavity in the ribs and long bones which are choked with hard bony tissue. This feature (pachyostosis) makes sirenian bone amongst the densest in the animal kingdom. The heavy ribs may have a ballast-like action, steadying the animal in the water and partly counteracting the high centre of gravity imparted by the lungs (fig.314).

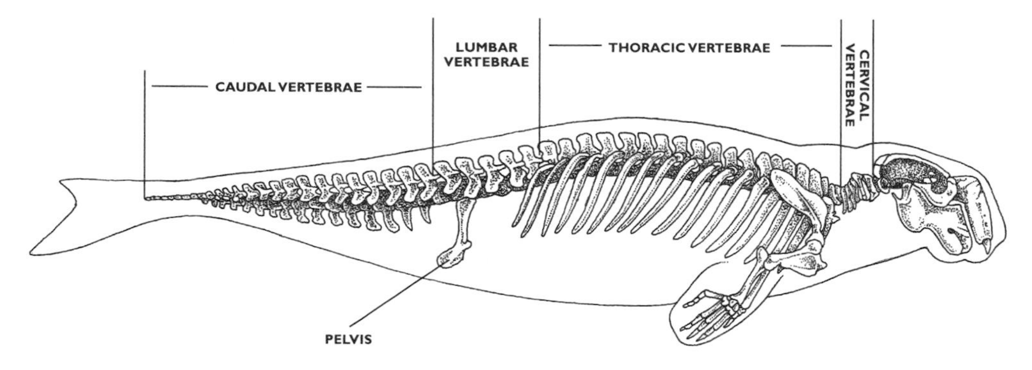

Figure 314. Skeleton of a Dugong. The rudimentary pelvis is present behind the last rib.

A side-by-side comparison of the skeletons of the Dugong and West Indian Manatee shows that the ribs are longer and wider in the manatee. The spine is also noticeably thicker and shoulder blades larger in the manatee. It is reasonable to assume that the Dugong, with its thin blubber layer and marine lifestyle in a more buoyant medium, has evolved to become less dependent on static buoyancy mechanisms than the manatee.

DIET

Sea cows eat a wide variety of aquatic plants. The West Indian Manatee is recorded to eat about sixty species of plants in Florida waters and seems to avoid those species that contain toxins (blue-green algae).

Manatees graze on floating plants at the water's surface (e.g. Water Hyacinth *Eichhornia crassipes*) and on underwater vegetation such as Hydrilla *Hydrilla verticillata*. The Dugong has a more downturned snout than the manatee, a feature that is more suited to its bottom-grazing diet of seagrasses. These include *Enhalus, Amphibolis, Halophila, Halodule, Cymodocea, Thalassia, Thalassodendron, Syringodium,* and *Zostera*. Dugongs leave distinct feeding trails visible at low tide when feeding on small, delicate sea grasses (e.g. *Halodule* and *Halophila*); they eat the roots and rhizomes as well as leaves and in this way the substrate becomes exposed. With tall-growing seagrasses (e.g. *Amphibolis antarctica*), only the leaves are cropped.

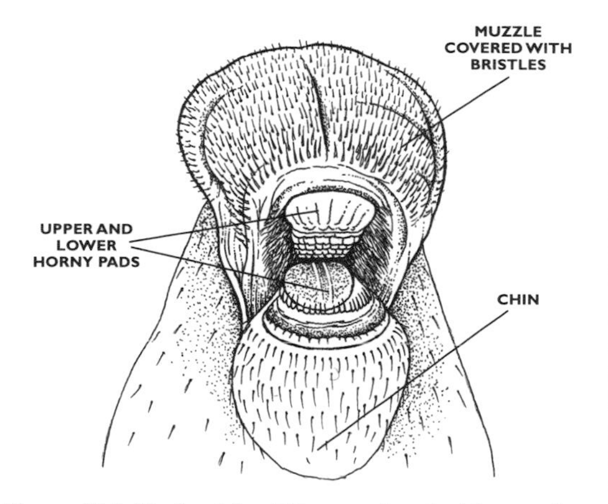

Figure 315. Underside of Dugong head with mouth opened to show horny mouth pads. The tongue is not visible since it lies behind the lower horny pad.

Seagrasses produce pollen and seeds and both are water borne. Seagrasses also reproduce under sediment by spreading rhizomes to produce dense stands or meadows. There are 49 species of seagrasses grouped in 12 genera; the majority of genera occur only in tropical waters and grow in soft substrates (muds to coarse coralline sand), often in shallow, sheltered reef areas where Dugongs are to be found. They are the only higher plants to have become adapted to a fully underwater lifestyle.

Sea cows use their paddles to assist with manipulating food. Manatees are seen to swish food within reach of their mouths with their paddles. Dugongs inhabiting the Red Sea use their paddles to dig sea grass from the sediment. The bristles around the snout (fig.315) help to deal with the slippery strands which are guided by the muscular lips into the mouth; the lips of the left and right sides can move independently. These structures have a comparable function to the cropping tongue and lips of cattle. A horny pad at the tip of upper and lower jaws

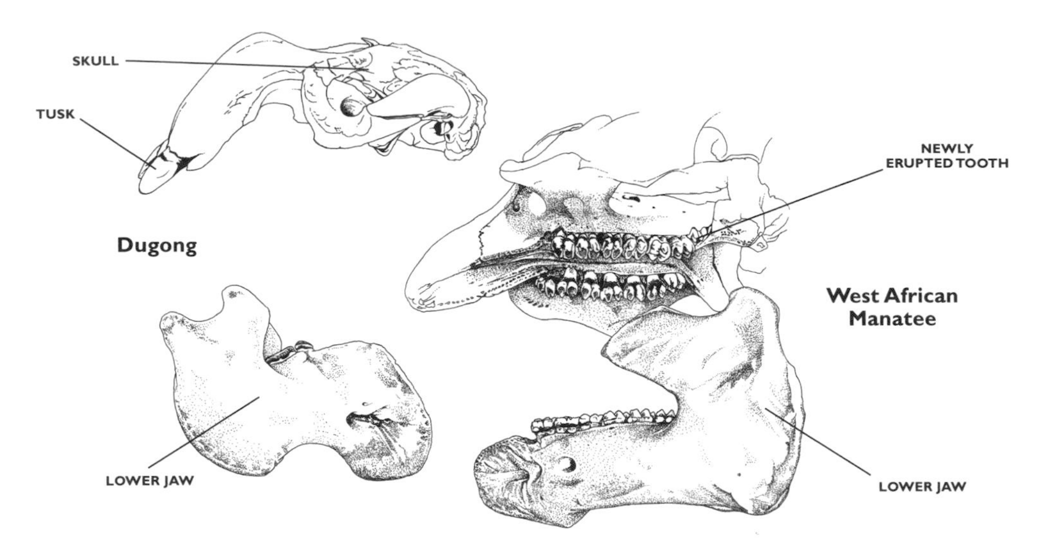

Figure 316. The teeth of manatees are curious and quite unlike those of the Dugong which, although reduced, fit more clearly into the general herbivore pattern of organisation and growth. Adult manatees are virtually toothless except for a small band of 4-7 molar teeth at the back of each half jaw. New teeth are formed throughout the manatee's life at the back of the jaws and these become pushed along the jaw during growth in a process called 'horizontal succession'. Molars are lost at the front end of the row after heavy wear. A manatee may form as many as eighty such teeth during its lifetime.

helps to grip and break off aquatic vegetation which is passed back into the mouth to be ground up by the teeth (fig.316).

Sea cows have a long digestive tract that is characteristic of herbivorous mammals. The large intestine is some 25m long in the Dugong, twice the length of the small intestine. The stomach itself is multi-chambered but relatively small in comparison to cattle, and in sea cows it is the large intestine that acts as the fermentation chamber during digestion. This type of digestive system is similar to land mammals such as the horse and elephant which are known as hind-gut ruminants. The large intestine of sea cows harbours a rich microflora that breaks down the tough cellulose (plant cell wall) component in the digesta. Digestion is a slow process. Food takes one week to pass through the gut in manatees although meals are enormous in bulk (up to 100kg wet weight per day in a large manatee).

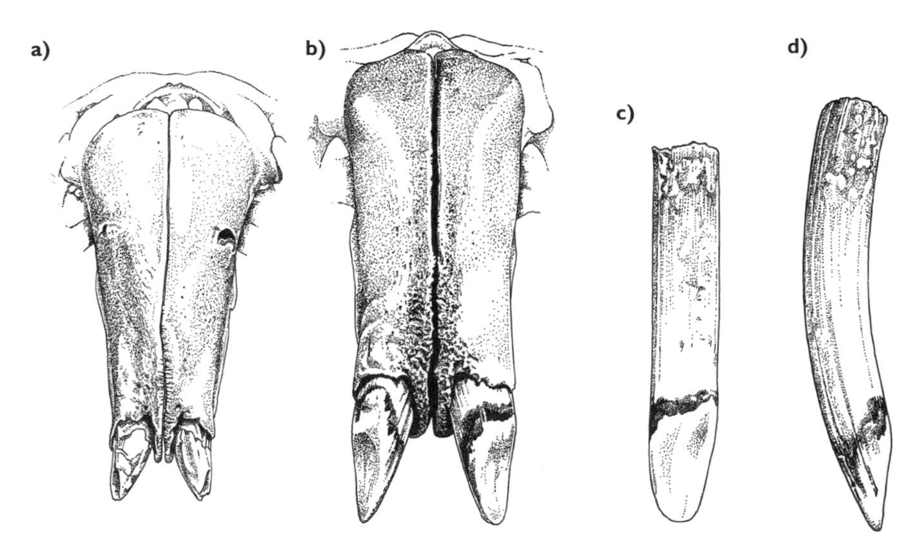

Figure 317. Anterior view of skull in: a) young male Dugong and b) adult male Dugong. c) Side view and d) top view of adult male Dugong tusk.

BEHAVIOUR AND ORIENTATION

Although large aggregations of both manatees and Dugongs are seen on good feeding grounds in some parts of their ranges, it is not yet known whether such groupings have any organised structure and, if this is the case, how it could relate to age, sex and reproductive condition. Careful observations on manatees in the wild suggest that there is little long term cohesion between individuals except that between mother and calf. When manatees encounter another individual there follows a period of snout-to-snout nuzzling suggesting an elaborate type of greeting ritual is taking place. At most other times, however, manatees are noticeably undemonstrative in their behaviour to each other.

Manatees and Dugongs produce chirp and squeak sounds of low frequency (less than 20kHz). It is not known what function these sounds could have. Manatee calves emit a call to which the mother will respond and this may be a type of distress signal; mothers and their calves are known to interact vocally, presumably to maintain and reinforce auditory contact between them. Although their eyes are small, direct observation suggests that manatees see well underwater. The eye has a shiny reflective layer behind the retina (tapetum lucidum) that is characteristic of animals that use vision at night. The tapetum produces a pinkish reflection or 'eye shine' like that of a cat at night. A tapetum is lacking in the Dugong eye, and little is known about vision in this species.

Manatees probably retain a sense of smell and the olfactory nerves are prominent on the base of the brain. There are also taste buds at the back of the tongue indicating that a taste sense is not absent in manatees.

REPRODUCTION

Manatees and the Dugong have a low reproductive rate. The Dugong does not reach puberty until 10 years of age (both sexes) and females have only one calf at a time with long rest periods between pregnancies. Calves suckle for at least 18 months in the Dugong and are born after a gestation period of 12 months. With a possible life span of about 70 years in the Dugong and an average calving interval of 5 years, a female could not expect to produce more than 12 young. This low fecundity increases the vulnerability of Dugong populations to the effects of over-exploitation and habitat destruction. Where hunting occurs, it has been estimated that the sustainable level of capture of female Dugongs is as low as 2% per year of all females in a population.

The tusks of the male Dugong erupt above the gum at about 12-15 years old and this coincides with the onset of sexual maturity. The tusk tip (fig.317) is worn to a chisel-like edge, the bevelled surface lying on the outer or lip-facing side. Although this indicates that the tusks are functional, the manner of their use remains speculative. The paired scars found on the backs of some Dugongs could have been produced by the tusks of rivals (fig.318); there is evidence that males fight for access to breeding females. Tusks may be used to grip the smooth back of the female during the process of turning the female on her back in the water prior to mating. However, tusks also erupt in old females.

Figure 318. Scarred back of Dugong. These weals were probably inflicted by the tusks of rivals.

Calving has been observed in captive manatees. Offspring can be born in either head-first or tail-first delivery. In Florida waters, the West Indian Manatee breeds year-round and both males and females are sexually active at all times of the year. Mating in the manatee differs from that in the Dugong. Copulation often occurs belly-to-belly with the male clasping the female with his paddles whilst positioned beneath her. The few observations of mating made in the wild suggest that manatees are promiscuous.

There is little information on birth in the Dugong; most specimens captured for display in aquaria have been immature. The few reports of calving in wild Dugongs suggest that the female beaches or drags herself onto a sandbar to give birth. There is one observation of a head-first delivery. If these reports are substantiated it would indicate that the female Dugong retains a residual attachment to the land in the reproductive cycle. Calves are about 1m in length at birth in the West Indian Manatee and 1.0 - 1.2 m long in the Dugong.

CONSERVATION

Exploitation of sea cows by native peoples has occurred through history and probably had little effect on sea cow populations until the introduction of modern hunting methods and equipment. In recent times, sea cows have been netted, trapped, harpooned and dynamited in many parts of their range for meat, skin and oil. The shallow riverine and tropical coastal habitats of sea cows are suitable as prime development sites and the future commercialisation of sea cow habitat will place further pressure on their populations. This has occurred in Florida with serious consequences for the local manatees; there is a high mortality rate in manatees from a variety of causes of which powerboat collisions are perhaps the most well known. Although the small manatee population of 1000-2000 animals is legally protected by the State (Florida Manatee Sanctuary Act of 1973), pressure on their habitat from boats continues to increase. Registrations of boats rose nearly seven-fold between 1960 and 1990 in Florida.

Until more is known about the sea cow populations in all areas of their range, the introduction of effective protection measures is difficult. Aerial survey is the technique of choice for locating widely distributed sea cows because of its potential to cover large areas in remote regions. However, it may not be available in all countries where occurrences are reported. The distribution of the Dugong encompasses the seas around more than 40 countries. Sea cows living in turbid coastal waters or in areas with overhanging vegetation are difficult to spot from aircraft.

Recent survey work in Australia has revealed that this country is the world's most important stronghold for the Dugong. The north coast of Australia has a population of up to 70,000 Dugongs with a year-round population of 17, 000 in the Gulf of Carpenteria. The Dugong is protected by law in Australia and it seems that here the future of this species can be safeguarded. The Dugong no longer occurs in the Laccadive, Maldive, Chagos, Nicobar, Barren, Narcondam, Cocos (Keeling), Christmas and Lesser Sunda Islands. In Micronesia, a relict population occurs in Palau Island where it is under threat of extirpation due to illegal poaching.

Steller's Sea Cow was hunted to extinction by fur traders in the short space of 27 years after its discovery in 1741 by the naturalist G. W. Steller on Bering Island. This loss to the world's marine mammal fauna is a reminder of the particular vulnerability of sea cows to hunting. It also stands as a testimony to the importance of developing effective protective measures for today's remaining populations of sea cows.

BIBLIOGRAPHY

Harrison, R. J. & King, J. E. 1980 (2nd ed.). *Marine Mammals.* Hutchinson, London.

Hartman, D. S. 1979. Ecology and Behavior of the Manatee (*Trichechus manatus*) in Florida. The American Society of Mammalogists, Special Publication. 5.

Marsh, H. (ed.) 1979. *The Dugong*. Proceedings of a Seminar/Workshop held at James Cook University 8-13 May 1979. James Cook University Publications, Townsville, Australia.

Reynolds, J. E. & Odell, D. K. 1991. *Manatees and Dugongs.* Facts on File, New York.

Pinnipeds

by

Kevin Morgan

INTRODUCTION

Figure 319. Two adult male Southern Elephant Seals *Mirounga leonina*, fighting for dominance of a beach.

Seals are carnivorous mammals that have adapted to a marine lifestyle. They are placed in the order Pinnipedia which literally means 'wing-footed', a reference to their webbed feet. The Pinnipedia divide into two main groups: the superfamily Phocoidea which contains only the family Phocidae (the true seals) and the superfamily Otarioidea which contains two families, the Odobenidae, whose only representative is the Walrus, and the Otariidae, the sea lions and fur seals (fig.320). Table 12 highlights the main differences between the Phocoidea ('the true seals' or phocids) and the Otarioidea ('the eared seals' and Walrus).

There are 33 species of pinniped.Pinnipeds match many terrestrial predators in their size and their array of formidable teeth. The Leopard Seal *Hydrurga leptonyx* can reach a length of 3.5m, whilst the Southern Elephant Seal *Mirounga leonina* can reach a length of 5m and weigh up to 3.9 tonnes.

Phocoidea	Otarioidea
No external ears	External ears (except the Walrus)
Hind flippers do not turn forward	Hind flippers can turn forward
Swim with hind flippers	Swim with fore flippers
Flippers completely furred	Flippers have naked 'soles'
Nails on 5 digits of hind flipper are the same size	Nails on digits 1 and 5 of hind flipper are minute
Have 2 or 4 teats	Have 4 teats
Marine and freshwater	Marine only

Table 12. The differences between phocids and otariids.

EVOLUTION

Pinnipeds evolved after the fully aquatic cetaceans (Cetacea) and sirenians (Sirenia) but before the sea otters (Mustelidae). There are two theories about their origins. The first is that the phocids and the otariids are monophyletic (i.e. they are descended from a common ancestor); the second is that they are biphyletic: the phocids are descended from an otter-like carnivore in the early Miocene while the otariids are descended from a bear or dog-like carnivore in the late Oligocene. The situation is unresolved and quite complex, but both theories agree that pinnipeds have evolved fairly recently.

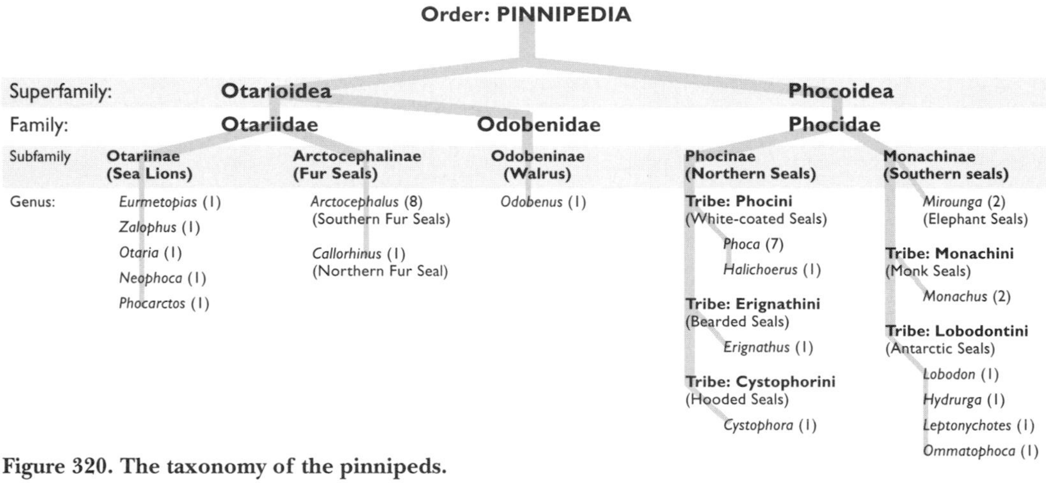

Order: PINNIPEDIA

	Otarioidea			Phocoidea	
Superfamily:	Otarioidea			Phocoidea	
Family:	Otariidae		Odobenidae	Phocidae	
Subfamily	**Otariinae (Sea Lions)**	**Arctocephalinae (Fur Seals)**	**Odobeninae (Walrus)**	**Phocinae (Northern Seals)**	**Monachinae (Southern seals)**
Genus:	*Eurmetopias* (1) *Zalophus* (1) *Otaria* (1) *Neophoca* (1) *Phocarctos* (1)	*Arctocephalus* (8) (Southern Fur Seals) *Callorhinus* (1) (Northern Fur Seal)	*Odobenus* (1)	**Tribe: Phocini** (White-coated Seals) *Phoca* (7) *Halichoerus* (1) **Tribe: Erignathini** (Bearded Seals) *Erignathus* (1) **Tribe: Cystophorini** (Hooded Seals) *Cystophora* (1)	*Mirounga* (2) (Elephant Seals) **Tribe: Monachini** (Monk Seals) *Monachus* (2) **Tribe: Lobodontini** (Antarctic Seals) *Lobodon* (1) *Hydrurga* (1) *Leptonychotes* (1) *Ommatophoca* (1)

Figure 320. The taxonomy of the pinnipeds.

NATURAL HISTORY

BLUBBER AND FUR

Since sea water is always colder than the mean body temperature of mammals (approximately 37°C), seals need to adapt to avoid excessive body-heat loss. Although many seals live in regions where the air temperature drops far below the minimum temperature of sea water (- 1.9°C), the problem is not so acute in still air. Air has a much lower thermal conductivity than water, so heat is lost from the body more slowly in air than in water. Air has a far lower heat capacity and it can be warmed up for a small expenditure of body heat loss (Bonner 1982). Pinnipeds conserve their body-heat in two ways: through their low surface area shape which effectively reduces heat loss, and through insulation. Most mammals insulate themselves with the hair covering their bodies and all seals have a hair coat which in some of the otariids is most luxuriant although, in the phocids, it is less dense. For example, the Northern Fur Seal *Callorhinus ursinus* may have up to 57,000 hair fibres per cm^2 (Scheffer 1962) while the Harp Seal *Phoca groenlandica* which lives in just as cold surroundings has only about 18,000 (Tarasoff 1972). The Fur Seals are further able to insulate themselves through the acquisition of two types of hair fibre, long thick guard hairs and much shorter underfur fibres (fig.321). Fur insulates the body by trapping air and sebaceous glands linked to the follicles help to keep the fur waterproof. The fur is at its thickest in pups which have yet to build up their layer of blubber and have to spend longer periods in very cold air temperatures. From a womb temperature of 37°C, a Weddell Seal pup can be ejected into conditions below -30°C. Fur insulation properties tend to decrease as seals dive to greater depths therefore otariids tend to spend much of their time near the surface. In phocids the main means of insulation is the blubber layer. In air, a blubber layer is about half as effective an insulator as an equal thickness of fur; but in water, unlike fur, it is incompressible and is unaffected by the depth to which the seal dives. Blubber is a layer of fat-filled cells lying beneath the skin (hypodermal adipose tissue) and in seals almost all the fatty tissue present in the animal is to be found in the blubber. Blubber, as well as acting as insulation, is also a food reserve. Most, but not all, seals inhabit cold seas but in hot, sunny climates they can easily overheat. Therefore, they will return to the water to cool down, or cover themselves with sand. Some otariids also urinate on their coats and spread their flippers to dissipate heat. The thick mane of hair and blubber around the neck in adult bulls is to provide extra protection during fights, to look imposing to other adult males and to attract females. For polar seals, adaptation to the transition from the extreme cold of the air to the relatively warmer conditions under water is complex and includes changes in the amount of blood that is supplied to areas close to the skin. Time spent under the ice varies between species.

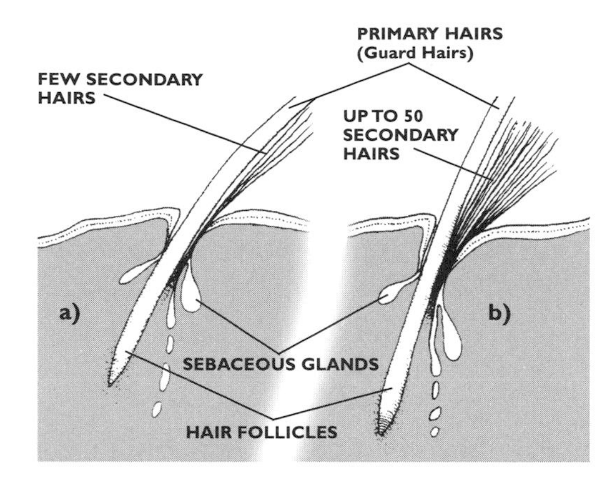

Figure 321. Section through a hair follicle of a) true seal (phocid) and b) fur seal (otariid).

All pinnipeds have to moult and most species do so some time after their breeding season has ended, usually in the warmer summer months. Otariids, because of their need to maintain their dense waterproof coat, have a more prolonged and more gradual moult than the phocids. In some species, notably Elephant and Monk Seals, the epidermis is shed with the hair. Elephant Seals remain out of water until their moult has been completed.

ADAPTATIONS FOR DIVING

Seals have proportionately a much greater blood volume per unit body weight than humans. Seal blood has a much higher haemoglobin concentration and their muscles contain large stores of myoglobin, a compound with an even greater oxygen-carrying capacity. Thus seals can carry large stores of oxygen in their blood and muscles when diving or swimming. While diving, seals can also restrict the blood flow to their vital organs such as the heart and brain, and can slow their heart rate down to only a few beats per minute.This helps reduce oxygen consumption and also prevents some physiological problems associated with prolonged deep diving. Seals also exhale when they take deep dives and the minimal air in the lungs helps to further prevent decompression sickness. In the relaxed position the nostrils are closed so that a resting seal will not suddenly breathe in underwater. The phocids are the deeper divers and spend longer underwater. The Weddell Seal *Leptonychotes weddellii* can reach a depth of 700m and range up to 12 kilometres from the breathing hole. They can remain underwater for up to 82 minutes, although dives are usually less than 30 minutes. The Southern Elephant Seal is another deep diver and the current record is to a remarkable depth of more than 1,700m, the longest duration being two hours. Eared seals can still reach a respectable depth of 190m.

HABITAT PREFERENCES

Seals generally leave the water to breed (although there is some evidence that the Common/Harbour Seal *Phoca vitulina* and the Walrus *Odobenus rosmarus* occasionally give birth in water) and most species spend longer periods out of water during their moult. Some species haul-out frequently throughout the year while others, such as Southern Elephant Seals, may spend up to 90% of their lives at sea. Many polar seals haul-out and breed on ice either in loose pack ice, at the ice edge or, in some cases, in fast ice many miles from open water. However, most species prefer to haul-out on remote or inaccessible coasts or islands away from disturbances (fig.304).

In the Antarctic, Weddell Seals can be found many miles from open water. They survive by using cracks in the ice and by maintaining breathing holes, gnawing the ice at the edges of the holes to prevent them from freezing over. Holes and cracks are used for breathing and for hauling-out onto the surface of the ice where they rest and where females give birth to their pups. Weddell Seals living on fast ice face a dilemma. If several seals share the same hole, less effort per individual is required to keep the hole from freezing over. This is important as canine teeth can become severely worn and seals may die from dental infections or from an inability to maintain their breathing holes - few Weddell Seals live beyond 20 years. If a number of seals use the same breathing hole there is an inevitable increase in competition for the food supplies within foraging range. During the breeding season, adult males compete for mates by defending breathing holes from other males, so restricting access to breeding females on the ice above.

Arctic seals that breed on fast ice (e.g. the Ringed Seal *Phoca hispida*) run the risk of predation by Polar Bears. These seals often congregate where cracks and gaps occur in the ice. Since Polar Bears often wait by these holes, the seals usually have a number of alternative holes which they can use. During their breeding season, Ringed Seal pups are protected in lairs or caves under ice ridges. By contrast, Hawaiian Monk Seals *Monachus schauinslandi* prefer to breed and haul-out on sandy beaches. Grey Seals *Halichoerus grypus* and Common/Harbour Seals haul-out on intertidal rocks, skerries and sandbanks. Grey Seals breed at traditional sites on uninhabited islands and frequently move off the beaches into less crowded grassy areas.

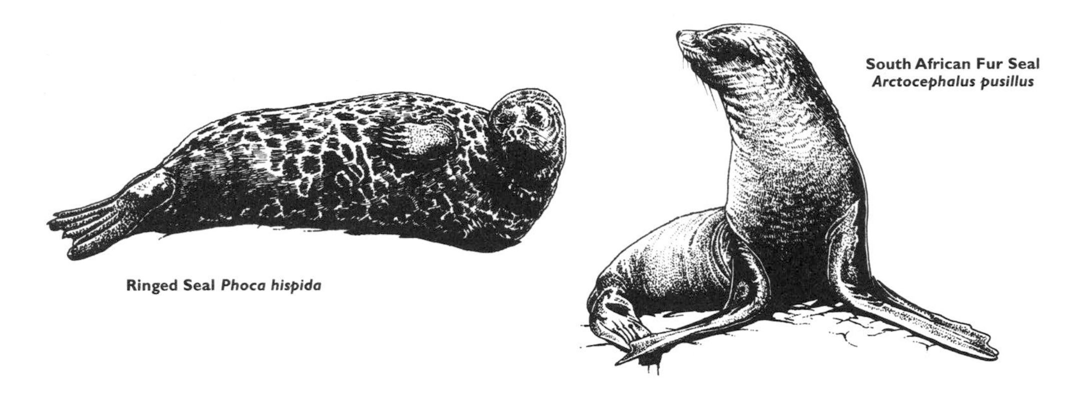

Figure 322. a) An adult Ringed Seal *Phoca hispida* and b) a South African Fur Seal *Arctocephalus pusillus*, showing their normal attentive posture when out of the water.

DISTRIBUTION AND STATUS

With the exception of India and south and west Asia, pinnipeds are widely distributed around the world. The ranges of some species overlap extensively. There are two landlocked species of seals: the Caspian Seal *Phoca caspica* and the Baikal Seal *P. sibirica* - both are phocids. Some species, such as the Harp Seal *P. groenlandica*, Hooded Seal *Cystophora cristata*, Northern Fur Seal *Callorhinus ursinus* and Antarctic Fur Seal *Arctocephalus gazella*, migrate between their wintering and summer breeding grounds. Individual animals may wander widely; for example the Walrus even occasionaly reaches British shores. Confusion can arise in the identification of certain species. The Fur Seals of the southern hemisphere do not differ greatly in appearance and where the ranges of two species overlap, care must be taken in identifying them. Extra care must be taken in those few locations where hybridisation occurs, for instance between the Subantarctic Fur Seal *Arctocephalus tropicalis* and the Antarctic Fur Seal *A. gazella* which are found together on a number of subantarctic islands.

Sightings of seals at sea are infrequent and any animals seen should be closely examined and a detailed description made, noting the species and location of the sighting. Photographs can be of considerable value.

The most abundant species of seal is the Crabeater Seal *Lobodon carcinophagus* with a population estimated to be between 11 and 12 million. It is one of the world's most abundant wild mammals. In the northern hemisphere the Ringed Seal is the most abundant seal with an estimated population of between 6 and 7 million. In contrast, the Hawaiian Monk Seal *M. schauinslandi* and the Mediterranean Monk Seal *M. monachus* only number a few hundreds each, while the West Indian Monk Seal *M. tropicalis* is now thought to be extinct.

Seals have been exploited by humans for centuries and a number of species were hunted almost to extinction in the 18th, 19th and 20th centuries. Many of these populations have made remarkable recoveries, especially the Northern Elephant Seal *Mirounga angustirostris*, and the Antarctic and Northern Fur Seals. Other species are recovering more slowly. The Northern Sea Lion *Eumetopias jubatus* population is declining at present for as yet unknown reasons. Curiously, the slaughter of whales in the southern oceans and the consequent rise in krill stocks may have benefited those seal species which feed on krill.

It is true that fish form a considerable part of a seal's diet and, this being the case, they are often blamed for reducing fish stocks. The culling of Harp Seal pups on the Canadian sea ice caused an international outcry. This was followed by a ban on the importation of sealskins into Europe imposed by the European

Economic Community (now the European Union). Grey Seals can take many valuable fish from fish farms and it is natural that fishermen should react to their behaviour, but Harp and Grey Seals are not globally threatened. Much more worrying is the reaction of fishermen to predation of fish stocks by the threatened Mediterranean Monk Seal. Many Mediterranean Monk Seal deaths result from the shooting of the animals by fishermen incensed by large sums of money being spent on their conservation while Mediterranean fish stocks continue to be depleted by over-exploitation. There has also been a reduction in haul-out sites that are free from human disturbance. Within the Mediterranean region, Monk Seals almost exclusively use caves as haul-out sites.

REPRODUCTION AND LIFE SPAN

In species that gather at traditional land sites to breed, males tend to be larger than females. This is because a bigger male can prevent smaller males from breeding with females which are in his territory or harem. Competition between males is intense and only the fittest males are able to breed. Males usually have a shorter lifespan than females. In species which breed on pack ice, females cannot group together to breed so the competition between males is not quite so severe. Consequently, the size difference between males and females is reduced and occasionally is reversed. Life is harsh for dominant bull seals and few bulls remain at the top for long. Bulls of many eared seals have a large lion-like mane, hence the name 'sea lion'. Bull elephant seals can inflate their snouts to impress the females (fig.319), whilst the Hooded Seal can inflate a cushion-like hood on its head and a red sac (the internasal septum), which protrudes out of its nostril .

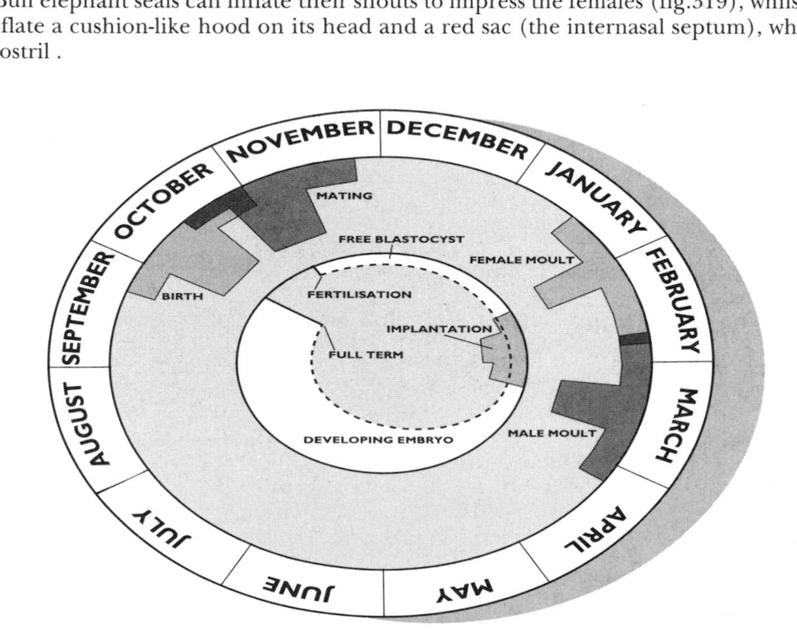

Figure 323. The breeding cycle of the Grey Seal *Halichoerus grypus* (after Bonner 1989).

In most polar seal species, pups are born in the spring but for species in warmer climes pups are born in the summer months. Mating occurs annually, generally soon after the females give birth. This suggests that the pregnancy takes a full year, but this is not so. In the case of the the Grey Seal, foetal development lasts about 250 days (fig.323). About ten days after fertilisation, development is suspended in the blastocyst stage for a period of about 100 days, followed by another 240 days of development, then around 15 days between birth and fertilisation, thus making a full year. It is thought that development is delayed to allow birth and mating to occur during one short period onshore. It is also believed that birth can be delayed if the mother is prevented from coming ashore by a disturbance on the beach. Once ashore, birth is rapid. Occasionally pups may be crushed to death on the breeding beaches. This mostly happens when rival bulls charge through the harems or when amorous bulls try to mate with receptive females which have new born pups beside them. Due to the richness of their mother's milk, pups quickly put on weight which minimises their exposure to predators and prepares them for the rigours of their new life.

All eared seals (otariids) breed on land and give birth to dark-coated pups. Phocid seals (e.g. elephant and monk seals) which have always bred on land also give birth to dark pups. Other phocids give birth to white-coated pups which are normally born on the ice. Grey and Common/Harbour Seals are exceptions as they usually breed on land but have white pups (a legacy of an earlier ice-breeding existence).

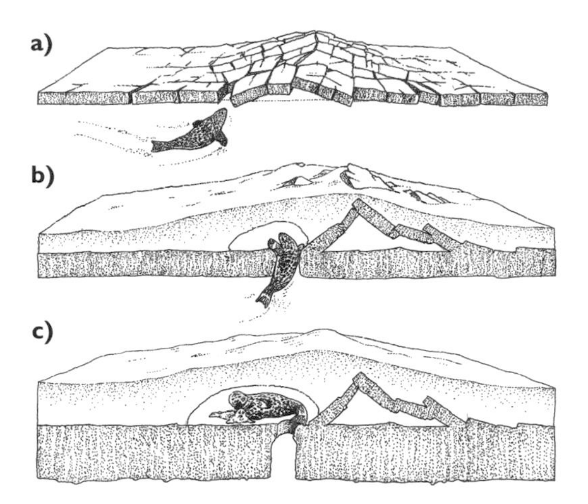

Figure 324. The breeding lair of a Ringed Seal: a), b) exploiting the buildup of snow around a pressure point in the ice sheet to hollow out a lair and c) female and pup within the lair.

Common/Harbour Seal pups moult their white coats before they are born. This is because mothers give birth in the intertidal zone and pups must have their first adult coat and be able to swim with the next high tide.

Most phocids, including Caspian and Baikal Seals, breed on the ice. Those in the Arctic have to adapt to the threat of Polar Bears and so wean their pups as quickly as possible. The Ringed Seal often has a breeding lair under the ice in the hollows formed by pressure ridges (fig.324). They are therefore able to avoid the attentions of Polar Bears and Arctic Foxes, and the lair also retains warmth.

Seals are at risk from predators when they go ashore to moult. Seals have a life span of from 10 to 35 years, with females usually living longer than males. There are records of Walruses living to at least 40 years, Ringed Seals to 43 years and Grey Seals to 46 years.

PREDATORS AND PREY

Seal diets are varied (fig.325). For most seals, fish and squid are the main food sources followed by crustaceans, bivalve molluscs and zooplankton; other warm-blooded animals such as birds and other seals may also be taken. Although Leopard Seals eat a large proportion of krill, they will take penguins (fig.326) and other seals when the opportunity arises.

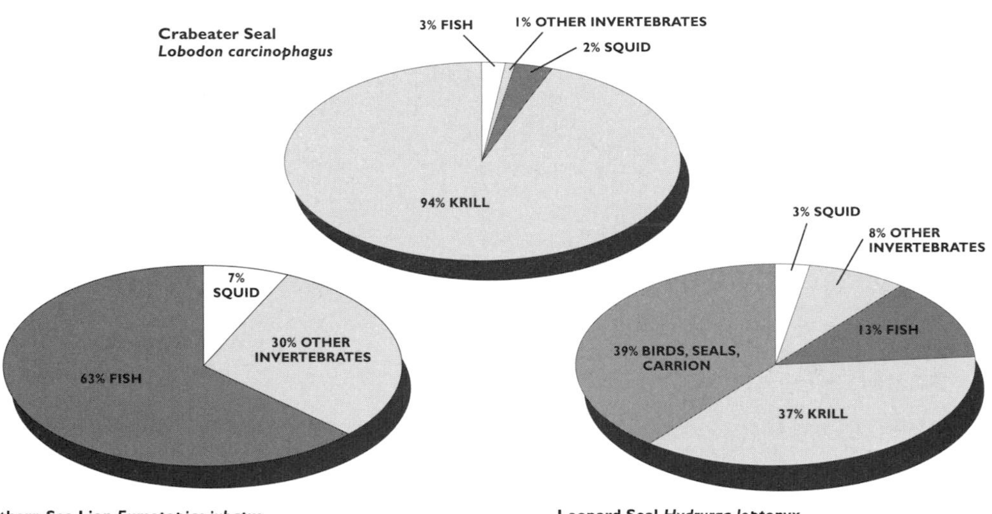

Figure 325. Average diets of an otariid and two phocids.

The Bearded Seal *Erignathus barbatus* feeds on the bottom and includes flatfish in its diet. The whiskers apparent on this seal, and to a lesser extent on other species, are important for probing nooks and crannies for food. The Walrus mainly feeds on shellfish and the Crabeater Seal, despite its name, feeds almost exclusively on krill. Where the ranges of seals overlap the feeding depth and preference for fish, squid, or other prey items appears to determine the allocation of resources between species.

Natural seal predators include sharks, the Killer Whale, and in the Arctic, the Polar Bear. The Great White Shark is a top predator around seal colonies (particularly in warm temperate seas). Seals and sea lions are attacked near the surface by the shark. The wounded animal is then carried underwater until it becomes moribund from blood loss. The Great White Shark has been observed swimming at high speed from

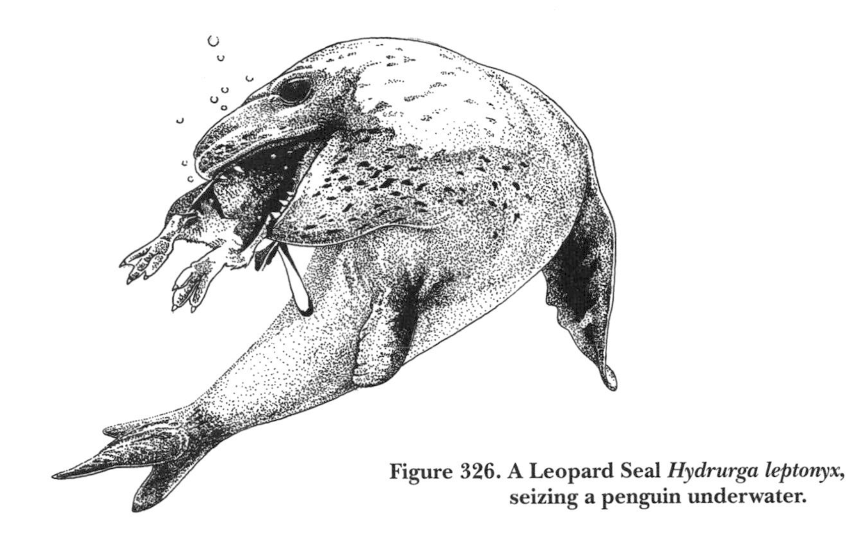

Figure 326. A Leopard Seal *Hydrurga leptonyx*, seizing a penguin underwater.

below to ram an unsuspecting seal swimming on the surface. The force of the impact can severely damage the seal and the injured animal is quickly dealt with by the shark. Killer Whales have been known to capture seals by tipping them off small ice floes and snatching sea lions from the beach close to the waters edge (fig.327).

Figure 327. Killer Whales *Orcinus orca* feeding on pinnipeds.

EFFECTS OF POLLUTION

Oil spills can have a direct effect on seal populations, particularly if a spill occurs close to a breeding colony. In 1988 at least 17,000 Common Seals died in the North Sea as a result of the phocine distemper virus (PDV) epidemic. This probably has occurred in the past in the North Sea, and huskies taken to Antarctica may have passed on distemper to seals there.

Toxic chemicals derived from industrial pollutants accumulate in the seal blubber as a consequence of eating contaminated prey. These pollutants appear to have a minimal effect if they are locked in the fat reserves; but under stress and during times of fasting such as occurs in the breeding season, the fat reserves are mobilised and toxins like PCBs are released and pass into the blood. Once in circulation, the toxins may depress the immune system thus making the seal more prone to disease. Seal mortalities are also caused by entanglement in fishing nets, lines and other human debris.

MOBILITY

All seals are expert swimmers. Walruses are not as fleet in water, probably because they do not have to chase after their primary food source of bottom-dwelling molluscs. Eared seals (e.g. sea lions) swim using their large paddle-like foreflippers for propulsion and their flexible hindflippers as stabilisers. They are quick and agile in the water and are quite nimble on land, walking or running on their foreflippers and forward-rotated hindflippers. Eared seals are innately curious and this, plus their agility, makes them popular animals in zoos and circuses. Phocids swim using their broad hindflippers for propulsion and their smaller foreflippers for low speed turning and manoeuvring. At high speeds the foreflippers are held against their sides. The fastest recorded swimming speed is 40km/hour (22 knots) by a Californian Sea Lion *Zalophus californianus,* while the land speed record of 19km/hour (11 knots) belongs to a phocid, the Crabeater Seal *Lobodon carcinophagus.*

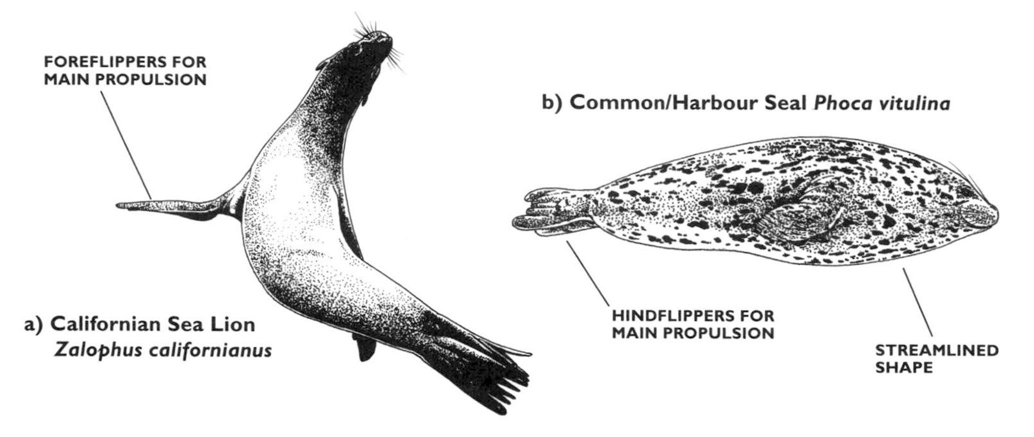

Figure 328. Swimming in a) a Californian Sea Lion *Zalophus californianus,* an eared seal (otariid), and b) a Common/Harbour Seal *Phoca vitulina,* a true seal (phocid).

IDENTIFICATION OF SEALS

The factsheet and plates 54-56 highlight the main field marks. Seals observed in the water can be very difficult to identify. Around breeding colonies identification can be assisted by observing a range of animals, including the bulls, but be aware that more than one species may be present.

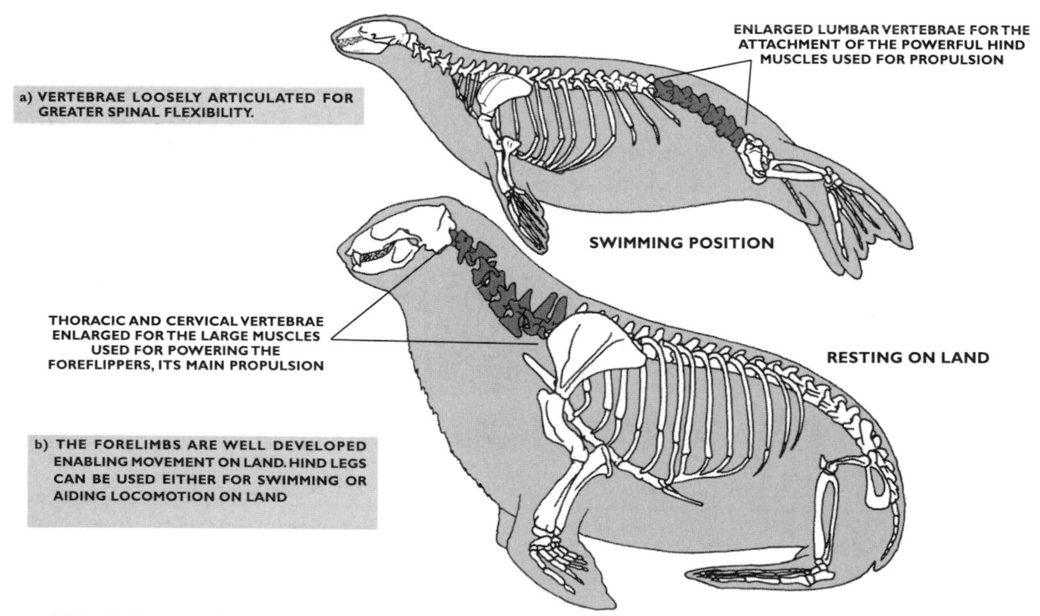

Figure 329. Skeletons of (a) phocid and (b) otariid.

BIBLIOGRAPHY

Arsenice, V. A. 1986. *Atlas of Marine Mammals.* TFH, New Jersey.

Bonner, N. 1982. *Seals and Man: A study of interactions.* University of Washington Press, Washington.

Bonner, N. 1989. *The Natural History of Seals.* Christopher Helm, London.

Godwin, S. 1990. *Seals: A Complete Photographic Survey.* Headline, London.

Jefferson, T. A., Leatherwood, S. & Webber, M. A. 1993. *Marine Mammals of the World.* FAO Identification Guide, Rome.

King, J. 1983. *Seals of the World.* (2nd Ed.) British Museum (Natural History), London.

Macdonald, D. (ed.).1989. *All the World's Animals: Sea Mammals.* Torstar Books, New York.

Macdonald, D. & Barrett, P. 1993. *Mammals of Britain & Europe,* HarperCollins, London.

Reeves, R., Stewart, B. S. & Leatherwood, S. 1991. *The Sierra Club Handbook of Seals and Sirenians.* Sierra Club Books, San Francisco.

Reinjders, P. *et al.* 1993. *Seals, Fur Seals, Sea Lions and Walrus (Status Survey and Conservation Action Plan)* IUCN/SSC Seal Specialist Group.

Ridgway, S. H. & Harrison, R. J. (eds). 1981. *Handbook of Marine Mammals: Vol.1. Walrus, Sea Lions, Fur Seals and Sea Otters,* Academic Press, London.

Ridgway, S. H. & Harrison, R. J. (eds). 1981. *Handbook of Marine Mammals: Vol.2. Seals,* Academic Press, London.

Riedman, M. 1990. *The Pinnipeds: Seals, Sea Lions and Walruses.* University of California Press, Berkeley.

Scheffer, V. B. 1962. *Seals, Sea Lions and Walruses: A review of the Pinnipedia.* Stanford University Press.

Tarasoff, F. O. 1972. *Comparative aspects of the hind limbs of the river otter, sea otter and seals.* In: *Functional Anatomy of Marine Mammals.* Vol.2. (ed. Harrison, R. J.), pp.333 - 359. Academic Press, London.

Seal Factsheet

TRUE SEALS

The 18 species of true seals, Phocidae, have no external ear flaps and short muzzles. Short body hair and fully furred flippers are also common features. When swimming they use undulations of their hindquarters and hindflippers, the foreflippers usually being tucked beside body and only used in tight manoeuvring. True seals are less playful and agile in water than eared seals (shown on plate 56), but still superb swimmers and deeper divers. Hindflippers cannot be rotated under body and with their relatively small foreflippers, they are only capable of slow progression on land by a belly-flopping type of motion. On ice they use rapid pulling motions of foreflippers to propel themselves along at a reasonable pace, or use a sculling motion of the hindflippers. True seals can be inquisitive on land or ice, often allowing approaches by humans; however, at sea they are quick to escape.

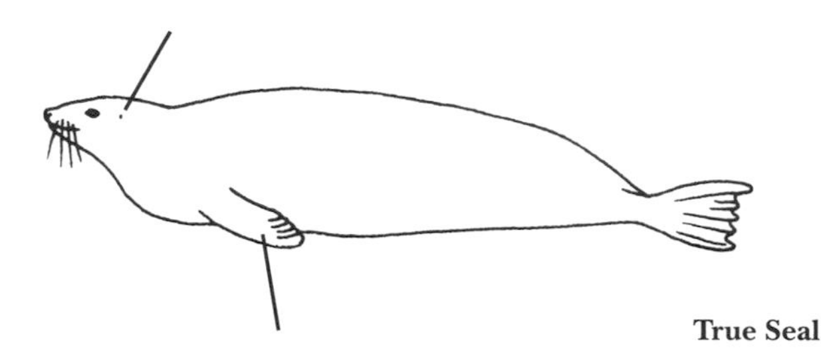

True Seal

EARED SEALS

The 14 species of eared seals, Otariidae, have small external ear flaps and use their large foreflippers for propulsion at sea and support on land. The hindflippers can be rotated forward, enabling them to walk on land. They are more agile and lithe at sea than phocids, often porpoising at the surface, but tend to be shallower divers. Playful and inquisitive, they are often curious near boats and people.

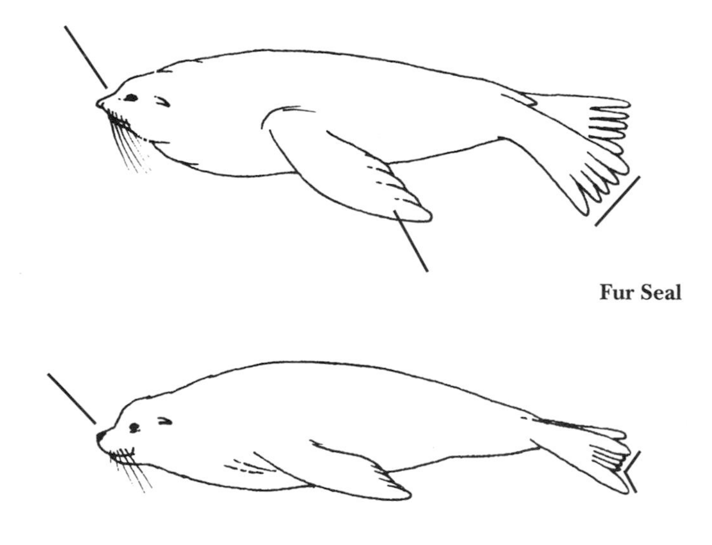

Fur Seal

Sea Lion

Marine Otters

by

Geoffrey Waller

Figure 330. A Sea Otter *Enhydra lutris* resting on its back, anchored by kelp fronds wound round its trunk. In some parts of Alaska, Sea Otters occur in areas without kelp where there is a rich bottom fauna.

The Sea Otter *Enhydra lutris* is adapted to living in the coastal waters of the north Pacific and can spend its whole life at sea, only coming ashore during storms. In this respect, the Sea Otter differs fundamentally from seals. The Sea Otter belongs to the group of carnivores called the Mustelidae (which includes weasels, skunks and badgers). It is one of the smaller species of marine mammal; males can reach a length of 1.48m and weigh 45kg (females are smaller and lighter). There is no blubber beneath the skin and therefore body weight is low relative to body size, in comparison to seals (in Harp Seals *Phoca groenlandica*, the blubber layer averages 40% of total body weight). Sea Otters are adept swimmers and divers as befits their fully marine lifestyle, and unlike the semi-aquatic river otters, they are ungainly on land, moving about with an odd hunched gait. Other important differences between the Sea Otter and river otters are given in Table 13.

SEA OTTER	RIVER OTTERS
Ears folded, valve-like	Ears cupped and projecting
Skin loose, dense fur	Skin sleek, thinner fur
Front paws with retractable claws	Claws fixed, reduced in some species
Back feet flipper-like; fully webbed and larger than front feet. Footpads lost except for vestiges on toe tips	Back feet not enlarged and only partly webbed. Obvious footpads
Tail short, straight, horizontally flattened with round tip	Tail long and pointed, only partly flattened

Table 13. External differences between the Sea Otter and river otters.

Sea Otters are dark brown in colour and older animals have a grizzled appearance with a distinctive lightening of the fur around the head and shoulders. There is a bushy mass of sensitive whiskers around the nose (fig.315). The coat, which has the highest density of hair of any mammal, is effective in trapping a layer of air between the sea water and the skin surface. The Sea Otter therefore has its own built-in wet suit which keeps it warm and increases its buoyancy at the surface. This layer of air is renewed during grooming by rearranging the folds of the coat to trap air, and also by air that the Sea Otter blows into its own fur. Grooming is important for waterproofing the coat, and an oily secretion from sebaceous glands in the skin is spread over the hairs.

Due to the need for continuous insulation, the moult is prolonged throughout the year rather than abrupt as in seals, and it reaches its peak in midsummer.

Sea Otters rest at the sea surface in a characteristic pose on their backs (fig.330). They are sociable animals and may raft together on a favourite kelp bed in numbers of up to two thousand animals. They entwine strands of kelp around themselves and, safely anchored in this way, float on their backs to and fro with the tide. At times they are inactive and have been observed with their front paws over their eyes in a pose much resembling sleep.

Figure 331. Sea Otters see well both in air and water. Their eyes have a focusing range over three times that of any terrestrial mammal.

The Sea Otter swims slowly (about 1 knot) on its back at the sea surface using a side-to-side sculling motion of its tail. Sea Otters will eat, rest, sleep and groom on their backs, and females will carry pups on their chests. When eating, the Sea Otter uses its chest as a table to prepare its food. When swimming on its back, the hind feet are used with alternate strokes and in this position, the outer ('little') toe is the first to enter the water during the propulsive stroke. This toe is curious in being the longest digit in its foot, and is an adaptation probably related to this particular habit of 'backstroke' swimming at the surface.

When swimming underwater, the sea otter swims on its front and undulates the posterior part of the body in a vertical plane (as in whales). The strongly webbed back feet are spread horizontally and together with the tail, provide a fluke-like propulsive surface for fast swimming (5 knots). The front feet are not used at all in swimming.

Unlike river otters which are primarily fish feeders, the Sea Otter has a specialised invertebrate diet, including sea urchins *Strongylocentrotus*, abalone *Haliotis*, clams *Gari*, mussels *Modiolus*, and crabs *Cancer*, *Pugettia*. The front paws are used to search around and beneath rocks and kelp holdfasts during underwater foraging. This is generally in the shallow sublittoral zone, but there is a report of a Sea Otter diving to 106m. With their thick pads, the paws can manipulate spiny sea urchins and the sharp-edged bivalve shells during feeding. A loose fold of skin under each armpit is used as a pouch to store food collected during dives since the Sea Otter must return to the surface to eat. Stone 'tools' used to crack open mollusc shells are also carried from the seabed in these pouches. The left pouch seems to be the favourite, Sea Otters being invariably right-handed.

The tough shells of these invertebrates are split open by the strong, flat-topped crushing teeth. Bivalves that are too large to be crushed with the teeth are broken open by pounding them against each other. Sea Otters also use a stone balanced on the chest in the same way as an anvil; hard-shelled prey is gripped in the paws and struck against the stone. Abalone shells are dislodged underwater by using a suitably shaped stone gripped between the paws to batter the mollusc. The abalone, however, is a tenacious mollusc and it may take the Sea Otter several dives before the shell releases its grip. An extra bone in the Sea Otter wrist may give the front paws additional leverage or dexterity in such circumstances. Among marine mammals such use of tools is unique to the Sea Otter. When fish are taken, it is usually by the larger males whose greater strength enables them to seize and pull apart such slippery prey.

The Sea Otter is a voracious feeder; it must eat 20-25% of its body weight in food every day. The liver is large (6% of total body weight, twice that of other marine mammals) to metabolise this rapid food intake. The gut, at ten times its body length, is greatly enlarged in comparison to similar-sized terrestrial carnivores (four times body length in the dog and cat). The Sea Otter must also drink sea water since not all of its water can be obtained from its food; the kidneys are enlarged (twice that of river otters) to remove the excess salt load. Sea Otters need to devote more time to foraging when prey of low nutritional value (e.g. sea urchins) forms much of their food intake.

In the wild, birth at sea has rarely been observed. The pups are precocious and of large size (60cm length) relative to the mother, with eyes open and milk teeth showing at birth. The young are able to swim at 4-5 weeks old, dive at 6 weeks and are independent at 6 months. The life span is commonly 10-15 years.

The Sea Otter has been heavily exploited in the past. This began shortly after 1741 when survivors of the Bering Expedition returned with Sea Otter furs. By 1911, when an estimated 500-1000 Sea Otters remained, hundreds of thousands had been slaughtered for their valuable pelts. In 1911, the Fur Seal Treaty was signed by USA, Britain, Russia and Japan and this included protective legislation for the Sea Otter. From 13 remnant colonies, the Sea Otter has recovered well with a population stronghold in the Alaskan peninsula (nearly 20,000 in 1986) and has returned (or been reintroduced) to parts of its historic range. This extended from Japan (Hokkaido) in the west, through the Aleutian Islands to the coast of the North American mainland south to Baja California.

The Sea Otter has few natural predators. Although the Bald Eagle *Haliaeetus leucocephalus*, the Great White Shark *Carcharodon carcharias* and the Killer Whale *Orcinus orca* are all thought to attack Sea Otters, the evidence indicates that predation is sporadic; eagles take only pups during the nesting season and Sea Otters have yet to be found in the stomachs of Great White Sharks and Killer Whales. On the rare occasions when Sea Otters come ashore, they may be susceptible to predation by bears and coyotes.

Local populations are vulnerable to oil pollution. When the Exxon Valdez oil tanker ran aground in Prince William Sound, Alaska in 1989, more than 1,000 oiled Sea Otter carcasses were retrieved. It is now known that the sticky crude oil causes waterlogging of the Sea Otter coat and the animal quickly chills in the cold sea water. Prolonged cooling is fatal. They may also succumb to toxic components in the oil itself.

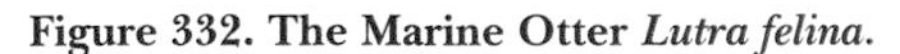

Figure 332. The Marine Otter *Lutra felina*.

Fossil evidence suggests that the first otter-like mammals appeared about 30 million years ago. During the evolutionary history of this group, there were probably marine otters living in the North Atlantic and North Pacific basins, although the genus *Enhydra* seems to have evolved in the last 2 million years. The present day Marine Otter *Lutra felina* is ecologically intermediate between true river otters and the Sea Otter. This species (fig.332) inhabits marine coastal waters of western South America (Peru and Chile) and, although known to science for over two centuries, little is recorded of its biology. It may be smaller than the Sea Otter with no size difference between the sexes. It swims on its back at the sea surface and comes ashore to shelter in caves on rocky coastlines. Due to hunting and a belief that the Marine Otter competes with local fishermen for shellfish, it has been heavily exploited and is now no longer found in Argentinian waters.

BIBLIOGRAPHY

Chanin, P. 1985. *The Natural History of Otters*. Croom Helm, London.

Kenyon, K. W. 1981. Sea Otter *Enhydra lutris* (Linnaeus 1758). In: *Handbook of Marine Mammals*. Vol. 1 (eds.
Ridgway S. H. & Harrison R. J.) pp. 209-223. Academic Press, London.

Love, J. A. 1990. *The Sea Otter*. Whittet Books, London.

Paine, S. & Foott, J. 1993. *The Nature of Sea Otters: a Story of Survival*. Greystone, Vancouver.

Polar Bear

by

Michael Burchett

Bears are thought to have evolved during the Miocene period about 20 million years ago from small, dog-like ancestors. The Polar Bear *Ursus maritimus* may have evolved from the closely related Brown Bear *Ursus arctos* that inhabits Siberia and North America and which is sometimes found along coastal areas.

The Polar Bear is well adapted for life in the Arctic where sea temperatures may fall below freezing and air temperatures can fall to -40°C. Polar Bears are the biggest of all the bears (Ursidae). Males can reach a length of 2.5m (tip of nose to tail) and weigh almost 800kg, while females are smaller with lengths reaching 2m and weights up to 300kg. The relatively large body size reduces the amount of heat loss compared to an animal of smaller size that loses heat more quickly. The Polar Bear has thick white fur that covers all parts of the body except for the foot pads and nose. There is a soft insulating underfur and overlying this, guard hairs of up to 15cm in length protect and maintain the integral nature of the underfur. Fur is present in other colours ranging from yellow, light brown and light grey. A greenish tint seen in the fur on the back of some bears is caused by microscopic green algae. The nose and skin are black in colour.

Large, partially webbed paws on the front limbs are used to assist swimming. Each paw has five digits, each with a non-retractile claw. Claws help to maintain a grip while walking on ice and snow. If the sea ice becomes thin and unstable the bear will spread its legs wide apart to distribute its body weight more evenly. Claws and feet can also used for digging, swimming, capturing and holding prey. Polar Bears are excellent swimmers and with their well insulated bodies they are able to spend long periods swimming in the freezing water. Bears have been observed swimming from island to island and between ice floes in search of food. They have also been seen swimming in open water many kilometres from the nearest land. During the summer months bears will often swim along the coast when overland travel may be difficult or time-consuming. The swimming action is similar to that of a dog.

The main food of the Polar Bear is Ringed Seal *Phoca hispida,* but they will also hunt Bearded Seal *Erignathus barbatus,* Harp Seal *Phoca groenlandica* and Hooded Seal *Cystophora cristata.* Occasionally they will take Walrus *Odobenus rosmarus* and Beluga *Delphinapterus leucas,* and they have also been seen eating Arctic Cod. When the Arctic sea is frozen, solitary bears or mothers with the previous year's young (cubs stay with their mother for two years) wander far and wide in search of seals. A bear will often wait next to a breathing hole in the ice and as the seal puts its head out of the water, the bear stuns the animal with a single blow from its powerful front leg. Bears have also been known to tip resting seals off small ice floes and break into the pupping dens of the Ringed Seal. During the spring months (March to May), seals are plentiful when females are hunting to feed their young. However, by late summer when the sea ice has mostly disappeared and the seals have dispersed, food becomes scarce for the bears. The bears may then come ashore and they are frequent visitors at the rubbish tips of coastal human settlements. Hunger may cause males to occassionally attack and kill bear cubs for food. However, females and their offspring generally avoid confrontation by maintaining a safe distance between themselves and the males.

Males and females come together to mate in late spring (April to June) and each male may mate with one or several females. Pregnant females excavate a snow den along the coastal fringes during November and December, and one to three cubs are born in December and January. When spring arrives the mother and her offspring leave the den in search of food.

Polar Bears have a northern circumpolar distribution. Their southern limit fluctuates with seasonal ice cover, ice movements and food (seal) availability. Solitary bears often move north and south with the ice edge and favour areas where currents and ice movements create open areas of water between ice floes. These are good areas for hunting seals. Large numbers of Polar Bears may congregate around seal colonies or when ice-free conditions cause seals to haul-out on shore, .

Polar Bears have relatively long lives and may live to be 30 years of age. Killer Whales *Orcinus orca* may take the occasional bear but humans are their only real threat. Bears are hunted commercially and for subsistence, mostly for hides and meat. There is active management and conservation in several areas but the Polar Bear is included on the IUCN vulnerable list. Population estimates range from 10,000 to 40,000 individuals.

BIBLIOGRAPHY

Jefferson, T. A., Leatherwood, S. & Webber, M. A. 1993. *Marine Mammals of the World.* FAO Species Identification Guide, Rome.

Stirling, I. & Guravich, D. 1988. *Polar Bears.* University of Michigan Press, Ann Arbor.

Reference Section

Observing, Recording and Sampling

by

Michael Burchett

INTRODUCTION

The right choice of equipment to carry out field studies is important and basic marine sampling and measuring equipment need not be complicated or expensive. Much of it is readily available in one form or another or can easily be made with a little practical experience and ingenuity. Much of the basic equipment can also be used for many other areas of outdoor study. However, if large sums of money are initially being spent on equipment, then a rethink of priorities may be in order.
There are four categories of basic equipment for:

1. Observing and recording
2. Sampling and collecting
3. Storage and preserving
4. Laboratory or home

A list of suitable equipment is given on p.449. However, this list is by no means comprehensive. Many of the pieces of equipment can be used for a number of tasks. Beware of the tendency to take too much equipment on a field trip. A good 'rule of thumb' is to leave equipment behind if it is not used on two or three successive outings.

Observing and Recording Equipment

BINOCULARS

Perhaps the three most important items of equipment are good binoculars, a field note book and pencil and relevant field guides. Care, time and advice should be taken when choosing a pair of binoculars. Generally buy the best quality that can be afforded and if properly cared for, they will give many years of useful service. For most bird observations, a reasonably small and light pair is required. Many bird-watchers choose 8x or 10x magnification with an object lens diameter of 40 or 50mm. The 'field of view' (as seen through the binoculars) and the image 'brightness' (light transmission) is extremely important and will vary according to the optical parameters and quality of lenses. It is important to try out several pairs and seek advice from people that regularly use binoculars outdoors.

PHOTOGRAPHIC EQUIPMENT AND CAMCORDERS

Photography is one of the most useful methods of recording events which provides reasonably accurate results. There are many aspects to photography and the level to which the naturalist wishes to go is a personal choice. Generally the most difficult subjects to photograph are the very small ones and fast moving animals.

At the most basic level, there are many good quality, 35mm automatic cameras on the market. Many of these have a zoom facility and some of the more expensive ones are also weatherproof. Many of the new generation cameras have automatic focus, light metering and wind-on. Underwater cameras are also available to the scuba diver, but these tend to be expensive and many of their features are adjusted for underwater conditions. A useful feature on many new cameras is the ability to produce the date on the exposure as it is taken. This makes subsequent cataloguing of photographs easier, especially if it is cross-referenced with field notes. There are three types of films (black and white, colour slide and colour negative), and they come in a variety of sizes, speeds (ASA) and exposures. Colour slide film is useful for giving lectures and slide shows, and many photographic agencies will use slide film for reference libraries. Prints can be obtained from colour slides, but this can be expensive, especially if large size prints are required. Most people find a 35mm, 36 exposure, colour print film of 100 ASA adequate.

Great care should be taken with camera equipment if it is to be used outdoors or in a wet environment. Padded waterproof bags are essential to keep film and camera isolated from damp, salty air which has the ability to ruin cameras and electronic equipment. Films should be changed in a dry, sheltered area away from direct sunlight and dampness. Always check that a film is properly loaded and there is actually a film in the camera. Ensure that enough film is taken on a field trip. When spare film is not used, it should be stored in a cool, dry place such as a plastic container in the fridge. When prints or slides are developed, catalogue them fairly quickly, as a fading memory can make one landscape or seascape look much like another. If a field note book is available when a photo is taken, write down the relevant time, date and location of the shot. This can then be linked to the individual exposure date at a later stage.

Video camcorders have become more popular, cheaper and reliable over recent years. They can be a valuable aid for recording natural events (e.g. whale sightings and large movements of birds). A video recording will also give a true feeling or representation of the areas in which the naturalist is travelling or working. Familiarity and ease of use is the key to competent photography. It is too easy to concentrate on using the equipment correctly and not to pay attention to the natural event which is taking place.

NATURE LOGS AND RECORDING SHEETS

Field note books are required if notes and data are to be recorded. The size will depend upon the circumstances. Sometimes a clipboard, pencil and pre-printed record sheet may be required for a specific task (see appendix for proformas, p.450). A hard, clear film of plastic or plastic bag over the top of the clipboard is useful to exclude the elements. On occasions, writing inside the plastic bag may have to be resorted to. Smaller A5 size, or pocket note books can be cheaply purchased overwhich a stiff waterproof sleeve may be slipped. There are also several types of waterproof notebooks on the market which have thin plastic sheets instead of the normal paper. This enables pencil notes to be taken in the wet, or even underwater. The pencil work can be erased and the pad reused. These pads are expensive, but a backup is useful if weather conditions become too severe for normal writing methods. Any field notes should be written up clearly into a logbook as soon as possible so that they may be understood, while also providing a second copy. An HB pencil with an eraser at the other end is all that is required for taking notes and producing field sketches. Make sure a spare pencil is available and that there is a means of sharpening it.

Another useful recording item is a pair of 'tally counters'. These are mechanical thumb-operated pieces that count from 0 to 9,999. The most robust ones are made of metal. They enable quick counting by the operator while ongoing observation is taking place.

Collecting and Sampling Equipment

Much of the collecting and sampling equipment can be bought locally or made from readily available materials. Many of the items of equipment are self-explanatory and a brief description of their use is given in the table of equipment (p.449). The following descriptions cover some of the more unusual pieces of equipment needing further explanation.

UNDERWATER VIEWER (POOL GOGGLER)

This is a simple plastic device that enables a person to look at life in the shallow rock pools or clear water without surface disturbance. It consists of a piece of plastic drainage tube about 300mm in length, with a diameter of at least 100mm. Set into one end is a round piece of 3-6mm clear acrylic or polycarbonate plastic (not glass). This can be held and sealed in place with appropriate sealant (fig.333).

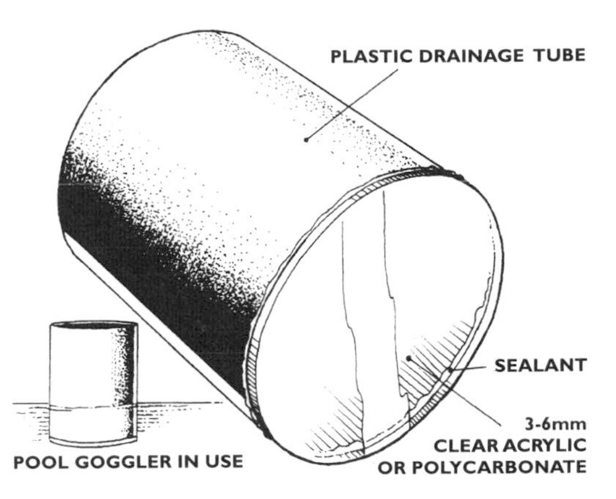

Figure 333. Pool goggler.

WEED DRAG

This item is a small type of grapnel hook. It is made from four pieces of mild steel, bent over at one end and joined to form a central shaft. The weed drag has a length of about 100mm. A strong nylon cord is attached to an 'eye ring' or more simply whipped around the central shaft. As its name implies, it is used for dragging up small samples of weed from shallow depths that cannot be reached by hand (fig.334).

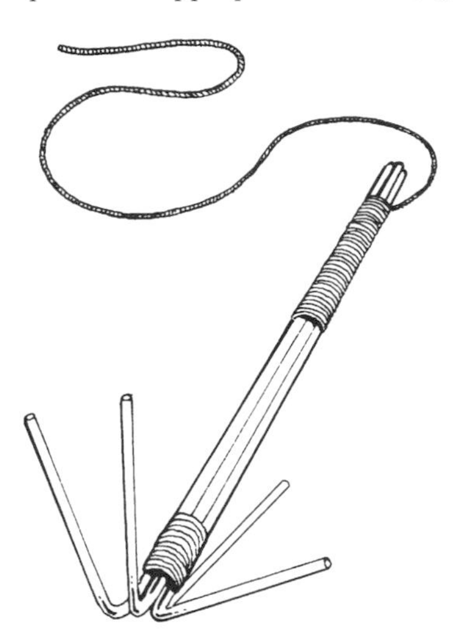

Figure 334. Weed drag.

VIEWING OR OBSERVATION CELL

This item of equipment can be dipped into the water and held to the light to view small organisms swimming inside. Scoop net contents can also be decanted into the cell. It is made from 3 pieces of clear plastic (acrylic or polycarbonate, 3mm in thickness) with the middle layer forming a U-shape. The viewing cell is 50-100mm square with an inside gap of about 6mm. The layers can be glued together or, alternatively, the middle layer can be substituted by a rubber gasket. The structure is then held together using two bulldog clips along the edge (fig.335).

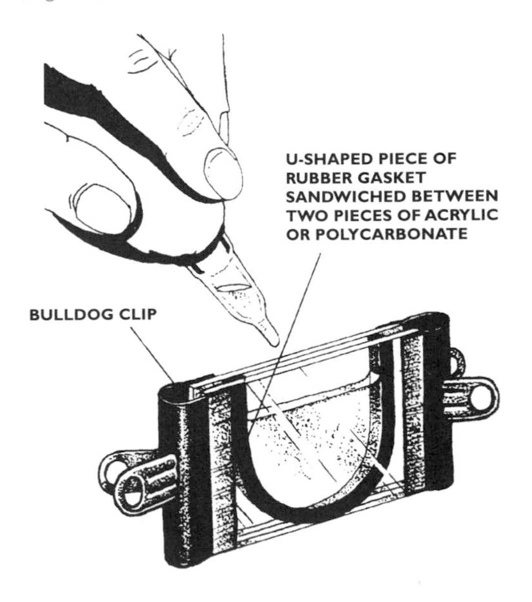

Figure 335. Viewing cell.

QUADRATS

A quadrat is a convenient apparatus for estimating the abundance of organisms within a small area. Quadrants come in two sizes with an area of 0.25 or 1.0 square metres. The numbers of individuals within the boundaries of the quadrat are counted and expressed as the number of individuals per square metre (n/m^2). If large numbers of animals are found, it is quicker to use a smaller quadrat, or divide up a larger quadrat, and multiply the final figure by the appropriate multiplication factor. Thus a 0.25 square metre quadrat needs to be multiplied by four to express the result as numbers of organisms per square metre. The easiest way to divide a large quadrat up is to use a cross-network of strings. If organisms can be removed from inside the quadrat as they are counted, this will prevent double counting. However, this is not always possible as many plants and animals are physically attached to the substrate.

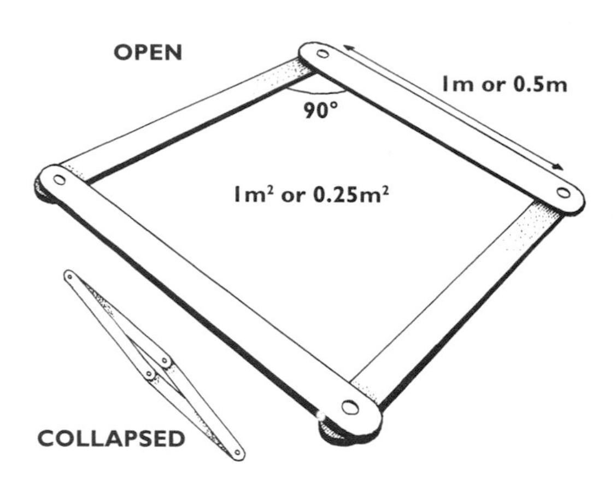

Figure 336. Quadrat.

Quadrats can be made from any thin lengths of material such as wood, plastic and alloy tubing. The equipment should form a square of four equal sides with a right angle at each corner. The quadrats are often made collapsible for ease of transportation and the corner joints are mobile to allow this (fig.336).

SAMPLING NETS

Towed Plankton Nets

For the most keen marine naturalists who wish to study the smaller plankton, some form of plankton net will be required. However, if a plankton net is purchased then appropriate microscopic equipment will also be required. Figure 337 shows the basic construction of a standard plankton sampling net. The mesh size of the net is critical as it determines the size of organisms caught. There a six standard size meshes available but the three most useful are as follows:

Mesh size	mpi	Aperture (mm)
Coarse	26	0.3240
Medium	49	0.0925
Fine	90	0.0630

A medium mesh size is probably the most useful for general boat use and should be used with a towing speed of less than three knots. Net cloth was once made from 'bolting silk' but modern plankton nets are mostly made of nylon, as this does not easily distort or rot. A plastic net bottle of about one litre, tied in at the far end, is generally adequate. Attachment of the plankton net to the leading metal hoop or ring may vary. Some nets may have a sleeved cuff for the hoop, while others may have a series of eyelets to lace the plankton net on.

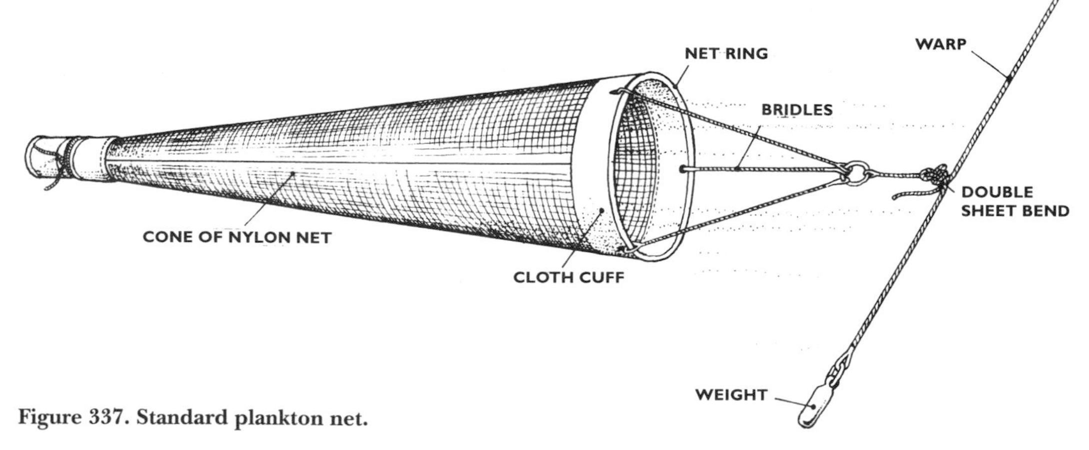

Figure 337. Standard plankton net.

Plankton nets can be set to be towed horizontally (fig.337) with enough warp paid out to ensure the net does not break the surface of the water. Alternatively it can be hauled up vertically from a predetermined depth using the boat winch or a hand winch. A lead weight of about 4kg is required to take the net down.

Hand nets

A small hand plankton net is a useful item of kit to dip into rock pools to capture the smaller organisms. It can easily be made from a bent hoop of metal, that is attached to a wooden dowel handle using whipping twine. The hand net should have a handle of about one metre in length (diameter approx. 25mm) and the plankton netting should be of fine mesh (90 mpi). If mesh cannot be obtained then a ladies stocking will do (fig.338).

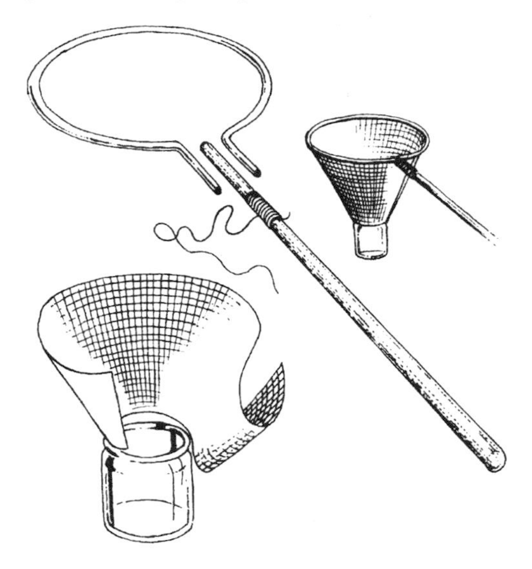

Figure 338. Hand plankton net.

Figure 339. D-shaped net.

D-shaped Collecting Net

This is one of the most useful nets for collecting larger rock pool animals such as shrimps, crabs and small fishes. It consists of a strong D-shaped metal hoop with a stout attachment for a wooden handle (see hand net). The netting is normally shrimp net which can be obtained from any good sea-angling shop (fig.339).

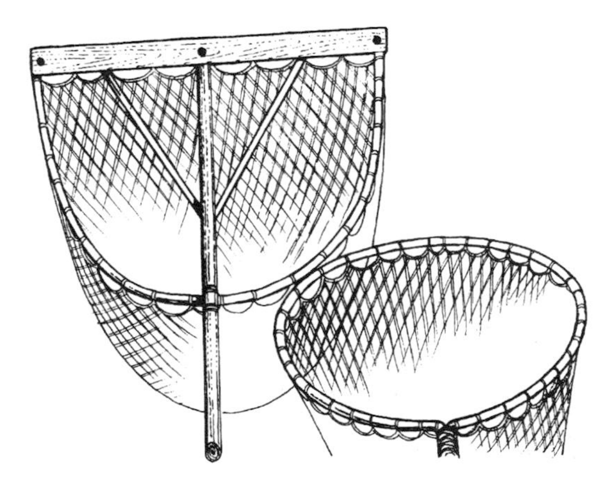

Figure 340. Shrimping nets.

Shrimping Nets

These come in various sizes and shapes, and most good sea-angling shops will stock a variety (fig.338). They are used by pushing the net along a sandy bottom just below the waterline. Generally it is more comfortable to push the net with the wind coming from behind the operator. The net should be lifted and the contents examined about every 30-50m. 'Shrimping' is probably the easiest way of catching shrimps, shore crabs and small fishes, especially during the warmer summer months when they move close to the shore.

COLLECTING AND SAMPLING TECHNIQUES

GENERAL SAFETY

On any excursion to a coastal area, shoreline, or by boat, safety must be a priority, especially where children are involved. Check weather forecasts, tide tables and prevailing weather conditions before setting out. Make sure adequate and appropriate clothing is provided to cope with the worst conditions that are likely to be encountered for that time of year. Wind chill can make any trip uncomfortable and in cold conditions it can be potentially dangerous, so take waterproof clothing and stout footwear when the conditions dictate.

Along sea cliff areas, where large tidal ranges may occur, tide tables should be checked carefully. If headlands or promontories have to be passed on foot, make sure the flooding tide does not cut off the return route. If this does happen try to reach a safe area above the high tide mark or strandline, but avoid climbing cliffs. It is always better to carry out shore explorations along cliff areas on an ebbing tide rather than the flooding tide. It may be a good idea to leave a message with someone who knows where you are going and at what time you expect to be back. If people are long overdue then the necessary alarm can be raised through the coastguard services.

Deep water is always one of the biggest dangers. On shores with large rock pools near the low tide region, the depth may be greater than the height of the person. On many headlands and cliff areas, rocks may enter straight into deep water. Slipping in can be quite easy, but getting out can become difficult on steep, weed-covered surfaces with waves surging back and forth. Young children who cannot swim should not be taken to potentially dangerous sites without adequate supervision, clothing or even a life jacket.

Sampling from boats can be even more hazardous and many fishermen will know of someone who has drowned or been carried overboard by equipment. At sea, the normal safety rules of boating and seamanship should be applied, especially if the craft is a small one which is more likely to be affected by adverse conditions. If any sampling is to be done over the side of a boat, adequate precautions should be taken to ensure a person's safety. The sampler should be wearing a life jacket and be securely attached to the boat by an appropriate means (e.g. safety harness). Beware of being caught in, or dragged over the side by heavy fishing or sampling equipment. Let winches and pulleys take the strain and not your body. A sharp serrated knife or hand axe should be available to cut free any entangled gear. Great care should be exercised when taking 'grab' samples from a moving boat. A full bucket of water exerts a lot of pull when the boat is underway. It is also highly dangerous to take a loop or 'turn' of rope around the wrist for any purpose whatsoever and this practice should be avoided at all costs. It is much easier and safer to have the other end of a rope attached to the boat.

Some form of medical kit should accompany a group of people on a coastal excursion or boat trip. Apart from the normal cuts, bruises and sunburn, the most likely accident is a bad fall, resulting in a damaged or broken limb. Proper first aid should be administered by a trained first-aider and help sought. People with leg injuries who cannot walk should not be carried any distance. It is better to seek medical help and obtain a stretcher.

Cliff areas must be approached with caution. They are ideal places for bird-watching and make excellent vantage points. Headlands will also allow a view along the coastline on either side. However, cliff tops are often windy, cold places, so adequate clothing and footwear should be worn, especially if cliff paths are narrow and precarious. Beware of venturing too close to cliff edges, as they are sometimes undercut and may easily give way. If cliffs are approached from the shore do not walk or sit directly underneath them as falling rock is a frequent hazard.

SAMPLING ETIQUETTE

It is essential to follow good codes of biological practice to safeguard the plants, animals and their environment. The following are some helpful guidelines:

1. If possible, examine all larger specimens where they are found and return them as soon as possible to the same place. Make field notes at the time when observations are fresh in the memory.

2. If some specimens are essential for further study at home, take no more than is necessary. While on the shoreline or out in a boat, temporary storage facilities will be needed for the specimens. Containers should be of adequate size for the specimens and they should be kept in cool, shady conditions with regular changes of water. Remember, temperature and salinity changes will 'stress' an organism quickly and if they do show signs of stress then return them immediately.

3. Handle marine animals with care and as little as possible with bare hands, as this can cause stress. In addition, many organisms bite or sting, especially tropical ones.

4. Approach habitats or areas of interest with a little stealth and caution. Patience and a slow approach will often reveal the more timid inhabitants. Avoid standing in rock pools or on their inhabitants.

5. There are now many coastal sites which have been designated conservation areas as part of coastal nature reserves; therefore refrain from removing organisms or damaging habitats in these areas.
6. Many marine animals are not protected by law. However, when birds or marine mammals do enter the territorial waters of a country they automatically come under its conservation laws. The taking of many of these animals for scientific purposes often requires permission and a licence. Bringing dead or live specimens into a country also requires permission from the appropriate authorities. So wherever possible, study and conserve living specimens.

SAMPLING METHODS

Rocky shores are a good place to start investigating marine life. Plants and animals are more apparent and easier to sample in these areas compared to sandy or muddy shores. Studying the zonation patterns with the help of a guide book is often the first step. Examining the contents of a rock pool can also be a rewarding experience if a net and 'pool viewer' are available. Shrimping nets pushed along the sandy shallows will also catch some of the shallow burrowing animals such as crabs, shrimps and fish. Animals that burrow deeper into the soft sediments need to be raked out, dug up and possibly sifted out. Near-surface animals such as the Common Cockle *Cerastoderma edule* can be removed with the 'cockle rake' (fig.341). Deep-burrowing animals such as the Lugworm *Arenicola marina*, ragworms *Nereis* and razor shells *Ensis* need to be dug up with a narrow fork or 'lug spade'. Sifting the excavated material through a 1mm mesh sieve will reveal many of the smaller animals.

The water's edge and strandline are other areas worthy of beach investigation. Many interesting artefacts from land and sea can be found here including dead animals, parts of animals, various rotting vegetable matter, shells, bones, egg cases and flotsam and jetsam. Care should be taken examining the contents of the strandline as sharp objects and pollutants are also common. Strandlines are well worth a visit after storms and gales. Large amounts of interesting debris may be washed up and the sand stripped away to reveal new items.

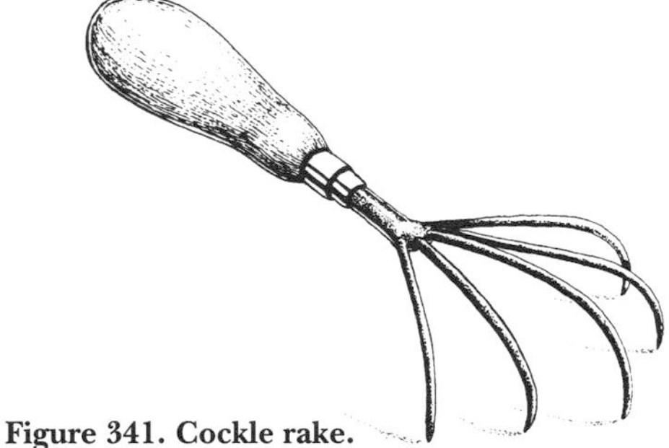

Figure 341. Cockle rake.

PRESERVING SPECIMENS

Any living tissue from marine organisms will quickly decay and smell once dead. If specimens are to be kept, they must be preserved by an appropriate method. Some of the smaller animals such as starfish, seahorses and brittle-stars can be air dried after they have been killed in 70% ethyl alcohol. Snails and bivalves can be killed in boiling water, then immersed into cold water so that the soft internal body of the animal can be easily removed from the shell. The empty shells can be gently wiped to remove algae, being careful not to damage the delicate outer surface. The two parts of the bivalve shell can be held together with an elastic band.

Soft-bodied animals such as jellyfish, sea anemones, polychaete worms, octopuses, squid, sea cucumbers, sea squirts and fish cannot be easily air dried and therefore require preserving in a liquid. Most soft-bodied animals need to be killed in 70% ethyl alcohol and then relaxed in a 7% solution of magnesium chloride and seawater for best results. After this process they can then be preserved. Crustaceans can be preserved in 70% ethyl alcohol. Soft-bodied animals will require initial preservation in a 5-10% solution of seawater formalin and then stored permanently in 70% ethyl alchohol after a few days.

Seaweeds need to be carefully handled as they readily fall apart. To preserve algae, first rinse them in sea water (not freshwater) to clean off any mud and sand. Cut off the parts to be preserved. Line a flat tray or dish with firm paper and use a small stone at each corner to hold the paper down. Fill the tray carefully to a depth of about one centimetre with clean sea water. Arrange the pieces of seaweed in the tray as desired, and lift the paper out slowly, allowing the surplus water to drain away. Place a new sheet of absorbent mounting paper on top of the seaweed and use a 'plant press' to store and dry them in the normal way. Do not over press. A plant press can easily be made from two pieces of galvanised weld-mesh tied together. A thin wooden board placed on top, with a weight (a brick will do) is all that is required. Preservatives such as ethyl alcohol (ethanol) and formalin can generally be purchased at chemists. Formalin is normally supplied as a 40% solution and it needs to be diluted to about 10% by adding nine parts water or sea water to one part of formalin. Formalin should be handled with care as it is a skin, eye and throat irritant. Overexposure can lead to sensitisation (allergy) to the chemical. Wash off any splashes from skin and storage jars, and wear disposable gloves when handing specimens that have been in the preservative. If possible, handle specimens with instruments, especially smaller animals or parts of animals. Carry out work in well ventilated areas and wash the specimens in sea water before examination as this will reduce the formalin fumes from the specimen.

SPECIMEN CONTAINERS AND LABELS

All specimens that are to be kept for long periods should be correctly stored and labelled as soon as possible. The container should be large enough to take the specimen without forcing it in. All containers must have sealable tops which are air tight and leak proof. Small 'vials' often have snap tops, while larger containers tend to have screw lids, or clamp lids with rubber or plastic gaskets. Glass containers are easy to see through but do break. To avoid this they may be rolled in paper or 'bubble-wrap' and placed in a larger container such as a milk crate. Plastic jars and containers for large specimens are probably better suited for boat use.

All specimens should be correctly labelled. Waterproof paper that will not disintegrate is required if the label is likely to become damp. Ready printed waterproof labels can be purchased, but they are sometimes difficult to obtain. All writing should be done in waterproof black, Indian ink or pencil. A basic label should have the following information on it: ▼

No.
Date:
Time:
Place:
Specimen:

Any additional information can be kept in a specimen log book, coded with the correct number corresponding to that on the specimen label. Sometimes a coded number on the label is all that is required as long as it can be cross-referenced with the log book information.

MEASURING AND TAGGING

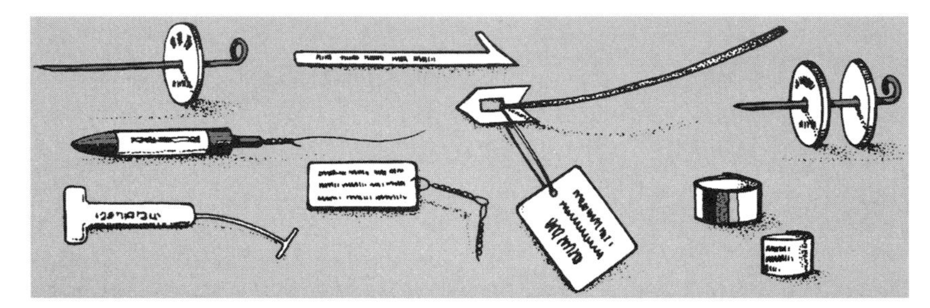

Figure 342. Tags for fish and birds.

Many plants and animals can be weighed and measured alive or dead. Many birds, seals and fish are routinely measured, tagged and released back into the environment for subsequent recapture. Using tag and recapture methods, much can be learnt about the growth and movement of animals. Measuring and tagging live animals is a skilled practice and should not be undertaken without training and permission. Tags have to be properly coded and authorised and the tagging programmes carried out by qualified biologists. Any tags or tagged birds that are found hold valuable information and should be returned to the appropriate authority. There is normally a small reward for each tag that is returned. If a tag is found, return it with as much information as possible. Many tags will not have an address so they should be returned to one of the major conservation societies. Anyone who is interested in helping with bird-tagging programmes and wishes to learn more should contact one of the various bird preservation societies.

There is no reason why a naturalist should not try taking some weights and measurements of animals to discover some of the basic growth relationships (e.g. length to weight relationship). However, the correct method for collecting raw data should be used and a relevant biological statistical book will be required to select the appropriate method of data analysis.

Bird-watching a popular pastime in many countries. In high latitudes, ideal times to observe seabirds and shorebirds are during the migration periods of spring and late summer. The higher latitude birds leave early and fly to the lower latitudes for the winter. Tidelines, estuary mudflats, saltmarshes, sand dunes and sea cliffs are all good locations for bird-watching. However, during the breeding seasons birds should be disturbed as little as possible. Walking through a nesting colony may cause parents to fly off, leaving the nest and eggs unguarded and exposed to predators. Late afternoon is often a good time to observe birds returning to the nesting areas for the night. Cliff tops and headlands are ideal vantage points to observe birds returning from sea. To learn more about birds there are several local clubs and societies willing to help new members. At sea the seafarer will have little choice but to use the appropriate guides for information and identification.

Observations and Record Keeping

Records of observations can be kept during sea passages or on a visit to a local shore. Keeping a record, or diary of observations, consolidates knowledge and increases a person's awareness and understanding. Record keeping should not be a tedious task but should be carried out quickly and simply. Appropriate and correct record sheets and methods should be used and only relevant information collected. Developing a keen sense of observation is one of the most important skills that a naturalist can learn. Often a seafarer may make some salient notes about wildlife in the ship's daily log. All good seafarers will keep a ship's log and on larger vessels it may be mandatory. However, it may be more appropriate to keep a separate 'nature log' to record any events or even the 'non-events'. Both environmental and biological data can be entered in the nature log. It should be remembered that what is not seen may be as important as what is seen, and therefore entering in the log book 'nothing seen' is a useful biological record.

ENVIRONMENTAL DATA

The following environmental data is worthwhile recording on a daily basis during a sea passage. It is often useful to take the readings at the same point of time each day. For example, a log taken on a daily basis at 1200hrs GMT (or close to it) may include the following information:

Yacht name: **Log. ref.:** **Date:**

Observer:

Local time:

GMT time:

Boat position:

Sea state:

Wind direction:

Pressure (mb): cloud cover: visibility: precipitation:

Air temperature (°C):

Surface sea temp. (°C):

Sea temp. at 10m (if possible):

Water colour:

Any biological notes (at specific time or over past 24 hours):

If a major biological event occurs, then it may be appropriate to record the above environmental data again and at the site of occurrence if this is possible. The extra biological data may include the following:

Yacht name: **Log. ref.**: **Date**:

Observer:

Local time:

GMT time:

Boat position:

Species name: 'common': **scientific**:

Animal numbers:	one: ☐	few (less 10): ☐	some (10-30): ☐	many (30+): ☐

Direction of animal travel:

Biological notes:

Time and location should be repeated in case there was no time to collect any relevant environmental data. For a typical record sheet see Recording Sheet A, p.450.

Temperature

This should be taken as degrees centigrade (°C) wherever possible. Many ocean yachts are now fitted with 'through the hull' temperature gauges which are connected to digital display units. This often gives a better reading as it is taken slightly below the surface. If temperatures are taken from a bucket of water the thermometer bulb should be held under water while the reading is taken.

Water Clarity

Water clarity can normally be estimated with a little experience. The following water classification may be useful:

VC	Very clear	e.g. 30m+ as in dark blue open ocean waters.
C	Clear	e.g. 10-30m as in coral reef areas.
FC	Fairly clear	e.g. 5-10m as in many coastal waters.
T	Turbid	e.g. 1-5m.
NV	Nil visibility	e.g. one metre or less (only an arm's length).

Sea State

This should be given in accordance with the internationally recognised Beaufort Scale (see Appendix III).

Cloud Cover

This should be given in the correct way using the standardised divisions of eigths (refer to a nautical almanac).

Air Visibility

The following meanings do not relate to the more complex or official versions but have been simplified for the purpose of biological observations. It should be noted that if a definite horizon can be seen, the visibility must be excellent.

Bad	**(B)**	Less than 0.5km or about 400yds.
Poor	**(P)**	0.5-1.5km or about 400yds to one mile.
Moderate	**(M)**	1.5-10km or about one to six miles.
Good	**(G)**	10-20km or about six to twelve miles.
Excellent	**(E)**	20km + or 12 miles +.

Directions

Directions can be given as a 'true' bearing in degrees for accuracy, or estimated using cardinal points. For most biological purposes an estimate of direction is good enough. It should be noted that wind direction is always given as 'coming from' while water currents are given as 'going to', thus, 'winds blow from while currents flow to'. Any animal movements should be given as the direction in which they are travelling.

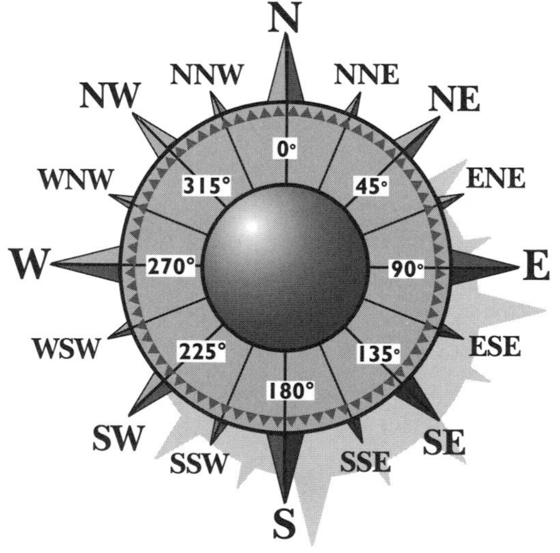

Figure 343. Compass rose.

BIOLUMINESCENCE

If bioluminescence is to be recorded then the following symbols may be inserted into the biological notes of a log:

Sparkles	**Sp.**	**Luminescent bands**	**Ba.**
Pulses and/or Blobs	**P/B.**	**Luminescent wheels**	**Wh.**
Erupting luminescence	**Er.**	**Milky seas**	**Mi.**

ANIMAL ABUNDANCE

Sometimes biological events need to have recording symbols instead of actual numbers, in situations where accurate counting is impossible or not necessary. The following may be useful to note in the log margins:

0	**None seen**	++	**Occasionally seen in large numbers**
+/-	**A few occasionally seen**	+++	**Regularly seen in large numbers**
+	**Regularly seen in small numbers**		

OTHER OBSERVATIONAL METHODS

SCUBA DIVING

Snorkelling (skin-diving) or scuba diving both require proper training and should only be undertaken after an appropriate training course has been completed. Snorkelling is often learnt first as it requires less training and less equipment (mask, fins, snorkel). When diving never go alone and always have adequate backup.

Scuba diving and snorkelling are not to everyone's liking or sense of adventure. However a great deal of fun can be had peering into the depths with a larger version of the 'pool goggler' (fig.333). In clear water regions it is also a useful way of checking a potential scuba diving site.

SKETCHING AND ILLUSTRATION

Anyone can learn to sketch to an understandable level and all it takes is a little care and practice. Before the invention of the camera, all naval midshipmen were taught to sketch and record views and profiles of coasts from the sea. This information helped assemble records for the formation of early charts, pilot books and anchorage information.

Sketching only requires an HB pencil, eraser and pencil sharpener. Drawings of specimens, parts of animals, plants and animals in their environment and coastal scenes can bring a dull log book to life. Two dimensional (flat) sketches are easier to learn before tackling three dimensional drawing techniques. Colour pencils and black Indian ink fineline pens can greatly enhance the pencil drawings. Drawings will allow details to be shown that cannot be easily achieved by photographic means. On long sea passages, sketching collected specimens can pass away the time. It also extends a person's familiarity with organisms and their various parts.

MICROSCOPY

Hand lenses are essential for the marine naturalist who is interested in plankton identification. The ideal hand lens is a dual one, with a 10x and 20x magnification. The 10x magnification is used for an overall scan of the plankton catch and the 20x for accurate identification.

The best instrument for viewing plankton, should expense and room allow, is a binocular microscope. These instruments are susceptible to damage and damp, and should be protected from both. Some smaller binocular microscopes can be bought with a watertight robust case, often attached to the microscope stand.

While viewing through a hand lens, or the binocular microscope, both eyes should be kept open. This may be difficult at first, but with practice and concentration the other eye's view will not interfere, and it is immeasurably more relaxing than squinting with the other eye, especially over long periods.

Sizing of the specimen under the binocular microscope can be facilitated by a micrometer-ruled eyepiece. This normally has to be purchased as an extra to the microscope, although some instruments may include this invaluable tool. You can, however, have a fine graded graph paper over which the slide or petri dish is placed. This is not as accurate as the first method, but it is the easiest method for the hand-held lens.

Small Petri dishes or viewing cells are useful for containing the specimens during viewing. Normally macroplanktonic animals such as small crustaceans and cnidarians are removed to a suitable container, and, after examination, placed back into the water. However, you may wish to prepare a temporary mounted slide. The easiest method is to use cavity or ring slides. The specimen sample of water is dropped into the 'bowl' via a teat pipette. Should you not possess these slides you can use a plain slide and a cover slip as shown in figure 344.

For permanent mounting, preserving microscopic specimens and further microscope techniques a laboratory guide should be consulted.

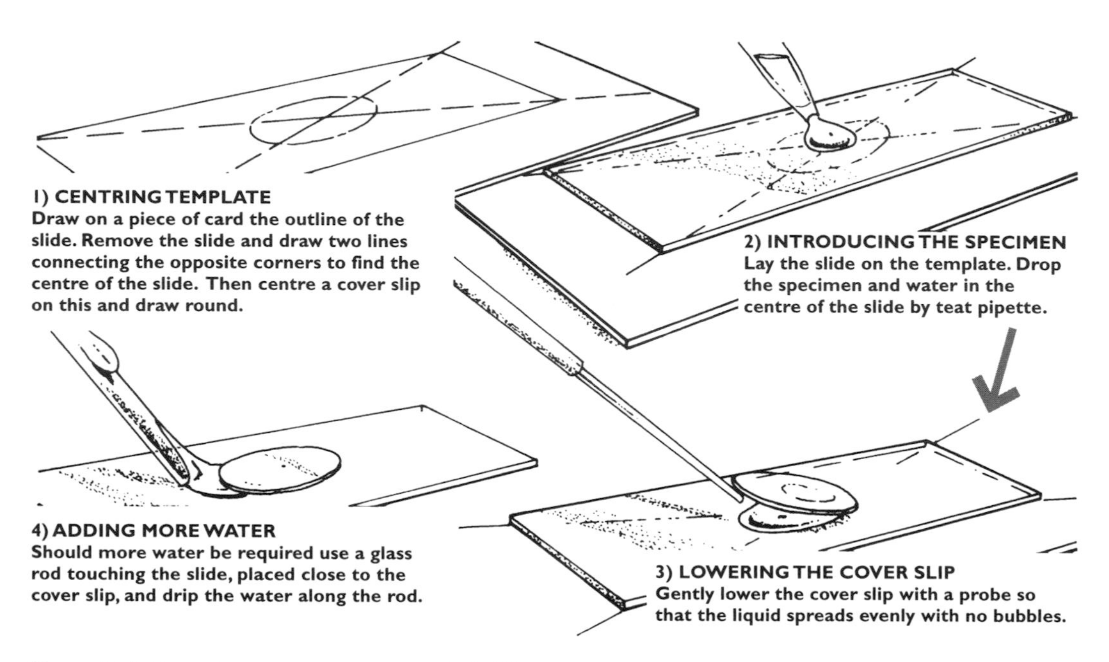

Figure 344. Preparing a temporary mounted slide.

EQUIPMENT LIST

General equipment to include:

Fisherman's scales (spring balance type).
Wrist watch and stop watch.
Tally counters from 0 to 9,999 (x2).
Maximum/minimum thermometer.
Digital electronic thermometer with long lead probe.
Vernier callipers (adjustable callipers for measuring parts of animals).
Rulers: plastic (150mm and 300mm).
Ruler: wooden type (one metre).
Sounding line (30m with 100mm markers for taking depth readings).
Tape measure (0-305cm with metric graduations in millimetres).
Measuring tape on wind-up reel (50m with metric graduations).
pH paper to measure pH and salinity.
Beakers: plastic (various sizes from 100-1,000ml).
Trays: plastic, photographic type (various sizes).
Pipettes (2-3ml capacity).
Buckets: plastic (10-15 litres).
Aquarium: clear plastic (320 x 300 x 230mm) with lid.
Aerator: 12v battery operated with one metre length of clear pvc hosing and diffuser block (air stone).
Alternative mains supply.
Various storage containers with sealable lids.
Petri dishes: round, clear plastic (90mm diameter) with lids.
Labels: waterproof paper and self-adhesive.
Trowel, narrow fork, cockle rake and lug-digging spade.
Knife: stainless steel with locking handle.
Pool goggler (useful for observing life in rock pools).
Specimen containers (various sizes from 25-100ml) with tops.
Weed drag (for dragging up weed samples from below the water).
Quadrat: folding type (0.25 and 1.0 square metres).
Sieve: round, galvanised metal type (for sieving animals from soft sediments).
Observation cell (to view small organisms when held up to the light).
Sampling nets: shell scoop; towed plankton net; hand plankton net; D-shaped hand net; shrimping net.
Hammer: geologists type (for collecting rock samples and fossils).
Bags: clear polythene (various sizes) with bag ties and labels.
Microscope and sundry microscope equipment.
Writing instruments including indian ink pen, HB pencils, waterproof felt marker pen, colour pencils and pencil sharpener.
Field note book, home note book, clipboard and 'nature log'.
Disposable gloves, rubber boots and adequate field clothing.
Compass: hillwalking type (useful for taking bearings).
Maps and charts (e.g. OS 1:50,000 Land Ranger; 1:25,000 for shore work; 'small boat charts' for coastal boat work).
Binoculars: compact, lightweight of about 8 x 40 magnification.
Photographic equipment (to suit the seafarer's interests).
Fishing equipment (to suit the seafarer's interests).
Preservation equipment: formalin, ethyl alcohol etc.

Dissection kit to include:

Brass awls (x10)
Wooden-handled seekers (angled, blunt, x2; straight, sharp x2)
Forceps (large and small)
Pin vice (up to 1mm capacity)
Tweezers (large and small); Reverse action tweezers (x1 straight, x1 angled)
Scissors (stainless surgical, open shank, one large, one small)
Scalpel handles (no.3 and no.4 with blades to fit)
Small, artist type paint brush
Hand lens loupes (triple element with 3x, 4x, 5x magnifications, glass lens types in fold-away metal case)
Alternatively, fixed hand lenses of 5x, 10x and 20x magnifications
Plastic nylon, disposable tweezers; dissection kit roll-bag

Form A	**Environmental Information**

Sheet No. [ ]

Ref. No. [ ]

REFERENCE DATA

Recorder's name: _ **Date:** _ _ _ _ _ _ _ _ _ _ _

Contact adddress: _

_ _

Boat name: _ _ _ _ _ _ _ _ _ _ _ _ _ _ **Boat position** (lat. & long.): _ _ _ _ _ _ _ _ _ _ _ _

or location: _

Observation time (local): _ _ _ _ _ _ _ _ _ _ **hrs** **(GMT):** _ _ _ _ _ _ _ _ _ _ **hrs**

ENVIRONMENTAL DATA

Sea conditions (Beaufort): _ _ _ _ _ _ **Wind direction:** _

Air temperature (°C): _ _ _ _ _ _ _ _ **Sea temperature** (°C) (if posssible): _ _ _ _ _ _ _ _ _

Air pressure (mb): _ _ _ _ _ _ **Cloud cover** (8ths): _ _ _ _ _ **Precipitation:** **Yes** [ ] **No** [ ]

Air visibility: Bad [ ] **Poor** [ ] **Moderate** [ ] **Good** [ ] **Excellent** [ ]

Water clarity: Nil visibility [ ] **Turbid** [ ] **Fairly clear** [ ] **Very clear** [ ]

Any other environmental remarks: _

_ _

_ _

_ _

_ _

_ _

_ _

_ _

Reference data and environmental data (this form) should accompany biological data (form B) if at all possible.

Form B — Biological Information

Sheet No. [ ]

Ref. No. [ ]

REFERENCE DATA

Recorder's name: _ **Date:** _ _ _ _ _ _ _ _ _ _

Contact adddress: _

_ _

Boat name: _ _ _ _ _ _ _ _ _ _ _ _ _ _ **Boat position** (lat. & long.): _ _ _ _ _ _ _ _ _ _ _

or location: _

Observation time (local): _ _ _ _ _ _ _ _ _ _ **hrs** **(GMT):** _ _ _ _ _ _ _ _ _ _ **hrs**

Do not fill in the above data if this form accompanies a form A with the same relevant information.

BIOLOGICAL DATA

Species name (common): _

(scientific if possible): _

Identification accuracy: Definite [ ] **Probable** [ ] **Possible** [ ] **Unknown** [ ]

Number of adults: _ _ _ _ _ _ _ _ _ _ _ _ _ _ **Estimated size** _ _ _ _ _ _ _ _ _ _ **cm/metres**

Number of juveniles: _ _ _ _ _ _ _ _ _ _ _ _ _ **Estimated size** _ _ _ _ _ _ _ _ _ _ **cm/metres**

OR

Abundance of animals: 0 [ ] **+/-** [ ] **+** [ ] **++** [ ] **+++** [ ]

Number of animals: 1 [ ] **2-10** [ ] **11-30** [ ] **31+** [ ]

Direction of travel of animal(s): _ _ _ **Nos. of species seen together:** _ _ _

Tags or rings observed: [ ] Yes [ ] No **if yes how many:** _ _ _ **any tag ref. nos.:** _ _ _ _ _

_ _ _ _ _ _ _ _ _ _ _ _ _ _ _ _ _ _ _ _

Photographs/video taken: [ ] [ ] **if yes ref no.:** _

Any other biological remarks: _

_ _

_ _

_ _

Biological data (this form) should accompany environmental data (form A) if at all possible.

Form C — Animal Measurement Information

Sheet No. [ ]

Ref. No. [ ]

REFERENCE DATA

Recorder's name: _ **Date:** _ _ _ _ _ _ _ _ _ _

Contact adddress: _

_ _

Boat name: _ _ _ _ _ _ _ _ _ _ _ _ _ **Boat position** (lat. & long.): _ _ _ _ _ _ _ _ _ _ _

or location: _

Observation time (local): _ _ _ _ _ _ _ _ _ **hrs** **(GMT):** _ _ _ _ _ _ _ _ _ **hrs**

Do not fill in the above data if this form accompanies a form A with the same relevant information.

SPECIES DATA

Species name (common): _ **Sex:** _ _ _ _ _ _ _

(scientific if possible): _

Animal condition: Alive [ ] **Freshly dead** [ ] **Dead** [ ]

Photographs/video taken: Yes [ ] No [ ] **if yes ref no.:** _

Animal type	Adult- A Juvenile-J	Measurements * (cms)			Fresh weight (whole) (kgs)
BONY FISH		TL	FL		
SHARK		TL	SL		
RAY		TL	DW		
TURTLE		L	Girth		
SEA SNAKE		TL			
CETACEAN		1 2 3 4	5 6 7 8	9 10 11 12	

Any other biological remarks _

_ _

_ _

Measurement data (this form) should accompany environmental data (form A) if at all possible.
* Refer to Factsheets for details of measurements.

GLOSSARY

Abductor. A muscle that draws a limb or a shell of an animal away from the centre of the body.

Abiotic. Absence of life; non-biological.

Aboral. Away from or opposite the mouth.

Abundance. The number of organisms present.

Abyssal zone. The deepest part of the sea floor below about 2000m.

Acoelomate. Lacking a fluid-filled body cavity (coelom).

Acron. Anterior nonsegmental part of the body of a segmented (metameric) animal.

Adaptive radiation. The varieties of adaptations that plants and animals have evolved to survive in different environments and habitats.

Adductor. A muscle that draws a limb or shell of an animal towards the centre of the body, e.g. the adductor muscles that bring the two halves of a bivalve shell together.

Aerobic. Living in oxygenated conditions or requiring oxygen.

Aerodynamic. The study of the interaction between air and solid bodies moving through it.

Aerofoil. A structure with a cambered surface that is designed to provide lift during flight.

Aftershaft. Small tuft of down that grows out from the base of some contour feathers.

Agnatha. Fish-like vertebrates that lack jaws (living lampreys and hagfishes, and other extinct groups).

Ahermatypic corals. Non-reef-building corals that lack zooxanthellae.

Air sac. Non-respiratory air bags that are connected to the lungs of a bird.

Alar bar. Contrasting mark on 'shoulders' of some cormorants formed by pale tips to scapulars.

Algae. Marine and aquatic plants that lack proper roots, stems and leaves. They are not vascular and can carry out photosynthesis.

Allopatric. Species or subspecies not occurring together, having discrete areas of distribution.

Alula. (Bastard wing). A group of feathers that provide a small aerofoil surface at the front edge of a birds wrist. The alula smooths air flow and prevents wing stalling at increased angles of attack.

Alveolus. A sac making the internal termination of a glandular duct (plural: alveoli).

Amino acid. Organic compound containing both basic amino (NH_2) and acid carboxyl (COOH) groups. The fundamental constituents of protein molecules.

Amino sugar. A monosaccharide which has at least one hydroxyl group (-OH) substituted with an amino group (commonly $-NH_2$).

Ammocoete. Larva of a lamprey, a jawless fish.

Amphioxus. Common name for small fish-like cephalochordates of the genera *Branchiostoma* and *Epigonichthyes*.

Ampullae of Lorenzini. Pores which form an electrical receptor system that is present in many cartilaginous fishes.

Anadromous. Refers to a fish that breeds in freshwater but spends most of its adult life in the marine environment.

Anatomy. The study of the structure of plants and animals.

Angiosperm. A plant that produces seeds in an ovary.

Angle of attack. The angle that an aerofoil wing meets the airflow.

Annulus. A growth ring, e.g. in the scales of fishes.

Antarctic circle. The latitude of 66.5° south.

Anticyclone. Pressure systems moving in a clockwise direction in the northern hemisphere or anticlockwise in the southern hemisphere.

Antifreeze. A substance that will lower the freezing point of a liquid.

Aphotic zone. The depth of oceans beyond which no light penetrates (1000m+).

Archaean eon. 4000 to 2000mya, during which life on this planet first began (formerly the second period of the Precambrian).

Arctic circle. The latitude of 66.5° north.

Artery. A vessel conveying oxygenated blood from the heart to the body.

Asexual reproduction. Reproduction without sex cell (gamete) formation, occurring by fission, budding or, in some invertebrates and vertebrates, parthogenesis (development of an egg without fertilisation).

Aspect ratio. A mathematical figure used in ornithology and obtained by dividing wing length by its breadth (span/chord).

Asthenosphere. The region of the Earth's upper mantle.

Atoll. A coral reef island that has grown around a subsided island (normally volcanic) and encloses a shallow lagoon.

Atrium. 1) the normally central chamber into which water is drawn in before being filtered; 2) the chamber in urochordates and cephalochordates which surrounds the pharyngeal region and opens to the exterior through the atrial opening.

Attenuation. A decrease in light energy through absorption and scattering in the water column.

Autotroph(ic). Refers to organisms that make their own organic materials from inorganic compounds (e.g. through photosynthesis).

Axilla. Armpit or area corresponding to the armpit. In elasmobranchs, the innermost corner of the pectoral fin.

Bacterioplankton. Planktonic bacteria.

Baleen. In Mysticeti, horny slats that hang transversely from the roof of the mouth with fringed inner edges used in filter feeding.

Bar. A linear ridge of sand in the nearshore zone that often runs parallel with the shore.

Barbel. A tactile (touch) process present on the head of fishes.

Bare parts. General term for non-feathered parts of a bird: bill, legs, facial skin etc.

Barrier island. A long narrow island of sand, built by wave action and separated from the mainland by a lagoon.

Barrier reef. A type of coral reef separated from the mainland by water.

Basal body. An organelle associated with the base of a cilium or flagellum, composed of microtubules, that is similar to a centriole.

Basalt. Dark volcanic rock found in the ocean crust with a high content of iron and magnesium.

Bathyal. The region of the sea floor from the shelf edge (200m) to the start of the abyssal zone (2000m).

Bathymetry. The study of the ocean depths.

Bathypelagic. Occurring in the water column below 1000m to the sea floor.

Bay. A semi-enclosed area of water open on one side to the open water.

Beach. A shoreline consisting mostly of loose sediment such as shingle, sand or mud that may be transported by wave action.

Beaufort scale. An internationally recognised scale of sea conditions related to the wind strength.

Berm. A flat area of sand that has accumulated above the high tide mark.

Benthos. Organisms living on the sea floor.

Bilateral symmetry (Bilateralism). Arrangement of body parts of an animal such that if it is divided along a mid-line the two halves are approximate mirror images of each other.

Bill. The beak of a bird, or jaws of a fish.

Billfishes. A group of fishes which includes sailfishes, spearfishes and marlins.

Bilobed. Consisting of two lobes.

Biological stress. The biological and environmental factors that adversely affect organisms.

Biomass. The total weight of a number of organisms or population.

Biota. The plants and animals of a region.

Biotic. Factors pertaining to life.

Biogeography. The study of global organism distribution.

Bioluminescence. Light produced by biochemical means in some organisms. The light produced is 'cold light'.

Biosphere. The envelope of the Earth that contains all living and nonliving things.

Bipolar. 1) occurring in both polar regions; 2) having two extremities (e.g. bipolar neurons with two long processes from the cell body).

Blastocyst. Mammalian embryo at the stage at which implantation in the wall of the uterus occurs. It consists of a thin-walled, hollow sphere of cells.

Blastopore. Transitory opening on the surface of an embryo in gastrula stage.

Blastozooid. Member of a colony of animals that is produced by asexual budding.

Blastula. Early embryonic stage, usually consisting of a hollow sphere of cells, prior to cleavage to form a gastrula (gastrulation).

Blow. Cloud of moisture-laden air exhaled by cetaceans at the surface which produces a characteristic sound as it is released.

Blowhole. Opening of nasal passage on top of the head in cetaceans; single in toothed whales, double in baleen whales.

Blubber. Layer of fatty tissue lying below epidermis (outer layer of skin) in marine mammals.

Bonnet. Largest callosity at the rostrum tip of Black Right Whale Balaena glacialis.

Braces. Contrasting pale stripes on the scapulars of a bird.

Bracket mark. Crescent-shaped marking behind eye in small sperm whales (Kogia).

Brackish water. Water that is a mixture of sea water and fresh water in differing amounts.

Branchial. Pertaining to the gills.

Branchial arch. A bony or cartilaginous arch on the side of the pharynx that supports a gill bar.

Breaching. A complete or almost complete leap above the water's surface accompanied by a highly audible splash.

Breeding plumage. Plumage of bird during nuptual or reproductive period.

Buccal. The cheek or mouth.

Budding. A form of asexual reproduction which adds new members to a colony.

Buoyancy. The ability to float or rise in a liquid.

Bursa. A saclike cavity or structure.

Byssal threads. Filamentous attachments, e.g. the threads that anchor a mussel (Mytilus) to a hard substrate.

Calcareous. Containing calcium carbonate (chalk).

Calcichordate. Extinct burrowing animals with external skeleton formed from plates; variously grouped with echinoderms or stem chordates.

Calcification. The process by which calcium and carbonate ions are combined to form calcareous skeletal materials.

Calcite. The type of calcium carbonate in the shells of most benthic molluscs.

Callosities. Discrete areas of roughened skin on the heads of Black Right Whales (Balaena glacialis) that have an orange colour due to the presence of numerous whale lice.

Calyx. A cup-shaped cavity or structure (e.g. the corallite exoskeleton of reef-building cnidarian polyps).

Canine tooth. Conical, pointed tooth between incisors and premolars. Prominent in carnivorous mammals where they are used to seize prey. In Pinnipedia, can be enlarged to form tusks, absent in Cetacea and Sirenia.

Cape. Darkly pigmented region on the backs of some dolphins, extending in front of and behind dorsal fin.

Carapace. 1) Domed top of a turtle or tortoise shell composed of two layers of interlocking plates and a covering of horny scutes; 2) scleratised plate covering the fused head and thorax of arthropods.

Carnivore. An animal that feeds exclusively on other animal tissue.

Carotene. Any one of a class of pigments (principally found in plants), derivatives of which are found as eye pigments of animals.

Carpal. Area of wrist; in birds the distal wing joint.

Cast. The non-digestible waste products that are regurgitated by birds as a ball of material (also known as a pellet).

Catadromous. Fishes that spawn in the sea whose young then migrate into freshwater to grow and mature.

Caudal. Tail (e.g. caudal fin).

Caudal peduncle. Laterally (side-to-side) compressed region between the dorsal fin and tail flukes of cetaceans. Region immediately anterior to the tail and behind the dorsal and anal fins in fish.

Cell. A single living unit of an organism that generally contains a nucleus.

Celsius. The temperature scale where 0°C (centigrade) is the freezing point of water and 100°C is the boiling point, at standard atmospheric pressure.

Centrifugal force. The force that moves an object away from the centre of a curved path which it is following.

Central nervous system. (CNS) The complex of nerve tissues that controls the activity of the body.

Centriole. Hollow, cylindrical structure, normally one of a pair composed of nine sets of three microtubules, which are thought to be organisers of the microtubular framework within animal cells and important in cell division.

Centripetal force. A counter-seeking force that makes a rotating body move towards the centre of rotation.

Cephalisation. Head development.

Cephalothorax. Fused head and thorax of arthropods.

Cerata. Projections from the body surface of nudibranchs.

Cerebroganglion. The ganglion that makes up the 'brain' of invertebrates (plural: cerebroganglia).

Chelate. Pincer-like or claw-like.

Cheliped. First pair of decapod crustacean legs that is large, heavy and chelate.

Chemoreceptor. A sensory receptor that is stimulated in response to contact with chemical substances.

Chemosynthesis. The fixation of carbon from carbon dioxide into organic compounds using energy derived from the oxidation of inorganic compounds such as ammonia, methane and sulphur.

Chlorophyll. A number of green plant pigments that capture the sun's energy and use it in the photosynthetic process.

Chlorophyta. Division of algae that contain chlorophyll.

Chondrichthyes. The cartilaginous fishes (chimaeras, sharks and rays).

Chordata. Phylum of animals with a notochord, hollow dorsal nerve cord, and pharyngeal (gill) slits. Comprises Urochordata (tunicates), Cephalochordata (lancelets) and Vertebrata (vertebrates).

Chromatophore. A pigment cell, or group of cells, that can be controlled by the nervous system to produce a colour or tone by alteration in shape.

Cilium. Hair-like structure which is used for locomotion or feeding (plural: cilia).

Claspers. Rod-like processes on pelvic fins of male elasmobranchs and chimaeras used for mating purposes.

Clay. Small particle sizes between silts and colloids.

Cnida. An eversible organelle found within a cnidocyte (plural: cnidae).

Cnidocyte. A specialised cell in cnidarians that is found throughout the epidermis and contains organelles capable of eversion (cnidae); they are especially abundant on the tentacles.

Coast. A strip of land that extends inland from the coastline as far as marine influences are still recognisable.

Coastline. Landward limit of an area affected by the highest wave action.

Coelom. Fluid-filled body cavity separating the body wall muscles from the gut.

Cohort. A group of organisms that are produced at the same time (e.g. a singe generation or year group).

Cold light. Bioluminescent light produced by some groups of animals.

Collagen. A fibrous scleroprotein of high tensile strength that is a major constituent of connective tissue.

Colonial organisms. Plants or animals that live together as a group and may be joined together.

Columella. A spiral rod-like structure that is formed by the fusion of the inner surfaces of gastropod mollusc shells that are tightly coiled.

Commensalism. A symbiotic relationship whereby one individual benefits at no expense or damage to the host.

Community. An ecological collection of different plant and animal populations within a given area or zone.

Compensation light intensity. The amount of light at which photosynthetic production just balances respiratory losses in plants.

Compound. A substance containing two or more elements combined in fixed proportions.

Compound eye. An eye that is made from many individual light sensitive units (ommatidia).

Condensation. The conversion of water vapour to a liquid that occurs through cooling.

Conspecific. Different forms belonging to the same species.

Consumer. A heterotrophic organism that ingests an external supply of food.

Continent. A large landmass exposed above sea level which is composed of granite and igneous rocks.

Continental drift. The process by which continents move around on the surface of the Earth, relative to each other.

Continental margin. The underwater edges of continents consisting of the continental shelf, the continental slope and the continental rise.

Continental rise. A gently sloping surface at the base of the continental slope consisting of deposited sediment.

Continental shelf. A gently sloping seabed extending from the low water mark of the shoreline to the shelf break at the upper continental slope.

Continental slope. The steep slope that starts at the shelf break (about 150-200m depth) and extends down to the continental rise of the deep ocean floor.

Contour feather. Predominant feather type of birds, characterised by the prescence of vanes.

Convection. The vertical transport of fluids or the transfer of heat within fluids.

Convergence. 1) evolution which produces similarity in some characteristic(s) between groups which were initially different; 2) the process by which different water masses come together which may often result in the sinking of surface water.

Coral. A group of benthic cnidarians belonging mainly to the class Anthozoa that often exist in large colonies and may secrete calcareous exoskeletons.

Coral head. A large piece of coral that grows to the surface of the sea.

Coral reef. A calcareous organic area composed of solid coral and coral sand, laid down by anthozoans and encrusting calcareous algae.

Core. The innermost centre of the Earth that begins at the base of the mantle and consists of high levels of iron and nickel.

Coriolis effect. An effect resulting from the Earth's rotation which causes particles in motion to be deflected to the right in the northern hemisphere and to the left in the southern hemisphere.

Cosmopolitan species. A species of plant or animal that is widely distributed throughout the world.

Cranium. The main part of the skull, excluding the lower jaw.

Crest. The highest point of a wave above the seabed.

Critical depth. The depth at which the total photosynthetic production is balanced by the total respiratory loss at any given point in the water column.

Crop. The enlarged oesophageal bag of a bird that is used as a temporary food store.

Crust. The thin outermost skin of the Earth that forms a solid layer of either basalt (oceanic crust) or granite (continental crust).

Current. A horizontal movement of water or air.

Cyclone. An mass of air moving in an anticlockwise direction in the northern hemisphere and clockwise in the southern hemisphere.

Cypris. The larval stage of barnacles that follows the nauplius stage.

Cytoplasmic streaming. Continuous movement of the cytoplasm within a cell.

Dactylozooid. A defensive or protective polyp within a colonial cnidarian.

Declination. The angular distance of the sun or moon above or below the plane of the Earth's equator.

Decomposer. An organism that breaks down dead organic matter into its inorganic components:

Delayed implantation. Period of suspended development in which the blastocyst lies free in the uterus prior to attachment to the uterine wall and placenta formation.

Delta. A low-lying deposit of soft sediment lying at the mouth of a river or estuary that typically consists of sand or mud.

Demersal. Living on or just above the seabed.

Demineralise. To remove salts.

Dendrite. Many-branched process arising from the cell body of a neuron that receives impulses from other neurons.

Density. In physical terms it is mass per unit volume, but in biological terms it is the number of individuals in a given area or volume.

Denticles. Small scale-like teeth (placoids) found on the skin of many elasmobranch fishes.

Deposition. The laying down of small particles to form accumulations of material.

Deposit feeder. An animal that feeds on organic particles found on, or in the sediments of, the seabed.

Dermis. The deeper layer of the skin which contains blood vessels, glands and connective tissue.

Desiccation. The process of drying out or detrimental water loss from an organism.

Detritus. Dead organic or inorganic material.

Detritivore. An animal that feeds on detritus.

Deuterostome. Member of one of several phyla of the animal kingdom in which the first embryonic opening (the blastopore) becomes the anus, and the second opening becomes the mouth.

Diffraction. The bending of a wave (light, water, radio etc.) around an object(cf. refraction and reflection).

Diffusion. The transfer of particles or materials by random molecular movement from an area of high concentration to an area of low concentration.

Dimorphism. Biological differences between two or more types (e.g. morphological differences between the males and females).

Dioecious. Having separate sexes (male and female) in unisexual individuals.

Diploblastic. Animals whose body is derived from only two embryonic cell layers and separated by a gelatinous layer (mesoglea).

Disphotic zone. The dimly lit 'twilight zone' where there is not sufficient light for the process of photosynthesis.

Diurnal. Occurring during the daytime.

Diurnal tide. A tide that has one high tide and one low tide in each tidal day of 24hr 50min.

Diversity. The richness of an area or the relative numbers of different types of organisms or groups of organisms in an area or region.

Divergence. A horizontal flow of water or air away from a central region. In water masses this often results in 'upwellings'.

Doldrums. A belt of light, variable winds occurring in a region 10° to 15° north and south of the equator. The Doldrums is the common name for the Intertropical Convergence Zone.

Dorsal. Upper surface or plane.

Downwelling. The sinking of water or water masses beneath the surface.

Drag. The force which tends to hold back or resist forward movement.

Driftfish. Oval-bodied, large-eyed marine fish (family Nomeidae) that are sometimes found accompanying jellyfish in the tropics and subtropics.

Drop-off. A steep cliff-like slope or underwater precipice.

Dune. A coastal deposition of sand resulting from onshore windblown action, driving sand inland from the beach area.

Ebbing. Receding or going out (e.g. a retreating tidal current).

Eccentric cell. A nerve cell found amongst the retinula cells of certain arthropod compound eyes. These neurons collect electrical information from surrounding retinula cells, which is then transmitted to the 'brain'.

Ecdysis. The periodic shedding of an exoskeleton or outer skin of animals.

Ecdysone. A hormone which brings about moulting in an arthropod, acting mainly on the epithelial cells.

Echolocation. The detection of an object by reflected sound, first described in bats. Also present in other mammals including shrews, some cetaceans, and in some birds.

Ecology. The study of organisms in relation to their surroundings.

Ecological efficiency. The efficiency of energy transfer from one trophic level to the next in a food chain or food web.

Ecosystem. All the living organisms in a given environment with its associated abiotic factors.

Ectoplasm. Outer gel-like layer of a protozoan that lies immediately below the cell membrane.

Eddy. A relatively small circular movement of water or air.

Ediacaran. Youngest time period of the Proterozoic period, about 600mya, with a marked increase in fossil remains. Named after the Ediacaran Hills, Australia.

Egestion. The removal of unwanted material from an organism (e.g. faeces).

Elasmobranch. Shark-like fish that have 5-7 gill slits, placoid scales and claspers.

Element. A substance composed entirely of the same atoms.

El Niño. A warm current flowing in a southerly direction which develops off the Ecuador coast just after Christmas. Sometimes it will move further south into Peruvian coastal waters and cause widespread death of plankton and fish which cannot tolerate its physical presence.

Ekman spiral. A wind-induced movement of water at an angle of 45° to the direction of the wind.

Ekman transport. The net flow of water to the right of the wind in the northern hemisphere and to the left of the wind in the southern hemisphere, caused by the deflection by the Coriolis effect.

Encrusting. Incorporating or covering with a hard material (e.g. the incorporation of calcium carbonate into the skeletons or tissues of some hydrozoans and red algae).

Endemic. Organisms which are only found in a specific area, habitat or region.

Endolithic. Growing or living between stones.

Endoplasm. Inner, more fluid layer of a protozoan in which the main organelles are embedded.

Endopodite. Inner appendage attached to the base of the outer appendage (exopodite) of a crustacean biramous limb.

Endostyle. A ciliated and glandular groove or pocket in the ventral wall of the pharynx used in suspension feeding.

Environment. The sum of biological and abiotic factors to which an organism, population or community is subjected.

Enzyme. A biological catalyst that helps a biochemical reaction to take place.

Ephyra. Immature medusa of a scyphozoan cnidarian (jellyfish) almost always microscopically small (plural: ephyrae).

Epibenthic. Refers to organisms living on the seabed (see demersal).

Epipelagic zone. The upper region of the sea from the surface to a depth of about 100m.

Epiphyte. Plants that grow on the surface of other plants without harming them.

Epizoic. Applies to non-parasitic animals that attach themselves to the outside of another normally larger animal.

Equinox. The period of time when the sun is over the equator, making day and night periods of equal length throughout the planet. The 'spring or vernal equinox' occurs about 21 March as the sun moves into the northern hemisphere; the 'autumn equinox' occurs about 21 September as the sun moves into the southern hemisphere.

Era. A short period of geological or biological history.

Erosion. The process by which rocks are broken down into smaller sizes by climatic, mechanical and chemical means.

Estuary. A semi-isolated coastal area of land that has a body of water which is a mixture of sea water and fresh water.

Euphotic zone. The well-lit surface layer of the ocean or sea that has enough light to support photosynthesis and primary production.

Euryhaline. An organism which is able to tolerate a wide range of salinities.

Eutrophic. An area that has high levels of nutrients and consequently has high biological productivity.

Evaporation. The physical process by which a liquid is converted to a gas at temperatures below the boiling point of the liquid.

Eviscerate. Disembowel.

Exopodite. See endopodite.

Exoskeleton. A skeleton of an animal which is external to the soft tissue of the body.

Excretion. The elimination of waste products from tissues or bodies. Excretion is often in the form of urea or ammonia.

Faeces. The solid waste products of animals.

Fahrenheit. Temperature scale (°F) in which the freezing point of water is 32° and the boiling point is 212°, at standard atmospheric pressure.

Falcate. Sickle-shaped.

Fauna. Animal life.

Fecundity. The level or rate of egg or offspring production.

Femtoplankton. The smallest size of classification of plankton (0.02 - 0.2μm in size).

Fetch. The distance and period over which the wind blows to create waves or wave systems.

Filoplumes. Type of feather resembling a thin hair.

Filter feeders. Animals that feed by sifting out small particles of organic material (see suspension feeders).

Finrays. Horny supporting structures inside the fins of fishes.

First-summer. Plumage stage acquired by moult replacing first-winter feathers (see below).

First-winter. Plumage stage replacing juvenile feathers in some birds.

Fjord. A narrow, deep, U-shaped inlet that has been submerged and forms part of a coastline.

Flagellum. A whip-like appendage used by some single-celled organisms and spermatozoa for locomotion.

Flank. The vertical sides of a body (e.g. the sides of a fish or bird).

Flexion. The bending of a limb.

Flotsam. Floating wreckage.

Flora. Plant life.

Fluke. 1) horizontally flattened and blade-shaped tail of cetaceans and Dugong. 2) parasitic flatworm.

Food chain. A linear progression of trophic (feeding) levels in which one organism is the food source for the next level.

Food web. A description of interactions or complex relationships between various trophic levels.

Foreflipper. The front flipper of a pinniped.

Foreshore. Another name for the intertidal zone which lies between the high and low tide marks.

Fossil. Any remains, trace or imprints of organisms that have been preserved in rock.

Freshwater. Water that has no salts and therefore is not saline.

Fringing reef. A reef of coral that is attached directly to the shore of a coastline or island and has no lagoon.

Frond. The leaflike structure of a plant.

Fucoids. Groups of large brown seaweeds (algae) belonging to the phylum Phaeophyta.

Fusiform. Spindle-shaped or torpedo-shaped and tapering at one or both ends.

Gadfly petrel. General term used to describe Pterodroma petrels.

Gamete. Sexual cell that fuses with another during fertilisation.

Gametogenesis. Formation of gametes from a cell that undergoes meiosis (fission of cell that halves its chromosome number).

Gametophyte. The gamete-forming phase in alternating plant generations.

Ganglion. Mass of nervous tissue that contains many neurons and synapses.

Gastrula. Developmental stage that succeeds the blastula stage in the embryonic development of an animal. Complex cell movements occur which are related to the formation of the adult body plan.

Genes. The unit hereditary factors in the chromosomes that carry the information influencing a set of characters in a particular way.

Genetic code. The sequence of codes on the genes that determines the organism and its characteristics.

Gill rakers. Small rod-like structures attached in a single or double row to branchial arches that are used to collect small particles of food.

Gizzard. A part of an animal's stomach where food is mechanically broken up.

Glacier. A large mass of slow-flowing ice that has formed on land by the recrystallisation of snow into ice by pressure.

Global plate tectonics. The process by which the lithospheric plates move across the surface of the Earth and alter the shape and size of continents and ocean basins.

Glycerol. A three carbon trihydroxyl (three -OH groups) alcohol, esters of which are important constituents of many lipids (e.g. fats that occur in cell membranes).

Gnathostomata. Fish with jaws (Elasmobranchii and Teleostomi).

Gravitational force. The force of attraction that exists between any two bodies. The level of the force will depend upon the mass of the two bodies and the distance between them.

Grazers. Animals that eat plant materials as their main source of energy.

Greenhouse effect. The warming of the Earth's atmosphere by terrestrially produced gases that are being absorbed into the atmosphere.

Guano. A phosphate deposit formed by the leaching of accumulated bird droppings in a dry, hot climate.

Gular pouch. Distensible pouch of skin under throat used in display (frigatebirds) or to hold food (pelicans).

Gulf stream. A fast moving western boundary current of the North Atlantic subtropical gyre, that flows northwards off the east coast of the United States. It brings warm water to northern Europe.

Guyot. A submerged flat-topped table mountain that is normally of volcanic origin.

Gyre. A large circulation of oceanic water which often has a stagnant central area. Gyres rotate clockwise in the northern hemisphere and anticlockwise in the southern hemisphere.

Hadal zone. The sea floor region associated with deep sea trenches (>4500m).

Haemoglobin. The red pigment of the blood that has an affinity for oxygen and transports it around the body.

Harem. A group of breeding females dominated by a single sexually mature male. In certain pinnipeds, harems are associated with territory formation by males (e.g. Northern Fur Seal Callorhinus ursinus).

Headland. A steep-faced promontory of rock that extends seawards from the coast.

Hemichordata. Phylum of animals with chordate features but lacking a true notochord; comprises worm-like Enteropneusta and stalked Pterobranchia.

Hemisphere. That half of the Earth which is either north of the equator (northern hemisphere), or south of the equator (southern hemisphere).

Herbivore. An animal that eats plants as its main source of energy.

Hermaphrodite. An animal that produces both male and female gametes

Heterocercal. A vertebral column that usually terminates in the upper lobe of the caudal fin of a fish. The upper lobe of the tail fin is usually larger the lower lobe.

Heterotroph(ic). Organisms that need organic material for food (which therefore includes most animals).

High water (HW). The highest level reached by a tide on a particular day before it starts to recede.

Hindflipper. The rear flipper of a pinniped.

Holdfast. A 'root-like' structure that anchors some large seaweeds to the bottom on hard substrates.

Holoplankton. Planktonic organisms that spend their entire lives in the water column and are permanent residents of plankton communities.

Homeothermic. An animal that is warm-blooded and is able to regulate its internal body temperature.

Homocercal. A tail with equal or nearly equal lobes and a vertebral axis ending about the middle of the base.

Homoeostasis. The tendency of a biological system to maintain itself in a state of equilibrium.

Homologous. Applies to an organ of an animal that is thought to have the same evolutionary origin as an organ of another animal.

Horse latitudes. The area between latitudes 30-35° north and south of the equator where the winds are light and variable.

Host. An organism having a parasite or commensal.

Hurricane. A violent, tropical, cyclonic storm where winds can reach speeds of over 120 miles per hour.

Hydrological cycle. The exchange of water among the oceans, atmosphere and the land through evaporation, rain, land run-off and ground water percolation.

Hydromedusa. The medusa (sexual stage) of hydroid cnidarians (plural: hydromedusae).

Hydrophilic. Attracting or liking water.

Hydrophobic. Repelling or disliking water.

Hydrosphere. The envelope of the planet containing all the gasses of the atmosphere and all the water.

Hyoid jaw suspension. A series of bones or cartilages evolved from the hyoid arch that forms the jaw suspension system in fishes.

Hypertonic. A fluid that has a higher osmotic pressure (therefore higher salinity) than fluid on the other side of a semipermeable membrane. Therefore the hypertonic solution will gain freshwater.

Hypotonic. The opposite of 'hypertonic'. The hypotonic solution will have a lower osmotic pressure and lose water to the solution of higher salinity.

Hypersaline. Water that has a high salinity of 40‰ or more.

Ichthyoplankton. Planktonic fish larvae and suspended fish eggs.

Igneous rock. One of the three main types of rocks which is crystalline in structure and formed from molten matter.

Immature. Arbitrary term used to denote avian plumage(s) between juvenile and adult.

Incisor tooth. Chisel-shaped tooth (primitively three on each side of each jaw) at the front of the mouth in mammals.

Indirect development. Having a larval stage in the course of development.

Infauna. Animals living in soft substrates like sand and mud.

Inshore. See nearshore.

Ingestion. The process by which food is taken in and swallowed.

Integument. A natural outer covering.

Interface. The surface or boundary separating two layers or substances that have different properties, i.e. air and water density, temperature, salinity etc.

Intermediate water. Water masses that normally form in polar regions and sink at the convergence zones to a depth of between 800-1,000m, before spreading out and flowing towards the equator.

Interstitial. Occurring in spaces between, e.g. interstitial organisms live between particles of sand.

Intertidal (littoral) zone. The zone between high and low water marks that is periodically exposed to the air.

Invagination. Infolding of the early embryo in which cells fold into the interior to form the primitive gut (gastrulation).

Invertebrate. Animal without a backbone.

Irides. Plural of iris; an anterior structure in the eye in front of the lens (colour is often identification feature).

Iridocytes. Cells or tissues containing reflective materials (e.g. mirror cells on the scales of fishes).

Isobar. A line on a map or chart joining places of equal atmospheric pressure.

Isohaline. Of the same salinity.

Isotonic. Of equal concentration on both sides of a membrane.

Isotherm. A line on a map joining places of equal temperature.

Jet stream. Easterly moving air mass at a height of about 10km, moving at speeds in excess of 300km per hour.

Jetsam. Floating debris washed ashore, often coming from ships.

Jizz. A combination of characters such as size, shape and flight action which assist with identification of birds in the field.

Junk. Fibrous, fat-filled layer of tissue lying between the spermaceti organ and cranium in the Sperm Whale.

Juvenile plumage. Plumage of young bird at point of fledging or departure from nest.

Keel (caudal). Surface projection generally of hydrodynamic importance on posterior region of body. Sideways paired projections in fish on peduncle and/or caudal fin, vertical projection(s) in cetaceans on peduncle.

Kelp. Giant brown algae that grow subtidally on hard substrates in the high and midlatitudes.

Keratin. A group of fibrous scleroproteins that usually contain large amounts of the amino acid cystine.

Knot. A unit of marine speed equal to one nautical mile per hour or about 51cm per second.

Krill. Shrimp-like crustaceans found in ocean waters (order Euphausiacea) which form an important food source for many of the larger marine animals.

Labyrinthodont tooth. Tooth with complex infoldings of the surface enamel layer found in primitive fossil amphibians (e.g.Ichthyostega) and certain rhipidistian fishes (e.g.Osteolepis). Provides important evidence for an evolutionary link between early tetrapods and lobe-finned fishes.

Lagoon. A relatively shallow stretch of water partly or completely separated from open water by a terrestrial barrier.

Lamella. Thin layer, membrane or plate-like structure 1) of gills in bivalve molluscs and fish or 2) with fine hair-like structures attached to the avian bill, enabling food particles to be filtered from water.

Lanugo. Birth coat of fur of pinniped pups. Shed in the uterus before birth in some species.

Larva. An immature stage of an animal which differs significantly from the mature adult. It normally follows the hatching of the egg.

Lateral line (acoustico-lateralis system). The sensory canal system running along the flanks of a fish that detects water movements, disturbances and vibrations.

Latitude. Location on the Earth's surface determined from the angular distance north and south of the equator, with the equator at 0° and the poles at 90°.

Lava. Fluid magma coming out of the surface of the Earth.

Lecithotrophic larvae. Meroplanktonic larval stages that do not feed on planktonic food.

Leeward. Direction towards which the wind and waves are blowing. The 'lee' of an island is the sheltered side. A 'lee shore' is an exposed shore.

Leptocephalus. The unusual looking larva of certain fishes such as eels (Anguillidae) and tarpons (Megalopidae).

Ligaments. Strong, fibrous bands of tissue connecting two or more moveable bones.

Limestone. Sedimentary rock mostly of calcium carbonate and magnesium carbonate, much of which has been formed from the shell debris and exoskeletons of animals.

Lithosphere. The outer layer of the Earth's structure including the crust and upper mantle.

Littoral zone. Another name for the intertidal zone.

Load. The amount of sediment that can be held in suspension by a moving body of water. When the water flow is reduced the load sinks to the bottom and is deposited.

Longitude. Location on the Earth's surface based on the angular distance east or west of the Greenwich Meridian (0° longitude). 180° longitude is the International Date Line.

Longshore drift. The transport of loose materials along a coastline by longshore currents and water movements.

Lophophore. The circular or horseshoe-shaped feeding organ of certain invertebrates, composed of ciliated tentacles surrounding the mouth.

Lug spade. A thin spade used for digging up lugworms (Arenicola) from soft sediments.

Macrobenthos. Another name for 'macrofauna' and includes all larger bottom-living animals.

Macrophagous. Collecting and ingesting of large food particles.

Macrophytes. Any large living marine plants such as algae, seagrasses and mangroves.

Macroplankton. Large plankton between 2 and 20cm in length.

Madreporite. Sieve-like button-shaped process on aboral surface of an echinoderm which forms the external opening of the water vascular system.

Mandible. 1) the lower jaw bone; 2) either part of the bill of a bird.

Mangroves. A variety of salt-tolerant trees and shrubs that inhabit the intertidal zones of tropical and subtropical regions.

Mantle. The area of the internal Earth extending from the base of the crust to the outside of the core.

Marginal sea. A semi-enclosed body of water adjacent to a large continent.

Mariculture. Marine agriculture and marine husbandry.

Marsh. An area of wetland that has been colonised by marsh grasses and other marsh plants, and is periodically flooded by water.

Mass stranding. Beaching of marine animals of the same species; in Cetacea involving three or more animals.

Mechanoreceptor. Sensory receptor that responds to mechanical stimuli such as touch and sound.

Medusa. The free-living body type of cnidarians resembling a gelatinous umbrella or bell with tentacles at its margins.

Megaplankton. Large planktonic animals between 20 and 200cm in length.

Meiobenthos. Small organisms that live between the particles of soft substrates.

Meiofauna. Refers to small animals, normally those living interstitially in sand.

Melanin. A dark brown pigment found in the skin (and feathers) of animals formed in melanoblast cells through the oxidation of aromatic compounds (hydrocarbons with aromatic properties).

Melon. Fatty cushion forming 'forehead' in toothed whales (anatomically homologous to highly modified upper lip of terrestrial mammals). Thought to act as sound lens.

Mermaid's purse. The leathery egg case laid by some cartilaginous fishes.

Meroplankton. Planktonic organisms that spend only part of their lifecycle in the water column, usually as an egg or larva.

Mesenchyme. A meshwork of loosely associated cells.

Mesentery. Vertical partitions of the body cavity of Anthozoa. In vertebrates, infoldings of the membrane lining the abdominal cavity that supports the abdominal organs.

Mesopelagic zone. The water column from the bottom of the epipelagic zone (100m) to about 1,000m in depth.

Mesoplankton. Plankton between 0.2 and 20mm in length.

Mesotrophic. An area or region that has moderate levels of nutrients in the sea water.

Metabolism. The chemical changes occurring in the bodies of organisms.

Metamorphosis. Abrupt change in body shape and structure that an animal goes through from embryo to adult stage.

Metameric segmentation. Repetition of organs and tissues at intervals along body of animal producing a linear series of segments.

Metaphyta. The kingdom of many-celled plants.

Metazoa. The kingdom of many-celled animals.

Microbenthos. Benthic animals that are small in size (e.g. many protozoans).

Microphagous. Collecting and ingesting of small food particles.

Microplankton. Plankton not easily seen by the naked eye (0.02- 0.2mm in length, or 20-200µm), but which can be caught in fine mesh plankton nets.

Migration. Long and often habitual journeys undertaken by animals normally for the purposes of feeding and reproduction.

Migratory circuit. The three-sided circuit of migration that animals undertake during their lifecycle from 'spawning grounds' to 'juvenile nursery areas' and finally to 'adult feeding grounds'.

Milt. The secretions from the male reproductive organs of fish.

Mimicry. Imitation by colour, resemblance or structure, of another object or organism for the purpose of self-protection or camouflage.

Mixed layer. The surface layer of oceanic water which is well mixed by wind and waves to produce relatively even physical and chemical conditions.

'M' mark. Diagonal line across the avian wing forming the shape of the letter 'M'.

Molar tooth. Posterior tooth in mammals used for crushing. Usually with several roots and biting surface formed from patterns of projections and ridges. A molar tooth is not preceded by a milk tooth.

Molecule. The smallest particle of a compound that retains the characteristics of it.

Mollymawk. General term for the smaller albatrosses.

Morph. Term to denote a colour form of a polymorphic species.

Morphology. The study of form and structure in plants and animals.

Mortality rate. The rate at which animals die.

Motive force. The force that moves an object in a given direction.

Moult. The periodic shedding of exoskeletons, hair, feathers or skin that allows new growth to take place underneath.

Mud. Sediments consisting of silt and clay particles which are smaller than 0.06mm in size.

Mudflats. Expanses of mud which are periodically exposed at low tide and are often found adjacent to saltmarshes.

Muzzle. Projecting part of the head in pinnipeds comprising mouth, nose and jaws.

Mutualism. A symbiotic relationship in which both partners benefit in some way or another.

Myomere. A muscle segment or muscle block divided off by connective tissue.

Mysticetes. Baleen whales of the order Cetacea.

Nannoplankton. Plankton too small to be caught in a fine mesh plankton net and which needs to be extracted by filter centrifuge. Includes many of the protozoans and phytoplankton (2 - 20µm in length).

Nares. Bony nasal openings of the skull.

Nasal plugs. Paired muscular flaps found at the base of the upper nasal passage of toothed whales that seal the airway during diving. Thought to be used (probably in combination with other nasal structures and the larynx) in the production of sounds.

Natural selection. The theory put forward by Charles Darwin by which the fittest, strongest and best adapted animals will survive.

Nauplius. The first free-swimming larval stage of many crustaceans including copepods, ostracods and decapods.

Neap tides. The smallest range of tides in a lunarcycle that occur when the moon is at the first and third quarters.

Nearshore zone. The shallow waters seaward of the line of breakers.

Necrotrophic. Applied to parasitic organisms that obtain nutrients from dead cells and tissues of their hosts.

Nekton. Pelagic animals with strong swimming capabilities that are able to swim against a current (e.g. adult fish, squid and marine mammals).

Nektobenthos. Active swimming animals that spend much of their time on or near the seabed.

Nematocyst. A stinging cell in cnidarians.

Neoteny. Persistence of larval characters in the adult form.

Neritic province. The area from the shoreline to a depth of about 200m that includes the continental shelf.

Neuromasts. Groups of sensory cells in the lateral line of fishes.

Neurophysiology. The science that studies structure and function of the nervous system.

Neuron. A nerve cell.

Neurotoxin. A toxin that affects the function of the nervous system.

Neuston. Organisms that inhabit the first few millimetres of the water surface.

Niche. The ecological role of an organism and its position in the ecosystem.

Nictitating membrane. A thin membranous eyelid that helps to protect the eyes and keep them clean. They are only found in some vertebrate groups including certain sharks.

Non-breeding plumage. Plumage of a bird during the winter or outside the breeding season.

Notochord. Rod-like supporting structure lying lengthwise between the central nervous system and gut. Present only in chordate animals.

Nutrients. Inorganic and organic compounds or ions that are used by plants in the production a new organic material (e.g. phosphates and nitrates).

Ocean. Usually refers to water well offshore beyond the 200m depth contour.

Oceanology. The study of the science of the oceans.

Oceanography. The study of the physical, chemical and living aspects of the seas and oceans and its interactions with the atmosphere.

Ocean basin. The seabed of an ocean lying between one or more continents.

Oceanic province. The pelagic ocean environment beyond the continental shelf.

Ocean ridge. A mid-ocean chain of mountains, mostly underwater, along which volcanic activities produce new material for the process of 'seafloor spreading' and therefore 'continental drift'.

Ocellus. 1) a simple eye which has a single, thickened cuticular lens; 2) eye-like marking in fish (plural: ocelli).

Odontocetes. Toothed whales of the order Cetacea.

Oligotrophic. Nutrient-poor areas that have low levels of biological production (e.g. the centres of many ocean gyres).

Olfaction. Sense of smell modulated by receptors which detect substances according to their chemical structure.

Olfactory. Pertaining to the sense of smell.

Ommatidium. A unit of the compound eye of an arthropod (plural: ommatidia).

Omnivore. An animal that feeds on both plant and animal tissues.

Ontogeny. Development of an animal from fertilisation of the egg to adult form.

Ooze. A marine sediment consisting of at least 30% animal skeletal remains (calcareous or siliceous) mixed with the inorganic clays.

Operculum. 1) a hard pad on the foot of some gastropod snails which is used to seal the opening of the shell. 2) the hard external cover of the gill opening of a fish (plural: opercula).

Osmosis. The movement of water through a semipermeable membrane that separates two solutions of differing solute concentrations. The passage of water across the membrane occurs to equalise to difference.

Osmotic pressure. The pressure required to counteract the passage of water through a semipermeable membrane.

Osmoregulation. The physiological mechanism used by an organism to regulate the salt and fluid balance of the body to an acceptable level.

Osphradium. Water-sampling organ, common to all neogastropod molluscs and also found in many other gastropod molluscs, used primarily for detecting prey.

Ossicles. 1) fenestrated calcareous plates, rods and crosses that are arranged to form a skeletal lattice; 2) small bones (e.g. in a vertebrate ear).

Ostracoderm. Extinct fish-like vertebrates lacking jaws. Ostracoderms, lampreys and hagfishes are grouped in the class Agnatha.

Otoliths. The calcareous ear stones located in the auditory/balance organs (inner ears) of many animals, including fishes.

Oviduct. A tube that carries eggs from ovary to the uterus or from the coelom (where the eggs are shed) to the exterior.

Oviparous. An animal that lays eggs and the young hatch outside the body of the female.

Ovoviviparous. An animal that produces membranous eggs that hatch in the maternal body and the young are later born alive.

Ozone (O_3). A tri-atomic form of oxygen that is formed and destroyed by ultraviolet radiation. Ozone cuts down the amount of harmful radiation reaching the surface of the Earth, thus protecting it.

Paddle. The front limb of a sirenian.

Palaeoceanography. A branch of oceanography that studies the biological and physical characters of ancient oceans.

Palaeogeography. A branch of geography that studies the shapes and positions of ancient continents and oceans.

Pangaea. A supercontinent of the Mesozoic era consisting of all the present-day continents put together.

Parenchyma. In platyhelminths the tissue that fills the interior of the body composed of cells and intercellular spaces.

Parthenogenesis. Reproduction by asexual means (without fertilisation) producing cloned offspring that are genetically identical to the parent.

Parasite. An organism that takes its nutrients from the tissues of another living organism at the host's expense.

Patchiness. Uneven distribution of organisms (e.g. in plankton).

PCB. Polychlorinated biphenyl, a broad-based organochloride pesticide that is also toxic to other animals.

Pectoral. Relating to the side of the body; in vertebrates the region supporting the forelimbs or fins.

Pedalium. The flattened, blade-like base of a tentacle in jellyfish.

Pedicellaria. Minute stalked organ in some invertebrates (e.g. echinoderms) used for grasping, defence and scavenging.

Peduncle. Invertebrate attachment stalk anchoring the body to the substrate. See caudal peduncle.

Pelagic. Pertaining to the open ocean away from the coasts and continental shelf areas.

Peristalsis. Rhythmic waves of muscular contraction.

Permeability. A condition that allows the passage of liquids through a substance or membrane.

Phagocytosis. A process in which a cell engulfs externally derived solid material within a vacuole without disrupting the continuity of the cell surface.

Pharynx. 1) part of the gut between the mouth and gullet (oesophagus) in vertebrates; 2) part of the gut into which slits open internally in tunicates, lancelets and hemichordates.

Phonation. The production of sound.

Phoresy. A method of dispersal when an animal clings to the body of a much larger animal of another species and is carried some distance before it releases itself.

Photic. Of or relating to light.

Photic zone. The well lit surface zone where photosynthesis can take place.

Photocyte. A special animal cell that produces cold, bioluminescent light.

Photophore. A complex animal organ in which cold bioluminescent light is produced.

Photosynthesis. The plant process by which water and carbon dioxide are used to manufacture energy-rich organic compounds in the presence of chlorophyll and energy from sunlight.

pH scale. A scale of 1 to 14 that represents the acidity or alkalinity of a solution, with pH7 being neutral.

Phylogeny. Relationships of groups of organisms as reflected by their evolutionary history.

Phylum. A primary division of taxonomic classification in which organisms are grouped according to their general similarities.

Physiology. Science of the functions of living organisms and their parts.

Physoclistous. The condition in which fishes do not have the swim-bladder connected by a 'pneumatic' duct to the digestive tract.

Physostomous. The condition in which fishes have the swim-bladder and digestive tract connected throughout life by a 'pneumatic' duct.

Phytoplankton. Microscopic planktonic plants (e.g. dinoflagellates).

Picoplankton. Small plankton measuring 0.2 - 2.0µm in length (e.g. bacteria).

Pinna. Fleshy flap forming outer ear.

Pitchpoling. Visual behaviour of the Sperm Whale where the head and eyes are raised vertically above the sea surface as the body is rotated slowly on its long axis.

Piscivores. Animals that eat fish.

Pitch. The 'up and down' movements of an object in the vertical plain.

Placoderm. Extinct class of fish-like vertebrates with heavily armoured head hinged posteriorly with the trunk shield.

Plankton. Organisms that are suspended in the water column and are not able to swim against the currents. Therefore they rely on water movements for distribution and transport.

Plankton bloom. A high concentration of phytoplankton resulting from the rapid reproductive multiplication of plant material in the presence of adequate sunlight and nutrients. The result may be seen as green, turbid waters.

Planktotrophic larvae. Meroplanktonic larvae that rely on planktonic food (e.g. bacteria and phytoplankton) for their growth.

Plasma. The 'liquid tissue' of the body fluids.

Pleopod. In crustaceans one of a number of abdominal appendages used for swimming, burrowing, producing water currents, gaseous exchange, obtaining food, carrying eggs, or (in males) as copulatory organs.

Pleuston. Animals that float passively at the surface of the sea with parts of their body projecting into the air. They are therefore moved by wind as well as by the water.

Pneumatic duct. A tube connecting the swim-bladder with the gut in fish.

Pneumatocyst (pneumatophore). A gas-filled cavity or float that keeps an organism buoyant in the water column. It can be found in animals (swim-bladders of fish) and plants (fucoid gas-bladders).

Pod. Collective name for a group of whales.

Poikilothermic. Having a body temperature that varies with the temperature of the surroundings.

Polar. Pertaining to the cold, high latitudes associated with the Arctic and Antarctic regions.

Pollution. Unusually high levels of substances or material that may cause damage to the environment.

Polychromatic. Many-coloured.

Polygynous. Animals that show a breeding behaviour where a male mates with more than one female in a breeding season.

Polymorphism. Distinct genetically determined colour variation within a species.

Polyp. A cnidarian individual forming part of a colony.

Polysaccharide. A linear or branched polymer composed of at least ten monosaccharide units (e.g. glucose and fructose).

Pool goggler. An underwater viewer for looking at organisms below the surface of rock pools.

Population density. The numbers of individuals of the same species expressed as a number per unit area or number per unit volume.

Postlarva. Larval stage in crustaceans, after the nauplius, when the full compliment of appendages and trunk segments appear. In bony fish, the stage between the disappearance of the yolk sac and the appearance of juvenile characteristics.

Post-canine tooth. Undifferentiated (molar-like) tooth behind the canines in pinnipeds, which together form a short series of cheek teeth.

Pre-adaptation. Possession of features by organisms that favour a new life style or habitat.

Precipitation. Rainfall, sleet, snow or hail derived from water that falls to the surface of the planet.

Predator. An animal which kills and feeds on other animals.

Prehensile. An appendage adapted for holding on to objects.

Premolar tooth. A bicuspid cheek tooth in front of the molars and behind the canines in mammals. Premolar teeth are preceded by milk teeth.

Primary production. New organic material synthesised by plants through the process of photosynthesis.

Priscoan. 4600 to 4000mya, from the formation of the Earth to the first appearance of life (formerly the first period of the Precambrian).

Proboscis. Tubular protrusion from the anterior of an animal; in pinnipeds an enlargement of the nose.

Productivity. The rate at which a quantity of organic material is produced by an organism.

Promontory. A high point of land that projects out into the sea.

Protandry. In animals, production of sperm in males before the female produces eggs.

Proterozoic. 2500 to 590mya, covering the first fossil period (formerly the youngest period of the Precambrian).

Protochordata. Small group of marine animals containing Urochordata (tunicates) and Cephalochordata (lancelets).

Protostome. Member of a major branch of the animal kingdom in which the mouth-like opening of the primitive gut (blastopore) in the early embryo contributes to the formation of the anterior mouth.

Protista. A collective name for bacteria, microscopic algae and protozoans.

Pycnocline. A depth zone in which water density changes rapidly.

Quadrat. An apparatus of fixed area that is used to help determine population numbers within its boundaries.

Quadruped. A four-footed animal.

Radial symmetry. The arrangement of the body of an animal such that parts are arranged symmetrically around a central axis.

Radiole. Ray or filament supported by a springy skeletal rod that forms part of the feathery crown of filter-feeding fanworms.

Reef. A consolidated mass of rock or coral often found in shallow waters or near the surface.

Reef slope. The steep gradient starting at the 'break of slope' which extends down to a more gentle gradient or the flat seabed.

Reef terrace. A flat platform of coral extending from the shore and which is often exposed at low tide.

Reflection. The process by which part of the energy of a wave at the sea surface is returned in the opposite direction, (e.g. a wave sent back offshore once it hits a cliff face).

Refraction. The process by which waves at the sea surface are slowed down and their direction altered by physical interference.

Relief. The difference in topographical elevation between the highest and lowest points of a particular area.

Remote sensing. Gathering information from a distance using automatic equipment.

Reproduction. The production of new living organisms.

Respiration. The metabolic process carried out by all living organisms in which organic substances are broken down and energy is released for utilisation. It results in the release of carbon dioxide and is the opposite biochemical reaction to photosynthesis.

Rete mirabile. Vascular bundle comprising exceptionally long, thin blood vessels with coiled or looped form. Characteristic of the blood vascular system of cetaceans and sirenians. In fish present in the swim-bladder, and in the swimming muscles of mackerel sharks and certain tuna species (plural: retia mirabilia).

Rip current. A narrow, fast-flowing current with a speed of about 4km per hour. It runs parallel to the shore and drains the surf zone water in a seaward direction, through channelled areas.

Rise. The long, broad elevation that rises gently and smoothly from the deep ocean floor.

Roll. The motion by which objects turn on their long axis.

Rooster tailing. Distinctive cone of spray formed during fast swimming at the surface in Dall's Porpoises, White-beaked Dolphins and Hourglass Dolphins.

Rorqual. Balaenopterids that have a grooved throat, dorsal fin, flat-snouted head and relatively short baleen plates. There are six species including the Blue Whale.

Rostrum. Upper jaw of cetaceans forming part of snout or beak. In fish, a forward projection of the snout. In crustaceans (prawns), a forward pointing spine on the carapace.

Sacoglossan. A herbivorous, opisthobranch gastropod named after the sac ('ascus sac') that contains the worn teeth after they are shed.

Saddle. Cape marking that dips below the dorsal fin to form a point in some dolphins. Also used to describe a darker back in certain seabirds (e.g. some terns).

Salinity. The level of dissolved mineral salts (mostly sodium chloride) in ocean water.

Saltmarshes. Flat areas of finely deposited silts that are occasionally flooded by sea water and dominated by communities of salt-tolerant plants.

Sand. Mineralised particles with a diameter of 0.062-2.0mm.

Sargasso Sea. The central area of the North Atlantic circulation gyre that has extensive beds of the floating brown alga Sargassum spp.

Scales. Plate-like structures produced by the skin that often overlap to produce an external protective covering (e.g. fish and reptile scales).

Schizocoel. Coelomic cavity derived from a splitting of the mesoderm.

School. A large number of fishes or cetaceans swimming together.

Scuba. 'Self-contained Underwater Breathing Apparatus' used by divers.

Scutes. Large thickened scales that are used for protection.

Sea. A subdivision of an ocean. Seas are smaller and often shallower bodies of water than oceans. They can be semi-enclosed (e.g. Black Sea) or marginal where they lie between land and the ocean (e.g. Caribbean).

Sea-floor spreading. The process producing the lithosphere where the addition of volcanic materials from ridges of undersea volcanoes causes the ocean basins to widen and therefore the continents to drift apart.

Seagrass. A collective term for the marine flowering (vascular) plants that grow in shallow waters on soft substrates.

Seamount. A peaked, underwater mountain that rises at least 1,000m above the ocean floor.

Sea pen. 1) Pennatulacea, an order of octocorals; 2) common name for the rudimentary shell of squid.

Sea state. The condition of the sea surface caused by a combination of wind and wave action (see Beaufort Scale).

Sea water. Oceanic water which contains dissolved salts and therefore has a salinity.

Seaweed. A collective term for all non-microscopic, non-vascular marine plants.

Secondary consumers. Carnivorous animals.

Secondary production. New animal tissue that is produced by consumers of primary producers.

Second-winter. Refers to intermediate plumage stage of some gulls.

Secondary symmetry. Symmetry that has developed in the course of evolution and was not of the original ancestral species.

Secretion. Substances produced from cells or glands.

Sediment. Particles of organic and inorganic material that loosely accumulate together.

Semicircular canal system. The canals and ducts of the inner ear that form part of the balance organs.

Semidiurnal tide. A tide that has two high waters and two low waters in each tidal day.

Seminal vesicle. Part of the male reproductive system that functions in the storage of sperm.

Septum. A partitioning wall of tissue that separates two areas (plural: septa).

Sessile. Animals that are permanently attached to a substrate and therefore cannot move to a different location.

Sexual dimorphism. Difference (e.g. size or plumage) between males and females of the same species.

Sexual reproduction. The production of offspring using sexual material from different parents (male and female) to produce individuals which are genetically non-identical to the previous generation.

Shelf break. The point where there is a change in the gradient of a slope from a gentle one to a steep one (e.g. the edge of an underwater cliff or the top of a coral slope).

Shoal. 1) a shallow area; 2) see school.

Shore. Another term given to the intertidal zone, but normally refers to the area that is exposed at low tide.

Shoreline (tideline). The water's edge which moves with the rise and fall of the tides.

Siliceous. Of or containing silicon dioxide (silica).

Sill. An underwater ridge or wall that divides two deeper areas over which the water can flow or spill.

Silts. Particles of sediment between the size of sands and clays with particle diameters of 4-62μm.

Slack water. The brief period between tides when there is little movement of water.

Slots. The gaps between the individual flight feathers of a bird.

Solute. A substance dissolved in a solution. Salts are the solute of sea water.

Solution. A mix of solutes in a liquid solvent.

Solvent. A liquid that has one or more solutes dissolved in it.

Species. A distinctive group of organisms that can successfully interbreed and produce viable offspring.

Species diversity. The numbers of different species and their relative abundance in a given area or region.

Spermaceti organ. Capsule of highly vascular tissue, filled with fatty esters (oils) and found in the snouts of sperm whales (Physeteridae).

Spermatophore. A capsule of fluid containing a number of sperms.

Spicule. A small, spiky skeletal element in sponges.

Spinal blaze. Light streak of colour jutting into the cape just below the dorsal fin in some dolphins.

Spiracles. A pair of dorsal openings found in many bottom-dwelling elasmobranch fishes that allows water to be drawn in and channelled to the gills.

Spiral valve. Special outfolds from the walls of the intestine that slow down the passage of food and increase the surface area for absorption. The spiral valve is present in most cartilaginous fishes.

Speculum. Contrasting patch of colour on inner secondaries of many species of duck.

Splash zone. Another term for the 'spray' or 'supralittoral zone', that is the area above the high tide mark that is affected by saltwater spray.

Sporophyte. A spore-producing phase in the life cycle of a plant that reproduces by asexual means or has alternating asexual and sexual generations.

Spring tide. A tide with a maximum range that occurs about every two weeks when the moon is new or full.

Spur. Plumage mark on a bird extending to a point, usually on the underwing or the sides of the breast.

Spyhop. Behaviour used by cetaceans when the head and eyes are raised vertically above the sea surface to visually observe their surroundings in air.

Stack. An isolated column of rock rising from the ocean, off the end of a headland or promontory. It has been isolated by the sea and is often the result of a collapsed arch.

Stall. The angle of the wing at a given air speed that does not provide lift to maintain a body at a set altitude.

Standing stock. The biomass or crop (total weight) of a population in a given area or volume.

Statocyst. A vesicle containing mineral grains that stimulate sensory cells as they move in response to the movement of the animal.

Stenohaline. Organisms that can only tolerate a narrow salinity range.

Storm surge. A rise of water above normal tidal levels caused by a combination of increased wave action and low atmospheric pressure.

Strandline. The line debris washed up along the shore.

Stridulation. Sound production through the plucking action of a specialised part of the body.

Subduction. The process by which one lithospheric plate on the Earth's surface slides beneath another at a plate boundary.

Sublittoral (subtidal) zone. The benthic zone extending from the low tide mark to the outer edge of the continental shelf (about 200m).

Suboceanic province. The benthic environment of the deeper ocean waters away from the continental shelf areas.

Subspecies. A form or forms distinguishable from other populations within a species.

Substrate. The seabed, on which or in which animals and plants may live. Substrates may be hard or soft depending on their nature.

Subtropical. The regions lying between the tropical and temperate latitudes.

Subsurface current. A slow, deep flowing current or watermass often found below the pycnocline, that may move in a different direction to surface waters.

Sulcus. Groove along the lower mandible holding coloured membrane, present in some Albatrosses.

Summer solstice. In the northern hemisphere, it is the point at which the sun moves north to the Tropic of Cancer before changing direction and moving southwards towards the equator, on 21 June.

Supercilium. Contrasting stripe above each eye of a bird.

Supralittoral zone. The narrow benthic zone on shores above the high water mark that is only wetted by storms and spray (see 'splash' or 'spray' zone).

Surge channel. A deep, narrow channel cut into a hard substrate through which water can enter or drain away from shallow areas.

Surf zone. The beach area on which waves are steepening and collapsing.

Swash zone. The narrow band of beach on which collapsed waves surge forwards and then drain backwards.

Swell. Large regular waves with an extensive 'wavelength' resulting from wave dispersion outside the 'fetch' area.

Swim-bladder. The gas-filled bladder of a fish that helps it maintain buoyancy .

Symbiont. A symbiotic organism.

Symbiosis. A linked association between two different organisms that is often of mutual benefit.

Sympatric. Species or subspecies occurring together, having overlapping areas of distribution.

Symphysis. Area of joining of bones, e.g. the two halves of the lower jaw (mandibular symphysis).

Synapse. The junction between two nerve cells across which the nerve impulse (pre-synaptic) crosses to stimulate the next nerve cell (post-synaptic).

Synaptic cleft. The very narrow gap across which a nerve impulse crosses using chemical messengers.

Syncitium. A cell or organism that has many nuclei which are not separated by cell membranes.

Systematics. The study of classification and nomenclature.

Tactile. Pertaining to the sense of touch.

Tail stock. Region just in front of the tail, connecting the tail to the rest of the body in cetaceans and sirenians.

Tally-counter. A piece of equipment operated by the thumb and used for counting organisms.

Tapetum lucidum. A reflective layer found in the eyes of some animals that double-reflects light back into the retina to make the best use of available light.

Taxonomy. The classification of plants and animals according to their biological and evolutionary characteristics.

Tectonics. Deformation of the Earth's crust by the heat generating forces of the interior of the planet (see global plate tectonics).

Teleostomes. 'True' fishes (i.e. those with a terminal mouth and bony skeleton) comprising the lobe-finned fishes and ray-finned fishes. The fossil acanthodian fishes may be teleostomes. The enigmatic fossil placoderm fishes are placed in a separate group.

Teleosts. Evolutionary advanced fishes with bony skeletons.

Telescoping. The progressive movement during evolution of the bones of the cetacean skull, particularly associated with the backward migration of the nostrils.

Temperate regions. The regions of the Earth that have pronounced seasonal changes and lie between 40-60° of latitude, north and south of the equator.

Temperature. A direct measurement of the average kinetic energy of the molecules of a substance.

Temperature gradient. The rate at which temperatures change over a given time and in a certain place.

Test. The hard outer covering of certain invertebrates.

Tetrapod. Four-limbed animals (amphibians, reptiles, birds and mammals).

Thermal. A rising current of heated air or liquid.

Thermocline. A water layer in which temperature changes rapidly with increasing depth.

Thermohaline circulation. The vertical movement of ocean water that is driven by density differences resulting from the combined effects of variations in salinity and temperature.

Threshold. A limit below which a stimulus causes no reaction.

Throat grooves. Regularly-spaced, channel-like infoldings of the skin behind the lower jaw found in most mysticetes and certain odontocetes (Ziphiidae).

Tide. The periodic rise and fall of oceanic water levels caused by the unequal gravitational attractions of the sun and moon acting on different parts of the rotating Earth.

Tideline. The level of water or the water's edge at any point during the tidal cycle.

Tidal range. The difference in height between consecutive high and low waters.

Tide tables. Prediction tables for the tidal ranges and the times of occurrence of high and low waters at any one place along a coastline.

Tissue. An aggregation of cells and their products produced by organisms to carry out a particular function.

Topography. The study of the shape of the Earth's surface including the geographical and physical features.

Torsion. Twisting especially of one end of a body while the other end is fixed.

Totipalmate. In certain birds, having webbing between all four toes.

Toxin. Any poisonous substance of plant or animal origin.

Trade Winds. Air masses that flow from the subtropical high pressure belts towards the equator. They are north-easterly in the northern hemisphere and south-easterly in the southern hemisphere.

Trench. A narrow, steep-sided underwater canyon associated with the abyssal sea floor region.

Triangulation. The engineering principle to gain strength and rigidity by using triangles in the design of a structure.

Trochophore. A free-swimming larval stage of polychaete worms and some molluscs that has bands of cilia around the margins of the body.

Trophic level. A feeding or consumer level in a food chain or food web.

Tropic of Cancer. The latitude 23.5° north.

Tropic of Capricorn. The latitude 23.5° south.

Tsunami. Large seismic oceanic waves generated by underwater earthquakes or volcanic events. They may occur in sequence and can reach great heights as they approach shallow coastal areas.

Tubenose. General term for albatrosses, petrels and shearwaters (which have tubular rather that open nostrils).

Tunic. The gelatinous or leathery body wall in tunicates (Urochordata).

Turbidity. The reduced visibility of water caused by the presence of suspended organic and inorganic material.

Turbulence. The physical and disorderly mixing of water.

Tympanic bulla. Enlarged, sometimes shell-like bone in the mammalian ear, supporting the tympanic membrane and derived from the angular bone of the lower jaw.

Typhoon. A severe tropical storm, especially in the western Pacific region.

Unicellular. Single-celled (e.g. organisms such as protozoans).

Ultraplankton. Plankton smaller than 5µm that are difficult to separate from the water.

Upwelling. Region of nutrient-rich waters that rises to the surface of the sea from deeper areas of the oceans.

Uropygial. The preen gland at the base of a bird's tail that produces an oily secretion used to waterproof feathers.

Vane. The flat unbroken surfaces of a feather either side of its midrib (rachis).

Vascular plants. Plants that have vessels to conduct and circulate fluid around their parts.

Veins. Branched vessels that convey de-oxygenated blood to the heart.

Veliger. A free-swimming larval stage of some molluscs.

Ventral. Lower surface or plane.

Vertebrate. Animal with a backbone.

Vertical migration. The periodic movement of animals up and down in the water column of the oceans.

Vertical zonation. The distinct bandings of groups, populations or organisms at different levels (e.g. on a shore or hard vertical substrate within intertidal areas).

Vial. A small glass cylinder with a plastic top.

Vibrissae. Whiskers.

Viewing cell. A piece of equipment for viewing small organisms when held up to the light.

Viscera. The organs within a body cavity, especially the abdominal organs.

Viviparous. Animals that give birth to live young after a period of nourishment inside the maternal body.

Vortex. Spiralling eddy of fluid that often forms at the trailing edge of a body moving through the fluid.

Water column. The vertical depth of water.

Water mass. A large body or volume of water that has a common origin and uniform characteristics.

Water vapour. The gaseous state of water.

Waterline. The surface level of water where it meets the atmosphere.

Wave. The disturbance that moves through a fluid, having a speed determined by the type of medium.

Wave frequency. The number of waves that pass a fixed point in a unit of time.

Wave height. The vertical distance between the 'crest' and the preceding 'trough'.

Wavelength. The horizontal distance between two similar points on successive waves, such as 'crest to crest', or 'trough to trough'.

Wave period. The elapsed time between two successive waves passing a fixed point.

Weathering. The process by which rocks are broken down into smaller pieces through physical and chemical actions.

Weed drag. An item of equipment used for dragging up samples of seaweed from below the surface.

Wetlands. Biologically productive areas bordering estuaries and other sheltered coastal regions that are occasionally flooded by tidal or river actions.

Whitecaps. Tops of waves that are 'blown off' by wind action in open waters.

Whorl. One of the coils in a molluscan shell.

Wind-driven circulation. Any movement of ocean water that is driven by the force of the wind blowing in a constant direction over a prolonged period of time. Most of these water movements are horizontal by nature.

Windward. The direction from which the wind is blowing.

Wingbar. Bar along the length of the avian wing, usually formed by contrasting tips to the wing-coverts.

Wing (span) loading. A mathematical figure obtained by dividing the weight of a body by the surface area of the wings.

Winter solstice. The point in time at which the southward moving sun reaches the Tropic of Cancer before changing direction and moving northwards, back towards the equator (22 December).

Wreck. Displacement of pelagic seabirds to inshore waters or even inland caused by storm conditions (sometimes resulting in mass mortality).

Yaw. Horizontal, side-to-side movements of an object.

Zoea. A planktonic larval stage in the life cycle of the class Malacostraca (shrimps, crabs and lobsters, and other crustaceans).

Zonation. Distinctive bands or levels of organisms related to variations in the physical environment.

Zoobenthos. Benthic animals living on the seabed.

Zooecium. The body wall of bryozoans.

Zoogeography. The geographical distribution of animals on the Earth.

Zooid. An individual member of a colony.

Zooplankton. Suspended planktonic animals of the water column.

Zooxanthellae. Photosynthetic micro-organisms (usually dinoflagellates) that live symbiotically in the tissues of organisms such as some corals and molluscs.

Zygote. A cell formed by the fusion of two gametes (one from a male and one from a female) through sexual reproduction.

APPENDIX I

The Metric System and Conversion Factors

Units of Area

1 square centimetre (cm^2)	= $100mm^2$	= 0.155 square inch
1 square metre (m^2)	= 10^4cm^2	= 10.8 square feet
1 square kilometre (km^2)	= 10^6m^2	= 247.1 acres
1 hectare (ha)	= $10{,}000m^2$	

Units of Concentration

molar concentration (M)	= gram molecular weight per litre	
µg $litre^{-1}$	= mg m^{-3}	
parts per million (ppm)	= mg $litre^{-1}$	
parts per billion (ppb)	= µg $litre^{-1}$	
µg $litre^{-1}$ ÷ molecular weight	= µM	= µmol $litre^{-1}$

Units of Length

1 angstrom (Å)	= 0.0001 micron	= 10^{-10} metres	
1 nanometre (nm)	= 10^{-9} metres		
1 micron (µ)	= 0.001 millimetre (or 10^{-3}mm)	= 10^{-6}metres	
1 millimetre (mm)	= 1000 microns	= 0.001 metre	
1 centimetre (cm)	= 10 millimetres	= 0.01 metre	= 0.394 inch
1 decimetre (dm)	= 10 centimetres	= 0.1 metre	
1 metre (m)	= 100 centimetres		= 3.28 feet
1 kilometre (km)	= 1000 metres	= 3280 feet	
1 kilometre (km)	= 0.62 statute mile	= 0.54 nautical mile	
1 inch (in)	= 2.54 centimetres		
1 foot (ft)	= 30.48 centimetres	= 0.3048 metre	
1 yard (yd)	= 3 feet	= 0.91 metre	
1 fathom	= 6 feet	= 1.83 metres	
1 statute mile	= 1.6 kilometres	= 0.87 nautical mile	
1 nautical mile	= 1.85 kilometres	= 1.15 miles	

Units of Mass

1 milligram (mg)	= 0.001 gram		
1 gram (g)	= 1000 milligrams		= 0.035 ounce
1 kiolgram (kg)	= 1000 grams		= 2.18 pounds
1 tonne (t)	= 1 metric ton (mt)	= 10^6grams	= 2205 pounds
1 pound (lb)	= 453.6 grams		

Units of Time

1 minute (min)	= 60 seconds (s) (mean solar)	
1 hour (h)	= 60 minutes = 3600 seconds	
1 day (d)	= 24 hours	= 86,400 seconds
1 year (yr)	= 365 days	= 31,536,000 seconds

Units of Velocity

1 centimetre per second (cm/s)	= 0.0328 feet per second (ft/s)	
1 metre per second (m/s)	= 2.24 statute miles per second (mi/s)	
1 kilometre per hour (km/h)	= 27.8 centimetres per second	= 0.55 knots
1 knot (kn) = 1 nautical mile per hour	= 51.5 centimetres per second	= 1.15 statute mph

Units of Volume

1 millilitre (ml)	= 0.001 litre	= 1 cubic centimetre (cm^3) (or 1cc)
1 litre (l)		= 1000 cubic centimetres
1 cubic metre (m^3)	= 1000 litres	

Units Used to measure Solar Radiation Energy

1 einstein (E) = 6.02×10^{23} quanta

Assuming an average wavelength of 550 nm,

$$1 \text{ einstein} = \frac{2.86 \times 10^7}{550} = 52 \times 10^3 \text{ calories}$$

1 watt m^{-2}	1.5×10^{-2} einsteins m^{-2} h^{-1}
1 joule m^{-2} s^{-1}	= 1 watt m^{-2}
1 calorie cm^{-2} min^{-1}	= 700 watts m^{-2}
1 langley	= 1 calorie cm^{-2}

Units Used in Production Studies

1 calorie (cal)	= 4.186 joules (J)	
1 kilocalorie (kcal)	= 1000 calories	
1 gram carbon (gC)	= 10 kilocalories	
1 gram carbon	2 grammes ash-free dr y weight (where ash free weight is dry weight less the weight of inorganic components such as shells)	
1 gram ash-free dry weight	21 kilojoules	= 21,000 joules
1 gram organic carbon	42 kilojoules	= 42,000 joules
1 gram ash-free dry weight	0.2 gram wet weight	
1 litre oxygen	= 4.825 kilocalories	
1 gram carbohydrate	4.1 kilocalories	
1 gram protein	5.65 kilocalories	
1 gram fat	9.45 kilocalories	

Units of Temperature

	Celsius (°C)	**Fahrenheit (°F)**	**Kelvin (K)**
Absolute zero (lowest possible temperature)	-273.2	-459.7	0.0
Freezing point of water	0.0	32.0	273.2
Boiling point of water	100.0	212.0	373.2

Conversions: °F = (1.8 x °C) + 32 [Roughly (2 x °C) + 30]
°C = (°F - 32)/1.8 [Roughly (°F - 30)/2]
K = °C + 273.2

Equivalence of Imperial and Metric Units

1oz	= 28.35g (approx. 30g)	100g	= 3.5oz	
1lb (16oz)	= 454g (approx. 450g)	1kg	= 2.18lb	= 35oz
1 fluid oz	= 28.35ml	1ml	= 0.035 fluid oz	
1 pint (20 fl oz)	= 568ml (approx. 570ml)	1l	= 35 fluid oz	= 1.75 pints
1g	= 0.035oz			

Imperial and Metric Equivalents of American Measures of Volume

American	**Imperial**	**Metric**
1 US gallon	6.6 imp. pints	3.78l
1 US quart	32 fl oz	945ml
1 US pint	16 fl oz	470ml

SI Units

In addition to the ordinary metric system there is a logical system of units, the Système Internationale d'Unités (SI units), which has been adopted by most scientific journals. This consists of seven basic units: length (metre), mass (gram), time (second), electric current (ampere), thermodynamic temperature (kelvin), luminous intensity (candela), and amount of substance (mole). Other units are derived from these basic seven.

Fractions and multiples of units are expressed as follows:

Fraction	Prefix	Symbol	Multiple	Prefix	Symbol
10^{-1}	deci	d	10	deca	da
10^{-2}	centi	c	10^{2}	hecto	h
10^{-3}	milli	m	10^{3}	kilo	k
10^{-6}	micro	μ	10^{6}	mega	M
10^{-9}	nano	n	10^{9}	giga	G
10^{-12}	pico	p	10^{12}	tera	T
10^{-15}	femto	f			

In the area of nutrition, the joule is a derived SI unit. 1kcal = 4186.8 J, expressed as 4.2kJ (see preceding tables). Note that megajoule has a capital M (MJ), while kilojoule has a small k (kJ). The SI unit for pressure is the pascal, with pressures normally measured in kilopascals (kPa).

Scientific Notation

To simplify writing very large and very small numbers, scientists indicate the number of zeros by scientific notation. One integer is placed to the left of the decimal, and multiplication by a power of 10 tells which direction and how far the decimal needs to be moved to write the number out in its long form. For example:

$$4.5 \times 10^{7} = 45,000,000$$

or

$$4.5 \times 10^{-7} = 0.00000045$$

Further examples showing numbers that are powers of 10 are:

1,000,000,000	= 1.0×10^{9},	or 10^{9}	giga
1,000,000	= 1.0×10^{6},	or 10^{6}	mega
1,000	= 1.0×10^{3},	or 10^{3}	kilo
100	= 1.0×10^{2},	or 10^{2}	hecto
10	= 1.0×10^{1},	or 10^{1}	deca
1	= 1.0×10^{0},	or 10^{0}	
0.1	= 1.0×10^{-1},	or 10^{-1}	deci
0.01	= 1.0×10^{-2},	or 10^{-2}	centi
0.001	= 1.0×10^{-3},	or 10^{-3}	milli
0.000001	= 1.0×10^{-6},	or 10^{-6}	micro
0.000000001	= 1.0×10^{-9},	or 10^{-9}	nano

To add or subtract numbers written as powers of 10, they must be converted to the same power:

Addition

$$2.1 \times 10^{4} + 2.0 \times 10^{6}$$
$$0.021 \times 10^{6} + 2.000 \times 10^{6}$$
$$= 2.021 \times 10^{6}$$

Subtraction

$$4.4 \times 10^{5} - 6.0 \times 10^{4}$$
$$4.4 \times 10^{5} - 0.6 \times 10^{5}$$
$$= 3.8 \times 10^{5}$$

To multiply or divide numbers written as powers of 10, the exponents are added or subtracted:

Multiplication

$$7.85 \times 10^{5} \times 3.1 \times 10^{2}$$
$$= 24.335 \times 10^{7}$$

Division

$$3.4 \times 10^{8} \div 1.7 \times 10^{5}$$
$$= 2.0 \times 10^{3}$$

APPENDIX II

Prefixes and Suffixes

The purpose of this appendix is to show the Latin or Greek derivations of some of the more common prefixes and suffixes likely to be encountered in the vocabulary of the marine environment. The prefix may vary slightly depending on whether or not it precedes a vowel or a consonant.

a- not, without (G)
ab-, abs- off, away, from (L)
abysso- deep (G)
acanth- thorn (G)
acro- top (G)
actin- a ray (G)
aero- air, atmosphere (G)
ad- toward, at, near (L)
al-, alula wing (L)
albi- white (L)
alga- seaweed (L)
alti- high, tall (L)
ampho- both, double (G)
anomal- irregular, uneven (G)
antho- flower (G)
apic- tip (L)
ap-, apo- away from (G)
apsid- arch, loop (G)
aqua- water (L)
arachno- spider (G)
arch- beginning, first in time (G)
arena- sand (L)
arthro- joint (G)
asthen- weak, feeble (G)
astro- star (G)
auto- self (G)
avi- bird (L)
balano- acorn (G)
bas- base, bottom (G)
batho- deep (G)
bentho- deep sea (G)
bio- living (G)
blasto- bad (G)
botryo- grape-like (G)
brachio- arm (G)
brachy- short (G)
branchi- gill-like (G)
broncho- windpipe (G)
bryo- moss (G)
bucc- cheek (L)
bysso- a fine thread (G)
calci- limestone (L)
calic- cup (L)
calori- heat (L)
capill- hair (L)
capit- head (L)
carno- flesh (L)
cartilagi- gristle (L)
cat- down, downward (G)
caud- tail (L)
cen-, ceno- recent (G)
cephalo- head (G)
cer-, cera- horn (G)
chaeto- bristle (G)
chir-, cheir- hand (G)
chiton- tunic (G)
chlor- green (G)
choano- funnel, collar (G)
chondri- cartilage (G)
chord- guts, string (G)
chorio- skin, membrane (G)
chrom- colour (G)
cilio- small hair (L)
cirri- hair (L)
clino- slope (G)
cloac- sewer (L)
cocco- berry (G)
coelo- hollow (G)
cope- oar (G)
cornu- horn (L)
cortic-, cortex- bark, rind (L)
costa- rib (L)
cran- skull (G)
crusta- shell (L)
cteno- comb (G)
cut-, cutis- the skin (L)
cyano- dark blue (G)
cyn- dog (G)
cypri- Venus-like (L)
-cyst bladder, bag (G)
-cyte cell (G)
dactyl- finger (G)
de- down, away from (L)
deca- ten (L)
delphi- dolphin
dent- tooth (L)
derm- skin (G)
di- two, double (G)
di-, dia- through, across (G)
dino- fearful (G)
diplo- double, two (G)
dolio- barrel (G)
duct- leading (L)
dur- hard (L)
e-, ex- out, without (L)
echino- spiny (G)
eco- house, abode (G)
ecto- outside, outer (G)
edrio- seat (G)
eid- form, appearance (G)
endo- inner, within (G)
entero- gut (G)
epi- upon, above (G)
erythr- red (G)
eu- good, well (G)
eury- broad (G)
exo- out, without
fec- excrement (L)
fecund- fruitful (L)
fer- carrier of (L)
fil- thread (L)
flacci- limp (L)
flagell- whip (L)
flora- flower (L)
fluvi- river (L)
gastro- belly (G)
geno- birth, race (G)
geo- Earth (G)
giga- very large (G)
globo- ball, globe (L)
glom-, glomer- ball of yarn (L)
gloss- tongue (G)
gnatho- jaw (G)
gracil- slender (L)
gul- throat (L)
gymno- naked, bare (G)
gyr- round, circle (G)
haem- blood (G)
halo- salt (G)
haplo- single (G)
helio- sun (G)
helminth- worm (G)
hemi- half (G)
hepat- liver (G)
herbi- plant (L)
herpeto- creeping (G)
hetero- different (G)
hexa- six (G)
hist- web, tissue (G)
holo- whole (G)
homo-, homeo- alike (G)
hydro- water (G)
hygro- wet (G)
hyper- over, above, excess ((G)
hypo- under, beneath (G)
ichthyo- fish (G)
-idae members of the animal family of (L)
infra- below, beneath (L)
insula- island (L)
inter- between (L)
intr- inside (L)
iso- equal (G)
-ite indicating a mineral or rock (G)
juven- young (L)
juxta- near to (L)
kera- horn (G)
kilo- one thousand (G)
kin- movement (G)
lacto- milk (L)
lamino- layer (L)
latero- side (L)
lati- broad, wide (L)
lecith- yolk (G)
lingu- tongue (L)
lipo- fat (G)
litho- stone (G)
lopho- tuft (G)
lorica- armour (L)
luci- light (L)
luna- moon (L)
lut- yellow (L)
luci- light (L)
macro- large (G)
magn- great, large (L)
mamilla- teat (L)
mari- sea (L)
mastigo- whip (G)
masto- breast, nipple (G)
maxillo- jaw (L)
medi- middle (L)

medull- marrow, pith (L)
medus- jellyfish (G)
mega- great, large (G)
meio- less (G)
melan- black (G)
mero- part (G)
meso- middle (G)
meta- after (G)
-meter measure (G)
-metry science of measuring (G)
micro- small (G)
milli- thousandth (L)
moll- soft (L)
mono- one, single (G)
-morph form (G)
myo- muscle (G)
myst- moustached (G)
nano- (10^{-9}) dwarf (G)
necto- swimming (G)
nemato- thread-like (G)
neo- new (G)
neph- cloud (G)
-nomy the science of (G)
noto- the back (G)
nutri- nourishing (L)
o-, oo- egg (G)
ob- opposite (L)
octo- eight (L/G)
oculo- eye (L)
odonto- teeth (G)
oiko- house, dwelling (G)
oligo- few, scant (G)
-ology science of (G)
omni- all (L)
ophi- serpent (G)
opisth- behind (G)
-opsis appearance (G)
opto- eye, vision (G)
ornitho- bird (G)
ortho- straight (G)
-osis indicating a process (G)
oste- bone (G)
oto- hear (G)
ovo- egg (L)
palaeo-, paleo- ancient (G)
pan- all (G)
para- beside, near (G)
pari- equal (L)
pecti- comb (L)
pedi- foot (L)
penta- five (G)
phag- eating (G)
phil- loving, friend (G)
pholado- lurking in a hole (G)
-phore carrier of (G)
photo- light (G)
phyl- tribe, race (G)
phyto- plant (G)
pisci- fish (L)
plano- flat, level (L)
platy- broad (G)
plankto- wandering (G)
pleisto- most (G)
pleuro- side (G)
plio- more (G)
pluri- several (L)
pneuma- air, breath (G)
pod- foot (G)
poikilo- variegated (G)
poly- many (G)
poro- channel (L)
post- behind, after (L)
-pous foot (G)
pre- before (L)
primo- first (L)
procto- anus (G)
proto- first (G)
pseudo- false (G)
ptero- wing (G)
pulmo- lung (L)
pycno- dense (G)
quadra- four (L)
quasi- almost (L)
ram- branch (L)
rept- crawl (L)
retro- backward (L)
rhin- nose (G)
rhizo- root (G)
rhodo- rose-coloured (G)
rhynch- beak, snout (G)
sali- salt (L)
sarc- flesh (G)
saur- lizard (G)
schizo- split, division (G)
scler- hard (G)
scyphi- cup (G)
semi- half (L)
septi- partition (L)
sessil- sedentary (L)
siphono- tube (G)
somato- the body (G)
sphen- wedge (G)
spiro- spiral, coil (G)
splanchn- viscera (G)
spondyl- vertebra (G)
squam- scale (L)
stego- roof (G)
sten- narrow, straight (G)
stoma- mouth (G)
strati- layer (L)
styl- style (G)
sub- below (L)
supra- above ((L)
syn-, sym- together (G)
taxo- arrangement (G)
tecto- covering (G)
tele- from afar (G)
terra- earth (L)
terti- third (L)
tetr- four (G)
thalasso- sea (G)
thec- case (G)
theri- wild animal (G)
therm- heat (G)
-tom cutting (G)
trem- hole (G)
tri- three (G)
trich- hair (G)
trocho- wheel (G)
trop- turn, change (G)
tropho- nourishment (G)
tunic- cloak, covering (L)
turbi- disturbed (L)
ultim- farthest, last (L)
vas- vessel (L)
vel- veil (L)
ventro- underside (L)
-vorous feeding on (L)
xantho- yellow (G)
xipho- sword (G)
zoo- animal (G)
zyg- pair (G)

APPENDIX III

Beaufort Wind Scale

Beaufort Number Force	Wind speed: knots	mph	metres per second	km per hour	World Meteorological Organisation (1964)	Estimating Wind Speed: Effects observed far from land	Effects observed near coast	Effects observed on land	Sea State: Terms and Height of Waves in metres	Code
0	<1	<1	0.0-0.2	<1	Calm	Sea like mirror.	Calm.	Calm; smoke rises vertically.	Calm, glassy, 0	0
1	1-3	1-3	0.3-1.5	1-5	Light air	Ripples with appearance of scales; no foam crests.	Fishing smack just has steerage way.	Smoke drift indicates wind direction; vanes do not move.		
2	4-6	4-7	1.6-3.3	6-11	Light breeze	Small wavelets; crests of glassy appearance, not breaking.	Wind fills the sails of smacks which then travel about 1-2mph.	Wind felt on face; leaves rustle; vanes begin to move.	Calm, rippled, 0-0.1	1
3	7-10	8-12	3.4-5.4	12-19	Gentle breeze	Large wavelets; crests begin to break; scattered whitecaps.	Smacks begin to careen and travel about 3-4mph.	Leaves, small twigs in constant motion; light flags extended	Smooth, wavelets, 0.1-0.5	2
4	11-16	13-18	5.5-7.9	20-28	Moderate breeze	Small waves, becoming longer; numerous whitecaps.	Good working breeze; smacks carry all canvas with good list.	Dust, leaves and loose paper raised up; small branches move.	Slight, 0.5-1.25	3
5	17-21	19-24	8.0-10.7	29-38	Fresh breeze	Moderate waves, taking longer form; many whitecaps; some spray.	Smacks shorten sail.	Small trees in leaf begin to sway.	Moderate, 1.25-2.5	4
6	22-27	25-31	10.8-13.8	39-49	Strong breeze	Larger waves forming whitecaps everywhere; more spray.	Smacks have doubled reef in mainsail; care required when fishing.	Larger branches of trees in motion; whistling heard in wires.	Rough, 2.5-4	5
7	28-33	32-38	13.9-17.1	50-61	Near gale	Sea heaps up, white foam from breaking waves begins to be blown in streaks.	Smacks remain in harbour and those at sea lie-to.	Whole trees in motion; resistance felt in walking against wind.		
8	34-40	39-46	17.2-20.7	62-74	Gale	Moderately high waves of greater length; edges of crests begin to break into spindrift; foam is blown in well-marked streaks.	All smacks make for harbour, if near.	Twigs and small branches broken off trees; progress generally impeded.	Very rough, 4-6	6
9	41-47	47-54	20.8-24.4	75-88	Strong gale	High waves; sea begins to roll; dense streaks of foam; spray may reduce visibility.		Slight structural damage occurs; slates blown from roofs.		
10	48-55	55-63	24.5-28.4	89-102	Storm	Very high waves with overhanging crests; sea takes white appearance as foam is blown in very dense streaks; rolling is heavy and visibility is reduced.		Seldom experienced on land; trees broken or uprooted; considerable structural damage occurs.	High, 6-9	7
11	56-63	64-72	28.5-32.6	103-117	Violent storm	Exceptionally high waves; sea covered with white foam patches; visibility still more reduced.			Very high, 9-14	8
12	64+	73+	32.7+	118+	Hurricane	Air filled with foam; sea completely white with driving spray; visibility greatly reduced.		Very rarely experienced on land; usually accompanied by widespread damage.	Phenomenal, 14+	9

Maps

The map section shows the geographical distribution of selected marine organisms. The range of occurrence of many marine animals is still poorly known and there are significant technical difficulties in accurately sampling the vast areas of ocean in which they occur. Further uncertainties are introduced by the effects of geographical variation, which are particularly pertinent in widespread species, compounding the difficulties of accurate identification. The distribution of marine organisms is probably influenced primarily by water temperature and, wherever possible in the relevant text, the geographical range of a species is qualified by sea temperature divisions as shown on page 10.

The maps that follow have been selected to provide assistance with identification. In cases where two or more species are similar in external appearance, but can be separated by partly or entirely different ranges, distribution maps can be of much value. For example, the three northern Pacific beaked whale species shown on page 479 are similar in external features, with adult males of all three possessing large tusk-like teeth. The maps show that their currently known ranges only partly overlap. The frequently confused Sei and Bryde's Whales (p.478) can be identified by differences in distribution, the range of Bryde's Whale being essentially tropical and less broad compared to the cosmopolitan Sei Whale. In the case of sea snakes (p.476), the Hydrophiinae have a distribution centred on coastal SE Asia although one widespread species (Pelagic Sea Snake *Pelamis platurus*) extends as far east as Panama and as far west as East Africa. This is the only sea snake species with such a broad area of distribution.

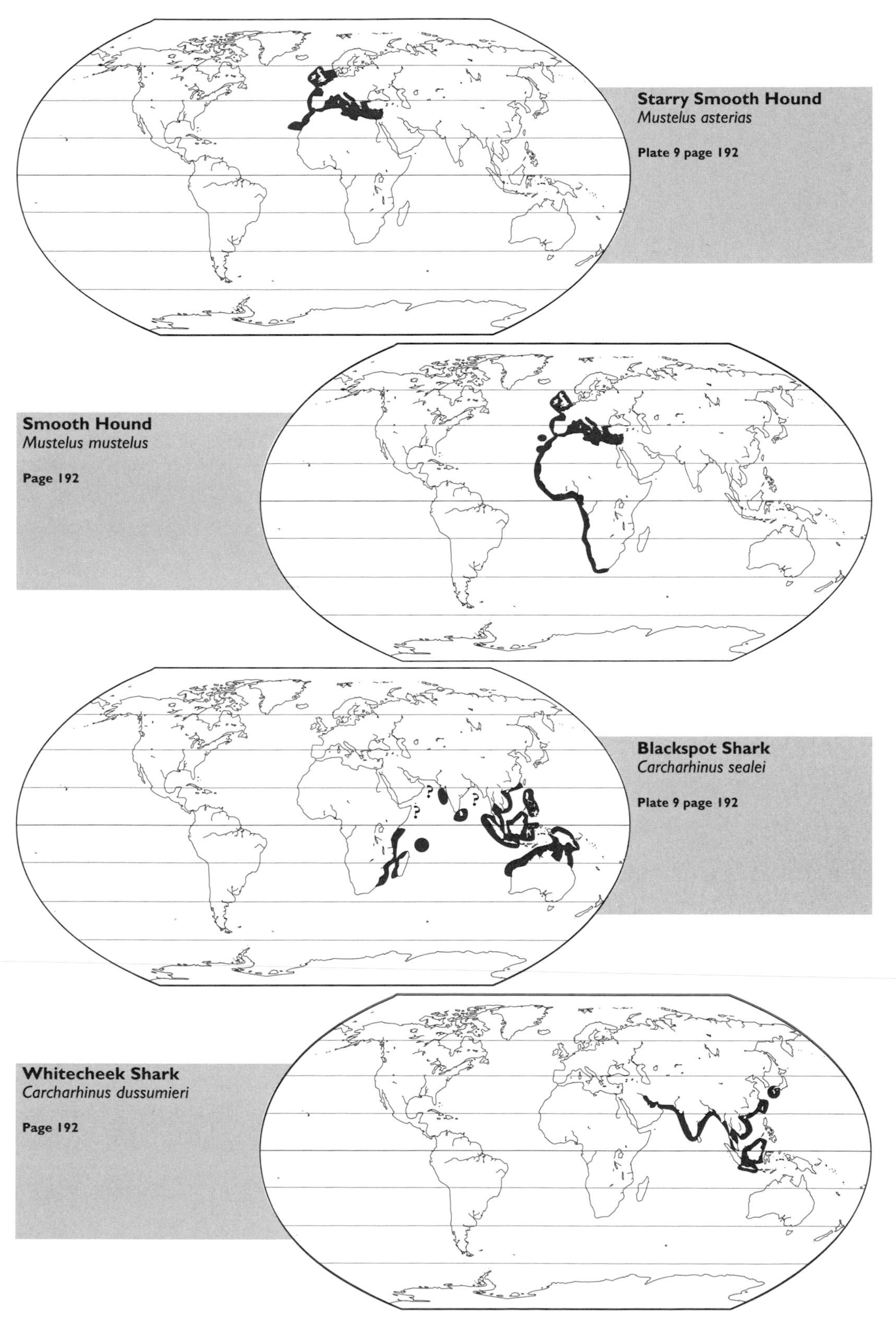

Starry Smooth Hound
Mustelus asterias

Plate 9 page 192

Smooth Hound
Mustelus mustelus

Page 192

Blackspot Shark
Carcharhinus sealei

Plate 9 page 192

Whitecheek Shark
Carcharhinus dussumieri

Page 192

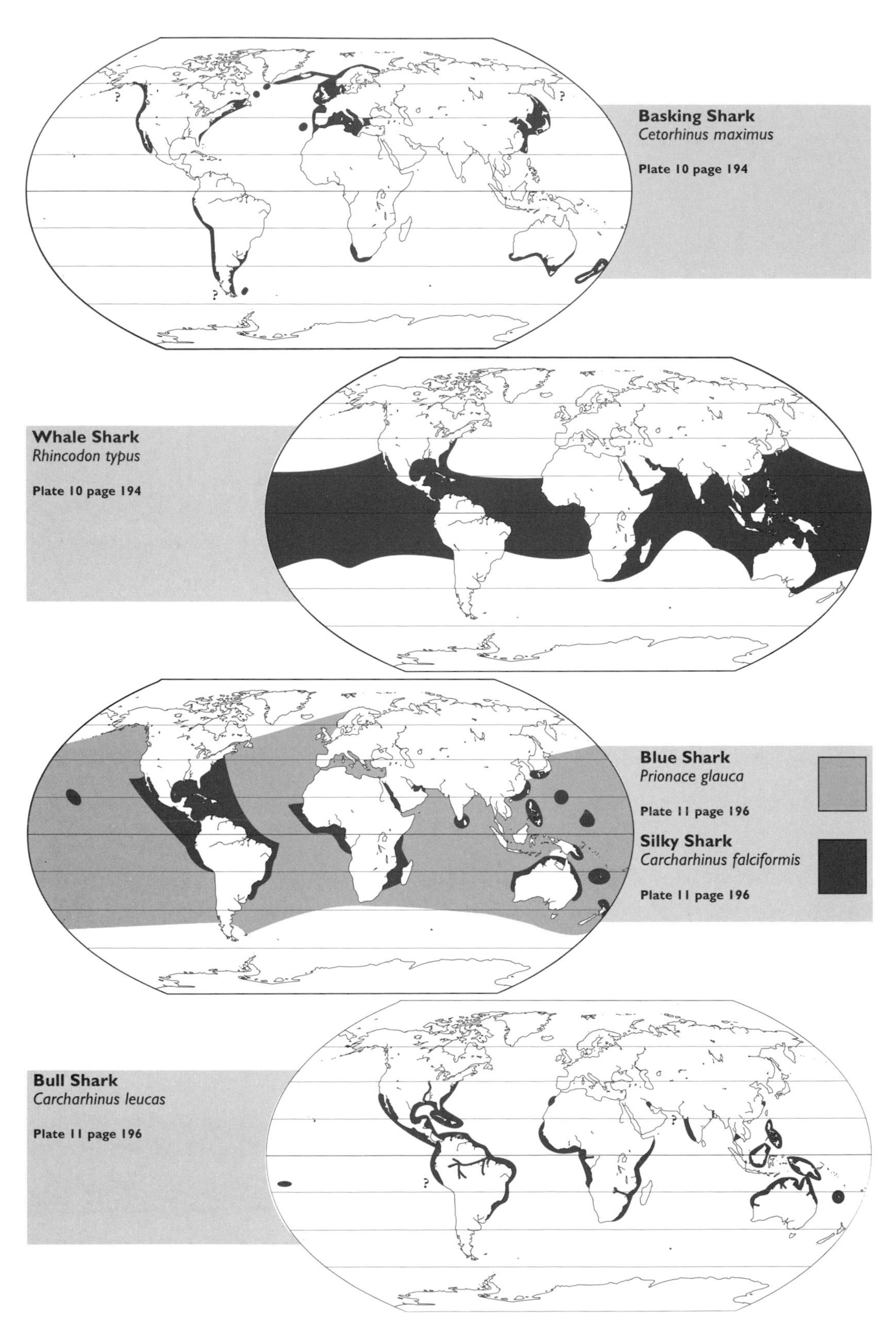
Basking Shark
Cetorhinus maximus
Plate 10 page 194
Whale Shark
Rhincodon typus
Plate 10 page 194
Blue Shark
Prionace glauca
Plate 11 page 196
Silky Shark
Carcharhinus falciformis
Plate 11 page 196
Bull Shark
Carcharhinus leucas
Plate 11 page 196

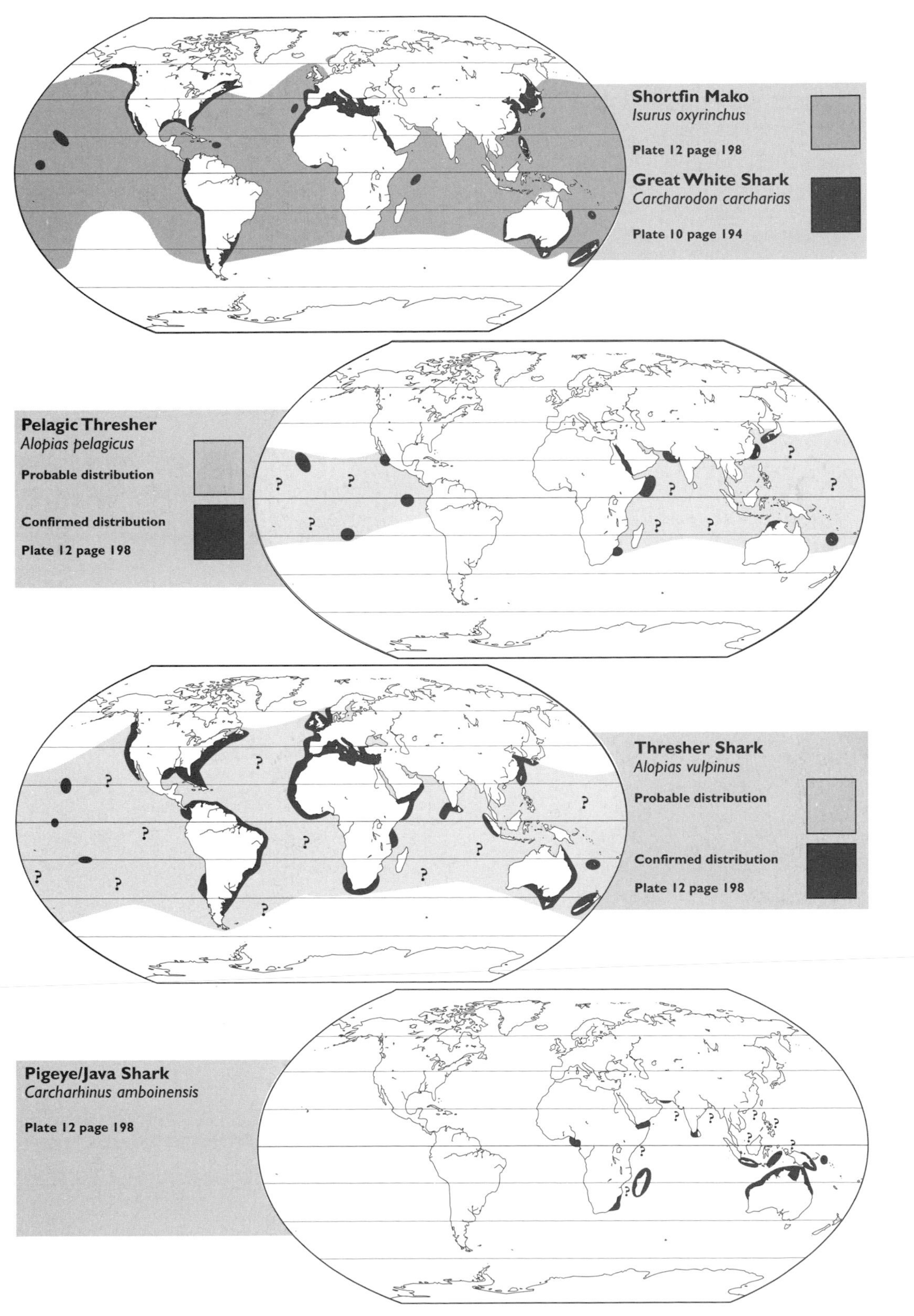
Shortfin Mako
Isurus oxyrinchus
Plate 12 page 198
Great White Shark
Carcharodon carcharias
Plate 10 page 194
Pelagic Thresher
Alopias pelagicus
Probable distribution
Confirmed distribution
Plate 12 page 198
Thresher Shark
Alopias vulpinus
Probable distribution
Confirmed distribution
Plate 12 page 198
Pigeye/Java Shark
Carcharhinus amboinensis
Plate 12 page 198

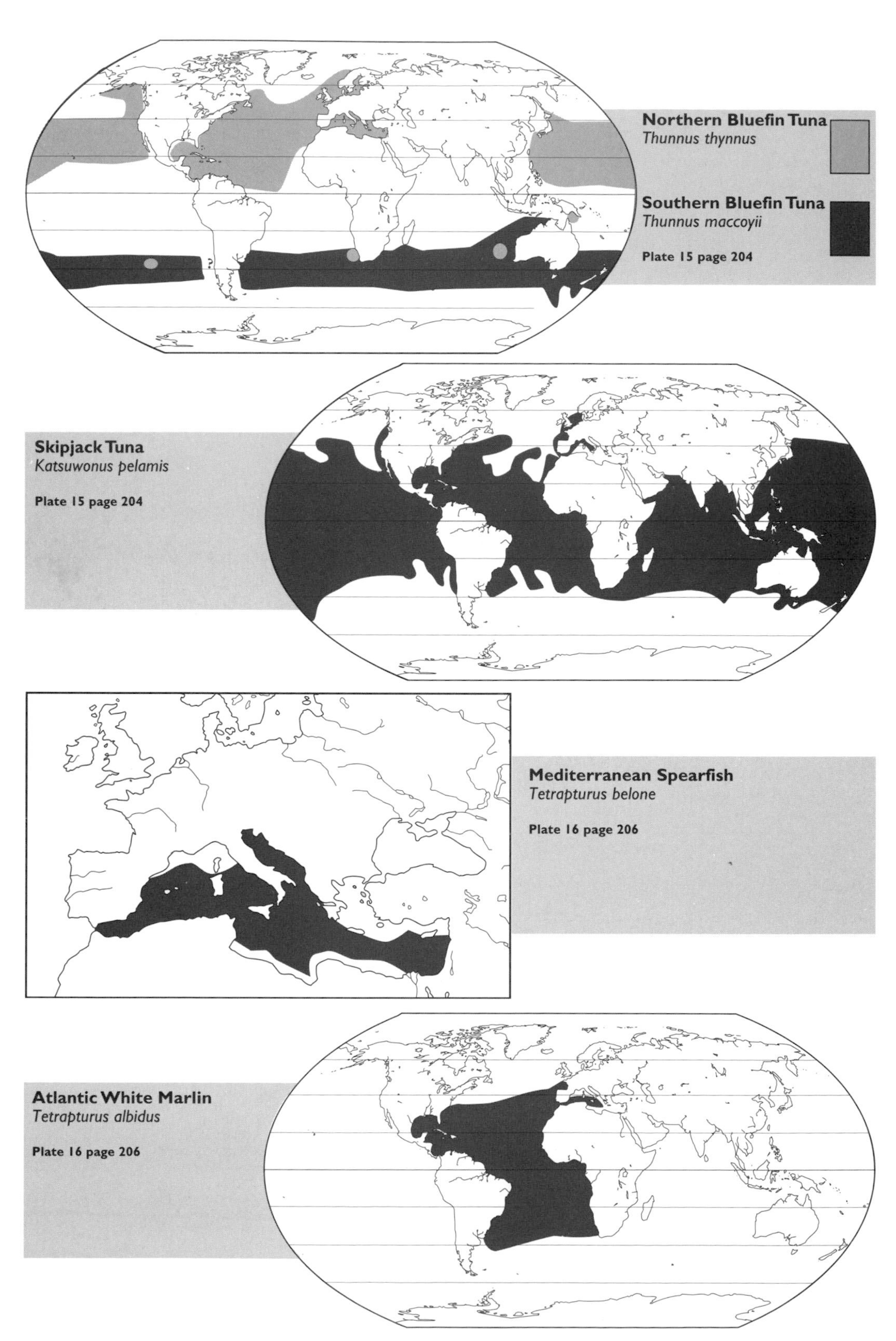
Northern Bluefin Tuna
Thunnus thynnus
Southern Bluefin Tuna
Thunnus maccoyii
Plate 15 page 204
?
Skipjack Tuna
Katsuwonus pelamis
Plate 15 page 204
Mediterranean Spearfish
Tetrapturus belone
Plate 16 page 206
Atlantic White Marlin
Tetrapturus albidus
Plate 16 page 206

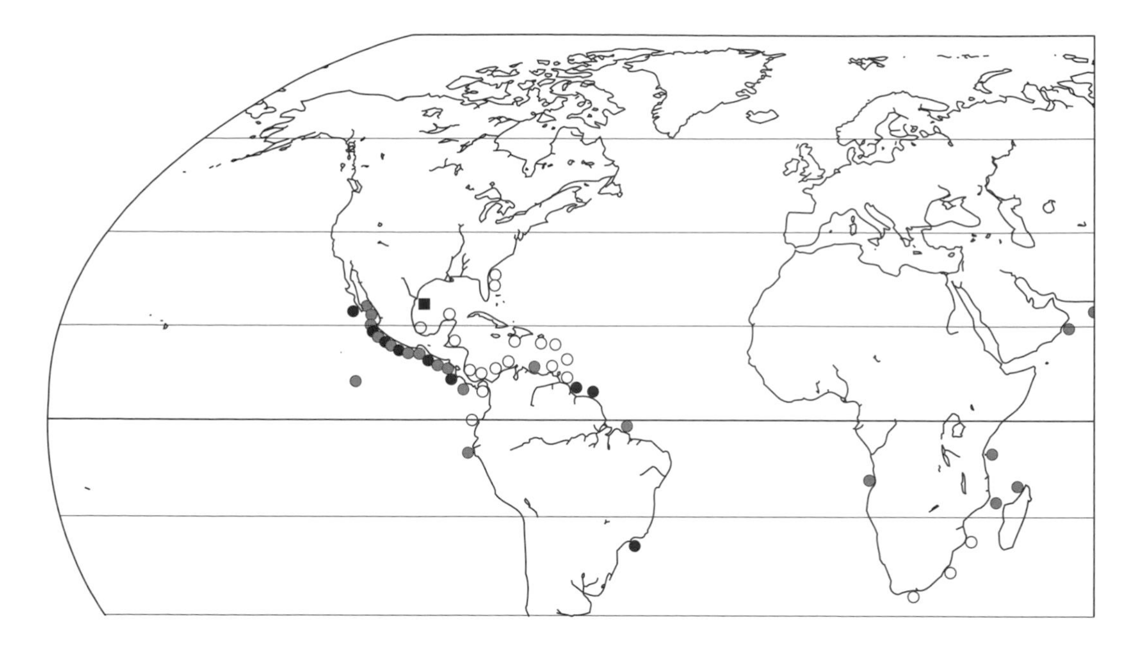

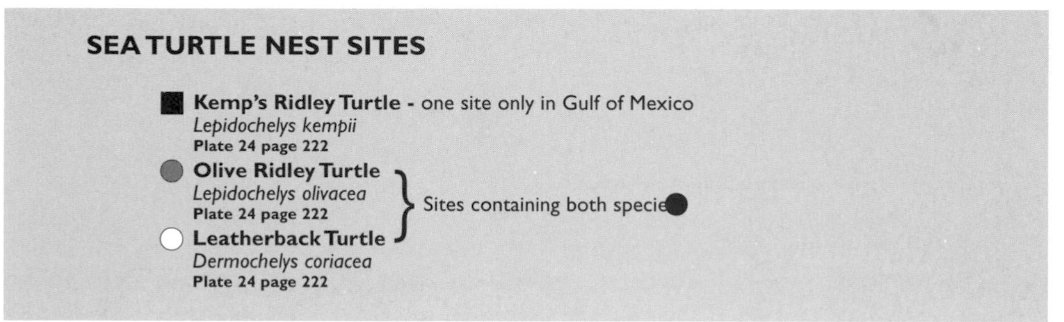

SEA TURTLE NEST SITES

■ **Kemp's Ridley Turtle -** one site only in Gulf of Mexico
Lepidochelys kempii
Plate 24 page 222

● **Olive Ridley Turtle**
Lepidochelys olivacea
Plate 24 page 222

○ **Leatherback Turtle**
Dermochelys coriacea
Plate 24 page 222

} Sites containing both species ●

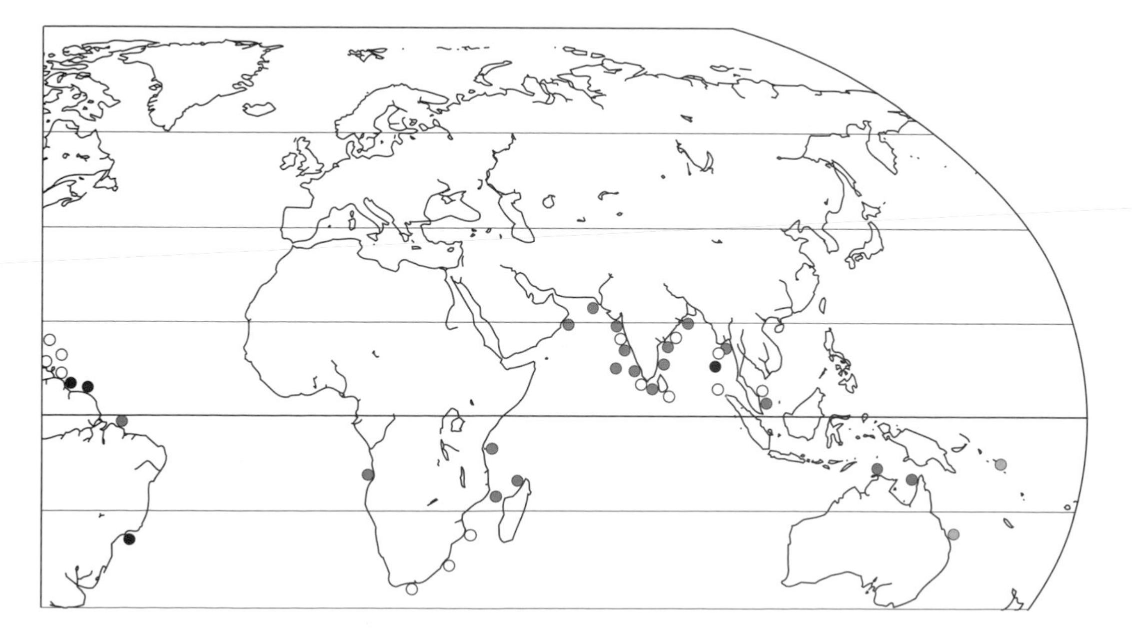

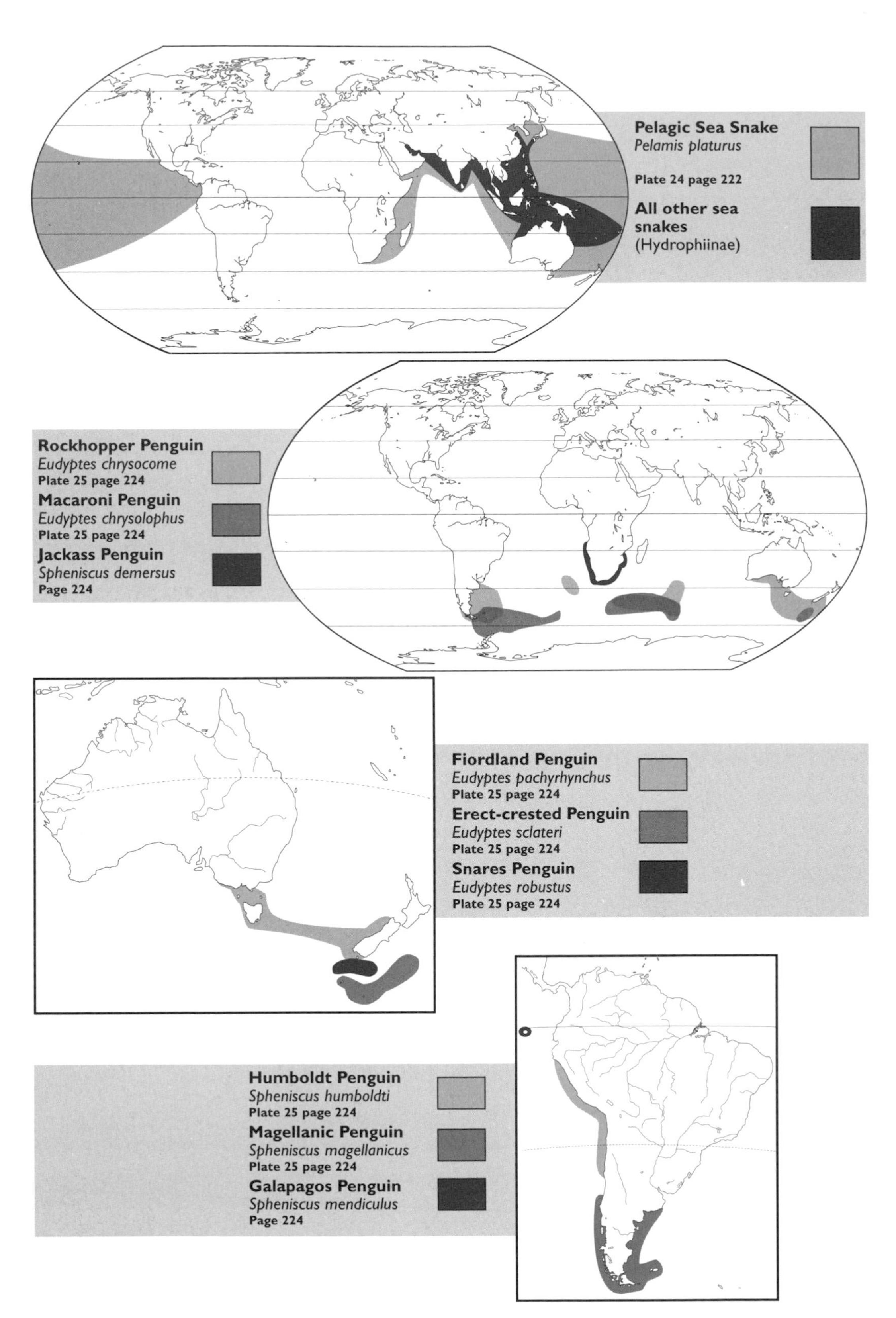

Pelagic Sea Snake
Pelamis platurus
Plate 24 page 222
All other sea snakes
(Hydrophiinae)
Rockhopper Penguin
Eudyptes chrysocome
Plate 25 page 224
Macaroni Penguin
Eudyptes chrysolophus
Plate 25 page 224
Jackass Penguin
Spheniscus demersus
Page 224
Fiordland Penguin
Eudyptes pachyrhynchus
Plate 25 page 224
Erect-crested Penguin
Eudyptes sclateri
Plate 25 page 224
Snares Penguin
Eudyptes robustus
Plate 25 page 224
Humboldt Penguin
Spheniscus humboldti
Plate 25 page 224
Magellanic Penguin
Spheniscus magellanicus
Plate 25 page 224
Galapagos Penguin
Spheniscus mendiculus
Page 224

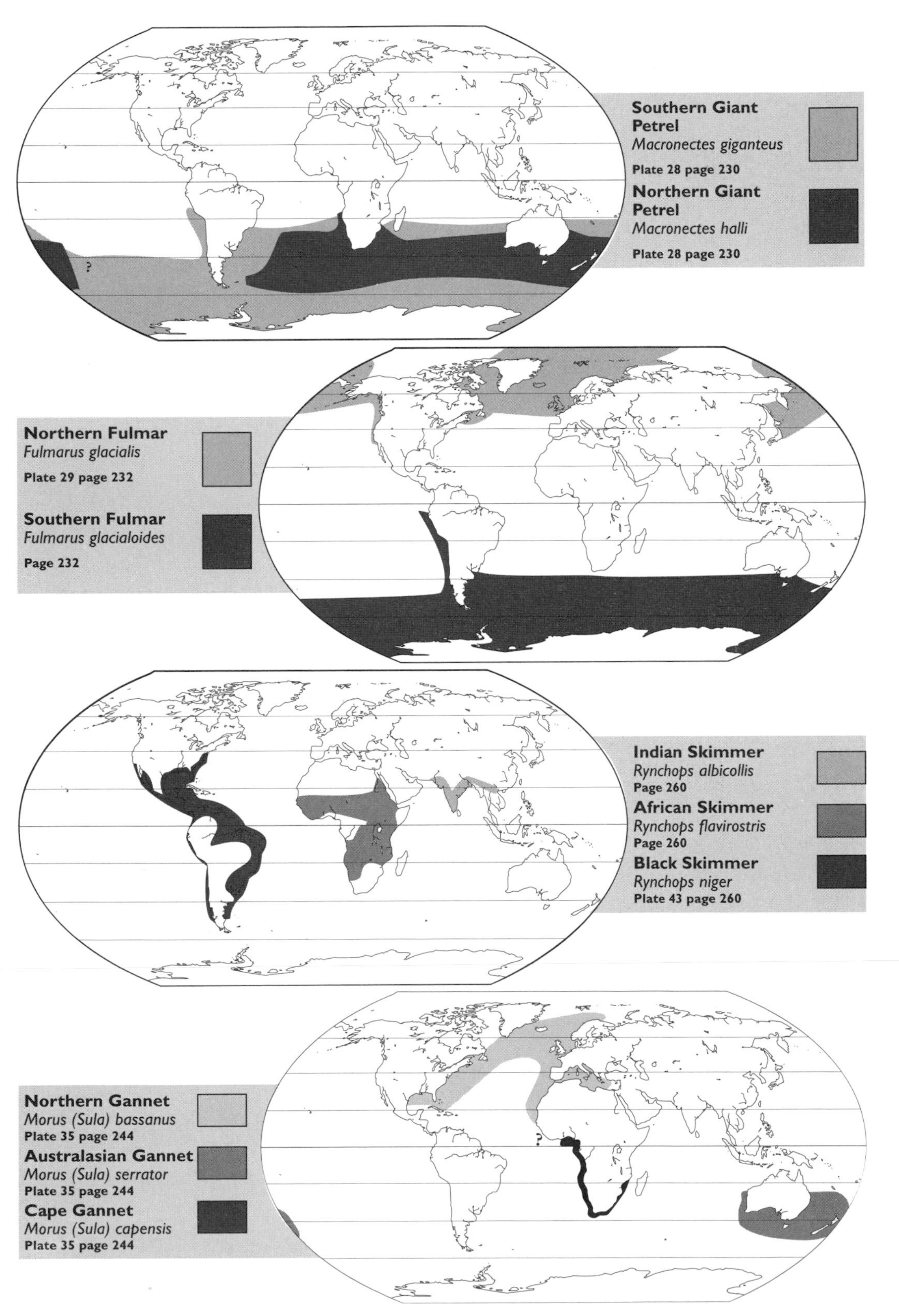
Southern Giant Petrel
Macronectes giganteus
Plate 28 page 230
Northern Giant Petrel
Macronectes halli
Plate 28 page 230
?
Northern Fulmar
Fulmarus glacialis
Plate 29 page 232
Southern Fulmar
Fulmarus glacialoides
Page 232
Indian Skimmer
Rynchops albicollis
Page 260
African Skimmer
Rynchops flavirostris
Page 260
Black Skimmer
Rynchops niger
Plate 43 page 260
Northern Gannet
Morus (Sula) bassanus
Plate 35 page 244
Australasian Gannet
Morus (Sula) serrator
Plate 35 page 244
Cape Gannet
Morus (Sula) capensis
Plate 35 page 244
?

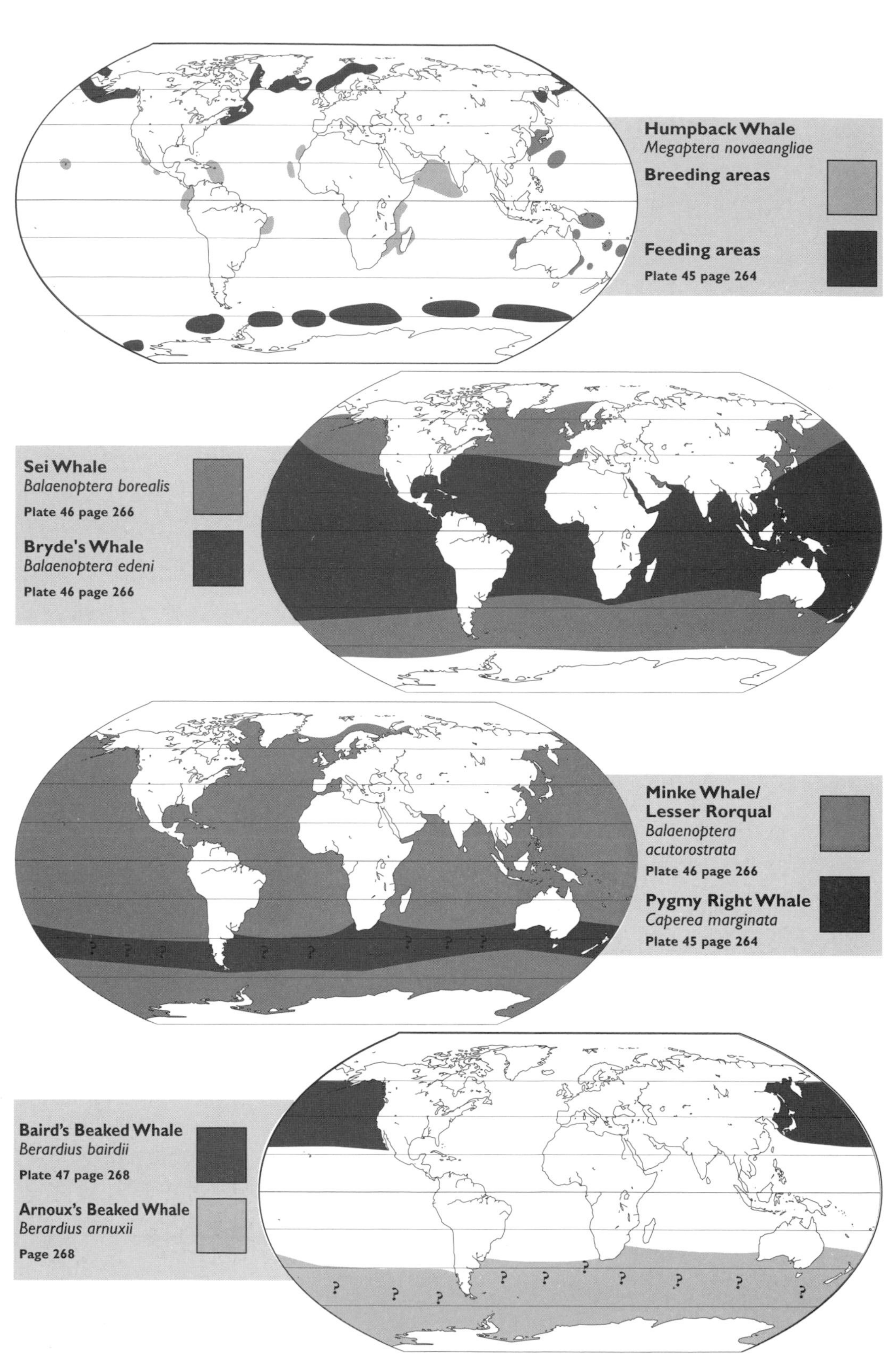
Humpback Whale
Megaptera novaeangliae
Breeding areas
Feeding areas
Plate 45 page 264
Sei Whale
Balaenoptera borealis
Plate 46 page 266
Bryde's Whale
Balaenoptera edeni
Plate 46 page 266
Minke Whale/
Lesser Rorqual
Balaenoptera
acutorostrata
Plate 46 page 266
Pygmy Right Whale
Caperea marginata
Plate 45 page 264
Baird's Beaked Whale
Berardius bairdii
Plate 47 page 268
Arnoux's Beaked Whale
Berardius arnuxii
Page 268

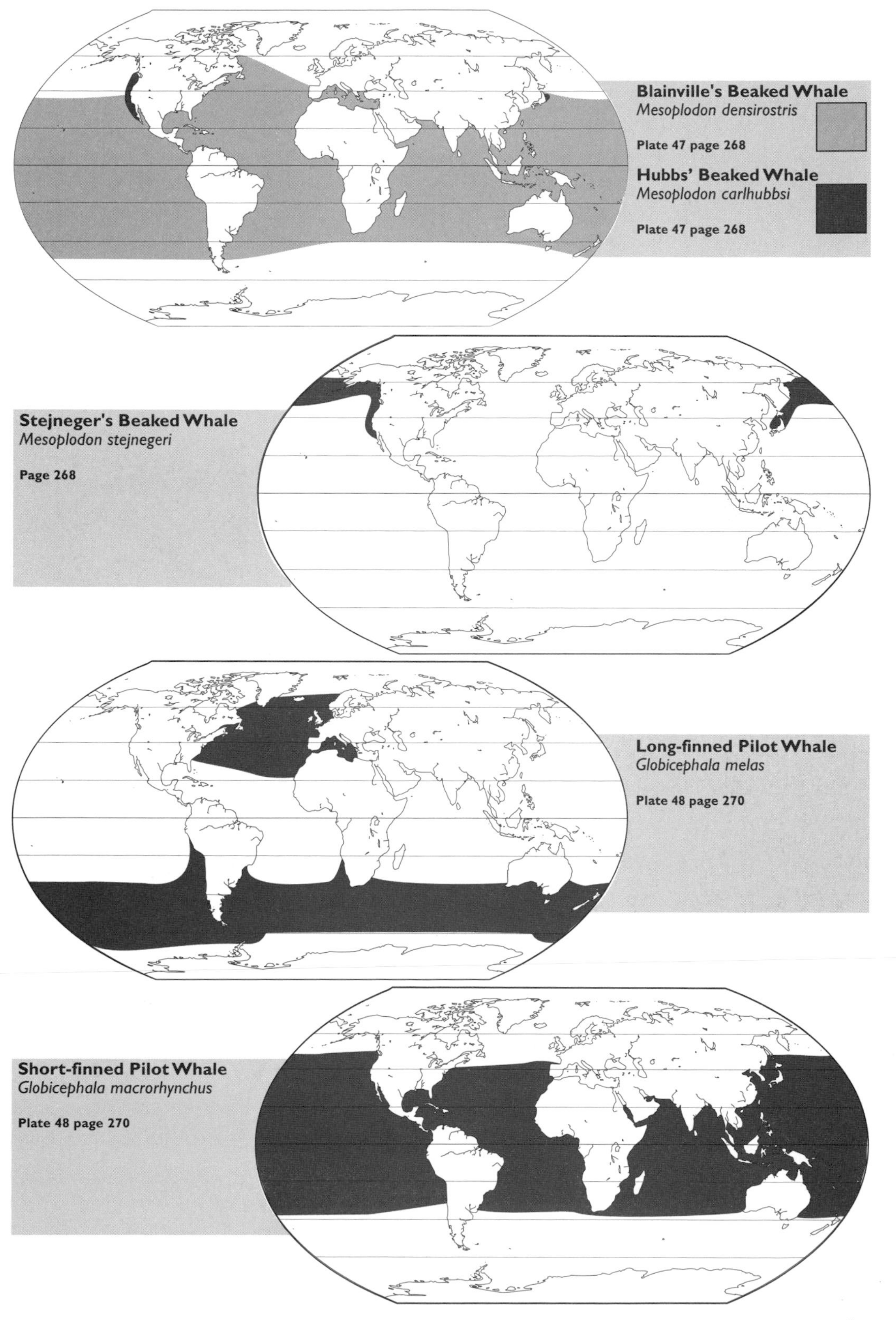
Blainville's Beaked Whale
Mesoplodon densirostris
Plate 47 page 268
Hubbs' Beaked Whale
Mesoplodon carlhubbsi
Plate 47 page 268
Stejneger's Beaked Whale
Mesoplodon stejnegeri
Page 268
Long-finned Pilot Whale
Globicephala melas
Plate 48 page 270
Short-finned Pilot Whale
Globicephala macrorhynchus
Plate 48 page 270

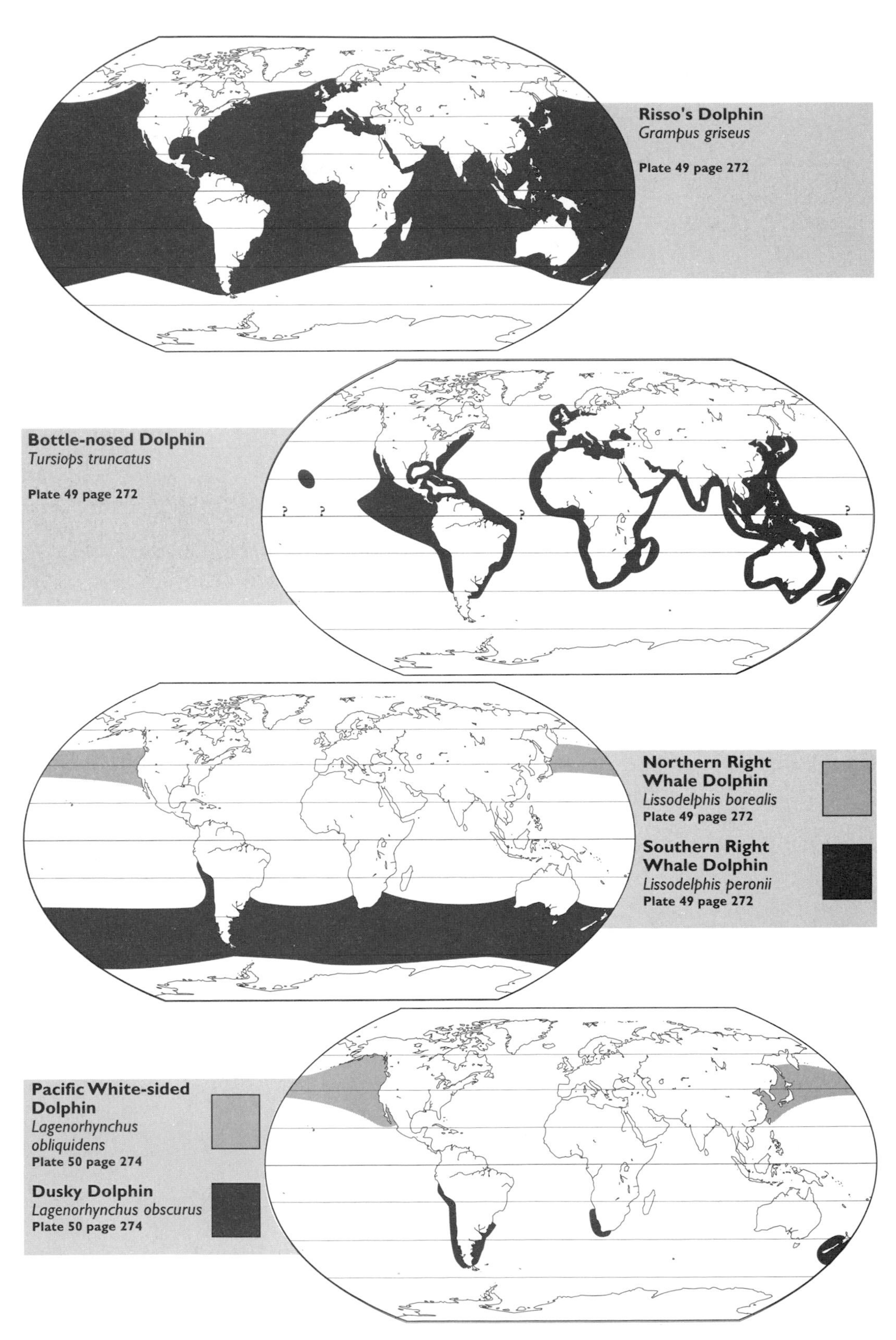

Risso's Dolphin
Grampus griseus
Plate 49 page 272
Bottle-nosed Dolphin
Tursiops truncatus
Plate 49 page 272
Northern Right Whale Dolphin
Lissodelphis borealis
Plate 49 page 272
Southern Right Whale Dolphin
Lissodelphis peronii
Plate 49 page 272
Pacific White-sided Dolphin
Lagenorhynchus obliquidens
Plate 50 page 274
Dusky Dolphin
Lagenorhynchus obscurus
Plate 50 page 274

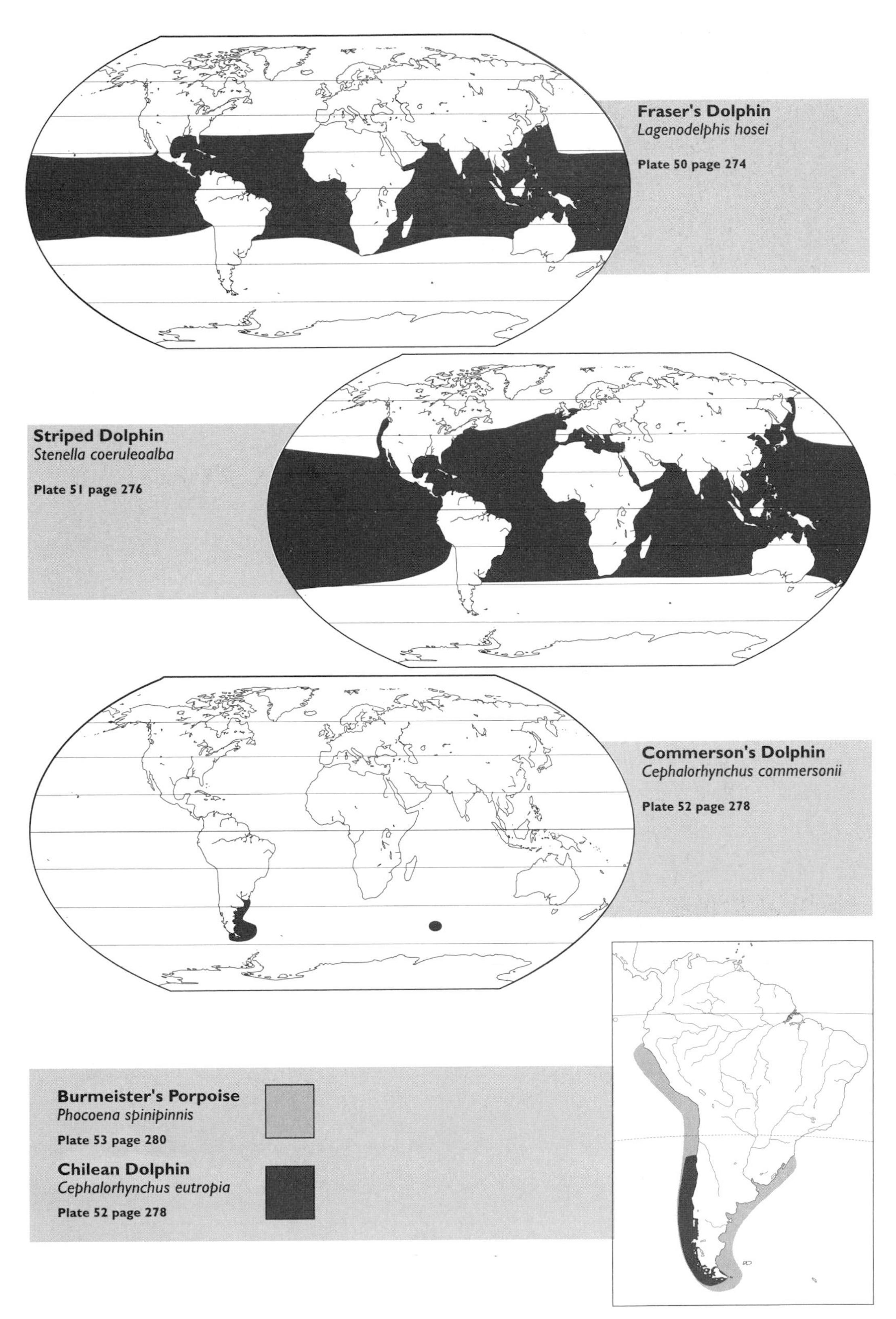
Fraser's Dolphin
Lagenodelphis hosei
Plate 50 page 274
Striped Dolphin
Stenella coeruleoalba
Plate 51 page 276
Commerson's Dolphin
Cephalorhynchus commersonii
Plate 52 page 278
Burmeister's Porpoise
Phocoena spinipinnis
Plate 53 page 280
Chilean Dolphin
Cephalorhynchus eutropia
Plate 52 page 278

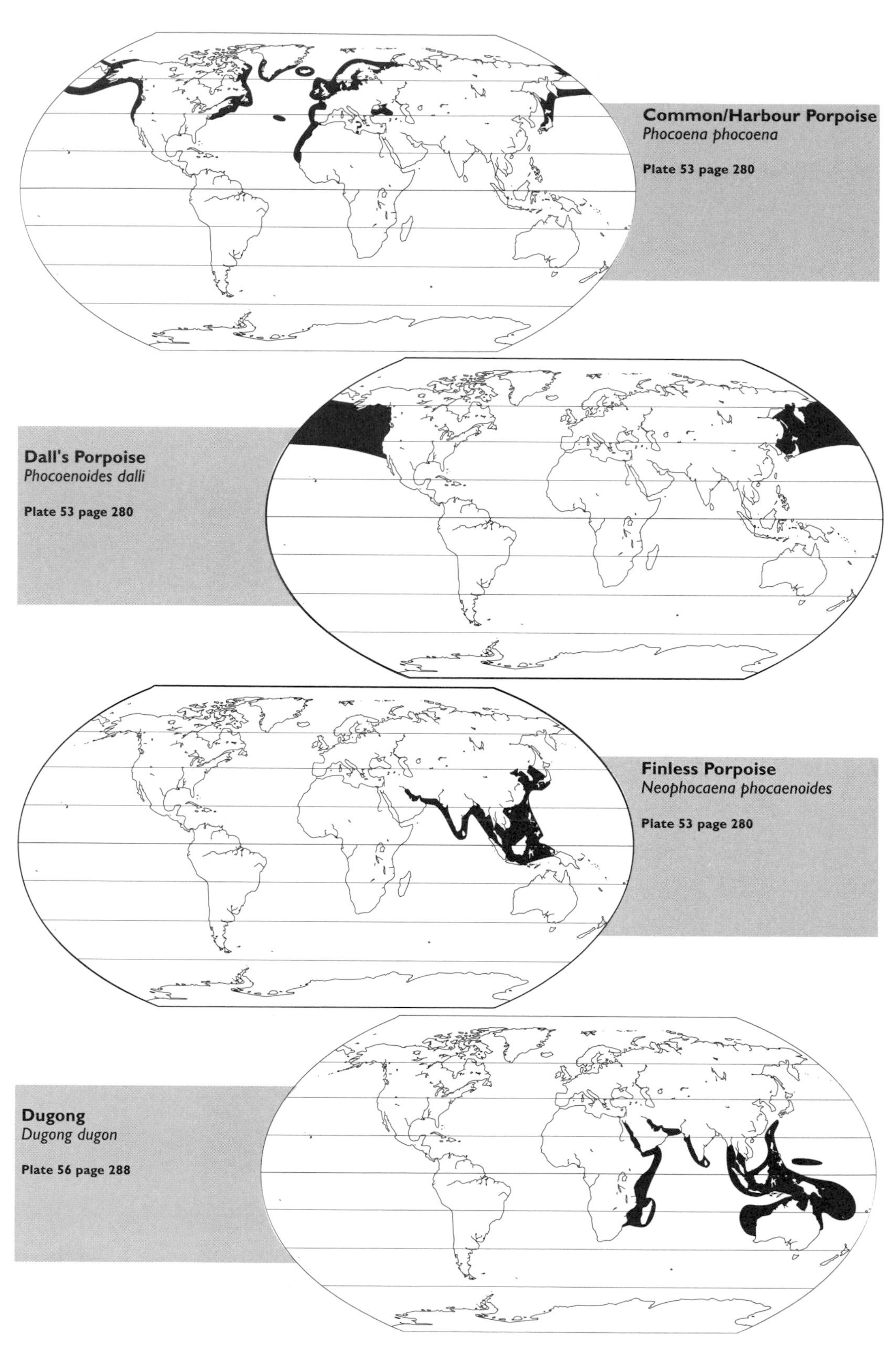
Common/Harbour Porpoise
Phocoena phocoena
Plate 53 page 280
Dall's Porpoise
Phocoenoides dalli
Plate 53 page 280
Finless Porpoise
Neophocaena phocaenoides
Plate 53 page 280
Dugong
Dugong dugon
Plate 56 page 288

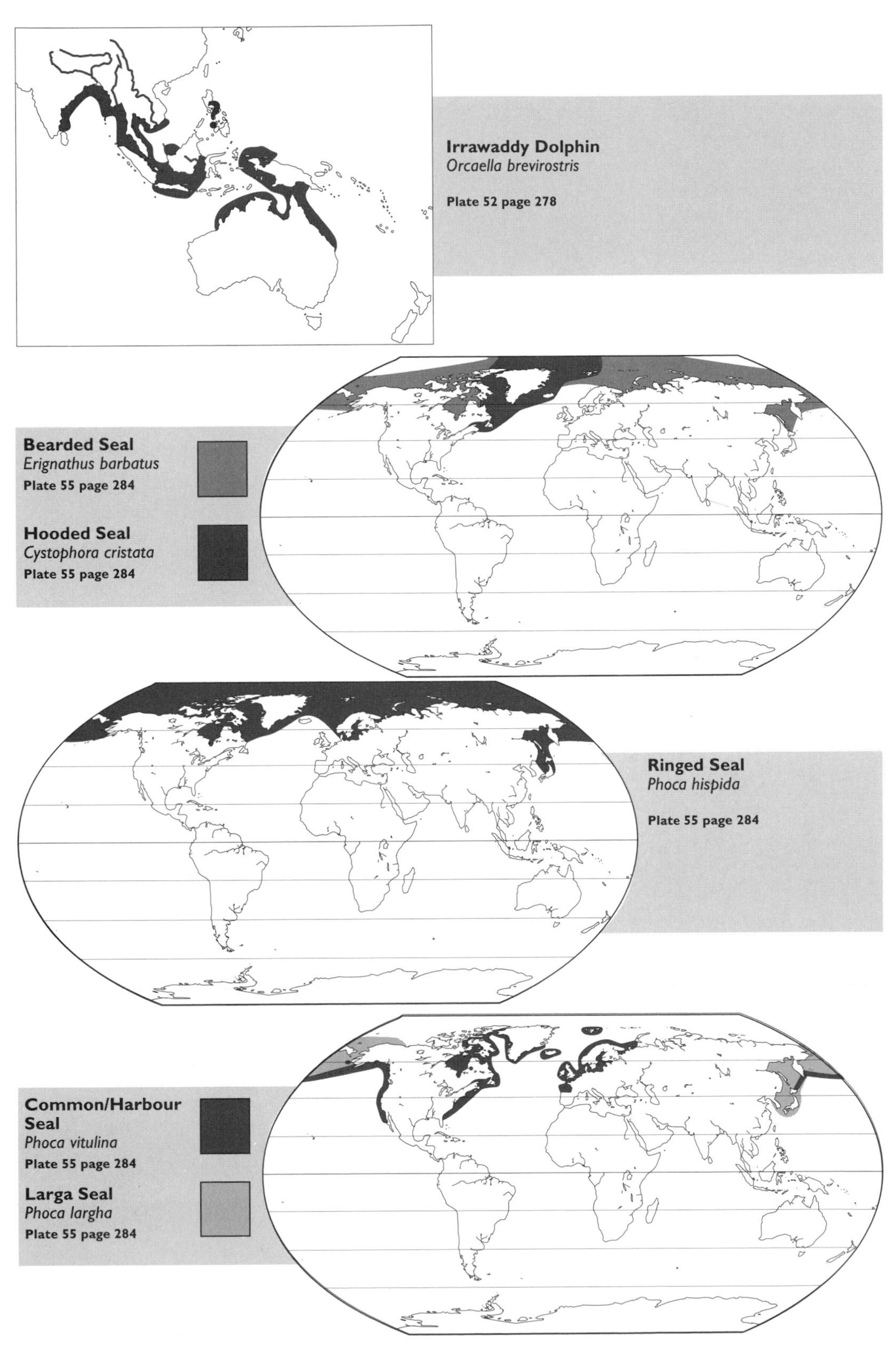
Irrawaddy Dolphin
Orcaella brevirostris
Plate 52 page 278
Bearded Seal
Erignathus barbatus
Plate 55 page 284
Hooded Seal
Cystophora cristata
Plate 55 page 284
Ringed Seal
Phoca hispida
Plate 55 page 284
Common/Harbour Seal
Phoca vitulina
Plate 55 page 284
Larga Seal
Phoca largha
Plate 55 page 284

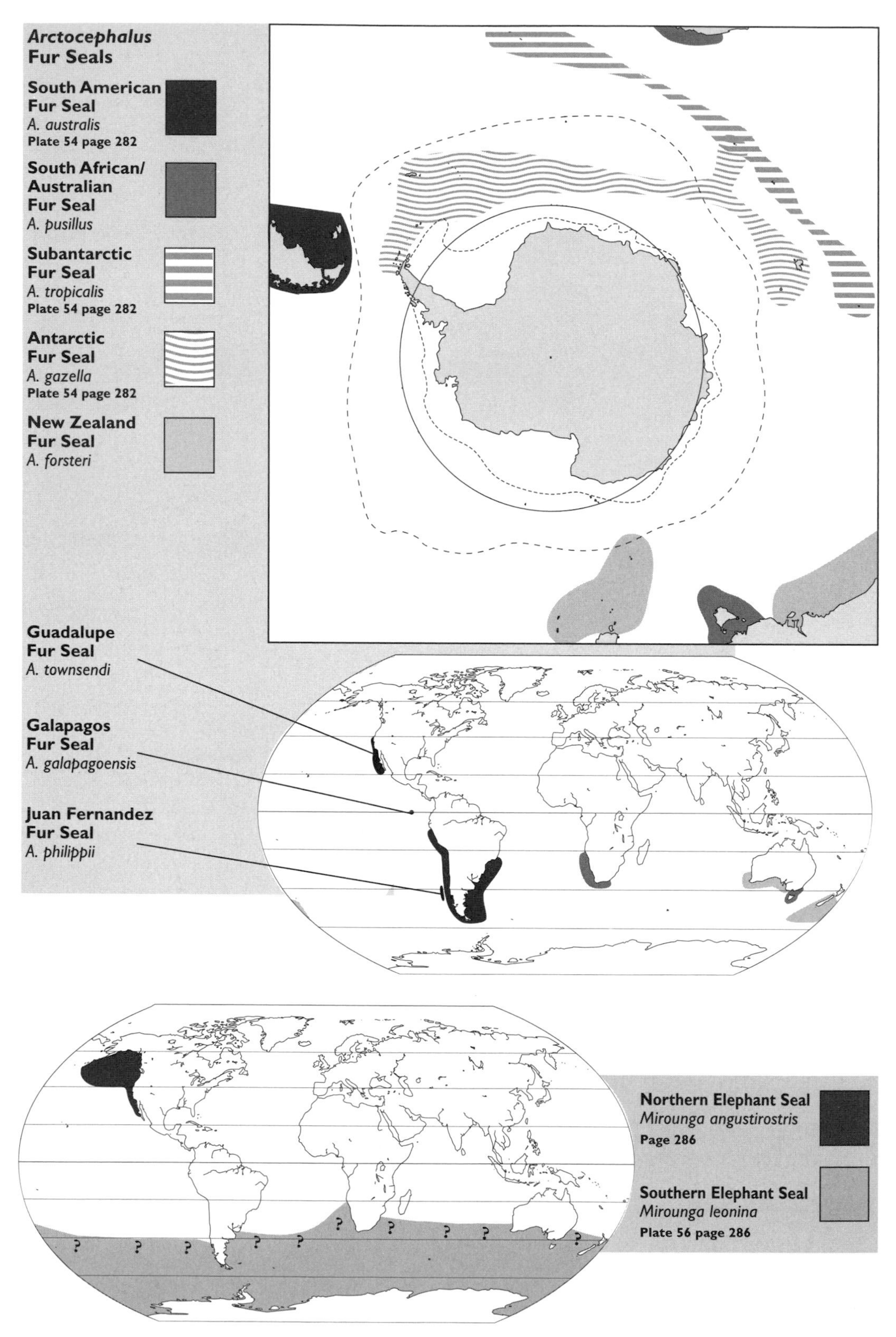
Arctocephalus
Fur Seals
South American
Fur Seal
A. australis
Plate 54 page 282
South African/
Australian
Fur Seal
A. pusillus
Subantarctic
Fur Seal
A. tropicalis
Plate 54 page 282
Antarctic
Fur Seal
A. gazella
Plate 54 page 282
New Zealand
Fur Seal
A. forsteri
Guadalupe
Fur Seal
A. townsendi
Galapagos
Fur Seal
A. galapagoensis
Juan Fernandez
Fur Seal
A. philippii
Northern Elephant Seal
Mirounga angustirostris
Page 286
Southern Elephant Seal
Mirounga leonina
Plate 56 page 286

INDEX

This index is both a subject index and a species index. Figures in plain type refer to a text entry, and in plain italics to a figure or table. Figures in bold type refer to a colour plate caption, and in bold italics to a map. All index entries refer to page numbers and not to plate or figure numbers. Glossary entries are not indexed.

C

D

G

H

I

M

Q

R

S

X

Y

Z